7th International Symposium on High-Temperature Metallurgical Processing

TMS2016

145th Annual Meeting & Exhibition

FEBRUARY 14-18 DOWNTOWN NASHVILLE, TENNESSEE MUSIC CITY CENTER

New proceedings volumes from the TMS2016 Annual Meeting:

- 7th International Symposium on High-Temperature Metallurgical Processing
- CFD Modeling and Simulation in Materials Processing 2016
- Characterization of Minerals, Metals, and Materials 2016
- Energy Technology 2016: Carbon Dioxide Management and Other Technologies
- EPD Congress 2016
- Light Metals 2016
- Magnesium Technology 2016
- Rare Metal Technology 2016
- REWAS 2016
- Shape Casting: 6th International Symposium
- TMS 2016 Supplemental Proceedings

7th International Symposium on High-Temperature Metallurgical Processing

Proceedings of a symposium sponsored by

the Pyrometallurgy Committee of the

Extraction and Processing Division (EPD) of

The Minerals, Metals & Materials Society (TMS)

held during

TMS2016

145th Annual Meeting & Exhibition

FEBRUARY 14-18 DOWNTOWN NASHVILLE, TENNESSEE MUSIC CITY CENTER

Edited by:

Jiann-Yang Hwang, Tao Jiang, P. Chris Pistorius, Gerardo R.F. Alvear F., Onuralp Yücel, Liyuan Cai, Baojun Zhao, Dean Gregurek, and Varadarajan Seshadri

Editors
Jiann-Yang Hwang
Tao Jiang
P. Chris Pistorius
Gerardo R.F. Alvear F.
Onuralp Yücel
Liyuan Cai
Baojun Zhao
Dean Gregurek
Varadarajan Seshadri

ISBN 978-3-319-48617-8
ISBN 978-3-319-48093-0 (eBook)
DOI 10.1007/978-3-319-48093-0

Chemistry and Materials Science: Professional

Published by Springer International Publishers, Switzerland, 2016
Reprint of the original edition published by John Wiley & Sons, Inc., 2016, 978-1-119-22575-1

Printed on acid-free paper

This Springer imprint is published by Springer Nature
The registered company is Springer International Publishing AG
The registered company address is: Gewerbestrasse 11, 6330 Cham, Switzerland

TABLE OF CONTENTS
7th International Symposium on High-Temperature Metallurgical Processing

Extraction and Recovery of Metals

Alloys and Materials Preparation

Fundamental Research of Metallurgical Process

Direct Reduction and Smelting Reduction

Sintering and Pelletizing of Iron Ores

Utilization of Complex Ores

Characterization and Simulation of High Temperature Process

Treatment and Recycling of Solid Slag/Wastes

Poster Session

PREFACE

This book collects selected papers presented at the 7th International Symposium on High-Temperature Metallurgical Processing organized in conjunction with the TMS 2016 Annual Meeting & Exhibition in Nashville, Tennessee, USA. More than 300 authors have contributed to the symposium with a total of 126 presentations. After reviewing the submitted manuscripts, 88 papers were accepted for publication on this book.

As the title of symposium suggests, the interests of the symposium is on thermal processing of minerals, metals, and materials that intends to promote physical and chemical transformations of materials to enable the extraction and production of valuable materials such as metals, alloys, ceramics and compounds.

The symposium was open to participants from both industry and academia and focused on innovative high-temperature technologies including those based on nontraditional heating methods as well as their environmental aspects such as handling and treatment of emission gases and by-products. Since high-temperature processes require high energy input to sustain the temperature at which the processes take place, the symposium intends to address the needs for sustainable technologies with reduced energy consumption and reduced emission of pollutants. The symposium also welcomed contributions on thermodynamics and kinetics of chemical reactions and phase transformations that take place at elevated temperatures.

This is the seventh book exclusively dedicated to this important and burgeoning topic. We hope this book will serve as a reference for both new and current metallurgists, particularly those who are actively engaged in exploring innovative technologies and routes that lead to more energy efficient and environmental sustainable solutions.

There could not be this book without contributions from the authors of included papers, time and effort that reviewers dedicated to the manuscripts, and help from the publisher. We thank them all! We also want to thank Miss Feng Chen and Mr. Zhixiong You for their assistance in collating the submitted abstracts and manuscripts.

Jiann-Yang (Jim) Hwang
Tao Jiang
P. Chris Pistorius
Gerardo R.F. Alvear F.
Onuralp Yücel
Liyuan Cai
Baojun Zhao
Dean Gregurek
Varadarajan Seshadri

EDITORS

Jiann-Yang (Jim) Hwang is a professor in the Department of Materials Science and Engineering at Michigan Technological University. He is also Chief Energy and Environment Advisor at the Wuhan Iron and Steel Group Company, a Fortune Global 500 company. He has been the Editor-in-Chief of the *Journal of Minerals and Materials Characterization and Engineering* since 2002. Dr. Hwang has founded several enterprises in areas including water desalination and treatment equipment, microwave steel production, chemicals, fly ash processing, antimicrobial materials, and plating wastes treatment. Several universities have honored him as a guest professor, including the Central South University, University of Science and Technology Beijing, Chongqing University, Kunming University of Science and Technology, and Hebei United University.

Dr. Hwang received his B.S. degree from National Cheng Kung University 1974, M.S. in 1980, and Ph.D. in 1982, both from Purdue University. He joined Michigan Technological University in 1984 and has served as Director of the Institute of Materials Processing from 1992 to 2011 and Chair of Mining Engineering Department in 1995. He has been a TMS member since 1985. His research interests include the characterization and processing of materials and their applications. He has been actively involved in the areas of separation technologies, pyrometallurgy, microwaves, hydrogen storages, ceramics, recycling, water treatment, environmental protection, biomaterials, and energy and fuels. He has more than 28 patents and has published more than 200 papers. He has chaired the Materials Characterization Committee and the Pyrometallurgy Committee of TMS and has organized several symposiums. He is the recipient of 2012 EPD Technology Award and the Michigan Tech Bhata Rath Research Award.

Tao Jiang received his M.S. in 1986 and Ph.D. in 1990, both from Central South University of Technology. He subsequently joined the university and served as an assistant professor (1990–1992) and full professor (1992–2000). From 2000 to 2003, he was a visiting scientist to the Department of Metallurgical Engineering at the University of Utah. Since 2003, Dr. Jiang has been a Professor in the School of Minerals Processing & Bioengineering at Central South University. He has been Specially-Appointed Professor of Chang Jiang Scholar Program of China since 2008 and dean of the school since 2010.

Dr. Jiang's research interests include agglomeration and direct reduction of iron ores, and extraction of refractory gold ores. He has undertaken more than 50 projects from the government and industry, including National Science Fund for Distinguished Young Scholars Program. He and co-workers invented the direct reduction process of composite binder pellets and three plants were set up in China based on the invention. He proposed the innovative composite agglomeration process of iron ore fines, which was put into production in Baotou Steel Company, China. He is actively involved in the areas of utilization of non-traditional ferrous resources such as complex ores and various solid wastes. Dr. Jiang has published 320 technical papers, six books including *Direct Reduction of Composite Binder Pellets and Use of DRI*, *Principle & Technology of Agglomeration of Iron Ores*, *Chemistry of Extractive Metallurgy of Gold*, *Electrochemistry and Technology of Catalytical Leaching of Gold*. He holds 35 patents and has more than 30 conference presentations.

P. Chris Pistorius is POSCO Professor and Co-Director of the Center for Iron and Steelmaking Research in the Department of Materials Science and Engineering at Carnegie Mellon University. His research focuses on pyrometallurgy, metals processing, and electrochemistry. He was an associate professor (1991–1996) and professor (1997–2008) in the Department of Materials Science and Metallurgical Engineering, University of Pretoria, South Africa, and served as the chair of that department from May 2002 to June 2008. Chris has a Master's degree in Metallurgical Engineering from the University of Pretoria, and a Ph.D. from the University of Cambridge, United Kingdom.

Gerardo R.F. Alvear F. is the Technology Manager for Pyrometallurgy & Refining at Glencore Technology. He is a chemical engineer and Dr. Engineer in Materials Science and Processing with over 25 years of experience in mining and metallurgy, with a focus in extractive metallurgy of copper, lead, zinc, and nickel. During his professional career, Dr. Alvear has been able to build a wide experience in operations, management, research and development, and technology innovation management in the base metals industry, participating in the generation of technology programs and in the execution of several projects across the world.

Industrial experience has also been balanced with academic work as a lecturer and assistant professor in Chile, Japan, and Canada teaching non-ferrous extractive metallurgy and metallurgical thermodynamics to undergraduate and graduate students and supervising master and doctor candidates in Japan, Chile, and Australia.

Dr. Alvear is also involved in professional societies, having currently the role of Chair of the Pyrometallurgy Committee of the Extractive and Processing Division of The Minerals, Metals & Materials Society (TMS); and also an active member of the Metallurgical Society of Canada, and member of the Copper Conference International Organizing Board, in charge of coordination of the Copper Conference Series.

Onuralp Yücel completed his technical education with a Ph.D. in Metallurgical Engineering from Istanbul Technical University (ITU) where he holds the post of Professor since 2002. He was a visiting scientist at Berlin Technical University between 1987 and 1988. He carried out postdoctoral studies at New Mexico Institute of Mining and Technology, Socorro, USA between 1993 and 1994. Prof. Yücel has more than 200 publications/presentations to his credit, which cover such topics as technological developments in the production of wide range of metals, ferroalloys, advanced ceramic powders and application of carbothermic and metalothermic processes among others. He was the vice chairman of ITU, Metallurgical and Materials Engineering Department between 2004 and 2007. He has been a director of ITU, Applied Research Center of Material Science & Production Technologies between 2006 and 2012.

Dr. Yücel's areas of interest include: Pyrometallurgy—Smelting and reduction of slags, production ferroalloys, alloys, and metals carbothermic and metalothermic processes in EAF or in ladle (copper, cobalt, vanadium, chromium, ferroboron, cobalt-boron, nickel-boron, ferromanganese, silicomanganese, ferrovanadium, ferrotungsten, ferrochromium, and aluminum-titanium-boron alloys);
Beneficiation of Industrial Wastes—Production of metals and compounds from galvanizing ash, brass production wastes, and vanadium sludges produced aluminum production; Grit production from aluminum, copper and steel slags.
Ceramic Powder Production and Processing—Production of carbide, nitride, boride powders and their processing by explosive consolidation or sintering techniques.

Liyuan Cai graduated from Central South University of Technology and received the doctoral degree of non-ferrous metallurgy in 1997. Chai is a professor at Central South University. He is the awardee of National Science Fund for Distinguished Young Scholars and Changjiang Distinguished Professor of the Ministry of Education. He is dean of the School of Metallurgy and Environment, director of the Chinese National Engineering Research Center for Control & Treatment of Heavy Metal Pollution, and director of the State Environmental Protection Engineer Center for Pollution Control of Nonferrous Metal Industry. He is also the Chairman of the International Organizing Committee of the 12th International Symposium on East Asian Resources Recycling Technology.

His research focuses on the heavy metal emission reduction and resource recycling. He led more than 50 major national research projects, published about 200 papers indexed by SCI/EI and 66 invention patents. The technology "Advanced Treatment and Reuse of Wastewater Containing Heavy Metals by Biologics" was awarded the Second Prize of State Technological Invention Award and the technology "Treatment and clean utilization of arsenic waste from non-ferrous metal smelting industry" was awarded the Second Prize of National Science and Technology Progress Award. These technologies are widely applied in non-ferrous metallurgical industries.

Baojun Zhao is Codelco-Fangyuan Professor in the School of Chemical Engineering at The University of Queensland, Brisbane, Australia. His primary fields of research are experimental investigation of physiochemical properties of high temperature melts, i.e., phase equilibria and viscosity of oxide systems. He has developed a number of novel research techniques to enable the phase equilibria and viscosity to be accurately determine for complex systems.

He has received the Spriggs Phase Equilibria Award from The American Ceramic Society (2014), Billiton Gold Medal from Institute of Materials, Minerals and Mining, U.K. (2009), and Non-Ferrous Pyrometallurgy Best Paper Award from Metallurgical Society of CIM, Canada (2009).

He has long-term close collaboration with international metallurgical companies including Baosteel, Codelco, Dongying Fangyuan Nonferrous, Rio Tinto Iron Ore, and Shougang.

Dean Gregurek has been a senior mineralogist in the RHI AG Technology Center Leoben, Austria since 2001. Dr. Gregurek received his master of science degree at the University of Graz in 1995 and his doctorate degree in Applied Mineralogy from the University of Leoben in 1999. Prior to RHI he worked two years for Luzenac Europe in the talc business. His current research interests and technical expertise are focused on chemical and mineralogical studies related to interactions between refractories, molten metals, and slags from pyrometallurgical furnaces. Dr. Gregurek has been a TMS member since 2012, currently is the *JOM* advisor for the Pyrometallurgy Committee, and a co-organizer for the 7th International Symposium on High-Temperature Metallurgical Processing (2016 TMS Annual Meeting & Exhibition, Nashville, Tennessee, USA).

Varadarajan Seshadri is a life member of the Indian Institute of Metals and other professional societies, such as the Brazilian Association of Metals and the American Iron and Steel Institute. Varadarajan Seshadri had his early education at Madras University, India, did post-graduation in metallurgy from Indian Institute of Science Bangalore in India, and received his doctorate from Technical University of Freiberg in Germany. He served in the Department of Metallurgical Engineering, Indian Institute of Technology, Kanpur and became Professor and Head of the Department of Metallurgical Engineering in MS University of Baroda. He was invited visiting professor in 1978 by the Department of Metallurgical Engineering and Materials, Universidade Federal de Minas Gerais, Belo Horizonte, Brazil, where he has continued ever since. He was also senior Humboldt Fellow in Max Planck Institut für Eisenforschung in Düsseldorf, Germany (1971–1973), Visiting Professor at Technical University, Aachen, Germany, University of Alberta, Canada, and University of Missouri, Rolla, USA, and has also served as a specialist sponsored by UNDP in India. He has won many awards for his publications in the fields of iron and steel making, transport phenomena, etc. He received the "Distinguished Professor" award from Universidade Federal de Minas Gerais, Brazil in 2006. He has served as organizing/advisory committee member of many international conferences. At present he is Prof. Emeritus in the Department of Metallurgical Engineering, Universidade Federal de Minas Gerais, Belo Horizonte, Brazil.

7th International Symposium on High-Temperature Metallurgical Processing

Energy Efficient Clean Metallurgical Technology

Session Chairs:

Jiann-Yang Hwang

Yousef Mohassab

7th International Symposium on High-Temperature Metallurgical Processing
Edited by: Jiann-Yang Hwang, Tao Jiang, P. Chris Pistorius, Gerardo R.F. Alvear F., Onuralp Yücel, Liyuan Cai, Baojun Zhao, Dean Gregurek, and Varadarajan Seshadri
TMS (The Minerals, Metals & Materials Society), 2016

FLASH REDUCTION OF MAGNETITE AND HEMATITE CONCENTRATES WITH HYDROGEN IN A LAB-SCALE REACTOR FOR A NOVEL IRONMAKING PROCESS

Yousef Mohassab, Mohamed Elzohiery, and Hong Yong Sohn

Department of Metallurgical Engineering, University of Utah, Salt Lake City, Utah 84112, USA

Keywords: Flash Reduction, Magnetite Concentrate, Hematite Concentrate, Hydrogen Reduction, Ironmaking

Abstract

As part of the development of a novel flash ironmaking process at the University of Utah, a lab-scale flash reactor was used to produce iron directly from iron oxide concentrate. This work tested the feasibility of the novel flash ironmaking process using the partial combustion of hydrogen as fuel and reductant. In this flash reactor, the energy required to reduce concentrate particles was obtained from the internal combustion of hydrogen and oxygen with complementary energy from electrical resistance heating to compensate for the heat loss. After the reaction shaft was electrically pre-heated, hydrogen and oxygen streams produced a non-premixed flame inside the reaction shaft. Various conditions such as flame configuration, flame power, positions of concentrate feeding ports, excess hydrogen amount and residence time were tested. More than 90% reduction of magnetite or hematite was achieved at temperatures as low as 1175 °C with < 100% excess hydrogen in <10 s of residence time.

Introduction

Recent developments in the ironmaking industry are mainly dedicated to reducing energy consumption and hazardous emissions. A novel flash ironmaking process is under development at the University of Utah[1-13] where hydrogen, natural gas, or coal gas is used as the reductant and fuel in a gas-solid flash reactor. In this process, iron is produced directly from iron oxides concentrate particles without the need of pelletization, sintering or coke reducing energy consumption and CO_2 emissions significantly. As part of the continuing effort to develop this process, a flash reactor was built and utilized to produce iron from iron oxides using the partial combustion of hydrogen as fuel and reductant. This was the first flash reactor used for the experimental realization of the flash ironmaking process where fuel/reductant is partially combusted in an oxy-fuel flame.

The factors affecting the extent of the reduction of iron oxide are mainly the nominal particle residence time and the reductant excess driving force. Also tested in this work, two different flame configurations two concentrate feeding ports. This work is important in understanding the process as well as developing pilot and industrial reactors.

Experimental Work

Two types of iron oxide concentrates were used in this work: magnetite and hematite ore. Both concentrate particles used were of irregular shapes or mean particle sizes of < 40 μm. Magnetite concentrate was from the Mesabi Range of the U.S. The hematite concentrate came from the Yuanjiacun Range, Shanxi Province, China.

Figure 1 shows the apparatus used in this work. The apparatus consisted of a vertical electrically heated furnace housing the reactor tube, an electric power control system, gas delivery lines, and pneumatic powder feeding system. The reactor tube was made of stainless steel 316 with 8.7" ID and 84" long. The furnace was electrically heated by six SiC heating elements grouped into two series and controlled by two SCR controllers with a maximum temperature of 1200 °C. Temperature was monitored inside the reactor tube at various locations and outside around the reactor body. Temperature and power were monitored and controlled with the aid of computer. The powder feeding system consisted of a syringe pump, vibrator, carrier gas line, powder container, and powder delivery line. The system was heated up under nitrogen flow until the temperature reached 1150 °C measured at depth 30" along the vertical axis of the reactor tube, nitrogen was switched to hydrogen with the predetermined flow rate then oxygen was introduced to the system to start the flame. Nitrogen was flowed through the powder feeder lines to carry the powder into the reactor. The flame ignition was indicated by an increase in the temperature measured by thermocouple placed near the burner nozzle. The system then was heated by both the flame and the electric power until the temperature at 30" is 1200 °C. The electric power was controlled to maintain the 30" measured temperature within 1205 ± 5 °C. Temperature profiles were measured for the reactor in order to define the reaction zone. The profiles were measured along the center of the reactor tube and close to the inner walls of the tube. All the profiles were measured while feeding powder under typical experimental conditions spanning wide range of flame power with and different flame configurations. The reaction zone was defined where the temperature is 1175 ± 25 °C and it was of 27"length. The radial variation in the temperature was negligible.

The powder was fed to the system at a constant rate of 0.12 - 0.24 ±0.006 kg/h. The particles were fed through the upper flange of the system either through the center of burner (namely burner feeding) or through two inlets on the side of the burner (namely two side feeding) as shown in Figure 2. When feeding was through the burner inlet, the feeding rate was 0.12 kg/h whereas in feeding through the two side inlets the feeding was 0.06 kg/h through each inlet with a total feeding rate of 0.12 kg/h.

Hydrogen flow rates were varied in the experiments to vary the nominal particles residence time inside the reaction zone. The amount of oxygen fed was varied to control the % excess driving force (EDF) for the reduction reaction [9, 13]. The flow rates of hydrogen and oxygen gases were controlled using accurate mass flow controller while the nitrogen gas flow rates were controlled using rotameters.

Reduced particles samples were collected at the end of the reactor tube where a stainless steel bowl was used for collection the powder. The samples were analyzed using ICP-OES to determine the reduction degree. The nominal residence time of the particles, excess driving force (% excess hydrogen) and the reduction degree calculations can be found elsewhere [9, 13].

The flame configuration inside the reactor was changed in some experiments by swapping the hydrogen and oxygen feeding ports in the burner nozzle. In the new flame configuration oxygen flow surrounds hydrogen feed stream. The old flame configuration had the two gas streams reversed. When the concentrate is fed through the center tube, the new configuration prevents the particles from being heated excessively and melting as shown by temperature contours in the two simulated configurations shown in Figure 3.

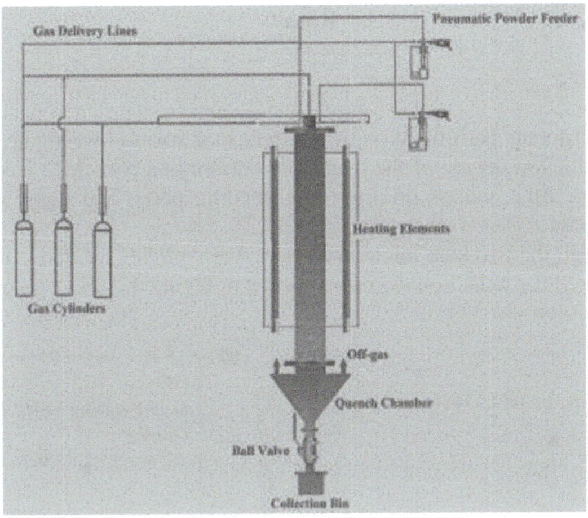

Figure 1. Schematic diagram for the lab-scale flash reactor used in this work.

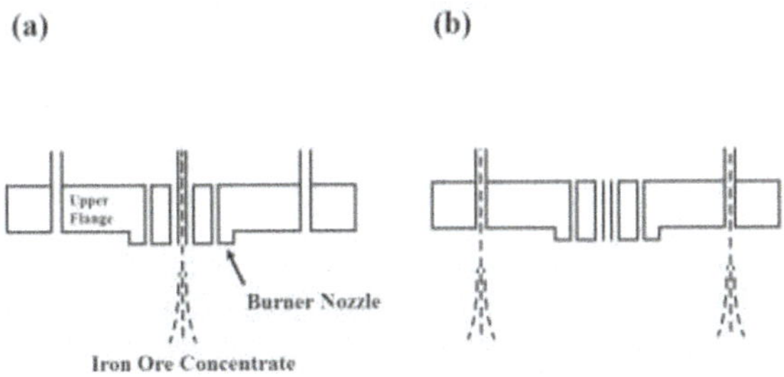

Figure 2. Powder feeding modes: (a) burner and (b) Two side-feeding.

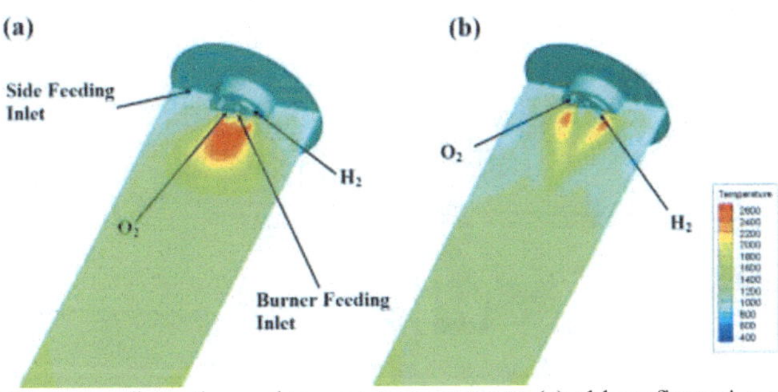

Figure 3. Flame configuration and temperature contours: (a) old configuration and (b) new configuration.

Results

Magnetite Concentrate

The aim of the experiments performed on the flash reactor was to investigate the conditions that yield the highest reduction degree of the magnetite concentrate particles. Various combinations of particles residence time, excess driving force, feeding ports, and flame configuration were studied for this purpose.

When feeding through the two side inlets, increasing the particles residence time and the excess driving force increased the reduction degree as shown in Figure 4.

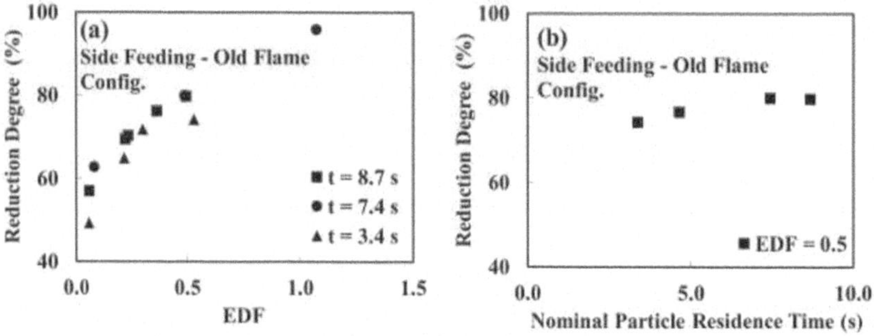

Figure 4. Feeding through the two side-inlets using the old flame configuration. (a) Effect of EDF on reduction degree (%) (b) Effect of nominal particle residence time.

Changing the feeding port from the two side inlets to the burner inlet resulted in lowering the reduction degree significantly due to the particles melting and becoming round as a result of exposure to the high temperature in the center of the flame, as shown in Figure 5. As shown in Figure 3(a), the temperature at the center of the flame where the particles are fed in the burner feeding mode using the old flame configuration reaches 2600 K. Melting and rounding reduces the active sites on the particles as compared with side feeding where the particles retain their irregular shape and reactivity that produced higher reduction degrees.

Figure 5. SEM micrograph for samples obtained with H_2 flow rate = 60 L/min and O_2 flow rate = 9.7 L/min and feeding through (a) side (RD = 63%) and (b) burner feeding (RD = 53%).

Under the same hydrogen flow rate, the flame power increases with oxygen flow rate. As mentioned earlier the particle nominal residence time is mainly controlled by the gas flow rates in this experiment. This means the power of the flame (directly linked to the flame temperature) increases with the decrease in residence time under the same excess driving force and temperature. For example, at 1175 °C and EDF of 0.5, flow rates of hydrogen 60 L/min and 9.7 L/min of oxygen (nominal particle residence time of ~ 1.9 s), respectively, generated higher flame power than that the hydrogen flow rates of 20 L/min and 2.9 L/min of oxygen (nominal particle residence time of ~ 4.6 s). In the former case, the high flame power results in increasing the particles temperature in the reaction zone and subsequently increased the reduction degree of the particles (RD = 60%, although shorter residence time, 1.9 s) as compared with latter case with the lower flame power (RD = 30%, although longer residence time, 4.6 s) as shown in Figure 6.

Changing the flame configuration by swapping the hydrogen and oxygen inlets ports resulted in changing the temperature profile at the top of the reactor. When feeding through the two side inlets, the particles experience higher temperature compared to the old flame configuration specially for higher flame power experiments where the H_2 flow rate is (40-60 l/min) and EDF = 0.5 as shown in Figure 7.

In burner feeding with the new flame configuration, the temperature in the center of the flame was lower (see Figure 3(b)) than in the old flame configuration (see Figure 3(a)). Thus, particle melting was not evident; rather the particles retained their irregular shape even when passing through the flame. Figure 7 shows the change in the reduction degree (%) with the residence time while Figure 8 shows the change in the particles shape with the flame configuration under otherwise the same conditions.

A reduction degree > 90% with % excess hydrogen driving force of less than 100 at temperature as low as 1175 °C and a feeding rate of ~ 0.2 kg/h in a few seconds of residence time was achieved.

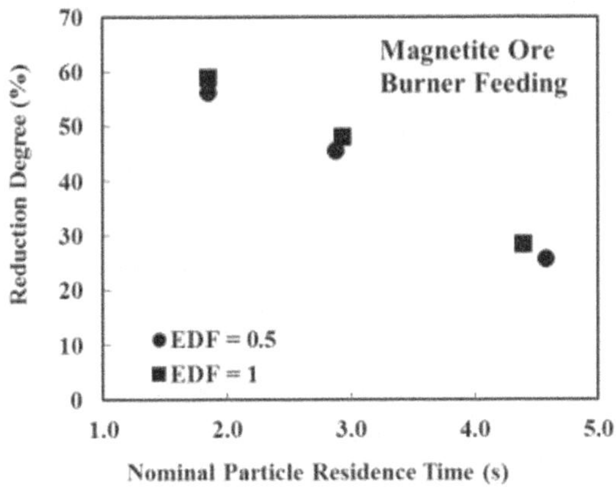

Figure 6. Reduction degree (%) vs. nominal particle residence time at different EDF when feeding through the burner inlet.

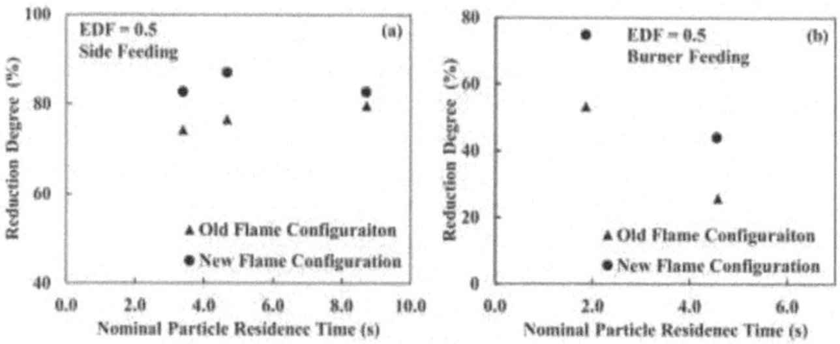

Figure 7. Reduction degree (%) vs. nominal particles residence time with EDF = 0.5 and different flame configurations for (a) side and (b) burner feeding.

Figure 8. SEM micrographs of the powders from the collection plate from experiment with EDF = 0.5, particle residence time = 1.9 s, and feeding through the burner using (a) old (RD = 59%) and (b) new flame configurations (RD = 75%).

Hematite Concentrate

Hematite concentrate was used in the flash reactor and experiments were performed with varying the EDF, residence time, feeding ports and flame configuration. Under certain conditions hematite reduction degree was found to be the same or higher than that of magnetite at the same conditions as shown in Figure 9. More experiments are being conducted to compare hematite flash reduction with hematite and magnetite concentrate particles. Hematite reduction degree > 90% with % excess hydrogen driving force of less than 100 at temperature as low as 1175 °C and a feeding rate of ~ 0.2 kg/h in a few seconds of residence time was achieved.

Conclusions

Hematite and magnetite concentrate particles were reduced in a gas-solid flash reactor heated by hydrogen/oxygen flame and electric power. Hydrogen gas was utilized as a reductant and fuel. Different conditions were tested such as the particle residence time, excess hydrogen driving force, feeding ports, and flame configuration. Hematite and magnetite reduction degree > 90% were

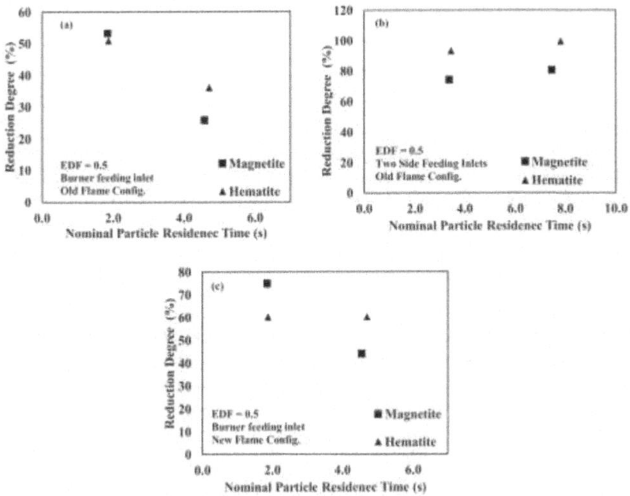

Figure 9. Comparison between magnetite and hematite showing the effect of nominal particle residence time on the reduction degree when EDF = 0.5 for (a) burner feeding with old flame configuration, (b) side feeding with old flame configuration, and (c) burner feeding with new flame configuration.

achieved at a temperature as low as 1175 °C with % excess hydrogen driving force of less than 100 % and a feeding rate of ~ 0.2 kg/h in a few seconds of residence time. Melting was found to decrease the reduction degree.

Acknowledgments

The authors thank Deqiu Fan, Brian Willhard, Andrew Laroche, Udo Fischer, Omar Kergaye, Osama Kergaye, Tuvshinbat Ganbat, and Jacqueline De Oliveira Cota for the help with the experimental runs and analytical work using ICP and. The authors acknowledge the financial support from the U.S. Department of Energy under Award Number DEEE0005751 with cost share by the American Iron and Steel Institute (AISI) and the University of Utah.

Disclaimer: This report was prepared as an account of work sponsored by an agency of the United States Government. Neither the United States Government nor any agency thereof, nor any of their employees, makes any warranty, express or implied, or assumes any legal liability or responsibility for the accuracy, completeness, or usefulness of any information, apparatus, product, or process disclosed, or represents that its use would not infringe privately owned rights. Reference herein to any specific commercial product, process, or service by trade name, trademark, manufacturer, or otherwise does not necessarily constitute or imply its endorsement, recommendation, or favoring by the United States Government or any agency thereof. The views and opinions of authors expressed herein do not necessarily state or reflect those of the United States Government or any agency thereof.

References

1. M.Y. Mohassab-Ahmed and H. Y. Sohn, *Method and Device for Digestion of Materials in a Microwave Oven*, 2012, US Patent App. 61/651.
2. M.Y. Mohassab-Ahmed, H. Y. Sohn, "Effect of Water Vapor Content in H_2–H_2O–CO–CO_2 Mixtures on the Equilibrium Distribution of Manganese between CaO–MgO_{sat}–SiO_2–Al_2O_3–FeO–P_2O_5 Slag and Molten Iron." *Steel Res. Int.*, 85 (2014), 875-884.
3. M.Y. Mohassab Ahmed, "*Phase Equilibria between Iron and Slag in CO/CO_2/H_2/H_2O Atmospheres Relevant to a Novel Flash Ironmaking Technology*" (PhD Dissertation, The University of Utah, 2013).
4. M.Y. Mohassab-Ahmed, H. Y. Sohn, "Effect of Water Vapor Content in H_2-H_2O-CO-CO_2 Mixtures on the Activity of Iron Oxide in Slags Relevant to a Novel Flash Ironmaking Technology." *Ironmaking Steelmaking*, 41(2014), 665 - 675.
5. Y. Mohassab and H. Sohn, "Effect of Water Vapour on Distribution of Phosphorus between Liquid Iron and MgO Saturated Slag Relevant to Flash Ironmaking Technology." *Ironmaking Steelmaking*, 41 (2014), 575-582.
6. Y. Mohassab and H.Y. Sohn, "Effect of Water Vapor on Sulfur Distribution between Liquid Fe and MgO-Saturated Slag Relevant to a Flash Ironmaking Technology." *Steel Res. Int.*, 86 (2014), 753–759.
7. Y. Mohassab and H.Y. Sohn, "Analysis of Slag Chemistry by FTIR-RAS and Raman Spectroscopy: Effect of Water Vapor Content in H_2-H_2O-CO-CO_2 Mixtures Relevant to a Novel Green Ironmaking Technology." *Steel Res. Int.*, 86 (2014), 740-752.
8. M.Y. Mohassab-Ahmed, H.Y. Sohn, and L. Zhu, "Effect of Water Vapour Content in H_2-H_2O-CO-CO_2 Mixtures on MgO Solubility in Slag under Conditions of Novel Flash Ironmaking Technology." *Ironmaking Steelmaking*, 41 (2014), 575-582.
9. F. Chen, Y. Mohassab, T. Jiang, and H.Y. Sohn, "Hydrogen Reduction Kinetics of Hematite Concentrate Particles Relevant to a Novel Flash Ironmaking Process." *Metall. Mater. Trans. B*, 46 (2015), 1133-1145.
10. F. Chen, Y. Mohassab, S. Zhang, and H.Y. Sohn, "Kinetics of the Reduction of Hematite Concentrate Particles by Carbon Monoxide Relevant to a Novel Flash Ironmaking Process." *Metall. Mater. Trans. B*, 46 (2015), 1716-1728.
11. H. Pinegar, M. Moats, and H. Sohn, "Flowsheet Development, Process Simulation and Economic Feasibility Analysis for Novel Suspension Ironmaking Technology Based on Natural Gas: Part 2 – Flowsheet and Simulation for Ironmaking Combined with Steam Methane Reforming." *Ironmaking Steelmaking*, 40 (2012), 44-49.
12. H. Pinegar, M. Moats, and H. Sohn, "Flowsheet Development, Process Simulation and Economic Feasibility Analysis for Novel Suspension Ironmaking Technology Based on Natural Gas: Part 1-Flowsheet and Simulation for Ironmaking with Reformerless Natural Gas." *Ironmaking Steelmaking*, 39 (2012), 398-408.
13. M.E. Choi and H.Y. Sohn, "Development of Green Suspension Ironmaking Technology Based on Hydrogen Reduction of Iron Oxide Concentrate: Rate Measurements." *Ironmaking Steelmaking*, 37 (2010), 81-88.

7th International Symposium on High-Temperature Metallurgical Processing
Edited by: Jiann-Yang Hwang, Tao Jiang, P. Chris Pistorius, Gerardo R.F. Alvear F., Onuralp Yücel, Liyuan Cai, Baojun Zhao, Dean Gregurek, and Varadarajan Seshadri
TMS (The Minerals, Metals & Materials Society), 2016

Investigation of Coal Tar Pitch Binder for the Production of Formed Coal Briquettes for COREX from High Volatile Coal Powder

Yongbin Yang[1], Yaxuan Wang[1,2], Qian Li[1], Bin Xu[1], Qiang Zhong[1], Wei Gao[1,3]

1.—School of Minerals Processing and Bioengineering, Central South University, Changsha, Hunan, 410083, China. 2.—e-mail: hey_yaxuan@163.com. 3.—e-mail:zhongqiang2008csu@163.com

Keywords: Formed Coal, Coal Tar Pitch, High volatile Coal, Carbonization

Abstract

Lump coal with high volatility is used for providing heat and reductive gas, and ensuring permeability in COREX. However, it is apt to be pulverized in transport and storage process. Therefore, comprehensive utilization of the pulverized coal to produce coal briquettes with appropriate binders is significant, among which the molasses binder cannot meet the technological requirements. In this paper, coal tar pitch binder and BJ additive were used in briquetting and carbonizing process of coal powder. The results showed that BJ effectively activated cold state cohesiveness of the pitch and make it bond the coke particles to a firm briquette at room temperature .Carbonizations of coal and coal tar pitch achieved obvious constriction and formed a carbonized skeleton. The drop resistance of wet briquettes reached over 50 times/1meter. When cured at 1000℃ for 1h under nitrogen atmosphere, the compressive strength of carbonized briquettes came to 6.43 Mpa.

Introduction

COREX, as a kind of new iron-making process, differs from the traditional blast furnace iron-making by its greatly reduced dependence of coke, thus the configuration of coke oven will not be needed. From the practical operation of existing COREX plants, the POSCO in South Korea has the lowest coke ratio, about 5%, while the coke ratio in others basically exceeds 20%[1]. Because of much higher fuel consumption in COREX than in blast furnace process, the important role of coal just ranks second to iron ore in raw materials.

Heat and reductive gas are provided through coal reaction, and necessary permeability can be realized from certain lumpiness range of coal as well[2].

As a result, in order to obtain stable furnace condition, COREX coal must have certain physical, chemical and high temperature performance[2].

As for the material requirement, COREX can only use 5~50mm nature lump coal and coke as fuel. However, the nature lump coal is limited both in varieties and quantity. In addition, it is inevitably pulverized in the process of transportation, transshipment and bin storage, which increases the investment and operating costs.

So adequately using the pulverized coal powders to produce coal briquettes plays a key role in practical production[3]. The COREX3000 in Baosteel adopts molasses briquetting technology from POSCO in South Korea to produce coal briquettes. But the high temperature behavior of molasses briquettes cannot reach the standards of lump coal.

Consequently, in this paper, coal tar pitch binder[4], used as an alternative to molasses, was studied in coal briquetting process systematically.

Compared with molasses briquetting technology, BJ, as an additive, was added into the coal tar pitch briquetting technology. The chemical ingredients of BJ was dimethylbenzene, it had been successfully applied in formed coke with coal tar pitch binder.

It had been shown that the coal and coal tar pitch could cause obvious constriction and form a carbonized skeleton when cured at high temperature, which got higher compressive strength of coal tar pitch briquettes and more ideal high temperature behavior than that of molasses coal briquettes[5].

Experimental

Materials

In this study, the coal, molasses and coal tar pitch were provided by Baosteel Corporation from China.

Table I. Proximate Analysis of Coal and Chemical Composition of Coal Ash/wt.%

Industrial analysis			Chemical composition				
FC_d	A_d	V_d	SiO_2	Al_2O_3	Fe_2O_3	CaO	MgO
62.35	6.18	31.47	25.21	5.71	10.75	25.19	6.33

Table II. AFT Results of Coal Ash Samples and Calorific Value of Coal

DT/℃	ST/℃	HT/℃	FT/℃	$Q_{gr,v,d}$/MJ.kg^{-1}	$Q_{net,v,ar}$/MJ.kg^{-1}
1180	1210	1220	1230	30.78	27.26

Materials used in this study include coal powder and coal tar pitch whose softening point was 80 ℃ [6]. Moreover, the main composition of coal tar pitch was hydrocarbons which will not add much ash content and affect slagging compared with molasses binder, also, using coal tar pitch as the binder meant improving the content of carbon in coal briquettes, so the calorific value of coal briquettes can be improved.

The results of industrial analysis of coal (see Table I) indicate that the volatile content of coal account for a high proportion, which can provide an enough reductive atmosphere. And the high calorific value of coal was 30.78MJ/kg (see

Table II), which can provide enough heat for reduction behavior.

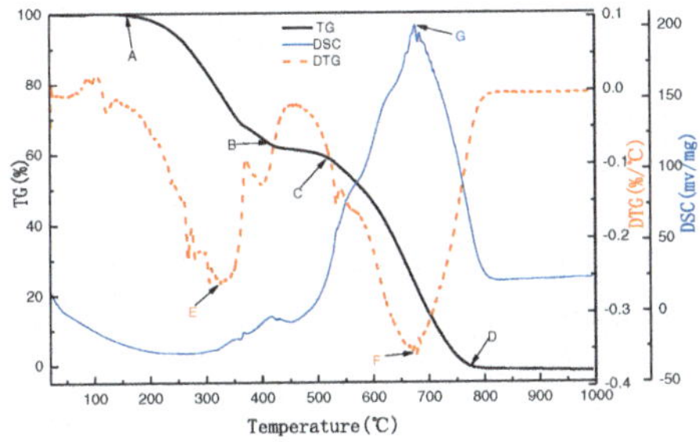

Fig 1. The thermogravimetric curve of coal tar pitch in nitrogen atmosphere

Figure 1 shows the thermogravimetric curve[7,8] of coal tar pitch in nitrogen atmosphere. We can see from the figure 1 that there are two points that the weight lose the most, one lies at 327℃ and the other at 668℃. Around 325℃, the decomposition speed of coal tar pitch is fast, the low molecular compounds of coal tar pitch evaporates under the effect of heat performance. After 668℃, the coal tar pitch has already turned into carbocoal and starts to change into bonding coke.

Experimental Method

The compression molding method was adopted to form coal briquettes, the material was compressed into a Φ20×20mm briquette through a Φ20×250mm cylindrical mold. Under certain pressure, displacement and deformation happened between material particles, and the briquette was compacted with the mechanical occlusion and interweave between the particles. Figure 2 was the schematic diagram of compression molding.

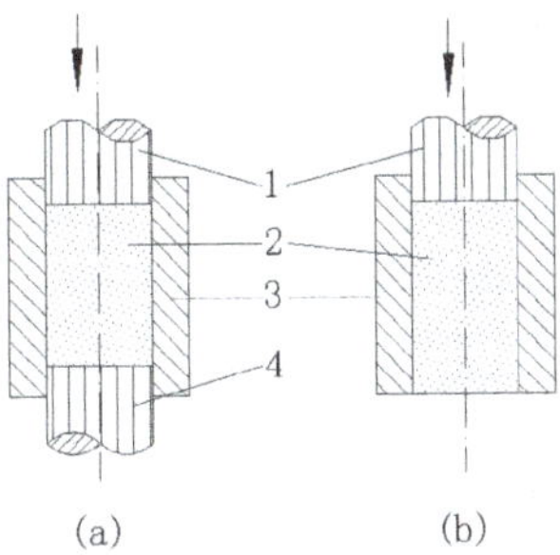

Fig 2. The schematic diagram of compression molding.
(a) One-way molded (b)two-way molded
1-top plunger; 2-material; 3-mould; 4-down plunger

The carbonization of coal briquettes was conducted in the shaft furnace with the protection of nitrogen atmosphere.

Results and discussions

Activation of BJ Additive in Briquetting Process

BJ was constantly used as solvent, the main effect of BJ was to activate cold state cohesiveness of the coal tar pitch and make it bond the coke particles to a firm briquette at room temperature .

When the briquetting pressure was 25MPa , the addition of water and coal tar pitch was 9% and 13% respectively. The effect of BJ on briquetting was shown in Table III.

Table III. Effect of BJ on Coal Briquettes

BJ / ml •100material^{-1}	Drop strength of wet briquettes/(1meter each)	Drop strength of dry briquettes/(2meters each)	Compressive strength of dry briquettes/MPa
0	0	0.58	1.06
1.2	0.14	2.67	2.28
1.8	20.00	5.00	3.66
2.4	37.29	6.13	4.71
3.0	38.36	6.43	4.56

It is clearly shown in Table III that the briquette strength increases with the increase of BJ. The strength increases fast during the period of 0-2.4ml • 100material^{-1}. When the BJ adding amount is 1.2 ml • 100material^{-1}, the drop strength of wet briquettes and dry briquettes is 0.14 times 1 meter each and 2.67 times 2 meter each, the compressive strength of dry briquettes is 2.28 MPa. After adding BJ to 2.4 ml • 100material^{-1}, the drop strength of wet briquettes and dry briquettes is 37.29 times 1 meter each and 6.13 times 2 meter each, the compressive strength of dry briquettes is 4.71 MPa. But there is no obvious increasing with more addition of BJ.

Effect of Coal Tar Pitch Binder on Briquetting

Coal tar pitch, as the binder, has the effect of infiltration and binding, so the strength of coal briquettes is closely related to the dosage of coal tar pitch.

When the briquetting pressure was 25MPa, the addition of water and BJ was 9% and 2.4ml.100material^{-1} respectively. The effect of coal tar pitch on briquetting was shown in Table IV.

It is obvious that with the addition of coal tar pitch increasing from 7% to 19 %, the drop strength of both wet and dry briquettes see a substantial upward trend, rising from 10.68 to over 50 times/1meter each and from 0.13 to 22.13 times/2meters each, respectively. Similarly, the compressive strength of dry briquettes increases steadily from 1.42 to 6.68 MPa, but when the addition of coal tar pitch amounts to 13%, this growth tendency declines gradually. In consideration of excessive binder being uneconomical and not environmentally-friendly, so the optimum addition of coal tar pitch is 13% and then the drop strength of wet and dry briquettes reaches 37.29 times/1meter each and 6.13 times/2meters each, respectively. At the same time, the compressive strength of dry briquettes arrives at 4.71MPa.

Table IV. Effect of Coal Tar Pitch Content on Coal Briquettes

Coal Tar Pitch /%	Drop strength of wet briquettes/(1meter each)	Drop strength of dry briquettes/(2meters each)	Compressive strength of dry briquettes/MPa
7	10.68	0.13	1.42
10	26.36	1.75	2.22
13	37.29	6.13	4.71
16	>50	11.75	5.86
19	>50	22.13	6.68

Effect of Time on Carbonization[9] Process

When the coal briquettes were carbonized, it was necessary to keep a certain amount of time under the carbonization temperature. After the pyrolysis condensation of coal and coal tar pitch, carbocoal was formed, and it still needs time to change into bonding coke.

After briquetting at optimum condition, the briquettes were carbonized at 700℃, the effect of time on carbonization was shown in Table V.

Table V. Effect of Time on the Strength of Carbonized Briquettes

Time/min	Drop strength of carbonized briquettes/(2meters each)	Compressive strength of carbonized briquettes/MPa
15	1.0	1.27
30	9.38	3.17
60	27.63	4.31
120	30.43	4.42
180	35.76	5.05
240	40.65	5.11

It is clearly shown in Table V that the strength carbonized briquettes increases when carbonization time is prolonged. The strength increases fast in the initial 15-60min, but after that this tendency declines obviously and then flattens out. When carbonized for 1h, the drop strength of briquettes is 27.63 times 2meters each and the compressive strength of briquettes is 4.31MPa. Considering that the extension of time will increase the cost, so carbonizing for 1h is the best.

Effect of Temperature on Carbonization Process

Due to the transformation of coal and coal tar pitch needs a certain temperature as a guarantee, so the carbonization temperature is also a very important condition for the carbonization behaviour.

When the carbonization time was fixed at 60min, the effect of temperature on carbonization was shown in Table VI.

Table VI. Effect of Temperature on the Strength of Carbonized Briquettes

Temperature/℃	Drop strength of carbonized briquettes/(2meters each)	Compressive strength of carbonized briquettes/MPa
500	13.5	2.22
600	27.6	5.09
700	20.5	5.36
800	16.2	5.10
900	20.0	5.19
1000	24.4	5.43

It is clearly shown in Table VI that when carbonized at 500℃, the drop strength of briquettes is 13.5 times 2meters each and the compressive strength of briquettes is 2.22 MPa, when the temperature rised to 600℃, the strength of briquettes is 27.6 times 2meters each and 5.09 MPa respectively. After that, the tendency flattens out, but the data shows a trend of fluctuations. As we can see, the drop strength of briquettes at 700℃ is 20.5 times 2meters each and the compressive strength of briquettes is 5.36 MPa, in order to be economical, the appropriate carbonizing temperature is 700℃.

Effect of Briquetting Pressure on Carbonization Process[10]

Table VII. Effect of Briquetting Pressure on the Strength of Carbonized Briquettes

Pressure/MPa	Drop strength of carbonized briquettes/(2meters each)	Compressive strength of carbonized briquettes/MPa
20	21.5	4.25
25	27.6	4.31
30	20.5	4.91
35	26.2	5.62
40	30.0	6.43
45	24.4	4.64
50	23.1	4.48

The briquetting pressure has a great impact on carbonization because the pressure is related with the tightness of the briquettes.

When the carbonization time and temperature was fixed at 60min and 700℃ respectively, the effect of briquetting pressure on carbonization was shown in Table VII.

As shown in the Table VII that the compressive strength of carbonized briquette increases when the briquetting pressure increases before 40MPa. But after that the carbonized briquette strength decreases with the increase of briquetting pressure. So for the carbonization behaviour, the optimum briquetting pressure is 40 MPa.

Conclusions

[1]. The additive BJ can activate cold state cohesiveness of the coal tar pitch and make it bond the coke particles to a firm briquette at room temperature. Adding BJ can make coal tar pitch better mixed with coal powder, and it is beneficial for the improvement of coal briquettes strength.

[2]. The optimum condition of briquetting is 9% water, 13% coal tar pitch, 2.4ml •100material^{-1} BJ, the briquetting pressure is 40MPa. The optimum condition of carbonization is carbonized at 700 ℃ for 1h in nitrogen atmosphere.

[3]. Under the optimum condition, the drop strength of briquettes can be 30.0 times 2meters each and the compressive strength of briquettes is 6.43 MPa

Acknowledgements

The authors gratefully acknowledge the financial supports from Baosteel Corporation of China as well as from post-doctoral workstation of Central South University.

References

[1]. Eberle, A. et al, "Start-up and Operating Results of POSCO COREX Plant," *Iron Steel Eng*, 1998, 75, (1), 25–28.

[2]. "Coals and Coal Requirements for the COREX Process," *Fuel and Energy Abstracts*, Volume 38, Issue 4, July 1997, Page 255.

[3]. Ayse Benk,Muzaffer Talu,"Phenolic Resin Binder for the Production of Metallurgical Quality Briquettes from Coke Breeze," *Fuel Processing Technology*.

[4]. Ayse Benk," Utilisation of the Binders Prepared from Coal Tar Pitch and Phenolic Resins for the Production Metallurgical Quality Briquettes from Coke Breeze and the Study of Their High Temperature Carbonization Behavior," *Fuel Processing Technology*.

[5]. Y.B. Yang et al., "The Preparation of High Strength Briquettes and Permeability Mechanism," *Proceedings of the 5th International Congress on the Science and Technology of Ironmaking*.

[6]. W.Zhang Jan T.Anderson,"Molecular Characterization of Large Polycyclic Aromatic Hydrocarbons in Solid Petroleum Pitch and Coal Tar Pitch by High Resolution MALDI ToF MS and Insights from Ion Mobility Separation," *Carbon,* Volume 95, December 2015, Pages 672-680.
[7]. J.Liu Y.F Pan," Determination of Ash Content and Concomitant Acquisition of Cell Compositions in Microalgae Via Thermogravimetric (TG) Analysis," *Algal Research,* Volume 12, November 2015, Pages 149-155.
[8]. F.J. Zhao, "Thermogravimetric Analysis and Application of Coal" *Fuxin, College of Resource and Environment Engineering*, 2005.
[9]. C.K Liu Y Feng," Effect of Carbonization Temperature on Properties of Aligned Electrospun Polyacrylonitrile Carbon Nanofibers," *Materials & Design,* Volume 85, 15 November 2015, Pages 483-486.
[10]. X.P. Xu, "Effect of Thickness of Plastic Layer and Final Shrinkage on Coke Quality," *Yangzhou Municipal Gas Company,* 1996.

7th International Symposium on High-Temperature Metallurgical Processing
Edited by: Jiann-Yang Hwang, Tao Jiang, P. Chris Pistorius, Gerardo R.F. Alvear F., Onuralp Yücel, Liyuan Cai, Baojun Zhao, Dean Gregurek, and Varadarajan Seshadri
TMS (The Minerals, Metals & Materials Society), 2016

Upgrading of Iron-Rich Titanium Minerals Using a Molten Salt Process

Farzin Fatollahi-Fard[1] and Petrus Christiaan Pistorius[1]

[1]Materials Science and Engineering, Carnegie Mellon University
5000 Forbes Avenue, Pittsburgh, PA 15213

Keywords: Titanium, Synthetic Rutile, Ilmenite, Molten Salt

Abstract

The production of synthetic rutile—a common starting material for most titanium production processes—from iron-rich titanium minerals such as pseudorutile and ilmenite is described. The process for "upgrading" the mineral to synthetic rutile is by carbothermal reduction of the mineral while immersed in a molten salt at high temperature, followed by magnetic separation of the reduced iron. A variety of factors are also investigated, such as particle size of the mineral, the type of mineral, and the molten salt used. The solubility of iron(III) oxide in different molten salts is also experimentally determined.

Introduction

There are many different types of titanium minerals, the most common of which is ilmenite ($FeTiO_3$). However, almost all processes which use titanium minerals require relatively high purity. Therefore, these minerals need to be "upgraded", to remove most of the iron. The most common processes used are smelting of the ilmenite to produce metallic iron and a titanium rich slag (the slag process), or carbothermal reduction followed by a chemical leach (the most common of this type is the Becher process).

The process for producing titanium slag involves melting the ilmenite coupled with reduction [1]. This yields molten iron, along with a titanium-rich slag containing 75-85% TiO_2. The resulting slag is often described as having the chemical formula of M_3O_5. The reaction is summarized in reaction (1) below. The resulting slag can then be used in the sulfate process for making pigments, or when sufficiently low in alkali earths can be used directly in the chloride process.

$$\mathrm{FeTiO_3} + x\,\mathrm{C} \longrightarrow \mathrm{M_3O_5} + \mathrm{Fe} + x\,\mathrm{CO}(\mathrm{M{=}Ti, Fe, Mn, ...}) \qquad (1)$$

Alternatively, the Becher process produces a product often referred to as synthetic rutile, as it often contains a similar impurity content to that of many natural sources of mineral rutile (TiO_2). This upgraded material can then be used directly in processes such as chlorination. To produce synthetic rutile, ilmenite is carbothermally reduced such that all of the iron oxide present in the ilmenite is reduced to metallic iron as in reaction (2). This iron can then be leached. The Becher process commonly uses a dilute solution of ammonium hydroxide in oxygenated water as the leaching agent, which causes the iron to be dissolved and form an oxide or hydroxide away from the titanium oxide, which can

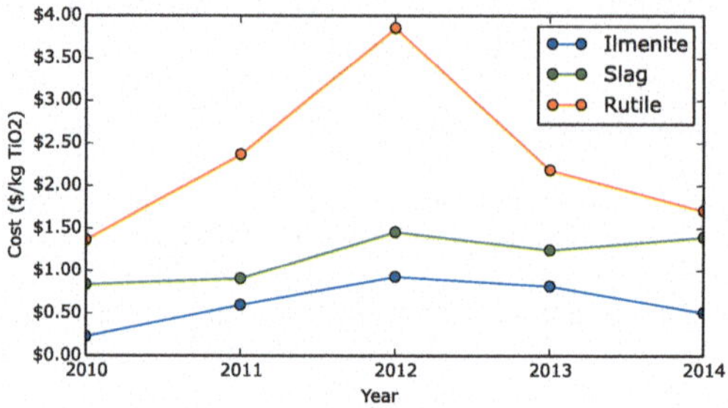

Figure 1: The cost of different titanium feedstocks from 2010-2014 [5].

then be removed, leaving behind porous synthetic rutile.

$$FeTiO_3 + C \longrightarrow TiO_2 + Fe + CO \tag{2}$$

Most processes for titanium mineral upgrading are one of the processes described previously or some variation thereof. However in the 1970s, Pollard and Stewart proposed a novel method for iron removal using a molten salt process [2, 3, 4]. This process involved using molten sodium chloride in conjunction with hydrogen chloride, carbon, and carbon monoxide to produce synthetic rutile and metallic iron. The belief of the authors was that the hydrogen chloride was responsible for the removal and reduction of the iron oxide from the ilmenite.

The difference in the prices of the different titanium feedstocks varies over time [5], as shown in Figure 1. However, the cost of ilmenite is always lower than that of slag and natural rutile. The price of synthetic rutile tends to be similar to that of natural rutile. If the cost of rutile—which may be readily used in the chloride process with few byproducts—can be lowered, this would be beneficial for producers of titanium tetrachloride (which is the feed for both pigment and titanium sponge production).

Methods

The raw materials used in this experiment were carbon black (Fisher Chemical) as the solid reductant, high-purity (99.995%) carbon monoxide and nitrogen gas for the atmosphere, iron(III) chloride tetrahydrate (Fisher Chemical) was used as the "HCl generator", and sodium chloride (Fisher Chemical) as the molten salt. We also used potassium chloride (Fisher Chemical) as a molten salt in some experiments when specified. A high-purity alumina crucible (Coorstek, 99.8%) was used in all experiments. The composition of the ilmenite materials used in these experiments is discussed in a later section.

In experiments which tested for particle size effects, the ore was ground to smaller particle sizes for 30 seconds using a "ring and puck" type tungsten carbide grinder (SPEX SamplePrep ShatterBox 8530). In experiments where the effect of oxidation state of the ore was tested, the ores were roasted in a box furnace for 24 hours in air at 850 °C.

Our initial experiments were meant to mirror those in the last patent by Pollard and Stewart [4]. Subsequent experiments were done without carbon monoxide and iron(III) chloride, to test the effect of removing these materials which were listed as necessary in the patents.

The standard procedure for leaching tests was to mix 1 gram of ilmenite, 1 gram of salt, 200 milligrams of carbon, and 30 milligrams of iron chloride (when applicable) in an alumina crucible. The crucible was then put into a vertical tube furnace, and taken to 900 °C at a rate of 15-20 °C/minute where it was held for 4 hours, and then cooled to room temperature at the same rate. Gas was flowed at low flow (300 mL/min) at a ratio of 2:1 nitrogen to carbon monoxide, or pure nitrogen.

For solubility tests, 50 milligrams of pseudorutile mineral sand was mixed with 50 grams of molten salt in an alumina crucible. The crucible was then put into a vertical tube furnace, and taken to 900 °C at a rate of 15-20 °C/minute where it was held for 1-24 hours, and then cooled to room temperature at the same rate. Nitrogen gas was flowed at low flow (200 mL/min) through the furnace throughout the course of the experiment.

To remove the sample from the solidified salt, the crucible was placed in a large bath of boiling water and mixed by magnetic stirring until the salt was fully dissolved. During this time, unreacted carbon could also be skimmed off the surface, when applicable. The sample was then filtered off for analysis. Optionally, if there was any material attached to the magnetic stir bar, this could also be kept for analysis.

Characterization of the samples was carried out by x-ray diffraction (XRD) using a PANalytical X'Pert Pro x-ray diffractometer. When unknown phases were encountered in the x-ray diffraction pattern, these were identified by QualX2 software [6]. Imaging was performed by an Aspex Explorer, Phillips XL30, or FEI Quanta 600 scanning electron microscope (SEM). Approximate compositions were determined by energy dispersive spectroscopy (EDS) on the SEM. In experiments which tested solubility, the SEM images were analyzed by ImageJ [7], where the leached and unleached areas were used to estimate the amount of iron dissolved.

Ore Materials

The ilmenites used in these experiments were a pseudorutile mineral sand from Florida, an ilmenite mineral sand and titaniferous slag of South African origin, and a synthetic ilmenite (Alfa Aesar, 99.8+% purity). The compositions are listed in Table I. The compositions were measured by x-ray florescence (XRF) spectrometry using a PANalytical Minipal 4, with the exception of the synthetic ilmenite, which was taken to be chemically pure.

Most of the experiments done were using the pseudorutile mineral sand. This was mostly due to the nature of the mineral sand itself, being a mixture of pseudorutile ($Fe_2Ti_3O_9$)

Table I: Compositions of all ores used in wt%.

	Pseudorutile (FL)	Ilmenite (SA)	Slag (SA)	Ilmenite (syn.)
TiO_2	80.5	47.0	81.8	52.6
FeO	15.9 (Fe_2O_3)	48.5	11.0	47.4
Al_2O_3	0.3	0.6	0.6	–
CaO	0.2	0.1	0.4	–
Cr_2O_3	0.4	0.4	0.3	–
MgO	<0.1	<0.1	0.1	–
MnO	0.9	1.1	1.9	–
SiO_2	1.6	2.2	3.7	–
ZrO_2	0.2	0.1	0.2	–

[8] and natural rutile. This allowed for easy identification of rutile relative to pseudorutile in SEM using a backscattered electron detector.

Results

Experiments with Pseudorutile Mineral Sand

Our initial experiment, which used the pseudorutile mineral sand as the starting material, along with all materials called for by Pollard and Stewart (carbon, carbon monoxide, iron(III) chloride tetrahydrate), showed that leaching did occur, as shown in Figure 2. The outer shell of the pseudorutile particle is porous, and has a lower contrast in backscatter imaging.

Subsequent experiments tested whether the presence of carbon monoxide has any effect on molten salt leaching, as indicated by the patent. These experiments showed no discernible difference to the case where carbon monoxide was present in the gas atmosphere.

The effect of particle size was also tested with ground pseudorutile (no carbon monoxide was used in the gas stream). The x-ray diffraction pattern of the resulting material is shown in Figure 3. The main peaks of rutile and pseudorutile are shown. The relative change in peak height indicates that the resulting material after leaching is more of a rutile than a pseudorutile. EDS measurements in SEM indicate that the overall iron content of the pseudorutile mineral sand has been reduced from 15.9 wt% to approximately 8 wt%.

Experiments with Ilmenite and Titaniferous Slags

Experiments were also performed on the South African ilmenite and titaniferous slag (with no CO in the atmosphere). However, they were entirely unaffected by the leaching conditions. This is likely because in these materials the iron is present mostly in divalent form, unlike the Fe(III) in pseudorutile.

The South African ilmenite mineral sand and the synthetic ilmenite powder were roasted

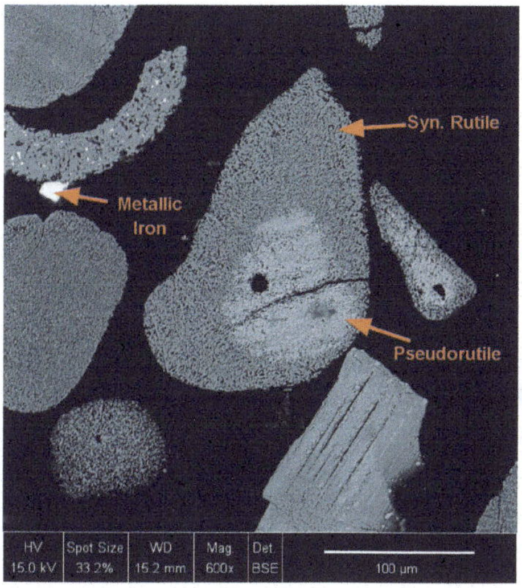

Figure 2: Backscattered SEM image of partially leached pseudorutile mineral sand.

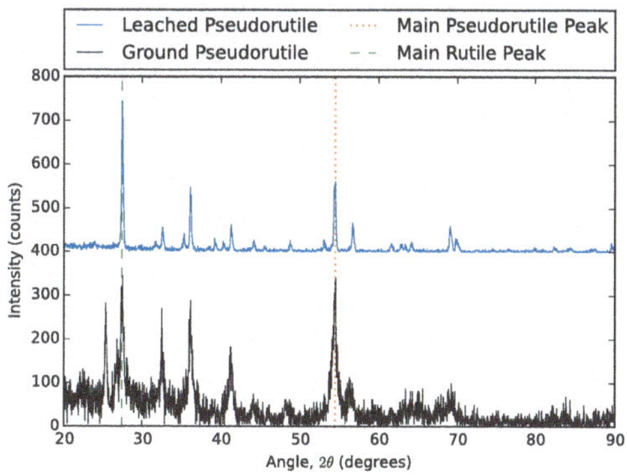

Figure 3: X-ray diffraction pattern of leached ground pseudorutile mineral sand.

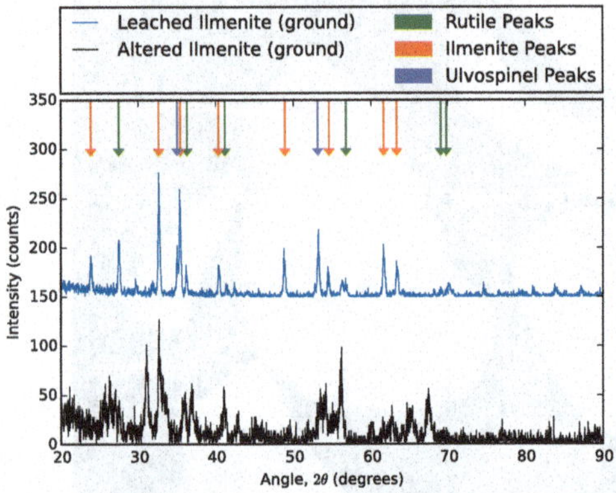

Figure 4: X-ray diffraction pattern of leached ground oxidized ilmenite.

Table II: Solubility of iron(III) oxide in different molten alkali chlorides at 900 °C.

Salt	Time (hours)	Solubility (ppm molar)
NaCl	1	17
NaCl	4	27
KCl	4	26
NaCl-KCl (eutectic)	24	43

(oxidized) in order to alter them into a higher oxidation state. In both cases, x-ray diffraction identified the predominant phases after roasting to be rutile (TiO_2), hematite (Fe_2O_3), and pseudobrookite (Fe_2TiO_5). These mineral sands were then ground to small particle sizes and used in the molten salt leaching conditions (no CO). X-ray diffraction of the resulting materials showed that both the altered natural and the synthetic ilmenite had formed ilmenite again, with a small amount of ulvospinel (Fe_2TiO_4). This is shown in Figure 4 for the synthetic ilmenite. It should however be noted that the back-reduction of ilmenite was by no means complete, and some synthetic rutile formed.

Solubility of Iron(III) Oxide in Molten Alkali Chlorides

In order to test the solubility of iron(III) oxide in chloride salts, a series of experiments were carried out to see how much iron(III) oxide would be leached under non-reducing conditions. This was done by immersing a small amount of pseudorutile mineral sand in a large volume of molten salts. The solubility was estimated based on the leached and unleached areas. Table II summarizes the results of these experiments.

Discussion

The experiments performed show that the process described by Pollard and Stewart is capable of leaching iron oxide from some iron-rich mineral sands. However, the experiments also showed that their proposed mechanism that a source of hydrogen chloride was necessary for this reaction to take place is not valid.

Based on the evidence from our experiments, we propose that the basis for the leaching is actually the fact that iron(III) oxide has some small solubility in both sodium and potassium chloride. As this iron is dissolved, it can be reduced by the carbon and precipitates as metallic iron, away from the titanium oxide particle. As this iron is no longer in solution, it allows for further dissolution of the iron(III) oxide from the mineral sand. Therefore—even though the solubility is low—given enough time synthetic rutile containing no iron(III) oxide will be produced.

The fact that the leaching had no effect on the South African ilmenite or titaniferous slags seems to suggest that iron(II) oxide in insoluble in sodium or potassium chloride or at least that the solubility is much lower than that of iron(III) oxide. However, this can be addressed by "roasting" the ores in air at high temperature. If the correct roasting temperature and time are used, then all of the iron(II) oxide should oxidize [9] to iron(III) oxide, which can then be leached by this process.

The behavior of the oxidized ilmenite in the molten salt leaching environment was initially surprising, but can be easily be explained by the highly reducing environment due to the amount of carbon present. Even in the pseudorutile, the reducing environment caused some very minor amount of metallic iron to form in the particle, as evidenced by the bright white spots in Figure 2. The ratio of carbon to mineral sands and molten salt was not varied from what was listed in the patent (10:2:10) [4]. However, that ratio of carbon to mineral sand is excessive. It is possible that a smaller amount of carbon, or a different kind of reductant other than solid carbon may make back-reduction to ilmenite less likely to occur.

In all cases the particles were never fully leached, although in the experiment with ground pseudorutile a reduction in iron content of approximately 50% was observed after 4 hours. We believe the diffusion of iron(III) oxide is limited by pore diffusion of dissolved iron oxide through the porous rutile product layer, which lowers the effective diffusivity. Therefore the leaching time (τ) can be reduced with an decreased particle size or higher solubility as given by equation 1 (where $C^* \approx 0.15$ kg/m^3 is the iron oxide concentration in the pseudorutile, $C \approx 1600$ kg/m^3 is maximum solubility or concentration in the salt, D_e is effective diffusivity, and $R \approx 150$ μm is particle radius). With our unground pseudorutile mineral sand, assuming a bulk diffusivity of $D = 1.8$ cm^2/s [10] and product porosity of $\epsilon = 0.15$ and $D_e = \epsilon^2 D$ we would estimate full dissolution to take approximately 72 hours.

$$\tau = \frac{C^* R^2}{6 D_e C} \tag{1}$$

Conclusions

It is possible to remove a significant amount of iron from an iron(III) oxide-containing titanium mineral sand using only a molten sodium or potassium chloride salt and carbon. This is mainly due to the (small) solubility of iron(III) oxide in those salts. Ilmenite can also be leached by this process, but only after roasting at a suitable temperature to oxidize the iron(II) oxide to iron(III) oxide. With these altered ilmenites, care should be taken to minimize the amount of carbon used in the molten salt as a highly reducing environment may cause the re-formation of ilmenite.

References

[1] G. Auer et al., "Pigments, Inorganic, 2. White Pigments," *Ullmann's Encyclopedia of Industrial Chemistry* (Weinheim: Wiley-VCH, 2012), 257–291.

[2] D. F. Stewart and L. J. Pollard, "Process of Beneficiating Titaniferous Ores in the Presence of Hydrogen Chloride," (Patent, US 3854929, 1974).

[3] D. F. Stewart and L. J. Pollard, "Process for Producing Metallic Iron Concentrates and Titanium Oxide Concentrates from Titaniferous Ores," (Patent, US 3816099, 1974).

[4] L. J. Pollard and D. F. Stewart, "Beneficiation of the Non-Ferrous Metal Values of Oxide-Containing Materials," (Patent, US 4047934, 1977).

[5] "USGS Mineral Commodity Summaries," (Report, Washington DC: U.S. Geological Survey, 2015).

[6] A. Altomare et al., "QUALX2.0 : a qualitative phase analysis software using the freely available database POW_COD," *Journal of Applied Crystallography*, 48 (2) (2015), 598–603.

[7] C. A. Schneider, W. S. Rasband, and K. W. Eliceiri, "NIH Image to ImageJ: 25 years of image analysis," *Nature Methods*, 9 (7) (2012), 671–675.

[8] I. E. Grey and J. A. Watts, "Mineralogical Nomenclature: Pseudorutile Revalidated and Neotype Given," *Mineralogical Magazine*, 58 (393) (1994), 597–600.

[9] X. Chen et al., "A simple oxidation route to prepare pseudobrookite from panzhihua raw ilmenite," *Journal of the American Ceramic Society*, 93 (10) (2010), 2968–2971.

[10] R. Brookes et al., "Diffusion Coefficients in Ionic Liquids: Relationship to the Viscosity," *The Journal of Physical Chemistry B*, 109 (14) (2005), 6485–6490.

7th International Symposium on High-Temperature Metallurgical Processing
Edited by: Jiann-Yang Hwang, Tao Jiang, P. Chris Pistorius, Gerardo R.F. Alvear F., Onuralp Yücel, Liyuan Cai, Baojun Zhao, Dean Gregurek, and Varadarajan Seshadri
TMS (The Minerals, Metals & Materials Society), 2016

DIRECT ELECTROLYTIC PRODUCTION OF Mo-Si-Ti-C COMPOSITES FROM THEIR OXIDES/SULFIDE/CARBON MIXTURE PRECURSOR IN MOLTEN SALT

Xingli Zou[1,*], Xionggang Lu[1,*], Qian Xu[1], Hongwei Cheng[1], Shuhua Geng[1], Zhongfu Zhou[1,2]

[1]State Key Laboratory of Advanced Special Steel, School of Materials Science and Engineering, Shanghai University, Shanghai 200072, P. R. China
[2]Institute of Mathematics and Physics, Aberystwyth University, Aberystwyth SY23 3BZ, UK
*Corresponding authors: Tel. & Fax: +86-21-56335768. E-mail addresses: xinglizou@shu.edu.cn (X. Zou); luxg@shu.edu.cn (X. Lu)

Keywords: Electroreduction, Molten salt, Composites, Sulfides, Oxides

Abstract

Herein we report an electrochemical route from oxides (SiO_2, TiO_2)/sulfide (MoS_2)/carbon mixture precursors to Mo-Si-Ti-C composites in molten salt. The electrochemical process was carried out in molten $CaCl_2$ at 800 °C with a potential of 3.1 V, and a graphite-based anode was used in the electroreduction process. The reaction mechanism and the influences of the stoichiometry of the initial mixture precursor on the morphology and phase composition of the final products were investigated. It has been found that Mo_5Si_3/$MoSi_2$/Mo_3Si/Mo, Mo_5Si_3/$MoSi_2$/Mo_2C/SiC, and Mo_5Si_3/Mo_2C/Ti_3SiC_2/TiC/Mo composites can be electrosynthesized directly from their oxides/sulfide/carbon mixture. The preliminary experimental results suggest that the electroreduction process can be used for the controllable electrolytic production of alloys/carbide-containing composites from oxides/sulfides/carbon precursors.

Introduction

Ti-Si and Mo-Si intermetallic compounds (such as Ti_5Si_3, $MoSi_2$, Mo_5Si_3, etc.) and their carbide-containing composites (such as Ti_5Si_3/TiC, $MoSi_2$/TiC, etc.) are regarded as high temperature structural materials with many unique properties such as low density, high melting temperature, and high oxidation resistance [1-13]. Especially, the carbides (e.g., SiC, TiC, etc.) reinforced Ti_5Si_3 and $MoSi_2$ matrix composites exhibit significant improved properties and have been intensively investigated [1-12]. Numerous approaches including spark plasma sintering [2], microwave reaction sintering [4], and combustion synthesis [5] have been applied to produce carbide-containing composites in recent years. However, these conventional methods typically involve ultra-pure Ti and Si elements, which are both generally produced by costly method.

Recently, direct electroreduction of metal oxides to metals/alloys in molten salts has been intensively investigated [13-16]. The direct molten salt electroreduction method offers a simple route to prepare metal carbide-containing composites directly from their corresponding metal oxides/carbon precursors [12]. More importantly, it has been proved that the direct molten salt electroreduction process can be used to reduce metal sulfide to produce metal in molten salt [17]. Therefore, the direct molten salt electroreduction process may also have potential to be used to reduce oxides/sulfide/carbon mixture precursors to prepare alloys/composites.

In this paper, sulfide (MoS_2), oxides (TiO_2 and SiO_2), and carbon have been used as starting materials to produce Mo-Ti-Si-C composites by using the molten salt electroreduction process. The variations of phase composition during electroreduction process and the microstructure of the final products have been investigated systematically.

Experimental

In this experiment, sulfide MoS_2, oxides TiO_2, SiO_2, and carbon powders were mixed and served as starting materials. These chemicals were analytical reagents and purchased from Sinopharm Chemical Reagent Co., Ltd (SCRC), China. All chemicals were used as received without further purification. To investigate the electrosynthesis of Mo-Si-Ti-C composites systematically, the electroreduction of MoS_2/SiO_2, $MoS_2/SiO_2/C$, and $MoS_2/SiO_2/TiO_2/C$ mixtures precursors was carried out in molten $CaCl_2$, respectively. These mixtures (MoS_2/SiO_2; $MoS_2/SiO_2/C$; and $MoS_2/SiO_2/TiO_2/C$) precursors with approximately 2 wt. % polyvinyl butyral binder were mixed at stoichiometric molar ratios ($MoS_2:SiO_2$ = 1:2; $MoS_2:SiO_2:C$ = 1:3:1; $MoS_2:SiO_2:TiO_2:C$ = 1:2:1:1) and homogenized by ball milling. The milled mixtures were pressed under 8-12 MPa to form cylindrical pellets and then assembled to form a cathode. A graphite-based anode (graphite rod) was used during the electroreduction process. The electroreduction experiments were carried out in molten $CaCl_2$ at 800 °C and 3.1 V. A Biologic HCP-803 electrochemical workstation was used to control the experiment and record current-time curve. The schematic of the electrolytic cell and more experimental details can be found in our previous work [12].
After experiment, the electroreduced products were washed with water and dried in air. The obtained products were characterized by X-ray diffraction (XRD) on a Rigaku D/Max-2550 diffractometer using Cu Kα radiation (λ = 0.15406 nm). The microstructure of the products was characterized using a JEOL JSM-6700F scanning electron microscope (SEM), and the elemental composition was analyzed by energy-dispersive X-ray spectroscopy (EDS) attached to the SEM.

Results and discussion

The current-time curves recorded during the electroreduction processes for MoS_2/SiO_2, $MoS_2/SiO_2/C$, and $MoS_2/SiO_2/TiO_2/C$ are similar, therefore, only a typical current-time curve of the electroreduction of MoS_2/SiO_2 in molten $CaCl_2$ at 800 °C and 3.1 V for 12 h is shown in Figure 1. It is observed that the variation trend of the current-time curve is well consistent with the explanation for the three-phase-interlines reaction mechanism during electroreduction process [15, 16]. The current-time curve shows that the residual current after 12 h electroreduction is approximately 0.5 A. After being electroreduced for 12 h, the XRD patterns (see Figure 2) show that MoS_2/SiO_2 precursor has been completely electroreduced to Mo-Si composites. The partial enlargements (Figure 2b and c) show that the intermediate products Ca_xMO_y (such as $CaSiO_3$) present during the electroreduction process. After being electroreduced for 6 h, the intermediate products disappear and Mo-Si composites ($Mo_5Si_3/MoSi_2/Mo_3Si/Mo$) products are formed. It should be noted that Mo_3Si occurs firstly during the electroreduction process and then converts to Mo_5Si_3 gradually. Mo_5Si_3 is the dominate phase of the final products. The morphology and elemental distribution of the final products are shown in Figure 3. It is obvious that the products show porous nodule structure with smooth particle surfaces. The elemental maps of the obtained final products (as shown in Figure 3c and d) suggest that Mo and Si elements distribute uniformly in the obtained Mo-Si composites.

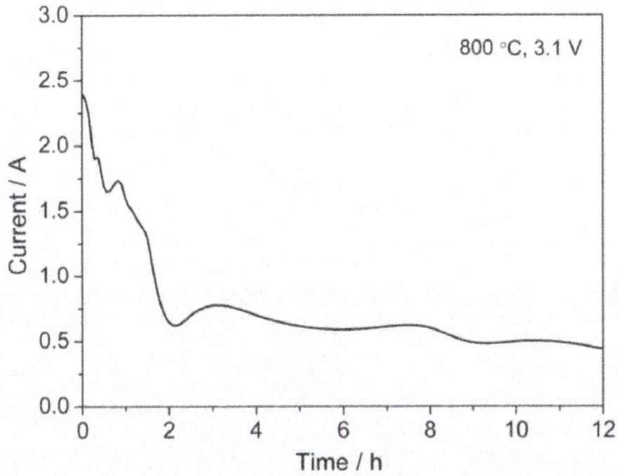

Figure 1. Typical current-time curve of the electroreduction of MoS_2/SiO_2 mixture precursor in molten salt.

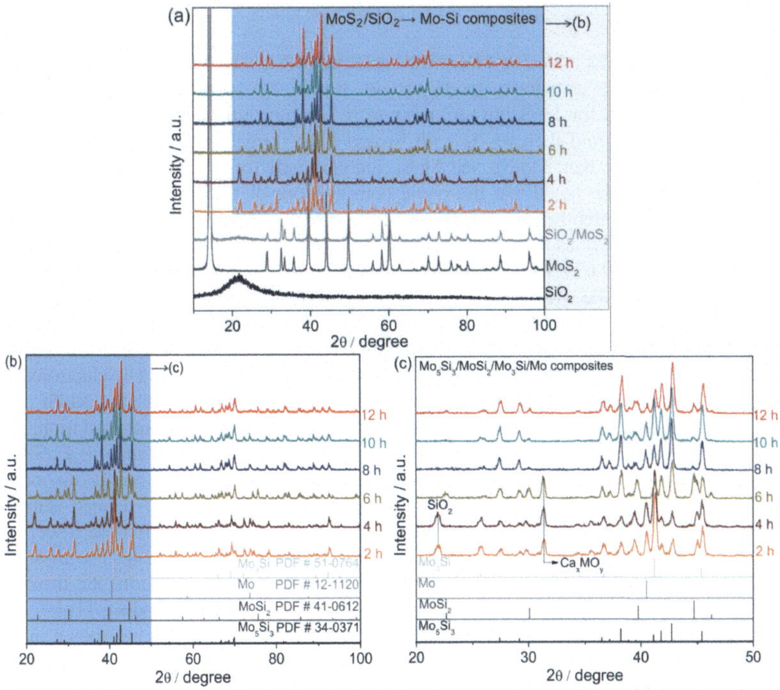

Figure 2. XRD patterns of SiO_2, MoS_2, SiO_2/MoS_2 precursors and the obtained products electroreduced at 800 °C and 3.1 V for different times in molten $CaCl_2$, (b) and (c) are the partial enlargements of (a).

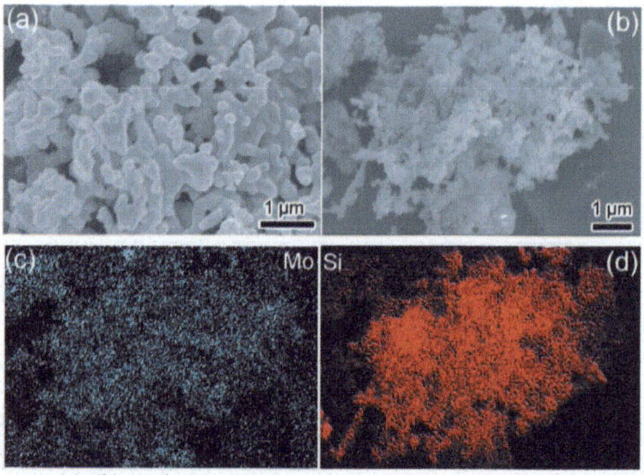

Figure 3. SEM images (a) (b) and elemental maps (c) (d) (corresponding to (b)) of the obtained final Mo-Si composites products.

While carbon powder was added in MoS_2/SiO_2 mixture precursors, the electroreduced products contain metal carbides, as shown in Figure 4. It can be seen from Figure 4 that the electroreduction process for $MoS_2/SiO_2/C$ mixture is similar with that for MoS_2/SiO_2 mixture. Mo is generated firstly and then reacts with Si and C to form Mo_3Si, Mo_2C, SiC during the electroreduction process, then Mo_5Si_3 and $MoSi_2$ are gradually formed as the dominate phases in the products. It should be noted that the phase compositions of the products obtained at 8, 10, and 12 h are similar, which mainly consist of Mo_5Si_3, $MoSi_2$, Mo_2C, and SiC phases. Notably, due to the Gibbs free energies for the reactions Si + C → SiC (ΔG = -63.208 kJ/mol) and 2Mo + C → Mo_2C (ΔG = -59.079 kJ/mol) at 800 °C are close, SiC and Mo_2C can be generated simultaneously during the electroreduction/carbonization processes. Actually, metal carbides such as HfC and carbide-containing composites such as Ti_5Si_3/TiC have been successfully electrosynthesized from their metal oxides precursors in previous work [12, 18]. It was suggested that metal carbides and composites can be synthesized using the molten salt electroreduction process. The final products obtained in this experiment typically contain Mo_5Si_3, $MoSi_2$, Mo_2C, and SiC phases, which imply that the electroreduction process should be controlled more strictly to produce $MoSi_2$/SiC composites. It should be pointed out that Ca_xMO_y such as $CaSiO_3$ and $CaTiO_3$ were usually observed as intermediate products during the electroreduction of metal oxides such as SiO_2 and TiO_2 [12, 13, 15, 16, 19]. However, $CaSiO_3$ formed during the electroreduction process can dissolve in molten $CaCl_2$, thus the dissolution-electrodeposition mechanism coexists with the solid-to-solid electroreduction mechanism [19]. The dissolution-electrodeposition process may cause a partial loss of element Si. In addition, the dissolution-electrodeposition mechanism can also influence the microstructure of the products [16].

SEM images and elemental maps of the final Mo-Si-C composites products are shown in Figure 5. It can be seen that that the morphology of the obtained Mo-Si-C composites products ($Mo_5Si_3/MoSi_2/Mo_2C$/SiC) is similar with that of the Mo-Si composites products (Figure 3). The particles exhibit smooth surfaces, and elements Mo, Si, and C distribute uniformly in the products. Compare to other conventional methods used to synthesize composites, the molten salt electroreduction process can produce composites with uniform elemental distribution.

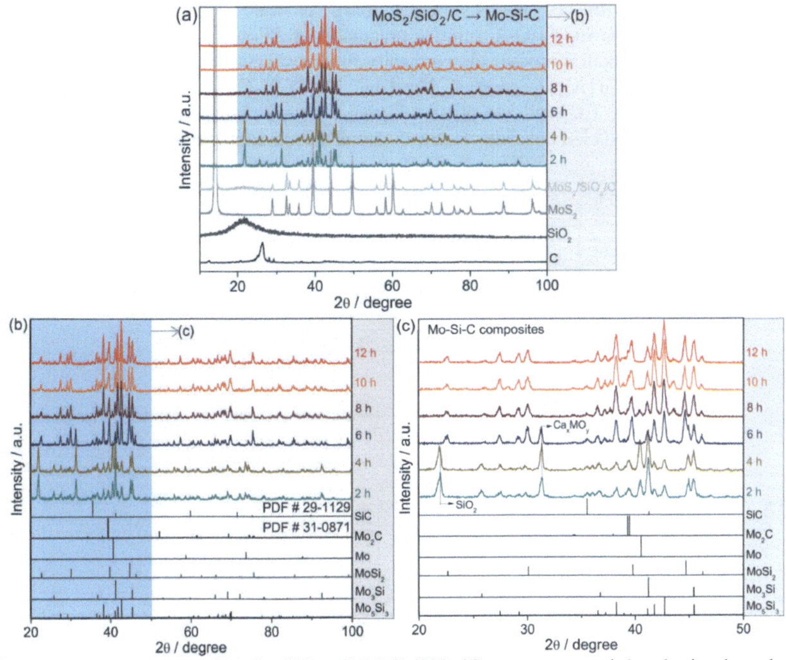

Figure 4. XRD patterns of MoS_2, SiO_2, C, $MoS_2/SiO_2/C$ precursors and the obtained products electroreduced at 800 °C and 3.1 V for different times in molten $CaCl_2$, (b) and (c) are the partial enlargements of (a).

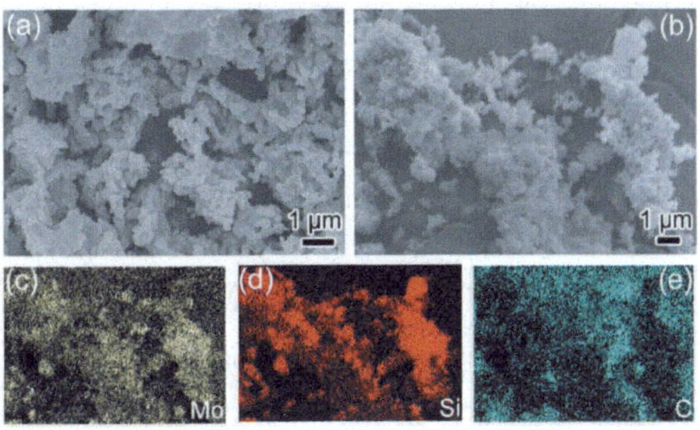

Figure 5. SEM images (a) (b) and elemental maps (c)-(e) (corresponding to (b)) of the obtained final Mo-Si-C composites products.

XRD patterns, SEM images, and elemental maps of the products obtained from the electroreduction of $MoS_2/SiO_2/TiO_2/C$ mixture precursors are shown in Figures 6 and 7. It can be seen that the electroreduction process also typically involves electroreduction and compounding (e.g., $TiO_2 + CaO = CaTiO_3$) processes. The final Mo-Si-Ti-C composites products contain Mo_5Si_3, Mo_2C, Ti_3SiC_2, TiC, and Mo phases. Figure 7 shows that the products is porous with micro/nanostructures and elements Mo, Ti, Si, C also distribute uniformly in the products.

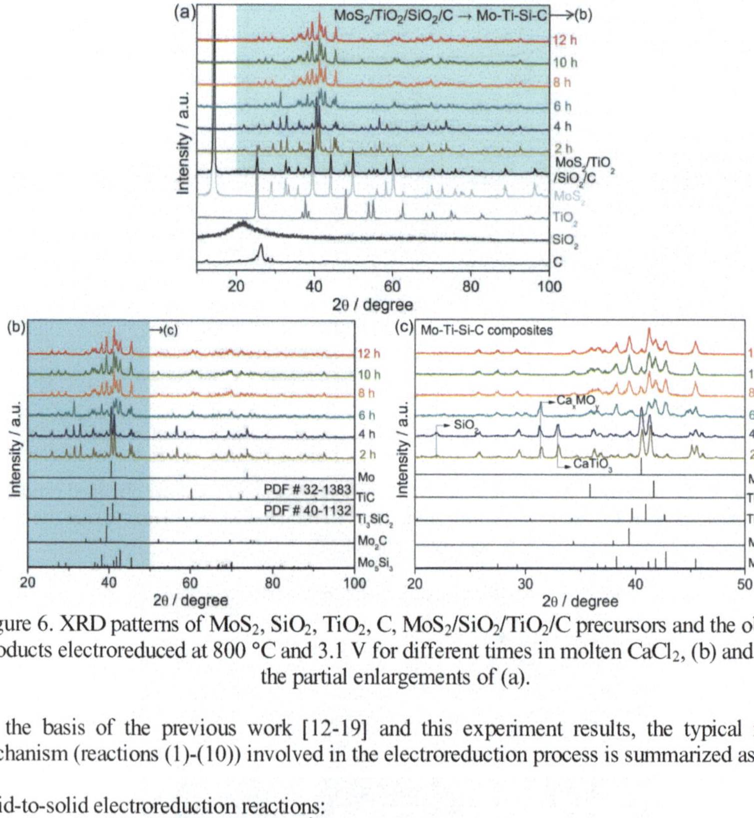

Figure 6. XRD patterns of MoS_2, SiO_2, TiO_2, C, $MoS_2/SiO_2/TiO_2/C$ precursors and the obtained products electroreduced at 800 °C and 3.1 V for different times in molten $CaCl_2$, (b) and (c) are the partial enlargements of (a).

On the basis of the previous work [12-19] and this experiment results, the typical reaction mechanism (reactions (1)-(10)) involved in the electroreduction process is summarized as follow.

Solid-to-solid electroreduction reactions:

$$MoS_2 + 4e^- \rightarrow Mo + 2S^{2-} \quad (1)$$

$$TiO_2 \,/\, CaTiO_3 + 4e^- \rightarrow Ti + 2O^{2-} \,/\, + CaO\,(Ca^{2+}, O^{2-}) \quad (2)$$

$$SiO_2 \,/\, CaSiO_3 + 4e^- \rightarrow Si + 2O^{2-} \,/\, + CaO\,(Ca^{2+}, O^{2-}) \quad (3)$$

Electroreduction/carbonization reactions:

$$2MoS_2 + C + 8e^- \rightarrow Mo_2C + 4S^{2-} \quad (4)$$

$$TiO_2 \,/\, CaTiO_3 + C + 4e^- \rightarrow TiC + 2O^{2-} \,/\, + CaO\,(Ca^{2+}, O^{2-}) \quad (5)$$

$$SiO_2 \,/\, CaSiO_3 + C + 4e^- \rightarrow SiC + 2O^{2-} \,/\, + CaO\,(Ca^{2+}, O^{2-}) \quad (6)$$

Carbonization/combination reactions:

$$Ti / Si + C \rightarrow TiC / SiC \tag{7}$$
$$3Ti + Si + 2C \rightarrow Ti_3SiC_2 \tag{8}$$
$$2Mo + C \rightarrow Mo_2C \tag{9}$$
$$xMo + ySi \rightarrow Mo_xSi_y \tag{10}$$

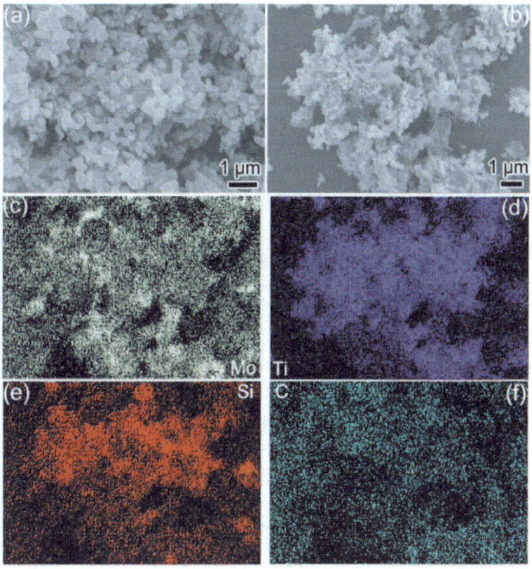

Figure 7. SEM images (a) (b) and elemental maps (c)-(f) (corresponding to (b)) of the obtained final Mo-Si-Ti-C composites products.

Conclusions

Direct electroreduction of MoS_2/SiO_2, $MoS_2/SiO_2/C$, and $MoS_2/SiO_2/TiO_2/C$ mixtures precursors to produce Mo-Si, Mo-Si-C, and Mo-Si-Ti-C composites in molten salt has been investigated in this work. The results show that the multicomponent $Mo_5Si_3/MoSi_2/Mo_3Si/Mo$, $Mo_5Si_3/MoSi_2/Mo_2C/SiC$, and $Mo_5Si_3/Mo_2C/Ti_3SiC_2/TiC/Mo$ composites can be obtained. The synthesized products typically exhibit porous micro/nanostructures with nodule particles, and elements Mo, Ti, Si, C distribute uniformly in the products. The molten salt electroreduction process has a promising to be extended for reduction of other sulfides/oxides mixtures.

Acknowledgments

We appreciate the financial support from the National Natural Science Foundation of China (Nos. 51304132, 51574164), China National Funds for Distinguished Young Scientists (No. 51225401), the National Basic Research Program of China (No. 2014CB643403), the Science and Technology Commissions of Shanghai Municipality (No. 14JC1491400) and Sichuan Province (No. 2013GZ0146).

References

1. X. Yang, et al., "Influence of preparation technology on the structure and phase composition of $MoSi_2$-Mo_5Si_3/SiC multi-coating for carbon/carbon composites," *J. Mater. Sci. Technol.*, 26(2) (2010), 106-112.
2. L.J. Wang, et al., "Microstructure of Ti_5Si_3-TiC-Ti_3SiC_2 and Ti_5Si_3-TiC nanocomposites in situ synthesized by spark plasma sintering," *J. Mater. Res.*, (19) (2004), 3004-3008.
3. J.F. Huang, et al., "A $MoSi_2$/SiC oxidation protective coating for carbon/carbon composites," *Corros. Sci.*, (53) (2011), 834-839.
4. J.G. Xu, F. Chen, and F. Tan, "In-situ preparation of SiC-$MoSi_2$ composite by microwave reaction sintering," *Ceram. Int.*, (38) (2012), 6895-6898.
5. Q.D. Hu, P. Luo, and Y.W. Yan, "Microstructures and densification of $MoSi_2$-SiC composites by field-activated and pressure-assisted combustion synthesis," *J. Alloys Compd.*, (468) (2009), 136-142.
6. M. Patel, J. Subramanyuam, and V.V. Bhanu Prasad, "Synthesis and mechanical properties of nanocrystalline $MoSi_2$-SiC composite," *Scripta Mater.*, (58) (2008), 211-214.
7. L. Sun, and J.S. Pan, "Fabrication and characterization of TiC_w/$MoSi_2$ and SiC_w/$MoSi_2$ composites," *Mater. Lett.*, (52) (2002), 223-228.
8. J.H. Meng, et al., "Preparation and properties of $MoSi_2$ composites reinforced by TiC, TiCN, and TiB_2," *Mater. Sci. Eng. A*, (396) (2005), 277-284.
9. L. Sun, and J.S. Pan, "Fabrication and characterization of TiC-particle-reinforced $MoSi_2$ composites," *J. Eur. Ceram. Soc.*, (22) (2002), 791-796.
10. Q. Zhu, et al., "Oxidation behavior of $Mo_{\leq 5}Si_3C_{\leq 1}$ and its composites," *J. Mater. Sci.*, (35) (2000), 863-872.
11. G.N. Komratov, and V.P. Kobyakov, "Oxidation of SHS Mo-Si-C materials in air," *Inorg. Mater.*, (36) (2000), 337-342.
12. X.L. Zou, et al., "Direct electrosynthesis of Ti_5Si_3/TiC composites from their oxides/C precursors in molten calcium chloride," *Electrochem. Commun.*, (21) (2012), 9-13.
13. X.L. Zou, et al., "Electrochemical extraction of Ti_5Si_3 silicide from multicomponent Ti/Si-containing metal oxide compounds in molten salt," *J. Mater. Chem. A*, (2) (2014), 7421-7430.
14. G.Z. Chen, D.J. Fray, and T.W. Farthing, "Direct electrochemical reduction of titanium dioxide to titanium in molten calcium chloride," *Nature*, (407) (2000), 361-364.
15. A.M. Abdelkader, et al., "DC voltammetry of electro-deoxidation of solid oxides," *Chem. Rev.*, (113) (2013), 2863-2886.
16. W. Xiao, and D.H. Wang, "The electrochemical reduction processes of solid compounds in high temperature molten salts," *Chem. Soc. Rev.*, (43) (2014), 3215-3228.
17. G.M. Li, et al., "Electrolysis of solid MoS_2 in molten $CaCl_2$ for Mo extraction without CO_2 emission," *Electrochem. Commun.*, (9) (2007), 1951-1957.
18. A.M. Abdelkader, and D.J. Fray, "Electrochemical synthesis of hafnium carbide powder in molten chloride bath and its densification," *J. Eur. Ceram. Soc.*, (32) (2012), 4481-4487.
19. W. Xiao, et al., "Verification and implications of the dissolution-electrodeposition process during the electro-reduction of solid silica in molten $CaCl_2$," *RSC Adv.*, (2) (2012), 7588-7593.

7th International Symposium on High-Temperature Metallurgical Processing
Edited by: Jiann-Yang Hwang, Tao Jiang, P. Chris Pistorius, Gerardo R.F. Alvear F., Onuralp Yücel, Liyuan Cai, Baojun Zhao, Dean Gregurek, and Varadarajan Seshadri
TMS (The Minerals, Metals & Materials Society), 2016

REDUCTION KINETICS OF MAGNETITE CONCENTRATE PARTICLES WITH H_2 + CO AT 1200 TO 1600 °C RELEVANT TO A NOVEL IRONMAKING PROCESS

Mohamed Elzohiery[1], Yousef Mohassab[1], Jagannath Pal[1,2], Shengqin Zhang[1,3], and Hong Yong Sohn[1]

[1] Department of Metallurgical Engineering, University of Utah, Salt Lake City, Utah 84112, USA

[2] National Metallurgical Laboratory, Council of Scientific and Industrial Research, Jamshedpur 831007, India

[3] School of Metallurgy and Materials Engineering, Chongqing University of Science and Technology, Chongqing 401331, China

Keywords: Reduction Kinetics, Magnetite Concentrate, Hydrogen Reduction, Ironmaking, Flash Reaction, Natural Gas

Abstract

The kinetics of magnetite reduction with mixtures of hydrogen and CO of various compositions has been investigated as part of the development of a flash ironmaking process at the University of Utah. This new process bypasses the cokemaking and pelletization or sintering steps required for the blast furnace. The particle kinetics were studied using magnetite concentrate particles of different sizes. Reduction degree of 60% was achieved in a few seconds in at 1350 °C using CO reductant alone and over 90% when H_2 + CO gas mixture was used. The effects of reductant partial pressures, temperature, and particle size on the reduction kinetics were studied.

Introduction

A flash ironmaking process is being developed at the University of Utah [1-13] to produce iron from iron ore concentrate particles in a flash reactor in few seconds of reaction time. Using hydrogen as the reductant/fuel, magnetite concentrate and hematite concentrate particles were more than 90% reduced at temperatures as low as 1250 °C within a few seconds of residence time and under hydrogen partial pressures as low as 0.3 atm. The eventual objective of this process is to replace the blast furnace. Natural gas represents an economically viable reductant/fuel for the flash ironmaking. For developing and designing the flash reactor running by natural gas, the rate equation for the reduction kinetics should be obtained. The rate equation for reduction reaction with H_2 gas was obtained and in this work, experiments were performed with a H_2 + CO gas mixture with a molar ratios H_2:CO close to the industrial syngas. Typical magnetite reduction curves using the gas mixture will be presented compared with those using only H_2 or CO gas as the reductant.

Experimental Work

Three different size fractions of magnetite concentrate particles were used in this work. The size fractions were 20-25, 32-38 and 45-53 μm which used for determining the effect of the particle

size on the reduction kinetics. Two drop tube reactors (DTRs) were used in this work allowing for wide ranges of operating temperature and residence times. The first reactor was used for the experiments performed in temperature range 1150-1350 °C and the other for 1400-1600 °C. The overall schematic diagram of the DTRs used in this study is shown in Figure 1.
The water cooled feeding line was inserted into a ceramic tube to avoid any soot formation due to the catalytic effect of the copper tubing that enhances the reduction of CO by H_2 gas. Negligible amounts of soot were noticed during the experiments and it was only on the upper part of the reactor away from the reaction zone.

The experimental conditions were designed in a way to resemble the most used industrial ratios of the syngas such as: H_2:CO = 2:1, 1:1, and 0.5:1 (molar ratio). The relevant equations for calculating the nominal particles residence time, % excess hydrogen, excess driving force (EDF), reduction degree (RD%),and the length of the reaction zone are described in Chen et al. work [10, 11].

Nitrogen gas was flowed in the reactor during heating up until the target temperature was reached then the reducing gases were introduced according to the experiment conditions. Nitrogen gas was used to adjust the partial pressure of the reducing gases and the particle residence time. Gases flow rates were controlled using rotameter flow meters. Once all the gases are introduced and the reactor temperature was stable, magnetite concentrate was fed into the reactor at the specified feeding rate which maintained the excess reducing gases (both H_2 and CO) for reduction more than 400% of each gas to ensure negligible change in the gas concentration along the reaction zone for accurate determination of the reaction kinetics.

The reduced powder sample was collected in a continuously heated collection bin to prevent any water condensation in the sample during the experiment. After the experiment, the collection bin was cooled by water under N_2 atmosphere to avoid any re-oxidation before opening and collecting the sample.

Samples collected were analyzed by ICP-OES for total iron content to calculate the reduction degree. Further analysis was performed on the samples as SEM. The reproducibility of the analysis method was about ± 1 wt%. The overall experimental precision was within ± 5%.

Results

Reduction Kinetics in Temperature Range 1200-1350 °C

The total iron fraction of the samples was determined using ICP-OES method and the reduction degree was calculated for each experiment. Some experiments were performed with conditions to mimic the complete reforming of natural gas yielding H_2:CO ratio = 2:1 (in molar ratio). These experiments were performed using CO and H_2 gas mixture with no nitrogen. Complete reduction was achieved at temperatures as low as 1200 °C and 3 s particles residence time (these results are not shown here).

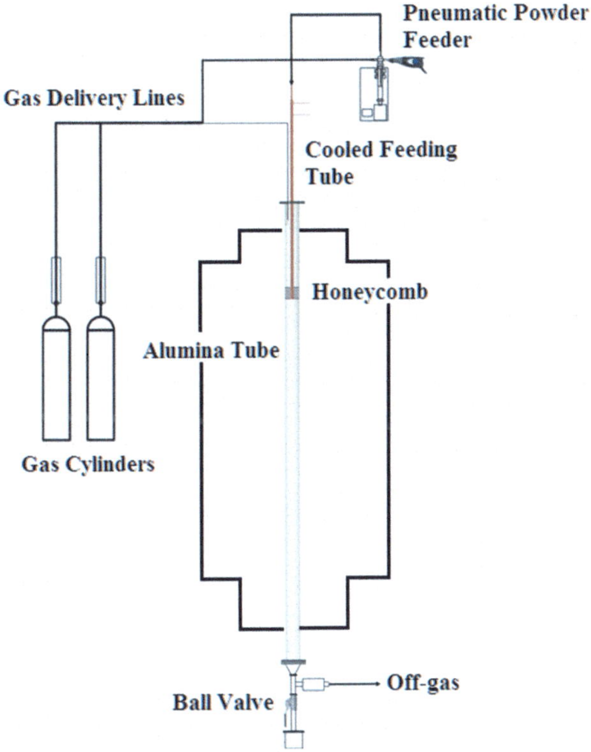

Figure 1. Schematic diagram for the drop tube reactor (DTR).

The results of the experiments were compared to the results obtained using only hydrogen as a reductant under otherwise similar conditions. More than 95% reduction degree was achieved at 1250 °C within 3 s of nominal particle residence time when using H_2 and CO partial pressures as low as 0.2 atm for each gas. Using only pure CO gas as the reductant, the maximum reduction degree achieved was 60% at 1350 °C within 6 s of reaction time. This implies the low contribution of CO to magnetite reduction. Adding CO to H_2 seemed to enhance the reduction kinetics. However at constant H_2 partial pressure and temperature, increasing CO partial pressure from 0.1 to 0.4 did not show any effect on the reduction kinetics. Although the overall contribution was not significant when comparing the results to the pure hydrogen experiments, as shown in Figure 2

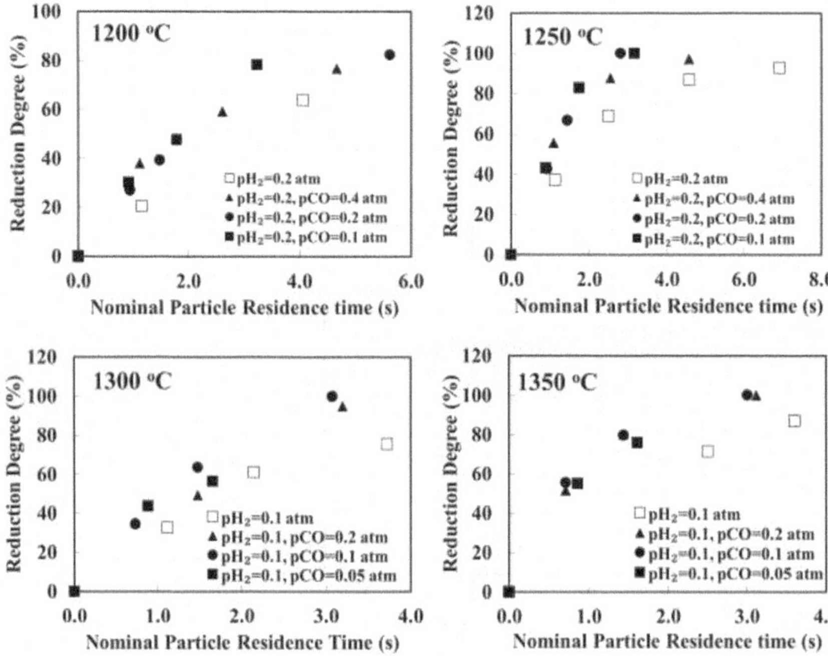

Figure 2. Effect of nominal particle residence time on the reduction degree of magnetite under different CO and H_2 partial pressures in the temperature range 1200-1350 °C.

Reduction Kinetics in Temperature Range 1400-1600 °C

At temperatures higher than 1350 °C, particles experience significant melting due to their gangue content. The experiments with pure reductants (no nitrogen was used) with the ratio H_2:CO = 2:1 showed complete reduction at 1400 °C within 2 s particles residence time. More than 95% could be achieved at 1400 °C, within 2 s using H_2 and CO partial pressures as low as 0.1 and 0.05 atm, respectively. The reduction degree for each sample was compared to the results of the pure hydrogen experiments. As shown in Figure 3, the contribution of CO to the reduction was more pronounced compared with its contribution in the temperature range 1200-1350 °C. The reduction degree for the mixture experiments were higher than that of the pure hydrogen experiments. The effect of increasing CO partial pressure at the same hydrogen partial pressure did not change much at the same hydrogen partial pressure as observed in the temperature range 1200-1350 °C.

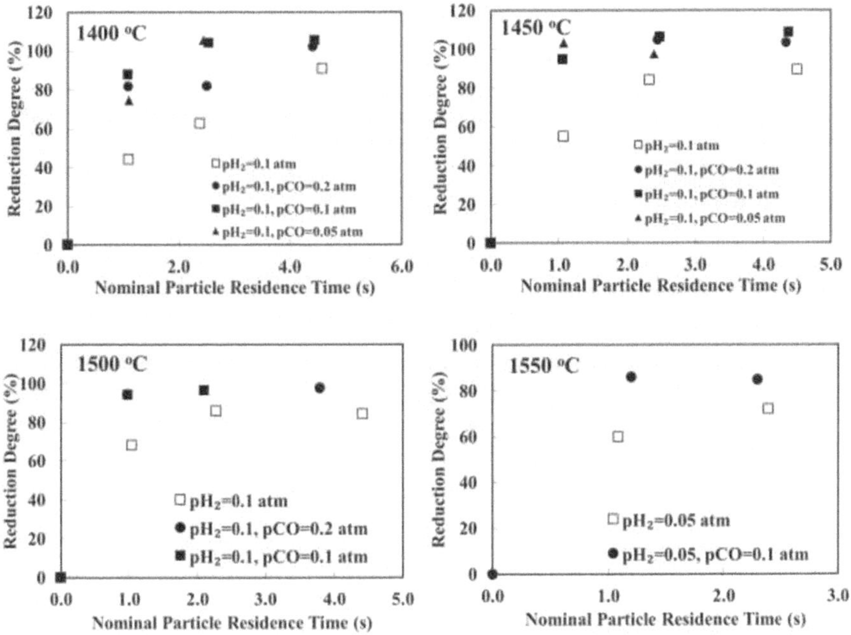

Figure 3. Effect of nominal particle residence time on the reduction degree of magnetite under different CO and H_2 partial pressures in the temperature range 1400-1550 °C.

Conclusions

The reduction kinetics of magnetite concentrate particles by H_2 and CO gas mixture was investigated in two temperature ranges: 1200-1350 °C and 1400-1600 °C. It was found that magnetite reaction with CO and H_2 gas mixture (with no nitrogen) with the molar ratio H_2:CO = 2:1 yielded complete reduction at temperatures 1200 and 1400 °C within 3 and 2 s of residence time, respectively. More than 95% reduction degree was achieved at 1250 °C within 3 s of nominal particle residence time when using H_2 and CO partial pressures as low as 0.2 atm for each gas. In the temperature range 1200-1350 °C, CO had low contribution compared to the H_2. The contribution of CO increased with increasing temperature. In temperature range 1400-1600 °C, CO had more significant contribution than at the lower temperature range. More than 95% could be achieved at 1400 °C, 2 s and H_2 and CO partial pressures as low as 0.1 and 0.05 atm, respectively. In both ranges, increasing the partial pressure of CO at the same H_2 partial pressure did not much affect the reduction degree. The overall enhancement in the reduction kinetics of magnetite by CO stayed small when compared with the reduction by hydrogen alone.

Acknowledgments

The authors thank Andrew Laroche, Omar Kergaye, Caio Melo, Jacqueline De Oliveira Cota, and Tuvshinbat Ganbat for the help with the experimental runs and the analytical work using ICP. The authors acknowledge the financial support from the U.S. Department of Energy under Award Number DEEE0005751 with cost share by the American Iron and Steel Institute (AISI) and the University of Utah.

Disclaimer: This report was prepared as an account of work sponsored by an agency of the United States Government. Neither the United States Government nor any agency thereof, nor any of their employees, makes any warranty, express or implied, or assumes any legal liability or responsibility for the accuracy, completeness, or usefulness of any information, apparatus, product, or process disclosed, or represents that its use would not infringe privately owned rights. Reference herein to any specific commercial product, process, or service by trade name, trademark, manufacturer, or otherwise does not necessarily constitute or imply its endorsement, recommendation, or favoring by the United States Government or any agency thereof. The views and opinions of authors expressed herein do not necessarily state or reflect those of the United States Government or any agency thereof.

References

1. M.E. Choi and H.Y. Sohn, "Development of Green Suspension Ironmaking Technology Based on Hydrogen Reduction of Iron Oxide Concentrate: Rate Measurements." *Ironmaking Steelmaking*, 37 (2010), 81-88.

2. M.Y. Mohassab-Ahmed and H. Y. Sohn, *Method and Device for Digestion of Materials in a Microwave Oven*, 2012, US Patent App. 61/651.

3. M.Y. Mohassab-Ahmed, H. Y. Sohn, "Effect of Water Vapor Content in H_2–H_2O–CO–CO_2 Mixtures on the Equilibrium Distribution of Manganese between CaO–MgO_{sat}–SiO_2–Al_2O_3–FeO–P_2O_5 Slag and Molten Iron." *Steel Res. Int.*, 85 (2014), 875-884.

4. M.Y. Mohassab Ahmed, "*Phase Equilibria between Iron and Slag in $CO/CO_2/H_2/H_2O$ Atmospheres Relevant to a Novel Flash Ironmaking Technology*" (PhD Dissertation, The University of Utah, 2013).

5. M.Y. Mohassab-Ahmed, H. Y. Sohn, "Effect of Water Vapor Content in H_2-H_2O-CO-CO_2 Mixtures on the Activity of Iron Oxide in Slags Relevant to a Novel Flash Ironmaking Technology." *Ironmaking Steelmaking*, 41(2014), 665 - 675.

6. Y. Mohassab and H. Sohn, "Effect of Water Vapour on Distribution of Phosphorus between Liquid Iron and MgO Saturated Slag Relevant to Flash Ironmaking Technology." *Ironmaking Steelmaking*, 41 (2014), 575-582.

7. Y. Mohassab and H.Y. Sohn, "Effect of Water Vapor on Sulfur Distribution between Liquid Fe and MgO-Saturated Slag Relevant to a Flash Ironmaking Technology." *Steel Res. Int.*, 86 (2014), 753–759.

8. Y. Mohassab and H.Y. Sohn, "Analysis of Slag Chemistry by FTIR-RAS and Raman Spectroscopy: Effect of Water Vapor Content in H_2-H_2O-CO-CO_2 Mixtures Relevant to a Novel Green Ironmaking Technology." *Steel Res. Int.*, 86 (2014), 740-752.

9. M.Y. Mohassab-Ahmed, H.Y. Sohn, and L. Zhu, "Effect of Water Vapour Content in H_2-H_2O-CO-CO_2 Mixtures on MgO Solubility in Slag under Conditions of Novel Flash Ironmaking Technology." *Ironmaking Steelmaking*, 41 (2014), 575-582.

10. F. Chen, Y. Mohassab, T. Jiang, and H.Y. Sohn, "Hydrogen Reduction Kinetics of Hematite Concentrate Particles Relevant to a Novel Flash Ironmaking Process." *Metall. Mater. Trans. B*, 46 (2015), 1133-1145.

11. F. Chen, Y. Mohassab, S. Zhang, and H.Y. Sohn, "Kinetics of the Reduction of Hematite Concentrate Particles by Carbon Monoxide Relevant to a Novel Flash Ironmaking Process." *Metall. Mater. Trans. B*, 46 (2015), 1716-1728.

12. H. Pinegar, M. Moats, and H. Sohn, "Flowsheet Development, Process Simulation and Economic Feasibility Analysis for Novel Suspension Ironmaking Technology Based on Natural Gas: Part 1-Flowsheet and Simulation for Ironmaking with Reformerless Natural Gas." *Ironmaking and Steelmaking*, 39 (2012), 398-408.

13. H. Pinegar, M. Moats, and H. Sohn, "Flowsheet Development, Process Simulation and Economic Feasibility Analysis for Novel Suspension Ironmaking Technology Based on Natural Gas: Part 2 – Flowsheet and Simulation for Ironmaking Combined with Steam Methane Reforming." *Ironmaking and Steelmaking*, 40 (2012), 44-49.

7th International Symposium on High-Temperature Metallurgical Processing
Edited by: Jiann-Yang Hwang, Tao Jiang, P. Chris Pistorius, Gerardo R.F. Alvear F., Onuralp Yücel, Liyuan Cai, Baojun Zhao, Dean Gregurek, and Varadarajan Seshadri
TMS (The Minerals, Metals & Materials Society), 2016

ADVANCED OXYGEN LANCES FOR SAFER FURNACE TAPPING OPERATIONS

Peter Sylvén[1], Darwin Morales[2]

[1]Envicom AB, Sweden, peter.sylven@envicomab.com
[2] Trefimet S.A., Chile, dmorales@trefimet.cl

Keywords: Taphole, Oxygen lance, Thermal lance, Oxygen cutting

Abstract

Despite being crucial for the smelting process, furnace tapping using oxygen lances has not been considered an operation that brings value. Therefore, these important tools have not seen a significant change in over 100 years. Problems related to the use of oxygen lances include:

- Inaccurate tool: causing damages shortening the life of the taphole refractories.
- Safety: extended exposure to the operator in an unsafe area of the furnace.
- Cost: high oxygen consumption & high lance consumption.
- Time consuming and frustrating to connect one lance to another.

This paper will suggest a very good solution; new concept thermal lances. We will show how these provide an outstanding performance for a more efficient, faster and cost-effective furnace tapping operation. These new lances are very accurate tools that enable the operator to perform a more controlled operation; resulting in longer taphole life, fewer repairs and less down-time.

Plain steel pipes vs. advanced lances

The basic concept of the thermal lance is that it generates a thermal energy flow from Fe combustion contained in the steel of the lance itself. In order for this combustion to take place, the Fe's ignition temperature must be reached and it must also be in contact with high purity oxygen. Once the steel reaches its ignition temperature, by applying oxygen, it reacts in combustion. However the kinetic energy of the oxygen drags the combustion forward leaving only the tip of the lance lit; its effective attack edge.

The plain steel pipe lance is a hollow tube, without inserts resulting in the steel coming into contact with oxygen only in the internal wall of the tube. The new advanced lances on the other hand, have several specially designed internal ducts which allow the oxygen to get in contact with both sides of its steel internal profiles. The result is a significantly increased steel/oxygen contact surface compared to the standard oxygen pipes.

The oxygen pipe ignites heating its tip with a solid combustible mass already ignited (i.e. burning coal), the oxygen increases the combustion, so that heat and oxygen bounce against the outer wall of the oxygen pipe (Figure 1 below).

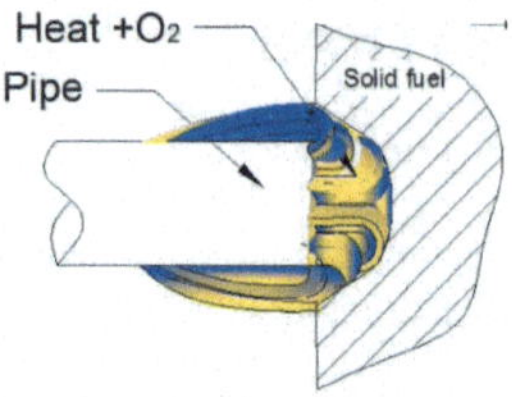

Figure 1. Oxygen lance tip combustion

Under this condition, any steel portion of the oxygen pipe, at Fe ignition temperature and in contact with oxygen, will react in combustion. However, as soon as the oxygen pipe is pulled back, the heat and oxygen bounce will end; extinguishing the Fe combustion and just a few sparks will remain. Many times it is hard to keep these sparks ignited until they reach the attack point, leaving the operator with a difficult and non-standard procedure to ignite the oxygen pipe.

The new advanced lances ignite just like standard steel oxygen pipes, by heating its tip and letting a bit of oxygen pass through. However, when the tip's combustion starts it will continue burning, even after the lance is pulled back from the burning mass. No matter the amount of oxygen, the lance will remain ignited thanks to its internal oxygen ducts that always keep Fe at ignition temperature and in contact with oxygen. This gives the operator an easy, safe and standard procedure to ignite the lance.

Since the standard oxygen pipe cannot keep being ignited by itself, it will only generate energy when the oxygen conducted through the pipe is bouncing against a surface. The bouncing oxygen ignites the external wall of the pipe, generating energy fluxes back around the pipe tip (usually the target material is in front), making it very inefficient (Figure 2 below).

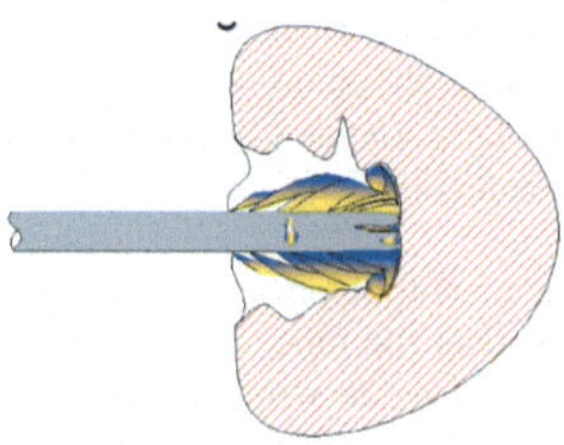

Figure 2. Inefficient cutting using standard lance

In contrast, thanks to their close to perfect steel/oxygen balance, the new advanced lances can keep being ignited without the need to bounce the oxygen through the lance against a surface. This will focus the energy forward and coaxially to the lance, concentrating the lance against the target material (Figure 3 below).

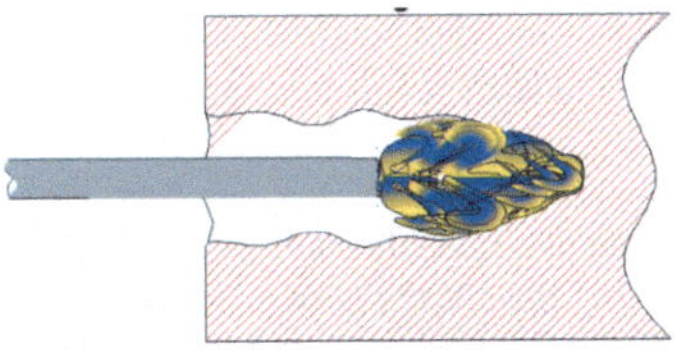

Figure 3. Efficient cutting using advanced lance

As previously stated, the standard oxygen pipe works only when the oxygen bounces, and it will ignite at random points behind the tip, sometimes several inches back. This random and unstable pipe burning causes energy loss, since a great deal of energy is used to heat and cut the pipe itself, focusing some of the energy backward and around the lance, lowering the effectiveness of the oxygen pipe energy generation.

Since the new advanced lance generates its energy forward, no points behind the tip of the lance are exposed to ignition temperatures: This fact dramatically lowers the energy wasted on heating and cutting the lance itself. This allows most of the energy to attack the target material (fig 3 above). The increased effectiveness of the energy generation of the new Trefimet lances makes them faster and more efficient than the standard oxygen pipes.

The standard oxygen pipes are completely hollow, thereby allowing the oxygen to pass through at its maximum flow for the given internal diameter. This could be harmful for some applications, such as furnace tapping, since oxygen excess can reach critical lining areas causing oxidation and deterioration. This is an especially critical issue for carbon and graphite refractories in the taphole area [1].

The new Trefimet lances are each specifically designed for different applications and conditions. Therefore, most of the oxygen is burned during the combustion process, and none of the oxygen excess will reach critical lining areas. In addition, since Trefimet lances operate faster, they will use oxygen during a fraction of the time compared to the standard oxygen pipe. This results in a significant saving in the oxygen consumption in order to get the same job done.

Wire filled steel pipes vs. advanced lances

There are several differences between the wire filled lance and the new Trefimet advanced lances. Here is a list of the most important:

- Oxygen/steel contact surfaces
- Oxygen Fluidity
- Constancy of the ducts geometry along the lance
- Oxygen flow direction
- Points/areas combustion in a given cross section
- Inserts retention

The wire filled lances have been on the market since the 1930s and they are made from a standard steel pipe filled with a bundle of inserted steel wires. These wires are retained inside the pipe through an external strangulation made by a forming press. This fact generates a number of inefficiencies, such as:

1. Since not enough oxygen can flow through the steel wires, most of the wire filled lance will be melted using the energy generated by the lance itself. In the Trefimet lance, the oxygen pass through the inner and the outer wall of the steel inserts (see Figure 4 below), making available more combustion surfaces, generating more energy and less energy is wasted melting the lance.

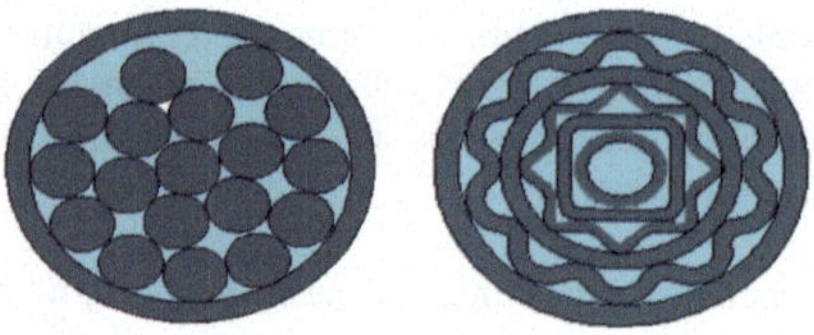

Figure 4. Wire filled lance The Trefimet TR-75 lance

2. When the bounce effect described earlier in this paper, burns the outer tube of the wire filled lance, it loses its oxygen conduction capacity, and at the attack edge, the wires are then uncovered, loosing contact with oxygen, thus ending the combustion process. After that occurs, it is very difficult to ignite the lance again.

3. Since the wires in the wire filled lance are locked inside the outer tubing through a strangulation made by a forming press, they frequently fall out (Figure 5 below). The inserts in the Trefimet lance are secured between themselves and to the outer tubing along its full length. This is done through measure interference from the smaller to the bigger inserts; from the internal to the external (the outer measure of the inner insert is greater than the inner measure of the outer insert). As the wires inside a wire filled lance are pressed very tightly together at the strangulation cross section (this is necessary to avoid the wires to be expulsed outside the lance by the oxygen flow), the oxygen flow is very restricted in that strangulation section, making the lance unable to operate under low oxygen pressure conditions.

   A Trefimet lance has several oxygen ducts configured between their inserts offering a total passage way along the whole length of the lance, no strangulation is required to maintain the inserts in place. The result is that the oxygen can "run freely" through the lance. That is the explanation to the fact that Trefimet Thermal Lances can operate well under low oxygen pressure conditions, using existing oxygen supply and plant equipment.

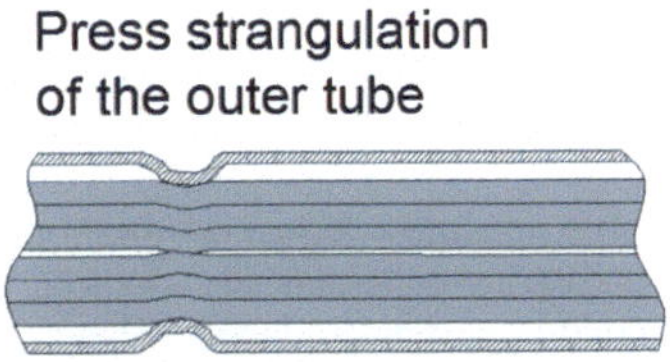

Figure 5. Wire filled lance strangulation

4. A randomized wire arrangement exists inside the wire filled lance, resulting in a variable cross section along the lance. This causes variations in oxygen speed and direction (multiple oxygen vectors along the lance), which in turn generate several random combustion points near its attack edge. This results in uncontrolled and inefficient lance combustion. Since the Trefimet lance oxygen ducts are regular and symmetrical along the full length of the lance, they contribute to more controlled oxygen vectors, resulting in a more focused and efficient lance combustion at its attack edge.

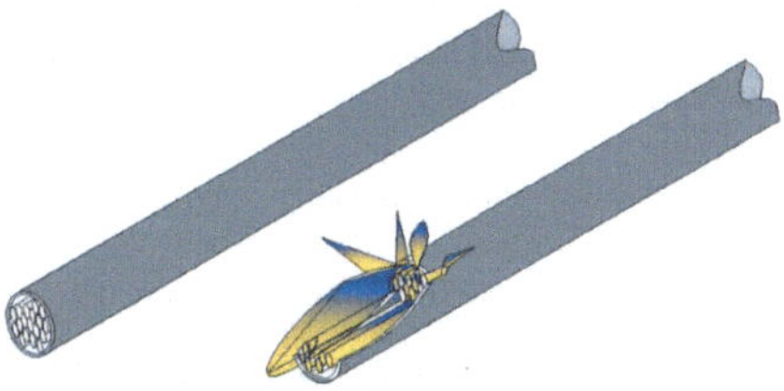

Figure 6. Wire filled lance with random combustion

5. Since it cannot operate after it has burned, the strangulated cross section (the wires will be completely loose), customers will always discard a portion of a wire filled lance. The Trefimet lance on the other hand, can be utilised to its full length. The cross section of these lances is the same along the lance and, furthermore, they are using the patented fast connection system called FCT that immediately and effortlessly connects a new lance to what remains of the used lance. This allows lance operators to achieve 100% lance consumption without any waste in addition to making very fast additions of new lances when needed.

The new advanced oxygen lance

Chilean based lance manufacturer Trefimet has over the last 15 years, together with the local copper producers, developed and patented an advanced thermal lance design and lance-to-lance connection system, which provides an outstanding performance for a faster, more efficient, and more cost-effective furnace tapping operation. These lances are supplied in several different models, each adapted to the exact application and conditions where it will be used.

The company has developed a perfect balance between speed and the available mass for the combustion process, creating the right exothermic reaction for each application.

These advanced oxygen lances offer an almost perfect combustion, resulting in an operation with minimum emissions of fugitive gaseous pollutants. In addition, it completes the operation in about one third of the time required by a standard lances. Therefore, it emits much less fugitive gases during operation. The reduced number of lances required, lower oxygen consumption and no left-overs (thanks to the special lance-to-lance quick connector system) are some of the benefits using these new lances.

Increased accuracy

These new lances are very accurate tools that enable the operator to perform a more controlled operation; resulting in longer taphole life, fewer repairs and less down-time [2]. Moreover, a lower projection of particles reduces the wear of PPE (personal protective equipment.) Increased safety is also an important consideration. Trefimet lances generate a more efficient and accurate energy which does not require pressing the lance against the target; it lowers the projection of incandescent particles and gas generation, giving greater control over the operation. The reduced operation time required decreases the time of exposure for the operator at high temperatures. Less stress, heat and sparks, coupled with greater cutting control, improve operations safety, cutting accuracy and working conditions.

The Trefimet TR model was developed specially for furnace tapping operations. It effectively eliminates so called Bounce Combustion as described above. This effect is predominant for standard lances, speeding up its consumption and making it less accurate. Without the Bounce Combustion effect, the Trefimet TR can focus all its power against the desired attack point. All these properties together improve operation safety, efficiency and working conditions, giving greater availability and asset optimization.

Suitable applications, in addition to tapholes on copper and ferroalloy furnaces, are cutting of accretions. In foundry processes, accretions of different types and sizes are generated, either in the smelters or by spills. Standard lances like steel pipes and burning bars, used in this application, do not achieve good results. They result in an increased requirement of time, effort, oxygen and lances. The Trefimet advanced thermal lance, with a concentrated power of penetration, is a perfect tool for cleaning and cutting accretions; to achieve results in the shortest possible time, with reductions in the consumption of oxygen, lances, safety equipment, and time of exposure of the operator, and as a result, an increase in safety.

Selecting the best lance design for the job

In order to choose the right lance, Trefimet reviews the key thermal properties of the material to be cut, including:

- Fusion Point
- Specific Heat

- Fusion Latent Heat
- Thermal conductivity
- Molecular mass
- Density

Based on the material properties listed above, it will be classified upon its thermal requirements. Trefimet uses the following classification, divided in High, Medium or Low thermal requirements.

Materials with High thermal requirements include:

- Graphite and all its alloys
- Carbides, titanium, tungsten, silicon, etc.
- Platinum
- Chrome
- Magnetite

Materials with Medium thermal requirements:

- **Copper**
- Alumina
- Titanium
- Refractory bricks

Materials with Low thermal requirements:

- Clay
- Calcium
- Aluminium
- Lead
- Zinc

Taking copper as an example, key properties deciding the thermal requirement rating above includes: Specific Heat Capacity = 0,385 KJ/(Kg °K), Fusion Latent Heat = 13,14 KJ/mol, Melting Point (Tm) = 1.356 °K and Molecular Weight (m)= 63,546 g/mol.

In addition to these parameters, and when selecting the appropriate Trefimet lance, the company will look at specific operating conditions such as: furnace type, type of cutting to be performed and distance from operator to object. And if used for tapping, crucial information includes: the taphole length and diameter, if drilling is used, drilling length, taphole clay used, taps per day, oxygen source, pressure and oxygen flow at generation point.

Conclusion

Standard oxygen lances have the following well known disadvantages:

- Ejects radial and axial energy
- Has an accelerated and costly consumption
- Makes irregular perforations
- Generates large amounts of fumes and particles
- Since it needs to be pressed forward, the lance bends

Trefimet has managed to develop a patented and advanced oxygen lance design that offers the following advantages:

- It focuses its energy on the attack point.
- It has lower consumption and is more cost efficient.
- It makes more accurate perforations.
- It generates less fumes and particles.
- Since it does not need to be pressed forward, it does not bend resulting in lighter and more accurate operation.

In addition to the above, the patented fast connection system (FCT) reduces waste and speeds up operations significantly.

As stated in the beginning of this paper, one of the most important tools for the metallurgical furnace industry and tapping operations in particular, has for over a century been the oxygen lance [3]. But this has not been considered an operation that brings added value, and, therefore, there has been practically no product development during all these years. Furnaces have been made more efficient and computer controlled, electrode systems have been modernized, taphole clays are today environmentally friendly as well as great for the tapping process, cooling systems have been improved, lance holders with built in safety valves are standard, refractory concepts and materials have been refined to offer improved life and safety [4], just to mention a few major developments that have taken place in recent years. And now, finally, the oxygen lance technology has reached a new level. These new lances are very accurate tools that enable the operator to perform a more controlled operation; resulting in longer taphole life, fewer repairs, higher safety and less down-time.

References

[1]C. Coetzee, P. Sylven, "No tap-hole – No furnace" (Paper presented at the SAIMM Refractories Conference, 2010), 55-66.

[2]C. Coetzee, P. Duncanson, P. Sylven, "Campaign extension for ferroalloy furnaces with improved taphole repair system" (Paper presented at INFACON XII, Helsinki, 2010), 857-865.

[3]L.R. Nelson, R.J. Hundermark, "The tap-hole – key to furnace performance" (Paper presented at the SAIMM Furnace tapping Conference,2014).

[4] P.L Duncanson, P. Sylven, "New refractory lining direction at Jindal stainless – first Indian FeCr producer to install UCAR Chillkote® linings" (Paper presented at the FAI Conference, Goa, India, 2011).

7th International Symposium on High-Temperature Metallurgical Processing
Edited by: Jiann-Yang Hwang, Tao Jiang, P. Chris Pistorius, Gerardo R.F. Alvear F., Onuralp Yücel, Liyuan Cai, Baojun Zhao, Dean Gregurek, and Varadarajan Seshadri
TMS (The Minerals, Metals & Materials Society), 2016

UTILIZATION OF PINE NUT SHELL FOR PREPARATION OF HIGH SURFACE AREA ACTIVATED CARBON BY MICROWAVE HEATING AND KOH ACTIVATION

Xuefeng Liao[1,2,3], Jinhui Peng[1,2,4], Shengzhou Zhang[2,3], Hongying Xia[2,3], Libo Zhang[2,3], Guo Chen[2,4], Tu Hu[2,3]

[1]State Key Laboratory of Complex Nonferrous Metal Resources Clean Utilization, Kunming University of Science and Technology, Kunming, Yunnan 650093, China;
[2]National Local Joint Engineering Laboratory of Engineering Applications of Microwave Energy and Equipment Technology, Kunming University of Science and Technology, Kunming, Yunnan 650093, China;
[3]Faculty of Metallurgical and Energy Engineering, Kunming University of Science and Technology, Kunming, Yunnan 650093, China;
[4]Key Laboratory of Resource Clean Conversion in Ethnic Regions, Education Department of Yunnan, Yunnan Minzu University, Kunming, Yunnan 650500, China;

Keywords: Pine nut shell; response surface methodology; microwave heating; high Surface area activated carbon

Abstract

Pine nut shell is as raw material for preparation of high surface area activated carbon (HSAAC) by microwave induced KOH activation. The effects of microwave power, activation duration and KOH/C mass rate (R) on the iodine adsorption capability and activated carbon yield were investigated. Additionally the surface characteristics of HSAAC were characterized by nitrogen adsorption isotherms and SEM. The operating variables were optimized utilizing the response surface methodology and were identified microwave power 738W, activation duration 17 min, and R 4, corresponding to a yield of 46.28 % and an iodine number of 2154 mg/g. The key pore structure parameters of HSAAC such as the Brunauer–Emmett–Teller (BET) surface area and total pore volume were estimated to be 3819 m^2/g and 2.09 mL/g, respectively. The findings strongly support the feasibility of microwave heating for preparation of HSAAC from spent pine nut shell by microwave induced KOH activation.

Introduction

Activated carbon is known as universal sorbent owing to its well-developed porous structure, high specific surface area and thermo stability [1-3]. It is being increasingly applied to different industries, including food, chemicals, national defense and environmental protection as adsorbent or catalyst [4, 5]. Demand for activated carbon was on the rise with the development of society and the improvement of environmental protection requirements in recent years [6-9]. Accordingly, the urgent need to improve the yield and quality of activated carbon, in particular, the high specific surface area activated carbon.

Basically, two activation methods for preparing activated carbon have been suggested which include physical and chemical activation [10, 11]. Physical activation requires a longer activation time and a higher activation temperature, it is gradually being replaced by chemical activation. Chemical activation refers to mixing the material with a suitable chemical agent, such as KOH, NaOH, K_2CO_3 and other alkali chemical reagents, then obtained activated carbon directly [12-15]. Many researchers have reported that KOH can provide activated carbon with a high specific surface area and pore volume through a series of crosslinking and polymerization reaction [16, 17]. In the process of preparing activated carbon, two kinds of heating methods, conventional heating and microwave heating, have been applied [18-20]. Recently, microwave heating is widely used in metallurgy and material industry, as heating is uniform, where the material can be heated at the same time inside and outside [21].

People have made a lot of research on activated carbon, preparation of raw materials also expanded from wood to nutshell, fruit skins, crop stalks and agricultural waste [22-24]. In this experiment, Pine nut shell was elected as the raw material since they are sustainable sources having high fixed carbon content, and the activated carbon was prepared from pine nut shell by KOH activation and microwave heating.

Experimental

Materials

Pine nut shell were obtained from Kunming city, Yunnan province of China. And it were crushed, sieved into a uniform size of 4-6 mm, then washed with deionized water and oven-dried in an electric oven at 105 °C and stored in a sealed containers for utilization in the experiment. Hydrochloric acid (HCl) and KOH, all of which are of analytical grade, were used for the experiments.

A sample of the pine nut shell were carried out by loading 1000 g into a resistance furnace, under the N_2 atmosphere (100 cc/min) and heating in to a carbonization temperature of 600 °C at a heating rate of 10 °C /min and was held for 90 min at 600 °C. The carbonized sample was crushed and washed sequentially with distilled water several times, then the washed sample was dried in an electric oven until the next experiment.

Preparation of High Specific Surface Area Activated Carbon

A tubular microwave furnace was made by the Key Laboratory of Unconventional Metallurgy, China Ministry of Education, Kunming University of Science and Technology, which was employed in the activation experiments, and the diagram of apparatus was shown in Figure. 1.

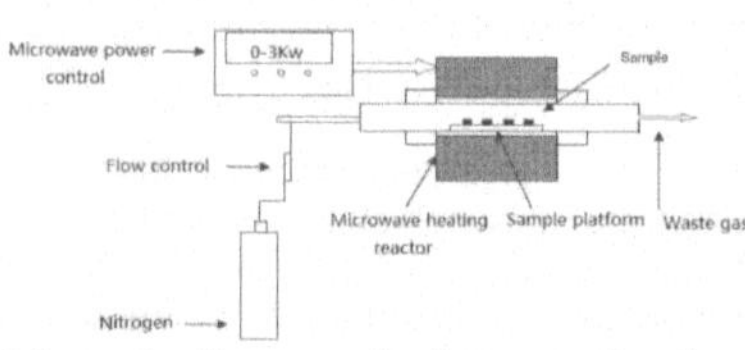

Figure 1. Schematic diagram of microwave heating apparatus

The microwave frequency was 2.45GHz, and the output power in the range of 0-3000W. The char was mixed with KOH at different KOH/C mass rate, placed in the tubular microwave furnace, and heated for different durations and at different microwave powers. The sample were allowed to cool down naturally inside the reactor to room temperature under nitrogen atmosphere. The resultant activated carbon was repeatedly washed with washed with 1:9 HCl (by volume), then washed with distilled water to remove the residual HCl until the pH was about 7, and then dried at 120 °C for 3 h and stored for further characterization.

Characterization of Activated Carbon

The yield of the activated carbon and the KOH/C mass rate are based on the following equation:

$$Y_2=W_A/W_B\times 100\% \quad (1)$$

$$IR=W_{KOH}/W_{char} \quad (2)$$

where Y_2 is the yield of the activated carbon; W_A is the weight of the activated carbon; W_B is the weight of the carbonized materials; IR is the KOH/C mass rate; W_{KOH} is the weight of the KOH; W_{char} is the weight of the carbonized.

The iodine number of the activated carbon was calculated using the standard testing methods of the People's Republic of China (GB/T12496.8-1999) [25]. The N_2 adsorption of the activated carbon was carried out at 77 K using an automatic adsorption apparatus (Autosorb-1-C, USA). The total pore volumes were estimated to be the equivalent liquid volume of the adsorbate (N_2) at a relative pressure of 0.99. The BET surface of the samples was calculated using the Brunauer–Emmett–Teller (BET) equation [26]. The Barrett-Joymer-Halenda (BJH) method was applied to calculate the micropore volume and external surface area [30]. Scanning electron microscopy (SEM, Philips XL30ESEM-TMP) analysis was used to assess the surface morphology.

Experimental Design

The response surface methodology (RSM), which is a popular statistical tool for modeling and analysis of multi parameter processes, and central composite design (CCD) was employed to design the activation experiments [27-29]. In this study, We consider the effects of three independent variables: X_1 (microwave power), X_2 (activation duration), and X_3 (KOH/C mass rate). The upper and lower limits of the process variables are shown in Table I. Iodine adsorption capacity (Y_1) and yield (Y_2) of the activated carbon were chosen as the response values. A total of 20 experiments consisted of 8 factorial points, 6 axial points and 6 replicates at the central points were employed. The experimental design matrix and results were listed in Table II. The experimental data were obtained by the Design Expert software version 7.1.5 (Stat-Ease Inc, Minneapolis, USA).

Table I Independent Variables and Their Levels for Central Composite Design

Factors	Code	Levels				
		-1.682	-1	0	-1	1.682
Microwave power	X_1	632.27	680	750	820	867.73
Activation time	X_2	9.27	12	16	20	22.73
KOH/C weight ratio	X_3	2.66	3	3.5	4	4.34

Table II Experimental Design Matrix and Results

Run	X_1	X_2	X_3	Y_1	Y_2	Run	X_1	X_2	X_3	Y_1	Y_2
1	632.27	16.0	3.50	1298	54.2	11	820.0	20.0	4.00	1543	38.1
2	750.00	16.0	3.50	1982	49.1	12	680.0	20.0	3.00	1371	53.9
3	750.00	16.0	4.34	2101	46.0	13	750.0	9.27	3.50	1326	59.5
4	750.00	16.0	3.50	1976	49.1	14	680.0	12.0	3.00	1164	60.5
5	680.00	20.0	4.00	2023	48.5	15	750.0	16.0	3.50	1965	49.1
6	750.00	16.0	3.50	1997	49.2	16	820.0	12.0	4.00	1655	51.9
7	820.00	12.0	3.00	1359	52.9	17	750.0	16.0	2.66	1509	58.1
8	680.00	12.0	4.00	1558	54.6	18	820.0	20.0	3.00	1223	46.6
9	750.00	22.7	3.50	1659	43.8	19	867.7	16.0	3.50	1309	40.1
10	750.00	16.0	3.50	1971	49.1	20	750.0	16.0	3.50	1967	49.1

Results and Discussion

Response Analysis and Verification of the Regression Model

The emphasis should be put on verifying the accuracy of the models for data analysis. And there will be a large discrepancy between the predicted and the actual results if the model accuracy is not high. The following Equations (1) and (2) were the final empirical models in terms of coded factors for iodine number (Y_1) and yield (Y_2), respectively:

$$Y_1=+1976.97-23.35X_1+72.05X_2+194.60X_3-115.00X_1X_2-53.75X_1X_3+35.25X_2X_3-242.06X_1^2-175.24X_2^2-64.76X_3^2 \quad (1)$$

$$Y_2=+49.13-3.80X_1-4.34X_2-3.02X_3-0.91X_1X_2+0.23X_1X_3-0.89X_2X_3-0.57X_1^2+1.02X_2^2+1.15X_3^2 \quad (2)$$

The correlation coefficient value was employed to assess the quality of the model equation, and the correlation coefficient value was 0.9867 in Equation (1) and 0.9771 in Equation (2), respectively. Both the correlation coefficient value for the iodine number and the activated carbon yield were of high proximity to unity, and the results demonstrating the suitability of the model equation and predicted data and the experimental data were in close agreement using the model. The comparison of the predicted iodine number and yield versus the experimental were shown in Figure 2 and Figure 3 respectively. As seen in the figure that the experimental data are distributed on both sides of the model prediction, indicating that the selected model accuracy was higher and the simulation results accurately, which simulate well activated carbon iodine number and yield results are accurate.

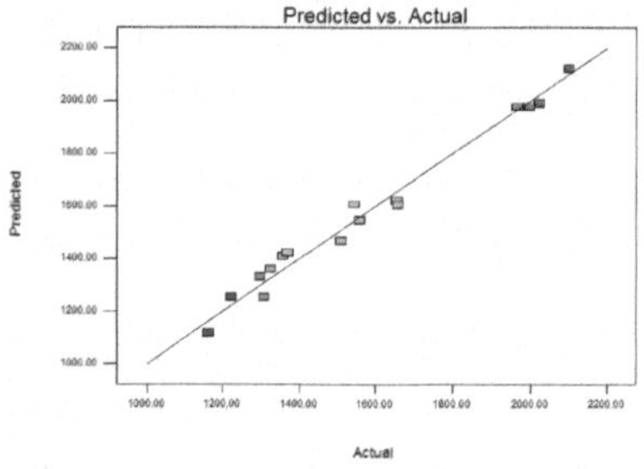

Figure 2. Predicted vs. experimental iodine number of activated carbon

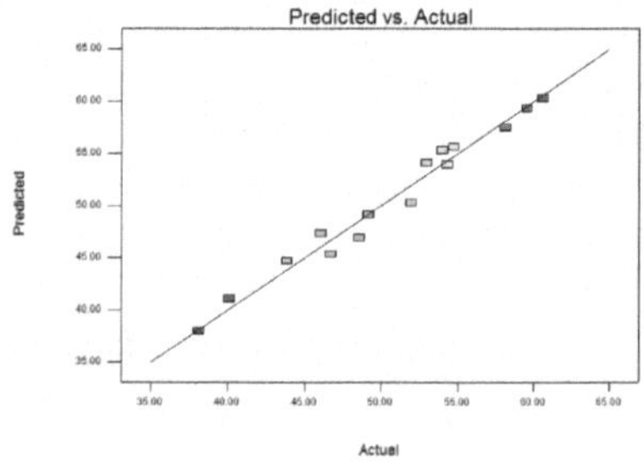

Figure 3. Predicted vs. experimental yield of activated carbon

Activated Carbon Iodine Number from Response Surface Analysis

Adsorption capacity is one of the most important characteristics of an activated carbon, which can characterized by pore volume, pore size, and surface area. Figure 4 shows the three-dimensional response surfaces of the combined influences of microwave power and KOH/C mass rate on the iodine number when the activation duration was 16 min. As seen in the figure the iodine numbers were 2101 mg/g and 1164 mg/g when the microwave power and KOH/C mass rate reached their maximum and minimum values, respectively. The increase of microwave power will lead to reduce the C-KOH activation energy of the reaction, thus increasing the rate of reaction. High reaction rates serve to increase the porosity of the activated carbon, namely the increase in iodine number. But the increase of microwave power also destroy the pores of activated carbon, so the iodine number appeared a drop process. Similarly the increase of KOH/C mass rate can lead to increase the extent of the C–KOH reaction, and it will form a correspondingly large number of pores. So, the adsorption capacity of the activated carbon is strengthened.

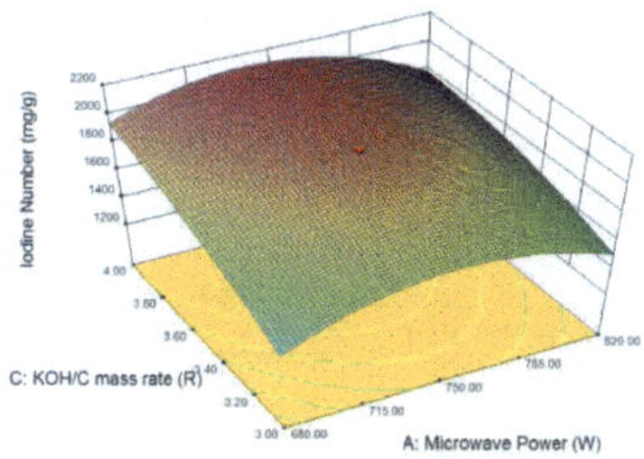

Figure 4. The three-dimensional response surfaces of the combined influences of microwave power and KOH/C mass rate on the iodine number (activation duration: 16 min)

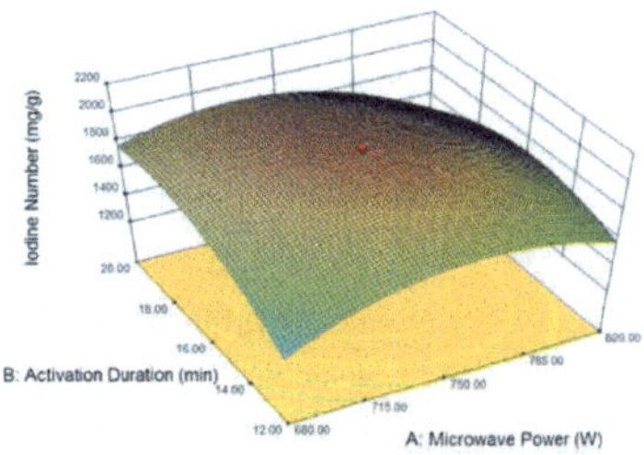

Figure 5. The three-dimensional response surface of the combined effect of microwave power and activation duration on the iodine number (R: 3.5)

Activated Carbon Yield from Response Surface Analysis

The yield of activated carbon is an important parameter and the three-dimensional response surface of the combined effect of microwave power and KOH/C mass rate on the yield of activated carbon when the activation duration was 16 min was shown in Figure 6. As seen in the figure, the yield of activated carbon was inversely proportional to microwave power and KOH/C mass rate. The maximum yield of the activated carbon was about 60.59%, corresponding to the minimum microwave power and KOH/C weight ratio. An increase in the microwave power and KOH/C weight ratio leads to a reduction in the activation energy of the carbon–KOH reaction, so the carbon–KOH reaction is strengthened, contributing to the reduction in yield of the activated carbon.

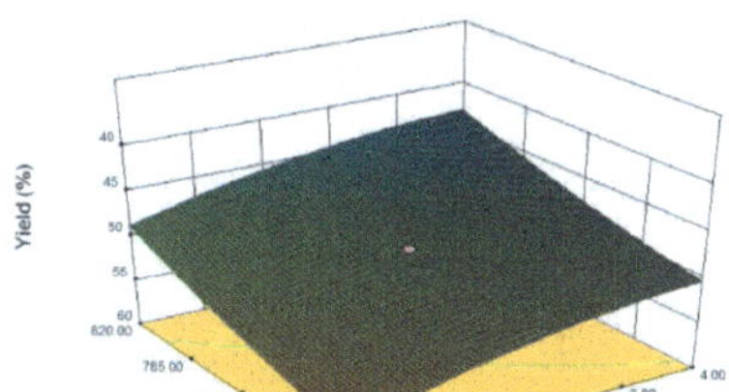

Figure 6. The three-dimensional response surface of the combined effect of microwave power and KOH/C mass rate on the yield of activated carbon (activation duration: 16 min)

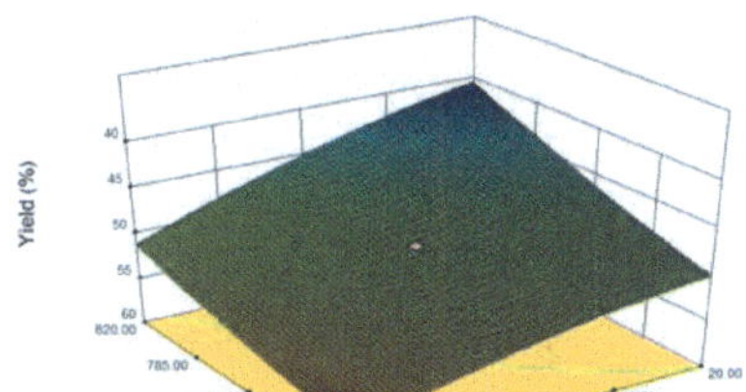

Figure 7. Three-dimensional response surface plot of activated carbon yield: effect of microwave power and activation time on the yield (R: 3.5)

Figure 7 was the influence of the microwave power and activation duration on the yield at a KOH/C mass rate of 3.5. The yield of activated carbon was inversely proportional to microwave power and activation duration. Figure 7 also shows that the maximum yield corresponded to the minimum microwave power and activation duration, whereas the minimum yield corresponded to the maximum microwave power and activation duration. An increase in the microwave power and KOH/C mass rate can speed up the carbon–KOH reaction. As a result, the yield of the activated carbon decreases.

Process Optimization

Generally, the industrial production of activated carbon augurs higher iodine number and yield, which contributes to improved economics of commercial manufacture. Therefore, industry needs a better activated carbon with high yield and adsorption performance, but the iodine number and yield affected by various process parameters.

Table III Validation of Process Optimization

Microwave	Activation	KOH/weight	Iodine number (mg/g)		Yield (%)	
X_1 (W)	X_2 (min)	X_3	Predicted	Experimental	Predicted	Experimental
738	17	4	2219	2154	44.88	46.28

The optimum combination of the parameters to maximize the yield and iodine number becomes important and the most desirable experimental parameters were identified by Design Expert software. The repeat experiments were carried out to verify the accuracy of the predicted result, and the experimental results were reported in Table III. It is obvious that the relative errors of iodine number and the yield were only 3.02% and 3.03%, respectively.

Pore Structure and Surface Area Analysis

The nitrogen adsorption isotherms of char and activated carbon under the optimum regeneration condition was shown in Figure 8. It can be seen that the shapes of these isotherms pertain to adsorption isotherm type II according to the IUPAC classification. By comparing the adsorption isotherms of activated carbon and char, the maximum absorption volume of the char was far less than the activated carbon, illustrating the number of pores in the activated carbon. The adsorption capacity continued to increase beyond of the relative pressure value of 0.1 that is typical for mesoporous materials. Figure 9 shows the cumulative pore volume of the activated carbon and char, and it can be seen that the cumulative pore volume of the activated carbon was far greater than the char.

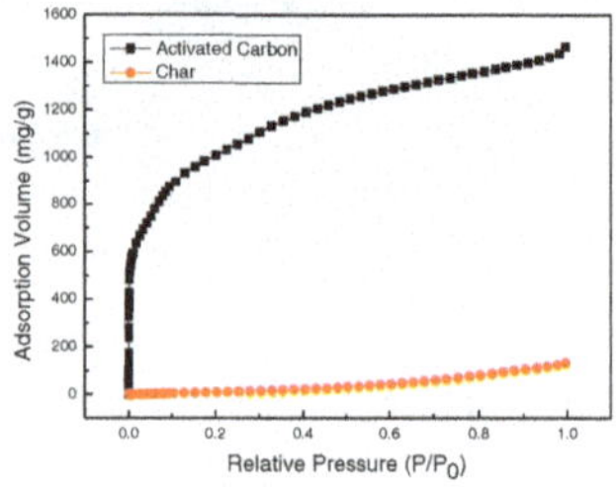

Figure 8. The nitrogen adsorption isotherms of the char and activated carbon

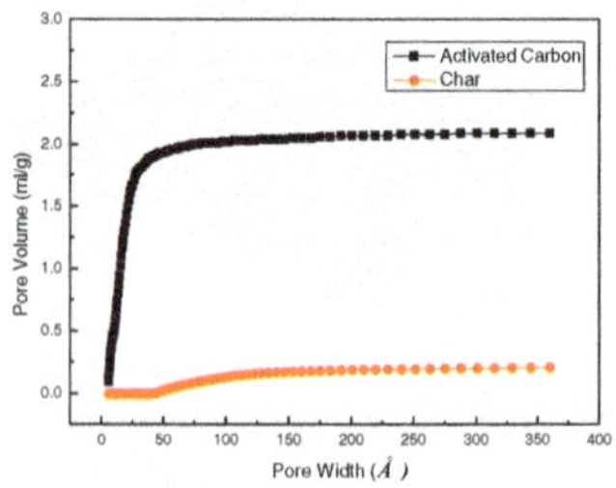

Figure 9. The cumulative pore volume of the char and activated carbon

Figure 10 shows the SEM microstructure of the char and activated carbon obtained under the optimum preparation conditions. It can be seen that the activated carbon has a relatively well-developed pore structure and a smooth surface while pore structure of the char was not very obvious. By comparison of the microstructure of the activated carbon with the char, a large number of micropores generated in the reaction of carbon–KOH. And those micropores significantly increases the surface area and pore volume.

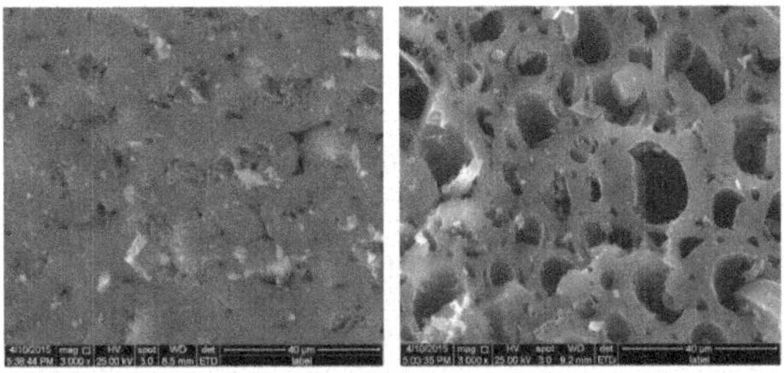

Figure 10. The SEM microstructure of the char and activated carbon ((a) char; (b) activated carbon)

Conclusion

Pine nut shell is utilized for preparing high surface area activated carbon with a well-developed pore structure by microwave heating induced KOH activation. In this research, the effects of three vital process parameters, microwave power, activation duration and KOH/C mass rate (R) on the iodine adsorption capability and activated carbon yield were investigated. The optimum conditions were: microwave power 738 W, activation time 17 min, and KOH/C weight ratio 4, with the resultant iodine number and yield being 2154 mg/g and 46.28 % respectively. The BET surface area and pore volume were evaluated using a nitrogen adsorption isotherm, and corresponded to 3819 m^2/g and 2.09 cm^3/g, respectively, under optimum conditions. The results illustrate that using pine nut shell for the preparation high specific surface area activated carbon is feasible by microwave heating induced KOH activation.

References

1. H.L. Mudoga, H. Yucel, and N.S. Kincal, "Decolorization of sugar syrups using commercial and sugar beet pulp based activated carbons," *Bioresource Technology*, 99 (9) (2008), 3528-3533.
2. S.L Liu, Y.N. Wang and K.T. Lu, "Preparation and pore characterization of activated carbon from Ma bamboo (Dendrocalamus latiflorus) by H_3PO_4 chemical activation," *Journal of Porous Materials*, 21 (4) (2014), 459-466.
3. A. Policicchio et al., "Higher methane storage at low pressure and room temperature in new easily scalable large-scale production activated carbon for static and vehicular applications," *Fuel*, 104 (2) (2013), 813-821.
4. G.M. WALKER and L.R. WEATHERLEY, "Textile Wastewater Treatment Using Granular Activated Carbon Adsorption in Fixed Beds," *Separation Science & Technology*, 35 (9) (2000), 1329-1341.
5. E. Kaçan and C. Kütahyal, "Adsorption of strontium from aqueous solution using activated carbon produced from textile sewage sludges," *Journal of Analytical & Applied Pyrolysis*, 97 (2012), 149-157.
6. Y.H Li et al., "Preparation of activated carbon from Enteromorpha prolifera and its use on cationic red X-GRL removal," Applied Surface Science, 257 (24) (2011), 10621-10627.
7. B. Kastening and M. Heins, "Properties of electrolytes in the micropores of activated carbon," *Electrochimica Acta*, 50 (12) (2005), 2487-2498.
8. Y.J. Kim et al., "Preparation and characterization of bamboo-based activated carbons as electrode materials for electric double layer capacitors," *Carbon*, 44 (8) (2006), 1592-1595.
9. C. Nieto-Delgado and J.R. Rangel-Mendez, "Production of activated carbon from organic by-products from the alcoholic beverage industry: Surface area and hardness optimization by using the response surface methodology," *Industrial Crops & Products*, 34 (3) (2011),

1528-1537.
10. Q.S Liu et al., "Preparation and characterization of activated carbon from bamboo by microwave-induced phosphoric acid activation," *Industrial Crops & Products*, 31 (2) (2010), 233-238.
11. A.L Cazetta et al., "Thermal regeneration study of high surface area activated carbon obtained from coconut shell: Characterization and application of response surface methodology," *Journal of Analytical & Applied Pyrolysis*, 101 (5) (2013), 53-60.
12. Y.P Guo et al., "Performance of electrical double layer capacitors with porous carbons derived from rice husk," *Materials Chemistry & Physics*, 80 (3) (2003), 704-709.
13. B. Karimi, M.H Ehrampoush, A. Ebrahimi et al., "The study of leachate treatment by using three advanced oxidation process based wet air oxidation," *Iranian Journal of Environmental Health Science & Engineering*, 10 (1) (2013), 119-124.
14. G. Hermosilla-Lara, G. Momen, P.H Marty et al., "Hydrogen storage by adsorption on activated carbon: Investigation of the thermal effects during the charging process," *International Journal of Hydrogen Energy*, 32 (2007), 1542-1553.
15. J.W Zondlo, and M.R. Velez, "Development of surface area and pore structure for activation of anthracite coal," *Fuel Processing Technology*, 88 (4) (2007), 369-374.
16. J.L Lim, and M. Okada, "Regeneration of granular activated carbon using ultrasound," *Ultrasonics Sonochemistry*, 12 (4) (2005), 277-82.
17. B. Kastening, and M. Heins, "Properties of electrolytes in the micropores of activated carbon," *Electrochimica Acta*, 50 (12) (2005), 2487-2498.
18. C.O Ania, J.B Parra, J.A. Menéndez et al., "Microwave-assisted regeneration of activated carbons loaded with pharmaceuticals," *Water Research*, 41 (15) (2007), 3299-3306.
19. M Kubota, A. Hata, and H. Matsuda, "Preparation of activated carbon from phenolic resin by KOH chemical activation under microwave heating," *Carbon*, 47 (12) (2009), 2805-2811.
20. X.H Duan, C. Srinivasakannan, J.H Peng et al., "Comparison of activated carbon prepared from Jatropha hull by conventional heating and microwave heating," *Biomass & Bioenergy*, 35 (9) (2011), 3920-3926.
21. X.H Duan, C. Srinivasakannan et al., "Regeneration of microwave assisted spent activated carbon: Process optimization, adsorption isotherms and kinetics," *Chemical Engineering & Processing*, 53 (2012), 53-62.
22. A.S Mestre, A.S Bexiga et al., "Activated carbons from sisal waste by chemical activation with K2CO3: Kinetics of paracetamol and ibuprofen removal from aqueous solution," *Bioresource Technology*, 102 (17) (2011), 8253-8260.
23. X.H Duan, C. Srinivasakannan, and J.S. Liang, "Process optimization of thermal regeneration of spent coal based activated carbon using steam and application to methylene blue dye adsorption," *Journal of the Taiwan Institute of Chemical Engineers*, 45 (2014), 1618-1627.
24. A.C Lua, and T. Yang, "Effect of activation temperature on the textural and chemical properties of02potassium hydroxide activated carbon prepared from pistachio-nut shell," *Journal of Colloid & Interface Science*, 274 (2) (2004), 594-601.
25. X.H Duan, Z.B. Zhang, C. Srinivasakannan, F. Wang et al., "Regeneration of spent catalyst from vinyl acetate synthesis as porous carbon: Process optimization using RSM," *Chemical Engineering Research & Design*, 92 (7) (2014), 1249-1256.
26. X.H Duan, C. Srinivasakannan, J.H Peng et al., "Preparation of activated carbon from Jatropha hull with microwave heating: Optimization using response surface methodology," *Fuel Processing Technology*, 92 (3) (2011), 394-400.
27. R.M Suzuki, A.D Andrade, J.C Sousa et al., "Preparation and characterization of activated carbon from rice bran," *Bioresource Technology*, 98 (10) (2007), 1985-1991.
28. K.Y Foo, and B.H. Hameed, "Adsorption characteristics of industrial solid waste derived activated carbon prepared by microwave heating for methylene blue," *Fuel Processing Technology*, 99 (7) (2012), 103-109.
29. K.Y Foo, and B.H. Hameed, "Porous structure and adsorptive properties of pineapple peel based activated carbons prepared via microwave assisted KOH and K_2CO_3 activation," *Microporous & Mesoporous Materials*, 148 (1) (2012), 191-195.

7th International Symposium on High-Temperature Metallurgical Processing
Edited by: Jiann-Yang Hwang, Tao Jiang, P. Chris Pistorius, Gerardo R.F. Alvear F., Onuralp Yücel, Liyuan Cai, Baojun Zhao, Dean Gregurek, and Varadarajan Seshadri
TMS (The Minerals, Metals & Materials Society), 2016

RESEARCH ON HIGH TEMPERATURE MECHANICAL PROPERTIES OF 50Cr5MoV ROLL STEEL

Dong-chang Huang, Guangliang Wu, Xin-bin LIU

(School of Minerals Processing and Bioengineering, Central South University, Changsha, Hunan, 410083, China)

Keywords: 50Cr5MoV roll steel; high temperature mechanical properties; Gleeble-1500 thermal simulator; brittle temperature zone

Abstract

High temperature mechanical properties of 50Cr5MoV roll steel was investigated by Gleeble 1500 thermal simulator testing machine. The morphology, composition and microstructure of fracture surface were observed and analyzed by OM, SEM and EDS. The results are indicative of that, there are two brittle temperature zones of 50Cr5MoV roll steel at the temperature of 600℃～950℃and 1300℃～1450℃, respectively. The section shrinkage rate is less than 60%. The fracture mode changes from mixed ones dominated by intergranular to toughness transgranular one with the increase of temperature at the range of 650℃～1300℃. The fracture is along the solid--liquid phase at the range of 1300℃～melting point. It is crucial to formulate the 50Cr5MoV casting technology, to prevent the cracks and improve the quality of 50Cr5MoV roll steel.

Introduction

With the continuous development of cold and hot rolling technology, backup roll, as one of the important parts on mill, its performance requirements become increasingly higher. The back-up roller is used to support the working roller to prevent Flexure deformation [1]. 50Cr5MoV, as the third generation in Cr series of forged backup roll steel, due to the pleasurable comprehensive mechanical properties, as an example, high impact toughness, wear resistance and resistance to high temperature, has been widely used in hot rolling bearing roller [1-2].

Because of its working condition, there is friction between the supporting roller and the working roller. The supporting roller needs to support high contact stress and shear stress. As a result, it would wear and deform. Therefore, the supporting roller should possess high abrasion performance, high compressive strength, accident resistance, fracture resistance, spalling resistance, hot-cracking resistance, excellent rigidity and toughness [2-4].

In the process of casting, the steel billet could be influenced by various stress factors.

Corresponding author: Guangliang Wu, Professor, Doctoral supervisor, E-mail: glwu_899@sina.com

The excess of high temperature ultimate strength and limiting strain of the billet, cause it to crack easily and makes it scrapped. To ensure the continuity of concreting and prevent cracking, the force must be lower than high temperature properties [5]. The former researches indicated that the mechanical properties of steel (thermoplastic and thermal strength) were the main index of the crack in continuous casting temperature range [6].

In view of the high temperature mechanical properties of 50Cr5MoV, It was tested on a Gleeble-1500 thermal simulator, and the mechanism of the fracture was discussed. It is crucial to formulate the 50Cr5MoV casting technology, to prevent the cracks and improve the quality of that.

Experimental

Raw Material

Specimen material is 50Cr5MoV roll steel, produced in a steel mill , whose composition is shown in Tab.I. The size of sample steel is Φ10mm×120mm, as shown in Fig.1.

Tab.I The chemical composition of 50Cr5MoV steel

C	Si	Mn	P	S	Cr	Mo	V
0.45~0.49	0.42~0.52	0.58~0.72	≤0.010	≤0.010	4.80~5.40	0.40~0.55	0.10~0.30

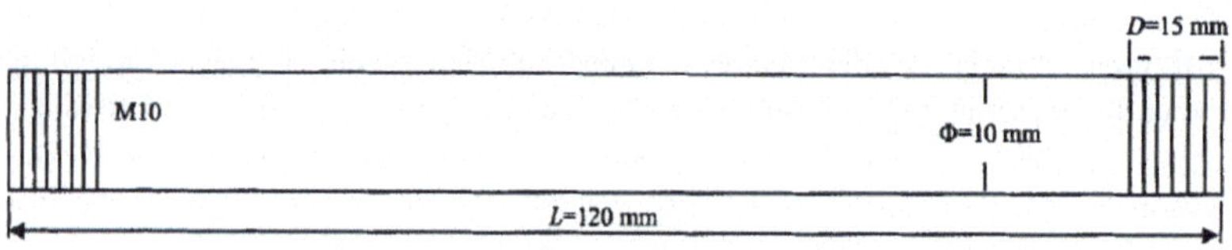

Fig.1 The sample size

Experimental Method

Based on the actual technological process parameters of 50Cr5MoV roll steel production line, Gleeble-1500 was adopted to test the reduction of area (R_A) and tensile strength. The samples were heated at the rate of 10℃/s in argon shielding gas. When the samples were heated up to the peak temperature 1465℃, and kept them heat preservation 5 min. Then the temperature was lowered to the experimental temperature at rate of 3℃/s, keeping the heat for 3min. If the experimental temperature ≤1250℃ (Casting mold temperature), high temperature tension experiments were carried out at strain-rate of 1.3×10^{-4}, if not, that were carried out at strain-rate of 6.46×10^{-3}. Fig.2 demonstrates the project of the experiment.

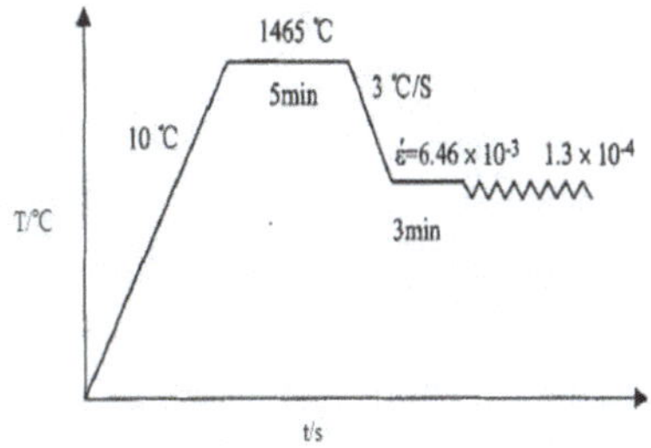

Fig.2 The project of the experiment

Results and discussion

High temperature tensile parameters

Those of experimental liquids temperature, cooling-rate and strain-rate parameter have a profound effect on high temperature mechanical properties. The experiment was simulated the solidification of continuous casting blank. In order to ensure the sample austenitized, the sample was heated up to liquids temperature, then cooled. Cooling-rate had corresponding relation with continuous casting cooling intensity, thus made an influence on the size of precipitate of steel and had effect on high temperature mechanical properties of steel. The influence on strain-rate to percentage reduction of area (RA) was not remarkable, on the condition of strain-rate below 10^{-2}. But the increase of strain-rate will made the tensile strength curve rise, and this enhanced with temperature rising, especially in the high temperature condition [7]. For an accurate simulation of the solidification process in casting, 50Cr5MoV liquids temperature, cooling-rate and strain-rate were calculated.

Combined the alloy steel liquids temperature formula with empirical revision, the liquids temperature formula of 50Cr5MoV was obtained,

$$T_L = 1538 - (\omega[\%C] + 6.2\omega[\%Si] + 1.7\omega[\%Mn] + 28\omega[\%P] + 40\omega[\%S] + 17\omega[\%Ti] + 2.6\omega[\%Cu]) - 1 \quad (1)$$

According to the solidification heat transfer principle, the cooling rate formula could be obtained:

$$C = (T_L - T_C) \times V_C \div L \div 60 \quad (2)$$

Where, T_L is liquids temperature of ingot; T_C is secondary cooling zone temperature;

V_C is casting speed, m/min; L is metallurgical liquid core length, m.

Mold strain rate can be calculated,

$$\dot{\varepsilon} = V_C |L - S| / SZ_0 \tag{3}$$

Where, L is mold transverse width, m; S is mold exit lateral arc length; Z_0 is mold height, m.

Correcting strain rate can be calculated,

$$\varepsilon = X \times (\frac{1}{R_1} - \frac{1}{R_2}) \times 100\% \tag{4}$$

Where, X is distances of a point to the center, $0 \leqslant X \leqslant D/2$, m; D is ingot thickness, m; R_1 is radius of the point before correcting; R_2 is radius of the point after correcting, m.

Combined the chemical composition of 50Cr5MoV steel with the actual technological process parameters and the Eq. (1)- Eq. (4), high temperature tensile parameters were obtained, as shown in Tab. II.

Tab. II High temperature tensile parameters

Liquids temperature/℃	Cooling rate/(℃.s^{-1})	Strain rate	Correcting strain rate
1465	3	6.46×10^{-3}	1.3×10^{-4}

Test result of high temperature mechanical properties

Fig.3 manifests the relationship between thermoplasticity and casting temperature. There are two low ductility regions, high temperature region (1300-1465℃), and low

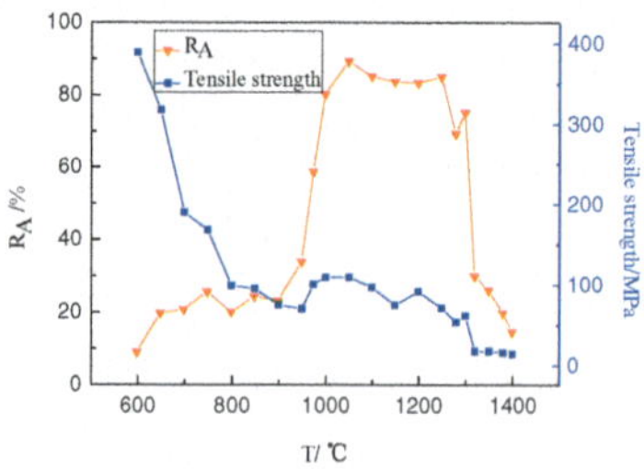

Fig.3 High temperature mechanical properties of 50Cr5MoV

temperature region (600-950℃), where the R_A is low (10%-40%) and changes dramatically. But during 950~1300℃ 50Cr5MoV has a favorable pyroplasticity and its R_A is above 60%. As the casting temperature increasing, the tensile strength shows a decreasing trend, but when the temperature is lower than 800℃, the decreasing range goes obvious slowly, that's because when the temperature increases, the form of dislocation movement has changed and a new mechanism arise, and function ability which strangle the dislocation weakens[8]. To improve the quality of 50Cr5MoV the reduction of area (RA) should be greater than 60% and avoid being correct in low ductility region, the correcting temperature should higher than 950℃.

Analysis of sample's fracture, micro structure and energy spectrum

High-temperature brittle region (1300–1465℃)

Fig.4 shows the specimen fracture morphology and fracture surface energy spectrum analysis at 1300℃. It can be seen from Fig.4a that, the grain interface is fuzzy, and part of the grain has been bonded by a film with thin liquid. From the 4b, the energy spectrum shows oxygen content 5.74%, nitrogen content 3.76%, carbon content 23.66%, sulfur content 0.16%, phosphorus content 0.11%. That is to say, harmful elements, including N、O、P、S are 10 times than elements in 50Cr5MoV matrix. At high-temperature brittle region, impurity elements gather at sample's fracture surface, causing the decrease of dendritic liquid membrane freezing point and the increase of liquid phase quantity. The brittle fracture is expanded along the branch boundary of the branch, and the impurity content of the ingot is increased. The brittle temperature range is enlarged and the zero strength temperature (ZDT) is increased [9].

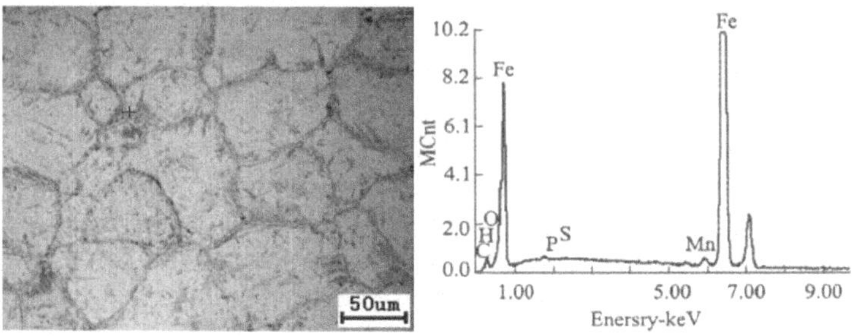

(a) (b)

Fig.4 The specimen fracture morphology and energy spectrum at 1300℃

High temperature plastic region(950–1300℃)

50Cr5MoV has a superior plasticity, and casting secondary cooling zone temperature needs to be controlled in this area, in order to effectively avoid cracking. Sample's fracture surface and fracture energy spectrum analysis at 1100℃, is shown in Fig.5. From the 5a, it is can be seen, there's a large tear dimple and precipitates in part of dimple at 1100℃. It has a good plasticity and demonstrates a transgranular fracture pattern. EDS (5b) analysis can be known, fracture has been oxidized, and the dimple has plenty of O(22.3%) and S(3.8%). they may exist as (Fe, Mn, A1)O 、 (Fe, Mn)S and etc. This kind of precipitates can decrease crystal boundary strength. The holes on the dimple ling may be caused by fracture carbides, inclusions and differences from matrix elastoplasticity, under the exogenous process the fracture deformed and holes appeared [10]。

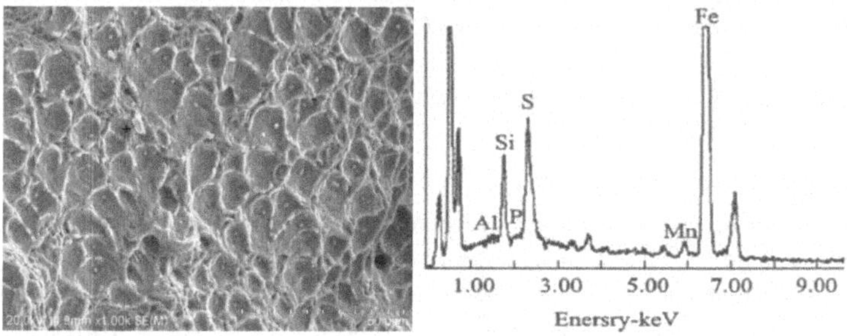

(a) (b)

Fig.5 The specimen fracture morphology and energy spectrum at 1100℃

Low-temperature brittle region (600-950℃)

Sample RA is 35% and its plasticity is unfavorable at 900℃. The average grain size is about 50um. From Fig.6a, fracture mainly made up by ductile intergranular fracture, brittle intergranular fracture as rock sugar and transcrystalline rupture partly. EDS spectrum analysis was carried out on the fracture cracks at grain boundaries (Fig.6b). The energy spectrum shows 30.35% carbon, 2.45% nitrogen and 23.39% oxygen. Precipitates of the fracture cracks might be tiny carbonitrides (Fe，Mn，A1)CN which are oxidized. From the 6c, fracture matrix microstructure is mainly gray martensite and partly ferrite. It suggest in 900℃ fracture matrix microstructure is made up by austenite and ferrite. The new generation of ferrite's strength is only austenite's 1/4 , brittle crack along the grain is expanded along the ferrite which separated out from the grain boundary of austenite. Transcrystalline rupture is caused by ferrite separated from austenite grain or compounds made by alloying elements in grain and harmful elements S、N、O and etc. Carbonitrides precipitates becomed the source of stress concentration at grain boundaries, and a collection of small voids caused cracks under the exogenous process.

Conclusions

Temperature of 50Cr5MoV casting ingot out crystallization should be lower than 1300℃, the correcting temperature higher than 950℃, Only which can effectively avoid crack.

The mechanism of crack formation demonstrates that, at low temperature brittleness region, the main crack way of casting ingot is brittle cleavage fracture (the intergranular and transgranular); At high temperature plastic region, the ingot is given priority to with the intergranular fracture toughness; At high temperature brittleness region, sample ruptures along the solid liquid two phase area.

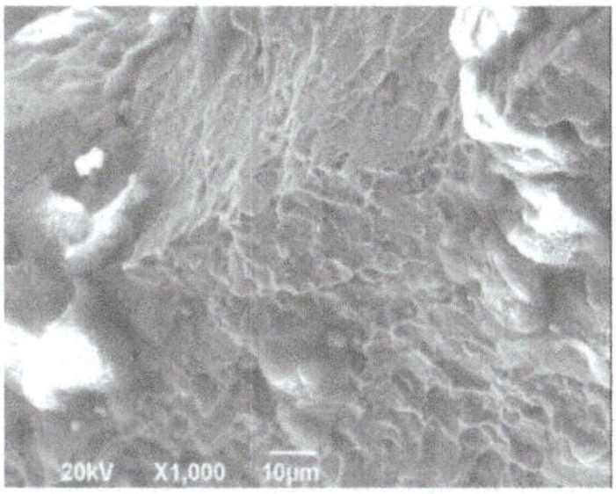

(a)

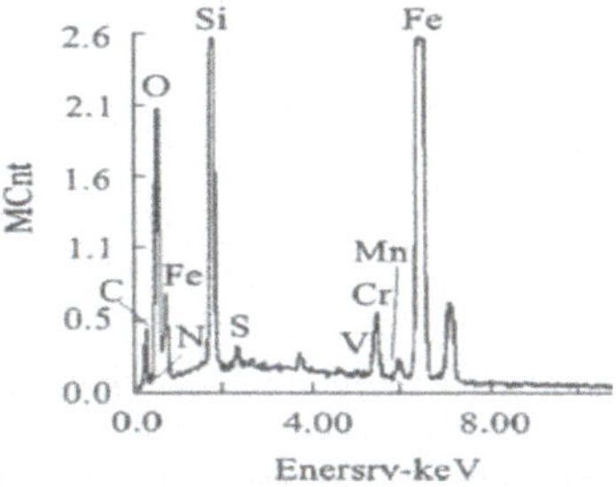

(b)

(c)

Fig.6 The specimen fracture morphology and energy spectrum at 900℃

References

[1] D. T. Kang, G. B. Ye, *Large Forgings Materials and Heat Treatment* (Bei Jing, BJ: Long Men Publishing Company1998), 454.

[2] Ohkomory Y, Kitagawa L, Shimozuka K, et al, "Study on Spalling of Hot Strip Mill Backup roll," *Journal of the Iron and Steel Institute of Japan*, 73 (1987), 691-698.

[3] Ohkomori Y, Kitagawa L, Shimozuka K, et al, "Cause and Prevention of Spalling of Backup Rolls for Hot Strip Mill," *Transactions of the Iron and Steel Institute of Japan*, 28 (1988), 68-74.

[4] Kapadia B M, Marsden K W, "Spalling Behavior of Back-up Roll Materials",(Paper Presented at the 39th Mechanical Working and Steel Processing Conference Proceedings, Indianapolis, IN, USA , Oct.19-22, 1997), 349-386.

[5] J. Wang, "Optimization of Continuous Casting Process for Improving The Quality of Medium Thin Slab" (M.Sc. thesis, Northeastern University 2006)

[6] Y. Gan, *Modern Continuous Casting Using Manual* (Bei Jing, BJ: Metallurgical Industry Press 2010), 100-125.

[7] Y. K. Yang, T. L. Wu, X. M. Jiang, *Metal High Temperature Strength and Test*

(Shang Hai, SH: Shanghai Science Press 1984), 33-37.
[8] J. S. Cheng, J. Liu, "Behavior and Analysis of High Temperature Thermoplastic of HG70 steel," *Research on Iron & Steel*, 38 (2010), 27-29.
[9] Y. H. Pan, "The Secondary Cooling Technology of Billet During High-efficiency Continuous Casting" (M.Sc. thesis, Chongqing University 2008), 15-20.
[10] Lankford W T, "Some Considerations of Strength and Ductility in the Continuous Casting Process," *Metallurgical Transactions*, 3 (1982),1331-1357.

7th International Symposium on High-Temperature Metallurgical Processing

Extraction and Recovery of Metals

Session Chairs:
Dean Gregurek
Ender Keskinkilic

7th International Symposium on High-Temperature Metallurgical Processing
Edited by: Jiann-Yang Hwang, Tao Jiang, P. Chris Pistorius, Gerardo R.F. Alvear F., Onuralp Yücel, Liyuan Cai, Baojun Zhao, Dean Gregurek, and Varadarajan Seshadri
TMS (The Minerals, Metals & Materials Society), 2016

SEPARATION OF RHENIUM AND MOLYBDENUM FROM MOLYBDENITE CONCENTRATE BY MICROWAVE-ASSISTED ROASTING

Tao Jiang, Linfeng Zhou, Zhiwei Peng, Guanghui Li, Rong Sun
(School of Minerals Pressing & Bioengineering, Central South University, Changsha, Hunan, Hunan 410083, China)

Keywords: Molybdenite concentrate, Rhenium, Molybdenum, Microwave, Roasting

Abstract

The effects of microwave roasting on separation of rhenium (Re) and molybdenum (Mo) from a molybdenite concentrate have been investigated based on evaluation of the heating characteristics of the concentrate in the microwave field. The results show that a variety of operating variables, namely microwave power, sample weight, sample geometry, roasting temperature, roasting time and atmosphere significantly impact on the volatilization rates of Re and Mo. The optimum conditions for Re-Mo separation are microwave power 1300 W, sample weight 20 g, sample geometry pellets, temperature 600 °C, time 1.5 h, and atmosphere 20%O_2. Under these experimental conditions, the volatilization rate of Re reaches 88.12%, while that of Mo is just 5.15%. Microwave-assisted roasting provides an alternative route to separating Re and Mo from molybdenite concentrate.

Introduction

Rhenium is a rare metal which has been widely used to produce reforming catalyst in oil industry and heat-resistant alloy. This metal is very rare in the earth's crust with content of just about 1×10^{-7}%. It is mainly distributed in molybdenite and copper sulphide ore, and sometimes found in lead, zinc, platinum, niobium and uranium ore.[1] The rhenium is mainly recovered from molybdenite concentrates by oxidative roasting and solidification roasting process.[2,3] During oxidative roasting molybdenum disulfide (MoS_2) changes to molybdenum trioxide (MoO_3) and rhenium sulfide (ReS_2) changes to rhenium heptoxide (Re_2O_7) which has a low melting point and a high solubility in water. The main chemical equations as follows:

$$MoS_2 + 7/2O_2 = MoO_3+2SO_2 \quad (1)$$

$$2ReS_2 + 15/2O_2 = Re_2O_7+4SO_2 \quad (2)$$

Because of these characteristics the rhenium heptoxide can be enriched in the dust while the molybdenum trioxide stays in calcine so that we can recover rhenium by washing dust with water or acid. During solidification roasting process molybdenum disulfide and rhenium sulfide reacts with calcium hydroxide ($Ca(OH)_2$) to generating calcium molybdate ($CaMoO_4$) and calcium perrhenate ($Ca(ReO_4)_2$) respectively. The main chemical equations as follows:

$$2MoS_2 + 6Ca(OH)_2 + 9O_2 = 2CaMoO_4 + 4CaSO_4 + 6H_2O \quad (3)$$

$$2ReS_2 + 5Ca(OH)_2 + 19/2O_2 = Ca(ReO_4)_2 + 4CaSO_4 + 5H_2O \quad (4)$$

The calcium molybdate and calcium perrhenate can be solubilized by leaching with dilute H_2SO_4, the Mo and Re can be separated by extraction and ion exchange. However, the operation conditions of traditional roasting processes mentioned above such as high roasting temperature and long roasting time are energy-intensive for today's society. Therefore, looking for a new kind of technology for Mo-Re separation is necessary.

Microwave heating is a unique and efficient technology for materials processing which was firstly used for treatment of iron ores in 1960s.[4] Around the same time, high-temperature microwave processing of oxide and sulfide minerals was reported.[5] Microwave heating has been extensively applied in various fields of materials processing because of its unique characteristics including volumetric and selective heating.[6] These characteristics contribute to many unexpected advantages including energy saving, improved product quality, faster processing and greater eco-friendliness when compared with conventional heating methods. While the application of mrciowave for roasting molybdenite concentrate have not been investigated. Microwave-assisted roasting may provide a new approach for separation of Re and Mo.

In this study, microwave roasting on separation of rhenium and molybdenum from a molybdenite concentrate is investigated. Through researching the heating characteristics of the concentrate in the microwave field the basic operation parameters of roasting were comfirmed. The major process of Re-Mo separation is volatilization of Re, therefore the volatilization rate of Re is the main evaluation index. Besides, the volatilization behavior of Mo and S were studied because of the significant influence to the volatilization behavior of Re. The investigation of microwave roasting may provide a potential high-efficiency method to separating Re and Mo.

Experimental methods

Materials

The molybdenite concentrate, having the chemical composition given in Table I, was supplied by Xi 'an Xincheng Investment co. LTD. The grade of Mo is 43.55%, it indicates that this kind of low grade molybdnite concentrate is suitable for roasting. The grade of Re is 321g/t which means the rhenium has great recovery value. The main mineral composition is shown in Figure 1. Molybdenum mainly exists in the form of molybdenite, The main gangue mineral is quartz and calcite.

Table I. Chemical Composition of Molybdenite Concentrate

Composition	Mo	Re	Fe	SiO_2	Al_2O_3	CaO	MgO	K_2O
Amount present	43.55	321 g/t	1.39	4.02	0.18	3.27	0.48	0.26
Composition	Na_2O	As	WO_3	Cu	Pb	Zn	P	S
Amount present	0.12	9.38g/t	1.66	0.026	4.52	0.06	0.016	32.46

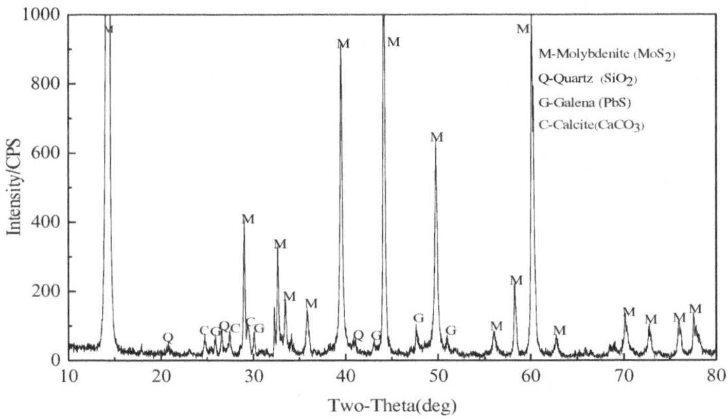

Figure 1. XRD Pattern of the Molybdenum Concentrate Sample

Experimental equipment

The microwave heating furnaces used in this study consists of a magnetron operating at 2.45 GHz with adjustable power within the range 0–1.5 kW. Figure 2 shows schematic of the furnace.

Procedure

At first, the heating characteristics of the concentrate in the microwave field was investigated. As microwave heating depends on sample geometry, molybdenum concentrate was made into pellets, briquettes and powder, respectively. Then the samples carried by the alundum crucible were put into the furnaces for heating until the temperature reached the highest value at the microwave power of 1200W. To evaluate the effect of microwave power on the separation of Re and Mo, the sample shape was kept unchanged while the microwave power was adjusted from 1000W to 1300W. After the basic parameters of the roasting experiment having been confirmed, the effect factors of roasting temperature, roasting time and atmosphere on the volatilization rates of Re and Mo were studied. First, the green pellets with the diameter ranging from 10mm to 12mm were dried in a baking oven at 80 ℃ for 4 h. The pellets were loaded into the alundum crucible, which was put into the quartz tube in the microwave heating furnace for heating at a set temperature, time and atmosphere. The volatilization rate is **α.** Equation is as follows:

$$\alpha = 1 - (M_2 \times \beta_2) / (M_1 \times \beta_1) \qquad (5)$$

where M_1 and M_2 denote the weights of pellets before and after heating, and β_1 and β_2 represent the grades of Re, Mo or S before and after heating. The content of rhenium was measured by ICP, while other species were determined by chemical analysis.

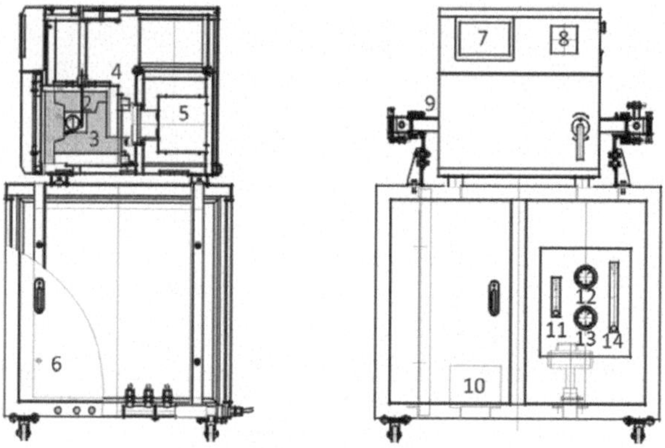

Figure 2. Schematic of Microwave Furnace
1–materials, 2–thermocouple, 3–thermal insulation material, 4–Furnace body, 5–magnetron, 6–workbench, 7–control panel, 8–vacuometer, 9–alundum tube, 10–vacuum pump, 11–flowmeter A, 12–water pressure gauge, 13–barometer, 14–flowmeter B

Result and discussion

Heating characteristics of the concentrate in the microwave field

The effect of sample shape on the heating rate was determined by fixing all other experimental variables, such as sample mass 20 g, microwave power 1300 W and gas flow rate 4 L/min. The effect of sample shape on the heating rate is given in Figure 3. The average heating rates for samples of different shapes are summarized in Table II.

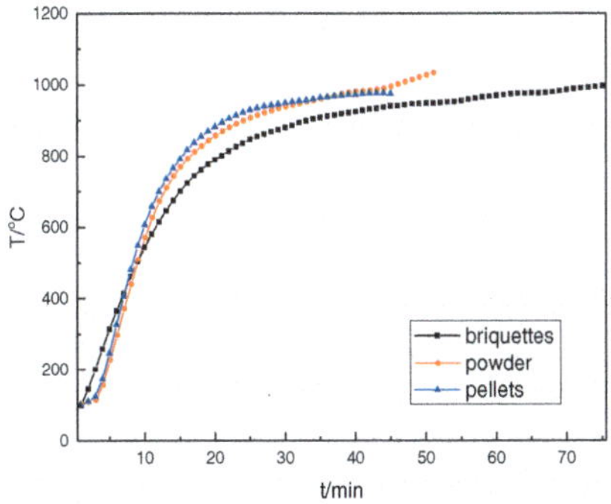

Figure 3. Effect of Sample Shape on the Heating Rate

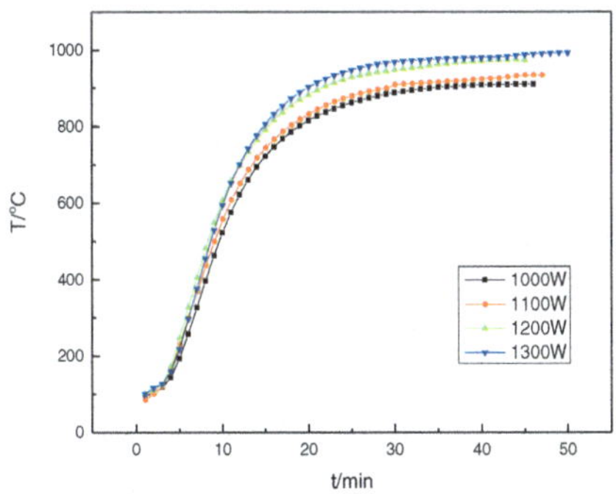

Figure 4. The Effect of Microwave Power on the Heating Rate

Table II. The Average Heating Rate of Different Sample Shapes

shape	briquettes	powder	pellets
heating rate（°C/min）	16.5	18.3	19.3

Figure 3 shows that the pellets have the fastest heating rate while the briquettes have the slowest heating rate because the compact structure of briquettes makes oxygen diffusion inside the sample difficult. On the contrary, the powder has large surface exposure to air, which would lead to fast oxidation reaction and sintering occurred between powder and the inner wall of alundum crucible. According to the above factors, pellet is the most suitable shape for microwave heating.

The effect of microwave power on the heating rate under air atmosphere, sample mass

20 g, sample shape pellets, gas flow rate 4 L/min is shown in Figure 4. The average heating rate was listed in Table III.

Table III. The Average Heating Rate of Different Microwave Power

power	1000W	1100W	1200W	1300W
heating rate（°C/min）	17.8	18.1	19.4	19.7

The heating rate of molybdenite concentrate can reach the highest value at around 600 °C and it becomes constant after 800 °C. According to the Table III， the most suitable microwave power is 1300 W.

The behavior of rhenium in the microwave oxidative roasting

Effect of roasting temperature The roasting temperature had significant effects on volatilization of Mo, Re and S as shown in Figure 5. The experiments were performed under air atmosphere roasting at different temperatures (450 °C—650 °C) for 120 min.

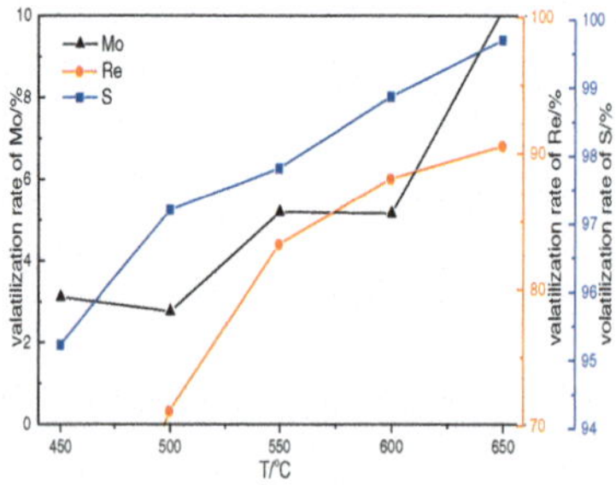

Figure 5. Effect of Roasting Temperature on the Volatilization Rates of Mo, Re and S

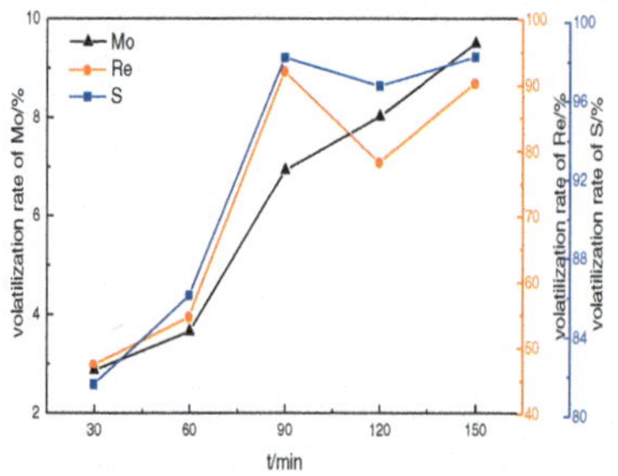

Figure 6. Effect of Roasting Time on the Volatilization Rates of Mo, Re and S

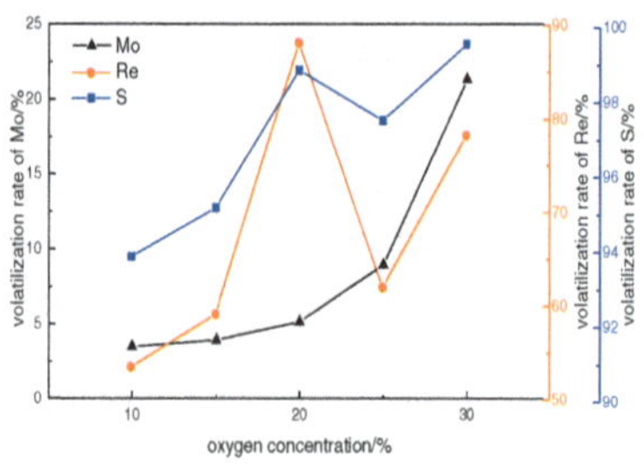

Figure 7. Effect of Roasting Atmosphere on the Volatilization Rates of Mo, Re and S

The volatilization rates of Mo, Re and S reach 3.11%, 47.82% and 95.22% at 450 °C after 120 min, respectively. It can be inferred that most molybdenum disulfide (MoS_2) has been oxidized while a bit of rhenium disulfide (ReS_2) oxidizes at this temperature. As the temperature increases from 450 °C to 600 °C, the volatilization rate of Re increases from 47.82% to 88.12%. Meanwhile, the volatilization rate of S increases from 95.22% to 98.87%, and the volatilization rate of Mo increases from 3.11% to 5.15%. Most of rhenium volatilizes to the air in the form of rhenium heptoxide ($Re_2 0_7$). However, as the temperature keeps increasing to 650 °C, the volatilization rate of Mo increases rapidly to 10.16% while the volatilization rate of Re remains relatively constant. More molybdenum trioxide (MoO_3) volatilization inhibits separating Re and Mo in the leaching processes. The optimal roasting temperature is 600 °C.

<u>Effect of roasting time</u> Figure 6 shows the effect of roasting time on the behavior of rhenium and the relationship of volatilization of Mo, Re and S in the molybdenum concentrate. Sample roasting was performed under air atmosphere at 600 °C for different durations (30 min—150 min). The volatilization rates of Mo, Re and S reach 2.87%, 47.52% and 81.66% after roasting at 600 °C for 30 min respectively. In such a short time both molybdenum disulfide and rhenium disulfide could not be oxidized completely. As the time increases from 30 min to 90 min, the volatilization rate of Re increases from 47.52% to 92.17% and the volatilization rate of S increases from 81.66 % to 98.23%. It can be deduced that the oxidation reaction of Re occurs after the oxidation reaction of Mo. Therefore, the volatilization rate of S is much larger than the volatilization rate of Re. As the roasting time reaching 150 min, the volatilization rates of Re and S keep constant where as that of Mo increases continuously. In conclusion, the optimal roasting time is 90 min.

<u>Effect of roasting atmosphere</u> Figure 7 shows the effect of roasting atmosphere on the volatilization rates of Mo, Re and S. The experiments were performed under different atmosphere with oxygen contents varying from 10% to 30% at 600 °C. Under low oxygen atmosphere, the molybdenite concentrate inside the pellets could not contact with oxygen sufficiently because the oxygen diffusion rate is relative slow. The oxidation reaction just occurs at the surface of pellets so that the volatilization rates of Mo, Re and S are just 3.50%, 53.56% and 93.90% respectively. Obviously, the volatilization rates of Re and S increase rapidly with the oxygen concentration increasing from 10% to 20%, reaching 88.12% and 98.87%, respectively. However, with the oxygen concentration increasing to 25% the huge quantity of heat generated from drastic oxidation reaction makes pellets melted, which lowers volatilization rates of Re and S to 62.02% and 97.54%. According to volatilization rate curve, high and low oxygen atmosphere is inappropriate for microwave roasting. The optimal roasting atmosphere is 20% O_2.

Conclusions

Microwave-assisted roasting has been used to separate Re and Mo from molybdenite concentrate. The results indicate this technique is an effective method for separation of the metals in molybdenite concentrate. The optimum conditions for Re-Mo separation are microwave power 1300 W, sample geometry pellets, temperature 600°C, time 1.5 h, and atmosphere 20%O_2. Under these experimental conditions, the volatilization rate of Re reaches 88.12%, while that of Mo is just 5.15%. Compared with the conventional roasting process, for the same indexes the roasting time can be cut at least half an hour with microwave irradiation. The heating efficiency is extremely high because of its unique characteristics including volumetric and selective heating. It can provides a new high-efficiency approach for Mo-Re separation through the investigation of microwave-assisted roasting.

References

1. Shunchang Wang, Shouzhi Qi, "Resource, Application and Market of Rhenium," *World Nonferrous Metals*, 02 (2001), 12-14.

2. P.R. Ammann, T.A. Loose, "Rhenium Volatilization during Molybdenite Roasting," *Metall. Trans*, 05 (1972), 1020–1021.

3. J.M. Juneja, Sohan Singh, D.K. Bose, "Investigations on the Extraction of Molybdenum and Rhenium Values from Low Grade Molybdenite Concentrate," *Hydrometallurgy*, 1996, 201-209.

4. L.H. Connell, L.A. Moe, "Apparatus for High Frequency Treatment of Ore," *US Patent*, (02) 1966.

5. J. D. Ford, D. C. T. Pei, "High Temperature Chemical Processing via Microwave Absorption," *Microw. Power*, 02 (1967), 61–64.

6. Zhiwei Peng, Jiann-Yang Hwang, "Microwave-assisted Metallurgy," *International Materials Reviews*, 1 (60) (2014), 30-63.

7th International Symposium on High-Temperature Metallurgical Processing
Edited by: Jiann-Yang Hwang, Tao Jiang, P. Chris Pistorius, Gerardo R.F. Alvear F., Onuralp Yücel, Liyuan Cai, Baojun Zhao, Dean Gregurek, and Varadarajan Seshadri
TMS (The Minerals, Metals & Materials Society), 2016

ACTIVE OXIDATION AND FUME FORMATION FROM LIQUID SiMn

Ida Kero[1] and Gabriella Tranell[2]

[1]SINTEF Materials and Chemistry; Alfred Getz vei 2, NO-7465 Trondheim, Norway
[2]Norwegian University of Science and Technology (NTNU); Alfred Getz vei 2, NO-7491 Trondheim, Norway

Keywords: Silicomanganese, oxidation, dust, fume

Abstract

Dust, or airborne particulate matter, from metal smelting can be either mechanically or thermally generated and it negatively affects the indoor air quality as well as the fugitive emissions from the plant. The thermally generated dust is often referred to as fume and it is typically generated whenever liquid metal comes into contact with air. For practical reasons, such as vehicle access, etc. it is often difficult to collect 100% of the fume. It is therefore highly desirable to develop processes and optimize operational procedures so that fuming can be kept to a minimum. In this series of experiments, the mechanism of active oxidation and fume formation from liquid silicomanganese under an impinging air jet in the temperature range 1400-1600°C has been explored. Characteristic properties of the fume have been established; particle diameter, shape, size distribution, and elemental composition are reported. The oxidation process is described in terms of vaporization, oxidation and agglomeration. In the boundary layer, two competing mechanisms seem to be operative in parallel: a direct oxidation of Mn vapor, and a two-step oxidation of silicon. The thermodynamic and kinetic conditions governing the two mechanisms are discussed.

Introduction

In the metallurgical industry today, extensive efforts and investments are directed towards improvements in occupational environment as well as emission control. This is typically done through large and sometimes expensive ventilation and filter systems, designed to collect and control the off-gases and fumes from the various process operations. From the corporate point of view, a process optimized to minimize fume formation would not only improve the indoor air quality and decrease the environmental impact of the plant; it would also minimize material and energy losses associated with the production. At the moment, however, fundamental knowledge about the origin and characteristics of most fumes is incomplete or lacking. The processes by which metallurgical fume forms are often vaguely understood, and rests heavily on relatively old studies on pure substances. As the chemical composition, phase composition and morphology of particles originating from different processes, conditions and alloys may vary greatly, it is also important to characterize the particles and link them to their origin.

This paper aims to describe the rate of oxidation and subsequent fume formation from a liquid SiMn alloy surface exposed to an impinging air jet. An additional objective is to provide basic characteristic information about the fume particles thus formed. The process of oxidation of silicon-rich alloys has been studied by several authors [1-3]. The oxidation and concurrent fuming from metallurgical grade silicon (MG-Si) have been described by Naess et al.[4-6] and the chemical composition of fumes from MG-Si and FeSi have been reported [7-9]. The properties and particle size distribution of industrial fumes from SiMn production sites have also been investigated [10-13].

Theoretical Background

Wagner's theory from 1958 [14] describes the fundamental principles of active oxidation of pure silicon. Since the development of the theory, the oxidation of silicon and silicon-rich alloys has been experimentally investigated by Hildal [2], Hinze and Graham [3] and Naess and co-workers [4, 5, 15]. Hinze and Graham verified Wagners theories through the finding that oxygen dissolved in the silicon alloy under oxygen exposure, thereby showing that oxygen can penetrate the boundary layer and form SiO gas. Naess and co-workers showed that oxygen diffusion through the gas boundary layer is rate limiting for the oxidation under both laboratory and industrial conditions and that the SiO to SiO_2 oxidation reaction takes place in the boundary layer at and above moderate gas flow rates. They also showed a correlation between SiO_2 particle size and gas flow rate

Turkdogan's theory from 1963 [16] describes a direct oxidation mechanism based on metal evaporation, enhanced by the formation of a so-called oxide mist above the boundary layer. Under this scheme, the metal vaporization is rate limiting. No, or very little oxygen reaches the metal surface as it is completely consumed in the boundary layer. Turkdogan and coworkers corroborated their theory experimentally by showing that it could be applied to a number of metals, including Mn and Fe. The direct oxidation mechanism seems quite adequate for metals with high vapor pressures, such as Mn. Lee & Kolbeinsen [17] suggested, however, that the rate limiting mechanism for FeMn fuming at temperatures above 1600°C is not the vaporization rate in itself, but rather the diffusion of the oxide mist through the gaseous boundary layer.

The temperature dependence of the vapor pressures of SiO (g) over Si (l) and Mn (g) over Mn (l) are shown in Figure 1. The vapor pressures were calculated for Si, SiO and Mn vapor, using the HSC software [18]. The vapor pressures of Mn and SiO are four orders of magnitude higher than that of Si (0,14 Pa at 1500°C).

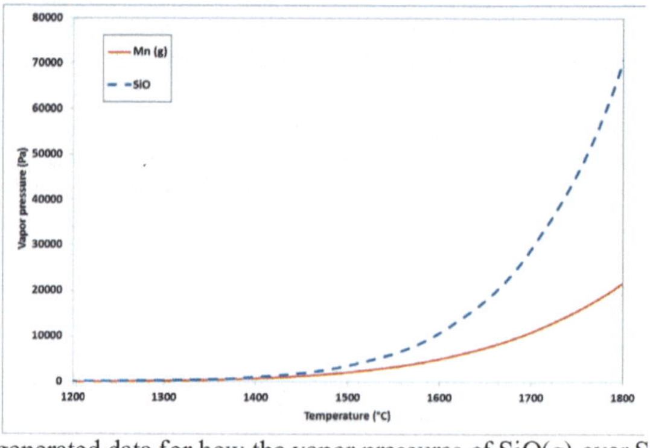

Figure 1. HSC-generated data for how the vapor pressures of SiO(g) over Si (l) and Mn (g) over Mn (l) vary with temperature.

Method

The term "fume," as used in this article, designates airborne particulate matter formed by the evaporation of liquid metal and subsequent oxidation of the metal vapor. The fuming experiments were carried out in an induction furnace with a graphite crucible, fitted with a graphite lid. The set-up is schematically illustrated in Figure 2. The lid has three orifices through which a graphite- and alumina-protected thermocouple, a graphite protected alumina lance and a steel exhaust pipe was mounted. The exhaust pipe is henceforth referred to as the "transition tube". It connects the cooler to the graphite lid. The cooler is a steel tube with a cooling water copper tube coil. The cooler is in turn connected to a dust collector, containing the filter. The filter fabric is industrial standard for SiO_2 capture in the silicon industry.

The induction heating of the furnace introduces a slow, vertical stirring movement in the alloy as it is melted; the alloy surface, however, remains visually motionless. The furnace was heated at a rate of approximately 30 degrees per minute and when the temperature was reached, technical (dry) air was introduced through the lance. The lance tip was positioned 2 cm above the liquid alloy surface and the air flow was regulated to 3 liters per minute. The experiments were carried out at temperatures between 1400 and 1600°C and with holding times of 10 to 20 minutes. Two reference experiments were carried out at two different temperatures, without gas being introduced through the lance – i.e. an atmosphere given by the crucible and metal present. These experiments confirmed that no detectable fume was produced without the air flow. In each experiment, 4 kg standard grade silicomanganese was heated and melted in the induction furnace. The alloy was of the size fraction 10-25 mm with approximately 1-2% C, 17-20 % Si, 65-70 % Mn and 7-10% Fe.

The fume collected from the various sites in the experimental set-up was analyzed with respect to particle shape, size, size distribution, elemental composition and phase composition by, respectively, scanning electron microscopy (SEM), Brunauer-Emmett-Teller surface adsorption (BET), inductively coupled plasma optical emission spectroscopy (ICP-OES) and x-ray diffractometry (XRD). The ICP-OES is not able to detect light elements such as oxygen and carbon. An estimate of the oxygen concentration was however calculated from the ICP-OES results by assuming that all the mass not accounted for by the elemental analysis was oxygen. The presence of agglomerates in ferroalloy fume is well documented. Dingsøyr et al. [19] applied the notions of "primary" and "secondary" agglomerates to describe the fume particles; the former being particles where the protoparticles are held together by material bridges and the latter being primary agglomerates attached to each other by weak van der Waals forces. Generally, such attachments are not visible on SEM and it is therefore only protoparticle size and size distribution which can be evaluated by SEM.

A flux was calculated from the amount of fume produced in each experiment according to:

$$J_m = \frac{m}{At^2} \qquad (1)$$

where J_m is the mass flux, m is the mass of fume produced, A is the surface area of the molten metal and t is the holding time of the experiment.

Results

The fume was multicolored and settled at different sites in the experimental equipment as illustrated in Figure 2.

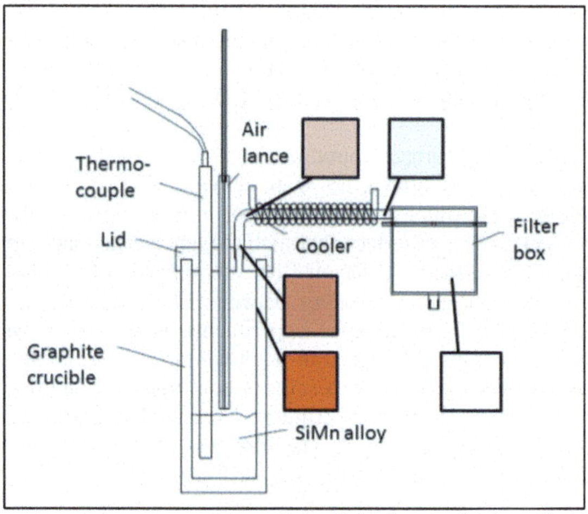

Figure 2. The experimental set-up and the colors of the fume which settled in different areas of the set-up.

The graph in Figure 3 shows the flux, plotted as a function of the temperature and Figure 4 shows scanning electron micrographs of some typical fume particles. The primary units, also known as protoparticles, are spheres of various sizes. The spheres appear to have very smooth surfaces and nano-sized protoparticles are clearly visible.

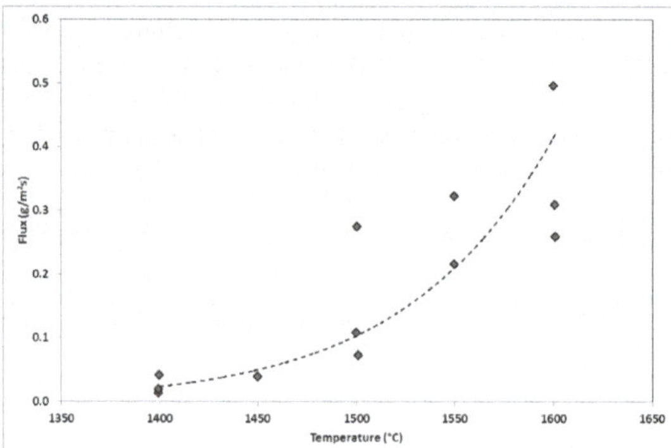

Figure 3. Measured mass flux over the SiMn alloy as a function of temperature

Figure 4. Scanning electron micrographs of some typical fume. Magnifications of a) 10 000 times and b) 50 000 times.

An average particle diameter of 0.18 µm was calculated from the BET surface area, assuming MnO density. This agrees well with the protoparticle diameters observed by SEM. The protoparticle size distributions of the filter fumes from three different temperatures, as obtained by visual evaluation of the SEM results, are shown in Figure 5.

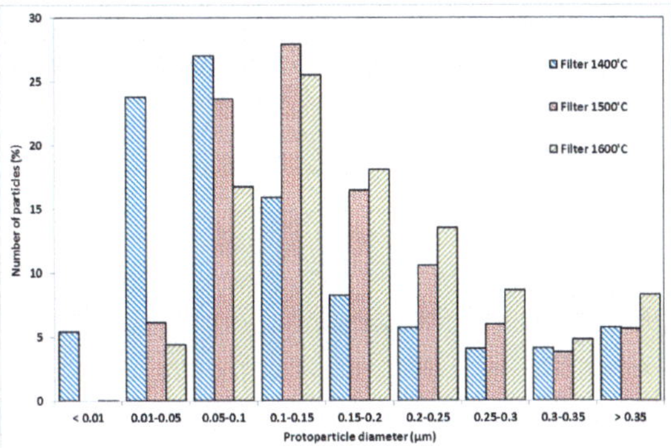

Figure 5. Protoparticle size distributions, as obtained by SEM, for filter fumes.

Discussion

During the oxidation and fuming from a liquid silicomanganese surface, two competing mechanisms may be operative in parallel: a direct oxidation of Mn vapor, similar to that described by Turkdogan [16] and a two-step oxidation of silicon in accordance with Wagner's theory [14]. This is schematically illustrated in Figure 6.

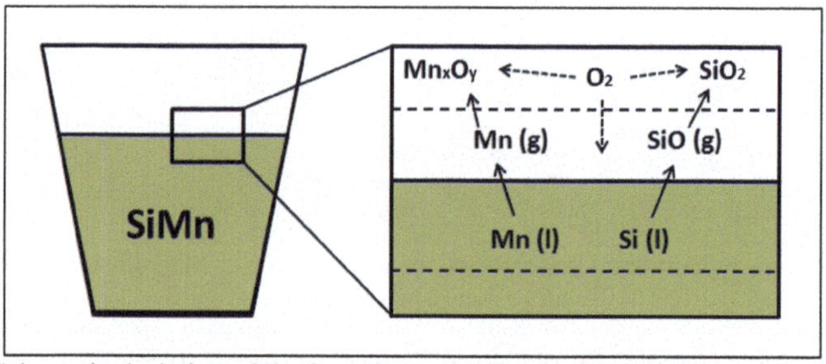

Figure 6. Schematic depiction of the fume formation from a liquid SiMn surface, illustrating the two competing oxidation processes of SiO and Mn.

At the lower temperatures, the vapor pressures of Mn and SiO are in the same order of magnitude but the vapor pressure of SiO increases much faster with temperature than that of Mn, this is shown in Figure 1. Therefore, the dominance of SiO oxidation becomes more pronounced with increasing temperature. Figure 7 shows that the molar Si/Mn ratio is linearly proportional to the vapor pressure of SiO (R^2=0.9994). Sun & Pehlke [20], observed that Si was oxidized before Mn during the oxidation of liquid iron. In our case, the ICP-OES analysis showed that the Mn concentration in the PM is decreasing with temperature while the reverse is true for Si.

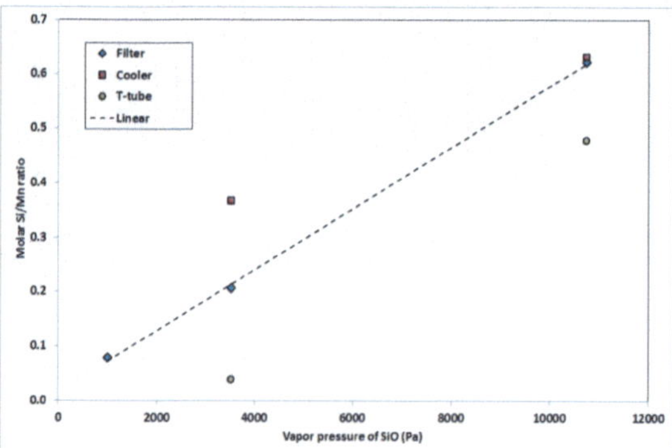

Figure 7. The molar Si/Mn ratio plotted against the vapor pressure of SiO (g).

The protoparticle diameter and size distribution are coarser than condensed silica fume from metallurgical grade silicon [5, 19, 21]. It should be kept in mind that the protoparticle size distribution is obtained by visual evaluation of the scanning electron micrographs and may be very different from an agglomerate size distribution.

These experiments have shown that the amount of fume produced by an impinging air jet on a liquid SiMn surface is of similar order of magnitude as the flux of condensed silica fume from MG-Si [21]. The particle shape is also very similar to results for MG-Si. Previously reported results from industrial measurements have shown that the fume collected in a SiMn plant was quite similar to that from a FeSi plant [12, 13].

Conclusions

The oxidation of liquid silicomanganese under an impinging air jet has been studied. Two oxidation mechanisms may be operating in parallel: a direct oxidation of Mn vapour and a two-step oxidation of Si, first to SiO and then SiO_2. The dominance of the SiO oxidation becomes more pronounced with increasing temperature. The vapor pressure of SiO increases much faster with temperature than that of Mn and the molar Si/Mn ratio is linearly proportional to the vapor pressure of SiO. Further studies are needed to illustrate the exact thermodynamic and kinetic mechanisms which govern the fuming of SiMn.

The characteristic properties of the fume generated by the active oxidation of SiMn have been established. Chemical composition and average particle diameter are reported as well as protoparticle shape and size distribution. The fume is largely amorphous and with an average particle diameter of approximately 0.18 μm.

Acknowledgements

These studies were funded by the Norwegian Ferroalloy Producers Research Association (FFF) and the Norwegian Research Council (NFR) through the project "Fugitive Emissions of Materials and Energy" (FUME). Irene Bragstad provided the BET analysis, Dmitry Slizovskiy carried out extensive laboratory work and Jan Inge Meling performed the protoparticle size assessment.

References

1. C. Wagner, "Passivity during the Oxidation of Silicon at Elevated Temperatures", *Journal of Applied Physics*, 29(9)(1958), 1295-1297.
2. K. Hildal, *Steam explosions during granulation of Si-rich alloys: Effect of Ca- and Al-additions*, in *Department of Materials Science and Engineering*. 2002, Norwegian University of Science and Technology (NTNU).
3. J.W. Hinze and H.C. Graham, "The active oxidation of Si and SiC in the viscous gas-flow regime", *Journal of the Electochemical Society*, 123(7)(1976), 1066-1073.
4. M. Næss, et al., "Active Oxidation of Liquid Silicon: Experimental Investigation of Kinetics", *Oxidation of metals*, 78(5-6)(2012), 363-376.
5. M.K. Næss, et al., "Mechanisms and kinetics of liquid silicon oxidation during industrial refining", *Oxidation of metals*, 78(3)(2012), 239-251.
6. M.K. Næss, et al., "Fugitive emissions related to oxidation of liquid silicon during ladle refining", *Oxidation of Metals*, DOI: 10.1002/9781118062081.ch11(2012.
7. E.H. Myrhaug and H. Tveit, *Material Balances of Trace Elements in the Ferrosilicon and Silicon Processes*, in *58th Electric Furnace Conference*. 2000: Orlando, FL.
8. M.K. Naess, et al., "A new method for estimation of emissions and sources of measurements error in the silicon refining process", *JOM*, 65(8)(2013), 997-1006.
9. M.K. Naess, et al., "Element distribution in silicon refining – Thermodynamic model and industrial measurements", *JOM*, 66(11)(2014), 2343-2354.
10. K. Gjønnes, et al., "Characterisation of workplace aerosols in the manganese alloy production industry by electron microscopy", *Analytical and Bioanalytical Chemistry*, 399(3)(2011), 1011-1020.
11. S. Gunst, et al., "Chemical composition of individual aerosol particles in workplace air during production of manganese alloys", *Journal of Environmental Monitoring*, 2(2000), 65-71.

12. I. Kero, et al. "Fume characterization in the ferroalloy industry" (INFACON XIII, Almaty, Kazakhstan,2013.
13. I. Kero, et al., "Particle size distributions of particulate emissions from the ferroalloy industry evaluated by electrical low pressure impactor (ELPI)", *Journal of Occupational and Environmental Hygiene*, 12(1)(2015), 37-44.
14. C. Wagner, "Passivity during the oxidation of silicon at elevated temperatures", *Journal of Applied Physics*, 29(1958), 1295-1297.
15. M.K. Næss, et al. "Fugitive emissions related to oxidation of liquid silicon during ladle refining" (2nd International Symposium on High-Temperature Metallurgical Processing, TMS Annual Meeting & Exhibition, San Diego, CA, USA,2011.
16. E.T. Turkdogan, et al., "Enhancement of diffusion-limited rates of vaporization of metals", *Journal of Physical Chemistry*, 67(1963), 1647-1653.
17. Y.E. Lee and L. Kolbeinsen, "Kinetics of oxygen refining process for ferromanganese alloys", *ISIJ International*, 45(9)(2005), 1282-1290.
18. *HSC Chemistry 7.1*, A. Roine, Editor. 2011, Outotec Research Center Oy: Helsinki, Finland.
19. E. Dingsøyr, et al., *Particle size and particle size distribution of Elkem Microsilica and its relevance to technical application*, in *The 5. European Symposium Particle Characterization*. 1992: Nurnberg, Germany.
20. H. Sun and R.D. Pehlke, "Modeling and experimental study of gaseous oxidation of liquid iron alloys", *Metallurgical and Materials Transactions B*, 27(1996), 854-864.
21. M.K. Næss, et al., *Experimental study of rate of silica fume formation from active oxidation of liquid silicon*, in *INFACON XIII - the 13th International Ferroalloys Congress*. 2013: Almaty, Kazakhstan. p. 957-966.

7th International Symposium on High-Temperature Metallurgical Processing
Edited by: Jiann-Yang Hwang, Tao Jiang, P. Chris Pistorius, Gerardo R.F. Alvear F., Onuralp Yücel, Liyuan Cai, Baojun Zhao, Dean Gregurek, and Varadarajan Seshadri
TMS (The Minerals, Metals & Materials Society), 2016

Research on Enrichment of MFe and RO Phase from Converter Steel Slag by Super Gravity

Chong Li, Jintao Gao, Zhancheng Guo
State Key Laboratory of Advanced Metallurgy, University of Science and Technology Beijing, 100083,
Beijing, P. R. China.
Keywords: Super gravity, Steel slag, Centrifugal enrichment, RO phase.

Abstract

A new approach to enrich MFe and RO phase from the converter slag by super gravity was investigated. The samples obtained by the gravity coefficient G=600 and cooling rate V=5K/min from 1623K to 1373K appeared significant stratification and the MFe gathered at the bottom of sample. With the help of metallographic microscopy and X-ray diffraction, it is found that large quantity of white RO phase gathered at the lower part of the sample as the concentrate, with volume fraction appearing gradient distribution along the direction of super gravity, while the upper part was mainly composed of dicalcium silicate as the tailing. The mass fractions of the FeO, MgO and MnO in the concentrate are up to 40.46%, 6.9% and 3.91%, while that in the tailing are just 11.65%, 2.15% and 1.55%. The recovery ratios of FeO, MgO and MnO in the concentrate are 60.35%, 58.45% and 72.24%, respectively.

Introduction

Steel slag is one of major by-products in steelmaking process, whose output is about 100-150kg for every ton of crude steel [1]. Except for limited utilization, most of them are disposed as waste, which leads to land resource waste and environment pollution. There are large quantities of metallic iron (MFe) and valuable metallic oxides such as FeO, MgO and MnO inlaying in steel slag, if they can be separated from slag, the separated high-grade metallic minerals could be reused into steelmaking process, and the tailing slag could be used as melting agent, concrete additive and steel slag cement etc [2,3].

The steel slag was mainly composed of three phases, namely dicalcium silicate (C2S), dicalcium ferrite (C2F) and solid solution of FeO, MgO and MnO (RO phase) [4,5]. C2S as Fe-poor phase will firstly precipitate from steel slag during solidification, and its density is lower than that of residual Fe-containing phases [6]. The RO phase will precipitate after C2S and its density is higher than that of residual melt. Up to date, magnetic separation is the main way to separate MFe and Fe-containing mineral from steel slag [7-9]. However, due to the small size and embedding distribution of MFe and Fe-containing mineral in steel slag, this method cannot recover MFe and Fe-containing mineral sufficiently. In addition, there are many other methods on separation of Fe from steel slag have also been studied, including gravity separation, floatation separation and carbon reduction etc [10]. However, those traditional methods cannot also separate Fe from the steel slag effectively. It is necessary, therefore, to find a novel method to separate Fe from steel slag in order to increase slag recycling. Considering the facts that super gravity has the traits to control the distribution of different phases and have successfully been applied in the separation and recovery of valuable elements from different kinds of slags [11-14] and removing impurities from alloy melt [15,16], it would be possible to recover MFe and RO phase from steel slag by super gravity.

In this work, the super gravity is used to enrich MFe and RO phase from converter steel slag. Simultaneously, microstructure of the sample and the recovery ratio of the sample after centrifugal enrichment have also been investigated.

Experimental

Experimental material and apparatus

The chemical composition of the steel slag from the Laiwu Steel Group Ltd in china was listed in table 1. In order to decrease melting point and viscosity of the steel slag, 10 percentage of CaF_2 needs to be added in the steel slag.

Table Ⅰ. The Chemical Composition of Steel Slag from Laiwu Steel Plant (mass,%)

composition	CaO	MgO	SiO_2	Al_2O_3	TFe
content	32. 88	5. 42	13. 04	0. 88	26. 49
composition	MFe	FeO	Fe_2O_3	MnO	P_2O_5
content	6. 36	21. 96	4. 37	2. 33	1.54

The super gravity field was generated by the centrifugal apparatus, as shown in figure 1. The heating furnace symmetrical with a counterweight about the rotation axis was fixed into a centrifugal rotor. It is heated with 0Cr27AlMo2 resistance wire whose highest temperature attains 1623K and controlled by a program controller with an R type thermocouple. The gravity coefficient, a ratio of super-gravitational acceleration to gravitational acceleration, was calculated by Eq. (1).

$$G = \frac{\sqrt{g^2 + (\omega^2 r)^2}}{g} = \frac{\sqrt{g^2 + (\frac{N^2\pi^2 r}{900})^2}}{g} \quad (1)$$

Where N is the rotating speed of the centrifugal, r/min; ω is the angular velocity, rad/s; r is the distance from the centrifugal axis to the sample, 0.25m; and g is normal gravitational acceleration, 9.8m/s^2. When N=0, G is equal to 1.

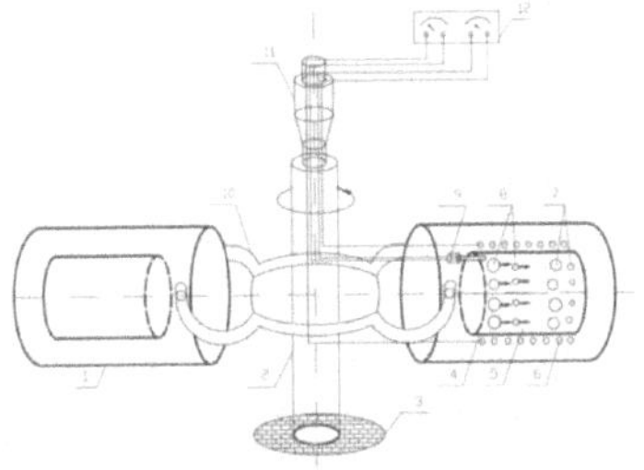

Figure 1. Schematic diagram of centrifugal separation apparatus;
1. Counterweight; 2. Centrifugal axis; 3. Base; 4. Magnesia crucible; 5. Slag melt; 6. Resistance coil;7. MFe and RO phase after centrifugal enrichment; 8. MFe and RO phase before centrifugal enrichment; 9. Thermocouple; 10. Horizontal rotor; 11.Conductive slipping; 12. Temperature controller.

Sample preparation and analysis

30g steel slag after completely mixed with 10% CaF_2 was put into a magnesia crucible and heated to the target temperature. The target temperature was kept for 10 min to ensure the slag fully melted, and then the centrifugal apparatus was started and adjusted to the specified angular velocity. The centrifugal apparatus was not turned off until the target time and then the sample was taken out to be quenched by water finally. The sample obtained by supper gravity was divided into halves along the center axis. One part was polished and investigated on the metallographic microscope and image analyzer software to gain the volume fraction of RO phase. The other one was cut along the interface between different colors into two parts and both of them were characterized by X-ray fluorescence and X-ray diffraction to determine the mineral composition and chemical component. At the same time, the parallel sample was prepared at same temperature and time without centrifugal treatment.

Results and Discussion

Macro and micro structure of the sample obtained by centrifugal enrichment

Figure 2 shows the macrostructure of the sample sectioned along the direction of super gravity after centrifugal enrichment with the gravity coefficient G=600 and cooling rate V=5 K/min from 1623 K to 1373 K, compared with the parallel sample without centrifugal enrichment. As shown in figure 2(b), all sizes of MFe gathered at the bottom of the sample after centrifugal enrichment and obvious stratification appeared at about 1cm position to the bottom of the sample. However, MFe in the parallel sample distributed uniformly and there was also no obvious stratification in the matrix, as shown in figure 2(a). In order to further investigate the microstructure of the sample, the positions of six areas characterized by the metallographic microscopy are shown in figure 2(b). The corresponding results are given in figure 3, wherein the white particles are RO phase. The volume fraction of RO phase presents gradient distribution along the direction of super gravity. It occupies just a small fraction from area a to area c, experiences a sharp increase in the stratified area d and achieves the maximum in the region f.

Under the normal gravity G=1, any sizes of MFe could not be separated from the steel slag because of the high viscosity, neither does the RO phase. However, super gravity as an intensified separation technique could affect the distribution of different phases in melt. The density of MFe is larger than that of steel slag and increasing super gravity coefficient equivalently improves the gravity acceleration of the MFe in the melt. As for the stratification, due to the adding of fluorine, dicalcium silicate-the first precipitated phase of high melting point-will partly react with fluorine to form low melting point eutectic, which reduce the melting point and viscosity of the steel slag. During the cooling process from 1623 K to 1373 K, the RO phase gradually precipitatd prior to the low melting point eutectic, and then was driven to the bottom to form the concentrate by super gravity for its larger density than residual melt.

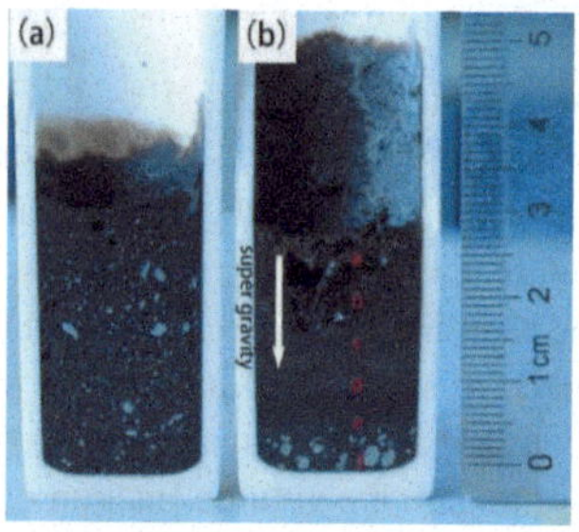

Figure 2. Effect of super gravity on macrostructure of samples: (a) G=1, V=5 K/min; (b) G=600, V=5 K/min.

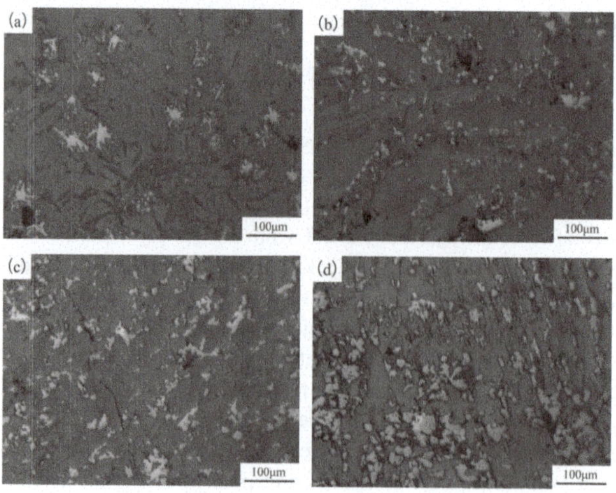

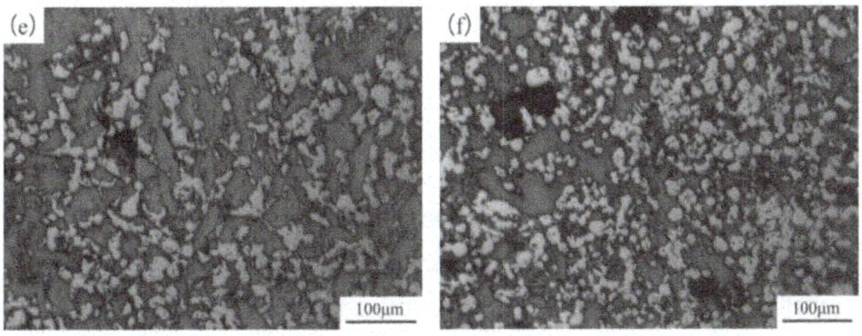

Figure 3. Micrographs of six areas of the sample obtained by centrifugal enrichment at G=600 and V=5 K/min from 1623 K to 1373 K.

Effect of gravity coefficient and cooling rate on volume fraction of RO phase

Figure 4 shows the variations of volume fraction of RO phase in different areas of the samples with different gravity coefficients at cooling rate V=5K/min from 1623 K to 1373 K and different cooling rates at G=600. The volume fraction of RO phase shows gradient distribution along the direction of super gravity and reach to the maximum value at the bottom of the sample. The slower cooling rate makes a larger gradient distribution of volume fraction under the same gravity coefficient, so does the larger gravity coefficient at the same cooling rate.

Increasing gravity coefficient means improving the driving force on the RO phase, and make it move faster under higher gravity coefficient. A slow cooling rate makes the melt viscosity increase slowly, which can reduce the movement resistance of RO phase and make it move faster at higher temperature. So the motion velocity of RO phase in the slag melt is proportional to the gravity coefficient and reciprocal of viscosity.

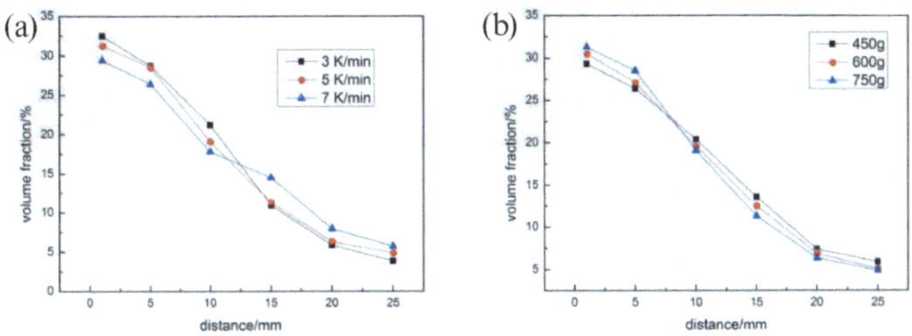

Figure 4. The variations of volume fraction of RO phase
a. different cooling rates from 1623 K to 1373 K at G=600
b. different gravity coefficients at cooling rate V=5 K/min from 1623 K to 1373 K

Components of stratified sample and recovery ratio of FeO, MgO and MnO after centrifugal enrichment

X-ray diffraction patterns indicate that the parallel sample mainly consist of dicalcium silicate, RO phase, calcium aluminoferrite, calcium silicate and calcium fluoride, as shown in figure 5. An overwhelming quantity of RO phase after centrifugal enrichment accumulates in the bottom of the sample as concentrate, while dicalcium silicate and calcium aluminoferrite occupy the upper region of the sample as tailing. Under the hypothesis that the iron, magnesium and manganese exist in the steel slag in the form of FeO, MgO and MnO, with the gravity coefficient G=600g and V=5 K/min from 1623 K to 1373 K, the mass fraction of the FeO, MgO and MnO in the concentrate are up to 40.46%, 6.9% and 3.91%, respectively, while that in the tailing are just 11.65%, 2.15% and 1.55%, as shown in the table 2. The recovery ratio of FeO, MgO and MnO in the concentrate are 60.35%, 58.45% and 72.24%, respectively, which were calculated via equation (2).

$$R_C = \frac{m_C \times \omega_C}{m_C \times m_C + \omega_t \times \omega_t} \times 100\% \quad (2)$$

Where R_c, m_c, m_t, ω_c and ω_t are recovery ratio in the concentrate, mass fraction of the concentrate and tailing, mass fraction of different oxides in the concentrate and tailing, respectively.

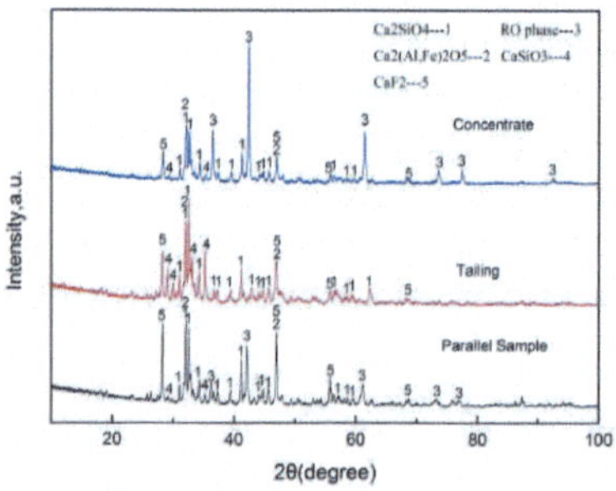

Figure 5. X-ray diffraction of the sample obtained by centrifugal enrichment and the parallel sample.

Table II XRF result of separation products after centrifugal enrichment (mass fraction, %)

products	CaO	FeO	SiO_2	MgO	MnO	P_2O_5	Al_2O_3	CaF_2	others
concentrate	31.77	40.46	8.38	6.9	3.91	1.92	1.41	3.69	1.55
tailing	49.71	11.65	17.86	2.15	1.55	2.79	1.49	9.83	2.97
parallel	43.64	21.30	15.08	3.75	2.19	2.5	1.48	8.46	1.6

Table III Recovery ratio of different oxids in the sample after centrifugal enrichment

products	Productivity/%	Recovery ratio of FeO/%	Recovery ratio of MgO /%	Recovery ratio of MnO /%
concentrate	32.28	60.35	58.45	72.24
tailing	67.72	39.65	41.55	27.76

Conclusions

It is experimentally confirmed that super gravity can be used to enrich MFe and RO phase in the steel slag. They are successfully driven to the bottom by super gravity and there is an obvious stratification appearing in the sample.

The volume fraction of RO phase shows gradient distribution and reach to the maximum value at the bottom of the sample. Larger gravity coefficient and lower cooling rate make the gradient distribution more obvious.

Under the gravity coefficient G=600 and V=5K/min from 1623K to 1373K, the mass fraction of the FeO, MgO and MnO in the concentrate are up to 40.46%, 6.9% and 3.91%, and the recovery ratio of FeO, MgO and MnO in the concentrate are 60.35%, 58.45% and 72.24%, respectively.

Acknowledgement

This work is supported by the National Natural Science Foundations of China (No.51234001 and No.51404025) and the Fundamental Research Funds for the Central Universities (FRF-TP-14-010A1).

References

1. R. Dippenaar, "Industrial Uses of Slag (the Use and Re-use of Iron and Steelmaking Slags)," Ironmaking and Steelmaking, 32 (1) (2005), 35-46.

2. A. S. Reddy, R. K. Pradhan, and S. Chandra, "Utilization of Basic Oxygen Furnace (BOF) Slag in the Production of a Hydraulic Cement Binder," International Journal of Mineral Processing, 79 (2) (2006), 98-105.

3. P. E. Tsakiridis, G. D. Papadimitriou, and S. Tsivilis, "Utilization of Steel Slag for Portland Cement Clinker Production," Journal of Hazardous Materials, 152 (2) (2008), 805-811.

4. D. Ouyang, Y. P. Xie, and J. Y. He, "A Study of the Mineral Phases of Oxygen Converter Slag and Their Cementious Properties," Journal of the Chinese Ceramic Society, 19 (6) (1991), 488–493.

5. G.H. Hou et al., "Microstructure and Mineral Phase of Converter Slag," Journal of the Chinese Ceramic Society, 36(4) (2008), 436-443.

6. H. Ono et al., "Removal of Phosphorus from LD Converter Slag by Floating Separation of Dicalcium Silicate during Solidification," ISIJ International, 21(1981) 135-144.

7. H. Kubo, K. Matsubae-Yokoyama, and T. Nagasaka, “Magnetic Separation of Phosphorus Enriched Phase from Multiphase Dephosphorization Slag,” ISIJ International, 50 (1) (2010), 59-64.

8. K. Matsubae-Yokoyama, H. Kubo, and T. Nagasaka, “Recycling Effects of Residual Slag after Magnetic Separation for Phosphorus Recovery from Hot Metal Dephosphorization Slag,” ISIJ International, 50 (1) (2010), 65-70.

9. K. Yokoyama K et al., “Separation and Recovery of Phosphorus from Steelmaking Slags with the Aid of a Strong Magnetic Field,” ISIJ International, 47 (10) (2007), 1541-1548.

10. K. Morita et al., “Resurrection of the iron and phosphorus resource in steel-making slag,” Journal of Material Cycles and Waste Management, 4(2)(2002), 93-101.

11. J.C. Li, Z.C. Guo, and J.T. Gao, “Isothermal Enriching Perovskite Phase from CaO-TiO_2-SiO_2-Al_2O_3-MgO Melt by Super Gravity,” ISIJ International, 54 (4) (2014), 743-749.

12. J.C. Li, Z.C. Guo, and J.T. Gao, “Laboratory Assessment of Isothermal Separation of V Containing Spinel Phase from Vanadium Slag by Centrifugal Casting,” Ironmaking and Steelmaking, 41 (9) (2014), 710-714.

13. J.C. Li et al., “Evaluation of Isothermal Separating Perovskite Phase from CaO-TiO_2-SiO_2-Al_2O_3-MgO Melt by Super Gravity,” Metallurgical and Materials Transactions B-Process Metallurgy and Materials Processing Science, 45 (4) (2014), 1171-1174.

14. J.C. Li J and Z.C. Guo, “Innovative Methodology to Enrich Britholite ($Ca_3Ce_2[(Si,P)O_4]3F$) Phase from Rare-Earth-Rich Slag by Super Gravity,” Metallurgical and Materials Transactions B-Process Metallurgy and Materials Processing Science, 45 (4) (2014), 1272-1280.

15. L. Hu et al., “Impurities Removal from Metallurgical-Grade Silicon by Combined Sn-Si and Al-Si Refining Processes,” Metallurgical and Materials Transactions B-Process Metallurgy and Materials Processing Science, 44 (4) (2013), 828-836.

16. L.X. Zhao et al., “Removal of Low-Content Impurities from Al By Super-Gravity,” Metallurgical and Materials Transactions B-Process Metallurgy and Materials Processing Science, 41 (3) (2010), 505-508.

7th International Symposium on High-Temperature Metallurgical Processing
Edited by: Jiann-Yang Hwang, Tao Jiang, P. Chris Pistorius, Gerardo R.F. Alvear F., Onuralp Yücel, Liyuan Cai, Baojun Zhao, Dean Gregurek, and Varadarajan Seshadri
TMS (The Minerals, Metals & Materials Society), 2016

VOLATILIZATION OF RHENIUM FROM MOLYBDENITE CONCENTRATE BY OXIDATIVE ROASTING

Guanghui Li, Rong Sun, Zhiwei Peng[1], Linfeng Zhou, Yuanbo Zhang

(School of Minerals Processing & Bioengineering, Central South University, Changsha, Hunan 410083, China)

Keywords: Rhenium, Molybdenite concentrate, Volatilization, Oxidative roasting

Abstract

The volatilization behavior of rhenium in the process of oxidative roasting of a molybdenite concentrate with 43.55 wt.% Mo and 321 g/t Re is explored based on identifying the relationship between volatilization of sulfur and rhenium. The results show that rhenium volatizes after sulfur and precise control of operating variables plays an important role in separating rhenium from molybdenite concentrate. When the roasting parameters are optimized as follows: roasting temperature 600 °C, roasting time 2 h, roasting atmosphere air, the volatilization rate of rhenium reaches 89.26%. In the process of solidification roasting, the volatilization rate of rhenium decreased almost to zero when the ratio of slaked lime/molybdenum concentrate was 10% excess to the stoichiometric requirement at 600 °C for 2 h.

Introduction

Rhenium is one kind of rare and refractory metal. Its melting point is 3180 °C, ranking second in all metals, behind tungsten. Because of its own catalysis effect, high temperature performance, corrosion resistance and other excellent characteristics, it is widely used in catalysis, aerospace, environmental protection and so on [1]. Rhenium is a rare metal with proven reserves of rhenium mineral only 7300-10300 t and the content of 1×10^{-7}% in the earth's crust. Rhenium is mainly distributed in molybdenite. It was also found in some copper mineral, and some niobium yttrium iron, platinum and uranium ore. There are no independent minerals of rhenium in nature as rhenium is highly dispersed. Rhenium minerals which have economic recovery value are mainly distributed in molybdenite and sulfide copper mines. In these mines, the existential form of rhenium is ReS_2 or Re_2S_7 [2].

Molybdenite concentrate is traditionally treated by oxidative roasting-ammonia leaching process and solidification roasting-acid leaching process [3]. The former process mainly considers the recovery of molybdenum. The contained rhenium is volatilized into dust in the

[1]Corresponding author: Dr Zhiwei Peng, Email: zwpeng@csu.edu.cn, Tel.: (+86) 731 88830542

form of oxides which are collected by further treatments [4, 5]. In the process of solidification roasting, the rhenium and molybdenum are retained in calcine and then recovered from leaching of calcine [6, 7]. Previous studies mainly studied the behavior of molybdenum in the roasting, the studies about the behavior of rhenium and the roasting parameters in favor of separating rhenium from molybdenum are rare. Therefore, the behavior of rhenium in the roasting of molybdenum concentrate requires further attention.

In this study, the behaviors of rhenium in roasting-ammonia leaching process and solidification roasting-acid leaching process were studied. The main evaluation of the behavior of rhenium is the volatilization rate in this thesis. Meanwhile, the relationship among volatilization of rhenium, sulfur and molybdenum was studied. Studying the volatilization rate of rhenium can provide guidance for Mo-Re separation to the resource of molybdenum concentrate containing rhenium.

Experimental

Materials

Molybdenum concentrate containing rhenium was sampled from a dressing plant in Shanxi Province. The chemical compositions are given in Table I. The main mineral composition is shown in Figure1, which reveals that molybdenite is the main molybdenum mineral in the ore.

Table I. Main Chemical Composition of Molybdenum Concentrate Sample /%

Component	Mo	Re	Fe	SiO_2	Al_2O_3	CaO	MgO	K_2O
wt%	43.55	321 g/t	1.39	4.02	0.18	3.27	0.48	0.26
Component	Na_2O	As	WO_3	Cu	Pb	Zn	P	S
wt%	0.12	9.38g/t	1.66	0.026	4.52	0.06	0.016	32.46

Methods

The molybdenum concentrate was ground to 100 wt. % undersize 0.074 mm. The sample was directly balled or mixed with a certain amount of additive and balled into green pellets with 10-14 mm in diameter. Green pellets were dried in a baking box at 80 °C for 4 h. The dried pellets were loaded into the reactor (a quartz cup 3 cm×6 cm), which was transferred into the heating zone of the quartz tube in the electric heating furnace Before each trial, nitrogen gas was flushed into the quartz tube until the temperature of the quartz tube reach the set temperature. Air was then blew into the tube for oxidative roasting. After roasting for a certain period, calcine was collected for subsequent analysis. The content of rhenium in the calcine was determined by ICP, while other species were determined by chemical analysis.

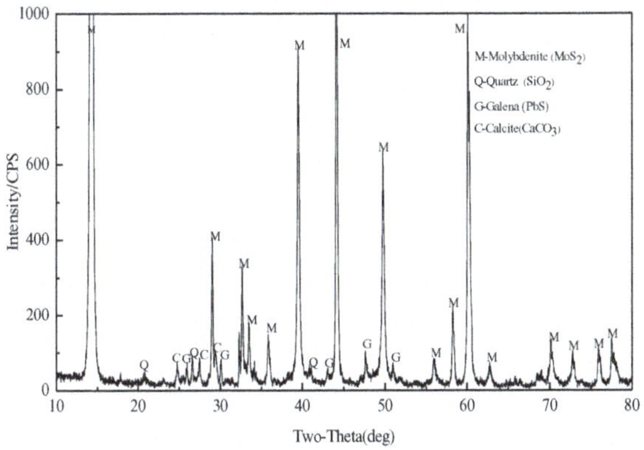

Figure 1. XRD pattern of the molybdenum concentrate sample.

Results and Discussion

The Volatilization of Rhenium In The Oxidative Roasting

The effects of roasting temperature Figure 2 shows the effect of the temperature on the volatilization of rhenium and the relationship of volatilization of Mo, Re and S in the molybdenum concentrate. The experiments were performed under air atmosphere at different temperature (450 °C-650 °C) for 120 min.

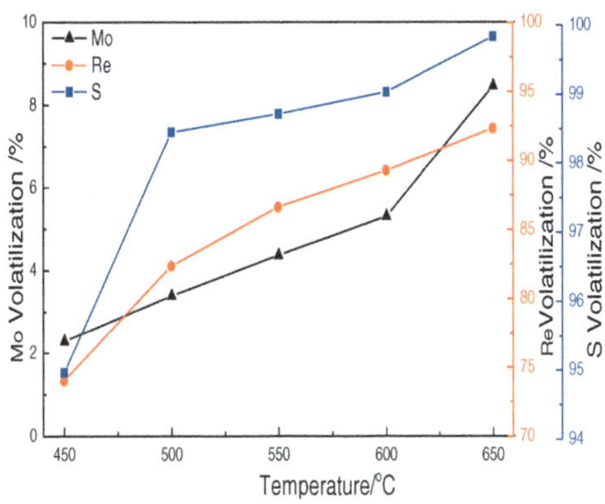

Figure 2. Effects of roasting temperature on the volatilization of rhenium.

The roasting temperature has a significant effect on volatilization of Mo, Re and S. The volatilization rate of Re increases from 74.05% to 92.34% and the volatilization rate of S increased from 94.97% to 99.84% with temperature increasing from 450 °C to 650 °C. However, as the temperature increases, the volatilization of Mo increased rapidly. The volatilization rate of Mo increases from 2.29% to 8.46% with the temperature increasing from 450 °C to 650 °C. When the temperature reaches 600 °C, the volatilization rate of S reaches 99.04%, indicating almost complete oxidation of molybdenite. Meanwhile, the volatilization rate of Re reaches 89.26%. It then remains nearly constant when the temperature is further increased conversely, the volatilization of Mo increases rapidly. Overall, the optimal roasting temperature is 600 °C.

The effects of roasting time Figure 3 shows the effect of the time on the volatilization of rhenium and the relationship of volatilization of Mo, Re and S in the molybdenum concentrate. The experiments were performed under air atmosphere for different time (20 min-150 min) at 600 °C.

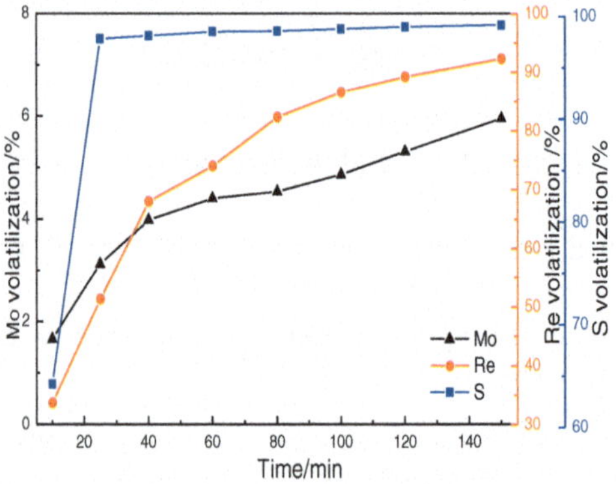

Figure 3. Effects of roasting time on the volatilization of rhenium.

Figure 3 shows molybdenum concentrate is oxidized completely in a relatively short period. The volatilization rate of S is 97.87% in 30 min. Further increase in the roasting time has little effect on the volatilization rate of S, but significantly influences the volatilization rate of Re. The rhenium oxide and sulfur dioxide formed during reduction reaction, which influences the volatilization rate of Re. It can be concluded that rhenium volatizes after sulfur during the process of the roasting of molybdenum concentrate. The volatilization rate of Re reaches 89.26% at the roasting time of 120 min. When the roasting time is further elongated, it remains relatively constant. On the contrary, the volatilization of Mo increases rapidly. The optimal roasting time is 120 min.

The effects of roasting atmosphere The influence of roasting atmosphere on the volatilization rates of Mo, Re and S is presented in Figure 4. The experiments were performed under different atmospheres with varying oxygen contents (15%-30%) at 600 °C for 2 h.

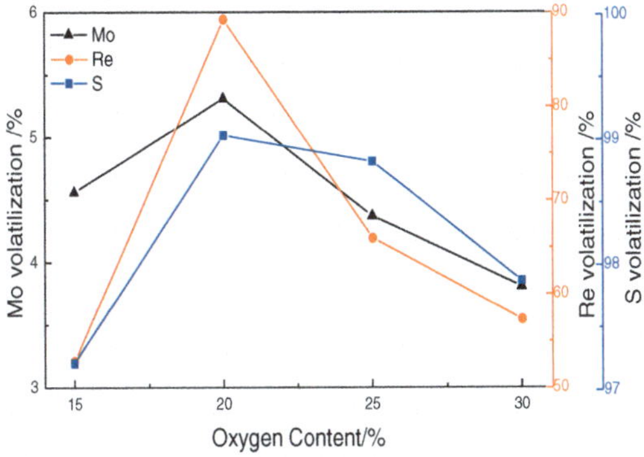

Figure 4. Effects of roasting atomsphere on the volatilization of rhenium.

Under low oxygen atmosphere, the molybdenum concentrate could not be completely oxidized, resulting in low volatilization rates of Re and S. Obviously, the volatilization rates of Re and S increase rapidly to 89.26% and 99.04% when the oxygen content increases from 10% to 20%(air atmosphere) respectively . However, as the oxygen content increases to 25% or more, the surface of pellets becomes oxidized quickly. During the process of the oxidation, the impurity elements (Pb, Cu, Fe, etc) in the molybdenum concentrate convert to the corresponding molybdate salts with low melting points, leading to serious sintering on the surface of pellets Obviously, the surface sintering inhibits the diffusion of gas that eventually causes lower volatilization of the metals. The optimal roasting atmosphere is 20%O_2.

The volatilization of rhenium in the solidification roasting

Slaked lime was used as additive in this study. The roasting of molybdenite in the presence of slaked lime results in the oxidation of molybdenum and conversion of associated rhenium values into calcium molybdate and perrhenate, which is called solidification roasting. The reaction process is as follows:

$$2MoS_2 + 6Ca(OH)_2 + 9O_2 = 2CaMoO_4 + 4CaSO_4 + 6H_2O \quad (1)$$

$$2ReS_2 + 5Ca(OH)_2 + 19/2O_2 = Ca(ReO_4)_2 + 4CaSO_4 + 5H_2O \quad (2)$$

The sulfur in the concentrate is converted into calcium sulfate, thereby making the process free from atmospheric pollution. The calcium molybdate and calcium perrhenate can be solubilized by leaching with dilute H_2SO_4 according to the reactions:

$$CaMoO_4 + H_2SO_4 = H_2MoO_4 + CaSO_4 \quad (3)$$

$$Ca(ReO_4)_2 + H_2SO_4 = 2HReO_4 + CaSO_4 \quad (4)$$

In order to characterize the degree of oxidation of the molybdenum concentrate, all roasted products were leached with 1 M H_2SO_4 at 80 °C for 2 h with constant stirring and the liquid-solid ratio of 10 to extract molybdenum and rhenium from the calcine. The calcium sulfate and the insoluble residue remained after leaching are removed by filtration, while molybdenum and rhenium are recovered from the filtrate by precipitation as ammonium molybdate and perrhenate, respectively.

The effects of the ration of slaked lime/molybdenum concentrate The influence of the ratio of slaked lime/molybdenum concentrate is presented in Figure5. The rhenium volatilization rate decreases from 71.69% to zero as the ratio of slaked lime/molybdenum concentrate increases from 0.5 to 1.2 at the roasting temperature of 600 °C for 2 h. Further increase in the ratio of slaked lime/molybdenum concentrate is unnecessary. From leaching results, we can also find that the leaching rates of Mo and Re reach maximum (98.63% and 84.93%) when the ratio is 1.2. It indicates that molybdenum and rhenium in the molybdenum concentrate are converted to calcium perrhenate and calcium molybdate by roasting with slaked lime. This ratio of slaked lime/molybdenum concentrate indicates a 10% excess of slaked lime over the stoichiometric requirement.

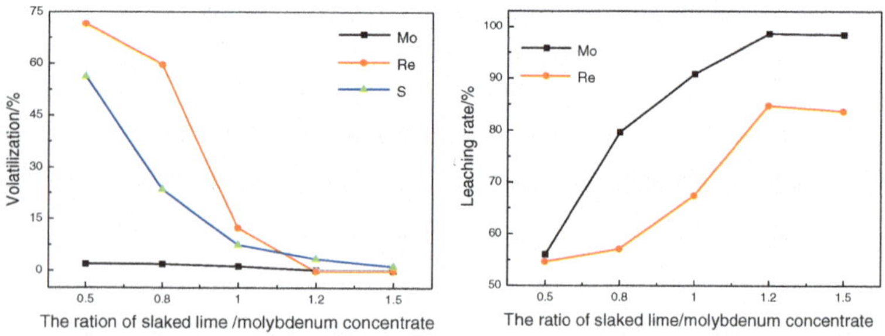

Figure 5. Effects of charge composition on the volatilization of rhenium.

The effects of roasting temperature The effect of the roasting temperature on the volatilization of rhenium is presented in Figure 6. The volatilization of Mo, Re and S has a little change with the variation of temperature. But the leaching rate of rhenium increases from 73.10% to 87.49% as the roasting temperature is raised from 450 °C to 600 °C and then remains constant beyond 600 °C at the ratio of slaked lime/molybdenum concentrate of 1.2 for 2 h. Meanwhile, the leaching rate of Mo has a little change with the variation of temperature. These results show molybdenum in the molybdenum concentrate can transform into calcium molybdate at a low temperature, but similar transformation of rhenium to calcium perrhenate requires a higher temperature. Thus, the suitable roasting temperature is 600 °C.

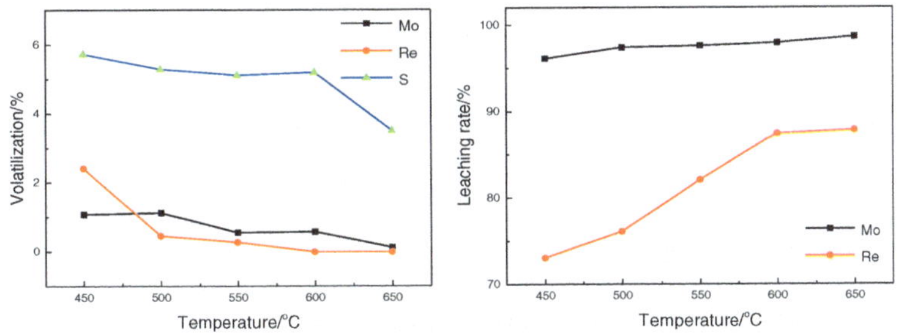

Figure 6. Effects of roasting temperature on the volatilization of rhenium.

The effects of roasting time The influence of the roasting time is presented in Figure 7. As discussed before, the volatilization of Mo and Re has a little change with the variation of time. The leaching results indicates that the leaching rate of rhenium increases from 80.61% to 87.49% as the roasting time is increased from 60 min to 120 min and then remains constant beyond 120min at the ratio of slaked lime/molybdenum concentrate of 1.2 and the roasting temperature of 600 °C. Meanwhile, the leaching rate of Mo has a little change with the variation of time. It is shown that calcium perrhenate forms later than calcium molybdate during the slaked lime cured roasting process. The suitable roasting time is 120 min.

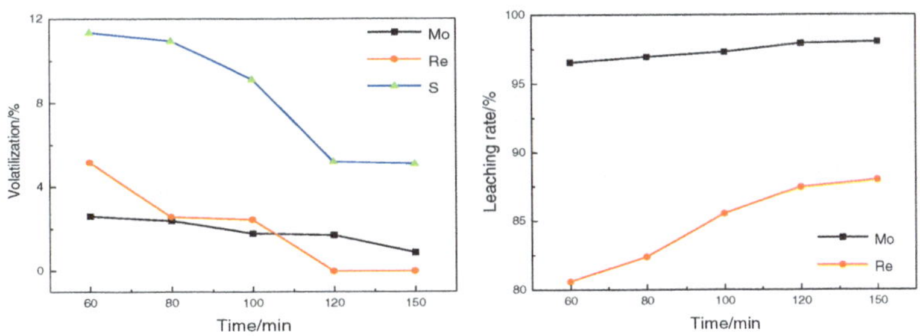

Figure 7. Effects of roasting time on the volatilization of rhenium.

Conclusions

During the process of oxidative roasting of molybdenum concentrate containing rhenium, the rhenium volatizes after sulfur. Meanwhile, optimum Re-Mo separation can be achieved when the sample is roasted at 600 °C, roasting time 120 min, roasting atmosphere air. Under these experimental conditions, the volatilization rate of Re reaches 89.26%, while that of Mo is just 5.31%. During the process of solidification roasting of molybdenum concentrate containing rhenium, molybdenum and rhenium in the molybdenum concentrate convert into calcium perrhenate and calcium molybdate when 10% excess of slaked lime over the stoichiometric requirement is used for roasting at 600 °C for 2 h. We found that molybdenum in the molybdenum concentrate can transform into calcium molybdate at a low temperature, but similar transformation of rhenium to calcium perrhenate requires a higher temperature. All these studies are helpful for the comprehensive utilization of molybdenum concentrate containing rhenium.

References

1. L.P. Li et al., "Extraction, and Research Status of Rhenium," *Express Inform Ataion of Mining Industry*, 2008, 10-13. (in Chinese)

2. S.C. Wang, S.Z. Qi, Resource, "Application and Market of Rhenium," *World Nonferrous Metals* 02 (2001), 12-14. (in Chinese)

3. Sutulov, *Molybdenum and Rhenium Recovery from Porphyry Coppers* (University of Concepcion, Chile, 1970).

4. L. C, I. Wilkomirsky, G Morizot, "Molybdenite Roasting and Rhenium Volatilization in a Multiple-hearth Furnace," *Trans. Inst. Min. Metall*, 1970, C34–C40.

5. P.R. Ammann, T.A. Loose, "Rhenium Volatilization During Molybdenite Roasting," *Metall. Trans. April*, 1972, 1020–1021.

6. M.H. Shariata, M. Hassani, "Rhenium Recovery from Sarcheshmeh Molybdenite Concentrate," *Journal of Materials Processing Technology*, 74 (1998), 243–250.

7. J.M. Juneja, Sohan Singh, D.K. Bose, "Investigations on the Extraction of Molybdenum and Rhenium Values from Low Grade Molybdenite Concentrate," *Hydrometallurgy*, 41 (1996), 201-209.

7th International Symposium on High-Temperature Metallurgical Processing
Edited by: Jiann-Yang Hwang, Tao Jiang, P. Chris Pistorius, Gerardo R.F. Alvear F., Onuralp Yücel, Liyuan Cai, Baojun Zhao, Dean Gregurek, and Varadarajan Seshadri
TMS (The Minerals, Metals & Materials Society), 2016

KINETIC INVESTIGATION OF THE ELECTRIC FURNACE COPPER SLAG TREATMENT

Stephan R. Steinacker[1], Juergen Antrekowitsch[1]

[1]Department of Nonferrous Metallurgy
Montanuniversitaet Leoben
Franz Josef-Straße 18, 8700 Leoben, Austria

Keywords: copper slag, electric furnace treatment, Boudouard equilibrium, kinetics

Abstract

The main aim of the electric furnace treatment in the primary copper industry is the reduction of valuable metals such as zinc, lead or copper itself. On the one hand legal restrictions for the further use of the treated slag are constantly getting stricter while on the other every type of loss in the oxidic phase leads to economic disadvantages. In order to fully understand and to improve the process, kinetics during the reduction play an essential role. In the first part, this paper highlights the theoretical background of the reduction process, while the following chapters present experimental aspects and results from laboratory scale trials at the Montanuniversitaet Leoben. In addition, the comparison between the chemical reaction and the diffusion model plays a continuously important role. By investigating the reducibility of electric furnace slag, process adaptions and ameliorations are developed.

Introduction

The treatment of the flash furnace slag constitutes a commonly performed process in the primary copper industry. In general, this procedure shows two main aims. On the one hand the contained copper oxides and sulfides should be reduced to metallic Cu, while on the other the particles are given a certain amount of time in order to coagulate and move to the bottom of the furnace for a better separation. While slag from the flash furnace typically displays copper contents between 1-2 %, slag from the copper converting step reaches 4-8 % and represents another possible input material for the electric furnace. The treatment time takes between 1-5 hours in order to guarantee a satisfactory reduction as well as a sufficient separation of the different phases. A slag low in copper which usually displays 0.7-1 % Cu and a copper matte with 65-70 % Cu represent the output materials.[1]

Copper losses in the slag can be divided into two general categories. On the one side there is the mechanical or physical fraction while on the other there is the chemical section. Latter one can subsequently be split into oxidic as well as sulfidic compounds while the physical losses usually consist of suspended droplets. The electric furnace's main advantage is represented by the excellent separation between the slag and the metal phase which derives from the long treatment time as well as the low turbulence. While mechanical losses sum up to 65-80 % of the total copper losses after the flash furnace, the electric furnace leads to a minimum amount of approximately 25 % in this section. Hence, the missing 75 % are represented by dissolved copper compounds.[2]

Beside the separation, the already mentioned reduction plays a highly important role. While the percentage of magnetite should be lowered to a value below 5 % in order to prevent a highly viscous and high-melting slag, the dissolved oxidic or sulfidic copper compounds have to be reduced. Typical reducing agents include coke, coal or pyrite. Introducing gaseous reduction agents such as methane shows a positive influence on the reaction kinetics. The deoxidation of the Cu_2O with carbon can either take place directly or indirectly as can be seen in equation 1 and equation 2.[1,3,4]

$$(Cu_2O)_{slag} + C\,(s) \leftrightarrow 2\,Cu\,(l) + CO\,(g) \tag{1}$$

$$(Cu_2O)_{slag} + CO\,(g) \leftrightarrow 2\,Cu\,(l) + CO_2\,(g) \tag{2}$$

During the electric furnace treatment several transport phenomena influence the process. On the one hand gas bubbles move up in the melt, while on the other metal and matte particles settle. In addition, microscopic processes such as the diffusion of the particles to the phase boundary surface as well as the contact interface between the reduction agent and the different compounds also influence the reduction.[1]

In order to keep the copper losses in the electric furnace slag at a minimum, optimum process conditions have to be adjusted. By adding a certain amount of SiO_2, fayalite forms and leads to a slag melting temperature of 1,100-1,300 °C – depending on the exact chemical composition of the input materials. An oxidizing atmosphere and the addition of carbon, however, lead to a higher melting point. The slag temperature should not exceed 1,250 °C as the copper solubility increases significantly at higher values. The ratio of Fe to SiO_2 has to stay below 1.5 and the total amount of magnetite must not surpass 7 % in order to achieve positive results. In addition, the process should be run close to the absolute copper saturation which means that the iron content of the matte must stay between a range of 6-9 %. As a result, the slag offers a minimum solubility for Cu as shown in Figure 1.[2]

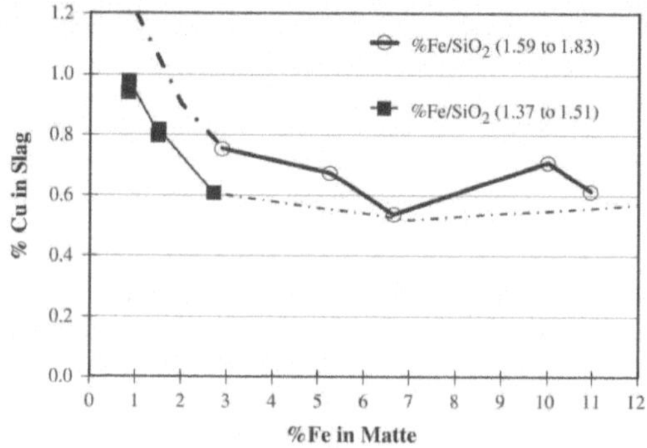

Figure 1: Solubility of copper in the slag depending on the iron content in the matte[2]

Reduction Kinetics Fundamentals

When aiming to achieve a better understanding for the reduction process, kinetics have to be considered. In general, four phases influence each other in a heterogenic system. These are the reduction agent – which is typically a solid – on the one hand and the liquid metal, the liquid slag as well as the gaseous phase on the other. The different reaction models offer explanations for the various process steps, while the actual reaction can usually be split into three main categories which include following:[5]

- Transport of the reactants to the phase boundary
- Chemical reaction at the phase boundary
- Transport of the products away from the phase boundary

The two-film model by Lewis and Whitman which can be seen in Figure 2 represents a basis for the mass transfer description.[5]

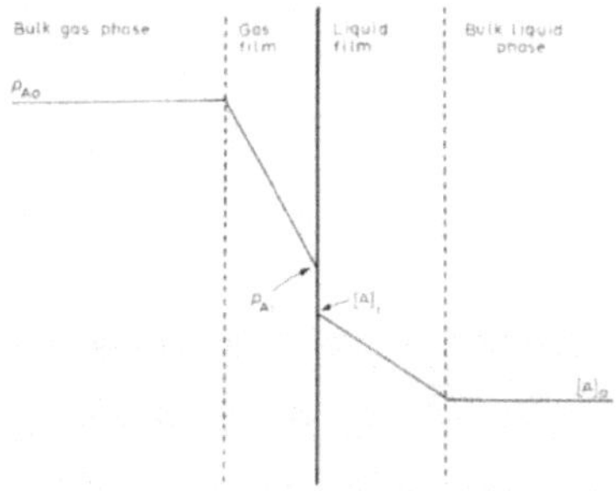

Figure 2: Two-film model by Lewis and Whitman for the mass transfer description[5]

A conductive boundary layer forms at both sides of the phase boundary through which the mass transfer takes place by diffusion. The concentration gradients can be described by Fick's first law, while a concentration leap can be detected at the phase boundary which occurs due to the different phase equilibrium conditions.[5] Fick's first law – as can be seen in equation 3 – states that the diffusion current density j_D is directly proportional to the concentration gradient dc/dt. The proportionality factor D represents the diffusion coefficient and is valid for the one-dimensional case.[6]

$$j_D = -D * dc/dt \qquad (3)$$

The reaction speed basically depends on the temperature, the concentration and possible catalysts. The order of the reaction derives from the sum of the partial reaction orders. In general, its determination is performed experimentally.[7]

Experimental

In order to investigate kinetic aspects and to draw conclusions from comparable research, several copper slag reduction trials were performed. As described above, a temperature of approximately 1,200-1,250 °C constituted the optimum value for the reduction of the contained copper. As a result and since no agitation within the bulk was required, a resistance furnace at the Montanuniversitaet Leoben represented the furnace of choice. Another advantage was offered by its high heating rate for which reason the desired temperature was reached within a minimum amount of time. The slag mainly consisted of fayalite – $2FeO.SiO_2$ – and also contained a significant amount of copper as well as other elements. Its general chemical composition can be seen in Table 1.

Table 1: Chemical composition of the fayalite slag from the primary copper industry

Compound	Al_2O_3	Cu	Fe	Fe_3O_4	S	SiO_2
[%]	3.7	1.5	39.3	8.2	0.2	32.4

Highly purified coke acted as reduction agent since it showed a carbon percentage over 99 % and only very low grades of impurities which was a fundamental condition for sulfur as a direct reduction of copper sulfide with carbon in general is not possible. The reducing agent was mixed with the slag in an over-stoichiometric ratio in order to guarantee a highly reducing atmosphere. Finally, the charge was put into a clay-graphite crucible, covered with a small additional amount of carbon to prevent reactions with the atmosphere and then inserted in the resistance furnace. The experimental setup while in operation can be seen in Figure 3.

Figure 3: Experimental setup of the reduction experiments

By performing several experiments with varying reduction times, the reducibility of the copper slag was investigated. Therefore, different periods from 40 to 70 minutes represented the experimental boundary conditions. After the treatment, the samples were instantly quenched in cold water in order to prevent the formation of new phases and to keep the reduced species at a maximum.

Results and Discussion

After the reduction trials in the resistance furnace as well as the quenching and drying of the slag samples, an X-ray fluorescence analysis took place in order to investigate the chemical composition and to determine the amount of the most important elements within the slag. A representative result from the analysis can be seen in Table 2.

Table 2: Distribution of the most important compounds of the treated slag

Compound	Cu	Fe_3O_4	S
[%]	0.62	5.43	0.20

An interesting fact is displayed by the sulfur which is still contained in the samples. As expected, the copper sulfide cannot be reduced with carbon, according to the equilibrium lines in the Richardson Ellingham diagram. Hence, a certain part of the Cu remains in the slag as sulfide – a fact which is underpinned by investigating the different phases in a scanning electron microscope. Figure 4 shows the overlay image of the contained copper and sulfur in the slag and therefore proves the existence of remaining sulfide. However, the majority of the contained copper in the slag is either reduced or already metallic for which reason a distinct separation takes place. A certain part of the magnetite is also reduced, hence promoting a successful reduction in the furnace.

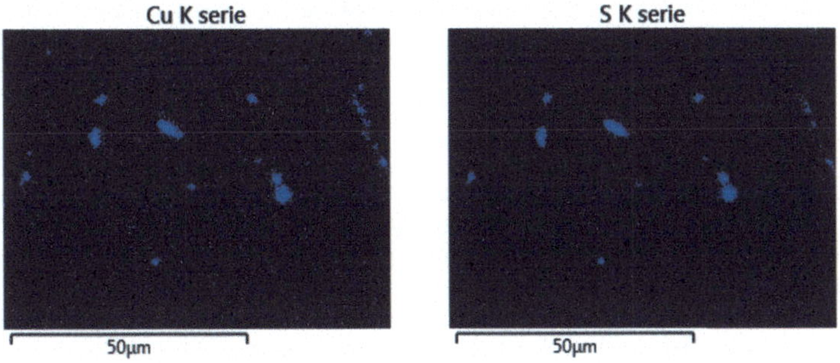

Figure 4: Correlation of Cu and S in a scanning electron microscope image

In order to investigate kinetic aspects of the trials, similar experiments performed by R. Reddy have to be considered. Reddy uses a similar slag with 1.625 % Cu and 6.34 % Fe_2O_3 and examines the reduction behavior of this material with pure graphite as reduction agent. The setup of the performed experiments is shown in Figure 5.[8]

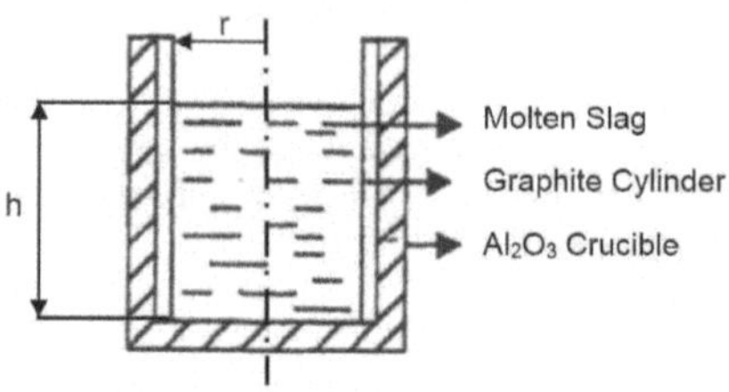

Figure 5: Experimental reduction setup by R. Reddy[8]

Since rising temperatures positively influence the reactions' kinetics, Reddy choses a high temperature of 1,300 °C for his reduction experiments. If the mass transfer of the copper oxide from the slag to the phase boundary of the graphite displays the rate-determining step, the entire process must be diffusion-controlled. However, as the forced convection by purging argon into the charge does not increase the reaction speed at all and hence does not enhance the reaction kinetics, the reduction of the copper oxide with carbon does not represent a diffusion-controlled process. As a result, the different chemical reactions during the reduction have to be investigated.[8]

The reduction of the copper oxide happens at the boundary layer between the slag and the carbon. Since sufficient C is available and CO_2 constitutes the main product, however, the Boudouard equilibrium – as described in equation 4 – has to be considered at elevated temperatures.[8]

$$C(s) + CO_2(g) \leftrightarrow 2\ CO(g) \quad (4)$$

In general, the Boudouard equilibrium represents an endothermic reaction. Therefore, the equilibrium moves to the products' side at a rising temperature. Since Reddy's experiments take place in the CO zone of the Boudouard diagram, an indirect reduction of the copper oxide with carbon monoxide has to be expected. As the general process is not diffusion-controlled, a chemical reaction must constitute the slowest partial step and hence the rate-determining factor of the entire process. The reaction of copper oxide with gaseous CO happens faster than with solid carbon for which reason the reduction with carbon monoxide is the most likely. Reddy compares his experiments with other systems where the Boudouard equilibrium displays the rate-determining step and at the end of his study detects an excellent correlation with his trials. Therefore the catalyzed reduction of copper oxide with CO must be the crucial step within the electric furnace process.[8]

Conclusion

The comparison of the experiments at the Montanuniversitaet Leoben and Reddy's reduction trials show an interesting correlation. The reducibility of the contained Cu can positively be demonstrated, while a certain amount of copper sulfide remains within the material. Settling processes also play an important role as the small droplets need a certain amount of time to coagulate and move down to the metallic phase. In general, diffusion does not influence the process significantly and hence does not constitute the limiting step. In contrast, the investigation of chemical reactions leads to the influence of the Boudouard reaction on the reduction. The use of a gaseous reduction agent could show positive results on the overall process.

The reduced and cleaned slag in general displays an excellent raw material for further applications. The main part of the slag goes to the road and construction industry for which reason the compliance of threshold values has to be severely observed. On the one hand the main problematic metals such as copper, zinc and lead have to be kept at a minimum level while on the other minor elements which can include arsenic or bismuth should entirely be avoided. Further application areas are represented by the cement and the glass industry.

To sum up, kinetic considerations play an essential role beside the thermodynamics during the reduction of copper slag. By determining the process's most important reactions and by understanding the key drivers, process adaptions can be formulated and therefore ameliorations be achieved.

Acknowledgement

The present work was written with the financial help of the Christian Doppler Forschungsgesellschaft, the Montanuniversitaet Leoben as well as the Austrian government. The authors also want to thank Stefan Wibner and Christian Wibner for their assistance and dedication during the experimental investigation.

References

1. M.E. Schlesinger: *Extractive metallurgy of copper*, Elsevier, Amsterdam, Boston, 2011.
2. P. Coursol, N. Cardona Valencia, P. Mackey, S. Bell, and B. Davis: *Minimization of Copper Losses in Copper Smelting Slag During Electric Furnace Treatment*, JOM vol. 64, pp. 1305–1513, 2012.
3. B. Friedrich, M. Zander, and C. Kemper: *Rückgewinnung von Kupfer und Kobalt aus schmelzflüssigen Schlacken der NE-Metallurgie*, Aschen, Schlacken, Stäube aus Abfallverbrennung und Metallurgie; Berliner Schlackenkonferenz, pp. 599–614, 2013.
4. B.E. Langner: *Understanding copper: Technologies, markets, business*, 1st ed., Winsen, Glockenheide, 2011.
5. C.H. Bamford, C.F.H Tipper and R.G. Compton: *Comprehensive chemical kinetics*, Elsevier, Amsterdam, Oxford, 1985.
6. G. Gottstein: *Physikalische Grundlagen der Materialkunde*, 3rd ed., Springer, Berlin, Germany, 2007.
7. P.W. Atkins and J. de Paula: *Physikalische Chemie*, 4th ed., Wiley-VCH, Weinheim, 2006.
8. R.G. Reddy, V.L. Prabhu, and D. Mantha: *Kinetics of Reduction of Copper Oxide from Liquid Slag Using Carbon*, High Temperature Materials and Processes, vol. 22, 2003.

7th International Symposium on High-Temperature Metallurgical Processing
Edited by: Jiann-Yang Hwang, Tao Jiang, P. Chris Pistorius, Gerardo R.F. Alvear F., Onuralp Yücel, Liyuan Cai, Baojun Zhao, Dean Gregurek, and Varadarajan Seshadri
TMS (The Minerals, Metals & Materials Society), 2016

THE EXTRACTION OF ZINC FROM WILLEMITE BY CALCIFIED-ROASTING AND AMMONIA-LEACHING PROCESS BASED ON PHASE RECONSTRUCTION

Wei Chen, Yufeng Guo*, Feng Chen, Tao Jiang, Xudong Liu
(School of Mineral Processing and Bioengineering, Central South University, Changsha, Hunan, 410083,China,)

Keywords: willemite, calcified-roasting, phase reconstruction, extraction of zinc

Abstract

Willemite is typically found in oxidized zinc ores and metallurgical secondary resources but without effective extraction method yet. Based on the ideas of zinc phase reconstruction in the willemite, the extraction of zinc from willemite by calcified-roasting using CaO as additive and ammonia-leaching process was studied in this paper. The influences of CaO dosage, roasting temperature, roasting time on the phase transformation and zinc leaching rate were studied, which were characterized by XRD. The results indicated that Zn_2SiO_4 was transformed into ZnO or $Zn_2CaSi_2O_7$ .After roasted with 1.4 times of theoretical CaO dosage at 1280 ℃for 40 minutes, willemite was thoroughly transformed into ZnO. The zinc leaching rate of the roasted sample was 97.45% in NH_4Cl-$NH_3 \cdot H_2O$ leaching system, which increased by 91.40 % compared with zinc leaching rate of 6.05% when willemite was directly leached in NH_4Cl-$NH_3 \cdot H_2O$ system.

Introduction

Several types of oxidized zinc ores have been found in nature so far, mainly include calamine ($ZnCO_3$), willemite (Zn_2SiO_4), hemimorphite ($Zn_4(Si_2O_7)(OH)_2 \cdot H_2O$) and zincite (ZnO) [1].China has an abundant reserve of oxidized zinc ores(more than 40 million tons) which is mainly found in the southwest and northwest areas like a world-class mine: Yunnan Lanping zinc ore. Oxidized zinc resources usually contain much silicon and alkaline gangue, belonging to a low-grade and refractory zinc resource. Among the oxidized zinc resources, hemimorphite and willemite are most refractory because of their stable structures and high silicon contents[2].

There are also a lot of oxidized zinc resources in lead (Pb) and zinc metallurgical secondary slag. The smelting zinc slag exhausted from blast furnace or direct reduction furnace contains 10 wt %. to 20 wt. % zinc. And 66.28wt.% of Zn in a typical lead smelting slag exists in the form of Zn_2SiO_4 and 31.63 wt.% in the form of $ZnFe_2O_4$[3]. In addition, many indigenous zinc smelting kilns had exhausted much smelting waste, in which zinc exists in the form of zinc metal and willemite [4].

Currently, the mainly processing methods of high-silicon contained oxidized zinc

*Corresponding author: Yufeng Guo, Email:guoyufengcsu@163.com

resource or smelting slag are: indirect smelting and direct smelting. Indirect smelting method enriches the zinc ores by floatation, followed by roasting, leaching and electrowinning (RLE) processes. Direct smelting method is processed without floatation and it includes two types: pyrometallurgical and hydrometallurgical method. Conventional flotation technique cannot effectively enrich the oxidized zinc resources because of the difficulty in separating zinc ores from gangue[5].Direct pyrometallurgical method usually refers to high temperature reducing-evaporation method which has various shortcomings, such as the stricter reducing conditions required for willemite and hemimorphite than zincite, more investment and the generation of harmful greenhouse gases[6]. So the dominant method for exacting zinc at present is hydrometallurgical process. Sulfuric acid leaching process is typically used in zinc producing process with zinc sulfide ore as the raw material. However when the process is applied to the utilization of high-silicon contained zinc resources, a large number of silica gel is formed due to the silicon's dissolution, resulting in difficulties in liquid-solid separation. Therefore, some measures have been studied to control the polymerization of silica gel, such as rapid leaching and flocculation [7]. The easy-separating ore pulp can be gained by these methods, but the leaching requires strict control of conditions with great technical difficulty. In addition, the acid consumption is too high and it is difficult to purify the leaching solution because of the high content of alkaline gangue and impurity elements. Thus, many scientists have studied the alkali leaching for high-silicon contained oxidized zinc resources[8]. Z.W. Zhao [9] used mechanical activation method to leach hemimorphite in NaOH solution. The results showed that silicon was dissolved into solution in the form of $H_2SiO_4^{2-}$ without the formation of silica gel while hemimorphite was dissolved in the form of $Zn(OH)_4^{2+}$. A zinc leaching rate of 93% was obtained by alkali leaching with the disadvantages of poor selectivity, poor quality of products and increased difficulty to the subsequent solution purification, solution regeneration and zinc recycling process[10]. Ammonia leaching oxidized zinc resources has the comprehensive advantages of high selectivity, fewer impurities and easy-to-purify. But the zinc leaching rate of willimete by ammonia leaching is much lower than that of other oxidized zinc ore[10]. Z.Y.Liu [11,14] used hemimorphite and willemite pure minerals to study the leaching mechanism in ammonia system. 90% of zinc in hemimorphite could be leached under the condition of high liquid-solid ratio while only 10% of zinc in willemite could be leached because it was limited by the low solubility of silica and precipitation rate of silica gel. Willemite seems to be most difficult to leach in ammonia system among all types of oxidized zinc ores.

Therefore, relying on previous studies, this paper proposed a calcified-roasting followed by ammonia leaching method to enhance the zinc leaching rate in ammonia system, based on the technical idea of reconstructing the existential form of zinc in willemite. Meanwhile, this paper takes examples of extracting vanadium by roasting and acid leaching process from vanadium slag[12], extracting vanadium by sodium-roasting or calcified-roasting process from stone coal[13] to study the willemite calcified-roasting and ammonia-leaching system .The mechanism of phase reconstruction during calcified-roasting process and the impact of reconstruction on zinc leaching efficiency are studied as well.

Experimental

Raw Materials

Willemite samples used in this study is obtained by roasting natural hemimorphite from Lanping Oxidized Zinc Ore, in Yunnan Province(China).Hemimorphite is crushed and milled to no large than 74μm, and then it is roasted at temperature of 1100 ℃ for 3 h. The roasting reaction equation is:

$Zn_4Si_2O_7(OH)_2(s) \rightarrow 2Zn_2SiO_4(s)+2H_2O(g)$

The XRD pattern of willemite sample is shown in Fig.1.

The main chemical composition of willemite sample is listed in Table 1 .

Table 1 The main chemical composition of willemite sample

Zn	Fe	SiO_2	CaO	MgO	Al_2O_3
57.89	0.34	0.02	0.10	0.033	0.095

Results in Figure 1 and Table 1 show that hemimorphite sample has completely dehydrated and transformed into willemite (JCPDF NO 83-2270) after roasting.The zinc element content of willemite sample is 57.89wt.%,which is very close to theoretical content (58.34 wt.%) of willemite .

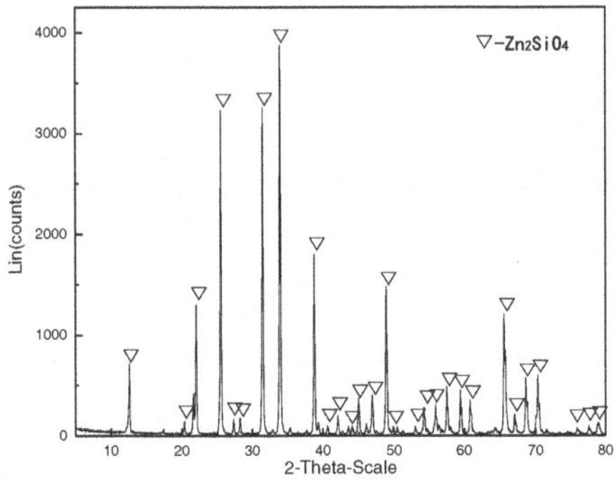

Fig.1 XRD pattern of willemite sample

Experimental Method

(1) Roasting Experiment: Willemite sample powder and analytical CaO powder are mixed by a certain percentage then the mixture is loaded in a corundum boat and roasted in a horizontal tube furnace at a target reaction temperature.

(2) Leaching Experiment: The roasted ore samples are ground into the particle with size <74μm.Leaching experiments was conducted in a250mL beaker heated in a water

bath at a target temperature. The leaching procedure initiated by pouring 100mL of required concentration ammonia leaching agent into a 250mL beaker with a plastic film covered. Then the leaching agent was heated to the experimental temperature with stirring. Time was counted when the willemite samples or calcified-roasting samples were added into the leaching beaker. When the required reaction time was up, the slurry was filtered, and then an appropriate dosage of dilution and leaching residue were taken for analysis. Each group of analysis is repeated for three times and the average value is taken as the final data.

Analysis Methods

Zinc content: EDTA Complexometric Titration (GB/T 8151.1-2012) .
Chemical phase analysis: X-ray Diffraction (Rigaku-TTRJII, Cu target, K_α, λ = 0 154 06 nm)

Results and Discussion

The main affecting factors of the calcified-roasting process are the dosage of CaO additive, roasting temperature and roasting time were studied. Willemite samples and calcified-roasting samples were leached in NH_4Cl-$NH_3 \cdot H_2O$ leaching system. Leaching conditions: the total ammonia concentration was 6mol/L, the mole ratio of NH_4Cl to $NH_3 \cdot H_2O$ was1:1,the leaching temperature was 40 ℃,the solid-liquid ratio was 20g / L, the stirring speed was 350 r / min and the leaching time was 3h. A blank comparative experiment referring to Dr. Z.Y.Liu's thesis[14] under the above leaching conditions was conducted and the zinc leaching rate of willemite directly leached in NH_4Cl - $NH_3 \cdot H_2O$ leaching system was 6.02%.

The Influence of the Dosage of CaO Additives on Zinc Leaching Rate

The influence of the dosage of CaO additive on zinc leaching rate is very significant as shown in Fig2. When the dosage of CaO additives increases from 0.5:1 (theoretical mole ratio) to 0.8:1, the zinc leaching rate increases from 69.19% to 98.22%.This is because willemite samples cannot fully contact with CaO when CaO is added in the theoretical amount. So since CaO dosage increases, willemite could become better contact with CaO which promotes the reaction and results in the increase of zinc leaching rate. Zinc leaching rate increases significantly, almost shows a linear relationship when the dosage of CaO additive increases from 0.5:1 to 0.7:1 (mole ratio).After that, a further increase of CaO dosage would not increase zinc leaching rate obviously, which indicates the CaO dosage has reached a sufficient dosage and it would no longer be a restriction factor of willemite roasting-transformation and leaching.

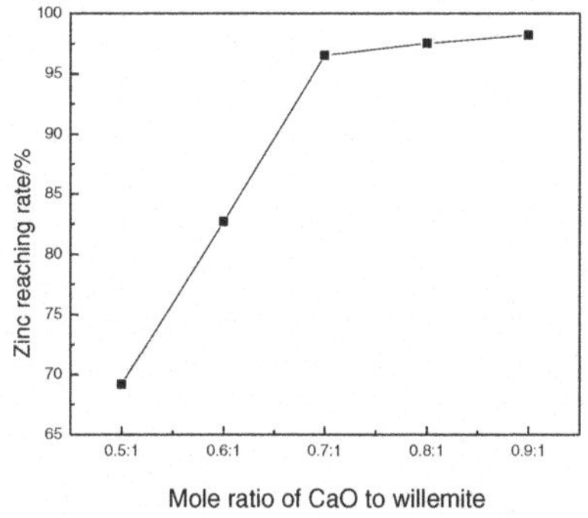

Fig. 2 Influence of CaO/ willemite(mole ratio) on zinc leaching rate

The Influence of The Roasting Temperature on Zinc Leaching Rate

The fully mixed willemite and CaO samples (CaO: willemite=0.7: 1 mole ration) were roasted at the temperature from 1200 ℃ to 1300 ℃ for 40min.The influence of roasting time on zinc leaching rate is shown in Fig.3.As seen from Fig.3, the impact of temperature on zinc leaching rate is obvious. As the roasting temperature increases, zinc leaching rate increases as well. The zinc leaching rate continues to rise sharply from 64% (at 1200 ℃) to 91.65% (at 1240℃) then the trend becomes slow. Zinc leaching rate almost does not continue to increase when the roasting temperature is above 1280 ℃.

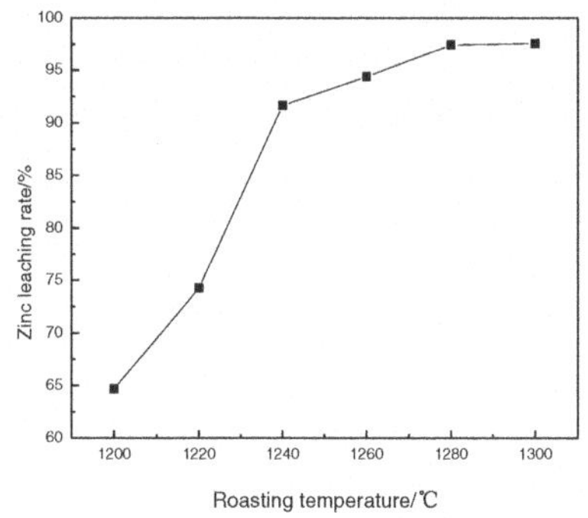

Fig.3 Influence of roasting temperature on zinc leaching rate

Influence of roasting time on zinc leaching rate

Influence of roasting time on zinc leaching rate is shown in Fig. 4. The XRD patterns of calcified-roasting sample (roasted at 1280 ℃for 10min or 40min) are shown in

Fig.5 and Fig.6.As shown in Fig.4, at lower temperatures, the influence of roasting time on zinc leaching rate is significant, when the temperature increases, roasting time has a weaker influence on the leaching rate. The appropriate roasting time at higher temperature is shorter than that of lower temperature. The appropriate roasting time at 1240℃ is 40min while the appropriate roasting time at 1300℃ is 30min with a higher zinc leaching rate. Zinc leaching rate does not increase obviously when continuing to prolong the roasting time longer than 40 min.

By analyzing Fig.5 and Fig.6,it can be seen that there are diffraction peaks of ZnO and $Zn_2CaSi_2O_7$ in roasted ore sample (roasted at 1280 ℃ for 10min) while diffraction peaks of $Zn_2CaSi_2O_7$ are not found in roasted ore sample(roasted at 1280 ℃ for 60min). It can be explained that most willemite has been transformed into ZnO after calcified-roasting for 60 min and that may account for the great increase of zinc leaching rate.

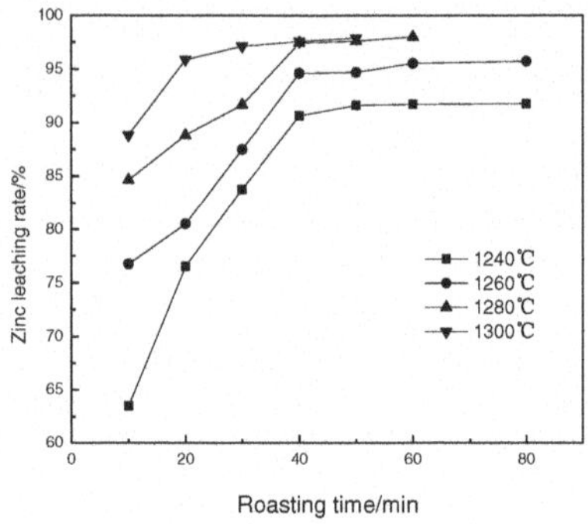

Fig. 4 Influence of roasting time on zinc leaching rate

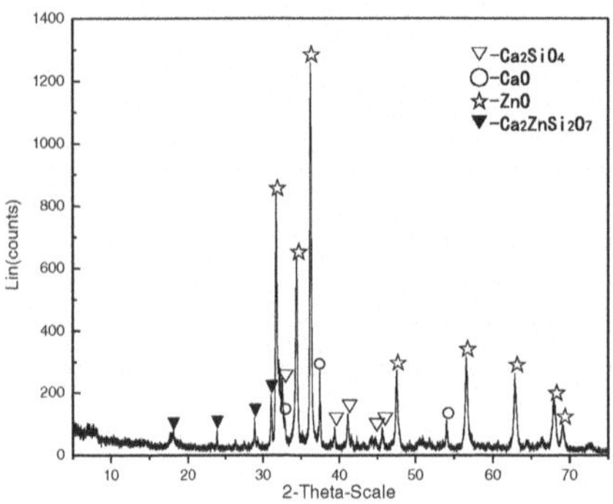

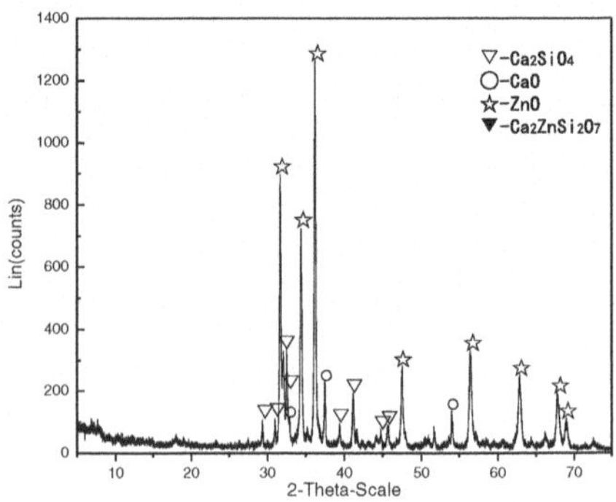

Fig. 5 XRD pattern of roasted sample (1280 ℃ for 10min)

Fig. 6 XRD pattern of roasted sample (1280 ℃ for 60min)

The chemical phase composition of leaching residue(roasted at 1280 ℃ for 60min followed by leaching in NH_4Cl - $NH_3{\cdot}H_2O$ agent system) is shown in Fig.7 and only the diffraction peaks of Ca_2SiO_4 can be found in the XRD pattern which proves that under adequate conditions, willemite can be fully transformed into ZnO during the calcified-roasting (phase-reconstruction) process and it means zinc element in reconstructed-roasting samples can be leached almost totally.

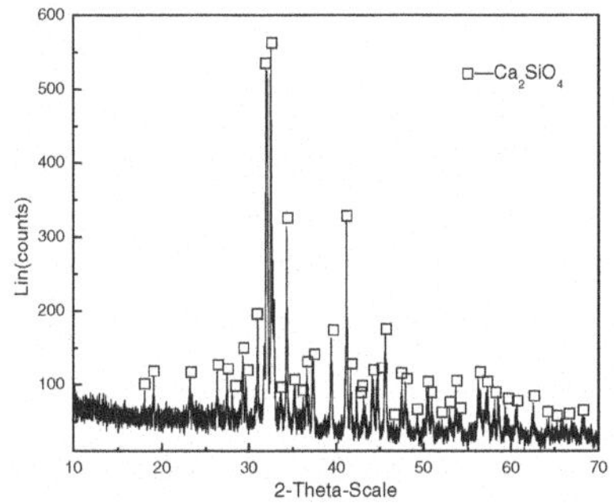

Fig.7 XRD pattern of leaching residue

Conclusions

(1) Within the roasting temperature range from 1200℃ to 1300℃,willemite (Zn_2SiO_4) can react with CaO and be transformed into $Zn_2CaSi_2O_7$ (hardystonite) or ZnO due to different CaO dosage, roasting temperature and roasting time.

(2)Compared with willemite, the calcified-roasted samples are much easier to leach in NH_4Cl-$NH_3{\cdot}H_2O$ leaching system. The zinc leaching rate of the roasted sample (roasted with CaO/ willemiteratio at 0.7:1, the roasting temperature at 1280 ℃for 40 min) is 97.45% while the zinc leaching rate of willemite is only 6.02% under the same leaching conditions .

(3)An appropriate increase of CaO dosage, roasting temperature or roasting time can promote the transformative reaction and lead to an obvious increase of zinc leaching rate.

(4)Under optimal roasting conditions, the calcified-roasting (transformative) reaction could react thoroughly, and most of willemite is transformed into ZnO, correspondingly, the zinc leaching rate is higher than 95%.

Acknowledgements

The authors are grateful to Major State Basic Research Development Program of China (973 Program Grant No. 2014CB643403);

References

1. Y.M. Zhang, J Li, "Influence of ultrasonic irradiation on ammonia leaching of zinc from low-grade oxide zinc ore," *The Chinese Journal of Nonferrous Metals*, 2009,19(5),961-966.
2. H.S. Xu, Research on sulfuric acid leaching of zinc silicate ore,(Ph.D thesis Kunming University of Science and Technology,2014).
3. W.Q. Qin,Mineralogical characterization of blast furnaces lag quenched by water,*The Chinese Journal of Nonferrous Metals*,2014,24(1),262-268.
4. P Wu, The mineralogical characteristics of Pb and Zn in zinc waste, *Acta mineralogica sinica,*2002,22(1),39-42.
5. A.L Chen, Z.W. Zhao,Comprehensive Utilization Status and Expectation of Zinc Oxide Ores, *Mining and metallurgical engineering*, 2008,28(6),62-66.
6. L.Z. Xiong, Q.Y. Chen, Preparation of metallic zinc from hemimorphite by vacuum carbothermic reduction with CaF_2 as catalyst. *Transaction of nonferrous metals society of China,*2011
7. Z.Y. Liu, Z.H.Liu, Leaching mechanism of willemite in $(NH_4)_2SO_4$-$NH_3 \cdot H_2O$ system, *Transaction of nonferrous metals society of China,*2011,21(11),2929-2935
8. H.S Xu, C Wei, Sulfuric acid leaching of zinc silicate ore under pressure, *Hydrometallurgy* ,2010,105, 186-190
9. Z.W. Zhao, X.J. Jia, Leachingzinc oxide ore in alkaline solution, *Journal of Central South University(Science and Technology),*2010,41(1),39-43
10. Z.W. Zhao, S Long, Alkali leaching of refractory high silica zinc oxide ore by mechanical activation, *Journal of Central South University(Science and Technology),*2010,41(4),1246-1250
11. Z.Y. Liu, Dissolution behavior of willemite in the $(NH_4)_2SO_4$-$NH_3 \cdot H_2O$ system, *Hydrometallurgy,*2012,125: 50-54
12. P Cao, Research on Vanadium Slag Roasted with Calcium Salt, *Iron steel vanadium titanium,*2012,33(1),30-34
13. S Bei, Z.J. Wang, Review of vanadium extraction from stone coal by roasting technique with sodium chloride and calcium oxide, *Chinese journal of rare metals,*2010,34(2),291-297
14. Z.Y. Liu, Leaching mechanism of zinc oxide ores in ammonia-ammonium aqueous solution, (Ph.D thesis ,Central South University,2012)

7th International Symposium on High-Temperature Metallurgical Processing
Edited by: Jiann-Yang Hwang, Tao Jiang, P. Chris Pistorius, Gerardo R.F. Alvear F., Onuralp Yücel, Liyuan Cai, Baojun Zhao, Dean Gregurek, and Varadarajan Seshadri
TMS (The Minerals, Metals & Materials Society), 2016

AN INVESTIGATION ON ANTIMONY PRODUCTION BY USING NIEDERSCHLAG PROCESS

Sedef Pinar Basag[1], Ahmet Turan[2], Onuralp Yucel[1]

[1]Istanbul Technical University, Faculty of Chemical and Metallurgical Engineering, Department of Metallurgical and Materials Engineering, 34469, Maslak, Istanbul, Turkey
[2]Yalova University, Faculty of Engineering, Department of Chemical and Process Engineering, 77100, Yalova, Turkey

Keywords: Antimony, Niederschlag process, Metallothermy

Abstract

Niederschlag process is commonly used to produce metallic antimony. The process is a metallothermic process which is conducted in one step for metallic antimony production. Using this method Sb_2S_3 is mixed with coal and metallic iron and subjected to reduction. During heating, antimony sulfide is decomposed and sulphur combines with iron to form the matte phase. Metallic antimony is collected at the bottom of the crucible. In this study, evaluation of Sb_2S_3 concentrate, extracted and obtained from West Anatolian Region of Turkey was investigated by using Niederschlag process. The reductant stoichiometry, reduction temperature and duration were carried out in the experiments. Chemical analysis methods by using atomic absorption spectrometry (AAS) and X-ray diffraction spectrometry (XRD) techniques were used for the characterization of the concentrate, metallic nuggets and matte phases.

Introduction

Antimony is a metalloid with atomic number 51. It is found in over 100 mineral species and found in nature mainly as a sulfide mineral stibnite (Sb_2S_3). Antimony is a lustrous grey metal that has a Mohs Scale hardness of three. This metalloid exists in two forms; metallic antimony is bright, silvery, hard and brittle, nonmetallic form is a grey powder. Antimony has poor electric and heat conduction. It is not attacked by dilute acid or by alkalis and stable in dry air [1].

Several production methods can be used for antimony production, such as; gasification and reduction method, electrolyte method and Niederschlag Process. Stibnite is totally oxidized at 350-400°C in reverber type furnaces for gasification and reduction method and it takes 10-12 hours approx. As electrolyte method; antimonite dissolves in hot sodium sulfide solution and then obtained sodium thio antimonite electrolyzed with steel anodes and cathodes. In Niederschlag Process, stibnite can directly reduced with coal and metallic iron in reverbetory type or blast furnaces [1].

$$Me_1X + Me_n = Me_1 + Me_nX \tag{1}$$

Niederschlag Process is the most economic production method for antimony production. In this method antimony can be reduced in one step. Sulphur is decomposed from stibnite while heating and interacts with iron then matte phase would obtain over the metallic antimony [2, 3].

$$Sb_2S_3 + 3Fe_{(k)} = 2Sb_{(s)} + 3FeS_{(s)} \quad (2)$$

Reduction reaction starts over 1100°C. Glauber salt or Na_2CO_3 can be used for avoid metal losses. System has FeS-Na_2S liquid matte phase and a metallic antimony phase underneath the matte phase [2].

In this study, the stibnite concentrate from Etibakir Halikoy mine be subjected to produce metallic antimony with Niederschlag Process. Dry concentrate mixed with borax, soda and SiO_2 as flux materials and iron nails were used as reductant. Experiments were conducted at various temperatures with various flux addition amounts.

Experimental Studies

In Niederschlag experiments, stibnite concentrate (particle size < 60.8 μm), iron source, glauber salt ($Na_2SO_4.10H_2O$), carbon black powders (99.8% pure and a specific surface area of 27 m^2/g), borax ($Na_2B_4O_7.10H_2O$), sodium carbonate (Na_2CO_3) and SiO_2 were used. The chemical analysis of the stibnite concentrate is given in Table I.

Table I. Chemical Analysis Results of Stibnite Concentrate (wt.%)

Sb	As	Pb	Fe	SiO_2	S	Al	Sn
68.67	0.08	0.14	1.08	0.14	27.11	0.08	0.12

The initial mixtures were prepared from dried powders at molar ratios Sb_2S_3:Fe as 1:1, 1:1.5, 1:2. The powder mixtures were charged into a cylindrical fireclay pots which have 58 mm inner diameter and 142 mm height. Reduction reactions took place in an electrical resistance furnace for one hour. Four different reaction temperatures (1273 K, 1373 K, 1473 K and 1573 K) were used for different experimental sets. The obtained phases were characterized by using XRF (ThermoScientific, w/ He tube) analyzer and chemical analysis methods. The flowchart of conducted experiments can be seen in Figure 1.

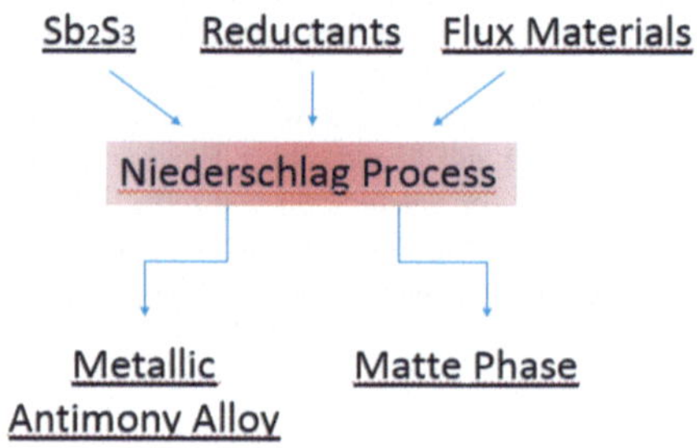

Figure 1. Flow chart of conducted experiments.

Results and Discussion

In the first experimental set, different Fe addition amounts were examined. There is a direct relation between Fe stoichiometry and obtained metallic phase. The effects of Fe stoichiometry are given in Table II. Metallic antimony phase was obtained in the experiment 100% reductant stoichiometry used and conducted for 80 minutes.

Table II. The Effects of Fe Stoichiometry

No	Raw Materials	Additives	Fe Sto.	Temp.	Duration	Metallic Phase	Matte-slag
1	Sb_2S_3: 33 g - Fe: 23.46 g	-	50%	1373 K	60 min.	-	-
2	Sb_2S_3: 45.3 g - Fe: 22.33 g	-	100%	1373 K	80 min.	17.4 g	12.2 g

The second experimental set, fixed Fe stoichiometry (100%) and increasing reaction temperature were carried out for 60 minutes process duration. The effect of temperature is given in Table III. Metallic phases were not obtained due to short process duration.

Table III. The Effects of Temperature

No	Raw Materials	Additives	Fe Sto.	Temp.	Duration	Metallic Phase	Matte-slag
3	Sb_2S_3: 45.3 g - Fe: 22.33 g	-	100%	1373 K	60 min.	-	-
4	Sb_2S_3: 45.3 g - Fe: 22.33 g	-	100%	1473 K	60 min.	-	-
5	Sb_2S_3:45.3 g - Fe: 22.33 g	-	100%	1573 K	60 min.	-	-

The last experimental set was carried out with different flux additions. The effect of flux additions are given in Table IV. It can be seen that flux addition and high reductant stoichiometries increased metal recovery rates.

Table IV. Effects of Additives, Fe Stoichiometry, Time and Temperature

No	Raw Materials	Additives	Fe Sto.	Temp.	Duration	Metallic Phase	Matte-slag
6	Sb_2S_3: 45.3 g - Fe: 22.3 g	Glau. S.: 5 g C: 5 g	100%	1373 K	80 min.	-	-
7	**Sb_2S_3: 45.3 g - Fe: 33.0 g**	**Borax: 15 g** **Soda: 30 g** **SiO_2: 10 g**	**200%**	**1373 K**	**60 min.**	**38.9**	**30.0**
8	Sb_2S_3: 45.3 g - Fe: 22.3 g	Borax: 15 g Soda: 30 g SiO_2: 10 g	100%	1373 K	60 min.	-	-
9	**Sb_2S_3: 50.0 g - Fe: 33.6 g**	**Borax: 5 g** **Soda: 10 g** **SiO_2: 5 g**	**200%**	**1373 K**	**60 min.**	**31.4**	**23.3**
10	Sb_2S_3: 45.3 g - Fe: 33.0 g	Borax: 15 g Soda: 30 g SiO_2: 10 g	200%	1273 K	60 min.	38.8	83.5

Chemical contents of metallic phases which were obtained in all experiments are given in Table V. Optical microscope micrographs of the metallic phases in question can be seen in Figure 2.

Table V. Chemical Analysis Results of Obtained Metallic Phases (wt.%)

No	Sb	Fe	S
2	55.15	35.72	8.24
7	68.61	26.16	4.89
9	41.55	37.74	6.92
10	69.72	20.46	8.57

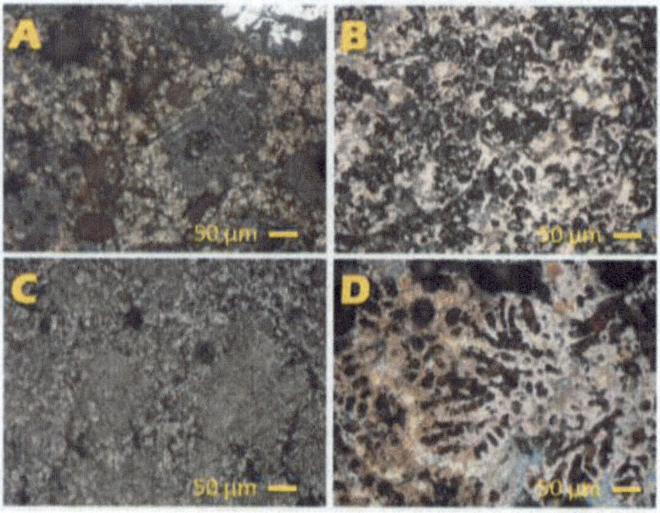

Figure 2. Optical microscope micrographs of metallic antimony phases after etching; (A) Exp. 2, (B) Exp. 7, (C) Exp. 9, (D) Exp. 10.

Conclusions

In this study several experimental works carried out for producing metallic antimony from stibnite concentrate with a metallothermic process which is named Niederschlag Process. During the experimental studies, reductant (Fe) amount, process temperature, duration and flux addition amounts were examined. According to the chemical analysis results, flux addition increased the metal recovery. The highest metallic phase weights were obtained in the Exp. 7 and Exp. 10. But, when we compare the reaction temperatures, Exp. 10 is technologically more efficient because of the obtaining of nearly same metal recovery rates at lower reaction temperatures. The experiment in question was conducted with 200% Fe stoichiometry, flux addition (Borax: 15 g,

Soda: 30 g, SiO_2: 10 g) at 1273 K. The content of metallic phase was determined as 69.72% Sb, 20.46% Fe and 8.57% S by using chemical analysis methods. In the further experiments, it is planned to conduct comprehensive experiments about the effects of flux materials.

References

1. Fathi Habashi, Handbook of Extractive Metallurgy, vol.2 (Heidelberg, Germany: WILEY-VCH, 1997).

2. M.E. Acma, "Behavior of Iron and Arsenic During Fire-Refining of Antimony" (Ph.D. thesis, Istanbul Technical University, 1988).

3. Fuat Y. Bor, Extractive Metallurgy Principles, vol.2 (Istanbul, Turkey: ITU press, 1989).

7th International Symposium on High-Temperature Metallurgical Processing
Edited by: Jiann-Yang Hwang, Tao Jiang, P. Chris Pistorius, Gerardo R.F. Alvear F., Onuralp Yücel, Liyuan Cai, Baojun Zhao, Dean Gregurek, and Varadarajan Seshadri
TMS (The Minerals, Metals & Materials Society), 2016

OXYGEN-RICH SIDE BLOW BATH SMELTING TECHNOLOGY – NEW DEVELOPMENTS IN CHINA

Lin Chen[1], Wei Chen[1], Hui Xiao[1,2], Tianzu Yang[1,*], Weifeng Liu[1], Duchao Zhang[1]

1 School of Metallurgy and Environment, Central South University, 410083, China
2 Hunan Jinwang Bismuth Industry Co., Ltd., Chenzhou 423000, China
* Corresponding Author, Tel: +86 13508482570, Email: tianzuyang@163.com

Keywords: Oxygen-rich side blow technology, bath smelting, pyrometallurgy

Abstract

Oxygen-rich side blow bath smelting (OSBS) technology was developed in 2000 based on Vanyukov process. This technology exceeds the traditional pyrometallurgical methods by its great processing capability, high metal recovery and environmental-friendly property. In the past 15 years, OSBS technology extended its application area from lead industry to bismuth, antimony and waste recycling industries, etc. This paper introduces the history and new developments of OSBS technology. The typical industrial applications examples of the OSBS technology in China, including high lead slag reduction, low grade bismuth-lead concentrate smelting and metallurgical waste recycling, are reviewed. Especially, the unique furnace properties and technical parameters are compared. Finally, the potential research directions of OSBS technology are summarized.

Introduction

Vanyukov process is a modern pyrometallurgy technology developed by Professor Vanyukov in the 1970s. By introducing oxygen-rich air from the tuyeres into the melt, autothermal smelting can be achieved for copper or copper-nickel concentrate smelting. Extensive efforts were then devoted to fundamental theory as well as engineering technology [1-3], however, its application area was mainly limited in Russia and Kazakstan. In 2001, Zhonglian corporation built the first experimental Vanyukov furnace (2.25 m^2) in Xinxiang (Henan, China) for lead smelting with the assistance of researchers from Russia [4]. Oxidative smelting of lead concentrate and reductive smelting of high lead slag in the furnace were investigated with intensive experiments and adjustments were done on the furnace structure and process operation parameters, steady operation were finally achieve in 2004. Based on the properties of the adjusted furnace structure, the technology was denominated as oxygen-rich side blow bath smelting (OSBS) technology.

Nowadays, more than 20 oxygen-rich air side blow furnaces (OSBF) were built in China and it is widely used for smelting a variety of non-ferrous metals, for example, lead, copper, bismuth, antimony, etc. This paper describes the OSBS technology and its key apparatus. Difference types of furnace structures are compared. The paper also reviews some of the typical applications of the OSBS technology in China. Finally, the advantages and future perspectives of the technology are summarized.

Description of the OSBS technology

Process description

Bath smelting is pyrometallurgy technology that metallurgical process, including melting, sulfurization, oxidation, reduction and fuming, etc., for the raw material containing metal values are conducted in a liquid bath formed by melted slag or matte. This technology strengthens the reactions by introducing materials (concentrate, matte or slag) with reactants (oxygen, coal or coke, sulfur concentrate, etc.) in a fiercely stirred liquid bath. In the OSBS technology, nonferrous metal minerals are usually charged from the feed inlet on the top of the furnace. Temperature of the mineral particles rises quickly in the reactor and melts to form metal (matte), slag and gas phases. These phases separate from each other due to their density difference, thus separation of metal values from the impurities is achieved. The metal (matte) phase usually possesses the largest density and precipitate from the slag phase into a hearth at the bottom of the reactor. The gas phase will go through the uptake into a waste heat boiler and the following purification facilities for heat recycling and dust removal. Oxygen-rich air is often used in the process, which not only act as the oxidant but also as the agitator for the bath. Therefore, heat and mass transfer between the gas-solid-liquid phases is greatly enhanced under the fierce agitation, leading to a chemical equilibrium state between the phases.

Reactor description

The key apparatus of the OSBS technology is oxygen-rich side blow furnace (OSBF). The furnace is widely used for liquid high lead slag reduction in China and two types of furnace structures were developed in industrial application. Fig. 1 shows the schematic diagram of the furnace most commonly used. The furnace is a rectangle shaft furnace. The hearth at the bottom of the furnace is built with magnesia-chrome firebricks, insulating bricks and refractories. The middle part of the furnace surrounded by copper water jackets and the cross sectional area is determined according to the processing capability of the furnace. The top part is constructed by firebricks, refractories and iron water jackets. The raw materials are added from the top of the furnace. The smelting products, including slag, matte and metal, can be released by a liquid sealed room, viz. siphon room. Oxygen rich gas is injected from the tuyeres on the copper water jackets. The schematic diagram of tuyere can be seen from Fig. 2 [5]. It can be used only for injecting oxygen-rich air rather than natural gas or other flammable gases. During production, tuyeres are used in turn to prolong their life time and those of the copper water jackets. Tuyeres can be shut down by inserting an iron rod thus melt can be prevented to enter the tuyeres.

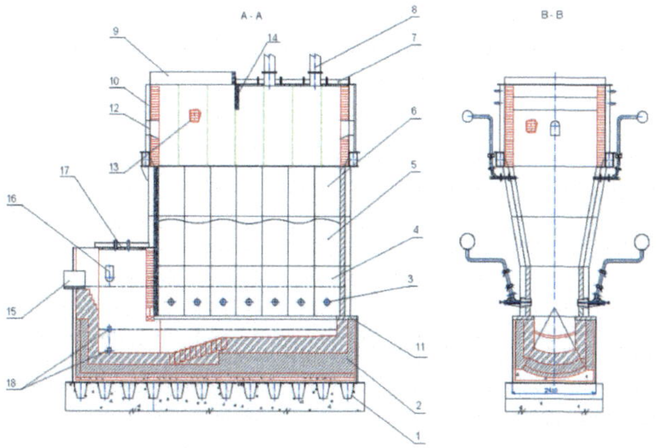

Fig. 1 Schematic diagram of OSBF. 1. Understructure, 2. Hearth, 3. Tuyere, 4. First floor copper water jacket, 5. second floor copper water jacket, 6. third floor copper water jacket, 7. Top iron water jacket, 8. Feed inlet, 9. Flue gas outlet, 10. Fourth floor iron water jacket, 11. Hearth copper water jacket, 12. Observation port, 13. Firebrick, 14. Separation wall iron water jacket, 15. Slag tapping hole, 16. Overflow, 17. Siphon room cover, 18. Satfy port.

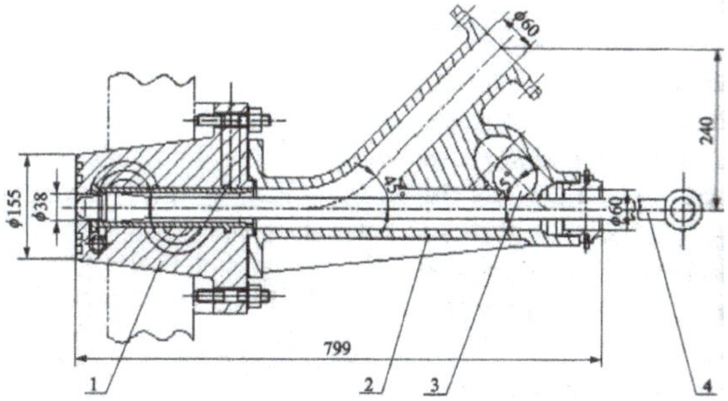

Fig. 2 Schematic diagram of the tuyere in OSBF. 1. Head, 2. Tube, 3. Seal ball, 4. Iron rod.

The other type of furnace is quite similar to the former one. However, its middle part is built with firebricks. Instead of contacting with molten slag directly, iron or copper water jackets are used on the outside to cool the firebricks. This type of furnace uses a nozzle similar to that used in the oxygen-rich gas bottom blow furnace (OBBF). The nozzle has a complicated structure consisted by several annular tubes, which enables the nozzle to inject both gas fuel and oxygen-rich air into the furnace at the same time. However, height of the molten bath should be lower than the nozzles before stop injecting air, since iron rod can not used to prevent the melt entering the nozzle.

These two types of OSBF both have applications in industry, but the former one is more widely accepted in China. The major differences between them are:

(1) Furnace wall structures

The former one uses copper water jackets to contain molten bath, water-cooled slag layer on the inside wall of the copper water jacket can effectively prevent the corrosion from the strongly agitated bath. On the other hand, inside lining is used as furnace wall for the latter furnace. This structure may experience serious corrosion in the molten bath.

(2) Reductants

The former one uses only coal as fuel as well as reductant. The latter one uses not only coal but also gaseous fuels.

(3) Gas injection

The former one injects oxygen-rich air with certain oxygen concentration into the melt from tuyeres. The tuyere can be easily opened and shut down with an iron rod. The latter one injects oxygen, air and gaseous fuels into the melt separately into the melt from nozzles, which is consisted by several annual tubes. The nozzle can directly raise the temperature of the melt, which is beneficial for steady operation of the furnace. However, the life time of the nozzles are quite short (20-30 d), molten slag should be discharged before replacing the broken nozzles, resulting in a low production rate.

Industrial applications

High lead slag reduction

In the 2000s, the most widely applied lead concentrate smelting technology in China is SKS process in which lead concentrate is first oxidized in an OBBF and then the high lead slag produced from the OBBF is reduced in a blast furnace [6]. The technology was developed jointly by China Shuikoushan mining administration and China Enfi engineering corporation in the 1980s. However, liquid high lead slag produced from OBBF requires casting before it can be further processed in blast furnace, which wastes the latent heat of the molten slag. In addition, dust emission in the casting process not only imposes threats to the environment but also causes unessential resource waste. Based on the OSBS technology, oxygen-rich side blow furnace (OSBF) was further developed for high lead slag reduction and was soon adopted by many lead companies in China [4, 7]. In a typical high lead slag reduction process, the liquid high lead slag is discharged directly from OBBF into OSBF. To meet the operation time of OBBF, an intermittent operation style is adopted for OSBF. The operation period is usually 120 min, among which 20-30 min are uses for charging molten high lead slag, 70-80 min are used for reducing high lead slag and the left 10-20 min are used for discharging the reduced slag. The lead produced form high lead slag reduction can be released from the siphon room, while reduced slag are quenched or sent for fuming. After heat recycling in the waste heat boiler, dust filtration in the bag filter and sulfide adsorption, the flue gas can be discharged [8]. The flow sheet of the process is shown in Fig. 3.

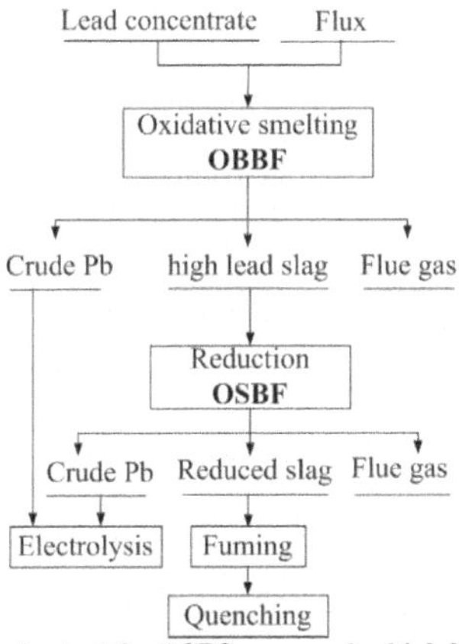

Fig. 3 Process flow sheet of the OSBS process for high lead slag reduction

Table 1 shows the technical index of SKS process and OSBS process. It can be inferred that lead recovery of OSBS is slight higher than that of SKS, while its comprehensive energy consumption is 70 kgce/t lower [9], indicating that OSBS exceeds the traditional SKS process by its high production capability, low energy consumption and strong metal-slag separation ability [8].

Table 1 Technical index of SKS and OSBS process

	unit	SKS	OSBS
Operation environment		Bad	Good
Production capability	$t/(m^2 \cdot d)$	50-65	50-80
Lead recovery	%	96.5-98	97-98
Dust rate	%	Oxidation 12-14 Reduction 6-7	Oxidation 12-14 Reduction 8-10
Oxygen consumption	m^3/t	270-280	320-330
Electricity consumption	kWh/t	115-125	60-80
Reductant consumption	%	14-18（coke）	8-10（coal）
Lead content in the reduced slag	%	2-3	<2
Comprehensive energy consumption	kgce/t	300	230

low grade bismuth-lead concentrate smelting

Bismuth is usually extracted from bismuth concentrate with high bismuth contents (> 15 wt.%) or secondary materials like copper smelting dust or lead anode slime [10]. During the past decades, precipitation smelting or reduction smelting have became the major treatment options in China [11]. However, the gradual decline in the quality of ores by the increase of other elements proposes challenge for these traditional methods which are criticized for their high energy consumption, low metal recovery and serious emission of dust and SO_2 [11, 12]. Recently, an industrial-scale OSBF system is built in Chenzhou (Hunan, China). Major components of the system contain two OSBFs which are used for the oxidative smelting of concentrate and reductive smelting of the oxidized slag from the preceding furnace, respectively.

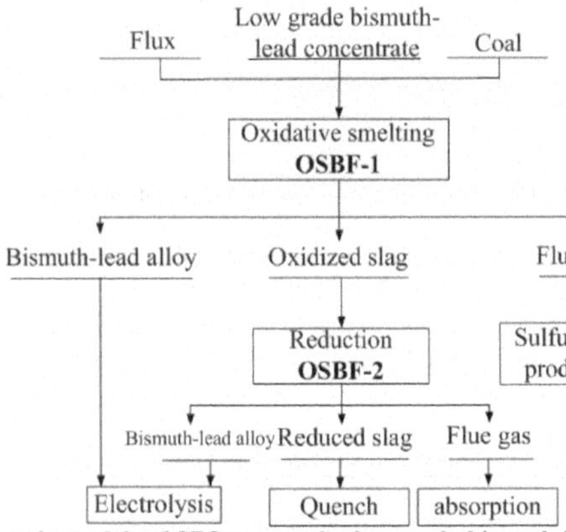

Fig. 4 Process flow sheet of the OSBS process for low grade bismuth-lead concentrate smelting.

Fig. 4 shows the process flow sheet of the OSBS process. The cross sectional areas of two OSBFs are 4.2 m^2 (OSBF-1) and 3.6 m^2 (OSBF-2), respectively. In the process, low grade Bi-Pb concentrate are continuously charged by belt conveyer with flux and coal for oxidative smelting in OSBF-1 to produce metal alloy and oxidized slag. Oxygen-rich gas is injected by means of tuyeres to create turbulent agitation in the foaming molten slag layer where rapid reaction between oxygen, coal, flux and concentrate happens [13]. Metal alloy in the bottom hearth is discharged by siphon to disk casting machine and the molten oxidized slag is tapped intermittently to OSBF-2. After reduction of the molten oxidized slag in OSBF-2, metal alloy and reduced slag are produced. Flue gases from OSBF-1 and OSBF-2 are sent for sulfuric acid production and alkaline absorption, respectively.

Advantages

The major advantage of the OSBS process is as below:

(1) Large production capability

Oxygen-rich air is injected into the molten bath with a pressure of 100 kPa, causing strong agitation and fierce bubbling. The heat and mass transfer in the furnace is greatly enhanced, dynamic barriers of the reactions are eliminated, thus different phases quickly reach a state near chemical equilibrium. The capacity of OSBF can reach as high as 120 t/(m^2·d).

(2) High metal recovery

The molten materials separate from each other according to their density differences in the OSBF. Oxygen-rich air is injected into the molten slag layer where a turbulent area is generated. In this turbulent area, metal or matte droplets quickly collide and agglomerate to bigger sizes, and then precipitate from the slag layer. Meantime, there is a relative peaceful area under the tuyeres, which serves effectively for metal-slag separation. Therefore, OSBF possesses excellent metal-slag separation ability. For example, lead content in the reduced slag can be reduced to < 2%, while those from blast furnace and OBBF are at ~3%.

(3) Low capital investment and production cost

The large production capability greatly reduced the furnace size required for processing large quantities of minerals. For example, an OSBF with a cross sectional area of 8.4 m^2 is usually used as the high lead slag reduction reactor for lead factories with capacities of 100 thousand tons crude lead per year. The capital investment for such an OSBF system cost less than 200 thousand U.S. dollars. Furthermore, the excellent metal-slag separation ability enables the technology to cancel the electric heating fore-well (EHFW) which is usually installed as a slag cleaning auxiliary equipment for blast furnace and OBBF. Therefore, the production cost of the OSBS process can be reduced to less than 100 U.S. dollar/$t_{\text{crude lead}}$.

Future perspectives

Beside high lead reduction and low grade bismuth-lead concentrate smelting, researcher in China are seeking to apply this technology to other areas, such as metallurgical waste cleaning, low grade antimony-lead concentrate smelting and waste electrical and electronic equipment recycling, etc. However, some questions should be clarified before industrial application:

(1) Automatic control system

OSBF belongs to the class of complex, multidimensional non-steady reactors [2]. In fact, to achieve steady operation, the oxygen concentration, gas volumetric velocity, coal ratio and mineral feeding should be adjusted frequently in a certain range. There is no quantitative standard, which proposes a big challenge to the engineers. On the other hand, as experienced engineers are hard to be trained and factories are not willing to share their technical secrets, thus it is difficult for new factories to find qualified engineers to operate the furnace, which hinders the wide application of the OSBS technology. Therefore, developing an automatic control system for OSBF is definitely beneficial for reducing production cost and extending the application of this technology to a wider area.

(2) New furnace structure

Furnace structure should be adjusted when it is applied for processing a new mineral. For example, when a low grade material is processed, volume of the hearth should be greatly reduced to ensure a quick replacement of liquid metal (matte) in it. When metallurgical wastes are processed, height of some tuyeres should be lowered to prevent the formation of matte layer between the slag and metal layers. Therefore, new furnace structure should be developed before applying the technology to other areas.

(3) Element distribution behavior

Although the OSBS technology has been successfully applied in lead industry, the element distribution behavior, especially behavior of some minor elements, is pretty vague. A viable thermodynamic model which connects process parameters and elements distribution behaviors is in urgent demand.

Acknowledgement

The financial support from National Natural Science Foundation of China (No. 21306231) and 2011 joint innovation center – "clean and effective utilization of strategic mineral resources" are gratefully acknowledged.

References

[1] S. Vaisburd, A. Berner, D. Brandon, S. Kozhakhmetov, E. Kenzhaliyev, and R. Zhalelev, "Slags and mattes in vanyukov's process for the extraction of copper", *Metallurgical and Materials Transactions B,* 33B (2002), 551-559.

[2] B.K. Mukhanov, A. Suleimenov, W. Wojcik, and K. Gromaszek, "Development of an optimal control system for smelting process in the molten-pool", *Przeglad Elektrotechniczny,* 11b (2012), 366-368.

[3] S. Vaisburd, D.G. Brandon, S. Kozhakhmetov, and E. Kenzhaliyev, "Physicochemical properties of matte-slag melts taken from vanyukov's furnace for copper extraction", *Metallurgical and Materials Transactions B,* 33B (2002), 561-564.

[4] L. Chen, W. Bin, T. Yang, W. Liu, and S. Bin, Research and industrial application of oxygen-rich side-blow bath smelting technology, in: J. Tao, J.-Y. Hwang, P.J. Mackey, O. Yucel, G. Zhou (Eds.) TMS Annual Meeting, TMS, San Antonio, 2013, pp. 49-55.

[5] L. Zhang, *Moden Lead Metallurgy*, (Changsha: Central South University Press, 2013), 165-215.

[6] D. Li, and Z. Zhang, "The application of new lead smelting technology of oxidizing in bottom blowing furnace and reduction smelting in blast furnace", *Nonferrous metals, Extractive Metallurgy,* 5 (2003), 12-13.

[7] L. Zhang, G. Lin, W. Bin, Y. Li, and X. Li, "Study on oxygen side-blown reducing furnace and smelting reduction process of high-lead slag", *China Nonferrous Metallurgy,* 2 (2012), 12-19.

[8] L. Chen, T. Yang, S. Bin, W. Liu, D. Zhang, W. Bin, and L. Zhang, "An efficient reactor for high-lead slag reduction process: Oxygen-rich side blow furnace", *JOM,* 66 (2014), 1664-1669.

[9] L. Chen, Z. Hao, T. Yang, W. Liu, D. Zhang, L. Zhang, S. Bin, and W. Bin, "A comparison study of the oxygen-rich side blow furnace and the oxygen-rich bottom blow furnace for liquid high lead slag reduction", *JOM,* 67 (2015), 1123-1129.

[10] L. Wang, *Bismuth Metallurgy*, (Beijing: Metallurgical industry press, 1986), 14-95.

[11] T.-z. Yang, J. Li, W.-f. Liu, L. Chen, and W.-d. Bin, Development of bismuth smelting technology in China, in: J. Tao, J.-Y. Hwang, P.J. Mackey, O. Yucel, G. Zhou (Eds.) TMS Annual meeting, TMS, San Antonio, 2013, pp. 631-642.

[12] L. Chen, Z. Hao, T. Yang, H. Xiao, W. Liu, D. Zhang, S. Bin, and W. Bin, "An efficient technology for smelting low grade bismuth-lead concentrate: oxygen-rich side blow process", *JOM,* 67 (2015), 1997-2004.

[13] R.R. Moskalyk, and A.M. Alfantazi, "Review of copper pyrometallurgical practice: today and tomorrow", *Minerals Engineering,* 16 (2003), 893-919.

7th International Symposium on High-Temperature Metallurgical Processing
Edited by: Jiann-Yang Hwang, Tao Jiang, P. Chris Pistorius, Gerardo R.F. Alvear F., Onuralp Yücel, Liyuan Cai, Baojun Zhao, Dean Gregurek, and Varadarajan Seshadri
TMS (The Minerals, Metals & Materials Society), 2016

CARBON REFRACTORIES IN AN OXIDIZING PROCESS? COPPER SMELTING IN AN OUTOTEC® AUSMELT TSL FURNACE WITH A UCAR® CHILLKOTE™ REFRACTORY SYSTEM

Jacob Wood[1], Stefanie Creedy[1], Peter Duncanson[2]

[1] Outotec® Pty Ltd, 12 Kitchen Rd, Dandenong, Australia, 3175
[2] GrafTech International, 791 Santa Fe Pike, Columbia, TN 38401 USA

Keywords: Cu Smelting, Ausmelt, UCAR, Carbon Refractories

Abstract

Carbonaceous refractory materials are widely accepted in many reducing smelting processes such as iron and ferroalloys. Oxidizing processes such as copper concentrate smelting traditionally use magnesia and/or alumina/silica refractory materials. The main objection to the use of carbon-based refractories as the exposed working lining in oxidizing applications is the potential for free oxygen to react with the carbon at high temperature, which could rapidly degrade the refractory lining.

In 2013, Outotec and GrafTech conducted joint trials of a thermally conductive, carbonaceous lining for smelting copper concentrates in the Outotec® pilot-scale top-submerged-lance (TSL) furnace in Dandenong, Australia. This testwork had two equally important goals; to demonstrate the ability of carbon refractories to withstand oxidizing process conditions, while ensuring that neither furnace operating conditions nor process stability were negatively affected by the refractory material.

The paper discusses the input parameters of the trials, key observations during the testwork program, and the results and conclusions from both process and refractory design standpoints.

1. Introduction

1.1. Outotec® Ausmelt TSL Technology

Outotec® Ausmelt Top Submerged Lancing (TSL) technology has been used for more than 30 years in a wide range of non-ferrous materials smelting applications. The technology utilises a vertically suspended lance for the injection of fuel and oxygen enriched air into the molten bath. The intense bath agitation achieved provides for highly efficient mass and energy transfer and thus a high specific smelting rate. Operation of the well sealed furnace under negative pressure ensures minimal fugitive dust and gas emissions and an offgas rich in sulphur dioxide (SO_2).

Historically, the Ausmelt furnace design comprised a water-cooled shell lined with conductive refractory [1] to promote formation of a protective lining of frozen slag. More recently, the typical furnace configuration has been a water-cooled, steel shell lined with chrome-magnesia, insulating refractories although in some process applications, water-cooled copper elements have also been incorporated within the furnace design.

1.2. UCAR® ChillKote™ Freeze Lining Technology

Carbon-based refractories have been standard technology for lining the crucible in blast furnaces for more than 50 years. In the last 20 years the technology has been adapted to submerged arc furnaces producing ferroalloys, mainly ferrochrome and ferromanganese. "Carbon refractory" commonly refers to materials with greater than 80% fixed carbon content, and can include various generic classifications such as carbon, semigraphite (or graphitic), and graphite. The different classifications have different ranges of physical properties (predominantly thermal conductivity), which are determined by raw materials and manufacturing processes. Compared with traditional alumina, silica, or magnesia refractory materials, carbon refractories generally have lower thermal expansion, lower elasticity, and higher thermal conductivity.

The UCAR® ChillKote™ System combines GrafTech's unique HotPressed Carbon and semigraphite bricks and other high conductivity materials with an active cooling system to create and sustain efficient heat transfer from the process to the cooling system. When properly configured, the hot face temperature of the refractory is low enough to freeze a protective layer of process material, usually slag, which protects and further insulates the refractory. When refractory temperatures are kept low, common temperature-dependent wear mechanisms such as erosion, thermal stress and chemical attack are prevented. Typical ChillKote System configurations are shown in Figure 1.

Iron and ferroalloys are produced via reduction processes, which leave virtually no oxygen available to react with carbon refractories under normal operating conditions. On the other hand, concerns about reactions between free oxygen and carbon refractories in an oxidizing process such as copper smelting have prevented the application of carbon refractories in such processes. As the reaction between carbon and oxygen requires high temperature, however, consideration was given to the possibility that a ChillKote lining may be able to maintain temperatures low enough to withstand the process.

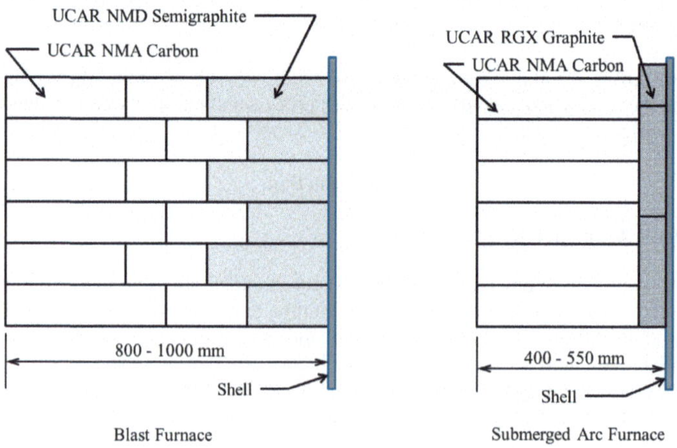

Figure 1: Typical UCAR® ChillKote™ Configurations

2. Pilot Scale Testwork Program

2.1. Testwork Preparation

To evaluate the performance of carbon refractories under oxidizing smelting conditions, a joint testwork program was conducted in 2013 by GrafTech and Outotec utilizing the Outotec pilot-scale Ausmelt TSL furnace in Dandenong, Australia. The testwork aimed to examine the performance of UCAR Grade NMD™ HotPressed™ semigraphite refractory bricks under the oxidizing process conditions in copper concentrate smelting to ensure neither the furnace operating conditions nor process stability were negatively affected by the refractory material.

The Outotec pilot-scale Ausmelt TSL furnace used in the testwork employed an enclosed water-jacket for cooling of the lower furnace walls along with a water-cooled, copper tapping block for removal of the molten furnace contents. The furnace hearth was lined with magnesia-chrome bricks sourced from RHI refractories, whilst UCAR Grade NMD HotPressed semigraphite bricks were used to line the furnace lower walls. UCAR Grade C46™ Cement was used as the mortar between adjacent bricks and to fill the small gap (<5mm) between the bricks and the furnace shell. In front of the semigraphite bricks, an approximately 50mm thick layer of magnesia-chrome, castable refractory was formed to act as a sacrificial refractory layer during the furnace heat-up. Steel formwork was also used to support this castable refractory. In total five (5) rows of NMD bricks with a total height of 960mm were used to line the lower furnace walls, with a small power pack being employed to ensure all bricks were tightly installed.

The upper furnace walls were lined with magnesia-chrome bricks, whilst the furnace roof and offtake were lined with castable magnesia-chrome refractory (Shiralite 140LW), sourced from RHI and Shinagawa Refractories Australia respectively. Photographs of the Ausmelt Pilot-Scale furnace and lower wall refractory installation are provided in Figure 2, Figure 3 and Figure 4.

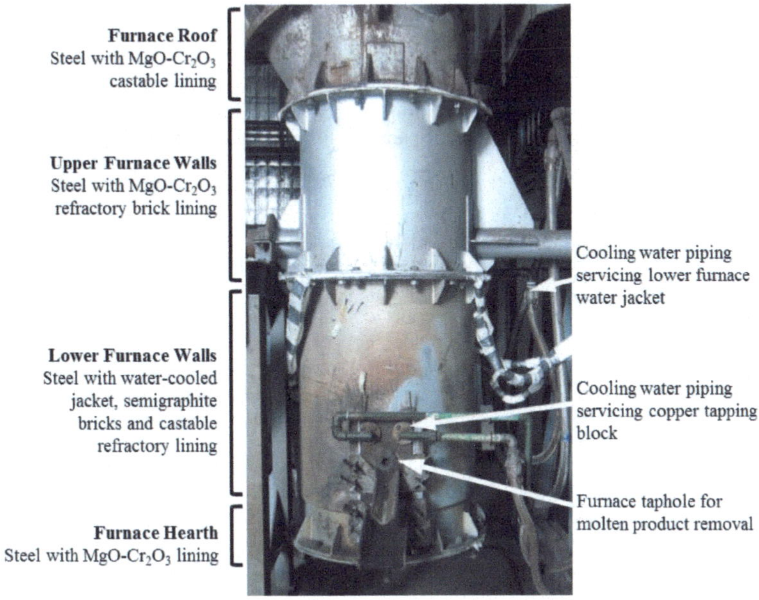

Figure 2: Outotec® Ausmelt TSL Pilot-Scale Testwork Furnace in Dandenong, Australia.

Eight (8) mineral insulated, metal sheath (MIMS) thermocouples were installed in the furnace lining to measure the temperature of the semigraphite refractory bricks. A small groove was cut into the back of selected bricks to ensure good contact between the semigraphite bricks and water-cooled furnace shell. Four (4) thermocouples were installed in the 2nd row of semigraphite bricks (designated T/C 1# - T/C #4), and four (4) thermocouples were installed in the 4th row of bricks (designated T/C 5# - T/C #8). Each row of four thermocouples was installed in 90 degree segments, offset from the taphole by 45 degrees.

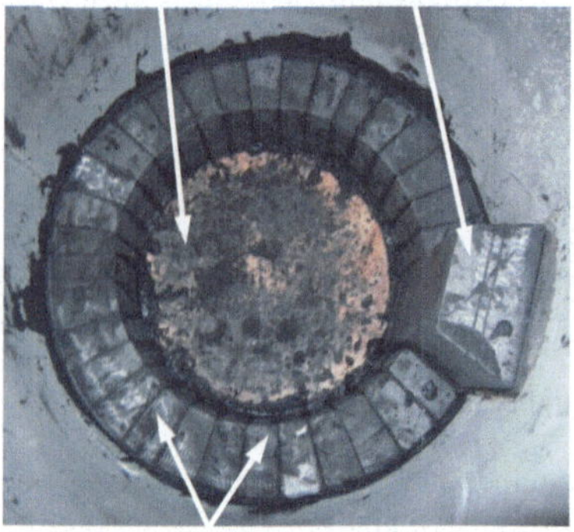

Figure 3: Hearth and Lower Wall Refractory

Figure 4: Top of Lower Wall Refractory

Prior to commencement of the testwork program, the furnace was heated over a period of approximately 6 hours, initially using a natural-gas fired burner installed via the furnace roof and subsequently using the Ausmelt lance. Heat-up of the furnace lining was done to remove any moisture present in the refractory materials, particularly the insulating $MgO-Cr_2O_3$ castable used in the upper furnace walls.

2.2. Testwork Execution

The testwork program examining the performance of GrafTech carbon refractories under oxidizing conditions comprised four individual (4) trials conducted on separate days, with each trial lasting 8 - 10 hours. In each trial, copper concentrate from Mt. Gordon copper mine in Queensland, Australia was smelted to produce copper matte and a low-copper content slag. Silica and limestone fluxes were also introduced to control the smelting slag composition. Target operating conditions for the testwork program are summarized in Table I, whilst details relating to material feedrates and lance flows in each of the trials are provided in Table II.

Table I: Target Operating Conditions for the Testwork Program

Bath Temperature (°C)	Slag Composition		Matte Grade (wt. % Cu)
	Fe/SiO2 ratio	wt. % CaO	
1210 - 1250	1.4	5	50

In the first trial (Trial 1) of the testwork program, slag from a previous set of trials was used for generation of a molten start-bath. During the start-bath formation, the steel formwork supporting the castable refractory in front of the semigraphite bricks was quickly dissolved, however the majority of the sacrificial layer of castable remained relatively intact. Copper concentrates were smelted continuously at a feedrate of 150kg/h for approximately 6 hours, with the tapping of slag and matte occurring three (3) times. Bath temperatures were controlled via adjustment of the lance natural gas flow and lance air oxygen enrichment. During the trial, none of the semigraphite bricks were directly exposed to the molten bath given that castable refractory remained in front of the bricks throughout. A frozen layer of slag was observed in front of the castable throughout the trial.

Table II: Lance Flows and Material Feedrates for each Trial

Trial no.	Trial Duration (hh:mm)	Natural Gas Fuel Rate (kg/h)	Lance Air O_2 Enrichment (%)	Concentrate Feedrate (kg/h)	Silica Flux Feedrate (kg/h)	Limestone Flux Feedrate (kg/h)
1	05:50	18 - 30	21 - 28	150	10 - 15	7
2	07:45	24 - 30	21 - 30	150	12 - 15	7
3	08:25	25 - 28	26 - 30	150	10 - 12	7
4	08:30	28 - 35	26 - 32	150	10 - 12	7

In Trial 2, slag from the previous trial was used to form a molten start-bath. Once again, copper concentrates were smelted continuously for nearly 8 hours, with tapping occurring four (4) times during the day. Whilst some localized thinning of the castable refractory layer was observed during this trial, again, none of the semigraphite bricks were directly exposed to the molten bath. Once again, a thin layer of frozen slag was present in front of the castable during Trial 2.

The testwork team agreed after the first two trials that the lining was "too stable" and further trials would not demonstrate the semigraphite brick's performance in direct contact with the molten bath. Prior to the third trial in the testwork program, therefore, a section of castable on the South-East side of the furnace was intentionally removed, exposing approximately 25% of the semigraphite bricks surface area (Figure 5).

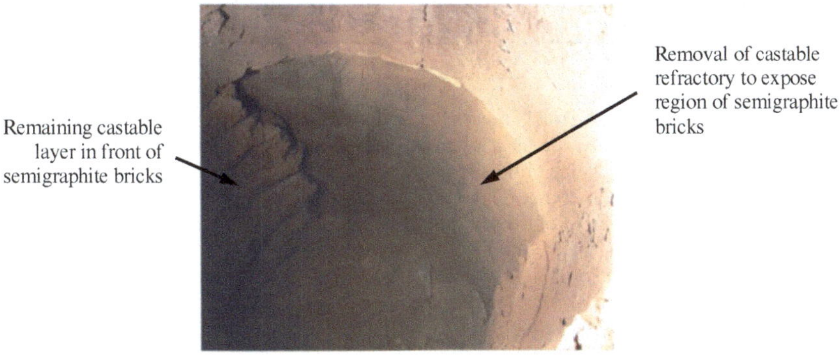

Figure 5: Section of Castable Refractory Removed Prior to Trial 3

During the trial smelting of copper concentrates was carried out for more than 8 hours. A higher lance natural gas flow and oxygen enrichment were required compared with previous trials as a result of the increased heat flux (losses) resulting from exposure of the semigraphite refractory directly to the molten bath. Tapping of slag and matte was completed 5 times in Trial 3. A frozen layer of slag was observed in front of both the remaining castable and directly on the face of the semigraphite bricks throughout the majority of the trial. During periods with a bath temperature of approximately 1300°C however, there appeared to be insufficient energy removal to maintain a stable layer of frozen slag and when molten slag impacted the semigraphite bricks, it would simply run down the brick 'hot-face' rather than forming a freeze lining.

Before the final trial in the testwork program, more of the castable refractory in front of the semigraphite bricks was removed from the South-West and Western regions of the furnace resulting in exposure of roughly 75% of the semigraphite refractory surface area. Copper concentrate smelting was completed for 8.5 hours, with tapping completed 6 times during the trial. Once again, the increased heat flux resulting from exposure of more of the semigraphite bricks resulted in an increased lance fuel and oxygen enrichment requirement to maintain the bath temperature than in previous trials. Once again, a layer of frozen slag was present in front of both the castable refractory and semigraphite bricks during Trial 4.

2.3. Testwork Results

Table III summarises the furnace operating conditions and composition of products in each trial. Application of a concentrate smelt air factor of 0.94Nm^3/kg early in Trial 1, resulted in generation of a high-grade matte and oxidation of a significant portion of iron into the slag meaning the slag Fe/SiO_2 ratio in this trial was higher than the target of 1.4. Application of a lower smelt air factor (0.8 Nm^3/kg) in subsequent trials meant matte and slag compositions were better controlled around the targets specified in Table I.

Table III: Furnace Operating Conditions and Product Compositions for each Trial

Trial no.	Bath Temp. (°C)	Concentrate Smelt Air Factory (Nm^3/kg)	Matte Grade (wt. % Cu)	Slag Composition	
				Fe/SiO_2 ratio	wt. % CaO
1	1210 - 1280	0.94	59 – 70	1.5 – 1.8	4 – 6
2	1180 - 1260	0.80	46 – 57	1.1 – 1.6	5 – 7
3	1150 - 1300	0.80	45 – 49	1.1 – 1.5	6 – 7
4	1180 - 1250	0.80	43 – 49	0.9 – 1.4	6 – 8

Following the completion of Trial 2, a sample of slag from in front of the castable refractory was taken and examined. The frozen slag layer was observed to be highly stable and difficult to remove, suggesting some degree of bonding with the castable refractory, whilst its thickness was measured to range from 10 – 20mm. This examination was repeated after Trial 4, where a sample of slag was taken from in front of the semigraphite refractories. The thickness of this slag varied between 5 and 10 mm (Figure 6), although in this case, there was noted to be no bonding of the slag directly with the refractory bricks.

Refractory brick temperatures measured in each of the trials indicated that with castable in front of the semigraphite bricks, brick temperatures were relatively low and unresponsive to changes in bath height and temperature. As such, minimal variations in individual brick temperatures were observed in Trial 1 and Trial 2.

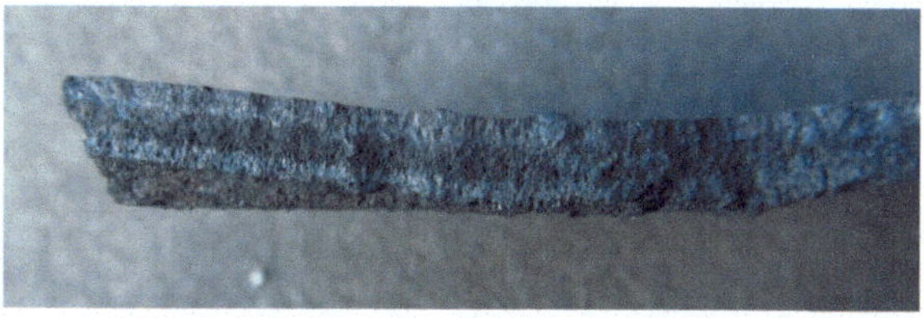

Figure 6: Frozen Slag Layer From in Front of Semigraphite Bricks

In Trial 3, removal of the castable refractory from in front of T/C #2 and T/C #6 resulted in significantly higher brick temperatures than in other regions of the furnace. Similarly, in Trial 4, measurements from T/C #2, T/C #3, T/C#6 and T/C #7 were consistently higher than for the other brick thermocouples. It was also observed that brick temperatures in Row 4 were higher than those measured lower in the furnace (Row 2), presumably due to differences in the heat flux between the furnace 'bath' and 'splash' zones. Additionally, brick temperatures in Row 2 were influenced to a greater extent by changes in bath height, as typically, after tapping of the furnace, the temperature of refractory bricks in this row would increase due to them being exposed to the furnace 'splash' zone.

Compared with the typical, pilot-plant furnace configuration utilizing insulating refractories, the furnace cooling water heat load (calculated from the measured furnace jacket cooling water flow and change in cooling water temperature across the furnace) in Trial 4 was roughly twice as high (Table IV).. As such, it would be expected that for full exposure of the semigraphite bricks (i.e. removal of all castable refractory), the heat load would likely be in the order of 3 times higher than for a furnace lined with insulating refractory.

Table IV: Average Furnace Cooling Water Heat Load for each Trial

Trial no.	1	2	3	4	*'Typical'*
Average Furnace Cooling Water Heat Load (kW)	36.0	35.3	49.5	69.2	30 - 40

2.4. Refractory Removal and Analysis

Following completion of the testwork program, the furnace lining was carefully removed to allow for inspection and sampling of the refractory materials. The condition of the semigraphite refractories was found to be almost new, with no significant wear of the lining observable. Figure 7 and Figure 8 depict the condition of the bricks during removal.

Eight bricks were retained from the lining for laboratory analysis: three from Ring 1, three from Ring 2, and two from Ring 3. Test results and comparisons with typical NMD brick properties are presented in Table V. Average CTE, Flex Strength and Young's Modulus showed somewhat higher values, whilst Specific Resistance illustrated lower values compared with typical NMD brick characteristics, which may suggest some absorption of molten furnace contents into the refractories. This conclusion is however contradicted by the slightly lower average Bulk Density compared with typical NMD brick properties. In any case, none of the measured properties indicated any form of refractory degradation.

Figure 7: Ring 5 With Portion of Frozen Slag Layer Removed

Figure 8: Hot Face of Lower Wall Bricks

Table V: Laboratory Analysis of Used NMD Bricks

Sample	CTE ($10^{-6}/°C$)	Bulk Density (g/cc)	Flex Strength (MPa)	Young's Modulus (GPa)	Specific Resistance (Ohm-m)
Average	2.02	1.872	8.93	9.00	18.21
Std. Dev.	0.18	0.01	0.23	0.22	2.15
Typical NMD	1.58	1.893	8.15	8.40	27.11

3. Conclusions

Pilot-scale testwork investigating the use of GrafTech carbonaceous refractories in the water-cooled Outotec Ausmelt furnace indicated these materials performed extremely well. The refractories demonstrated an ability to withstand the oxidizing process conditions in copper concentrate smelting during the timeframe of exposure in the trials (typically 8-10 hours per trial). The semigraphite bricks demonstrated no measureable wear and a stable layer of frozen slag was easily established both in front of the sacrificial castable refractory and directly on the bricks themselves. It was also observed that the molten slag did not wet the surface of the bricks.

The furnace heat flux (loss) during the testwork was significantly higher than typically experienced for a furnace lined with insulating refractories. As such, furnace fuel and oxygen consumptions were also higher than in previous in the testwork programs. Based on the data obtained, it would be expected that without any castable refractory in front of the semigraphite bricks, the furnace heat flux would likely be in the order of 3 times higher. Control of matte and slag compositions in each of the trials was relatively good, with the furnace refractory configuration (i.e. use of carbonaceous refractories) not having any impact on metallurgical process control and stability.

References

1. B. Chitambira and P. Markham, *Submerged Lance Smelting of Nickel – Copper Leach Residues – Fundamental Aspects and Development of the Process by Rio Tinto Zimbabwe at Empress Nickel Refinery.* (Copper, Cobalt, Nickel and Zinc Recovery Conference, Victoria Falls, Zimbabwe, 16 – 18 July, 2001, The South African Institute of Mining and Metallurgy, Johannesburg, 2001)

7th International Symposium on High-Temperature Metallurgical Processing
Edited by: Jiann-Yang Hwang, Tao Jiang, P. Chris Pistorius, Gerardo R.F. Alvear F., Onuralp Yücel, Liyuan Cai, Baojun Zhao, Dean Gregurek, and Varadarajan Seshadri
TMS (The Minerals, Metals & Materials Society), 2016

ENRICHMENT OF GOLD IN LOW GRADE COPPER MATTE FROM ARSENICAL REFRACTORY GOLD CONCENTRATE VIA MATTE SMELTING METHOD

Duchao Zhang, Qingkai Xiao, Tianzu Yang, Weifeng Liu, Lin Chen
School of Metallurgy and Environment, Central South University, Changsha 410083, China

Keywords: Refractory gold concentrate, Matte smelting, FeO-SiO_2-CaO ternary slag type system, Recovery ratio

Abstract

To solve the problem of low leaching ratio of direct gold cyanide leaching from arsenical refractory gold concentrate, this paper proposed a matte smelting process for gold enrichment in low grade copper matte. The effects of copper content in the mixed material, ratio change in slag type, smelting temperature and reaction time on recovery rate of gold and copper were investigated. The experimental results indicated that the recovery rate of gold increased with the increase of copper content in the mixed materials, temperature and reaction time. It was also noted that the recovery rate of gold and copper were respectively 98.16% and 96.39% when the addition of copper content in the mixed material was 3%, slag type of CaO/SiO_2 and FeO/SiO_2 were 0.6 and 0.9 in 1250°C and the time of reaction was 40min. Most of the arsenic was oxidized into As_2O_3 existed in the dust. The results showed that the matte smelting method can achieve efficient recovery of gold from arsenical refractory gold concentrate.

1. Introduction

Refractory gold ore, also known as refractory gold mine or refractory gold deposit, refers to the golden ore whose cyanide leaching rate is less than 80% even though it is finely grounded [1]. Among them, the arsenic refractory gold reserves belongs to the largest and difficult process refractory gold ores, mainly because the gold occurs in the lattice of pyrite, pyrrhotite and arsenopyrite in microscopic state or disseminations. For this kind of refractory gold ore, gold cannot connect with cyanide solution by conventional leaching technology, which results in the leaching rate of gold is lower than 50%. Hence, arsenic-bearing refractory gold ore must be pretreated before leaching in order to expose the gold which is wrapped up by pyrite, pyrrhotite and arsenopyrite.

In recent years, there are many studies about pretreatment of refractory gold ore, such as oxidizing roasting [2-4], pressure oxidation [5], bacterial oxidation [6-7], arsenic-fixing roasting [8], microwave roasting [9], chemical oxidation [10], nitric acid catalyzing and oxidizing [11], et al. However, only three of them have been applied in industry, namely, oxidizing roasting, pressure oxidation and bacterial oxidation. The gold recovery rate is greatly improved after the pretreatment of arsenic-bearing refractory gold ore by using the above-mentioned methods. Despite their potential in industry, certain constraints hinder the extension and application of these pretreatment technologies, for instance, low recovery rate of gold and associated valuable metals, high equipment requirements, huge investment, long pretreatment cycle, poor adaptability to different feed stocks and high environmental cost, et al [12-13].

During pretreatment by oxidizing roasting, arsenic-bearing refractory gold ore was roasted in a two-stage furnace in fluid bed. After arsenic and sulphur was oxidated to volatile As_2O_3 and SO_2, the ore turned out to be porous calcining [12]. This method has been widely used because of its great advantages in simple process and low cost. However, during roasting, the leaching rate of gold varied greatly due to the wrapping of gold by Fe_2O_3. Besides, generation of SO_2 and As_2O_3 will pollute environment. These defects block the further application of oxidizing roasting [14].

Pressure oxidation refers to that the coated gold is exposed by oxidative decomposition of pyrite and arsenopyrite under the condition of high temperature, high pressure, acidic and oxygen atmosphere [13]. This method is of great advantage in short reaction time, high leaching rate of gold, low sensitivity to poisonous metal. The disadvantages of the approach are high cost and low recovery of silver. Pyrite and arsenopyrite are removed through the action of bacteria to expose the gold, which is known as bacterial oxidation [15-17]. Arsenic-bearing refractory gold ore was pretreated through biological oxidation by Langhans et al. (1995). The results showed that the leaching rate of gold increased from 49% to 80% after the pretreatment. Bacterial oxidation is characterized by simple operation, long period and high cost in environment protection [16].

In this work, a pyrometallurgy method was proposed to overcome the drawbacks of traditional pretreatment methods and bath smelting. Arsenic-bearing refractory gold ore and copper-contained materials were smelted together to produce low-grade copper matte as enrichment carrier of gold. Based on this, enrichment of gold and valuable metals with high recovery rate and low cost could be realized. The study in this paper provided an effective method for pretreatment of arsenic-bearing refractory gold ore.

2. Experimental

1.1 Materials and reagents

Arsenic-bearing refractory gold ore whose granularity was less than -150 were used, and its chemical components were listed in Table I. Copper concentrate with the granularity of less than -150 was supplied by Jiangxi Copper Corporation. Chemical components of the copper concentrate were listed in Table II.

Table I. Chemical composition of the gold ore samples by XRF (mass fraction, %)

Elements	Content	Elements	Content	Elements	Content
As	5.58	Fe	20.28	Cu	0.21
S	16.74	Sb	0.31	Pb	0.22
K	0.71	Ca	1.92	Al_2O_3	9.27
SiO_2	30.32	Au*	72.07	Ag*	39.37

Notes: * correspond to g/t

Table II. Chemical composition of the copper concentrate by XRF (mass fraction, %)

Elements	Content	Elements	Content	Elements	Content
Cu	20.87	Fe	24.28	S	28.33
Pb	2.13	Zn	15.68	As	0.28
Al	0.66	Mg	0.26	K	0.17
Ca	0.31	O	5.51	Si	1.39

1.2 Experiment procedure

The solvent was added into 150g gold concentrate according to certain ratio of iron-to-silicon and ratio of calcium-to-silicate. In order to ensure the content of copper, copper-contained materials were added. The material was poured into a 600mL clay crucible after being uniformly mixed, and then weighed them all. The crucible was put into a pit furnace. Open the dust collection and processing equipment, warm the pit furnace and flow the gas after the preset temperature has been reached. Taken out the crucible after a certain time and weighed them all after the crucibles were cooled to room temperature. The slag phase and matte phase were collected and weighed separately. Then mash the slag phase and matte phase. Analysis of element content and characterization of element were followed.

3. Results and discussion

2.1 Effect of melting temperature

The Effect of melting temperature on the content of gold and copper in smelting slag is shown in Fig.1. From Fig.1, it can be seen that the content of gold in the slag decreases with the increase of smelting temperature. When the temperature goes up to 1200°C, the content of gold and copper in the slag are 10.5g/t and 0.16%, respectively. At the temperature of 1250°C, the content of gold and copper in the slag reduce to 0g/t and 0.06%. Beyond this point, the content of gold in the slag remains almost unchanged, meanwhile, the content of copper in the slag continuously increases with the increase of temperature. With all factors considered, 1250°C is suitable for smelting.

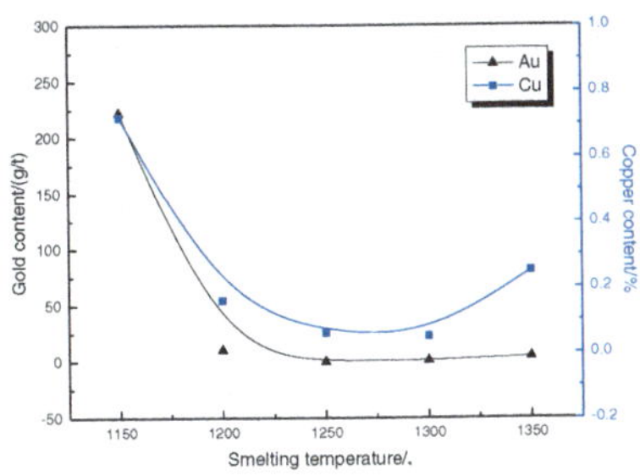

Fig.1 Effect of melting temperature on the content of gold and copper in smelting slag. Conditions: FeO/SiO_2=1.2, CaO/SiO_2 =0.4, 5% copper content, pump the oxygen-rich air as the flow of 0.8L/min for 10min, and keep the crucibles at the constant temperature for 30min.

2.2 Effect of the settling time of heat preservation

The effect of the settling time of heat preservation on the content of gold and copper in smelting slag is shown in Fig.2. From Fig.2, it can be seen that as the settling time is extended, the content of gold in the slag get down. When the settling time is 30min, the content of gold and copper in the slag are 1.25g/t and 0.17%, respectively. However, the content of gold and copper in the slag remain stable when the settling time is over 30min. The matte phase and the slag phase cannot be separated completely within a shorter settling time, resulting in low degree of gold enrichment. In addition, energy consumption will increase with the increase of settling time. Therefore, the optimum settling time is 30min.

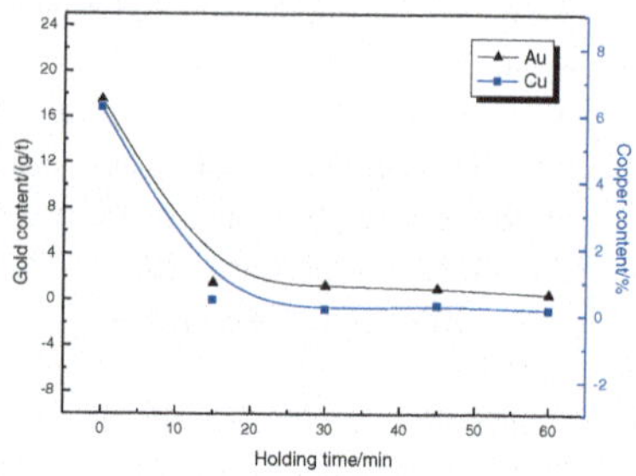

Fig.2 Effect of the settling time of heat preservation on the content of gold and copper in smelting slag. Conditions: FeO/SiO_2=1.2, CaO/SiO_2 = 0.4, 5% copper content, pump the oxygen-rich air as the flow of 0.8L/min for 10min, and 1250°C.

2.3 Effect of FeO/SiO_2 ratio

The effect of FeO/SiO2 ratio on the content of gold and copper in smelting slag is shown in Fig.3. From Fig.3, it can be seen that gold and copper content in slag are 0.5g/t and 0.23% at the FeO/SiO2 of 1.1. When FeO/SiO2 goes up to 1.3, gold and copper content rise up to 1.0g/t and 0.65% respectively. If FeO/SiO2 keeps increasing, content of Fe3O4 that have high melting point and large specific gravity will rise up, which could decrease the flow ability of slag, go against the department of slag and matte, and lead to loss of gold and copper in slag. So it can be concluded that the best FeO/SiO2 is 1.1.

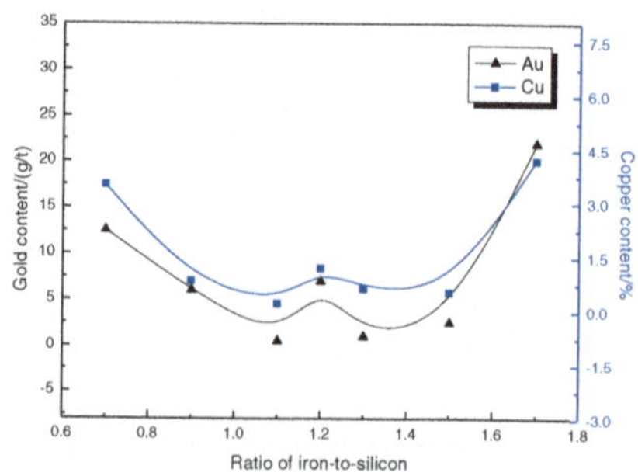

Fig.3 Effect of FeO/SiO_2 ratio on the content of gold and copper in smelting slag. Conditions: $CaO/SiO_2 = 0.4$, 5% copper content, pump the oxygen-rich air as the flow of 0.8L/min for 10min, 1250°C and keep the crucibles at the constant temperature for 30min.

2.4 Effect of CaO/SiO_2 ratio

The effect of CaO/SiO_2 ratio on the content of gold and copper in smelting slag is shown in Fig.4. From Fig.4, it can be seen that gold and copper content in slag are 1g/t and 0.34% at the CaO/SiO_2 of 1.1 respectively. When CaO/SiO_2 gets up to 0.4, gold content hold the line and copper content rise up to 0.65%. If CaO/SiO_2 keeps increasing, the flow ability of slag decrease as follow, making the loss of gold and copper in slag increase. Consequently, the better CaO/SiO_2 is 0.4.

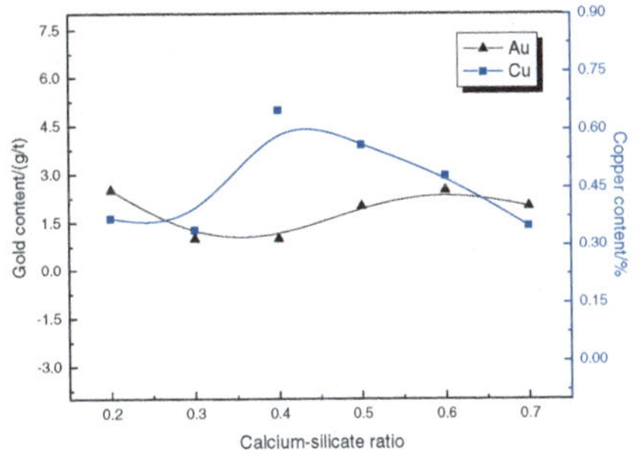

Fig.4 Effect of CaO/SiO_2 ratio on the content of gold and copper in smelting slag. Conditions: $FeO/SiO_2 = 1.2$, 5% copper content, pump the oxygen-rich air as the flow of 0.8L/min for 10min, 1250°C and keep the crucibles at the constant temperature for 30min.

2.5 Effect of copper content

The effect of copper content on the content of gold and copper in smelting slag is shown in Fig.5. From Fig.5, it can be seen that when the content of copper range from 1% to 5%, low content of gold in the slag can be observed. The contents of gold and copper in the slag are 0.5g/t and 0.16% when the copper content in the material is 3%. Recovery rate of gold is increasing with the increase of copper content in the material. However, the larger content of slag and the high content of copper in matte phase are unfavorable for the following leaching of copper and gold. As a result, the content of copper in the material should remain at 2-5%.

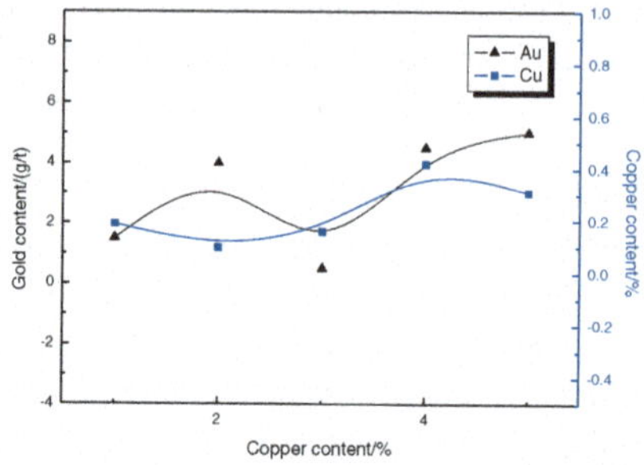

Fig.5 Effect of copper content on the content of gold and copper in smelting slag. Conditions: $FeO/SiO_2 = 1.2$, $CaO/SiO_2 = 0.4$, pump the oxygen-rich air as the flow of 0.8L/min for 10min, 1250°C and keep the crucibles at the constant temperature for 30min.

2.6 The comprehensive experiment

The optimal condition was chosen as follows: 1.1 of FeO/SiO_2, 0.4 of CaO/SiO_2, 5% of copper content, 1250°C, pumping the oxygen-rich air as the flow of 0.8L/min for 10min and keeping the crucibles at the constant temperature for 30min before taking them out. The weight of the mixture is 510g and the contents of Au, S and As are 41.95g/t、16.51% and 3.34%, respectively.

Smelted slag of 268.04g was obtained, in which content of copper and gold were 0.34% and 1.8g/t, respectively. The weight of matte phase was 164.34g. And the compositions are shown in Table3.

Table III. The chemical composition of the smelted low grade copper matte by XRF (mass fraction, %)

Elements	Content	Elements	Content	Elements	Content
Fe	39.74	S	30.04	Cu	14.67
As	2.03	Pb	1.27	Au*	127.25

Notes: * correspond to g/t

From Table III., it can be seen that the content of copper and gold in the smelted low grade copper matte are 14.67% and 127.25g/t, while the content of copper and gold in the smelted slag are 0.34% and 1.8g/t. Based on the data, it can be calculated that recovery rate of copper and gold are separately 96.4% and 97.8%. At the same time, compared to the mixed material, 3 times enrichment of gold is achieved through matte smelting. Besides, the content of copper in the produced low grade copper matte is 14.67%. It can be inferred that high grade copper matte will be produced by continuous converting of the low grade copper matte, which may be beneficial to the following recovery of copper and gold.

4. Conclusions

(1) The optimum condition was chosen as follows: 0.4 of CaO/SiO_2, 1.1 of FeO/SiO_2, 3% of copper content, 1250°C of melting temperature, 10min of aeration time and 30min of settling time. Under the optimum conditions, the content of gold and copper

in the slag are 3.0g/t and 0.83%, respectively. The results showed that over 98% of gold can be enriched in the low grade copper matte.

(2) The recovery rate of gold from arsenical refractory gold concentrate increases with the increase of copper content in the mixed materials, temperature and reaction time.

(3)The matte smelting method can achieve efficient enrichment of gold in low grade copper matte from arsenical refractory gold concentrate, therefore is beneficial to the following recovery of gold and copper.

5. References

[1] T.Z. Yang. *Metallurgy and deep production of noble metal* (Changsha: Central south university press, 2005).

[2] T.N. Angelidis and K.A. Kydros, "Selective gold dissolution from a roasted auriferous pyrite-arsenopyrite concentrate," Hydrometallurgy, 37 (1) (1995),75-88.

[3] Q.C. Li, D.X. Li, and F.J. Qian, "Pre-oxidation of high-sulfur and high-arsenic refractory gold concentrate by ozone and ferric ions in acidic media," Hydrometallurgy, 97 (1–2) (2009), 61-66.

[4] J.G.Dunn and A.C. Chamberlain, "The recovery of gold from refractory arsenopyrite concentrates by pyrolysis–oxidation," Minerals Engineering, 10(9) (1997), 919-928.

[5] F.P.Gudyanga et al., "An acidic pressure oxidation pretreatment of refractory gold concentrates from the KweKwe roasting plant, Zimbabwe," Minerals Engineering, 12 (8) (1999), 863-875.

[6] T.L. Deng and M.X. Liao, "Gold recovery enhancement from a refractory flotation concentrate by sequential bioleaching and thiourea leach," Hydrometallurgy, 63 (3) (2002), 249-255.

[7] A.Hol et al., "The effect of anaerobic processes on the leach ability of an arsenopyrit refractory ore," Minerals Engineering, 24 (2011), 535-540.

[8] J.Liu et al., "Selective arsenic-fixing roast of refractory gold concentrate," Metallurgical and Materials Transactions B, 31B (2000), 1163-168.

[9] P.K.Amankwah and C.A.Pickles, "Microwave roasting of a carbonaceous sulphidic gold concentrate," Minerals Engineering, 22 (2009), 1095-1101.

[10] M.Saba et al., "Diagnostic pretreatment procedure for simultaneous cyanide leaching of gold and silver from a refractory gold silver ore," Minerals Engineering, 24 (2011), 1703-1709.

[11] G.L.Gao et al., "Kinetics of high-sulfur and high-arsenic refractory gold concentrate oxidation by dilute nitric acid under mild conditions," Minerals Engineering, 22 (2009), 111-115.

[12] K.B. Shen, "Research on two stages roast for pretreatment of sulfur & arsenal enriched refractory golden ore," Yunnan Chemical Technology, 34(5) (2007), 26-29.

[13] T.S. Qiu, G.H.Nie, and Q. Zhang, "Present situation and development of pretreatment and leaching technology for refractory copper-bearing golden ore," Gold, 26(8) (2005), 30-34.

[14] B.S.Kafui Nyavor and Nosa O. Egiebor, "Controlling SO_2 emissions in the roasting of gold concentrate," JOM, 43 (12) (1991), 32-34.

[15] R.C. Cui et al., “Bacterial oxidation-cyanidation leaching for varied arsenal-bearing golden ore,” Chinese Journal of Nonferrous Metals, 21(3)(2011), 694-699.
[16] D. Langhans et al, “Biooxidation of an arsenic-bearing refractory gold ore,” Minerals Engineering, 8 (1-2) (1995), 147-158.
[17] S. Ubaldini et al., “Combined bio-hydrometallurgical process for gold recovery from refractory stibnite,” Minerals Engineering, 13 (14-15) (2000), 1641-1646.

7th International Symposium on High-Temperature Metallurgical Processing

Alloys and Materials Preparation

Session Chairs:

P. Chris Pistorius

Merete Tangstad

7th International Symposium on High-Temperature Metallurgical Processing
Edited by: Jiann-Yang Hwang, Tao Jiang, P. Chris Pistorius, Gerardo R.F. Alvear F., Onuralp Yücel, Liyuan Cai, Baojun Zhao, Dean Gregurek, and Varadarajan Seshadri
TMS (The Minerals, Metals & Materials Society), 2016

ZINC AND REFRACTORIES – A NASTY RELATION

D. Gregurek[1], S. Redik[1], C. Wenzl[2], A. Spanring[2]

[1]RHI AG, TC Leoben, Magnesitstrasse 2, Leoben, A-8700, Austria
[2]RHI AG, Wienerbergstrasse 9, Vienna, A-1100, Austria

Keywords: Non-ferrous metal production, RHI, Zinc, Refractories

Abstract

Zinc is a component of many input materials used in various pyrometallurgical processes, for example primary lead and zinc production from Pb/Zn ores and recycling processes dealing with zinc-containing residues (e.g., copper recycling, WAELZ process). Hence, zinc and its compounds are present in the respective metallurgical vessels and interact with the refractory lining. In the present work the zinc attack on the refractories out of various primary and secondary furnaces is briefly introduced and discussed. At the prevailing processing temperatures in the metallurgical vessels, zinc oxide is highly corrosive for the brick components. The knowledge of the wear behavior is based not only on a detailed chemical and mineralogical characterization carried out on provided post mortem samples, but also on FactSage calculations. This together with results obtained by practical testing in the RHI pilot plant represents an important prerequisite for product development and brick selection for the individual customer application.

Introduction

Zinc is normally mined together with other metals, especially Pb, but also Ag and Cu [1, 2]. Zinc enters other metal production processes via:

- Copper: ores, (brass) scrap, residue recycling
- Lead: ores, residue recycling
- Silver: ores, residue recycling
- Steel – EAF route: recycling of galvanized steel

Hence, zinc is an important by-product in many metallurgical processes, e.g., flue dusts from copper, lead and steel production (evaporation during "normal" process or separate Zn slag fuming stage). Pyrometallurgical zinc production is mainly done using the WAELZ process, the most important input material being EAF dust (Zn coming from galvanization layers on recycled steel scrap) and Zn-containing residues from hydrometallurgical Zn production.
Zn is a relatively ignoble metal (standard electrode potential of -0,76 V) and also has a relatively low melting and evaporation point (Tm = 420 °C; Tv = 907 °C) [3]. This leads to the following behaviour:

- Reductive effect for oxides of more noble elements
- Easy evaporation of metallic Zn in reduction processes

While these effects can be exploited in the metallurgical processes, they also have to be considered when choosing refractory materials for the respective applications.

Evaporation

Metallic zinc is easily evaporated under the prevailing process temperatures in nonferrous metals production and depending on the process conditions, namely available oxygen in the furnace atmosphere, either stays in the metallic form as Zn vapour or gets re-oxidized. The latter formation of solid ZnO is an exothermic reaction and hence also has to be considered in the thermal balance of the furnace, especially if larger amounts of Zn are present. Depending on the offgas flow conditions, the Zn or ZnO in the gas phase will contact the furnace refractory lining in a certain area and will show according interactions with the refractories.

Reductive Effect

Due to the position of zinc in the Ellingham diagram, it is apparent that zinc has a tendency to act as reductant for oxides of more noble metals. In the case of refractories, this is especially important for iron-oxides, as present in different amounts in most refractory materials, but also for chromites.

The presence of Fe facilitates the reduction of Cr oxide, as the activity of Cr is reduced by alloy formation. In case of reduction of Fe oxide (to metallic Fe) from a mixed Fe-Cr oxide, active centers are formed at the Cr oxide where the reduction reaction can proceed. This effect, that is exploited in FeCr production, also occurs in the case of Zn contacting Fe-Cr oxides in the refractory materials, as the Cr oxides used in refractory production also contain a certain amount of Fe oxides [3, 4].

Similarity of ZnO and MgO

As ZnO normally is less noble than the metals it is co-produced with, a certain amount of ZnO found in the slag. Due to the similar valence and size of Mg and Zn cations, ZnO gets incorporated in the magnesia of basic bricks and reacts with the alumina in non-basic bricks [5, 6].

Hence, the main reaction mechanisms are:

- ZnO: incorporation in MgO of basic materials & spinel formation with alumina in non-basic materials
- Zn: reductive effect on certain refractory oxides

These two above mentioned effects are the basis for refractory attack by Zn.

Post Mortem Studies

The wear mechanisms described in the following sections are representative case studies of investigations carried out at the laboratories in the RHI Technology Center Leoben and aim to highlight the complexity of in service wear processes.

In the Figure 1 (left) a massive (22 tons) build-up had formed in a WAELZ kiln processing EAF dust and residues from Zn leaching, namely in the area where the magnesia-chromite bricks are installed. The build-up formation however is not actually a Zn-related wear effect, but more due to the prevailing process conditions, namely co-reduction of Fe oxides to metallic iron and re-oxidation to magnetite [7].

Figure 1. Massive build-up formed in a WAELZ kiln in the installation area of the magnesia-chromite brick (left). Metal penetration between the brick joints (rectangulars) (right). Metal coating on the immediate brick hot face.

Generally, every single post mortem study starts with the visual inspection which is carried out on the brick cross section followed by selection of samples for chemical analysis and mineralogical investigation.

The received and investigated sample shown in Figure 1 (right) represents a broken piece which is partly covered with a thin metal coating on the immediate brick hot face. On the cross section cracks parallel to the brick hot face are visible. These cracks are partly filled with metal. Slag and metal also penetrated between brick joints. The brick is completely infiltrated with complex metallic sulphides.

ZnO attack on basic and non-basic refractories

In addition to the slag corrosion affecting mainly magnesia, followed by formation of different silicates such as monticellite ($CaMgSiO_4$), forsterite (Mg_2SiO_4), merwinite ($Ca_3MgSi_2O_8$) [8]. etc., a high ZnO supply to basic brick results in a complementary refractory corrosion by incorporation of ZnO in MgO. The latter can frequently be observed in the roof lining of various furnaces in nonferrous metallurgy where process conditions favour Zn evaporation and reoxidation in the furnace atmosphere.

The supplied ZnO reacts with the magnesia mainly due to diffusion processes. This was confirmed by scanning electron microscope investigations which showed an increased ZnO content in the magnesia. The second brick component chromite is usually also enriched with ZnO but in comparison to magnesia less corroded (Figure 2 left).

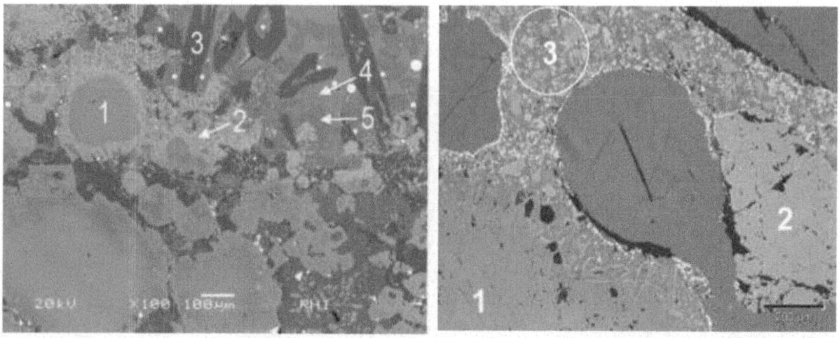

Figure 2. Mineralogical investigation by reflected light microscopy and scanning electron microscope. Magnesia-chromite brick out of the roof of copper production furnace (left). Completely corroded magnesia. Chromite relics (1) highly enriched with ZnO. Zn-Sn-(Cu)-Ni-Mg-Fe-Al-Cr-oxide of spinel type (2). Forsterite (3) enriched with NiO and ZnO. (Sn)-(P)-(Zn)-(Cu)-Na-Ca-Mg-Fe-Al-silicates (4 and 5). Alumina-silica brick out of the same furnace (right). The main brick component andalusite (1) is more corroded than the fused alumina (2). Reaction zone (3).

The severest degradation of the basic refractory in industrial furnaces was observed in a furnace for silver production. In this particular case the magnesia shows numerous newly formed (Zn)-Mg-oxide precipitations in the matrix. At the brick hot face an additional corrosion of chromite under the formation of Zn-Mg-(Mn)-Fe-Al-Cr-spinel could be observed.

Also in case of the non-basic refractories (alumina-silica brick) a strong chemical attack due to ZnO supply was observed. The refractory hot face is enriched with up to 11 wt% of ZnO (Table I).

Table I. Chemical analysis of a alumina-silica brick out of a copper production furnace

wt%	MgO	Al_2O_3	SiO_2	P_2O_5	K_2O	CaO	TiO_2	Cr_2O_3	Fe_2O_3	NiO	CuO	PbO	ZnO
Hot face	0.2	32.0	23.0	1.0	1.0	0.5	0.4	2.0	10.0	1.6	9.4	5.1	11.0
Cold end		69.0	27.1	2.3	0.3	0.1	0.2		0.7				

The microstructure of the refractory hot face is changed, showing a chemothermally changed microstructure with a high pre-wear of the matrix fines. The coarse grains of andalusite (Al_2SiO_5) and fused alumina (Al_2O_3) are "swimming" without bonding in the reaction product. In this particular case the main brick component andalusite is more corroded than the fused alumina (Figure 2, right).

Zinc metal and sulphide infiltration

As already macroscopically observed in Figure 1 (right), there is nearly pore-filling infiltration of the brick microstructure with complex metallic sulphides (FeS, PbS, Fe-Cu-sulphide, Zn-Fe-Mn-sulphide, Ba-Cu-Fe-Pb-sulphide).

Infiltration is generally combined with filling of the pore space within the refractory, without any corrosion of the single refractory components. Nevertheless, such deeply-infiltrated brick microstructure has an increased thermal conductivity and is therefore more susceptible to crack formation and spalling especially in batch-wise processes. The metal or sulphide infiltration is depending on several parameters such as porosity of the brick, bath density, temperature, surface tension of bath and the wetting angle [9].

Exemplary metallic zinc penetration in a non-basic brick (SiC) up to the cold face is shown in Figure 3. Not only infiltrated metallic zinc was determined within the brick joints, but also ZnO. The latter was formed by reaction with the bonding component of the SiC-bearing mortar (phosphate bonded) and subsequently formed Zn-silicate.

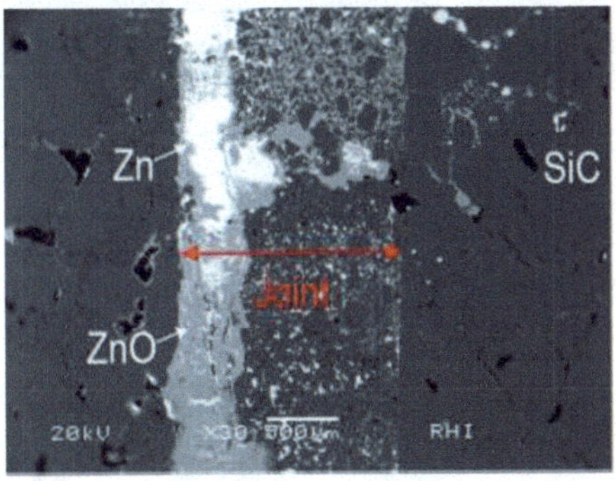

Figure 3. Mineralogical investigation by scanning electron microscope. Metallic zinc penetrated in the brick joints. The ZnO was formed by reaction with the bonding component of the SiC-bearing mortar and subsequently formed Zn-silicate.

Laboratory Testwork

To better understand the wear phenomena in the industrial furnaces, a static cup test was performed in the pilot plant of the RHI Technology Center Leoben. For that purpose a sample with dimension 80x80x70 mm was cut out of the brick. The dimensions of the inner cylindrical hole were 40 mm diameter and 45 mm height (see Figure 4, left).

For the testwork, which was performed in a laboratory electric furnace, 250 grams of customer-provided zinc were used. The filled crucible was sealed at the top with the refractory material. Testing was performed on basic (magnesia-chromite) and non-basic (alumina-chrome and alumina-silica) bricks at a temperature of 1100°C for 10 hours. The cross-section of a magnesia-chromite brick is shown in Figure 4 (right).

The first results showed that almost all tested magnesia-chromite bricks show severe reduction of Fe-oxides, chromite, chromite precipitations to metallic Fe(Cr) (Figure 5 left). Only one tested magnesia-chromite brick (4,1 wt% Fe_2O_3 content) was not affected by the reduction of chromite. The highest reduction was observed for the bricks with a high Fe_2O_3 and Cr_2O_3 content (14 wt% and 19 wt%, respectively). This confirms the theoretical considerations regarding Fe-Cr co-reduction (see above).

Figure 4. Cup sample before testing (left) and after testing (cross sectional view, right). Sample for mineralogical investigation (rectangular).

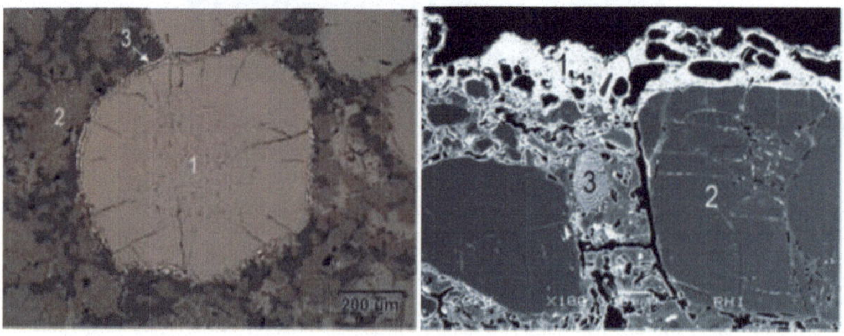

Figure 5. Mineralogical investigation by reflected light microscopy and scanning electron microscope. Magnesia-chromite brick (left). Severe reduction of chromite (1) to metallic Fe(Cr) (3). Magnesia (2). Alumina-silica brick (right). Zn-Ca-Al-phosphate (1) formed after corrosion of andalusite (2). Zirkonmullite (3).

Based on the practical results and FactSage calculations, the possible explanation for the above phenomena in basic bricks is as follows:

- Infiltration of molten metallic zinc into the brick microstructure

- Co-reduction of Fe oxides and chromite by metallic zinc and simultaneous oxidation of zinc to ZnO
- ZnO-enrichment of the magnesia/chromite by diffusion

In case of the alumina-chrome brick no reduction of the brick components was visible. In the first few mm of the cup bottom (i.e., metal contact zone) the ZnO-reacted with fused alumina respectively Cr-alumina.
The alumina-silica brick showed an enrichment of the brick microstructure with ZnO over the whole polished section (0-25 mm). The andalusite and the brick bonding phase (phosphate bonded brick) reacted with ZnO, resulting in formation of Zn-Ca-Al-phosphate (Figure 5 right).

These findings agree with the results of the FactSage calculations: Figure 6 shows the general formation of spinel phase and inclusion in MgO phase in non-basic and basic bricks.

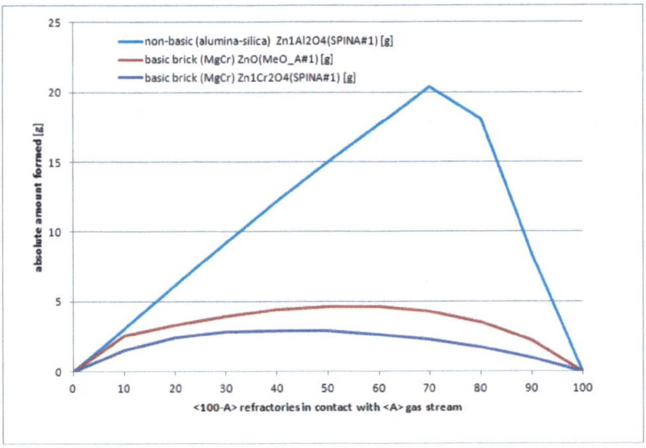

Figure 6. Effect of Zn in non-basic and basic refractories – formation of spinel phase and inlusion in MgO phase (x-axis: 100% refractories at <A> = 0, 100% gas phase at <A> = 100)

Conclusions

Zinc in the form of Zn metal, oxide or sulphide affects both basic and non-basic refractory materials. The main mechanisms are redox reactions with refractory oxides, dissolution/incorporation in brick components and infiltration.

To better understand the wear phenomena in the industrial furnaces cup tests were performed in the pilot plant of the RHI Technology Center Leoben. Based on the first results it can be concluded, that in case of basic bricks (magnesia-chromite) the reduction potential is obviously related to Fe_2O_3 content of the brick, as well as to content of chromium ore which is high in Fe_2O_3 content. Unlike basic bricks, the non-basic bricks (i.e., alumina-chrome and alumina-silica) do not show reduction of single brick components – nevertheless reactions of the main refractory components with ZnO were detected. Further trials and comparison with FactSage calculations are currently ongoing.

References

1. C. Hassall et al, "Changing Dynamics of Long-Term Zinc and Lead Supplies". *Erzmetall* 68, (4), (2015), 218-224.

2. B. Elvers, S. Hawkins, and W. Russey, *Ullmann's Encyclopedia of Industrial Chemistry*, (John Wiley & Sons Inc., 1990), 632.

3. M. Gasik, *Handbook of Ferroalloys* (Kidlington, Oxford OX5 1GB, UK: Elsevier Science Ltd., 2013), 509.

4. G. Routschka, "Handbook of Refractory Materials", (Essen: Vulkan-Verlag GmbH, 2012), 86-92.

5. Chen et al., "The influence of ZnO in fayalite slag on the degradation of magnesia-chromite refractories during secondary Cu smelting", *Journal of European Ceramic Society*, 35 (2015), 2641-2650.

6. E. Riedel, "Anorganische Chemie", (Berlin: Walter de Gruyter GmbH & Co.KG, 2004), 75.

7. P.A. Kozlov, *The Waelz Process, Ore and metals* (PH, Moscow, 2003), 160.

8. D. Gregurek et al., "Refractory Wear Mechanisms in the Non-Ferrous Metal Industry – Testing and Modelling Results", *Journal of Metals*, 65 (11) (2013), 1622-1630.

9. D. Gregurek et al., "Overview of Wear Phenomena in Lead Processing Furnaces", *Journal of European Ceramic Society*, 35 (2015). 1683-1698.

7th International Symposium on High-Temperature Metallurgical Processing
Edited by: Jiann-Yang Hwang, Tao Jiang, P. Chris Pistorius, Gerardo R.F. Alvear F., Onuralp Yücel, Liyuan Cai, Baojun Zhao, Dean Gregurek, and Varadarajan Seshadri
TMS (The Minerals, Metals & Materials Society), 2016

Preliminary Study on Preparation of Al-Sc Master Alloy in Na_3AlF_6-K_3AlF_6-AlF_3 Melt

Zhongliang Tian, Yanqing Lai, Kai Zhang, Xun Hu, Hongliang Zhang, Jie Li

School of Metallurgy and Environment, Central South University,
932 Lushan South Road, Changsha, Hunan, 410083, China

Keywords: Al-Sc master alloy, Na_3AlF_6-K_3AlF_6-AlF_3 melt, Electrolysis

Abstract

The Al-Sc Alloys show many advanced properties for a wide range of engineering applications. But the metal Sc must be added into the metal Al in the form of the Al-Sc master alloy, which is usually produced by the thermal reduction process. In this paper, the decomposition voltages of Sc_2O_3 and Al_2O_3 in Na_3AlF_6-K_3AlF_6-AlF_3 melt were calculated based on the Nernst equation. And the Al-Sc master alloy was prepared in the melt at 950℃ by electrolysis. The results show that the Al-Sc master alloy can be prepared by the co-deposition of Sc^{3+} and Al^{3+} at the cathode if the concentration ratio of Sc_2O_3 to Al_2O_3 in the melt is more than 2:1 under the condition of the aluminum electrolysis. SEM and ICP analysis of the product indicates that the Al-Sc master alloy is made of the phase Al and the phase Al_3Sc.

Introduction

Scandium is one of the most efficient hardening additives in the aluminum alloys, due to the formation of coherent Al_3Sc phase precipitates with the size less than 10 nm. And it has been investigated as an alloying element for aluminum alloys in recent years[1,2]. For example, the introduction of 0.3% scandium can increase the ultimate rupture strength of annealed aluminum sheets from 55 to 240 MPa, that of the Al-1%Mg alloy from 110 to 300 MPa, and that of the Al-5%Mg alloy from 260 to 400 MPa[3]. However, to produce the Al-Sc alloy, the metal Sc must be added into the metal Al in the form of the Al-Sc master alloy because of the great melting point difference between Sc and Al. From the binary alloy phase diagram, the melting point of the Al-2wt.%Sc is close to that of the metal Al[4]. Therefore, the Al-2wt.%Sc master-alloy is the key raw material for preparing the Al-Sc alloys[5].

The current major method of preparing the Al-Sc master alloy is to mix Sc into Al by melting. This process has several problems, such as easy segregating and heavy burning loss, etc. The molten salt electrolysis is alternative method to prepare the Al-Sc master-alloy [6,7]. Sc_2O_3 of low price can be used as raw material, and the segregation of alloying elements also can be reduced. Thus, the cost of production can be decreased. In addition, it is very convenient to realize the large-scale continuous production. The method of preparing the Al-Sc master alloy by the molten salt electrolysis has been reported to be successful in NaF/AlF_3-ScF_3-Sc_2O_3 or KF-AlF_3-Sc_2O_3 melts [2,8]. But the content of Sc in the alloy collected at the cathode is low because of the low solubility of Sc_2O_3 in these melts. To improve the content of Sc, it is beneficial to increase the solubility of Sc_2O_3 in the melt.

The solubility of Sc_2O_3 in Na_3AlF_6-K_3AlF_6-AlF_3 melt has been determined based on the previous study. And the solubility of Sc_2O_3 could reach 7.058 wt.%, even if the concentration of Al_2O_3 is 5.0 wt.% in the melt[9]. Obviously, it is more than that in Na_3AlF_6-AlF_3 melt[10]. Therefore, the Al-Sc master alloy with the high content of Sc may be prepared in Na_3AlF_6-K_3AlF_6-AlF_3 melt by the electrochemical method. In this work, the decomposition voltages of Sc_2O_3 and Al_2O_3 in the melt were calculated based on the Nernst equation in the electrochemical process. The Al-Sc master alloy was prepared in the laboratory electrolysis cell by adding Sc_2O_3 and Al_2O_3 into the melt of Na_3AlF_6-K_3AlF_6-AlF_3 at 1223K and the cathode current density of 1.5 A/cm^2. The morphology and composition of the electrolyzed sample was also discussed.

Experimental

Chemicals and materials

The melt was made up of raw materials Na_3AlF_6, K_3AlF_6, AlF_3, Al_2O_3 and Sc_2O_3. Among them, Na_3AlF_6, K_3AlF_6 and Al_2O_3 were reagent grade. AlF_3 was purified by sublimating three times and its purity was greater than 99.5 wt.%. The main composition of Sc_2O_3 was listed in table Ⅰ. All the components were dried at 473 K for at least 48 hours in the vacuum oven to remove the water before being used.

Table Ⅰ Composition of scandium oxide

Composition	Sc_2O_3	SiO_2	Fe_2O_3	CaO
Content/wt.%	>99.99	$<5\times10^{-3}$	$<5\times10^{-2}$	$<1.2\times10^{-2}$

Cell design and electrolysis procedure

A sketch of the experimental cell is presented in Fig.1. Alumina crucible was set in the graphite crucible and about 500g electrolyte was contained. A hole was drilled at the bottom of the crucible to collect conveniently the product. Under the operating conditions of the laboratory test, the cell could not be thermally self-sustaining. It was necessary to provide the extra heat by placing the experimental cell in a vertical laboratory furnace and the furnace was heated to the desired temperature. The temperature of the melt was measured with Pt/Pt-10%Rh thermocouple and the uncertainty of temperature was about ±1℃. The laboratory electrolysis process was carried out at the constant current with the cathode current density of 1.5A/cm^2 and duration of 2h at 1223 K.

After electrolysis, the cell was left to cool with the furnace. And then, the product collected at the cathode was removed from the crucible, and the solid electrolyte on the surface of the Al-Sc master alloy was cleaned carefully. The content of Sc in the Al-Sc master alloy was analyzed by the inductively coupled plasma (ICP). The phase composition of the sample was investigated by the X-ray diffraction (XRD) and the scanning electron microscopy-energy dispersive spectroscopy (SEM-EDS).

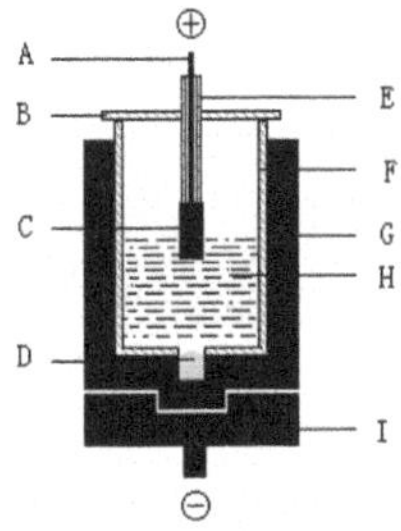

Fig.1. Experimental cell
A-Stainless steel anode rod, B-Alumina cover, C-Graphite anode, D-Cathode product, E-Alumina tube, F-Alumina crucible, G- Graphite crucible, H- Electrolyte, I-Graphite mechanical support

Result and Discussion

Theoretical analysis of decomposition voltage

Sc_2O_3 and Al_2O_3 dissolved in Na_3AlF_6-K_3AlF_6-AlF_3 melt will be decomposed into the metal Sc and the metal Al by the electrolysis according to the reactions (1) and (2).

$$Sc_2O_3 — 2Sc + 3/2O_2 \qquad (1)$$

$$Al_2O_3 — 2Al + 3/2O_2 \qquad (2)$$

For the oxide in the melt, its decomposition voltage can be calculated based on the Nernst equation as the following Eq. (3).

$$E = E^0 + \frac{RT}{nF}\ln\frac{a_{Me_2O_3}}{a_{Me}} \qquad (3)$$

Here, E is the calculated decomposition voltage, E^0 is the theoretic decomposition voltage for the oxide studied, n is the number of electrons transferred, F is the Faraday constant, R is the ideal gas constant, T is the temperature, $a_{Me_2O_3}$ and a_{Me} are the chemical activities, and the activities can be described by the following relation (4).

$$a_i = r_i \times c_i \qquad (4)$$

Here, r_i and c_i are respectively the activity coefficient and the mole fraction for the component i.

According to the reference [11], the chemical activities of Sc_2O_3 and Al_2O_3 can be replaced by $c_{Sc_2O_3}^{4.3}$ and $c_{Al_2O_3}^{3}$, where $c_{Sc_2O_3}$ and $c_{Al_2O_3}$ are respectively the concentrations of Sc_2O_3 and Al_2O_3 in the melt. At the beginning, the content of Sc in the Al-Sc master alloy is very low, and the main composition of the alloy is the liquid Al. Thus, the value of a_{Al} is generally considered as 1 while a_{Sc} is regarded as c_{Sc}, which is the mole fraction of Sc in the Al-Sc master alloy. The decomposition voltages of the oxides are listed in table II.

Table II Decomposition voltage of the oxide in Na_3AlF_6-K_3AlF_6-AlF_3 melt(1223K)

$c_{Sc_2O_3}$/wt.%	$E_{Sc}=E_{Al}$/V	$c_{Al_2O_3}$/wt.%	$c_{Sc_2O_3}/c_{Al_2O_3}$
2.50	2.464	1.05	2.39
3.00	2.451	1.36	2.21
3.50	2.439	1.69	2.07
4.00	2.429	2.05	1.95
4.50	2.420	2.43	1.85
5.00	2.412	2.82	1.77
5.50	2.405	3.24	1.70

As shown in table II, the decomposition voltages of Sc_2O_3 and Al_2O_3 will decrease if the concentrations of the oxides in Na_3AlF_6-K_3AlF_6-AlF_3 melt increase. For example, when the concentration of Sc_2O_3 increases from 2.50wt.% to 5.50wt.%, its decomposition voltage decreases from 2.464V to 2.405V. For Al_2O_3, its decomposition voltage will decreases from 2.464V to 2.405V if the concentration increases from 1.05wt.% to 3.24wt.%.

It also can be concluded that the concentration ratio of Sc_2O_3 to Al_2O_3 decreases when the decomposition voltages of Sc_2O_3 and Al_2O_3 are kept equally and the concentration of Sc_2O_3 in the melt increases. This means that a high concentration of Al_2O_3 should be maintained in Na_3AlF_6-K_3AlF_6-AlF_3 melt when the concentration of Sc_2O_3 is high in order to make the metal scandium deposit by the electrolysis prior to the metal aluminum at the cathode. For instance, when the concentration of Sc_2O_3 is 4.50wt.%, the concentration of Al_2O_3 should be lower than 2.43wt.% to make the metal Sc a priority to produce at the cathode. However, when the concentration of Sc_2O_3 is 2.50wt.%, the concentration of Al_2O_3 cann't be more than 1.05wt.%, or the metal Al will be produced preferentially.

As we all know, the concentration of Al_2O_3 is kept about 2.0wt.% in the modern aluminum electrolysis[12]. So, to obtain the Al-Sc master alloy by the electrolysis in Na_3AlF_6-K_3AlF_6-AlF_3 melt, the concentration of Sc_2O_3 should be more than 4.00wt.% and the concentration ratio of Sc_2O_3 to Al_2O_3 is more than 2:1.

Analysis of deposition

After electrolysis, some metal balls are recovered at the cathode (Fig.2). The size of these metal balls is no more than 12mm in diameter. At the same time, there are some holes in the metal balls. On one hand, the wettability of the Al-Sc master-alloy for the graphite cathode is very poor, and it is advantageous to collect the Al-Sc master-alloy particles. On the other hand, the spherical shape of the metal acquired results from the principle of energy minimization in the grain growth process.

Fig. 2 Photograph of metal balls recovered at the cathode

Among them, a large metal ball is selected and sectioned to analysis further. From XRD pattern of the metal ball recovered at the cathode (Fig.3), there are the phase peaks of Al and Al_3Sc in the production. It means that the constituent elements of the metal balls are aluminum and scandium, and the element of scandium exists in the form of Al_3Sc. It shows that the Al-Sc master alloy can be prepared by the electrolysis in Na_3AlF_6-K_3AlF_6-AlF_3 melt under the experimental conditions.

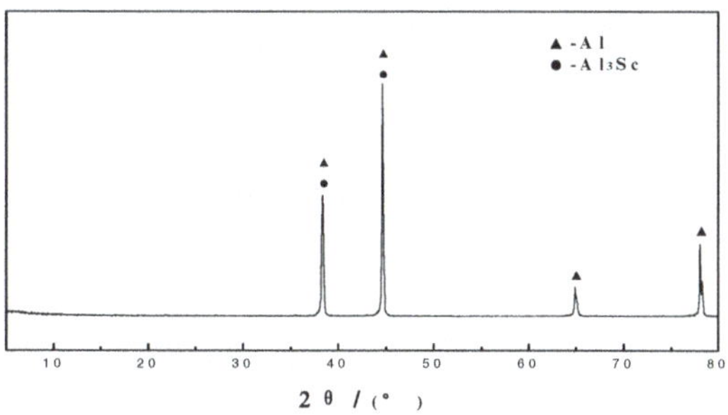

Fig. 3 XRD pattern of the metal ball recovered at the cathode

To obtain the content of Sc in the metal ball, four samples are taken from the different positions (Fig.4) and are analyzed by ICP, and the results are listed in table III. The results show that the content of Sc nearby the edge of the metal ball sample is almost the same, and the content of Sc for the three different positions are 1.18wt.%, 1.23wt.% and 1.22wt.% respectively. However, for the position A where is the center of the metal ball sample, the content of Sc is only 0.26wt.%. It is lower than that in the other three positions tested. This can be attributed to the melting points of the metal of scandium and aluminum, which are 1541℃ and 660℃ respectively. When cooling with the furnace, there is the serious segregation phenomenon for the Al-Sc master alloy.

Table III Content of Sc in the different positions

Position	A	B	C	D
Content of Sc/wt.%	0.26	1.18	1.23	1.22

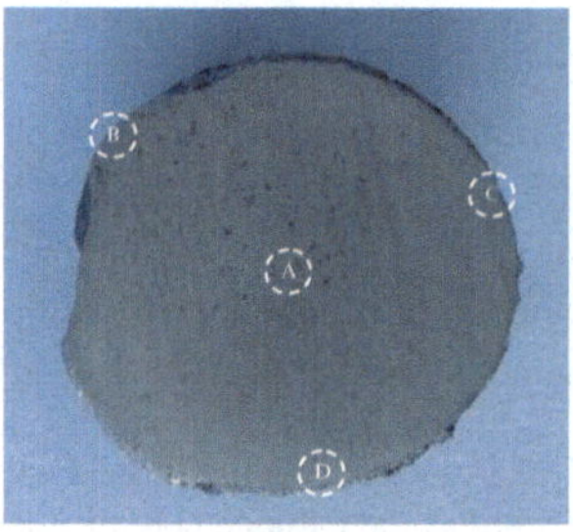

Fig. 4 Sectional view of metal ball recovered at the cathode

Fig. 5 shows the SEM micrograph for the edge of the Al-Sc master alloy ball. It can be seen that there are the metal phase α-Al matrix (gray) and many regular shapes of second phase (white) in the sample. And the results from EDS for the white phase shows that there are two elements of Al and Sc, and the ratio of the atom Al to Sc is close to 3:1, which indicates that it is the Al_3Sc particles.

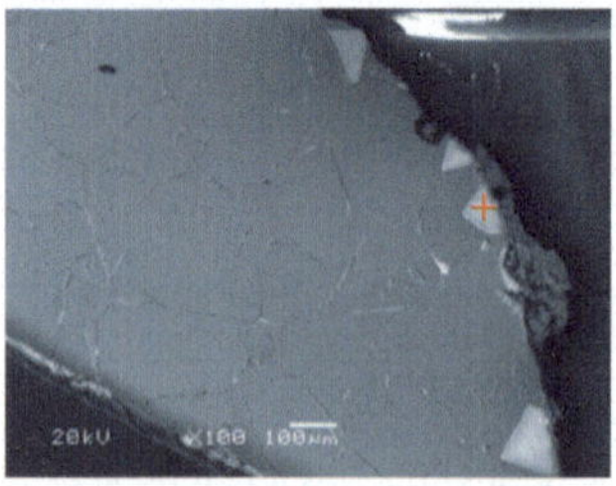

Fig. 5 SEM micrograph for the edge of metal ball

It can be seen in Fig.5 that there are more Al_3Sc particles at the edge of the meal ball than that in the center. It confirmed further that the segregation phenomenon appears when the molten Al-Sc master alloy cools.

Conclusions

The decomposition voltage of Sc_2O_3 and Al_2O_3 based on the Nernst equation shows that the Al-Sc master alloy can be prepared by the electrochemical co-deposition of Sc^{3+} and Al^{3+} at the cathode. Under the condition of the modern aluminum electrolysis, the concentration ratio of Sc_2O_3 to Al_2O_3 should be controlled more than 2:1 to obtain the Al-Sc master alloy by the electrolysis in Na_3AlF_6-K_3AlF_6-AlF_3 melt.

The Al-Sc master alloy can be prepared by the electrolysis in Na_3AlF_6-K_3AlF_6-AlF_3 melt under the experimental conditions. There exists the Al_3Sc phase and the Al phase in the production. However, the Al_3Sc phase is unevenly dispersed in the matrix of Al because of the melting point difference of metals between scandium and aluminum.

Acknowledgement

The authors gratefully acknowledge the financial supports from the State Natural Science Funds (No. 51474238 and No. 51334002).

References

1. N.A. Belov et al., "Effect of scandium on structure and hardening of Al-Ca eutectic alloys," *Journal of Alloys and Compounds*, 646(2015), 741-747.
2. C.Y. Guan et al., "Preparing aluminum-scandium alloys using direct hall reduction process," *3rd International Symposium on High-Temperature Metallurgical Processing*, ed. T. Jiang et al(Orlando, FL, USA: TMS, 2012), 243-250.
3. V.V. Zakharov, "Effect of scandium on the structure and properties of aluminum alloys," *Metal Science and Heat Treatment*, 45(7)(2003),246-253.
4. Z.M. Ying et al., *Scandium and its alloys* (Changsha: Central South University Press, 2007),265. (In chinese)
5. G. C. Teng, Y. C. Chai, and Y. Yang, "Preparation technology of aluminum-scandium master-alloy," *World Nonferrous Metals*, 1(2010), 70-71. (In chinese)
6. Y.W. Tian, "Preparation of Al-Sc Alloy in molten salt," *Material Science and Technology*, 15(2)(1998),626-631.
7. Z.X. Qiu and M.J. Zhang, "Preparation of aluminum master alloy by electrolysis in molten cryolite," *Aluminum*, 66(6) (1990),560-564.
8. Q.C. Liu et al., "Processing Al-Sc alloys at liquid aluminum cathode in KF-AlF_3 molten salt," *ECS Transactions*, 50(11)(2012),483-489.
9. Z.L. Tian et al., "Solubility of Sc_2O_3 in Na_3AlF_6-K_3AlF_6-AlF_3 melts," *3rd International Symposium on High-Temperature Metallurgical Processing*, ed. T. Jiang et al(Orlando, FL, USA: TMS, 2012), 105-112.
10. G.M. Lu and X.S. Liu, "Dissolution of Sc_2O_3 in fluoride molten salt," *The Chinese Journal of Nonferrous Metals*, 9(3)(1999),624-626. (In chinese)
11. S. Yang, Q. Li, and S.Q. Gu, "The thermodynamics of electrolysis Al-Sc alloy," *Nonferrous Metals (Extractive Metallurgy)*, 2(2003),26-29. (In chinese)
12. Y.X. Liu and J. Li., *Modern Aluminum Electrolysis* (Beijing: Metallurgical Industry Press, 2008),395. (In chinese)

7th International Symposium on High-Temperature Metallurgical Processing
Edited by: Jiann-Yang Hwang, Tao Jiang, P. Chris Pistorius, Gerardo R.F. Alvear F., Onuralp Yücel, Liyuan Cai, Baojun Zhao, Dean Gregurek, and Varadarajan Seshadri
TMS (The Minerals, Metals & Materials Society), 2016

EFFECT OF THE REDUCTANTS ON THE PRODUCTION OF IRON BASED ALLOYS FROM MILL SCALE BY METALLOTHERMIC PROCESS

Mehmet Buğdaycı[1, 2], Ahmet Turan[2], Murat Alkan[1], Onuralp Yücel[1]

[1]Istanbul Technical University, Faculty of Chemistry & Metallurgy, Metallurgical and Materials Eng. Dep., 34469, İstanbul, Turkey
[2]Yalova University, Chemical and Process Engineering Department, Yalova, 77100, Turkey

Keywords: Metallothermic process, Mill-scale, Effect of reductants

Abstract

In this study, iron based alloys (Fe, Fe-Cr-Ni-Mo, etc.) were produced by using metallothermic methods starting from Mill Scales obtained continuous casting processes. During the process, the combustion was realized by ignition and the produced wave propagates throughout the reactant mixture yielding the desired product. In this work, Mill scale used as an iron source, which includes +2, +3 values iron oxides. Aluminum, Magnesium and Silicon were used as reductants. In these metallothermic experiments, different ratios of reductants were examined, and their effects on the metal recovery and alloy compositions were investigated. The samples were characterized by using chemical analysis (AAS), XRD and XRF techniques.

Introduction

Metallothermic process is an alternative method for producing Iron based alloys such as FeCr, Fe Mo, Fe Ni, Fe-Cr-Ni-Mo etc. Process has several stages this are:

1- Reaction is initiated from sample surface with heat flux

2- After initiation reaction becomes self-sustaining and propagates

3- All of the mixture was reacted and metallothermic products obtained

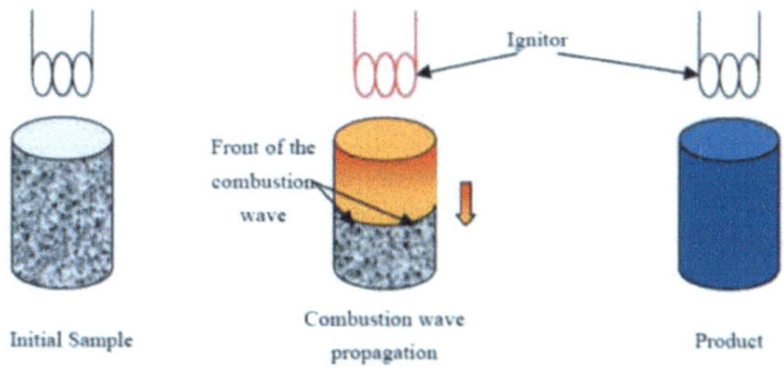

There are some advantages in this technique such as, less energy requirement, very short processing time, simple process, sub-micron sized particles, high purity products. Although there are many advantages, this technique have some defects. These are controlling reactions, cost of reductants, and controlling quality of final products. For this reason thermodynamic data's "Specific heat and adiabatic temperature" have very important role for predict reaction circumstances.

Specific heat; determining the heat evolved during the reaction and to estimate if the temperature achieved is sufficient to smelt the metal and the slag and to separate them due to the different density.

Calculated by dividing the enthalpy of the reaction at 25 °C by the sum of the molecular weights(*MW*) of the reaction products.

$$\mathbf{Me_IO + Me_{II} \rightarrow Me_I + Me_{II}O \qquad \Delta G_R = \Delta GMe_{II}O - \Delta GMe_IO < 0}$$

Specific heat < 2250 J/g, insufficient to melt the charge and to separate metal/slag

Specific heat > 4500 J/g, violent reaction and may even be explosive.

Specific heat : 2250 – 4500 J/g, a controlled and self-sustaining metallothermic reaction

Adiabatic Temperature must be higher than 1527 °C to start metallothermic reaction.This study aims to produce Iron based alloy from mill scale via aluminothermic method and investigate effect of Si and Mg addition.

Experimental Procedure

The mill scale utilized in this study is formed in a continuous casting plant and is of non-oily nature. Other than iron oxides of different valence states, it contains metallic iron and impurities. Table I, displays the chemical analysis of the mill scale.

Table I. Chemical Analysis of the Dried Mill Scale

Component	Ratio, wt. %
Total Iron	70.95
Fe^{2+}	24.59
Fe^{3+}	42.80
Fe^0	3.65
Mn	0.14
Cu	0.75
SiO_2	0.39

In the metallothermic experiments, a mixture of Mill-scale, NiO, Cr_2O_3, MoO_3 and Al powders were used in order to produce metallic, Fe-Ni, Fe-Ni-Cr, and Fe-Ni-Cr-Mo alloys. After the first experimental set low Chromium efficiency detected than to improve Chromium efficiency, new reductants (Magnesium and Silicon) used with Aluminum. The metal oxide powders have over 96 % purity and 150 μm average grain sizes. Table II, displays purity and grain size of powders.

Table II. Purity and Grain Size of Raw Materials

Raw Materials	Purity, %	Grain Size
Al	>96.0	<150 micron
Cr_2O_3	>99.0	<150 micron
NiO	>99.0	<150 micron
Fe	>98.0	<150 micron
Mill Scale	>70.95	<150 micron

The reaction mixtures were mixed thoroughly 15 minutes in a turbula mixer and powder mixtures (approximately 150 g) were charged into Cu crucible and compacted. W (tungsten) wire was placed at the top of copper crucible and the reaction realized by passing current through the wire. After initiation, a highly exothermic reaction became self-sustaining and propagated throughout the SHS mixture. The obtained SHS products were discharged from the crucible after cooling.

Results and Discussions

In metallothermic experiments, productions of Iron based alloys were investigated by using a mixture of Mill scale + NiO + Cr_2O_3 + MoO_3 + Al as a reductant with using Cu crucible. The list of experiment parameters, weight of the initial mixtures and final products were given in Table III. NiO addition into the green mixtures decreased the total metal recovery. Because Ni firstly uses main energy of the process and other components cannot find enough energy for reduction. On the other hand, Cr_2O_3 addition into the green mixtures decreased the scattered ratio due to the lower exothermicity of Cr_2O_3 reduction but Cr efficiency presents the worst results of the all experiments.

Table III. Weight of the Initial Mixtures and Final Products (Stainless Steel "S.S.").

Experiment Parameters	Initial Mixture, (g)				Final Products		Total Metal Recovery, %	Scattered Ratio, %
	Mill Scale	Al	NiO	Cr_2O_3	Alloy	Slag		
Fe Ni	100	37,7	25,4	x	73,2	68,8	78	10,9
18-8 S.S.	83,6	34,1	10,5	21,7	65,2	81,4	79	6,7
18-8 S.S.	100	40,8	12,5	25,9	88	86,6	89	3,0
305 S.S.	100	41,8	15,4	26,6	89,7	80,1	88	1,7
305 S.S.	50	20,9	8	13,3	46,4	43,3	91	1,8
201 S.S.	100	38	5,7	22,5	80,2	85	88	6,8
201 S.S.	50	19	2,9	11,3	44,9	21,1	98	3,5
301 S.S.	100	38,8	8,3	23,2	81,5	88,7	87	2,4
316 S.S.	100	42,2	14,2	25,2	88,5	82,3	87	2,8
317 S.S.	100	44,7	16,6	28,6	94	86,5	87	4,2

Metallic distributions into alloy, slag and scattered parts obtained after metallothermic experiments were given in Table IV. Fe recovery was reduced with the NiO addition into the green mixture, but it was ascended with the Cr_2O_3 addition. Highest metal recoveries were measured as 100% Fe, 99.75 % Ni, 75 % Cr, 91 % Mo respectively. Figure 1. presents distribution into the alloys, slags and scattered parts

Table IV. Metallothermic Process Products, Metal Recoveries

Experiment Parameters	Fe, %	Al, %	Cr, %	Ni, %	Mo, %
Fe Ni	93.80	x	X	86.30	x
18-8 S.S.	100.0	x	67.20	95.00	x
18-8 S.S.	100.0	x	72.00	100.0	x
305 S.S.	94.30	x	59.40	99.75	x
305 S.S.	99.00	x	74.30	92.16	x
201 S.S.	95.70	x	75.00	94.80	x
201 S.S.	96.00	x	72.30	86.30	x
301 S.S.	97.00	x	66.80	100.0	x
316 S.S.	95.00	x	61.50	99.00	91.00
317 S.S.	95.70	x	62.45	88.00	78.12

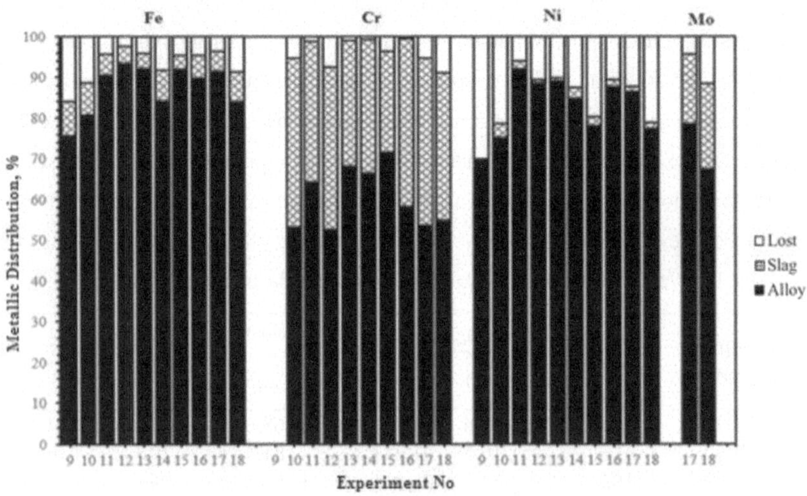

Figure 1. Distribution into the alloys, slags and scattered parts for Iron based alloy experiments.

In the last series of metallothermic process experiments, reductant effect has been investigated by adding Mg and Si. 316 and 201 stainless steel standards selected for this experiments, because 316 series Cr recovery were the worst of experiments and 201 series had the best Cr recoveries in the same series. Both Mg and Si additives positive effected to the Cr recovery the best result have been detected in % 100 Millscale + %95 Al +%5 Si system. The Cr recovery increased from % 61,5 to % 69,4 but the same effect did not seen in 201 stainless steel standard experiment. The list of experiment parameters, weight of the initial mixtures and final products were given in Table V.

Table V. Weight of the Initial Mixtures and Final Products

Initial Mixture, (g)							Final Products		Total Metal Recovery, %	Scattered Ratio, %	Additives Stoichiometric Ratio, %
Mill Scale	Al (g)	Mg (g)	Si (g)	NiO (g)	Cr_2O_3 (g)	MoO_3 (g)	Alloy	Slag			
100	42,2	2,1	x	14,2	25,2	3,3	88,1	87	87	2,6	100 Al, 5 Mg
100	40,1	2,2	x	14,2	25,2	3,3	89,2	73,7	88	1,8	95 Al, 5 Mg
100	38,0	4,2	x	14,2	25,2	3,3	88	70,3	87	4,4	90 Al, 10 Mg
100	38,0	4,2	x	14,2	27,7	3,3	94,5	86,5	92	0,3	110 Cr
100	42,2	x	2,1	14,2	25,2	3,3	89,5	83	89	2,9	100 Al, 5 Si
100	40,1	x	2,2	14,2	25,2	3,3	90,3	83	89	4,9	95 Al, 5 Si
100	40,1	1,0	1,0	14,2	25,2	3,3	83,8	84,3	85	8,7	95 Al, 5 Mg-Si
100	36,1	1,9	x	5,7	22,5	x	85,6	12,3	94	0,8	95 Al, 5 Mg
50	18	1	x	2,9	11,3	x	41,8	31,4	92	1,5	95 Al, 5 Mg

Metallic distributions into alloy, slag and scattered parts obtained after SHS experiments were given in Table VI. Fe recovery was reduced with the NiO addition into the green mixture, but it was ascended with the Cr_2O_3 addition. Highest metal recoveries were measured as 99% Fe, 98 % Ni, 72 % Cr, 95,4 % Mo respectively. Figure 2 presents Distribution into the alloys, slags and scattered parts.

Table VI. Metallothermic Process Products, Metal Recoveries

Experiment No	Additives Stoichiometric Ratio, %	Fe, %	Al, %	Cr, %	Ni,%	Mo, %	Alloy
19	100 Al, 5 Mg	95.90	x	64.70	98.10	95.40	316 S.S.
20	95 Al, 5 Mg	95.80	x	68.80	96.50	93.20	316 S.S.
21	90 Al, 10 Mg	95.60	x	63.00	92.70	91.30	316 S.S.
22	110 Cr	90.50	x	54.00	92.00	80.00	316 S.S.
23	100 Al, 5 Si	96.00	x	69.41	94.00	92.70	316 S.S.
24	95 Al, 5 Si	99.00	x	67.20	83.60	78.10	316 S.S.
25	95 Al, 5 Mg-Si	99.00	x	68.70	99.50	92.20	316 S.S.
26	95 Al, 5 Mg	96.00	x	71.00	98.00	x	201 S.S.
27	95 Al, 5 Mg	97.00	x	72.00	96.00	x	201 S.S.

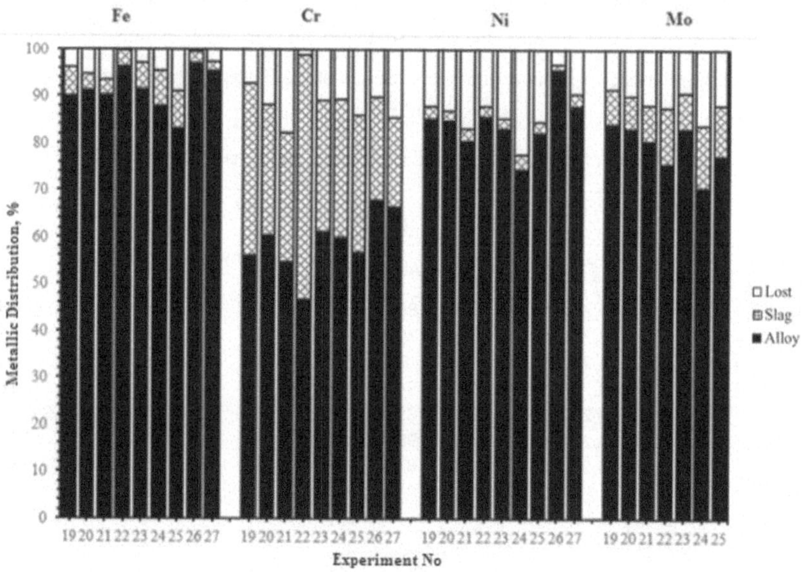

Figure 2. Distribution into the alloys, slags and scattered parts for additional reductant experiments.

Conclusions

In this study, Fe-based alloys were produced by using mill scale formed in a continuous casting plant via metallothermic process.. In metallothermic experiments, Mill scale-NiO-Cr_2O_3-Al system, NiO addition hasn't got any positive effects on the metal recovery; however NiO+Cr_2O_3 addition made an increase in the metal recoveries. In reductant effect SHS series, Mg and Si additives increased Cr and Metal recoveries, stoichiometric 95 % Al, 5 % Mg, experiment was given the best Cr recovery. This mixture also was given the best metal recovery.

References

1. O. Yücel, F.C. Sahin, A. Tekin, "The Preparation of Ferroboron and Ferrovanadium by Aluminathermic Reduction," High Temperature materials and Processes, 15 (1-2) (1996) 103–106.

2. A.G. Merzhanov, "Self-propagating High-temperature Synthesis (SHS)," (ISMAN, Russia, 2002).

3. I.P. Borovinskaya, "Chemical Classes of the SHS Processes and Materials," Pure and Applied Chemistry, 64 (1992) 919-940.

4. V.I. Yukhvid, "Modifications of SHS Processes," Pure and Applied Chemistry, 64 (1992) 977-988.

5. F.C. Demirci, O. Yücel, "Research on the Carbothermic Reduction Conditions of Mill Scale from Continous Casting Processes," (Paper presented at the 140th TMS Annual Meeting, San Diego, California, 27 February-3 March 2011), 651-659.

6. Bugdayci M., Turan A., Alkan M., Demirci F., and Yucel O., A Comparative Study on the Reduction of Mill Scale from Continuous Casting Processes. (Paper presented 143nd TMS Annual Meeting, February 16-20, 2014 San Diego, California, USA), 499-508.

7th International Symposium on High-Temperature Metallurgical Processing
Edited by: Jiann-Yang Hwang, Tao Jiang, P. Chris Pistorius, Gerardo R.F. Alvear F., Onuralp Yücel, Liyuan Cai, Baojun Zhao, Dean Gregurek, and Varadarajan Seshadri
TMS (The Minerals, Metals & Materials Society), 2016

EXPERIMENTAL STUDY ON IRON-BASED ALLOY AS CLADDING LAYER—IMPROVING HIGH TEMPERATURE OXIDATION RESISTANCE OF FURNACE ALLOY

Wang Yanze[1], Chen Chen[1], Hong Xin[1]

[1]Shanghai Key Laboratory of Modern Metallurgy & Materials processing, Shanghai University, shanghai 200072, China

Keywords: Iron-based alloy, plasma powder surfacing, Oxidation resistance

Abstract

Furnace superalloy was applied in harsh environment within the erosion of high temperature and oxidizing atmosphere. With plasma powder surfacing technology, iron-based alloy powder was sprayed on the surface of 310S stainless steel to form oxidation layer that increases the oxidation resistance and reduces the cost of materials. With optimum technological parameters, in this paper, oxidation test of 100 hours on iron-based cladding alloy and furnace alloy were carried out at 1200℃. The technique of XRF, XRD and EDS were used to characterize composition, oxide morphologies and surface mechanical properties. The results showed that: Element composition of cladding layer formed by Plasma powder surfacing was well distributed. Its oxidation weight gain rate of 100 hours was $0.4552 g/m^2h$, belonged to antioxidant level. The oxidation weight gain rate of furnace alloy was $1.4617 g/m^2h$, belonged time-antioxidant level. And on the mechanical comparison test, surfacing alloy is relatively excellent than furnace alloy.

Introduction

With the development of modern industrial technology, a large number of engineering components applies in complex working conditions. The productivity of mechanical parts is seriously affected by oxidation and corrosion of the material surface itself. Surface Technology is helpful to improve the corrosion resistance, wear resistance, fatigue resistance and other properties while increasing the working life of materials, reducing the waste of resources and production costs. Currently, the surface treatment welding technology is not only applied in manufacturing equipment, but also in the repair field of mechanical components.

In this paper, Plasma powder surfacing technology is used in improving the oxidation resistance of furnace alloy though spraying iron-based alloy powder on the surface of 310S stainless steel. In this way, the service life of components is significantly improved and broken surface of the original parts is repaired. Therefore, the application of Plasma powder surfacing technology to improve high temperature oxidation resistance has good prospects [1-4].

Experimental

Materials and Characterization

In this experiment, metal powder used by plasma spray welding machine come from Tai Xin Long materials company in Beijing and alloy powder mainly contains iron, chromium, nickel, aluminum and silicon, their purity are: ≥99.0%. Table I shows the chemical compositions of iron-based alloy powder. Matrix material of this test is 310S stainless steel (2520 steel) and furnace alloy comes from a small factory in Shanghai as a benchmark to measure properties of iron-based high temperature alloy cladding layer.

Table I. Alloy powder composition after correction, wt. %

element	Fe	Cr	Ni	Al	Si
Contents (%)	62	27.1	4.4	3.6	2.9

The character methods used in this experiment include: XRF, X-ray diffraction (XRD), scanning electron microscopy (SEM) coupled with energy dispersive spectroscopy (EDS). Experimental equipment include: Plasma powder surfacing machine, Trough type heating furnace. Experimental method contains orthogonal experiment method and Pattern recognition method.

Table II shows the compositions of furnace alloy by using XRF method. It is clear that this is a kind of iron-based high temperature alloy. The high content of Cr make it has good oxidation resistance, and Ni make the stability of austenite strong. But the lack of Al makes furnace alloy show poor oxidation resistance at high temperature of 1200℃. At the same time, the lower content of Si decreases the content of SiO_2, the adhesion between oxidation film and matrix and anti-stripping capacity.

Table II. The composition of super alloy from the heating furnace/*wt. %*

Fe	Cr	Ni	Si	Co	Cu	Mo	V	P	S
59.89	25.06	13.31	0.84	0.16	0.08	0.04	0.03	0.03	0.02

Experimental processes

This test uses the previous iron-based superalloy component(Cr27%, Ni8%, Al3%, Si2%, Fe balance) that makes an optimal oxidation at 1200℃ as surfacing layer to spray on 310S stainless steel by Plasma powder surfacing. According to the appropriate process parameters, the thickness of the cladding layer is 2mm, length is 30mm, and the width is above 15mm. By the size of the test standard, cladding layers are cut into the sample, and the test method of HB5258-2000 steel and high temperature alloy is used. High temperature oxidation test and wear test were performed by HB5258-2000 and GB/T12444-2006.

Main experimental contents:

(1) Due to the different alloy powder component and the substrate 310S stainless steel components, changes of the alloy powder after surfacing cladding composition occur. Therefore, the superalloy component must be changed to get the proper surface composition of the cladding layer by detection result, and then the required cladding layer is obtained.

(2)With the optimum superalloy composition, alloy powder are sprayed on the substrate surface of 310S stainless steel and the oxidation test was carried out under the condition of 1200℃.

(3)Using the pattern recognition method to optimize the oxidation resistance of iron-based high temperature alloy composition in the early stage of the research group, select the optimal area and design better test points, and then carry out the plasma surfacing test to form cladding layers, and carry out the test to explore the antioxidant capacity of the alloy cladding layer at a higher temperature by 1300℃.

(4)The properties of cladding layers are tested though Rockwell hardness and the wear testing, which must be contrasted with the alloy used in the factory.

Results and Discussion

The correction of Surface composition and element distribution

Table III shows the primitive component of surfacing layers. From this table, Nickel content is higher but other elements are lower. Because of the high Nickel content of 310S stainless steel, matrix composition diluted composition of surfacing layers. So the composition of cladding alloy powder should be corrected according to the composition of matrix material. According to table III, Nickel content of alloy powder should be reduced and other element should be increased appropriately.

Table III. The composition of surfacing layer, wt. %

element	Fe	Cr	Ni	Al	Si	others
Design composition	60	27	8	3	2	/
Actual composition	57.95	26.30	11.54	2.49	1.34	0.38

After correction, Table IV shows the optimum composition of alloy powder. According to table IV, alloy powder after correction is sprayed on the surface of 310S stainless steel and alloy cladding layer is formed.

Table IV. Alloy powder composition after correction, wt. %

element	Fe	Cr	Ni V	Al	Si
Composition	62	27.1	4.4	3.6	2.9

From table V, we can see that the composition on the surface of surfacing layers is close to the design composition.

Table V. The composition of surfacing layer after correction, wt. %

element	Fe	Cr	Ni	Al	Si	others
Composition	59.03	27.35	8.67	2.90	1.81	0.24

The oxidation resistance of high temperature alloy is directly related to the distribution of its elements. If the distribution of elements is not uniform, especially the distribution of antioxidant elements, it is easy to cause the oxide layer to be oxidized from the position of its relatively low concentration, and then further into the inner layer of the alloy. The distribution of elements on the surface of cladding layers is detected by EDS, and it is verified, as shown in Figure1.

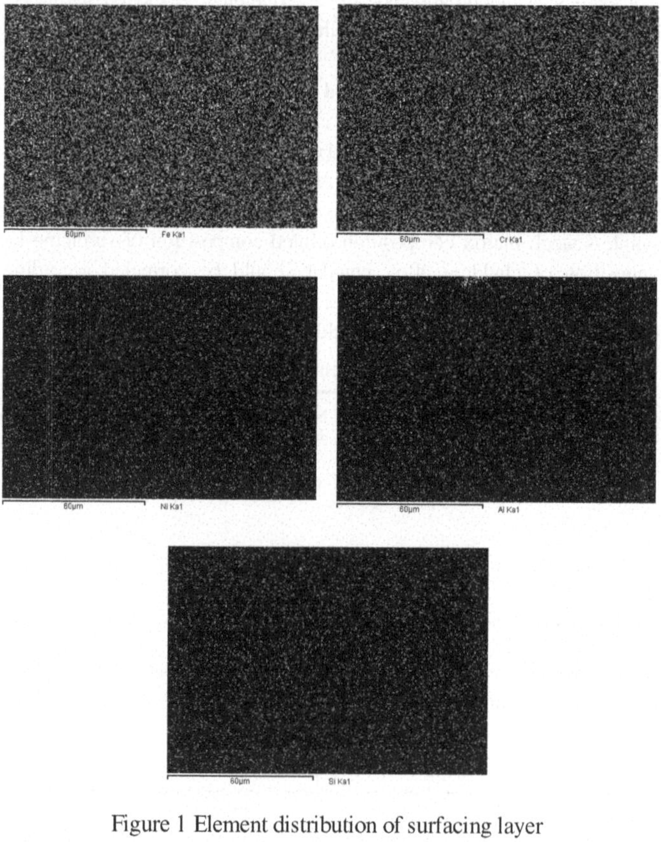

Figure 1 Element distribution of surfacing layer

Determination of antioxidant activity

According to HB5258-2000, High-temperature oxidation test of cladding layers is carried out by oxidation weight increase method. The sample is cut into 20x10x4mm specimen by linear cutting. Then the surface of these samples should be polished by polishing machine, while all edges and corners should be polished smooth. After that, high temperature oxidation test started in the trough type furnace.

The whole antioxidant test lasts 100 hours. Every 25 hours is a cycle and these samples should be weighed after each cycle. By computing weighing results of these samples, high-temperature weight gain rate of oxidation K can be obtained, and the quality of the antioxidant properties of these samples can be defined. TableVI shows the weight gain rate of the oxidation and classification of the oxidation resistance.

Table VI. The determination of antioxidant properties

level	K g/(m^2h)	classification
1	<0.1	complete-antioxidant
2	0.1~1.0	antioxidant
3	1.0~3.0	time-antioxidant
4	3.0~10.0	Weak antioxidant
5	>10.0	No antioxidant

In the antioxidant ratings of 100 hours antioxidant test, the oxidation rate K of samples is calculated according to the following formula:

$$K = (m_1 - m_2)/50S$$

Of which m_1 is the weight of the samples and container after 100 hours, m_2 is the weight of the samples and container after 50 hours, and S is the surface area of the samples.

Analysis on oxidized weight gain rate of samples

Figure 2 shows oxidation weight gain rate of each sample at 1200℃. The oxidation weight gain rate of iron-based alloy sample is 0.4552g/m^2h, it belongs antioxidant level because the oxidation weight gain rate is between 0.1 g/m^2h to 1.0 g/m^2h. The oxidation weight gain rate of furnace alloy sample is 1.4617g/m^2h, belongs time-antioxidant level. The oxidation weight gain rate of 310S stainless steel is 1.5427 g/m2h, belongs time-antioxidant level. The results show that compared with 310S stainless steel substrate, iron-based alloy can significantly improve the high temperature oxidation resistance by plasma spray welding technology. And the oxidation resistance of surfacing layers is much better than furnace alloy at 1200℃.

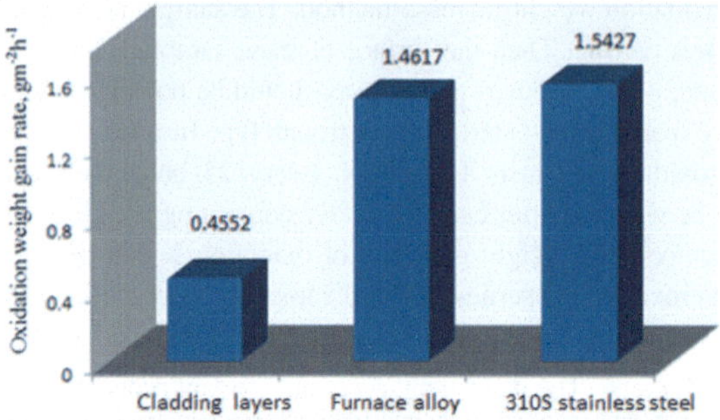

Figure 2 The oxidation weight gain rate of specimens, 1200℃, 100h

Average oxidation weight gain rate of each sample is listed in tableⅦ.

Table Ⅶ. Average oxidation weight gain rate of specimens, g/m^2h

Classification	0-25h	25-50h	50-75h	75-100h
Surfacing layers	0.2699	0.5800	0.3436	0.5667
Furnace alloy	1.2099	1.6589	1.4008	1.5226
310S stainless steel	1.3108	1.7213	1.4224	1.6631

Morphology analysis of surfacing layers

Figure 3 shows SEM microscopic detection of welding samples after oxidation test of 100 hours. There are two different regions on the surface of surfacing layers. Bright region and dark region are both dense, not undulating. Elemental compositions of two areas are swept by EDS in table Ⅷ .

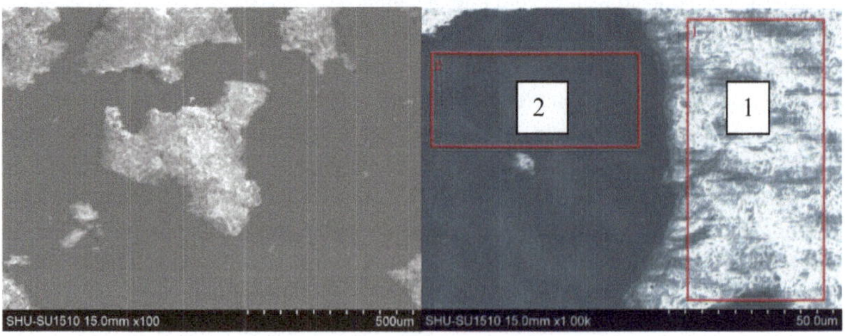

Figure 3 The SEM images of the surfacing sample after oxidation

Dark region is unoxidized metal surface because oxygen element is not detected in this area. Bright parts are mainly composite oxide film of Cr_2O_3 and Al_2O_3, in which the content of Fe is low and oxide film combines with the substrate tightly. Because of the existence of Al and Si elements, the adhesion between oxide film and substrate is strong and the oxide film is dense.

Table Ⅷ. The areas analysis results of the EDS, wt. %

region	O	Fe	Cr	Ni	Al	Si
1	46.20	2.67	19.87	/	31.27	/
2	/	48.73	27.10	18.93	4.41	0.82

Phase analysis of oxide film

Figure 9 shows the XRD test result on surface of surfacing layers. It can be seen that the main component on the surface of the sample is $NiFe_2O_4$, Cr_2O_3, Al_2O_3 and SiO_2. The surface oxide film of the surfacing samples is a variety of composite materials. SiO_2, Al_2O_3 and Cr_2O_3 are attached to each other to increase the adhesion and density of the oxide film and the substrate. In the oxidation experiment of 100 hours at 1200℃, the composite oxide film have a good protective effect on the substrate, which shows good antioxidant activity.

The composition of oxide film is a decisive factor to determine the antioxidant properties of the substrate, and the oxidation resistance of the alloy is actually controlled by the composition of the oxide film. If Al is added to the sample, Al_2O_3 and Cr_2O_3 composites effectively, oxidation resistance of the coating will be improved. After increasing the content of Si, oxide dispersion can be formed in the oxide film, which can effectively fill the vacancy between the oxide film and substrate, increase the density of the oxide film, and improve the adhesion and spalling resistance of the oxide film and substrate.

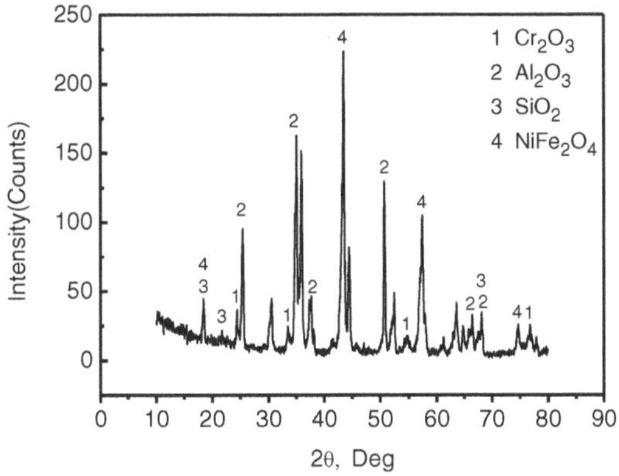

Figure 9 The XRD pattern of the surfacing sample after oxidation

Conclusion

An experimental study to improve oxidation resistance of furnace alloy has been carried out in this paper. The optimum composition(Fe62%, Cr27.1%, Ni4.4%, Al3.6%, Si2.9%)and technological parameters were obtained, with which an antioxidant layer was cladded on the surface of relatively cheap alloy substrate through the plasma spray welding, thus effectively improving the oxidation resistance of the matrix. The above mentioned research provided a technical route and experiment basis for the improvement of workpiece used in furnace for better oxidation resistance, lower manufacturing costs and longer the service life.

References

[1] LI Min, LI Hui dong, LI Hui qi, etc. Study on surface metallurgical technology by Plasma jet [J].Mining & Processing Equipment. 2004, 32(5): 23-25.

[2] LU Jian bo, YAO Sun, LOU Song nian, etc. Coat components optimization and wearability for plasma transferred arc processing [J].Materials for Mechanical Engineering.2006, 30(3):35-37.

[3] Hasan Jamshidi, Sayed Ali Sadough Vanini, Aliraza Attari. Investigation of Creep Feed Grinding Parameters and Heat treatment Effects on the Nickel-base Superalloys[J]. Transactions of material and heat treatment. 2004, 5(25): 92-96.

[4] Qin X, Guo J, Yuan C, etc. Evolutions of microstructures and mechanical properties of two cast Ni-based superalloys during long-term thermal exposure[J]. Acta Metall. Sin. 2010, 46: 213-219.

7th International Symposium on High-Temperature Metallurgical Processing
Edited by: Jiann-Yang Hwang, Tao Jiang, P. Chris Pistorius, Gerardo R.F. Alvear F., Onuralp Yücel, Liyuan Cai, Baojun Zhao, Dean Gregurek, and Varadarajan Seshadri
TMS (The Minerals, Metals & Materials Society), 2016

PRODUCTION OF FeMn ALLOYS WITH HEAT TREATED Mn-NODULES

Merete Tangstad[1], Eli Ringdalen[2], Edmundo Manilla[3], Daniel Davila[3]

[1] Norwegian University of Science and Technology, 7491 Trondheim, Norway
[2]SINTEF, Trondheim, Norway
[3]Autlan, Ejido Tamos C.P. 92018, Veracruz, Mexico

Keywords: Manganese ferroalloys, manganese ores, ore properties, pilot scale experiment

Abstract

FeMn alloys are typically produced with lumpy ore or sinters. Minara Autlan, the Mexican FeMn producer, is however making Mn-nodules as a raw material for Mn-alloy production. The Mn nodules have been characterized by XRF, XRD, by wet chemical methods and by EPMA. The Mn/Fe ratio is somewhat lower compared to commercial Mn ore on the marked. The basicity of the ore is however close to 1, and additional fluxes may not be used. The Mn-nodules will be in a state of manganosite (MnO), tephroite ($(Mn, Ca)_2SiO_4$) and galaxite ($(Mn,Mg)Al_2O_4$). This leads to a low melting temperature area. A low porosity and low amount of higher Mn-oxides gives a high thermal strength to the nodules.

Introduction

Manganese ferroalloys are typically produced in submerged arc furnaces; however some is still produced in blast furnaces. The main reaction zones in the furnace are however quite similar in both events, and in this paper the submerged arc furnace process will be used as an example. The process may be divided according to the temperature: In the upper part of the furnace the temperature is quite low, less than 1000-1200 °C. The raw materials will be solid, and hence all the reactions in this zone are gas-solid reactions. As the charge is being heated the higher manganese oxides will be reduced to MnO and the iron to FeO or metallic iron. The carbonates, like limestone, will decompose. If the higher manganese oxides, and the decomposition of the limestone, is occurring at relatively high temperatures (assumed to be above 800°C), this CO_2 will react with C according to the Boudouard reaction ($C+CO_2=2CO$). The reaction is endothermic and hence, it will lead to a higher carbon and energy consumption. The prereduction rate of the raw materials is hence important. Reduction of higher manganese oxides is exothermic, and hence an increased amount of higher manganese oxides decreases the total energy consumption. In the high temperature zone, from 1250 – 1600 °C, the raw materials will start to melt. This is also called the cokebed zone, where liquid slag is reduced, and liquid slag and metal flows through the cokebed. In this zone the main reaction is MnO reduction with solid carbon to Mn alloyed with iron and dissolved carbon.

The Mn-sources may be lumpy ore or heat treated ore, like sinter, pellets and nodules. This paper will show the main properties of Mn-nodules produced by Autlan, and how these nodules are compared to other commercial Mn-raw materials. Autlan has a large deposit of manganese in the state of Hidalgo, Mexico. The mineral is manganese carbonate and is processed for use as manganese nodules in the ferroalloy industry with a rotary kiln, where a thermal conversion is carried out.

Characterization of Mn nodules

The chemical analysis of the typical Mn nodule is shown in Table I. Compared to other commercial ores, shown in Table II, the chemical composition is similar to the some of the South African ores. The Mn/Fe ratio is around 5 and the overall basicity $(MgO+CaO)/(SiO_2+Al_2O_3)$ is close to 1. This means that the addition of basic fluxes like dolomite or limestone is not necessary when Mn-nodules are used for HC FeMn production.

Table I. Chemical analyses of Mn nodule (dry basis).

	Mn	Fe	SiO_2	Al_2O_3	CaO	MgO	S	P	MnO_2
%	37.71	7.94	14.88	3.90	11.06	8.13	0.14	0.085	0.00

Table II Chemical composition of some commercial Mn-ores[6].

Manganese ore	Mn/Fe	H_2O	SiO_2	Al_2O_3	MgO	CaO	K_2O	P	CO_2
Comilog MMA	18.5	8.7	4	5.5	0.3	0.2	0.7	0.11	0.1
Comilog MMD	9.5	9	7.7	7.2	0	0.1	0.8	0.09	0.1
Comilog MMR	13.9	9	5	6.1	0.1	0.1	0.8	0.11	0.1
Comilog MMS	12	9	7.7	7.5	0	0	0.8	0.09	0.1
Comilog Sinter	16.7	1.5	7	6.5	0	0.1	0.75	0.12	0
Asman 48	5.1	0.9	5.5	0.4	0.7	4.3	0	0.04	0.8
Amapa Sinter	5.1	1	7.6	7.6	0.5	0.8	0.3	0.1	0
Amapa Miudo 40	3.3	10	5.9	8.1	0.1	0.3	0.8	0.11	3.5
Mamatwan	8.2	1	4	0.5	3.5	14.7	0	0.02	17
Gloria	7.8	0.4	5.7	0.3	3.8	12.7	0	0.02	15.4
Groote Eylandt	11.6	2.7	6.9	4.2	0.1	0.1	2	0.09	0.5
CVRD sinter	11.5	0.6	5.4	8.7	0.5	1.9	1.4	0.11	0.2
Wessel 38%	3.2	1.2	4.9	2.5	1	6	0.1	0.04	3.6
Wessel 50%	5	0.9	3.6	0.4	1	5.6	0.1	0.04	2.6

As the ore is heat treated, all CO_2 is removed and all manganese oxides are present as MnO. This is also verified by XRD measurements. The XRD results show that the nodules has a mineralogical composition of 33% manganosite (MnO), 63% tephroite (Mn_2SiO_4) and 4 % galaxite ($(Mn,Mg)Al_2O_4$). This will probably give the nodules quite low melting temperatures as the melting point of Mn_2SiO_4 is 1347 °C while MnO and $MnAl_2O_4$ melts at 1842 and 1745 °C (incongruently) respectively. The tephroite phase is not a 100% match, which may be an indication of the solid solution of Ca in the tephroite, as $(Mn,Ca)_2SiO_4$ (olivine). This microstructure is similar to the phases found in sinters. In the sinters, the manganese rich phase is however closer to Mn_3O_4 than MnO, as seen in the Mn-nodules.

Particles of about 2-3 mm were investigated by EPMA (microprobe). The edge and center of each particle was studied. The phase distribution of the nodules is similar to all heat treated manganese ores. There is a spherical phase distributed in a matrix. This phase has the highest manganese content and is the manganosite phase. The analysis of this phase is 70-80%MnO with about 20%FeO and 5-10%MgO dissolved. The main matrix phase is the tephroite phase with about 30%SiO_2, where the 10%MgO and 20-30% CaO is dissolved in 30%MnO. A minor phase in-between in the grey matrix is galaxite with 50% Al_2O_3 and a solution of 20-25%MnO, 10%MgO and 12%FeO.

The lower oxygen content in the Mn-nodules has two competing effects on the energy consumption. Higher manganese oxides will be reduced exothermically down to MnO with CO gas. The developed CO_2 may however react according to the Boudouard reaction, which is endothermic.

$$\tfrac{1}{3}\,Mn_3O_4 + \tfrac{1}{3}\,CO(g) = MnO + \tfrac{1}{3}\,CO_2(g) \qquad \Delta H^\circ_{298} = -16.9\ kJ$$
$$\tfrac{1}{3}\,C + \tfrac{1}{3}\,CO_2(g) = \tfrac{2}{3}\,CO(g) \qquad \Delta H^\circ_{298} = +57.5\ kJ$$
$$\tfrac{1}{3}\,Mn_3O_4 + \tfrac{1}{3}\,C = MnO + \tfrac{1}{3}\,CO(g) \qquad \Delta H^\circ_{298} = +40.5\ kJ$$

Figure 1 shows the energy consumption as a function of oxygencontent in the ores. It is seen that the Mn-nodules, who is containing only MnO oxides, will have about the same energy consumption as an Mn_2O_3 ore. It will have no exothermic reduction of higher manganese oxides, but also no extent of the endothermic Boudouard reaction.

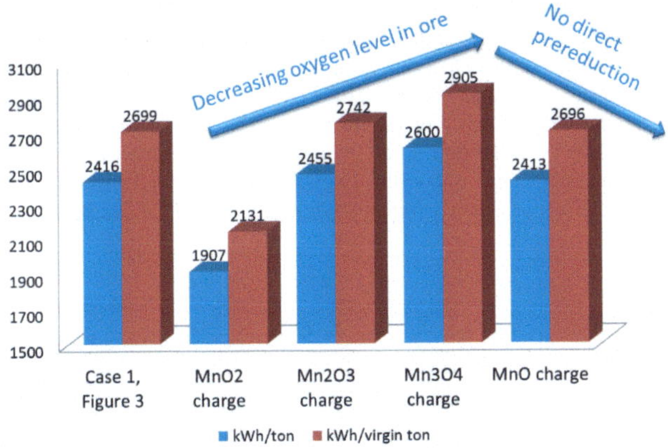

Figure 1. Electrical power consumption with decreasing oxygen content in the ore. All other parameters are assumed constant[11].

The density is in the area of 3.8-4.0 and the porosity of the nodules is between 3 and 9%. Compared to other commercial ores, the porosity of the Mn nodules are in the low area and comparable to the South African ores and Groote Eylandt ore[2,9,8].

Warm strength

To summarize the method, the ore is exposed to mixture of CO and CO_2 - gas during heating. After being heated to 1100 °C, the ore is cooled and then sieved for size distribution. The sample is then tumbled and then sieved again for size distribution.

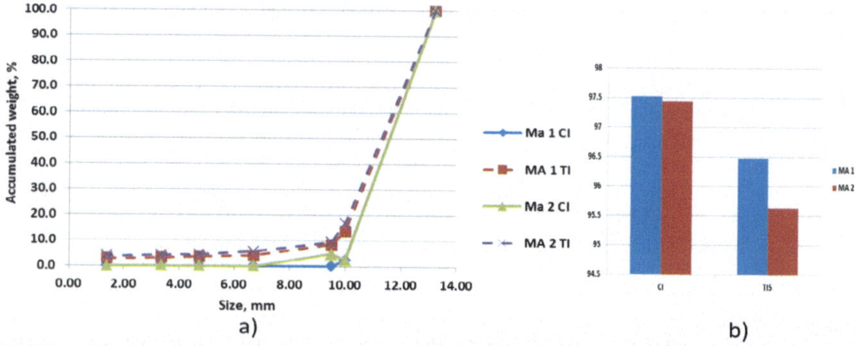

Figure 2. a) Size distribution of Mn nodules before (CI) and after (TI) tumbling and b)calculated CI and TI.

The size distribution after heating, before and after tumbling, is shown Figure 2. This can be expressed as the Cohesion Index (CI) and the Tumbler Index (TI). The CI is the amount > 10 mm after heating and TI is the amount > 6 mm after heating and tumbling. It is seen that the Mn nodules are very strong, and even after tumbling, they have less than 5% -5mm particles. This is much higher strength compared to previous published ore data, as shown in Figure 3. While the Mn nodules have a CI and TI in the mid 90ies, typically ores and sinter are in the range 40 to 80% and even lower. The only material, previously published, that can compare in strength is the pellets made during oxidizing condition, which also show a CI and TI above 90%.

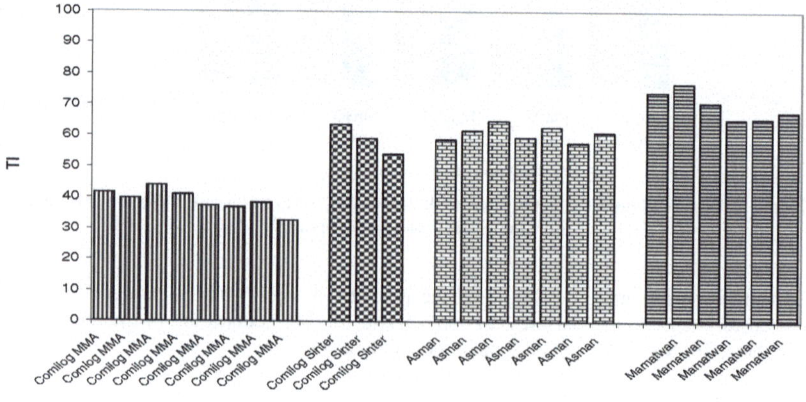

Figure 3. Tumbler index after heating in CO and CO2 gas mixture followed by tumbling[5].

As previously mentioned, the reduction rate is dependent on when the slag phase becomes liquid. This slag phase will be in coexistence with a solid MnO phase. As the reduction of MnO dissolved in the liquid slag phase is reduced with carbon, the solid MnO phase will dissolve in the liquid slag phase. To investigate these phenomenon two different types of experiments has been done in the Mn nodules; the sessile drop method and the slumping test. These are done in two different scales; <1g and about in the order of 1 kg respectively.

In the sessile drop a small amount of material is investigated. To get a representative sample, the ore is crushed down and then formed to small pellets. As the pellets are visualized one can determine:

- Start of melting - when the shape of the particle has lost the sharp edges
- Final melting - when the droplet is round, with no artefacts
- Start of reduction - the first bubble is visible

In addition other phenomenon like shrinking is seen. The results are shown in Table **III** and Figure 4 for Mn nodules. Around 1300 °C a visual acknowledgement of the start of melting and reduction occur. The ore will however at this temperature, visually behave like a solid as it contain a lot of solid phase. At temperatures between 1400 and 1500 °C, all the solid phase is reduced and dissolved, and hence the slag is now completely liquid and could drain through the cokebed and be tapped, if it was in an industrial furnace. These temperatures are compared to other published data in Table IV.

Table III Results for sessile drop test.

	Test	Initial melting	Complete melting	start of reduction	shrinking
Minera Autlan 1	1	1329	1521	1292	1240
Minera Autlan 2	1	1314	1401	1284	

Ore	Initial melting	Start Reduction	Complete melting
Minera Autlan 1	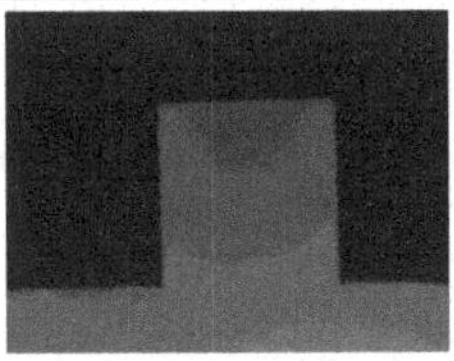 **1240°C**	 **1292°C**	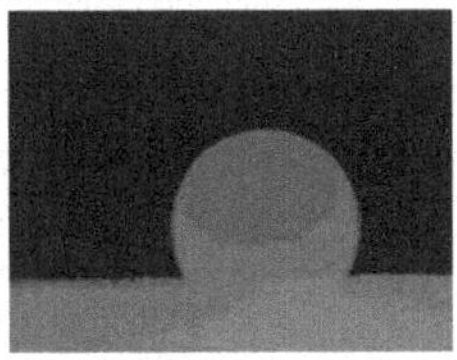 **1521°C**
Minera Autlan 2	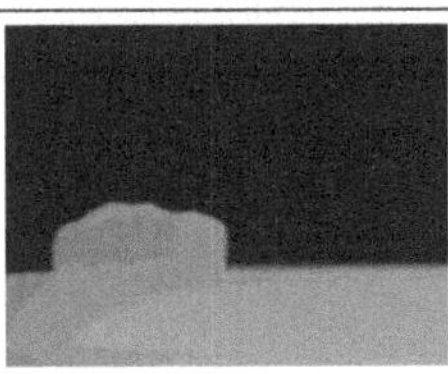 **1314°C**	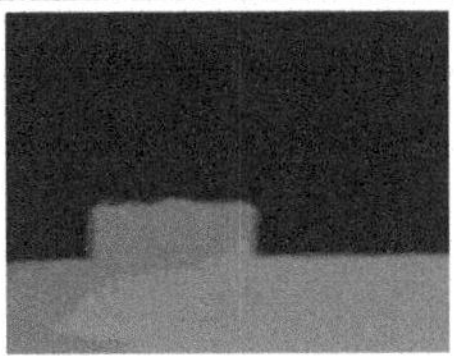 **1284°C**	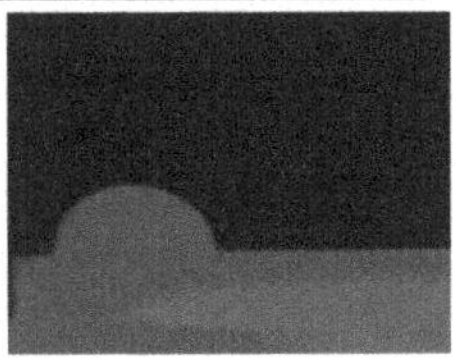 **1401°C**

Figure 4 Pictures from sessile drop measurements.

Table IV. Summary of melting and reduction temperatures measured by sessile drop by other authors (after review of Ringdalen et al, 2015[10]).

Material	Initial melting °C	Complete melting °C	Material	Initial melting °C	Complete melting °C
Assmang ore	1450	1510	Comilog ore	1490	1540
Assmang Nchwaneng ore	1420	1624	Gabonese ore	1485±11	1538±9
Assmang ore	1446±70	1513±57	Gabonese ore	1442	1531
Nchwaneng A	1478	1600	Gabonese sinter	1339	1498
Nchwaneng B	1519	1628	Gabonese pellets	1358	1516
Nchwaneng C	1521	1550	CVRD sinter	1350	1410
Nchwaneng D	1450	1560	HCFeMn slag	1220	1230
Assmang Gloria pellets	1622	1683	CVRD ore MF15	1385	1505
Gloria A	1588	1702	CVRD ore	1461±13	1494±23
Gloria B	1512	1674	CVRD sinter	1395±85	1489±132
CVRD pellets MF15	1322	1421	CVRD sinter	1228	1305

Compared to other Mn sources used in FeMn production, the Mn nodules is melted and reduced at a lower temperature. The closest one is CVRD sinter, which melts around 1350 °C. Comilog ore, CVRD ore and Assmang ore melts at quite higher temperatures.

This is in agreement with the slumping experiments where about 1 kg of Mn ore was heated between two layers of coke in an induction furnace. The height of the ore layer will decrease with melting, reduction and flow into the coke layer.

The test was performed in a graphite crucible, 110 mm ID, 150 mm OD and 400 mm high. A graphite block on the top of the material in the crucible was used to measure changes of the charge level. The crucible was first charged with 100 mm of coke, +9.5 – 16 mm. A 70 mm thick layer of manganese ore was added. Finally a 100 mm thick layer of coke was added to the top of the manganese ore, and on the top of this graphite block to monitor changes in charge level. The set up for the crucible is shown in Figure 5. The filled crucible was loaded into the induction furnace, and then slowly heated at 10-15°C/min until the height of the ore bed had decreased 40-50 mm, measured by the relative height change in the graphite block on top of the charge. The crucible was cooled down and then filled with epoxy. When the epoxy had hardened, the crucible was cut in half with a diamond saw. The melted and reacted phases were then studied.

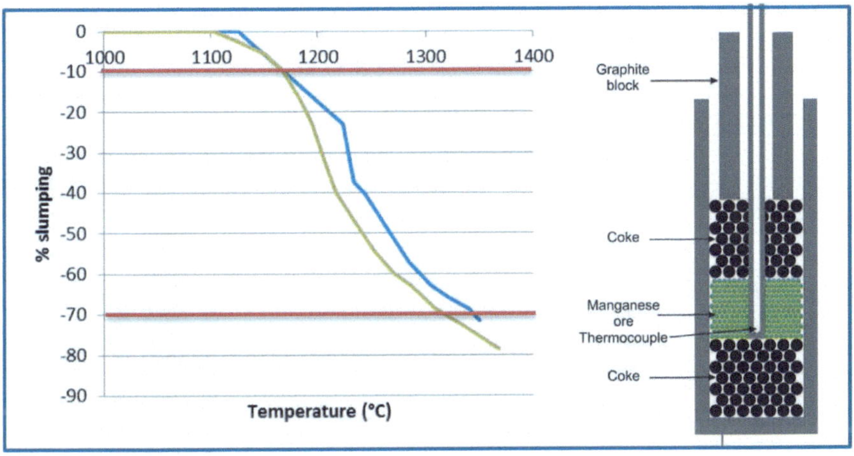

Figure 5. Percentage slumping of ore layer as it is melted, reduced and flowing into the coke layer and the sketch of crucible for bulk softening and melting measurements.

By definition it is determined that initial melting is at 10% slumping and final reduction at 70% slumping. This means that the initial melting for Mn nodules are less than 1200 °C (1168 °C) and the final reduction is a bit higher than 1300 °C (1345 and 1322 °C). This is a bit lower than the sessile drop apparatus, as it is actually measuring a bit different phenomenon. However, the main results here are that Mn melts and is reduced at quite low temperatures.

The electrical resistivity of the Mn-nodules is also studied. The equipment for resistivity measurements consist of power supply, data logging system and crucible with electrical connections and thermocouples. The dimensions of the crucible are 115 mm ID, 150 mm OD and 400 mm high. The top of the crucible has six M8 threads tapped into the wall, to attach a top plate 110 mm ID, 200 mm OD and 20 mm thick. The electrical connection is attached to the top plate with a stainless steel plate 50 mm wide and 3 mm thick. The top electrical connection was made with a graphite electrode 72 mm OD and 300 mm long, which rested on top of the charge. Graphite paper is used between all electrical connections to reduce the contact resistance.

In the experiment, there was a bed of SSAB coke +10 –16 mm in the bottom, up to the level of the bottom wire, with the ore up to the second wire and then 70 mm of coke above the ore.
The power of the induction furnace was controlled to maintain a heating rate of 10 °C, although this was not always possible. As the sample was heated, the height of the top electrode was measured to correlate with the slumping of the ore sample. The goal was to continue the experiment until the ore bed had collapsed to 80% of the original height. As the sample was heated, a series of electrical pulses was placed across the two electrodes.

The specific resistivity of the ore is shown in Figure 6 . The resistivity decreases drastically in the area when the nodules melt, as in accordance with results from Miyauchi[3,4]. Below 1200 °C, where the ore is mainly solid, the specific resistivity is above 10 ohm•m. At the highest temperatures where the ore is assumed to be mostly liquid phase, the specific resistivity is in the area of 0.5 ohm•m. However, this is still in the order of 10 to 100 times higher than a cokebed without slag as shown by Eidem[7]. Hence, based on this, it is still believed that the major part of the electrical current will go through the cokebed, and not the half-melted ore.

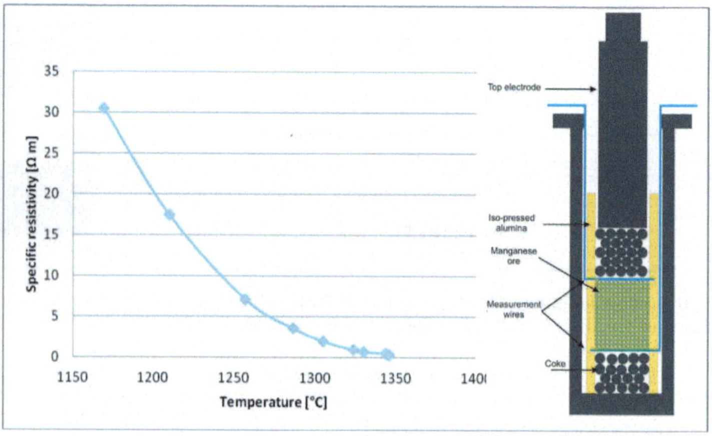

Figure 6. Specific resistivity versus temperature and sketch of crucible setup.

REFERENCES

1.A.Muan, E.F.Osborne (1966), *Phase equilibria among oxides in steelmaking*, Addison-Wesley

2.L.S.Todd and D.S Martin (1979), *Solid state reactivity of manganese ores – job no 528-50534*, Union Carbide report, 1979

3.Miyauchi, Y., M. Mochida and Y. Fuchi, (2001) *High Thermal Electrical Property of Manganese Ore in Production of High Carbon Ferromanganese*. INFACON 9, pp. 236-243.

4.Miyauchi, Y., T. Nishi, K. Saito and Y. Kizu, (2004), *Improvement of High-Temperature Electric Characteristics of Manganese Ores*. INFACON X, Cape Town, South Africa, pp. 155-162,

5.M.Tangstad, P. Calvert, H.Brun, (2004), A-G.Lindseth, *Use of Comilog ore in FeMn production*, Infacon X, Cape Town.

6.S.Olsen, M.Tangstad, T.Lindstad, (2007) *Manganese ferroalloy production*, Tapir

7.P.A.Eidem (2008), *Electrical resistivity of cokebeds*, PhD Thesis, IMT, NTNU, Norway

8.M.Tangstad, D.Leroy, E.Ringdalen, (2010) *Behaviour of agglomerates in ferromanganese furnace*, INFACONXII, Helsinki, Finland, 2010

9.K.Turkova, D.Slizovskiy, M.Tangstad, (2014), *CO reactivity and porosity of manganese materials*, ISIJ International. vol. 54 (6).

10.E. Ringdalen, M.Tangstad, T.Brynjulfsen (2015), *Melting behavior of Mn-sources – Effect on furnace performance*, Infacon XIV, June 2015, Kiev, Ukrain

11.M.Tangstad, K.Ichihara, E.Ringdalen, (2015), *Pretreatment unit in ferromanganese production*, Infacon XIV, June 2015, Kiev, Ukrain

7th International Symposium on High-Temperature Metallurgical Processing
Edited by: Jiann-Yang Hwang, Tao Jiang, P. Chris Pistorius, Gerardo R.F. Alvear F., Onuralp Yücel, Liyuan Cai, Baojun Zhao, Dean Gregurek, and Varadarajan Seshadri
TMS (The Minerals, Metals & Materials Society), 2016

THERMODYNAMIC ANALYSIS AND EXPERIMENTS ON VACUUM SEPARATION OF Sn-Sb ALLOY

Junjie Xu [1, 2, 3, 4] Lingxin Kong [1, 2, 3, 4] Yifu Li [1, 2, 3] Bin Yang [1, 2, 3, 4, *] Yongnian Dai [1, 2, 3]
Kunhua Wu [1, 2] Anxiang Wang [1, 2, 3]

[1]National Engineering Laboratory for Vacuum Metallurgy, Kunming 650093, PR China
[2]Key Laboratory of Vacuum Metallurgy for Nonferrous Metal of Yunnan Province, Kunming 650093, PR China
[3]Faculty of Metallurgical and Energy Engineering, Kunming University of Science and Technology, Kunming 650093, PR China
[4]State Key Laboratory of Complex Nonferrous Metal Resources Clean Utilization, Kunming 650093, PR China

Keywords: Sn-Sb Alloy; Vacuum distillation; Distillation parameters; Separation

Abstract

In this study, the saturated vapor pressures of tin (Sn) and antimony (Sb), the separation coefficient β and the vapor-liquid phase equilibrium of Sn-Sb alloy were theoretically analyzed, which demonstrate that it is possible to separate Sn and Sb by vacuum distillation. The process parameters of vacuum distillation, including the distillation temperature, distillation time and alloy mass (thickness of raw materials) on the direct yield of Sn and the content of Sn in liquid phase were investigated by using single factor experiments. The preliminary results show that the direct yield and the content of Sn are 98.77 wt.% and 96.01% with the optimized distillation conditions of a distillation temperature of 1473 K, a distillation time for 45 min and a Sn-Sb alloy mass of 125 g (thickness of 8mm). The distillation parameters in the study provide effective and convenient conditions on separation of Sn-Sb alloy.

Introduction

As one of the earliest metals that human discovered and used, Sn is widely used and plays an indispensable key role in military industry and modern cutting-edge technology areas [1, 2]. Based on the unique physical and chemical properties, Sn and Sb were involved in a national strategy aiming at accelerating the cultivation and development of strategic emerging industries including the energy conservation and environmental protection, next generation information technology, biotechnology, new energy, new energy vehicles, high-end equipment manufacturing, and new materials industries [3]. However, the contradictions between the remaining productive life and demands of Sn and Sb resourses are becoming increasingly acute.

Large numbers of waste Sn-based alloys will be recycled from electroplates, solders and other various industries in China and elsewhere due to Sn was usually used to produce alloys with other metals, which will causes serious resources waste and environment pollution if the

* **Corresponding author. Tel.:** +86 871 65161583; fax: +86 871 65161583.
E-mail address: kgyb2005@126.com (B. Yang) ; xjjkm@sina.com (J.J. Xu)

alloys cannot be recycled cleanly and efficiently. The common refining techniques to remove Sb from Sn-Sb alloy are pyrometallurgy and electrolytic method. There are some serious problems in conventional processs. The conventional process has some disadvantages such as procedure complex, lower direct rate of tin, bad state of operation, evident environment pollution [4-6]. Over past decades, vacuum distillation has been studied and successfully used in refining and recovery of various nonferrous crude metals and alloys by Kato [7], Gopala [8], Zhan [9], and Ali [10, 11] etc., and the area of its application is being extended rapidly.

The separation of Sn-Bi, Sn-Pb, Sn-Sb, Ag-Pb-Sb alloys by vacuum distillation has been studied in China and elsewhere for a long time, which has been described in numerous works [12-16]. The purpose of this study, therefore, was to obtain more obvious and optimum process parameters of Sn-Sb alloy during vacuum distillation, which will provide experience used for industrial production.

Theoretical analysis

Saturated vapor pressure

Crude metal can be separated from impurities by vacuum distillation due to the different properties of components contained when vaporizing and condensing. The basic principle for vacuum separation of crude metal is the difference in saturated vapor pressure of each element in a finite temperature. The saturated vapor pressure of pure components of these two alloys can be calculated from the following equation [17]

$$\lg P^* = AT^{-1} + B\lg T + CT + D \quad (1)$$

Where P^* represents the saturated vapor pressure of pure components; A, B, C and D are the evaporation constants of components of alloys, which were available in Ref. [17].

Separation coefficient

Dai [17] introduced the separation coefficient, viz. β, to determine whether two components can be separated from each other by vacuum distillation according to the composition difference between vapor phase and liquid phase. To estimate the feasibility of separation of alloys by vacuum distillation, the following equation from theoretical derivation for *i–j* binary alloy was derived, that is

$$\beta = (\gamma_i / \gamma_j) \cdot (P_i^* / P_j^*) \quad (2)$$

Where γ_i and γ_j are activity coefficient of *i* and *j* components; P_i^* and P_j^* are saturated vapor pressure of *i* and *j* in pure state, respectively. With a known value of β, the ratio of vapor densities of *i* and *j* can transform into

$$\frac{\rho_i}{\rho_j} = \beta \frac{\omega_i}{\omega_j} \quad (3)$$

Where ρ_i/ρ_j is the mass ratio of component *i* in the vapor phase to that of *j*, and ω_i/ω_j is the mass ratio of component *i* in the liquid phase to that of *j*, respectively. The separation of *i* and *j* could happen while $\beta > 1$ or $\beta < 1$, but it could not happen while $\beta = 1$.

After substituting activity coefficients γ_i and γ_j at different temperatures into Eq. (2), the separation coefficients can be easily calculated as shown in Fig 1., which indicates the content of Sb in gas is much higher than those in liquid. Therefore, Sb can be effectively concentrated in the gas phase by vacuum distillation.

Vapor-liquid phase equilibrium in vacuum distillation

For *i-j* alloy, the relationship between the mass fraction of *i* and *j* components in vapor and liquid phase can be respectively expressed as,

$$\omega_{i,g} + \omega_{j,g} = 1 \tag{4}$$

$$\omega_{i,l} + \omega_{j,l} = 1 \tag{5}$$

Where $\omega_{i,g}$, $\omega_{j,g}$ are mass fraction of *i* and *j* components in the vapor phase, respectively; $\omega_{i,l}$, $\omega_{j,l}$ are mass fraction of *i* and *j* components in the liquid phase, respectively.
When the two phases are in equilibrium, the mass fraction of component *i* in the vapor phase is related to the vapor densities of *i*, *j* components as follows [17].

$$\omega_{i,g} = \frac{\rho_i}{\rho_i + \rho_j} = \frac{1}{1 + (\rho_j / \rho_i)} \tag{6}$$

Substituting Eqs. (2) and (3) into Eq. (6), the mass fraction of component *i* in the vapor phase can be expressed as,

$$\omega_{i,g} = \left[1 + \left(\frac{\omega_{j,l}}{\omega_{i,l}}\right) \cdot \left(\frac{\gamma_j}{\gamma_i}\right) \cdot \left(\frac{P_j^*}{P_i^*}\right)\right]^{-1} \tag{7}$$

Where ω, γ, P^*, β are the mass fraction, activity coefficient, saturated vapor pressure, and separation coefficient, respectively. The relationship diagram of $\omega_{i,g} - \omega_{i,l}$ can be calculated by γ, P^* and a series of $\omega_{j,l} / \omega_{i,l}$ at required temperatures, that is the vapor–liquid phase equilibrium diagram for *i–j* alloy system.

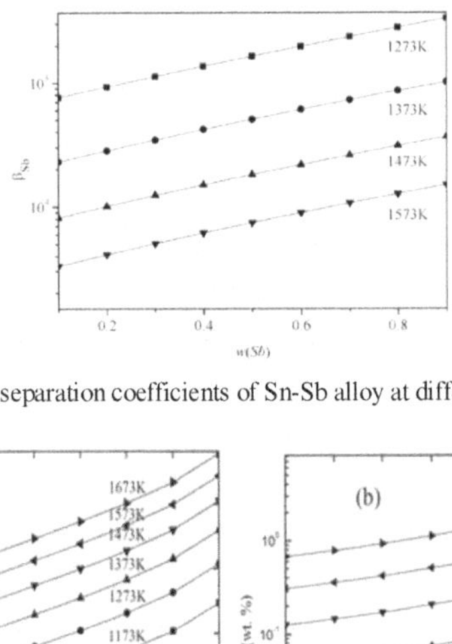

Fig 1. The separation coefficients of Sn-Sb alloy at different temperatures

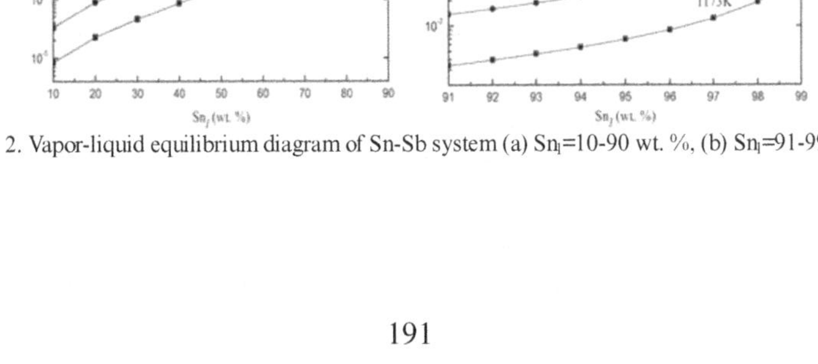

Fig 2. Vapor-liquid equilibrium diagram of Sn-Sb system (a) Sn_l=10-90 wt. %, (b) Sn_l=91-99 wt. %.

Based on Tao's the molecular interaction volume model (MIVM), a reliable and stable model in predicting thermodynamic properties of liquid alloys [18], the vapor-liquid equilibrium diagram of Sn-Sb system is shown in Fig 2. As it shows that the content of Sn in vapor phase, Sn(g), increases with increasing the distillation temperature and the content of Sn in liquid phase, Sn(l), which theoretically demonstrates the different content of Sn between Sn(g) and Sn(l) in a various range of distillation temperature.

Experimental

Raw materials

For all alloys (Sn 51.3 wt.%, Sb 48.7 wt.%), the tin (99.995 wt.%, YT Co., Ltd) and antimony (99.999 wt.%, YT Co., Ltd) were weighed out in the analytical balance, then arc melted under a high purity argon (99.999 wt.%, Messer®) atmosphere.

Equipment

The internal structure schematic diagram of the vacuum furnace used is shown in Fig.3. The vacuum system consists of mainly (1) Pit furnace, (2) Evaporator-Collector-Condenser set, (3) Retort-flange assembly, (4) Vacuum system using an oil pump, (5) Circulating water control system, and (6) Vacuum resistance furnace with two electrodes. A cylindrical crucible (ID × height = 60 × 100 mm) which consists of isostatic fine grain high density graphite (R-7340 of SGL Carbon, Germany) was used as the evaporator. The collector is a circular cold plate (surface area = 314 cm^2) made up of stainless steel.

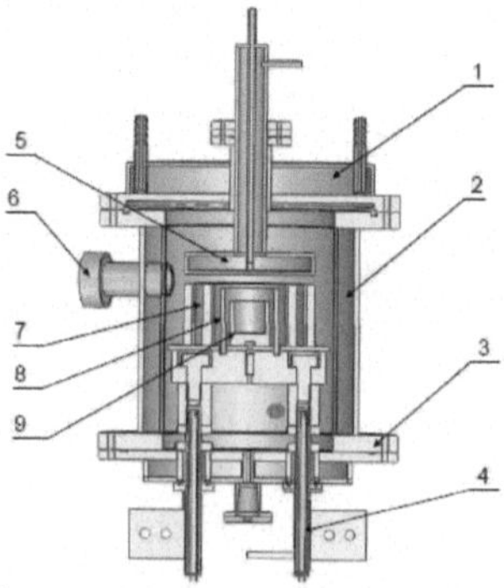

Fig 3. Internal structure schematic diagram of the vacuum furnace. 1. furnace lid; 2. furnace body; 3. furnace bottom; 4. electrode; 5. cold plate; 6. observation door; 7. heat holding cover; 8. heating unit; 9. graphite evaporator.

Experiments design

Learning from industrial practice, the major effects on this study include distillation temperature, distillation time, alloy mass (thickness of raw materials) and vacuum degree. With a stable equipment of 5 Pa below, however, the vacuum degree causes little effect. Therefore, the vacuum distillation was carried out for Sn-Sb alloy at the distillation temperature range of 1173-1673 K, distillation time range of 15-65 min and alloy mass range of 85g (4mm)–185g (14mm).

The temperature error of ± 3 K and the alloy mass error of ± 0.3 g were found due to device limitation and operating error.

After each distillation experiment, the component content in the residual and volatile were analyzed by ICP-AES combined with chemical analysis method. The content of Sn in liquid phase (the component content in the residual) and the direct yield of Sn were taken as evaluation indexes. And the direct yield of Sn was defined as following equation,

$$\text{The direct yield of Sn} = (m_0 \times x)(m_1 \times y) \tag{8}$$

Where m_0 and m_1 are the raw materials and the mass of residuals, respectively; x and y are the content of Sn in liquid phase and the content of Sn in raw materials, respectively.

Results and discussion

Effect of distillation temperature

The vacuum distillation was carried out at the distillation temperature range of 1173 - 1673 K with a gradient of 100 K, distillation time of 35 min, and alloy mass of 125 g (8 mm). The relationship between components contents of products and temperature is shown in Fig 4..

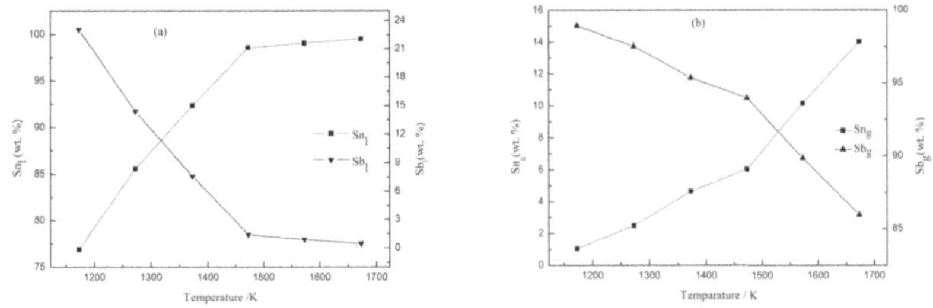

Fig 4. Effect of temperature on the component of (a) residue, (b) volatile

It can be seen from Fig 4. that the content of Sn and Sb in residue were76.89 wt. % and 23.05%, while in vapor phase, they were 1.05 wt. % and 98.95 wt. % at 1173 K. Owing to a quicker growth rate of the statured vapor pressure of Sb with the increasing temperature, the volatilizing quantity of Sb increases constantly, leading the concentrating of Sn in liquid phase. However, the growth rate of Sn in residue slows down when the distillation temperature is higher than 1473 K. This phenomenon shows that part of Sn is wasted under a higher distillation temperature.

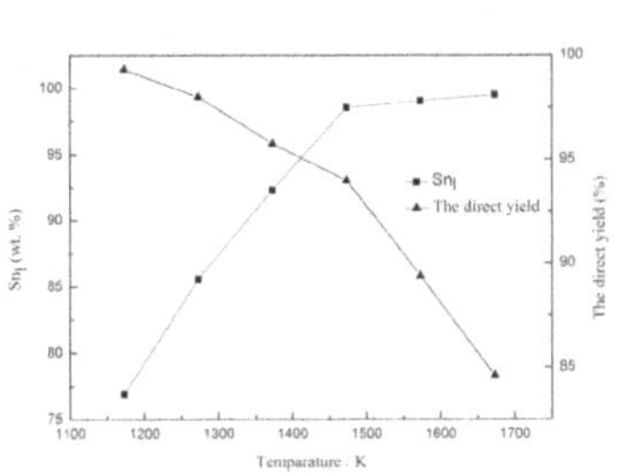

Fig 5. Effect of temperature on the the content of Sn in liquid phase and the direct yield of Sn

As can be seen in Fig 5., the direct yield of Sn decreases with the increase of distillation temperature while the content of Sn in liquid keeps raising until at 1473 K which shows the turning point for a quicker decreasing rate of the direct yield of Sn. The proper distillation temperature in this study presents to 1473 K.

Effect of distillation time

Based on the proper distillation temperature of 1473 K, the following experiments aim to investigate optimum distillation time of vacuum separation of Sn-Sb alloy. The results at conditions of distillation time range of 15 - 65 min with a gradient of 10 minutes and alloy mass of 125 g (8 mm) were shown at Table 1..

Table 1. Effect of Distillation Time on the Separation of Sn-Sb Alloy

Distillation time/ min	Alloy mass/ g	Residue			Volatile			The direct yield/ %
		Sn wt. %	Sb wt. %	Mass/ g	Sn wt. %	Sb wt. %	Mass/ g	
15	125.25	97.13	2.85	62.74	5.29	94.71	62.5	99.31
25	125.11	97.69	2.27	61.83	5.96	94.04	63.27	98.00
35	124.89	98.44	1.53	61.13	6.10	93.90	63.76	95.76
45	124.96	98.79	1.15	60.88	6.18	93.82	64.08	93.99
55	124.88	98.35	1.64	60.83	6.62	93.38	64.05	89.38
65	125.04	98.83	1.15	60.54	6.69	93.31	64.5	84.61

The content of tin in the vapor phase increased with the increase of distillation time resulting in continuous decreasing of the direct yield of Sn. The content of Sn in liquid phase increased to 98.44 wt. % at the distillation time of 35 min. However, there was not obvious change for Sn in the residue after extending the distillation time to 65 minutes. The content of tin in the liquid phase was 98.44 wt pct, while extending distillation time at a gradient of 10 minutes, it had a range of ± 0.4 wt. %. Based on the fact of the ceaseless volatility of Sn in liquid phase, there is no need to set a longer distillation time.

Effect of alloy mass

According to the recommended parameters in this study, the distillation time and temperature were selected at 1473 K and 35minutes. The results under conditions of alloy mass range of 85g-185g with a gradient of 20 g were shown in Fig 6..

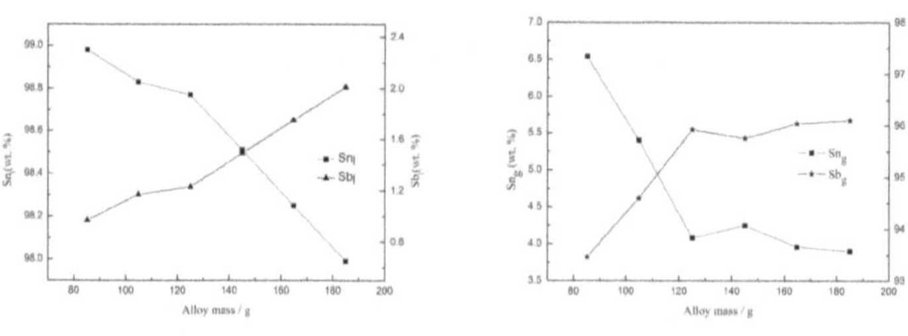

Fig 6. Effect of alloy mass on the component of (a) residue, (b) volatile

It can be seen from Fig 6. (a) that the content of tin in the liquid phase sostenuto decreased with the increase of alloy mass, especially, it declined rapidly after the mass at 125 g. Fig 6. (b) shows that the content of Sn in the vapor phase decreased slowly when the alloy mass was 125 g. It delicates that the alloy mass has little effect on the volatilizing of Sn and Sb. However, the production of vacuum distillation will decrease with the increase of alloy mass (thickness of raw materials).

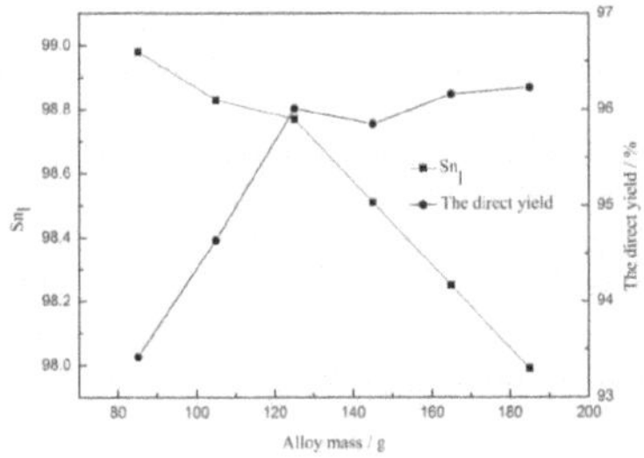

Fig 7. Effect of alloy mass on the the content of Sn in liquid phase and the direct yield of Sn

After increasing to a peak at alloy mass of 125 g (thickness of 8 mm), as can be seen in Fig 7., the direct yield of Sn decreased with the incremental alloy mass , which shows a turning point for obtaining the Sn production. And the content of Sn in liquid phase reached a stable level when the raw materials of 125 g. It demonstrates that the optimal treatment amount for vacuum distillation of Sn-Sb alloy is 125 g (thickness of 8 mm).

Conclusions

The results validate the probability of the separation of Sn-Sb alloy by vacuum distillation. After the vacuum processing, the content of Sb in Sn-Sb alloy and the direct yield of Sn can attain levels of 2 wt. % below and 95 % above, respectively. The distillation parameters, such as distillation temperature, feeding materials (thickness of 8mm) and distillation time, have effect on the direct yield of Sn and the content of Sn in both liquid and vapor phases. In this study, optimized distillation parameters of a distillation temperature of 1473 K, a distillation time for 45 min and a Sn-Sb alloy mass of 125 g (thickness of 8mm) were obtained, which provide referential experience for the distillation separation of Sn-Sb alloys in cleaning industry production.

Acknowledgements

This research work is supported by the General Program of Natural Science Foundation of China (Grant No. 51474116), the Program for Innovative Research Team in University of Ministry of Education of China (Grant No. IRT1250), the Free Exploration Program for Academician of Yunnan Province (Grant No. 2014HA011) and the Cultivating Program for Scientists Leader of Yunnan Province (Grant No. 2014HA003).

References

1. M.F. Liu, "The 'Sn' dream of China," *China Metal Bulletin*, 2011, No. 32:16-18. (in Chinese)
2. National Development and Reform Commission, Ministry of Land and Resources, Commerce Department, State Environmental Protection Administration, State Administration for Industry and Commerce, State Quality Inspection Administration, General Administration of China Customs, "The advice on strengthening management of W, Sn and Sb industry," *Advanced Materials Industry*, 2005, No. 8:30-32. (in Chinese)
3. The State Council of the People's Republic of China, "Speeding up the cultivating and developing strategic emerging industry," *New Industrialization*, 2011, No. 1:79-84. (in Chinese)
4. H.W. Yang, et al, "Application of molecular interaction volume model in vacuum distillation of Pb-based alloys," *Vacuum*, 86 (9) (2012), 1296-1299.
5. Y.N. Qin, "Making high quality tin by electrolysis from high-antimony tin alloy" *China Nonferrous Metallurgy*, 2009, No. 4:74-76. (in Chinese)
6. C.G. Wei, "The distribution and comprehensive recovery of antimony in tin smelting process" *Tin Technology*, 4 (3) (2002), 39-44. (in Chinese)
7. T.Kato, et al, "Distillation of cadmium from uranium–plutonium–cadmium alloy," *J. Nucl. Mater.*, 340 (2005), 259-265.
8. A. Gopala, et al, "Process methodology for the small scale production of m6N5 purity zinc using a resistance heated vacuum distillation system," *Mater. Chem. Phys.*, 122 (2010), 151-155.
9. L. Zhan, et al, "Separating zinc from copper and zinc mixed particles using vacuum sublimation," *Sep. Purif. Technol.*, 68 (2009), 397-402.
10. S.T. Ali, et al, "Reduction of trace oxygen by hydrogen leaking during selective vaporization to produce ultra-pure cadmium for electronic applications," *Mater. Lett.*, 61 (2007), 1512-1516.
11. S.T. Ali, et al, "Preparation of high pure zinc for electronic applications using selective evaporation under vacuum," *Sep. Purif. Technol.*, 85 (2012), 178-182.
12. H.W. Yang, et al, "Calculation of phase equilibrium in vacuum distillation by molecular interaction volume model," *Fluid phase Equilib.*, 314 (2011), 78-81.
13. W. Chen, et al, "Research on the law of vacuum distillation of bismuth," J. Vac. Sci. Technol., 14 (4) (1994) 254-259. (in Chinese)
14. L.X. Kong, et al, "Application of MIVM for Pb-Sn system in vacuum distillation," *Metall. Mater. Trans. B*, 43 (2012), 1649-1656.
15. A.X. Wang, et al, "Process optimization for vacuum distillation of Sn-Sb alloy by response surface methodology ," *Vacuum*, 109 (2014) 127-134.
16. L. Li, et al, "Volatilization behavior of metals from silver-contained lead-antimony multi-component alloys by vacuum distillation" *Journal of Central South University: Science and Technology*, 43 (7) (2012), 2489-2494. (in Chinese)
17. Yongnian Dai, Bin Yang, *Vacuum Metallurgy for Non-Ferrous Metals and Materials* (Beijing, Metallurgical Industry Press, 2006), 21.
18. D.P. Tao, "A new model of thermodynamics of liquid mixtures and its application to liquid alloys," *Thermochim. Acta*, 363 (2000), 105-113.

7th International Symposium on High-Temperature Metallurgical Processing
Edited by: Jiann-Yang Hwang, Tao Jiang, P. Chris Pistorius, Gerardo R.F. Alvear F., Onuralp Yücel, Liyuan Cai, Baojun Zhao, Dean Gregurek, and Varadarajan Seshadri
TMS (The Minerals, Metals & Materials Society), 2016

Simulation of Solidification Microstructure of 30Cr2Ni4MoV Steel Ingot under Different Intensities of Mechanical Oscillation Condition

ShuangYu Du, JieYu Zhang*, Bo Wang, SenYang Qian, Jian Zhao

(State Key Laboratory of Advanced Special Steel, School of Materials Science and Engineering, Shanghai University, Shanghai 200072, China)

Keywords: Mechanical oscillation, Solidification microstructure, Numerical simulation

Abstract

During the solidification process, the flow pattern of the molten steel has significant influence on the formation of solidification microstructure. In present study, the solidification microstructure of 30Cr2Ni4MoV steel ingot under different intensities of mechanical oscillation(forced convection) was simulated by using ProCAST®software. The results showed that the increase of the intensity of oscillation could restrain the development of the columnar grains and refine the grains. As the intensity of oscillation increased, the proportion of columnar grains decreased and the proportion of equiaxed grains increased. When the intensity of oscillation was at a low level, the effect of refining grains could be improved obviously by increasing the intensity. However, as the intensity increased to over 0.833, the effect of refining grains would not be improved.

1 Introduction

In the situations of solidification process, the flow pattern of the molten steel has significant effect on the formation of solidification microstructure. As the forced convection is produced by adding mechanical oscillation, The fluid flow will change the temperature field and the solute distribution in front of the solidifying interface, which influences the final solidification microstructure. In this paper, to analyze how oscillation influences the final microstructure, the solidification microstructure of 30Cr2Ni4MoV steel ingot under different intensities of mechanical oscillation was simulated using ProCAST®software.

The ProCAST® is a software which could simulate the solidification process. The basis of the ProCAST® model is the Navier-Stokes equations for a Newtonian fluid. An enthalpy method is adopted to handle the phase transition problem during the solidification[1,2]. The ProCAST® provides a parameters setting of adding a rotation to the steel ingot as mechanical oscillation to simulation the casting under the forced convection condition. In ProCAST®, microstructure can be calculated by CAFE model which couples the finite element method (FEM) on the micro-scale of casting and

*Corresponding author.
E-mail address: zjy6162@staff.shu.edu.cn (Jieyu Zhang)

cellular automaton (CA) method on the micro-scale of the grains structure [3].

2 simulation parameters

2.1 Simulation boundary conditions

In this paper, the model of simulation was a steel ingot(100×150×180mm) made of 30Cr2Ni4MoV ultra-super-critical rotor steel (Tab.2-1) using ProCAST® software. The mold was made of 6mm thickness refractory alumina. With a consideration of calculation precision and calculation time, the mesh generation of computation domain was divided into 78006 nodes and 396206 elements.

Table.2-1 Chemical element content of low pressure $30Cr_2Ni_4MoV$

Composition	C	Si	Mn	S	P	Cr	Ni	Mo	V
Mass fraction,%	0.20	0.38	0.25	0.017	0.019	1.63	3.34	0.49	0.094

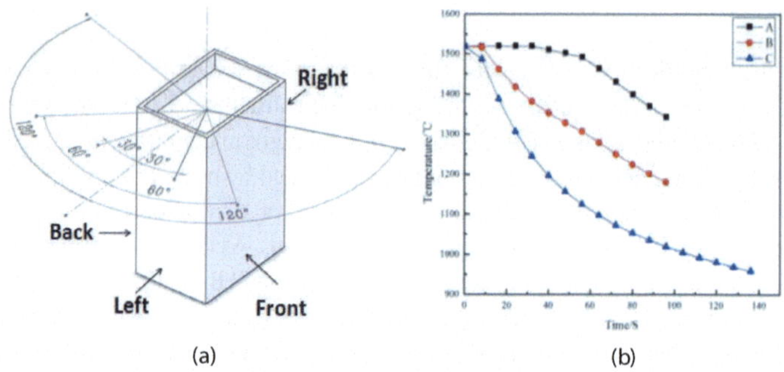

Fig.2-2 (a) Schematic of model used for simulation; (b) Temperature as a function of time;

Fig.2-2 is the schematic of model, temperature as a function of time. The cooling condition was slow cooling with an initial temperature of 1520℃. Each side of the mold was given temperature boundary condition in Fig.2-2(b). The left side of mold was assigned temperature curve A. The front side and the back side were assigned temperature curve B and the right side was curve C. Interfacial convective heat transfer coefficient between mold and sides of ingot was $200w/(m^2k)$, while heat transfer coefficient between mold and bottom of the ingot was $500w/(m^2k)$.

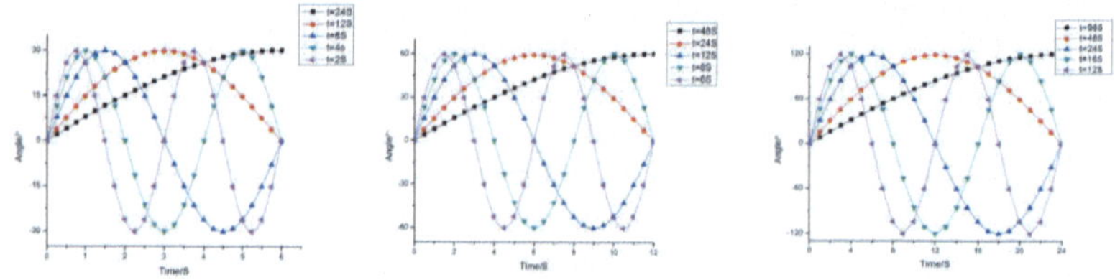

(a)A=30° (b)A=60° (c)A=120°

Fig.2-3 Rotation curves of the model

Table.2-4 parameters of mechanical oscillation intensities

test	A(°)	ω/(°/s)	T(s)	f(Hz)	f^2*A
01	120	5	96	0.0104	0.0130
02	60	5	48	0.0208	0.0260
03	30	5	24	0.0417	0.0521
04	120	10	48	0.0208	0.0521
05	60	10	24	0.0417	0.1042
06	30	10	12	0.0833	0.2083
07	120	20	24	0.0417	0.2083
08	60	20	12	0.0833	0.4167
09	120	30	16	0.0625	0.4688
10	30	20	6	0.1667	0.8333
11	120	40	12	0.0833	0.8333
12	60	30	8	0.1250	0.9375
13	60	40	6	0.1667	1.6667
14	30	30	4	0.2500	1.8750
15	30	40	3	0.3333	3.3333

The ingot and mold was simulated under the forced convection condition (addition of rotation)which were rotated by axle wire reciprocating motion as schematic shown in Fig.2-2(a) and the rotation curves were sine curves with three kinds of amplitude shown in Fig.2-3(a)(b)(c) but didn't stop rotating until the end of the solidification process. As J. Campbell [4] studied, f^2A was introduced to characterize the oscillation intensities. For the purpose of comparing the influence of oscillation intensity on solidification microstructure, the tests were listed in an ascending order of oscillation intensities(f^2A) in Table.2-4.

Tab.2-5 Nucleation parameters used in microstructure simulation

Nucleation parameter	$n_{s,max}$ /m^{-2}	$\Delta T_{s,max}$ /℃	$\Delta T_{s,\sigma}$ /℃	$n_{v,max}$ /m^{-3}	$\Delta T_{v,max}$ /℃	$\Delta T_{v,\sigma}$ /℃
Value	5×10^7	0.5	0.1	7×10^8	2.5	1

The Gibbs-Thompson coefficient is $\Gamma=3\times10^{-7}m\cdot k$. The coefficients of dendrite tip growth kinetics were calculated to be $\alpha=8.9\times10\text{-}6m\cdot s^{-1}k^{-2}$ and $\beta=4.5\times10\text{-}6m\cdot s^{-1}k^{-3}$ by ProCAST® software. Nucleation parameters are listed in Table.2-5.

2.2 Simulation domains of microstructure

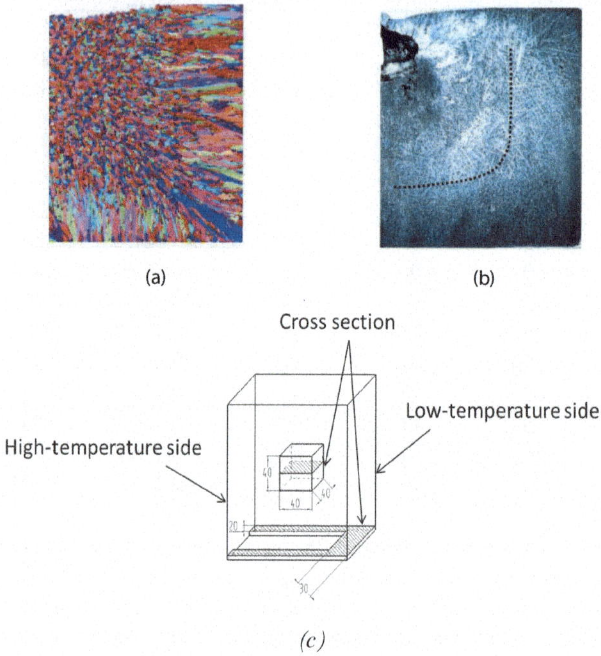

Fig.2-6 (a) Experimental result (b) Simulated result of Stationary(c)Simulation donmin

The microstructure and morphology predicted by a static model in Fig.2-6(a) could be compared with the features of the experiment in Fig.2-6(b). It was found that columnar grains zones were mostly grouped on the right wall surface and the lower bottom edges, While the equiaxed grains zones were mostly grouped in the central and the left top portions of the ingot. To better understand the influence of oscillation intensity on solidification microstructure, A 40*40*40mm cubic simulation domain shown in Fig.2-6(c) was selected from the ingot center as the equiaxed grains simulation domain, and a central cross-section of domain was taken for observing and analyzing. A U-zone, as shown in Fig.2-6(c), was selected from the bottom as the computation domain of columnar grains, and a cross-section 1mm away from the bottom surface was taken.

3 Results and discussion

3.1 Influence of oscillation intensities on microstructure and morphology

At initial solidification stage, a large undercooling was developed because of the contact between the molten steel and the coarse mold wall, resulting in quick formation of fine equiaxed grains on the surface. Due to the existence of a temperature gradient in parallel with the mold wall, those grains oriented perpendicularly to the mold wall grew the fastest, forming columnar grains zones near the wall. As the columnar grains growth proceeded, the solid-liquid interface front released crystallization latent heat, heating the temperature up; meanwhile, the heat dissipation along the direction perpendicular to the mold wall slowed down and the central temperature gradient decreased ,as a result, making the particles ready for nucleation. Columnar grains stopped growing when they came in contact with the central grains, hence an equiaxed grains zone was formed in the center of the ingot.

(1)(A=120°, ω=5°/s)

(2)(A=60°, ω=5°/s)

(3)(A=30°, ω=5°/s)

(4)(A=120°, ω=10°/s)

(5)(A=60°, ω=5°/s)

(6)(A=30°, ω=5°/s)

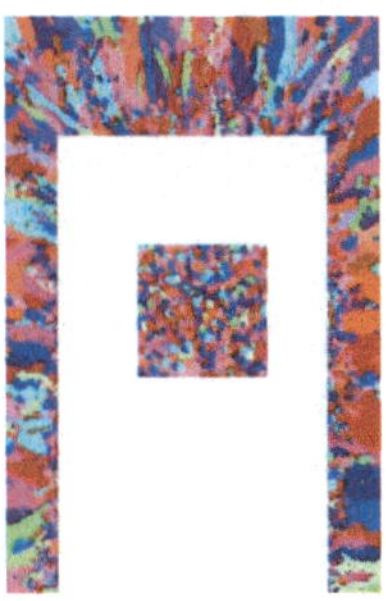

(7)(A=120°, ω=20°/s)

(8)(A=60°, ω=20°/s)

(9)(A=120°, ω=30°/s)

Fig.3-1 microstructure and morphology under different oscillation intensities

Fig.3-1 provides the 30Cr2Ni4MoV solidification microstructure and morphology under different oscillation intensities. It is clear from the Fig.3-1 that after introduction of forced convection, the bottom were still dominated by columnar grains whereas the central portion was an equiaxed grains growth zone. With the use of oscillation, the temperature fluctuation and breaking effect restrain the growth of columnar grains. Such breaking effect generated a lot of nuclei in the front of the columnar grains, as a result, equiaxed grains grew on the columnar grains zone front. As the oscillation intensity increased, the effect of restraining columnar grains became more appreciable and more equiaxed grains were formed in the columnar grains zone at the bottom. In test 11, most of the bottom columnar grains zone was occupied by equiaxed grains.

when the oscillation intensified further to over 0.833, the rotation frequency became excessively high and the molten steel collided frequently with the wall. The temperature fluctuation and the breaking effect introduced by forced convection was consequently lessened, the nucleus number on front of the columnar grains decreased and the equiaxed grains growth was subdued; in terms of microstructure and morphology, all this translated into a larger columnar grains zone and less equiaxed grains in the zone.

3.2 Influence of microstructure shape factor

We introduced φ (1)as a parameter that describes quantitatively the grains morphology. It plays an important role in exploring into the influence of different oscillation intensities on solidification structures.

$$\varphi = \frac{4\pi S}{P^2} \tag{1}$$

Where S is the area of the gains; P is the circumference; φ ranges between 0 and 1, when the closer the particles is with a circular, φ is more close to 1. When the φ is 1,it indicates that particle is circular.

Table.3-2 Shape factor φ_1: equiaxed grains; φ_2: columnar grains.

A(°)	ω/(°/s)	T(s)	f(Hz)	f^2 A	φ_1	φ_2
120	5	96	0.0104	0.0130	0.5291	0.4589
60	5	48	0.0208	0.0260	0.5333	0.4596
30	5	24	0.0417	0.0521	0.5350	0.4606
120	10	48	0.0208	0.0521	0.5388	0.4719
60	10	24	0.0417	0.1042	0.5516	0.4746
30	10	12	0.0833	0.2083	0.5531	0.4751
120	20	24	0.0417	0.2083	0.5540	0.4773
60	20	12	0.0833	0.4167	0.5543	0.4780
120	30	16	0.0625	0.4688	0.5587	0.4782
30	20	6	0.1667	0.8333	0.5627	0.4815
120	40	12	0.0833	0.8333	0.5855	0.4888
60	30	8	0.1250	0.9375	0.5834	0.4871
60	40	6	0.1667	1.6667	0.5734	0.4864
30	30	4	0.2500	1.8750	0.5679	0.4847
30	40	3	0.3333	3.3333	0.5634	0.4829

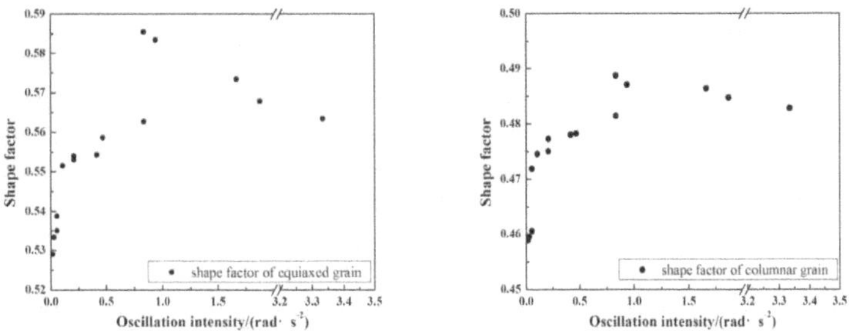

(a) (b)

Fig.3-3 shape factor (a) equiaxed grains (b) columnar grains

The Fig.3-3 is the shape factor of the equiaxed grains and columnar grains. It showed that oscillation intensities altered the shape factors of columnar and equiaxed grains, and the columnar grain shape factor coefficient was smaller than the shape factor coefficient of the equiaxed grains. At a weaker oscillation intensity, the oscillation intensity had a large influence on the shape factor, and the shape factor increased with an increasing oscillation intensity, indicating that the forced convection had a stronger stirring effect on the molten steel, as a result, the temperature fluctuation, breaking effect became more effective, which produced shorter columnar grains shown as a greater shape factor.

As the oscillation intensity was increased to over 0.833, the growth trend of the columnar grain with an increasing oscillation intensities became less noticeable, and the equiaxed grain shape factor dropped with an increasing oscillation intensities. This is because once the oscillation factor reached over 0.833, a lot of dynamic energy was dissipated due to frequent collisions between the molten steel and the steel ingot wall. the forced convection had a weak stirring on the molten steel, resulting in a smaller shape factor. it is clear from the Table.3-2 that for given a same oscillation intensities, the larger the oscillation amplitude (A) was given, the better stirring effect the forced convection had on the molten steel and the larger the shape factor was .

4 Conclusions

(1) The simulation using ProCAST®software can predict the microstructure evolution and the changes of grain morphology in solidification process under different intensities of mechanical oscillation.

(2) The increase of the of oscillation intensity can restrain the development of the columnar grains and refine the grains. With the increase of the intensities of oscillation, the proportion of columnar grains decrease while the proportion of equiaxed grains increase.

(3) Under the same intensity of mechanical oscillation, the refine effect is better when the rotation amplitude is bigger. When the intensity of oscillation was at a low level, the effect of refining grains can be improved obviously by increasing the intensity. However, when the intensity increase to over 0.833, the increase of the intensity of oscillation couldn't make the effect of refining grain better.

References

1. Incropera F P. *Fundamentals of Heat and Mass Transfer* (Beijing: Chemical Industry Press, 2007.)

2. Wenquan Tao. *Numerical Heat Transfer* (Xi'an: Xi'an Jiaotong University Press, 2001.)

3. LI Xiang-mei, ZHANG Jie-yu, WANG Bo, REN Zhong-ming, ZHOU Guo-zhi. "Simulation of stray grain formation during unidirectional solidification of IN738LC superalloy,"*Cent. South Univ. Technol.* (2011) 18: 23–28.

4. Campbell J. "Effects of vibration during solidification." *International Metals Reviews*, 1981, 26(1): 71-108.

7th International Symposium on High-Temperature Metallurgical Processing
Edited by: Jiann-Yang Hwang, Tao Jiang, P. Chris Pistorius, Gerardo R.F. Alvear F., Onuralp Yücel, Liyuan Cai, Baojun Zhao, Dean Gregurek, and Varadarajan Seshadri
TMS (The Minerals, Metals & Materials Society), 2016

PRODUCTION OF ZrB_2-B_4C COMPOSITE MATERIALS VIA SHS PROCESS

Kağan Benzeşik[1], Mehmet Buğdaycı[1,2], Ahmet Turan[2], Onuralp Yücel[1]

kaganbenzesik@gmail.com, mbugdayci@itu.edu.tr, yucel@itu.edu.tr

[1]Istanbul Technical University, Metallurgical and Materials Engineering, Maslak, Istanbul, 34469, Turkey

[2]Yalova University, Chemical and Process Engineering Department, Yalova, 77100, Turkey

Abstract

ZrB_2 and B_4C are one of the most significant advanced ceramics. They are being used in various industrial areas from space technology to nuclear industry owing to their physical properties such as high melting point, high thermal/electrical conductivity and low density. Present study was conducted in two main stages: Self-propagating high-temperature synthesis (SHS) reactions and leaching. ZrO_2, Carbon black, B_2O_3 were used as starting material to produce ZrB_2-B_4C powders by SHS. Also Mg powder was used as the reductant. The reaction mixtures (approximately 100 g) were mixed thoroughly 15 minutes in a turbula mixer and charged into Cu crucible and compacted. W (tungsten) wire was placed at the top of copper crucible and the reaction realized by passing current through the wire. After initiation, a highly exothermic reaction became in a self-sustaining mode and propagated throughout the SHS mixture. The samples were characterized by using AAS and X-Ray Diffraction.

Introduction

Self-propagating high temperature synthesis process (SHS) is a combustion synthesis process which presents some benefits such as high quality of production, low cost, low processing temperatures, low energy requirement, very short processing time and simple operation [1, 2]. In a SHS process, the ignition begins the combustion and it propagates throughout the reactant mixture yielding the desired product. However, the disadvantages of the process such as unreacted products due to undesirable reaction rates needs to be overcome by changing some parameters such as ignition temperature, particle size, additive, atmosphere etc. [3, 4]. Aim of the study is the production of ZrB_2-B_4C composite materials. Present study was conducted in two main stages: self-propagating high-temperature synthesis (SHS) reactions and leaching process. ZrO_2, B_2O_3, carbon black were used as starting materials to produce ZrB_2-B_4C composite materials via SHS.

Experimental Procedure

In the SHS experiments, a mixture of ZrO_2, B_2O_3 and Mg powders were used in order to produce high technology ceramic composite powders such as ZrB_2, B_4C and ZrB_2- B_4C. The metal oxide powders have over 97% purity and 150 μm average grain sizes.

Table I displays purity and grain size of powders. The advanced thermodynamical simulations of the reactions were investigated including different ratios of initial mixtures as well as different initial temperatures in order to reduce the number of experiments.

Table I. Purity and Grain Size of Raw Materials

Raw Materials	Purity, wt.%	Particle Size, μm
Mg	99.7	<150
H_3BO_3	99.5	-
B_2O_3	97.0	<53
ZrO_2	99.0	-
C	98.0	-

The reaction mixtures were mixed thoroughly 15 minutes in a turbula mixer and powder mixtures (approximately 100 g) were charged into Cu crucible and compacted. W (tungsten) wire was placed at the top of copper crucible and the reaction realized by passing current through the wire. After initiation, a highly exothermic reaction became in a self-sustaining mode and propagated throughout the SHS mixture. The obtained SHS products were discharged from the crucible after cooling. Experimental set-up of the SHS synthesis was illustrated schematically in Figure 1. After the SHS process the leaching step was started to eliminate Mg-containing-phases. First, ZrB_2-B_4C composite materials were produced via SHS, then optimum leaching conditions were defined. Experimental set-up of the leaching step was illustrated schematically in Figure 2.

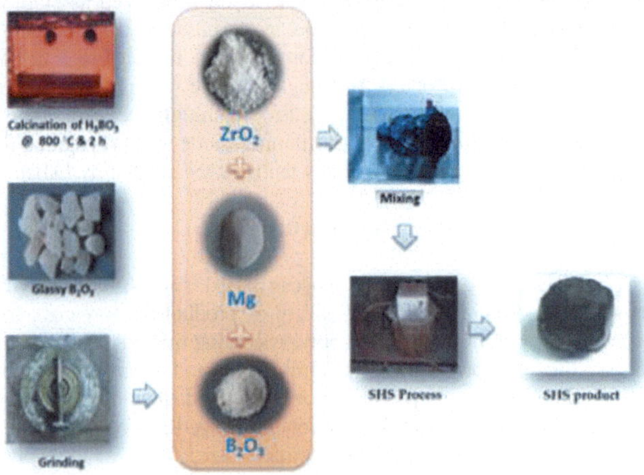

Figure 1. Flowchart of SHS experiments.

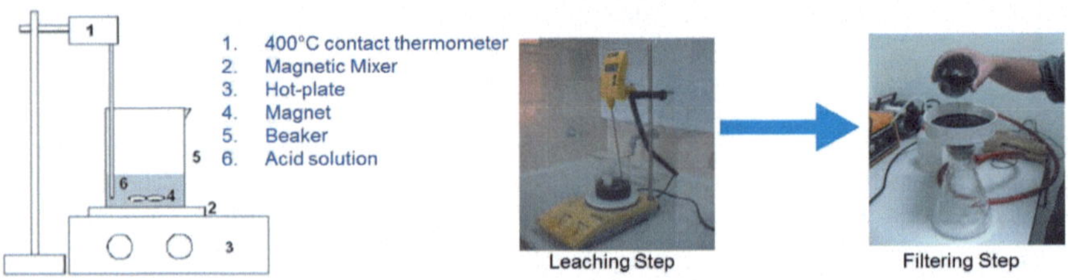

Figure 2. Flowchart of leaching experiments.

Results and Discussions

The first experimental set, ZrB_2-B_4C composite materials were produced via SHS. After the SHS process optimum leaching parameters were determined. Table II shows the initial conditions of powder mixtures and Figure 3 shows the XRD results of the SHS products.

Table II. Initial Conditions of Powder Mixtures

EXP. NO.	STOICHIOMETRY	INITIAL CONDITIONS			
		ZrO_2 [g]	B_2O_3 [g]	C [g]	Mg [g]
1	100% B_4C-0% ZrB_2	0.000	45.74	6.30	47.92
2	90% B_4C -10% ZrB_2	3.630	43.21	5.67	47.42
3	80% B_4C -20% ZrB_2	7.260	40.69	5.04	46.93
4	70% B_4C -30% ZrB_2	10.89	38.16	4.41	46.44
5	60% B_4C -40% ZrB_2	14.52	35.64	3.80	45.95
6	50% B_4C -50% ZrB_2	18.00	33.12	3.15	45.46
7	40% B_4C -60% ZrB_2	21.80	30.60	2.52	44.96
8	30% B_4C -70% ZrB_2	25.41	28.07	1.90	44.97
9	20% B_4C -80% ZrB_2	29.04	25.54	1.26	44.00
10	10% B_4C -90% ZrB_2	32.70	23.02	0.63	43.49
11	0% B_4C -100% ZrB_2	36.34	20.56	0.00	43.03

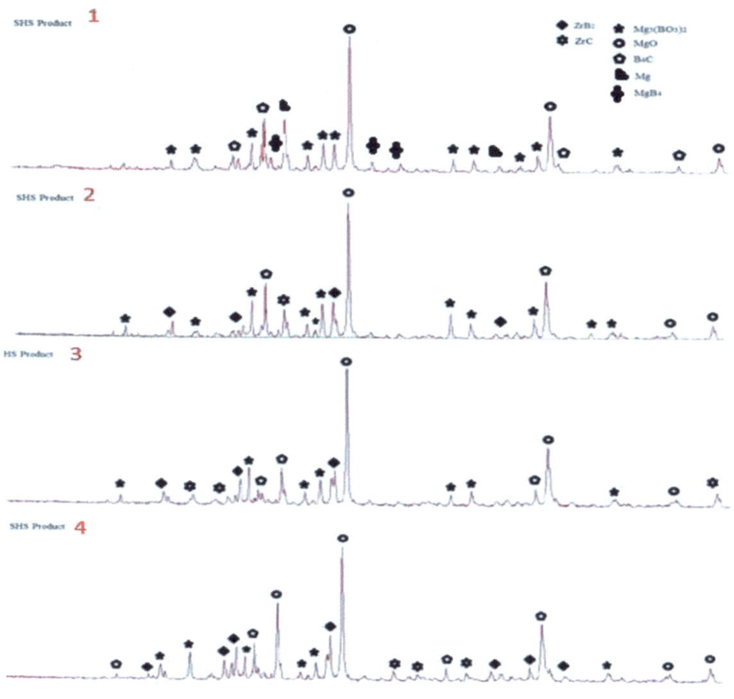

Figure 3. XRD Patterns of SHS Products

XRD results show many Mg phases, to eliminate these impurities two different leaching processes were carried out. In the leaching step, the SHS products were leached by HCl solution. Table III shows initial conditions of first type of leaching experiments, Table IV shows initial conditions of second type of leaching experiments, Figure 4 presents XRD results of leached products and Figure 5 presents comparison of XRD results between first type leached product and second type leached product.

Table III. Initial Conditions of First Type of Leaching Experiments

S/L ratio	1/5
Leaching temperature	Room Temperature
Leaching Time	60 Minutes
HCl Concentration	37% HCl

Table IV. Initial Conditions of Second Type of Leaching Experiments

S/L ratio	1/5
Leaching temperature	85 °C
Leaching Time	60 Minutes
HCl Concentration	37% HCl

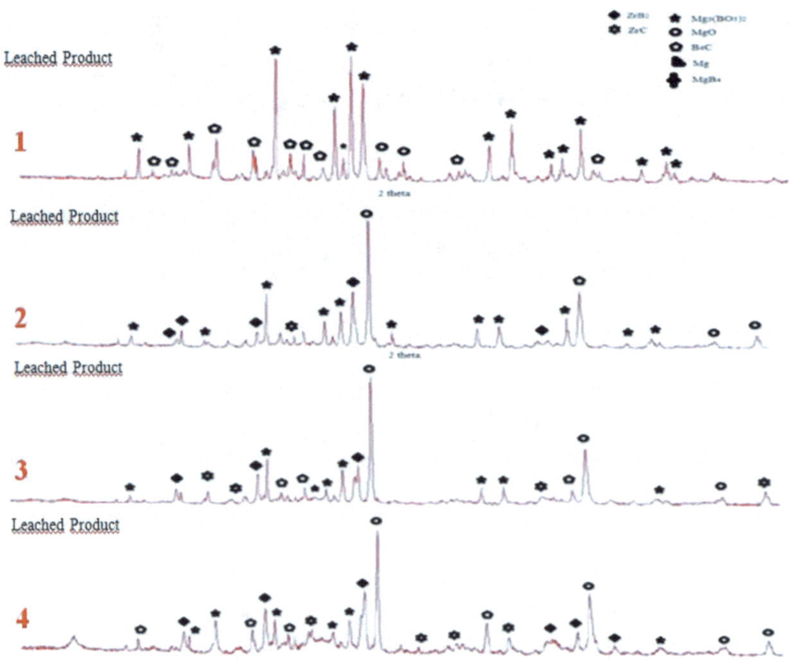

Figure 4. XRD Results of Leached Products

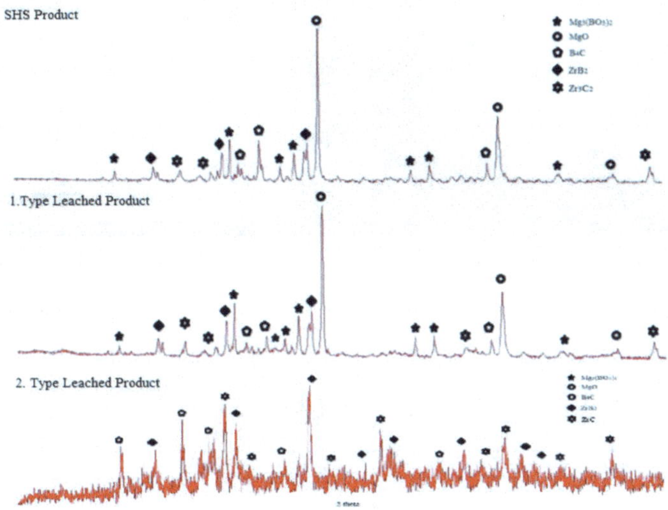

Figure 5. Comparison of XRD Results between First Type Leached Product and Second Type Leached Product

After SHS and leaching processes, all the products were analyzed chemically. Figure 6 presents Mg ratio of the products during the leaching step. Figure 7 shows the SEM images of the SHS product and leached product. Table V presents BET analysis of SHS product and leached product.

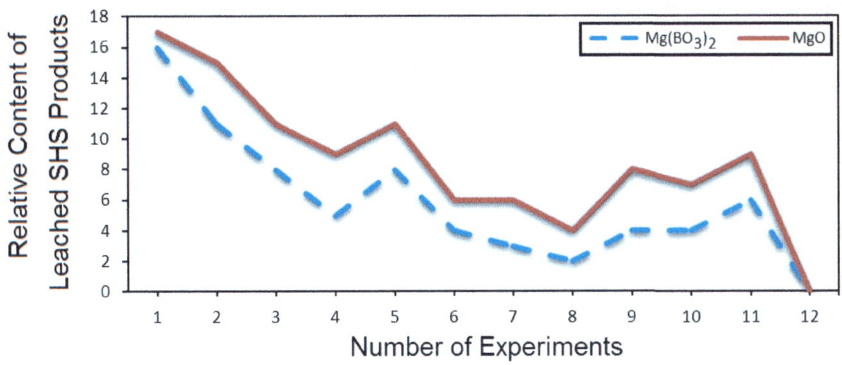

Figure 6. Mg Ratio of the Products

Figure 7. SEM Images of SHS Product (Left) and Leached Product (Right)

Table V. BET analysis of SHS Product and Leached product

BET Analysis of SHS Product	BET Analysis of Leached Product
1,04 m^2/g	2,20 m^2/g

Conclusions

SHS process was carried out to synthesis B_4C from B_2O_3-C, ZrB_2 from ZrO_2-B_2O_3, with magnesiothermic reduction under atmospheric atmosphere followed by HCl acid leaching. The XRD analysis showed that SHS product was composed of MgO, $Mg_3B_2O_6$ phases along with B_4C, or ZrB_2 phase. MgO and $Mg_3B_2O_6$ phases were successfully eliminated after the second type of leaching step.

References

1. O. Yücel, F. C. Sahin, A. Tekin, "The Preparation of Ferroboron and Ferrovanadium by Aluminothermic Reduction", High Temperature Materials and Processes, 15 (1-2) (1996) 103–106.

2. A. G. Merzhanov, "Self-propagating High-temperature Synthesis (SHS)", (ISMAN, Russia, 2002).

3. I. P. Borovinskaya, "Chemical Classes of the SHS Processes and Materials", Pure and Applied Chemistry, 64 (1992) 919-940.

4. V. I. Yukhvid, "Modifications of SHS Processes", Pure and Applied Chemistry, 64 (1992) 977-988.

7th International Symposium on High-Temperature Metallurgical Processing
Edited by: Jiann-Yang Hwang, Tao Jiang, P. Chris Pistorius, Gerardo R.F. Alvear F., Onuralp Yücel, Liyuan Cai, Baojun Zhao, Dean Gregurek, and Varadarajan Seshadri
TMS (The Minerals, Metals & Materials Society), 2016

RESEARCH ON MICROWAVE ROASTING OF ZnO AND APPLICATION IN PHOTOCATALYSIS

Qin Guo [1,2,3], Linqing Dai * [1,2,3], Shenghui Guo [1,2,3], Libo Zhang [1,2,3], Jinhui Peng [1,2,3]

[1]Kunming University of Science and Technology; Kunming 650093, China

[2]Key Laboratory of Unconventional Metallurgy, Ministry of Education; Kunming 650093, China

[3]Yunnan Provincial Key Laboratory of Intensification Metallurgy, Kunming 650093, China

Keywords: microwave roasting, ZnO, Photocatalytic, Cellulose

Abstract

This paper adopts a simple and novel preparation. Taking cellulose as a template, a tubular morphology structure of micro/nano ZnO crystals was obtained by microwave roasting at high temperature. The higher the microwave roasting temperature, the larger the degree of crystallization. The experiments indicate that the best photo catalytic effect can be obtained at 850℃: At this temperature, the oxidation of cellulose is complete. Large quantities of hexagonal shaped ZnO crystal can be observed, specifically the (0001) crystal face. Under ultraviolet-visible irradiation, taking long arc mercury lamp as light source, the material had good photocatalytic activity. The decomposing efficiency of methylene blue reaches 97% in 60min, which is superior to that of preparations without templates at the same temperature.

1. Introduction

Photocatalyst becomes important not only due to use in solar energy water decomposition hydrogen energy storage, but also in environmental pollution problems. They are used for development in many fields, such as photocatalytic hydrogen production[1], degradation of organic dyes[2], sterilization[3], the heavy metal ions in waste water treatment[4] and air purification[5] etc. The current mainstream of photocatalysts include titanium dioxide, zinc oxide, tin oxide, zirconium dioxide, cadmium sulfide and various oxide and sulfide semiconductor. Nitrides, phosphide and boride catalyst are reported, but they have less research. In the light of the existing catalyst, TiO_2 with its unique and excellent performance, is the most common green catalyst. In practice, the application of TiO_2, is nevertheless, not perfect. Main problem of restricting its development is its wide band gap, low efficiency of the use of solar energy, the high rate of light carrier recombination rate and the low quantum efficiency etc. Hence, it is imperative to carry on the positive exploration and research on the whole photocatalytic material system.

At present, ZnO and TiO_2 are comparable semiconductor photocatalytic materials.

Some studies showed that ZnO is more effective than TiO_2 in certain areas of photocatalytic degradation, such as: photocatalytic degradation of bleached pulp waste water [6], photosensitive oxidation of phenol [7] etc. Researchers have found that different exposure ratio (0001) of the micro/nano ZnO crystal is obtained by mild deposition method, which has good catalytic activity for thermal decomposition of ammonium perchlorate [9]. This preparation method, however, requires close control of the crystal growth.

2. Experiment

Experiment Method and Raw Material

This experiment adopts a novel preparation. Using cellulose as a template with zinc acetate, (0001) surface micro/nano ZnO was obtained by microwave roasting at high temperature. Not only is the method simple, but also it can yield the tubular structure of cellulose. Catalytic degradation of methylene blue, with long arc mercury lamp as light source, showed that with this method of synthesis of ZnO its photocatalytic performance is better than that of commercial ZnO and of ZnO prepared without cellulose template.

Raw Material: absorbent cotton, zinc acetate, ZnO, anhydrous ethanol, methylene blue.

Specific experiments were designed as follows, first, immerse 15g cotton, washed with anhydrous ethanol and dried, in 2mol/l zinc acetate, leave 12 hours, take out it and put into a corundum crucible , put into a drying oven, dried 12h at 50℃, take 7 dry samples and 5g each, microwaved in the microwave high temperature equipment, conduct XRD and SEM analysis of the product after it cools.

The calcination temperature set to 700℃, 750℃, 800℃, 850℃, 900℃, 950℃ and 1000℃, respectively. To partition the process of microwave roasting, calcination temperature can be divided into three stages. The first stage, before the T(Final Temperature)-200, of the heating rate is 50℃/min, when reach the 200℃ away from final temperature, and keep 30 min. The second stage, between T(Final Temperature)-200 and T(Final Temperature)-100, of the heating rate is 20 ℃/min, and keep 30 min. when reach the final temperayure with the rate of 20 ℃/min , keeping 60 min with furnace cooling to room temperature.

3. Results and Discussion

XRD Pattern of Roasted Product

This paper reports the XRD pattern of microwave roasted product at different temperatures, result shown in Figure 1.

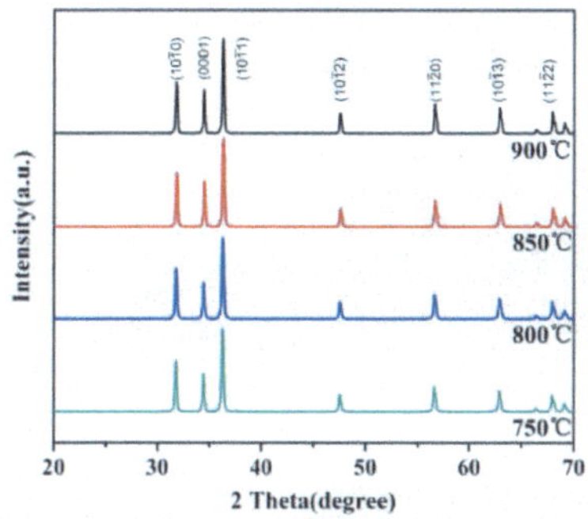

Figure 1. XRD Pattern of Roasted Product

We can see from Figure 1, the different position of the diffraction peaks is intactly match the lattice constant of a = 0.3249 nm, c = 0.5206 nm hexagonal wurtzite, comparing with the standard spectra of JCPDS 36-1451. The diffraction peak is very sharp, which shows that the sample of ZnO has a high crystallinity. As showed in Figure 1, once its microwave roasted temperature becomes higher, its diffraction intensity of (1011) increase, and in each individual pattern (1011) diffraction peak intensity is the strongest. At 850℃ the diffraction peak of (0001) is slighter higher than other temperature, which shows that the exposure quantity in this temperature is more than at other temperatures.

SEM Image of Roasted Product

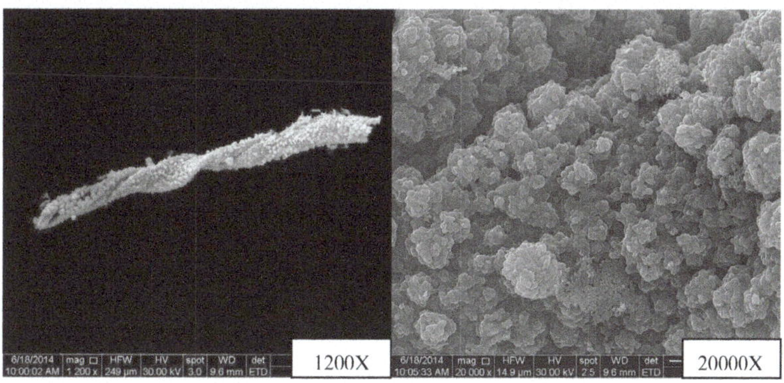

Figure 2. SEM Image of ZnO Prepared at 750℃

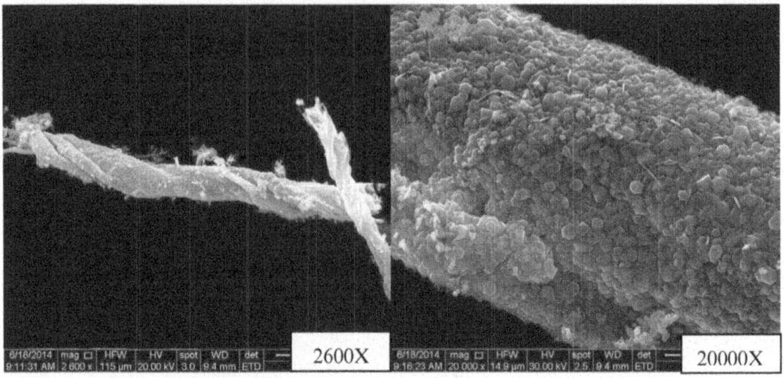

Figure 3. SEM Image of ZnO Prepared at 850℃

Figure 4. SEM Image of ZnO Prepared at 900℃

We can see in Figure 2, Figure 3 and Figure 4, when the microwave calcination temperature is 750 ℃, 850 ℃, 900 ℃ respectively, taking cotton as template, SEM images of the prepared ZnO. The images at left show lower magnification. It is confirmed that the preparation of ZnO with tubular, absorbent cotton as template, is feasible. The higher magnification images at right show that, when the calcination temperature is 850 ℃, ZnO surface crystallization is significantly better than at 750 ℃ and 900 ℃. The ZnO clearly shows hexagonal wurtzite crystal for 850 ℃, with (0001) facets. Others researchers have found that the ZnO crystals with (0001) facets can improve the photocatalytic properties of ZnO photocatalytic material [10].

<u>The Test of Photocatalytic Activity</u>

A long arc mercury lamp, main wavelength 365 nm, as catalytic light, was used to degrade methylene blue (MB), to evaluate the photocatalytic properties. Specific experiments are designed as follows. First, add 0.048g catalyst in 25mg/l methylene blue solution, after taking the Eeo sample, a long arc mercury lamp light source, for 100 min, taking 5 ml solution of the reaction of catalytic and methylene blue every

10mins. The concentration of methylene blue solution was determined by spectrophotometry.

The physical figures and degradation curves, showing the effect of ZnO from different calcination temperatures on the degradation of MB, are shown in Figure 5 and Figure 6 respectively.

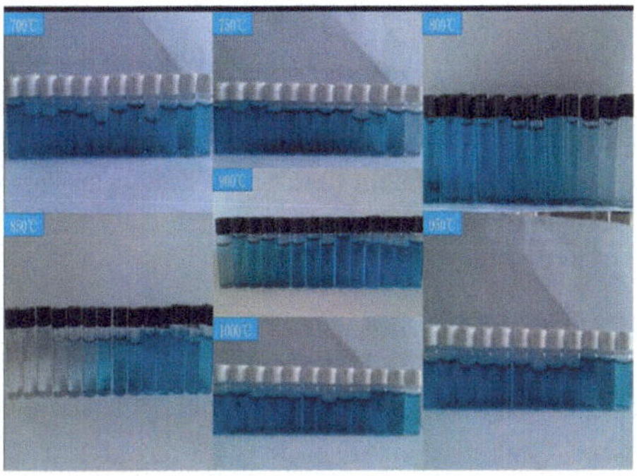

Figure 5. Physical figures of ZnO from different calcination temperature on the degradation of MB

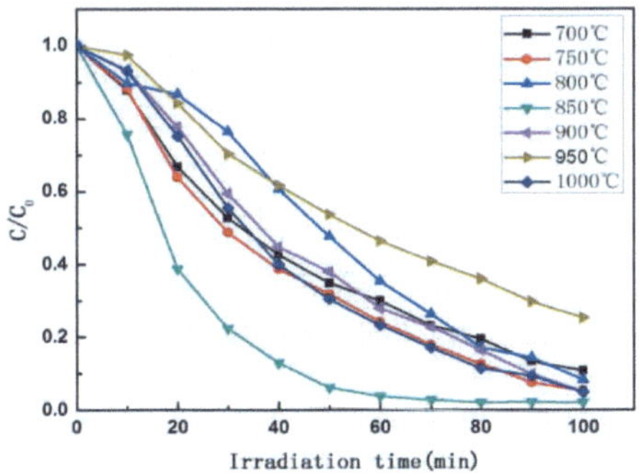

Figure 6. Degradation curves of ZnO from different calcination temperature on the degradation of MB

From figure 5 and 6 we can see that, in the dark reaction stage, material for methylene blue (MB) has no chemical reaction, also does not have the adsorption stripping effect. The material itself does not adsorb methylene blue. Moreover, the pores in the absorbent cotton template are too large for significant adsorption capacity. Microwave calcined material heated to 850 ℃ has better catalytic activity. The degradation of methylene blue reached 97% in 60mins. The characterization results of the analysis is consistent with previous results. It appears that exposure of (0001) facets plays a role. It is confirmed that ZnO (0001) crystal facets really can improve the photocatalytic activity of ZnO photocatalytic materials.

4. Conclusion

This work, with absorbent cotton as template, used high temperature calcination to produce ZnO with cotton fiber structure and tested its photocatalytic performance. After microwave calcination at 850℃, the catalytic activity for photocatalytic degradation of methylene blue is best. Absorbent cotton as a template for ZnO photocatalytic performance played a positive role. After 850 ℃ microwave calcination treatment, ZnO crystals on the surface of the tubular structure exposed a number of (0001) facets, which play a vital role to promote the photocatalytic activity of ZnO. This method is more convenient to improve ZnO catalytic activity.

References

[1] Fujishima A, Hond K. Electrochemical Photolysis of Water at a Semiconductor Electrode [J]. Nature, 1972, 238: 37-38.

[2] Denis D, Nikodem T, Han M Y, et al. Reversible Phase Transfer of (CdSe/ZnS) Quantum Dots between Organic and Aqueous Solutions [J]. ACS Nano, 2009, 3(3): 661-667.

[3] Han Z, Ren L, Cui Z, et al. Ag/ZnO flower heterostructures as a visible-light driven photocatalyst via surface plasmon resonance [J]. Applied Catalysis B : Environmenta, 2012, 126(25): 298-305.

[4] Kikuchi Y, Qian Q R, Machida M, et al. Effect of ZnO loading to activated carbon on Pb(II)adsorption from aqueous solution [J]. Carbon, 2006, 44: 195-202.

[5] Park N K, Lee Y J, Han G B, et al. Synthesis of various zinc oxide nanostructures with zinc acetate and activated carbon by a matrix-assisted method [J]. Colloids and Surfaces A–Physicochemical and Engineering Aspects, 2008, 66: 313-314.

[6] Li Y Z, Xie W, Hu X L, et al. Comparison of Dye Photodegradation and its Coupling with Light-to-Electricity Conversion over TiO2 and ZnO [J]. Langmuir, 2010, 26(1): 591–597.

[7] Khodja A A , Sheili T, Pihichowski J F, et al. Photocatalytic degradation of 2–phenylphenol on TiO2 and ZnO in aqueous suspensions [J]. Journal of Photochemistry and Photobiology A: Chemistry, 2001, 141: 231-239.

[8] Serpone N, Maruthamuthu P, Pichat P, et al. Exploiting the interparticle electron transfer process in the photocatalysed oxidation of phenol, 2–chlorophenol and pentachlorophenol: chemical evidence for electron and hole transfer between coupled semiconductors [J]. Journal of Photochemistry and Photobiology A: Chemistry, 1995, 85: 247-255.

[9] Tang G, Tian S, Zhou Z, et al. ZnO Micro/Nanocrystals with Tunable Exposed (0001) Facets for Enhanced Catalytic Activity on the Thermal Decomposition of Ammonium Perchlorate [J]. The Journal of Physical Chemistry C, 2014, 118(22): 11833-11841.

[10] Zhu Z, Zhou J. Rapid growth of ZnO hexagonal tubes by direct microwave heating [J]. International Journal of Minerals, Metallurgy, and Materials, 2010, 17(1): 80-85.

7th International Symposium on High-Temperature Metallurgical Processing

Fundamental Research of Metallurgical Process

Session Chairs:
Gerardo R.F. Alvear F.
Lifeng Zhang

7th International Symposium on High-Temperature Metallurgical Processing
Edited by: Jiann-Yang Hwang, Tao Jiang, P. Chris Pistorius, Gerardo R.F. Alvear F., Onuralp Yücel, Liyuan Cai, Baojun Zhao, Dean Gregurek, and Varadarajan Seshadri
TMS (The Minerals, Metals & Materials Society), 2016

REDUCTION KINETICS OF HEMATITE CONCENTRATE PARTICLES BY CO+H_2 MIXTURE RELEVANT TO A NOVEL FLASH IRONMAKING PROCESS

Yousef Mohassab[1], Feng Chen[1,2], Mohamed Elzohiery[1], Amr Abdelghany[1], Shengqin Zhang[1,3], and Hong Yong Sohn[1]

[1] Department of Metallurgical Engineering, University of Utah, Salt Lake City, Utah 84112, USA
[2] School of Minerals Processing and Bioengineering, Central South University, Changsha, Hunan 410083, China
[3] School of Metallurgy and Materials Engineering, Chongqing University of Science and Technology, Chongqing 401331, China

Keywords: Hematite, CO+H_2 mixtures, Reduction kinetics, Flash ironmaking

Abstract

The kinetics of hematite concentrate reduction by mixtures of hydrogen and CO of various compositions has been investigated as part of the development of a flash ironmaking process at the University of Utah. This process produces iron directly from iron oxides concentrates by the gas-solid flash reaction based on the partial oxidation of natural gas, resulting in a significant reduction in energy consumption and greenhouse gas emission. The reduction kinetics of hematite concentrate of an average particle size 21.3 μm by the above mentioned gases in the temperature range 1423 to 1623 K (1150 to 1350 °C) was investigated. Hematite concentrate particles can be reduced to > 90% by any of these reductants in several seconds of residence time typically available in a flash reactor. The activation energy ranged from 214 kJ/mol for hydrogen to 231 kJ/mol for CO.

Introduction

The high energy consumption and CO_2 emissions in the conventional blast furnace process have given rise to the need for alternative ironmaking technologies. Although the blast furnace has high production rates among other advantages, it faces problems arising from high energy consumption and CO_2 emissions due to the use of coke and the need for sintering, pelletization and cokemaking. At the University of Utah, a novel flash ironmaking process is under development [1-10], in which iron ore concentrate is flash-reduced by gaseous reductants at temperatures above 1473 K (1200 °C). This flash ironmaking is more energy efficient and drastically reduces the emissions of carbon dioxide based on the elimination of sintering, pelletization and cokemaking steps required for the blast furnace process. Although hydrogen is the cleanest and most efficient reductant, other gaseous reductants can be used as well in the flash ironmaking process. Reformed natural gas and coal gas, which are mainly composed of CO and H_2, are also possible reductants in this novel process.

Most of the previous work [11-17] on the reduction of iron oxides was conducted at temperatures lower than the minimum temperature, 1473 K (1200 °C), expected for the new flash process. Furthermore, the majority focused on the reduction of hematite samples with particle size larger than 100 μm, which is significantly larger than the concentrate particles for the novel flash process. Additionally, in the rapid reduction of fine hematite particles at high temperatures, the reduction does not proceed by forming the next lower oxides in order; rather, a combination of hematite, magnetite, wustite and iron is present at any given moment.

There had been little work on the gaseous reduction of iron oxide concentrates before the conception of the new flash ironmaking technology. As an integral part of the novel flash ironmaking process, the goal of this research was to perform a systematic measurement of the hematite concentrates particles reduction kinetics by mixtures of H_2 and CO of various compositions and derive the rate expressions in the temperature range 1423 to 1623 K (1150 °C to 1350 °C). Rate equations that contain the effects of process variables in a flash ironmaking reactor, including temperature and partial pressures of reductant gases, were developed.

Experimental Work

The experimental apparatus, the materials, and the detailed procedures have been described elsewhere [9, 10], the summary of which is presented in this paper. Hematite concentrate used in this study was from the Yuanjiacun Range, Shanxi Province, China, with average particle size of 21.3 m. About 30 mass pct magnetite was originally contained in the ore and then removed by magnetic separation to separate hematite particles. The chemical composition of the magnetically concentrated hematite sample (about 88 wt pct hematite and 3.4 wt pct magnetite) is presented in Table I.

Table I. Chemical Composition of Hematite Concentrate Used in This Work

Component	Total Fe	FeO in Fe_3O_4	SiO_2	Al_2O_3	K_2O	Na_2O	P	S
wt pct	63.98	1.05	5.53	0.82	0.06	0.02	0.03	0.01

A drop-tube reactor system consisted of a vertical tubular furnace housing an alumina tube, a pneumatic powder feeder, gas delivery lines, a powder cooling and collecting system, and an off-gas burning system, was used in this work to reduce hematite concentrate particles. The isothermal zone (L_{iso}) where the temperature was maintained within ±20 K started at the bottom of cylindrical honeycomb and extended 91 cm downward. When the target temperature was reached, reductant gas (hydrogen/carbon monoxide mixtures) and nitrogen were charged into the reactor at predetermined flow rates after the system was appropriately purged by nitrogen. After a thorough reductant gas leak test, the hematite particles were fed into the reactor tube. After a certain amount of the concentrate particles were supplied into the reactor, the syringe pump was stopped and the vibrator was kept running for 3 minutes longer to discharge the powder in the delivery lines. Then, the flow of the reductant gases was stopped and the flow rate of nitrogen increased to purge the whole system for 15 to 20 minutes. The valve of the powder collection bin was closed and the collector was disconnected from the reactor and quenched by water to room temperature to cool the reduced powder to avoid the re-oxidation when the fine particles contact with air. The reduced sample was transferred into a vial sealed with a cap for further analysis by ICP, SEM, and XRD.

Previous work [9, 10] has described the definition of parameters, including the degree of reduction (X), the residence time (τ) and pct excess reductant gas. It is noted that the barometric pressure at Salt Lake City is 0.85 *atm* (1 *atm* = 101.32 *kPa*), which was used in all the experimental design and analysis in this work. In addition, pct excess H_2 or CO is > 500 in the reduction experiments by single gaseous reductant, and > 300 pct in the reduction experiments by $CO+H_2$ mixture.

Results

Effect of Experimental Variables

Reduction Temperature: Previous work [18, 19] indicated that the magnetite reduction rate by hydrogen was fast enough in this novel flash system when the temperature was higher than 1373 K (1100 °C). The experiments were carried out under 1423 K (1150 °C) to 1623 K (1350 °C) in this work to examine the effect of reduction temperature on the reduction degree of hematite concentrate particles.

Reductant Gas Partial Pressure: It was concluded from previous work [19, 20] that high reduction degree of magnetite concentrate particles could be obtained within a short residence time in this novel flash reduction system. N_2 was added to change the partial pressure of reductant gas.

Hematite Reduction Kinetics by Single Reductant Gas

Hematite concentrate reduction kinetics by CO in the temperature range 1473 to 1623 K (1200 to 1350 °C) and that by H_2 in the temperature range 1423 to 1623 K (1150 to 1350 °C) were investigated by Sohn and coworkers [10-11]. It was found that within a few seconds of residence time, a reduction degree of over 90 pct was achieved using either CO or H_2 at temperatures above 1573 K (1300 °C), as shown in Figures 1 and 2.
The nucleation and growth rate equation with the Avrami parameter n = 1.0 adequately described both the carbon monoxide and hydrogen reduction kinetics of hematite concentrate particles, as detailed elsewhere [9-10]. In addition, the dependence with respect to gas partial pressure was 1st-order for both CO and H_2. The activation energy ranged from 214 kJ/mol for hydrogen to 231 kJ/mol for CO. The following complete rate equations (Eq. [1] and [2]) are, respectively, the hematite kinetics rate expressions separately by CO and H_2:

$$\left(\frac{dX}{dt}\right)_{CO} = 1.91\times10^{7}\times e^{\frac{-231000}{RT}}\times\left(pCO-\frac{pCO_2}{K}\right)\times(1-X) \qquad [1]$$

$$\left(\frac{dX}{dt}\right)_{H_2} = 4.41\times10^{7}\times e^{\frac{-214000}{RT}}\times\left(pH_2-\frac{pH_2O}{K}\right)\times(1-X) \qquad [2]$$

where X is the fraction of oxygen removed from iron oxide (degree of reduction), R is 8.314 J/mol·K, T is in K, p is in atm, and t is in seconds.
The agreement between the calculated reduction degrees vs time using the developed rate equation and the experimental results is satisfactory, as shown in Figures 3 and 4.

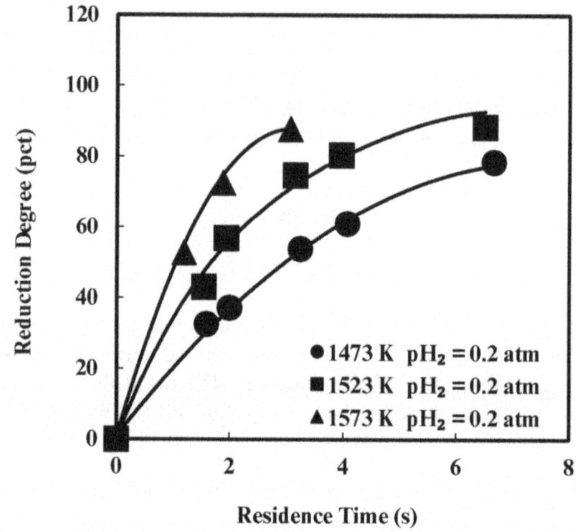

Figure 1. Reduction degree of hematite by H_2 *vs* residence time. The solid lines represent the calculated reduction degree using the developed rate expression, Eq. [1]. (pct excess H_2 > 500 in all experiments.)

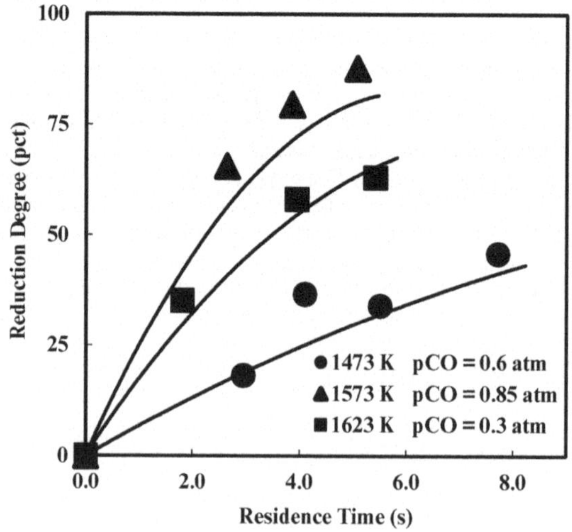

Figure 2. Reduction degree of hematite by CO *vs* residence time. The solid lines represent the calculated reduction degree using the developed rate expression, Eq. [2]. (pct excess CO > 500 in all experiments.)

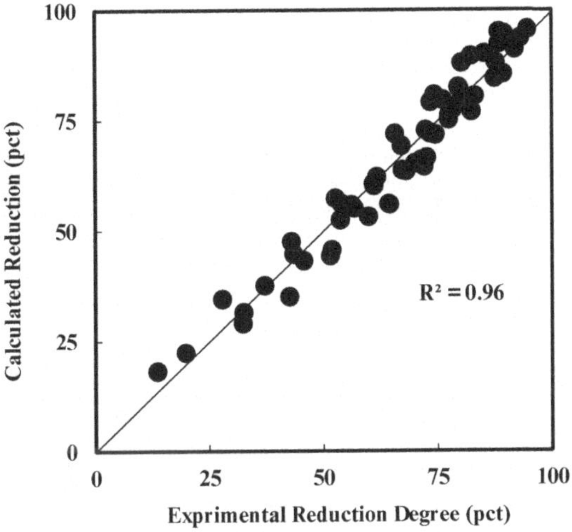

Figure 3. Comparison between the calculated reduction degree (using Eq. [1]) and the experimental results for all the runs made in the reduction by H_2.

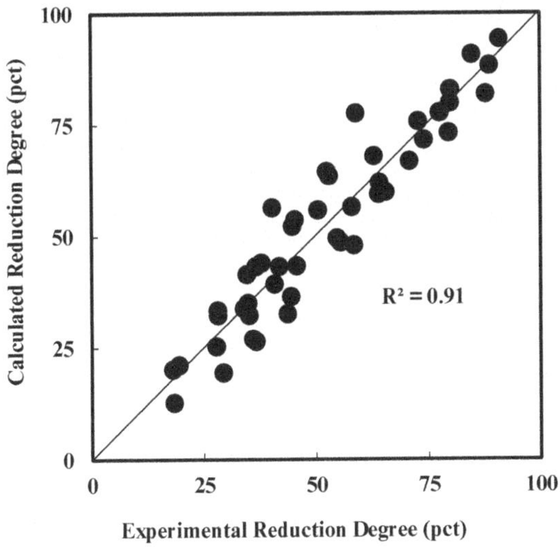

Figure 4. Comparison between the calculated reduction degree (using Eq. [2]) and the experimental results for all the runs in the reduction by CO.

Hematite Reduction Kinetics by $CO+H_2$ Mixtures

Hematite reduction kinetics by $CO+H_2$ mixtures, was conducted to represent the case of using natural gas or coal gas as the reductant/fuel in the flash ironmaking process. The hematite reduction rate by $CO+H_2$ mixtures is complicated due to the simultaneous reduction by the two reductants. It was found that within a few seconds of residence time, a reduction degree of over 90 pct was achieved using either $CO+H_2$ at temperatures above 1573 K (1300 °C), as shown in Figure 5. It is also clear that adding CO to the H_2 boosts the reduction kinetics as compared with the reduction by a single gas H_2 or CO. It was found, however, that increasing CO partial pressure from 0.05 to 0.2 atm while holding hydrogen partial pressure at 0.2 atm did not affect the reduction kinetics of hematite concentrate, as Figure 5 shows. More reduction experiments by $CO+H_2$ mixtures are being conducted.

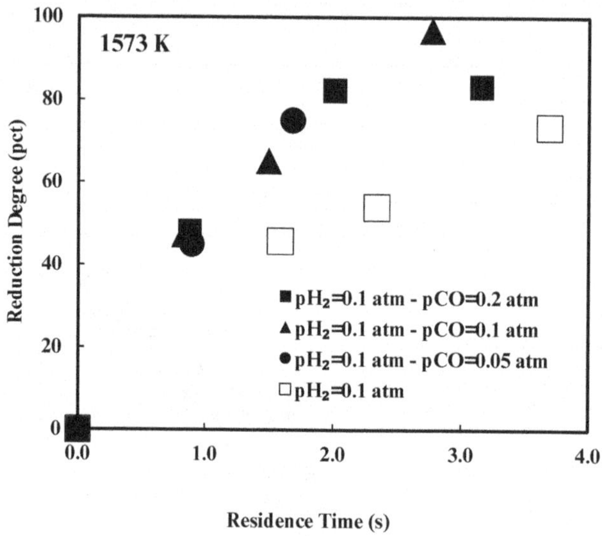

Figure 5. Reduction degree comparisons between single reductant gas (H_2/CO) and $CO+H_2$ mixtures at 1573 K (1300 °C).

Conclusions

The reduction kinetics of hematite concentrate particles of an average particles size 21.3 μm by gaseous reductants, including H_2, CO and their mixtures, were investigated. The dependences of rate on partial pressure of gaseous reductant, residence time and temperature were determined. The reduction temperature ranges of hematite concentrate particles by H_2, CO and their mixtures were 1423 to 1623 K (1150 to 1350 °C).

The results clearly indicated that the hematite concentrate can be reduced to greater than 90 pct by H_2, CO and their mixtures in the several seconds of residence time typically available in a flash reactor. The nucleation and growth rate equations with an Avrami parameter $n = 1$ well describe the kinetics of hematite reduction by H_2 and CO. The reduction rate has a 1st-order dependence on the partial pressure of reductant gas, including H_2 and CO. The activation energy of carbon monoxide reduction of hematite concentrate was 231 kJ/mol in contrast to 214 kJ/mol for the hydrogen reduction. Complete rate equations were developed respectively to satisfactorily represent the reduction kinetics of hematite concentrate particles by H_2 and CO and are suitable for the design of a flash reactor.

Acknowledgments

The authors thank Omar Kergaye, Andrew Laroche, Caio Melo, and Tuvshinbat Ganbat for the help with the experimental runs and the analytical work using ICP. The authors acknowledge the financial support from the U.S. Department of Energy under Award Number DEEE0005751 with cost share by the American Iron and Steel Institute (AISI) and the University of Utah.
Disclaimer: This report was prepared as an account of work sponsored by an agency of the United States Government. Neither the United States Government nor any agency thereof, nor any of their employees, makes any warranty, express or implied, or assumes any legal liability or responsibility for the accuracy, completeness, or usefulness of any information, apparatus, product, or process disclosed, or represents that its use would not infringe privately owned rights. Reference herein to any specific commercial product, process, or service by trade name, trademark, manufacturer, or otherwise does not necessarily constitute or imply its endorsement, recommendation, or favoring by the United States Government or any agency thereof. The views and opinions of authors expressed herein do not necessarily state or reflect those of the United States Government or any agency thereof.

References

1. M.Y. Mohassab-Ahmed and H. Y. Sohn, *Method and Device for Digestion of Materials in a Microwave Oven*, 2012, US Patent App. 61/651.
2. M.Y. Mohassab-Ahmed, H. Y. Sohn, "Effect of Water Vapor Content in H_2–H_2O–CO–CO_2 Mixtures on the Equilibrium Distribution of Manganese between CaO–MgO_{sat}–SiO_2–Al_2O_3–FeO–P_2O_5 Slag and Molten Iron." *Steel Res. Int.*, 85 (2014), 875-884.
3. M.Y. Mohassab Ahmed, "*Phase Equilibria between Iron and Slag in CO/CO_2/H_2/H_2O Atmospheres Relevant to a Novel Flash Ironmaking Technology*" (PhD Dissertation, The University of Utah, 2013).
4. M.Y. Mohassab-Ahmed, H. Y. Sohn, "Effect of Water Vapor Content in H_2-H_2O-CO-CO_2 Mixtures on the Activity of Iron Oxide in Slags Relevant to a Novel Flash Ironmaking Technology." *Ironmaking Steelmaking*, 41(2014), 665 - 675.
5. Y. Mohassab and H. Sohn, "Effect of Water Vapour on Distribution of Phosphorus between Liquid Iron and MgO Saturated Slag Relevant to Flash Ironmaking Technology." *Ironmaking Steelmaking*, 41 (2014), 575-582.

6. Y. Mohassab and H.Y. Sohn, "Effect of Water Vapor on Sulfur Distribution between Liquid Fe and MgO-Saturated Slag Relevant to a Flash Ironmaking Technology." *Steel Res. Int.*, 86 (2014), 753–759.
7. Y. Mohassab and H.Y. Sohn, "Analysis of Slag Chemistry by FTIR-RAS and Raman Spectroscopy: Effect of Water Vapor Content in H_2-H_2O-CO-CO_2 Mixtures Relevant to a Novel Green Ironmaking Technology." *Steel Res. Int.*, 86 (2014), 740-752.
8. M.Y. Mohassab-Ahmed, H.Y. Sohn, and L. Zhu, "Effect of Water Vapour Content in H_2-H_2O-CO-CO_2 Mixtures on MgO Solubility in Slag under Conditions of Novel Flash Ironmaking Technology." *Ironmaking Steelmaking*, 41 (2014), 575-582.
9. F. Chen, Y. Mohassab, T. Jiang, and H.Y. Sohn, "Hydrogen Reduction Kinetics of Hematite Concentrate Particles Relevant to a Novel Flash Ironmaking Process." *Metall. Mater. Trans. B*, 46 (2015), 1133-1145.
10. F. Chen, Y. Mohassab, S. Zhang, and H.Y. Sohn, "Kinetics of the Reduction of Hematite Concentrate Particles by Carbon Monoxide Relevant to a Novel Flash Ironmaking Process." *Metall. Mater. Trans. B*, 46 (2015), 1716-1728.
11. I.-J. Moon, C.-H. Rhee, and D.-J. Min, "Reduction of Hematite Compacts by H_2-CO Gas Mixtures." *Steel Res.*, 69 (1998), 302-306.
12. A. Bonalde, A. Henriquez, and M. Manrique, "Kinetic Analysis of the Iron Oxide Reduction Using Hydrogen-Carbon Monoxide Mixtures as Reducing Agent." *ISIJ Int.*, 45 (2005), 1255-1260.
13. L.-y. Yi, Z.-c. Huang, H. Peng, and T. Jiang, "Action Rules of H_2 and CO in Gas-Based Direct Reduction of Iron Ore Pellets." *J. of Central South University*, 19 (2012), 2291-2296.
14. M. Kazemi, B. Glaser, and D. Sichen, "Study on Direct Reduction of Hematite Pellets Using a New Tg Setup." *Steel Res. Int.*, 85 (2014), 718-728.
15. E.R. Monazam, R.W. Breault, and R. Siriwardane, "Kinetics of Hematite to Wüstite by Hydrogen for Chemical Looping Combustion." *Energy Fuels*, 28 (2014), 5406-5414.
16. K. Piotrowski, K. Mondal, T. Wiltowski, P. Dydo, and G. Rizeg, "Topochemical Approach of Kinetics of the Reduction of Hematite to Wüstite." *Chem. Eng. J.*, 131 (2007), 73-82.
17. H.S. Ray and N. Kundu, "Thermal Analysis Studies on the Initial Stages of Iron Oxide Reduction." *Thermochim. Acta*, 101 (1986), 107-118.
18. M.E. Choi and H.Y. Sohn, "Development of Green Suspension Ironmaking Technology Based on Hydrogen Reduction of Iron Oxide Concentrate: Rate Measurements." *Ironmaking Steelmaking*, 37 (2010), 81-88.
19. M.E. Choi, "*Suspension Hydrogen Reduction of Iron Ore Concentrate*" (PhD Dissertation, The University of Utah, 2013).
20. J. Vehec, H.Y. Sohn, H.G. Kim and M.E. Choi, *Hydrogen Flash Smelting*. Session 13: 1-6, Proceedings of EECRsteel-METEC, Dusseldorf, Germany, June 26-July 1, 2011.

7th International Symposium on High-Temperature Metallurgical Processing
Edited by: Jiann-Yang Hwang, Tao Jiang, P. Chris Pistorius, Gerardo R.F. Alvear F., Onuralp Yücel, Liyuan Cai, Baojun Zhao, Dean Gregurek, and Varadarajan Seshadri
TMS (The Minerals, Metals & Materials Society), 2016

SO_3 FORMATION IN COPPER SMELTING PROCESS: THERMODYNAMIC CONSIDERATION

Mao Chen[1], Zhixiang Cui[2], Leonel Contreras[3], Chuanbing Wei[2], Baojun Zhao[1]

[1]The University of Queensland, Brisbane, Australia
[2]Dongying Fangyuan Nonferrous Metals, Dongying, China
[3]National Copper Corporation of Chile, Santiago, Chile

Keywords: SO_3 formation, Copper smelting, Temperature, P_{O_2}

ABSTRACT

Bottom blown oxygen copper smelting process developed at Dongying Fangyuan Nonferrous Metals shows many advantages such as flexible feed requirement, high productivity, low energy consumption and high direct copper recovery. However, the weak acid generated during the process is not stable which influences the waste management and continuous operation. The amount of the weak acid is determined by SO_3 formed in copper smelting process. SO_3 can be formed inside copper smelting furnace and boiler at different conditions. Effects of temperature and oxygen partial pressure on the formation of SO_3 during copper smelting process have been discussed thermodynamically. Understanding of the mechanisms for SO_3 formation will enable the operation of bottom blown copper smelting furnace to be optimized and generation of the weak acid in the process can be managed. The fundamentals developed in this study can also be applied to other copper smelting processes.

Introduction

The bottom blown copper smelting furnace has been operated in Dongying Fangyuan Nonferrous Metals for 7 years and proved to be successful in treatment of different kinds of feeds including low grade complex copper concentrate, gold- and silver-containing minerals. The production has been also shown good performances of high copper recovery rate and energy efficiency. [1-2]

Off-gas handling is always one of the key processes in the overall production. In the copper smelting conditions SO_2 is usually a stable phase for sulphuric acid production. However, SO_3 could be formed in the smelting furnace and boiler. The formation of SO_3 during the copper smelting process will results in weak acid which cannot be recycled directly and significantly reduces the recovery of sulphur.

Oxidation of SO_2 to SO_3 can be divided into two stages: 1) low temperature (< 700 °C), thermodynamically favored to form SO_3 but limited by the reaction rate; 2) high temperature, controlled by the oxygen partial pressure of the reaction. [3] The study of the coal combustions in the power plant demonstrates that the formation of SO_3 strongly depends on resident time at low temperature and amount of excess air at high temperature [4]. The tendencies from the study [4] show that excess air increases the SO_3 concentration, and low temperature helps to form SO_3. The formation of SO_3 is necessary to be studied separately in different part of the copper smelting process.

Although the kinetics also plays an important role on the formation of SO_3, the present study is focused on thermodynamic aspect. The formations of SO_3 in copper smelting process is discussed for the bottom blown furnace but the theory can be applied to other copper smelting processes. Due to the practical difficulties to measure the actual SO_3 concentration in different positions of the process, the effects of temperature and oxygen partial pressure on the formation of SO_3 are discussed with the assistance of FactSage computational software [5].

Thermodynamic calculations and discussions

The thermodynamic calculations were undertaken using FactSage 6.4 [5]. Based on the differences in the temperatures and oxygen partial pressures at different positions of the process, four situations can be identified for generating SO_3 during the bottom blown smelting process. The formation of SO_3 was traced back from the lance of the furnace until the off-gas goes to the boiler and shown in the schematic diagram (Figure 1):

Case 1: SO_3 forms directly at the flame of lance where oxygen is injected into matte layer
Case 2: SO_3 is converted from SO_2 gas above the bath
Case 3: SO_3 is formed in the off-gas hood
Case 4: SO_3 is formed in the boiler

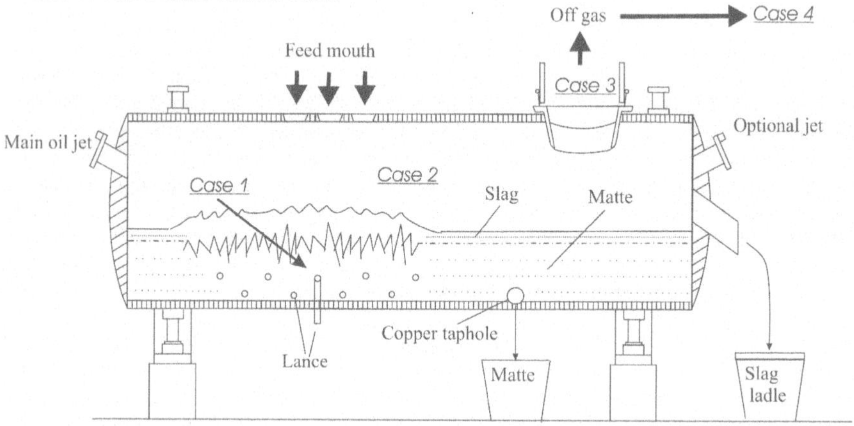

Figure 1. Schematic diagram of BBF copper smelting furnace showing the possible formation of SO_3 at four different locations

The SO_3 formations due to the thermodynamic equilibrium in different parts of the process are therefore discussed separately.

Case 1: SO_3 formation on the flame above the lance where oxygen is injected into matte layer.

$$CuFeS_2 + O_2 \rightarrow (Cu,Fe)O + SO_3 + SO_2 + S_2$$

In the bottom blown furnace, oxygen and air are blown through the lance into the matte layer. The oxygen partial pressure close to the lance is high. The exothermic reaction between sulphide and oxygen generates large amount of heat and the temperature will be much higher than the bath temperature. In the present calculations, in order to closely simulate the actual conditions within the bottom blown furnace, three levels of oxygen partial pressures (0.99, 0.7, and 0.5 atm) were assumed. The temperature ranges were set to be 1200 to 1800 °C to cover the possible situations. The volume ratio of SO_3 among all the S-containing gases were predicted and shown in Figure 2. As it can be seen from Figure 2 that up to 4% SO_3 could be generated on the burning zone at high oxygen partial pressure. The formation of SO_3 is favorable at high oxygen partial pressure and low temperature. Effect of temperature is more significant on the formation of SO_3. In BBF operation oxygen and air are injected separately and the temperature could be very high. The proportion of SO_3 formed would be lower than other processes. It can be seen from Figure 2 that the total SO_3 vol% is less than 1 % when temperature is over 1500 °C which is most likely true in BBF.

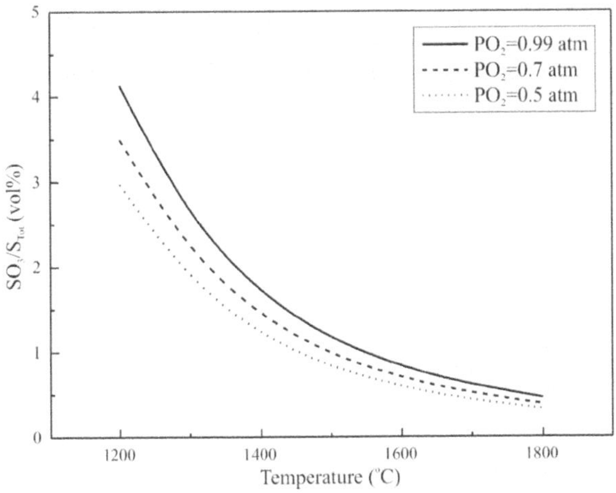

Figure 2. Proportion of SO_3 in total S-containing gases (S_{tot}: SO_3, SO_2 and S_2) on the flame above the oxygen lance

Case 2: SO_3 is converted from SO_2 gas above the bath.
In the smelting process, large amount of SO_2 gas can be generated from the reaction of sulphide and oxygen at high temperature. Before entering the off-gas hood, SO_2 gas has the possibility to be converted to SO_3 and S_2 gases by the reactions below:

$$2SO_2 + O_2 \rightleftharpoons 2SO_3$$

$$2SO_2 \rightleftharpoons S_2 + 2O_2$$

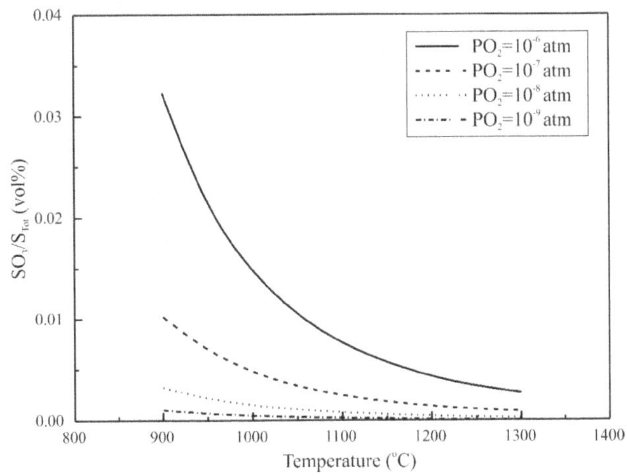

Figure 3. Proportion of SO_3 gas in total S-containing gases above the bath

The temperature above the bath in BBF was assumed to be lower than the bath, and the oxygen partial pressure is low (~10^{-8} atm) as injected oxygen from the bottom has been almost consumed. The calculations of equilibrium were carried out at temperature range 900 – 1300 °C and P_{O_2} range 10^{-6} – 10^{-9} atm. The proportion of SO_3 among all SO_3, SO_2 and S_2 gases and its tendencies as functions of temperature and oxygen partial pressure are shown in Figures 3 and 4. It can be seen from Figure 3 that negligible amount of SO_3 is generated above the bath in the range of temperature and oxygen partial pressure discussed. The proportions of SO_3 decrease with increasing temperature (Figure 3) and decreasing Po_2 (Figure 4). It can be seen from Figure 4 that formation of SO_3 is more sensitive at lower temperature and higher oxygen partial pressure. For example, it can be seen from Figure 4 that at 1100 °C, increase of Po_2 from 10^{-8} to 10^{-7} atm only increases the proportion of SO_3 by 1.5 times. Further increase of Po_2 from 10^{-7} to 10^{-6} atm will increase the proportion of SO_3 by over 3 times.

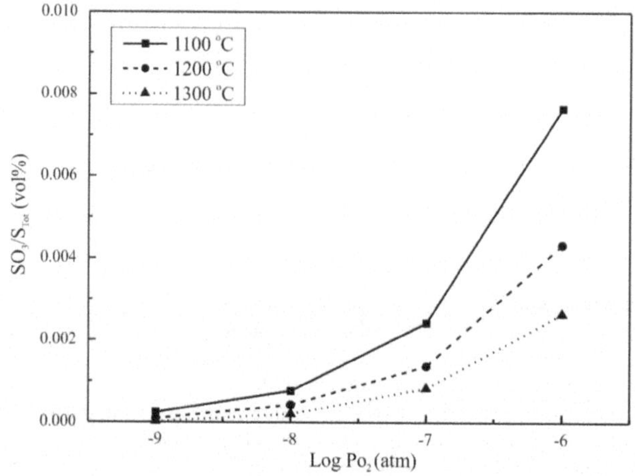

Figure 4. Proportion of SO_3 gas volume ratio as a function of Po_2 at fixed temperatures

SO_3, SO_2 and S_2 are the major S-containing phases formed in copper smelting process. It can be seen from Figure 5 that, at fixed Po_2 10^{-8} atm, element sulphur can be as high as 10% and it increases with increasing temperature. If the sulphur gas is cooled at the off-gas hood, the liquid and solid sulphur will be formed and stuck to the hood wall. This has been observed in BBF operations and certain amount of oxygen is necessary in the gas mixture to avoid the formation of liquid and solid sulphur in the off-gas hood.

The feeding materials may also affect the SO_3 formation. The feeds containing $CuSO_4$ could be used in the smelting operations and it is argued that decomposition of $CuSO_4$ would form SO_3. Figures 3 - 5 show that the proportion of SO_3 in the gas mixture is only functions of temperature and oxygen partial pressure. At the given ranges of temperature and oxygen partial pressure, the SO_3 released from $CuSO_4$ will be converted to SO_2 and S_2. The SO_3 came from the feeds is not necessary the source of the SO_3 in the off-gas.

Up to 10% H_2O could be present in the copper concentrates which was claimed to affect the formation of SO_3. FactSage calculations show that thermodynamically, at fixed 1200 °C and Po_2 10^{-8} atm, presence of H_2O has little effect on the formation of SO_3. SO_3 concentrations are less than 0.001% and they are almost independent of H_2O concentrations. It is possible that presence of H_2O could accelerate the oxidation reaction $2SO_2 + O_2 \rightarrow 2SO_3$. Kinetic rather than thermodynamic may play important role in the formation of SO_3 when H_2O is present.

From the discussions above it can be seen that, in the bottom blown copper smelting furnace, limited SO_3 can be formed above the bath due to the oxidation of the SO_2 as most of oxygen has been consumed in the bath. The SO_3 gas from concentrate or flame above the oxygen lance will be converted to SO_2 above the bath as a result of the equilibrium reaction $2SO_3 \rightarrow 2SO_2 + O_2$.

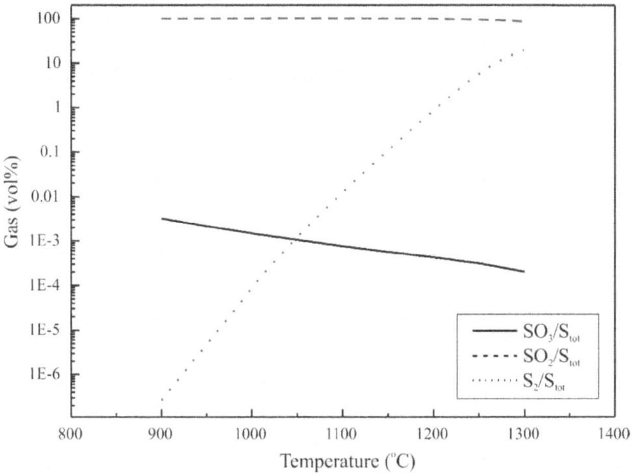

Figure 5. Distribution of S-containing gases (SO_3, SO_2 and S_2) at fixed Po_2 10^{-8} atm

Case 3: SO_3 is formed in the off-gas hood

The off-gas produced from the smelting process enters the off-gas hood where the temperature is decreased significantly. It has been shown in Figure 5 that, at fixed Po_2 10^{-8} atm, element sulphur can be formed and it may be converted to liquid and solid in the off-gas hood. This has been observed in BBF operations and it caused great difficulty to the operation. In practice a certain amount of air is introduced from the off-gas hood to reoxidise the element sulphur to form SO_2 $S_2 + 2O_2 \rightarrow 2SO_2$. If the amount of air entering into the off-gas is controlled accurately in a narrow range, it is possible to only oxidise the element sulphur to SO_2. However, excessive air could be introduced into the off-gas which would increase the oxygen partial pressure significantly and result in the formation of SO_3. If the air leakage rate is 30% the oxygen partial pressure will be approximately 0.05 atm. It can be seen from Figure 6 that in the temperature range 1000 – 1200 °C, significant amount of SO_3 can be formed. The volume ratio of SO_3 increases with increasing oxygen partial pressure and with decreasing temperature. Air leakage on the off-gas hold will also decrease the gas temperature. It can be seen from the figure that, at a given oxygen partial pressure, the amount of SO_3 formed at 1000 °C can be 1-2 times higher than that formed at 1100 °C.

The discussions above show that it is important to accurately control the air leakage rate. In BBF operation the efficiency of oxygen consumption is very high and the oxygen partial pressure above the bath is very low. Significant amount of element sulphur can be formed inside the furnace and condensed on the wall of the off-gas hood. It is necessary to have air leakage to oxidise the element sulphur. If the air leakage rate is controlled properly, the oxygen partial pressure and temperature of the off-gas will enable the levels of element sulphur and SO_3 to be manageable. The oxidization of element sulphur will be exothermic and the local temperature could be raised if only a small amount of air is introduced into the system. A higher temperature would be favorable for SO_2. On the other hand, if excessive air is introduced at off-gas hood, the oxygen partial pressure will be high and the local temperature will be lower that are all favorable for the formation of SO_3.

If the air leakage rate is very high, higher oxygen partial pressure and lower temperature are thermodynamically favorable for the formation of SO_3. However, diluted SO_2 concentration and very low temperature will result in lower reaction rate. The concentration of SO_3 in the off-gas is actually lower than the theoretical level. Higher air leakage rate may also increase the gas flowrate and therefore decrease the residence time of the gas mixture in the boiler. The SO_3 formed in the boiler which requires catalysis reaction could be significantly decreased.

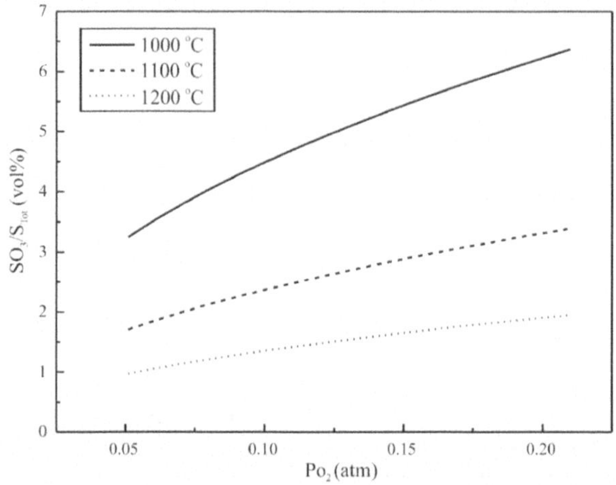

Figure 6. Proportions of SO_3 gas volume ratio in the off-gas hood calculated by FactSage

Case 4: SO_3 is formed in the boiler

The off-gas flows through the exhaust pipe to the boiler for heat exchange. The temperature of the gas mixture keeps decreasing during the whole path. Thermodynamic calculations were carried out between 200 and 900 °C at two oxygen partial pressure ranges Po_2 10^{-6} - 10^{-8} atm and 0.001 - 0.21 atm. The predictions are shown in Figures 7 and 8 respectively. Both figures show that sulphur will be thermodynamically stable in the form of SO_3 at low temperatures in wide Po_2 range. It can be seen from the figures that in these ranges of oxygen partial pressure and temperature, the SO_3 concentration is very sensitive to the oxygen partial pressure and temperature. For example, at Po_2 10^{-6} atm as shown in Figure 7, the SO_3 in all S-containing gases is increased from 10% to 50% when the temperature is decreased from 450 to 400 °C. On the other hand, if the gas temperature is fixed at 400 °C, the SO_3 in all S-containing gases will be increased from 10% to 50% when Po_2 is increased from 10^{-8} atm to 10^{-6} atm.

When reasonable air is introduced the Po_2 range 0.001 - 0.21 atm given in Figure 8 would be more appropriate to describe the real operating conditions. In this Po_2 range SO_3 is the dominant phase at temperatures below 450 °C. The amount of SO_3 formed is more sensitive to oxygen partial pressure and temperature. If the heat exchange is not sufficient and the gas temperature entering into the boiler is relatively higher, the formation of SO_3 will be significant as the reaction rate is higher. Comparing the conditions given in Figures 7 and 8 it can be seen that, at 600 °C, the ratio of SO_3 is below 2% in Po_2 ranges 10^{-6} - 10^{-8} atm. At the same temperature the ratio of SO_3 will be 20 - 80% in Po_2 ranges 0.001 - 0.1 atm.

In actual industrial practices, kinetics should be considered in addition to the thermodynamic aspect as the production is continuous and the equilibrium is hard to be achieved, in particular at low temperatures. A number of parameters such as concentrations of SO_2 and O_2, temperature, gas residence time and presence of catalysts all can have significant effects on the reaction rate. Unfortunately the kinetics of the reaction for oxidation of SO_2 cannot be theoretically predicted. Systematic experiments are required in the conditions close to copper smelting process to accurately characterize the reaction mechanisms for SO_3 formation.

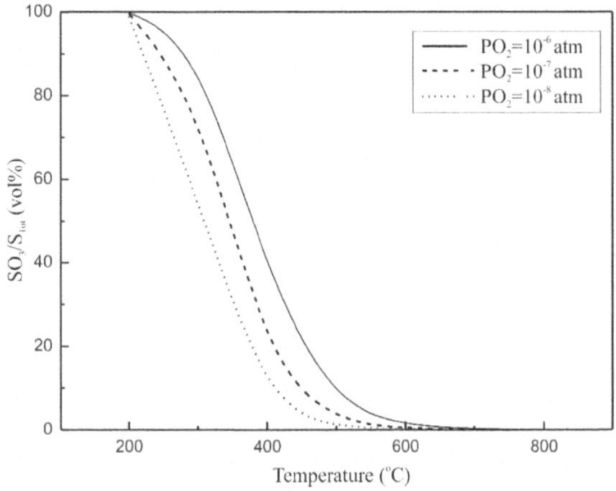

Figure 7. Proportions of SO_3 gas volume ratio in the boiler at low air leakage rate

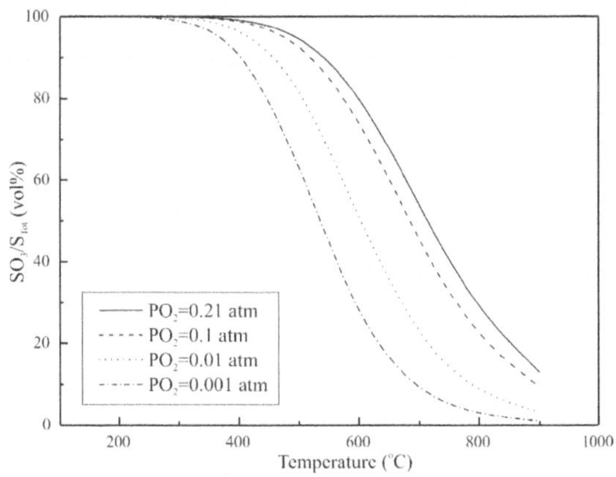

Figure 8. Proportions of SO_3 gas volume ratio in the boiler at high air leakage rate

Conclusions

Thermodynamic calculations were performed to predict the formation of SO_3 at different locations of the bottom blown smelting furnace and its follow-up processes. It was found from the calculations that negligible amount of SO_3 can be generated within the furnace under smelting conditions. The main risk of forming SO_3 is due to the excess air in the off-gas hood and the boiler.

In order to minimize the formation of SO_3, it is suggested that air leakage rate at off-gas hood has to be accurately controlled so that oxygen partial pressure and temperature can be adjusted properly to manage the formations of both SO_3 and S_2 gases. Proper control of the air leakage rate at off-gas hood affects the formation of SO_3 not only in off-gas hood but also in the boiler. The present study will help researchers and engineers to gain deeper understanding of the SO_3 formation in copper smelting process. The thermodynamic fundamentals discussed in this study is based on BBF copper smelting furnace but they will be applicable to other copper smelting technologies.

Acknowledgements

The authors would like to thanks National Copper Corporation of Chile (Codelco), Dongying Fangyuan Nonferrous Metals (Fangyuan) and The University of Queensland for financial support through the "Codelco-Fangyuan Professorial Funding".

References

1. B. Zhao, Z. Cui, and Z. Wang. "A New Copper Smelting Technology – Bottom Blown Oxygen Furnace Developed at Dongying Fangyuan Nonferrous Metals," (4th International Symposium on High-Temperature Metallurgical Processing, San Antonio, US, 2013), 3-10.
2. M. Chen, Z. Cui and B. Zhao, "Slag Chemistry of Bottom Blown Copper Smelting Furnace at Dongying Fangyuan," (6th International Symposium on High Temperature Metallurgical Processing, Florida, USA, 2015).
3. E. Herrera and L. Mariscal, "Controlling SO_3 formation in the off-gases process from the Isasmelt™ furnace at southern Peru Ilo smelter," (Copper 2013, Santiago, Chile, 2013), 61 – 74.
4. R. Srivastava, C. Miller, C. Erickson and R. Jambhekar, "Emissions of Sulfur Trioxide from Coal-fired Power Plants," *J. Air & Waste Manage. Assoc.*, 54(2004), 750-762.
5. C. Bale, E. Bélisle, P. Chartrand, S. Degterov, G. Eriksson, K. Hack, I. H. Jung, Y. B. Kang, J. Melançon, and A. Pelton, "FactSage thermochemical software and databases--recent developments," *Calphad*, 33(2009), 295-311.

7th International Symposium on High-Temperature Metallurgical Processing
Edited by: Jiann-Yang Hwang, Tao Jiang, P. Chris Pistorius, Gerardo R.F. Alvear F., Onuralp Yücel, Liyuan Cai, Baojun Zhao, Dean Gregurek, and Varadarajan Seshadri
TMS (The Minerals, Metals & Materials Society), 2016

EFFECT OF OXIDATION ON WETTING BEHAVIOR BETWEEN SILICON AND SILICON CARBIDE

Yaqiong Li, Lifeng Zhang, Zineb Benouahmane

School of Metallurgical and Ecological Engineering,
University of Science and Technology Beijing, Beijing 100083, China
Email: zhanglifeng@ustb.edu.cn

Keywords: Oxidation, Wettability, Silicon, SiC

Abstract

Experimental oxidation tests at high temperature (1000-1200 °C) on SiC samples have been respectively performed under wet and dry air atmosphere for different holding times. The thickness of thermally grown oxide was experimentally measured by DEKTAK Stylus Profilometry. The results showed that the oxidation rate was increased with temperature and holding time, corresponding to a passive oxidation regime which led to the formation of SiO_2 layer. The sessile drop method was employed in order to measure the contact angles between Si and SiC at high temperature. Contact angle can be varied from 39.6° to 75.6° by controlling the atmosphere. The deoxidation of the oxide layer and infiltration of liquid silicon caused a decrease in the contact angle.

Introduction

Silicon carbide is an attractive candidate for electrical, electronic, structural, and high-temperature applications because of its outstanding chemical and physical stability and mechanical property. In view of this considerable interest in SiC, the understanding of interfacial liquid metal/solid SiC is of enormous importance, and several studies have been conducted to investigate the wettability of SiC using many metals in various atmosphere [1-3]. Notable that, the oxygen partial pressure (P_{O2}) will influence the surface chemistry of SiC and the applied metals, and thus result in different the wettability of SiC under various conditions [4, 5].

The aim of this present work is to investigate the effect oxidation on the wetting behavior in the system of Si-SiC, and several factors, namely temperature, holding times, and atmosphere, are considered here.

Experimental Procedure

In order to study the effect of oxidation on wetting behavior between Si and SiC, the thermal oxidation process was carried out in the temperature range from 1000 °C to 1200 °C for different oxidation time i.e. 2, 4, 6, 8, 10, and 12 h. The wet and dry oxidizing ambient was respectively applied in order to analyze the exact thermal oxidation behavior of SiC. After the experiment, the thickness of thermally grown oxide was analyzed by DEKTAK Stylus Profilometry. Wetting was studied by the sessile drop method in Ar and air atmosphere, respectively. A $SiMo_2$ electric furnace was used in this experiment. Approximately 100 mg of pure Si rod (7 N) was placed on a SiC plate, and moved into the furnace. Thereafter, the temperature was increased up to a temperature higher than the melting point of Si at 5 °C/min. The furnace is equipped with windows, through which the contact angle can be determined by a camera.

Oxidation of SiC

For both wet and dry oxidation processes, passive oxidation occurs involving the reaction between the base material SiC and oxygen to form a dense SiO_2 layer, which is described by the following reaction [6]. Theoretically, the mass increased by 50wt% due to the addition of O and evaporation of CO.

$$SiC(s) + \frac{3}{2}O_2(g) = SiO_2(s) + CO(g) \qquad [1]$$

The thickness of the oxidized layer was determined by the Stylus Profilometry, and plotted against the holding time in **Figure 1**. As shown in this figure, the thickness of SiO_2 layer increased with increase in temperature and holding time. When the SiC oxidized under wet air at 1200 °C for 12 h, it grew about a length of 4.5 nm. Notable that, there is a difference in the SiO_2 thickness between oxidizing in wet and dry air, and the wet air atmosphere is found to prefer and promote the growth of SiO_2 layer, resulting in a high growth rate.

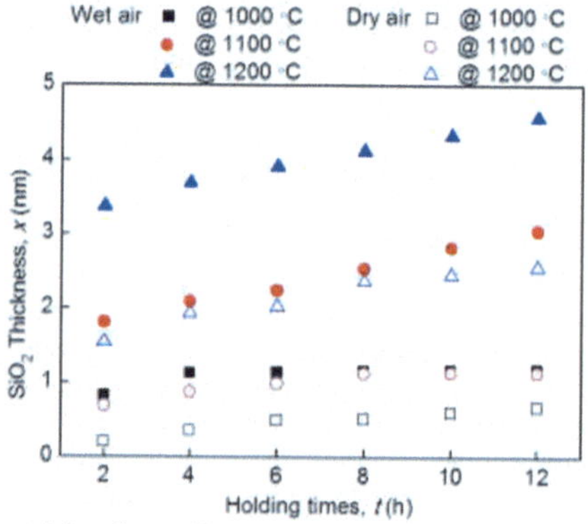

Figure 1 Plots of wet and dry thermally grown oxide thickness at 1000, 1100, and 1200 °C

According to **Ref.[7]**, **Eq.[2]** is applied to calculate the average growth rate of SiO_2, which is based on the Deal and Grove oxidation model [8] to investigate the relations between the thickness of the oxidized layer (x) and oxidation time (t). The results for *dx/dt* is plotted in **Figure 2**.

$$\left.\frac{dx}{dt}\right|_{sample} \Rightarrow \text{Average} \left.\frac{dx}{dt}\right|_{sample} \qquad [2]$$

As shown in this figure, the average oxide growth rate of SiO_2 is not constant, which attains the highest value at the beginning of the oxidation process at any temperature but decreases with increasing the oxide thickness. That is because the diffusion rate of oxygen in SiO_2 is quite low, and the formation of SiO_2 can thus inhibit the further oxidation of SiC and work as a protective mask.

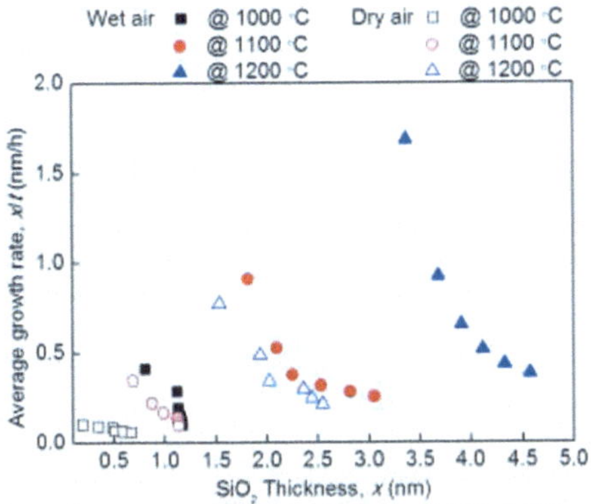

Figure 2 Plots of average oxide growth rate at 1000, 1100, and 1200 °C

Additionally, the effect of oxidation on the morphology of SiC was studied below. Two SiC plates were respectively heated at 1300 °C for 2 h under Ar (P_{O2}=10^{-15} Pa) and air atmosphere. After thermal treatment, the macro- and microscopic surface characteristics were observed and showed in **Figure 3**, together with the SiC raw plate for comparison. In **Figure 3** (a), the OM image of the raw plate shows a rough surface, and contains a large number of holes. In **Figure 3 (b)**, the morphology of sample obtained under Ar gas looks similar to the SiC raw plate, but become a little smooth. In **Figure 3 (c)**, by the thermal treatment in air atmosphere, the third sample is covered a white thin layer on its surface without any large holes, exhibiting a smooth surface.

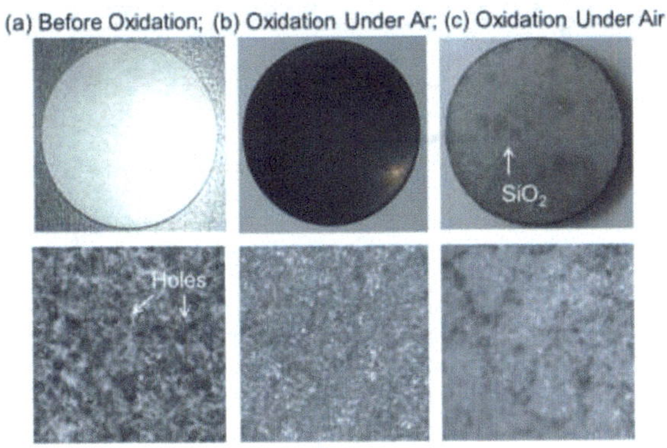

Figure 3 Macro- and microscopic surface characteristics of SiC plates

It is evident from the above discussions that the oxidation of SiC occurred easily and quickly when it exposed to the oxidizing atmosphere at high temperature, causing a smooth morphology and consequently, influenced the wetting behavior of SiC. The effect of oxidation on the wetting behavior between Si and SiC is discussed below.

Wetting Hehavior

The contact angles for the Si-SiC system in both atmospheres (Ar and air) are shown in **Figure 4**. The time $t = 0$ stands for the complete melting of Si, and at this time, the contact angle is determined and named as initial contact angle. Under Ar atmosphere, the P_{O2} is quite low, and the Si-SiC system was considered to be a non-reactive metal/solid system, revealing a fresh SiC surface. While under air atmosphere, a protective oxide layer formed and covered on the top of SiC plate with a thickness of several nanometers, as determined in previous work. The initial contact angle obtained under air atmosphere is 75.6°, large than that measured under Ar atmosphere (i.e. 64.1°), but close to the contact value of 83° in the Si-SiO_2 system [9]. Therefore, it can be inferred that the initial value obtained under air atmosphere is determined by the SiO_2, not SiC. When the Si is completely melted, the initial contact angle decreases rapidly to a final value, and attains an equilibrium state. As shown in **Figure 4**, under Ar atmosphere, the contact angle decreased from an initial value of 64.1° to a steady contact angle of 44.9° in about 60 s, while in the case of air atmosphere is about 360 s. In this duration, the following reaction happened and consumed SiO_2. When the SiO_2 layer exhausted, the fresh SiC exposed and showed a rough surface which adhered to the Si melt and resulted in a smaller contact angle as shown in **Figure 5**.

$$SiO_2(s) + Si(l) = 2SiO(g) \quad [3]$$

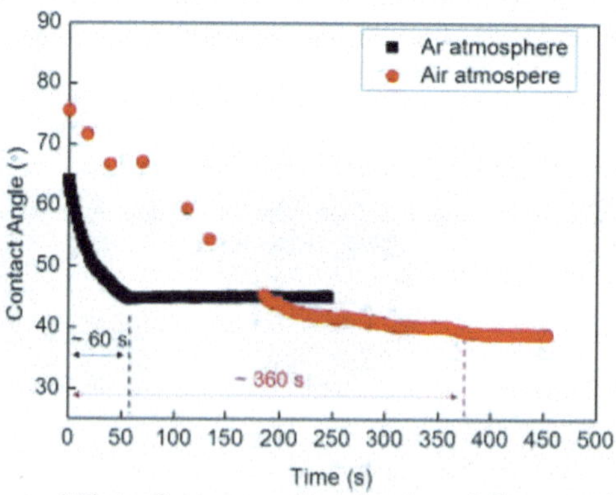

Figure 4 Plots of contact angle under Ar and air atmosphere

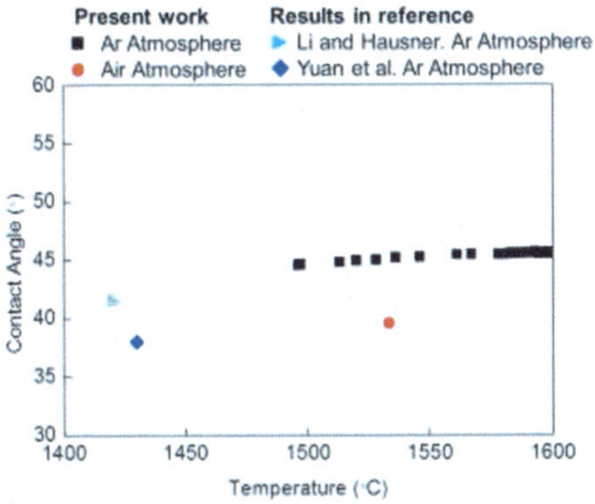

Figure 5 Plots of final contact angle under Ar and air atmosphere [10, 11]

Based on the above discussion, the effect of oxidation on the wetting behavior between Si and SiC can be illustrated by **Figure 6**. The formation of SiO_2 layer was found to be a key to effective modify the SiC, which provided a large initial contact angle for the Si-SiO_2 system. When the reaction between O and Si happened, the consumption of SiO_2 made the SiC rough and gave rise to the infiltration of Si, and finally resulted in a small contact angle for the Si-SiC system.

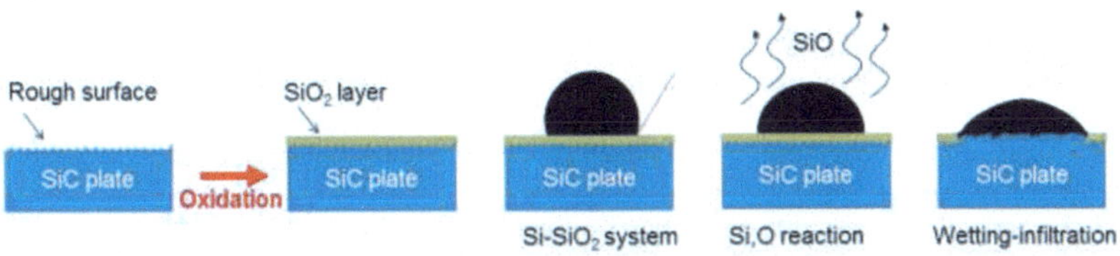

Figure 6 Effect of oxidation on the wetting behavior between Si and SiC

Summary

The oxidation behavior of SiC has been investigated for both oxidizing ambient (dry and wet) at high temperature (1000-1200 °C). It has been concluded that the oxidation of SiC occurred easily and quickly when it exposed to the oxidizing atmosphere at high temperature, causing a smooth morphology without large holes. Due to the formation of SiO_2 under air atmosphere, the initial contact angle attained a large value (75.6°), close to the contact value of 83° measured in the Si-SiO_2 system, but decreased rapidly to an equilibrium value of 39.6° due to the deoxidation of the oxide layer and infiltration of liquid silicon

Acknowledgements

The authors are grateful for support from the National Science Foundation China (Grant No. 51334002), State Key Laboratory of Advanced Metallurgy, Beijing Key Laboratory of Green Recycling and Extraction of Metals (GREM), the Laboratory of Green Process Metallurgy and

Modeling (GPM[2]) and the High Quality Steel Consortium (HQSC) at the School of Metallurgical and Ecological Engineering at University of Science and Technology Beijing (USTB), China.
Correspondence author: Lifeng Zhang, Email: zhanglifeng@ustb.edu.cn .

References

[1]. H. Jones and H. Atkinson, "The wettability of silicon carbide by liquid aluminium: the effect of free silicon in the carbide and of magnesium, silicon and copper alloy additions to the aluminium", *Journal of Materials Science*, 28(10) (1993), 2654-2658.

[2]. M. Pech-Canul, et al, "Optimum parameters for wetting silicon carbide by aluminum alloys", *Metallurgical and Materials Transactions A*, 31(2) (2000), 565-573.

[3]. J. Rocher, et al., "Wetting improvement of carbon or silicon carbide by aluminium alloys based on a K_2ZrF_6 surface treatment: application to composite material casting", *Journal of Materials Science*, 24(8) (1989), 2697-2703.

[4]. Dezellus, et al., "Wetting and infiltration of carbon by liquid silicon", *Journal of Materials Science*, 40(9-10) (2005), 2307-2311.

[5]. Mailliart, et al., "Influence of oxygen partial pressure on the wetting of SiC by a Co-Si alloy", *Materials Science and Engineering: A*, 495(1) (2008), 174-180.

[6]. T. Narushima, et al., "High-Temperature Passive Oxidation of Chemically Vapor Deposited Silicon Carbide", *Journal of the American Ceramic Society*, 72(8) (1989), 1386-1390.

[7]. S. K. Gupta and J. Akhtar, "Silicon carbide-materials, processing and applications in electronic devices", Facts2011: INTECH Open Access Publisher.

[8]. B. E. Deal and A. Grove, "General relationship for the thermal oxidation of silicon", *Journal of Applied Physics*, 36(12) (1965), 3770-3778.

[9]. B. Drevet, et al., "Wetting and adhesion of Si on Si3N4 and BN substrates", *Journal of the European Ceramic Society,* 29(11) (2009), 2363-2367.

[10].Z. Yuan, et al., "Wettability and reactivity of molten silicon with various substrates", *Applied Physics A*, 78(4) (2004), 617-622.

[11].J. Li and H. Hausner, "Wettability of silicon carbide by gold, germanium and silicon", *Journal of materials science letters*, 10(21) (1991), 1275-1276.

7th International Symposium on High-Temperature Metallurgical Processing
Edited by: Jiann-Yang Hwang, Tao Jiang, P. Chris Pistorius, Gerardo R.F. Alvear F., Onuralp Yücel, Liyuan Cai, Baojun Zhao, Dean Gregurek, and Varadarajan Seshadri
TMS (The Minerals, Metals & Materials Society), 2016

EFFECT OF CaO/SiO_2 AND P_2O_5 ON THE VISCOSITY OF $FeO\text{-}SiO_2\text{-}V_2O_3\text{-}CaO\text{-}P_2O_5$ SLAGS

Zhen Zhang[1], Bing Xie[1], Pan Gu[1], Jiang Diao[1], Hongyi Li[1]

[1]College of Materials Science and Engineering, Chongqing University, Chongqing, 400044, China

Keywords: CaO/SiO_2, P_2O_5, viscosity, basic vanadium slag, network ploymerization.

Abstract

Viscosities of $FeO\text{-}SiO_2\text{-}V_2O_3\text{-}CaO\text{-}P_2O_5$ slags have been measured at the temperature range of 1673 K to 1773 K (1400 ℃ to 1500 ℃). This investigated system represents the major components of basic vanadium slag that is an important intermediate product of a novel vanadium extracting process, in which removal of phosphorus and extraction of vanadium can be achieved from the vanadium-containing hot metal simultaneously. It was found that higher CaO/SiO_2 ratio resulted in lower viscosity in the composition range of interest. Also, the viscosities increase with the increasing P_2O_5 concentration. The varying viscosities can be attributed to the modification in complexity and polymerization of the slag network. The variation of apparent activation energies are in tune with the viscosity measurements. The results of this work revealed the effect of CaO/SiO_2 and P_2O_5 on the viscous properties of $FeO\text{-}SiO_2\text{-}V_2O_3\text{-}CaO\text{-}P_2O_5$ slags and therefore provide the theoretical foundation for the control of the coupling de-P and V-extraction process.

Introduction

Vanadium-titanium magnetites (VTM) are comparatively abundant in China especially the Panxi region. Traditionally, a complete set of blast furnace-LD converter process is adopted to extract metal from the VTM ore. In this process, the VTM ore is primarily smelted and reduced in blast furnace to produce vanadium-containing hot metal which is then oxidized in LD converter to obtain vanadium slag and semi-steel. Vanadium slag is traditionally sodium roasted followed by water leaching to extract vanadium. However, the sodium roasting technique has some shortages in terms of environmental protection [1].

Recently, a novel method that is named "direct roast-selective leach" have been proposed, in which adopt calcium roasting followed by acidic leaching to extract vanadium from vanadium slag [2]. This approach allows a relatively high content of CaO in the vanadium slag. By adding CaO into LD-converter, de-P and V-extraction can be achieved simultaneously during the oxidation of vanadium-containing hot metal. In this way, the burden of de-P of semi-steel during steelmaking can be relieved. The schematic diagram of the proposed coupling de-P and V-extraction process is illustrated in Figure 1[2].

During the vanadium extraction process, vanadium slag is a crucial intermediate product. It is thus worthwhile to investigate the physical-chemical properties of this slag system. Previous studies on vanadium slag mainly focused on the crystallization behavior and mineralogical characterization under different compositions and processing conditions [2-4]. However, to the knowledge of the authors, the researches on vanadium slag containing both P_2O_5 and CaO have not been reported yet. Based on previous researches [5-7], the viscosity of slag generally increases with the increasing content of P_2O_5 due to the polymerization of the slag melt networks. In addition, it is well-established that higher basicity gives lower viscosity since the basic oxide will depolymerize the slag structure [8-10].

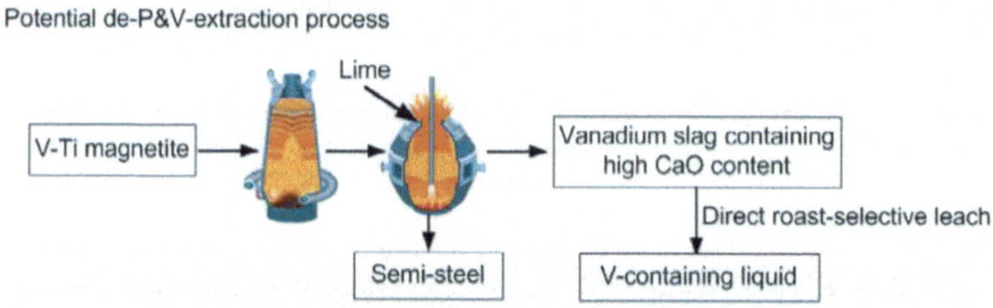

Fugure 1. Flow diagram of the novel De-P&V-extraction process [2].

Considering the viscous properties of slag is of great importance since it governs the mass transfer and the crystallization behavior of the molten slag. In this study, we measured the viscosity of the CaO-SiO_2-40 wt pct FeO-15 wt pct V_2O_3- P_2O_5-based slags to clarify the effect of CaO/SiO_2 and P_2O_5 on the viscous flow of molten vanadium slags at high temperatures. This may provide the clues how to control and facilitate coupling of V-extraction and De-P during the oxidation process of V-contained hot metal in converter.

Experimental method

The designed compositions of two series (namely, R-serie and P-serie) slag samples are listed in Table I. Analytical-grade CaO, SiO_2, V_2O_3, $Ca_3(PO_4)_2$ were used as raw materials. FeO was prepared by using appropriate ratio of Fe powder and reagent grade Fe_2O_3. Those original chemicals were dried and thoroughly mixed before melting.

The viscosity measurements were carried out by using rotating cylinder method. The working bob, crucible and the connecting shaft were all made of molybdenum. A vertical tubular furnace with six U-shape $MoSi_2$ heating elements was employed. A Pt-10 pct Rh/Pt thermocouple protected by alumina tube was placed beneath the bottom of the crucible to monitor the experimental temperature. Before measuring the viscosity of slags, the viscometer was calibrated using silicone oil with known viscosities at different temperature.

The Mo crucible filled with about 150 g slag sample was placed at the constant temperature zone of the furnace. Purified Ar gas was inlet into the reaction chamber at a flow rate of 300mL/min throughout the experimental process to prevent the oxidation of the slags. The furnace was initially heated up to 1773 K (1500℃) and held for 2 hour to guarantee completely molten and homogenized slags. Then, the bob was carefully immersed into the slag melt and maintained at 10 mm above the crucible base. Before each measurement, the molten slag was preserved at the corresponding measuring temperature for half an hour to ensure thermal equilibrium.

After finishing the measurement of viscosity, the slags were reheated to 1773 K and held for 1 hour, and then they were quenched by ice water to obtain the glassy slag samples. The compositions of the quenched slags were analyzed using X-ray fluorescence (XRF) spectroscopy, and no apparent variation was observed from the initial weighed mass.

Table I. Compositions of experimental slags.

serie	No.	R	FeO	SiO_2	CaO	V_2O_3	P_2O_5
R-serie	R1	0.5	40.00	27.33	13.67	15.00	4.00
	R2	1.0	40.00	20.50	20.50	15.00	4.00
	R3	1.5	40.00	16.40	24.60	15.00	4.00
	R4	2.0	40.00	13.67	27.33	15.00	4.00
P-serie	P1	1.5	40.00	18.00	27.00	15.00	0.00
	P2	1.5	40.00	17.20	25.80	15.00	2.00
	P3	1.5	40.00	16.40	24.60	15.00	4.00
	P4	1.5	40.00	15.60	23.40	15.00	6.00

*The content of all components of sample P3 are the same as that of sample R3.

Results and discussion

Effect of CaO/SiO_2 and P_2O_5 on the slag viscosity

The effect of basicity (CaO/SiO_2) on the viscosity of the CaO-SiO_2-40 wt pct FeO-15 wt pct V_2O_3-4 wt pct P_2O_5 system is shown in Figure 2. It is clear that increasing the basicity of the slag will result in decrease in the viscosity of slags, which shows a similar variation tendency with the previous result [10].

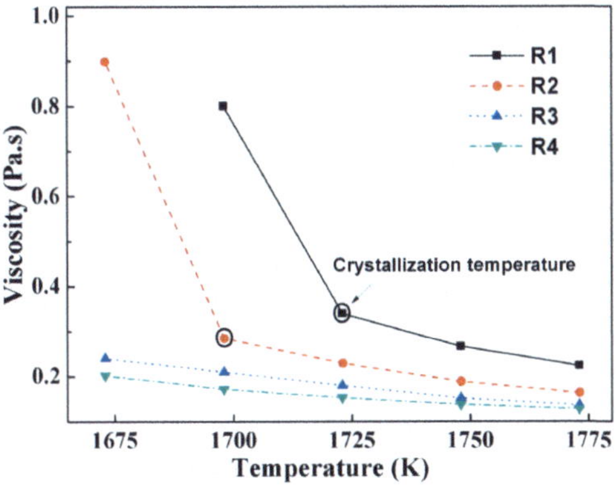

Figure 2. The viscosity of R-serie slags as a function of temperature. Circles mark the inflection points and lines are guides to the eyes.

According to Sridhar et al. [11], the appearance of this inflection point in viscosity can be ascribe to the initial precipitation of crystal phases in the molten slag, and the corresponding temperature is referred to as crystallization temperature or break temperature of the melt, below which the slag turns into non-Newtonian fluid. The inflection points of samples R1 and R2 were detected and are marked as circles in Figure 2. The corresponding crystallization temperatures of samples Rl and R2 are 1723 (1450℃) and 1698 K (1425℃) respectively. In contrast, the crystalline behaviors are not observed in sample R3 and R4, implying the increase in basicity will decrease the crystallization temperature and therefore weaken the crystallization capacity of the slags at certain temperature.

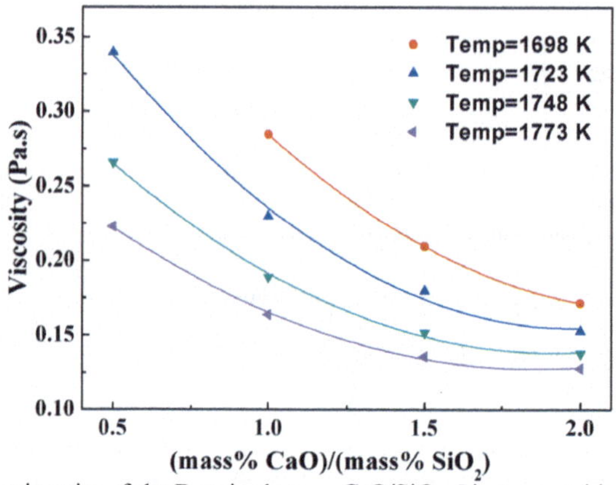

Figure 3. The viscosity of the R-serie slags *vs* CaO/SiO_2. Lines are guides to the eyes.

The viscosity of the CaO-SiO_2-40 wt pct FeO-15 wt pct V_2O_3-4 wt pct P_2O_5 slags as a function of basicity is also shown in Figure 3. Increasing basicity from 0.5 to 1.5 exhibits a considerable decrease in the viscosity, while a subsequent increase in the basicity from 1.5 to 2.0 was less effective in lowering the viscosity, especially at higher temperature of 1773 K (1500℃) where the viscosity shows only slightly decrease when the basicity higher than 1.5. Similarly, Kim et al. [10], investigated the influence of CaO/SiO_2 on the viscous behavior CaO-SiO_2-$Al2O_3$-MgO slags, and it was found that the effect of basicity variation on the viscosity is more prominent when the basicity is lower than 1.0.

Based on previous work done on relating the viscosity with slag structure [12-14], it can be inferred that the decrease in viscosity can be contributed to the depolymerization of the slag structure. With the increase in basicity, more free oxygen ions (O^{2-}) will be released to the slag due to the dissociation of CaO. The released O^{2-} reacts with the bridged oxygen (O^0) and therefore fragments the network structure of the slag resulting in lower viscosity. However, when excess O^{2-} is provided to the slag with the increasing basicity but only relatively few complex network structures are available, then further depolymerization of the slag structures is impossible and thus the effect on the viscosity can become negligible as seen from the results in Figure 3.

In addition, the depolymerization of the slag can also occur with higher temperatures, where higher thermal energy can provide sufficient energy to break the existing network structures. Thus at higher temperature such as 1773 K (1500℃), higher basicity and subsequently higher O2- can have comparatively less effect on the viscosity of the slag since more existing complex structures have already been broken at the temperature of 1773 K (1500 ℃) compared to lower temperatures due to the higher temperature induced higher thermal energy.

The effect of P_2O_5 on the viscous behavior of the CaO-SiO_2-40 wt pct FeO-15 wt pct V_2O_3-P_2O_5 system at a fixed basicity (CaO/SiO_2) of 1.5 is illustrated in Figure 4. The viscosities, as expected, continuously increased as the P_2O_5 content increasing at a given temperature. It was also noted that the effect of P_2O_5 addition on the viscosity increase was more pronounced at the lower temperatures, whereas P_2O_5 addition did not significantly affect the viscosity of slags at higher temperatures. The viscosity of P-bearing slags measured by other researchers is also labeled in Figure 4 for comparison [5, 7]. It can be seen that those reported viscosity values showed a similar trend with the present measured results, despite the slag compositions were not the same.

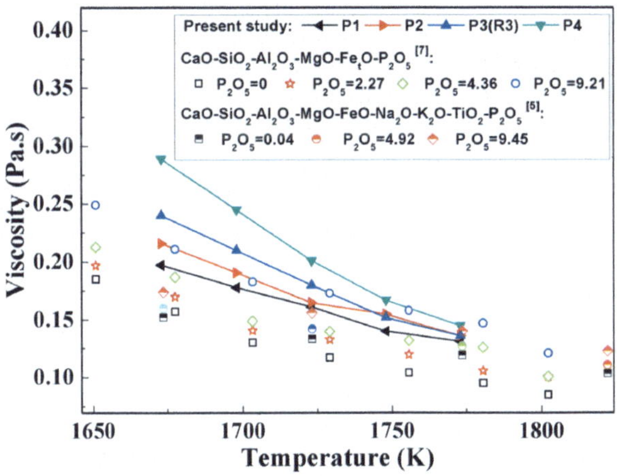

Figure 4. The viscosity of the slags with different P_2O_5 contents.

Apparent activation energy of the slags

The temperature dependence of the viscosity for the present study is shown in Figure 5. The activation energy (E_η) for the viscous flow of the silicate melts is obtained from the Arrhenius-type equation:

$$\eta = A\exp\left(-E_\eta / RT\right) \tag{1}$$

Where η (Pa.s) is the viscosity, A is a constant, E_η (kJ/mol) is the activation energy of viscous flow, R is the gas constant, and T(K) is the absolute temperature. E_η represents the energy barrier that the cohesive flow units in slags have to overcome when the units move between different equilibrium states [15]. The variation of E_η suggests structural changes in slags, and its value is expected to be constant for a certain molten slag system before precipitation occurs [16]. The activation energies can be evaluated from slope of linear fitting of the natural logarithm of the viscosity ($\ln\eta$) with reciprocal temperature ($1/T$) and the calculated results are provided in Table II.

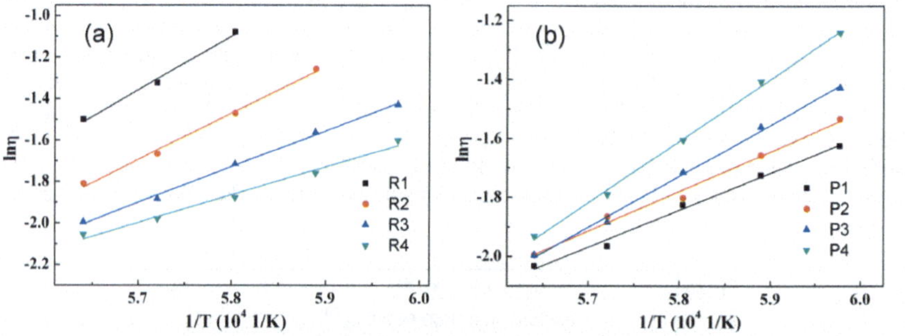

Figure 5. Natural logarithm of the viscosity in fully liquid region for (a) R-serie and (b) P-serie slags *vs* reciprocal temperature.

Table II. Activation energy for the various slag composition in the present study.

Sample	R1	R2	R3(P3)	R4	P1	P2	P4
E(KJ.mol^{-1})	214.3	185.8	144.0	112.0	104.2	112.0	173.9

As expected, the addition of P_2O_5 to the slags results in an increase of the activation energy. This indicates the rise of energy barrier for viscous flow and some more complex structural units formed in the slag melts which agrees with the measured viscosity shown in Figure 4. The effect of P_2O_5 on the viscosity increase can be interpreted based on the basis of an increase in the degree of polymerization (DOP). A comparison of the activation energies at fixed P_2O_5 of 4 mass pct and variable basicity shows a decrease in the activation energy with higher CaO/SiO_2, which is in accordance with the decrease in the viscosity and depolymerization of slag structures with increasing CaO content.

Conclusions

The effect of P_2O_5 and CaO/SiO_2 on the viscous flow of the CaO-SiO_2-FeO-V_2O_3-P_2O_5 slags was studied by measurements of viscosity. The conclusions are as follows:

(1). Slag viscosity decreases with increasing basicity, indicating the formation of some fragmented structural units. CaO/SiO_2 of 1.5 seems to be a critical basicity for control the viscous properties of the slags since slight decrease in viscosity can be achieved with basicity higher than 1.5.

(2). The viscosity of slag melts increases with increasing P_2O_5 content. This is ascribed to that P5+ act as network former copolymerized with Si-O based network therefore forms more complexed structure.

(3). The activation energies of slags decrease with higher CaO/SiO_2 and increase with the addition of P_2O_5 under a constant basicity, which correlated well with the results of the viscosity measurements.

Acknowledgements

This work was financially supported by National Basic Research Program of China (973 Program, grant no.2013CB632604) and National Natural Science Foundation of China (grant nos. 51090382, 51474041).

References

1. D. X. Huang, *Vanadium Extraction and Steelmaking* (Beijing, Metallurgical Industry Press, 2000), 178.
2. H.X. Fang et al., "Influence of CaO on Existence form of Vanadium-containing Phase in Vanadium Slag," *ISIJ International*, 55 (1) (2015), 200-206.
3. X. Zhang et al., "Nucleation and growth kinetics of spinel crystals in vanadium slag," *Ironmaking & Steelmaking*, 39 (2) (2012), 147-154.
4. J. Diao et al., "Mineralogical characterisation of vanadium slag under different treatment conditions," *Ironmaking & Steelmaking*, 36 (6) (2009), 476-480.
5. M.J. Toplis, D.B. Dingwell, and G. Libourel, "The effect of phosphorus on the iron redox ratio, viscosity, and density of an evolved ferro-basalt," *Contributions to Mineralogy and Petrology*, 17 (3) (1994), 293-304.
6. C. Mercier et al., "Influence of P_2O_5 content on the structure of SiO_2-Na_2O-CaO-P_2O_5 bioglasses by ^{29}Si and ^{31}P MAS-NMR," *Journal of Non-Crystalline Solids*, 357 (24) (2011), 3901-3909.
7. Z.J. Wang et al., "Effect of P_2O_5 and Fe_tO on the Viscosity and Slag Structure in Steelmaking Slags," *Metallurgical and Materials Transactions B*, 46 (2) (2015), 758-765.
8. J.S. Machin and D.L. Hanna, "VISCOSITY STUDIES OF SYSTEM CaO–MgO–Al2O3–SiO2: 1, 40% SiO2*," *Journal of the American Ceramic Society*, 28 (11) (1945), 310-316.
9. Y.S. Lee et al., "Influence of basicity and FeO content on viscosity of blast furnace type slags containing FeO," *ISIJ international*, 44 (8) (2004), 1283-1290.
10. H. Kim et al., "Effect of Al_2O_3 and CaO/SiO_2 on the Viscosity of Calcium-Silicate–Based Slags Containing 10 Mass Pct MgO," *Metallurgical and Materials Transactions B*, 44 (1) (2013), 5-12.
11. S. Sridhar et al., "Break temperatures of mould fluxes and their relevance to continuous casting," *Ironmaking & steelmaking*, 27 (3) (2000), 238-242.
12. H.S. Park, H. Kim, and I. Sohn, "Influence of CaF_2 and Li_2O on the viscous behavior of calcium silicate melts containing 12 wt pct Na_2O," *Metallurgical and Materials Transactions B*, 42 (2) (2011), 324-330.
13. G.H. Kim and I. Sohn, "Influence of Li_2O on the Viscous Behavior of CaO-Al_2O_3-12 mass% Na_2O-12 mass% CaF_2 Based Slags," *ISIJ international*, 52 (1) (2012), 68-73.
14. J.H. Park, D.J. Min, and H.S. Song, "FT-IR Spectroscopic Study on Structure of CaO-SiO_2 and CaO-SiO_2-CaF_2 Slags," *ISIJ international*, 42 (4) (2002), 344-351.
15. G. Urbain, Y. Bottinga, and P. Richet, "Viscosity of liquid silica, silicates and alumino-silicates," *Geochimica et Cosmochimica Acta*, 46 (6) (1982), 1061-1072.
16. K. Zheng et al., "Investigation of the Viscosity and Structural Properties of CaO-SiO_2-TiO_2 Slags," *Metallurgical and Materials Transactions B*, 45 (4) (2014), 1389-1397.

7th International Symposium on High-Temperature Metallurgical Processing
Edited by: Jiann-Yang Hwang, Tao Jiang, P. Chris Pistorius, Gerardo R.F. Alvear F., Onuralp Yücel, Liyuan Cai, Baojun Zhao, Dean Gregurek, and Varadarajan Seshadri
TMS (The Minerals, Metals & Materials Society), 2016

HEAT LOSSES TO FURNACE COOLERS AS A FUNCTION OF PROCESS INTENSITY

M.W. Kennedy[1,2], A. MacRae[3], H. Haaland[4]

[1]Department of Materials Science and Engineering, Norwegian University of Science and Technology, N-7491 Trondheim, Norway.
[2]ProVal Partners, SA, Avenue de Sévelin 6 b, 1007 Lausanne, Switzerland.
(*Corresponding author: mark.w.kennedy@ntnu.no, m.kennedy@provalp.com)
[3]MacRae Technologies, Inc., 1000 Silver Maple Lane, Hayward, CA 94544-6681, USA.
[4]Elkem Technology, Drammensveien 169-171, 0277 Oslo, Norway.

Keywords: Furnace, Refractory, Cooler, Heat flux

Abstract

Furnace refractories are in most cases chemically incompatible with smelter slag, which leads to a steady erosion in their thickness over time. Once too thin, refractory walls become mechanically unstable and catastrophic failure can result. Historically, external shell cooling was applied to generate a freeze lining of slag and thus prevent refractory erosion. Many modern high intensity smelting furnaces instead maintain their physical integrity by the use of internally cooled wall panels, plates or finger coolers. An initial refractory lining is often installed inside of the coolers, such that the furnace originally operates with a temperature (insulated) rather than a heat flux (freeze lined) boundary condition. This paper examines the change in slag-wall heat transfer coefficient and slag superheat as a function of process intensity. The implications of the changes in heat transfer on residual brick thickness are explored using analytical modelling.

Introduction

Modern designs for both arc and slag resistance furnaces often incorporate extensive use of copper coolers [1-3], as shown in Figure 1 for an Electric Arc Furnace (EAF). Intensive cooling allows for long campaign life due to lower refractory wear, as well as the use of increased furnace power intensity and hence greater productivity [4].

Many existing electric furnaces have been 'upgraded' over time by operating at higher power with their present equipment (e.g. moving to brush arc) or adding both larger transformer capacity and steadily improved cooling and wall/hearth binding systems [5-10]. Increases in furnace power intensity (particularly so-called 'bath' power [11], i.e. Joule heating due to current conduction in the slag) can lead to greater slag superheat and produce higher sidewall heat fluxes increasing the difficulty of maintaining furnace vessel integrity even with the modern refractory, coolers and binding systems.

As furnace power intensities have increased, the sidewall heat losses have tended to rise and therefore the intensity of sidewall cooling technology has also increased, as indicated in Table I. Modern coolers such as those shown in Figure 1 can be designed to have limiting heat fluxes in excess of 1000 kW/m^2 [3].

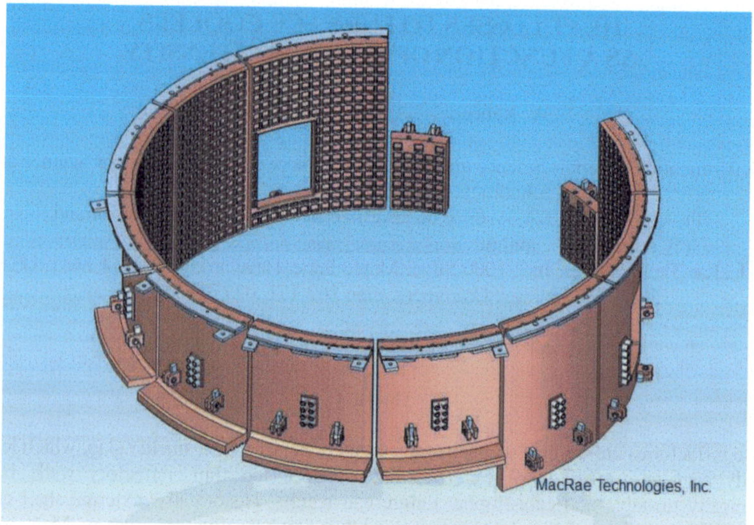

Figure 1: A modern Electric Arc Furnace (EAF) body design with extensive use of copper coolers [reproduced with permission of MacRae Technologies Inc.].

Table I: Typical Operational Heat Fluxes and System Design Limits for Various Furnace Sidewall Cooling Technologies

Cooling Technology	Typical kW/m^2	Design kW/m^2	References
Natural air convection	1	4.5	[12]
Forced air	2	6	[12]
Ripple cooling	3	10	[12]
Spray cooling	5	20	[12, 13]
Vertical chamber (atm)	5	20	[12]
Horizontal channels (2-4 bar)	10	40	[12]
Vertical channels (2-4 bar)	20	80	[12]
Finger coolers	15	40	[3]
Copper plates	20	100	[3, 12]
Vertical copper coolers with a hot-face pattern	30	>1000	[3]

Actual system limits can vary greatly, due to the use of high velocity or advanced technology.

This paper will explore the reaction of slag containing furnaces to changes in process operating intensity by way of an example using a generic $CaO{\cdot}SiO_2$ slag (48.3 wt.% CaO and 51.7 wt.% SiO_2). The implications of changes in furnace operating power with respect to bulk slag physical properties, sidewall heat losses, and equilibrium refractory thickness will be discussed.

Theory

Electric furnaces using immersed electrodes have been commercially applied to produce molten slag for the 'blowing' of rock wool [13, 14]. Such slag melting furnaces have operated using oil-cooled sidewalls to maintain a stable furnace lining at power intensities as high as ~360 kW/m^2 [15] as shown in Figure 2.

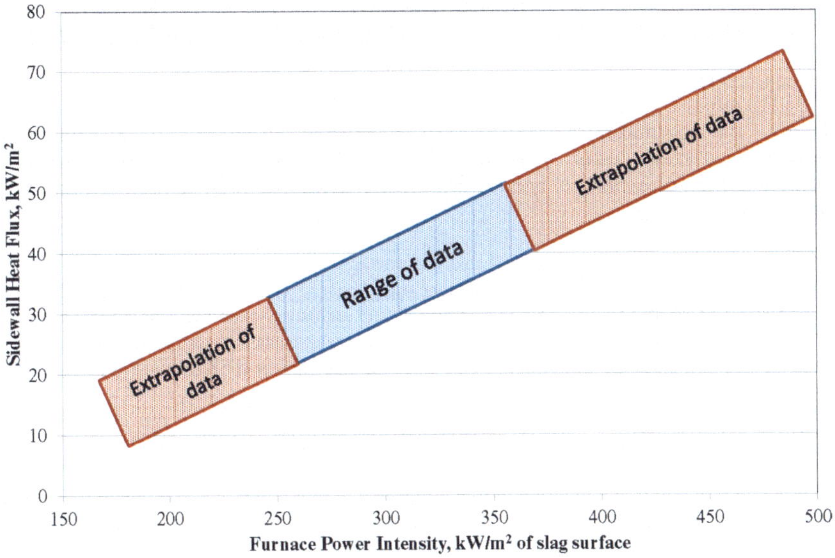

Figure 2. Approximate change in industrial furnace sidewall heat flux with furnace power intensity based on reference melting plants [15, 16], including linear extrapolation marked in red.

In the case of a slag melter, negligible endothermic reduction reactions take place. The supplied energy is consumed primarily by the sensible heat and net heat of fusion/mixing of the feed materials, and a smaller portion is lost through the walls, hearth and roof. Heat is generated close to the immersed electrodes warming the slag, which flows upwards primarily due to liquid buoyancy (in the absence of gas bubbles, and ignoring magneto-hydrodynamic effects [17]). The slag subsequently transfers sensible heat to the solid charge, due to the slag superheat as shown by Equation (1):

$$Q_{slag\text{-}charge} = A_{slag\text{-}charge}\, h_{slag\text{-}charge}\,(T_{slag} - T_{liquidus}) \tag{1}$$

where: $Q_{slag\text{-}charge}$ is the amount of thermal energy per unit time that must be transferred to heat, and melt the charge [W], $A_{slag\text{-}charge}$ the interfacial area between slag and charge [m^2], $h_{slag\text{-}charge}$

the overall slag-to-charge heat transfer coefficient (W/m^2·K), T_{slag} the average temperature of the slag under the charge [K], and $T_{liquidus}$ the liquidus temperature of the slag [K].

The temperature required to melt or dissolve the charge, is not strictly defined. For example, the charge in this simple case could be a mixture of pure CaO and SiO_2, a blend of CaO/SiO_2 containing materials or Wollastonite. During melting, the surface of the charge (interphase between solid and liquid) must adapt towards the local liquidus temperature, which can be either higher or lower than that of the bulk slag. However, the overall heat-transfer driving force remains that of Equation (1), as below the bulk slag liquidus, the bulk slag will begin to solidify.

From Equation (1) it is apparent that the average bulk slag superheat ($T_{bulkslag} - T_{liquidus}$), should also tend to increase with smelting rate, i.e. with greater required heat transfer to the charge, $Q_{slag\text{-}charge}$. The magnitude of the increase depends on: (i) the change of $h_{slag\text{-}charge}$ with power intensity; related to the degree of agitation, slag viscosity, thermal conductivity, density and coefficient of thermal expansion, and (ii) heat transfer area; related to the feed size and bath coverage.

To simplify the discussion, it is assumed that the average slag superheat increases linearly with furnace power intensity. An arbitrary initial slag superheat (85 K at 100 kW/m^2) was selected to closely match this study's results with the magnitudes of industrial example heat fluxes shown in Figure 2. Industrial benchmark data on the change in slag superheat with furnace intensity are lacking in the literature and would therefore be a very useful subject for future in-furnace studies.

The hot slag flows along the charge-slag interface transferring energy as it moves towards the sidewall where it encounters a relatively stagnant zone of slag and the furnace lining [17]. At thermal equilibrium an incipient freeze lining is created protecting the refractory from further chemical attack. The maintenance of this protective freeze lining requires that all energy arriving at the slag-wall interface be removed, which necessitates a heat flux boundary condition that can be described in a linear fashion by Equation (2):

$$\frac{Q_c}{A_c} = h_{slag-wall}\left(T_{bath} - T_{hotface}\right) \tag{2}$$

where Q_c is the heat flow [W], A_c is the cooler area [m^2], $h_{slag\text{-}wall}$ is the heat transfer coefficient between the molten slag and the hot-face of the freeze lining [W/m^2·K], T_{bath} is the temperature of the bath material arriving into the natural convection zone near the wall [K] and $T_{hotface}$ is the hot face temperature of the furnace lining in front of the cooler [K].

T_{bath} is assumed to be approximately equal to the average bulk temperature, $T_{bulkslag}$. At thermal equilibrium between the furnace lining and slag, $T_{hotface}$ is often taken as the liquidus temperature [18], and sometimes the average of the liquidus and solidus [19], but in the case of a compound like CaO·SiO_2, it is simply the slag melting point, 1813 K (1540°C) [20]. Equation (2) can be re-written specifically for the example used in this paper:

$$\frac{Q_c}{A_c} = h_{slag-wall}\left(T_{bulkslag} - 1813\right) \tag{3}$$

The sidewall has been described by Bendzsak and Baines as "a surface with approximately constant temperature next to a virtually stagnant slag bath" [17]. The conditions at the sidewall are therefore approximately those of a vertical plate undergoing natural convection heating. Kang and Robertson have published results on the heat transfer modelling of furnace copper coolers [21, 22]:

$$Nu = 0.32\left(GrPr\right)^{0.3} \quad (4)$$

where Nu is the Nusselt number at the slag-furnace lining interface [unitless], Gr is the Grashof number [unitless], and Pr is the Prandtl number [unitless], defined as:

$$Nu = \frac{h_{slag-wall}L}{k_{eff}}, \ (5) \quad Gr = \frac{g\rho_m^2\beta\left(T_{bath} - T_{hotface}\right)L^3}{\mu_m^2}, \ (6) \quad Pr = \frac{C_p\mu_m}{k_m}, \quad (7)$$

and $h_{slag-wall}$ is the natural convection heat transfer coefficient at the molten slag-furnace lining interface [$W/m^2 \cdot K$], L a characteristic length, i.e. the slag height along the cooler, which is assigned a maximum value equal to 1 in the present study [m], k_{eff} the effective thermal conductivity of the molten slag [$W/m \cdot K$] is taken as 0.25 at the liquidus and not adjusted for temperature, due to the uncertainty in literature values versus temperature for different measuring methods [23, 24], g the acceleration due to gravity, 9.81 [m/s^2], ρ_m the molten slag density ~2600 [kg/m^3], β the slag volumetric expansion coefficient ~9.9×10^{-5} [$m^3/m^3 \cdot K$], $_m$ the viscosity [Pa·s] and C_p the specific heat capacity of the liquid slag 1445 [J/kg·K].

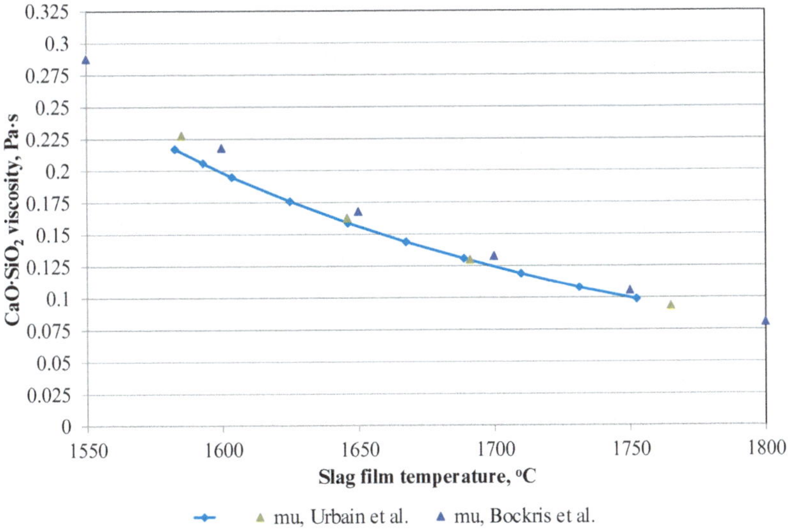

Figure 3. $CaO \cdot SiO_2$ modelled slag viscosity and comparison with literature values versus temperature [34].

High temperature slag properties are extremely difficult to measure and published data are often not in good agreement. Some physical and thermodynamic property data are available for $CaO{\cdot}SiO_2$ [20, 23-33]; however, the Ken Mills Excel "slag properties" spreadsheet [34] was used to estimate ρ_m, β, C_p, and $_m$ using the Urbain method [35]. All the properties were evaluated at the 'film' temperature of the slag at the sidewall (i.e. average of the slag bulk and liquidus temperatures or in this case the melting point). The slag property estimation methods described by Ken Mills *et al.* [36] provide an easy to apply set of temperature dependent equations. The Urbain method was selected for viscosity, being both 'simple' and in good agreement with published viscosity values for $CaO{\cdot}SiO_2$ [32, 33] as shown in Figure 3.

There is a general lack of liquid slag thermal conductivity data, k_{eff}, in the literature, and significant disagreement in conductivity values obtained by different methodologies (e.g. transient hot wire versus laser flash) at any given temperature [23, 24]. The correct value of k_{eff} versus temperature; therefore, introduces the most significant uncertainty in this analysis. The relative contributions of radiation, convection, and conduction to k_{eff} are also difficult to quantify, and are clearly composition, temperature and system geometry dependent.

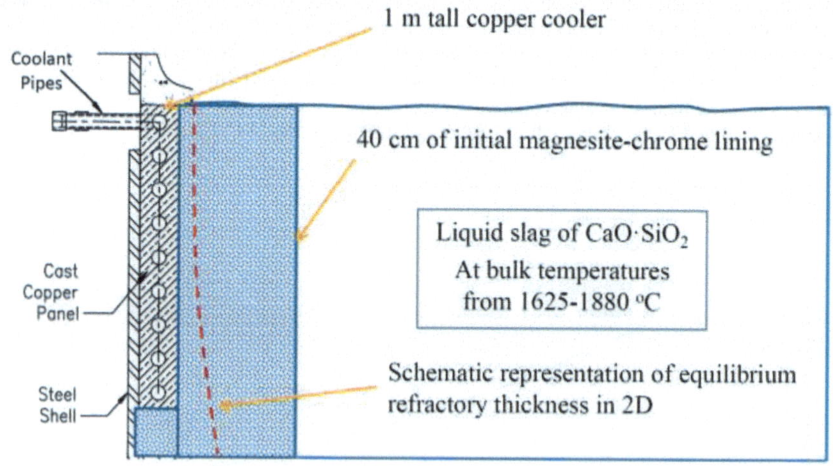

Figure 4. Schematic diagram of the system being modelled.

The 1D heat flux through the refractory lining and cooler, assuming negligible thickness of freeze lining and equal heat transfer areas can also be estimated by:

$$\frac{Q_c}{A_c} = \frac{(T_{hotface} - T_{coolant})}{\frac{d_{refractory}}{k_{refractory}} + \frac{d_{copper}}{k_{copper}} + \frac{1}{h}} \tag{8}$$

where $d_{refractory}$ is the thickness of the refractory assigned an initial value of 0.4 [m], $k_{refractory}$ the thermal conductivity of the refractory with a representative value of 2.23 [W/m·K] for a fired magnesite-chrome lining [37], d_{copper} the thickness of the copper between the cooling channels and the hot-face, assigned a value of 0.05 [m], k_{copper} the thermal conductivity of warm impure copper with a representative value of 334 [W/m·K] consistent with 85% of the theoretical conductivity [38], h the copper-coolant heat transfer coefficient assumed to have a value of 2000 [W/m²·K] representative of an effective oil based coolant, such as Mono-Ethylene-Glycol (MEG) and $T_{coolant}$ is assumed to be 60 °C, which is a typical for oil cooled systems [39]. For round furnaces with very thick walls or relatively small diameter, Equation (8) should be modified to account for the average heat transfer area through each resistive layer.

Rearranging Equation (8) to solve for the equilibrium refractory thickness yields:

$$d_{refractory} = k_{refractory}\left[\frac{(T_{hotface} - T_{coolant})}{\frac{Q_c}{A_c}} - \frac{d_{copper}}{k_{copper}} - \frac{1}{h}\right] \tag{9}$$

Equation (9) is solved by first determining $h_{slag\text{-}wall}$ using Equations (4-7) and then finding Q_c / A_c using Equation (3). A schematic of the system being studied is shown in Figure 4.

Results and Discussion

Equations (3) to (9) have been solved for the values of viscosity shown in Figure 3 for a variety of film temperatures (average of bulk temperature and wall hot-face temperatures) representative of different furnace power intensities. The 1D analytical results are given in Figure 5.

Figure 5 indicates that for a $CaO{\cdot}SiO_2$ slag, the assumption of linearly increasing slag superheat with furnace power intensity also produces a nearly linear increase in $h_{slag\text{-}wall}$ rising from 90 $W/m^2{\cdot}K$ at 100 kW/m^2 to 162 $W/m^2{\cdot}K$ at 400 kW/m^2. The overall wall heat loss also increases in an approximately linear fashion from 7.7 to 55 kW/m^2 over the same range of applied furnace power intensities. The roughly linear relationship is in agreement with the limited available industrial reference data summarized previously in Figure 2. Careful examination of Figure 5 shows that the combined effect of increases in both heat transfer coefficient and superheat with power intensity is not truly linear (with an exponent of ~1.4 against power intensity). This deviation from linearity is typically not observable with actual operational data, due to the lack of steady state conditions (high variability with time) and limited range in the changes in furnace operating intensity.

At a furnace bath power intensity >200 kW/m^2, a cooling methodology better than shell spray cooling (e.g. finger coolers) would be necessary, as indicated previously in Table I to prevent the internal refractory from becoming dangerously thin. Copper plates or vertical coolers with a hot face pattern would be necessary at applied power intensities beyond ~325 kW/m^2.

It can be observed that for the chosen model parameters that a freeze lining would be produced below about 110 kW/m^2 of applied power for the initial refractory lining of 400 mm of fired magnesite-chrome. At >110 kW/m^2 the original lining will be insulated until such time as the slag has consumed enough refractory to reach thermal equilibrium and produce a minimal initial freeze lining as indicated by the red-line with square symbols on Figure 5. There is currently no accepted method to accurately estimate the transition time from the initial insulated/temperature to the equilibrium freeze-lined/heat flux boundary conditions.

At furnace 'bath' power intensities >175 kW/m^2, the equilibrium refractory thickness is <50% of original and potentially mechanically unstable. At >300 kW/m^2, <25% of the initial refractory remains and the installation will be critically unstable. Results of a 1D model only give an indication of average equilibrium thickness. In actual fact there exists a 2D refractory profile, which varies from top (thin) to bottom (thick). Ultimately the wall stability will be determined by the thinnest part of the refractory. In order to obtain a 2D profile it is necessary to perform either 2D, 2D-axial symmetric or 3D CFD and FEM modelling, which necessitates quite accurate estimation of both liquid and solid slag properties over a wide range of temperatures.

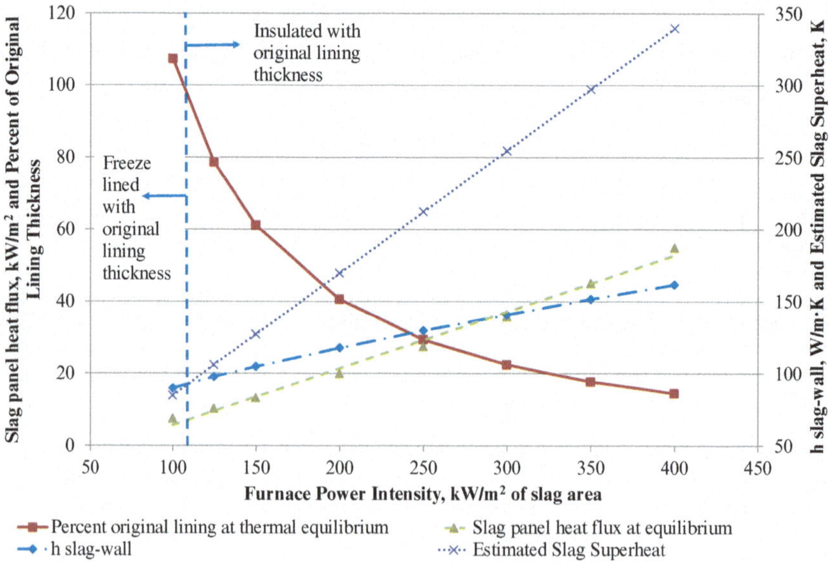

Figure 5. 1D analytical results for: (i) slag panel heat flux [kW/m^2], (ii) lining thickness at thermal equilibrium [m], (iii) slag-wall heat transfer coefficient $h_{slag\text{-}wall}$ [W/m·K], and (iv) estimated slag superheat [K], as a function of furnace power intensity, [kW/m^2].

It has been noted that slag melters reach a 'viscosity balance' [40] in operation, i.e. that the slag must achieve a certain fluidity in order to transfer heat effectively from the slag to the charge and disadvantageously from the slag to the wall. The effect is observed in actual operation when minor deviations in slag chemistry cause the liquidus of the slag to increase and the operating temperature in the furnace 'naturally' rises at a more or less constant applied power to achieve a superheat comparable to previous values, and correspondingly similar viscosity, melting rate and heat loss, i.e. more or less the same thermal efficiency as previously.

The 'viscosity balance' concept can be explored mathematically for furnaces operating with very different slag chemistries, e.g. over a large variation in basicity (from say 0.5 to 2.0). It must be noted that slag flow is induced at the electrode due to buoyancy and that greater upward velocity increases $h_{slag\text{-}charge}$. Examination of Equations (1-2) and (4-7) suggest that slags with low viscosity at a given operating temperature (high Rayleigh number) will therefore require a lower superheat to achieve a given melting duty; however, this low viscosity will also result in correspondingly higher heat losses at the wall for the same superheat. The result of the correlation between $h_{slag\text{-}wall}$ and $h_{slag\text{-}charge}$ is that slag furnaces working with different slags can operate at quite different superheats for the same power intensity, but achieve similar heat losses and therefore thermal efficiencies as a function of applied power intensity.

Conclusions and Recommendations

Using the modelling methodology of Mills [34] and assuming a linear increase of slag superheat produces a roughly linear increase in both slag-wall heat transfer coefficient, $h_{slag\text{-}wall}$, and also in slag-wall heat losses, Q_c/A_c, which is in agreement with industrial benchmark information over a large range of furnace power intensities.

In order to allow for better thermal analyses of furnaces, the actual change in slag superheat with furnace power intensity should be studied. The effective heat transfer conductivities of liquid slags (including conduction, radiation and convective components), k_{eff}, must also be studied more intensively and verified preferably by means representative of real slag-wall geometries, i.e. relatively thick liquid slag and a freeze lining on a cooled substrate.

Furnaces with low viscosity slags should operate with lower superheat at any given power input and melting rate in comparison with those using higher viscosity slags; however, both should operate with similar overall heat losses and thermal efficiencies.

References

1. J. Kunze and R. Degel, "New Trends in Submerged Arc Furnace Technology," in *Proceedings: Tenth International Ferroalloys Congress*, INFACON X, (2004), 444-454.
2. P. Argenta, C. Oertel, and B. Nourse, "Recent Development in Submerged Arc Smelting Technology," *Steel and Metallurgy*, November, (2008), 22-24.
3. J. Sarvinis, N. Voermann, C. Crowe, J. Bianchini, and B. Wasmund, "Furnace Design for Modern, High-Intensity Pyrometallurgical Processes," *Metallurgical Plant Design and Operating Strategies*, Sydney, NSW, 15-16 April, 2002, AusIMM, 318-331.
4. M. W. Kennedy, H. Haaland, J. A. Aune, "High Intensity Slag Resistance Furnace Design," *Proceedings of the Conference of Metallurgists, Nickel-Cobalt 2009*, CIM, Sudbury, Ontario, (2009) 101-110.

5. A. Daenuwy, A. Dalvi, M. Solar, and B. Wasmund, "Development of Electric Furnace Design and Operation at PT Inco (Indonesia)," in *International Symposium on Trace Metals and Furnace Practices in Non-Ferrous Pyrometallurgy,* Met. Soc. of CIM, Edmonton, Alberta, (1992) 1-23.
6. A. Matyas, R. Francki, K. Donaldson, and B. Wasmund, "Application of New Technology in the Design of High-Power Electric Smelting Furnaces," *CIM bulletin,* vol. 86, (1993), 92-99.
7. N. Voermann, V. Vaculik, T. Ma, C. Nichols, G. Roset, and W. Thurman, "Improvements to Stillwater Mining Company's Smelting Furnace Yielding Increased Capacity and Productivity," *Sulfide Smelting'98: Current and Future Practices,* (1998), 503-518.
8. J. Sarvinis, S. De Vries, K. Joiner, C. Van Mierlo, N. Voermann, F. Stober, C. Rule, and P. Majoko, "Improvements to BHP Hartley Platinum's Smelting Furnace," *Copper 99-Cobre 99,* (1999), 613-628.
9. L. Nelson, J. Geldenhuis, B. Emery, M. de Vries, K. Joiner, T. Ma, J. Sarvinis, F. Stober, R. Sullivan, and N. Voermann, "Hatch Developments in Furnace Design in Conjunction with Smelting Plants in Africa," *Southern African Pyrometallurgy 2006,* (2006), 417-436.
10. F. Stober, T. Miraza, A. T. Hidyat, I. Jauhari, K. Belanger, D. Fowler, T. Gerritsen, A. Matyas, C. Nichols, and N. Voermann, "Furnace Upgrade with Hatch Technology at PT Antam FeNi-II in Pomalaa, Indonesia," *INFACON XI, New Delhi, India,* (2007), 638-653.
11. N. Voermann, T. Gerritsen, I. Candy, F. Stober, and A. Matyas, "Furnace Technology for Ferro-Nickel Production - an Update," *International Laterite Nickel Symposium 2004, TMS Annual Meeting*, (2004), 563-577.
12. GDMB, "Heft 78, Feuerfestwesen in Metallhütten," *Schriftenreihe der GDMB,* (1997), 97-110.
13. G. Ellefsen and Ø. Hallquist, "The Elkem Multi-Purpose Furnace® Used for Slag Production, Technical Description with Plant Features," presented at the Elkem seminar, Bhubaneswar, India, (1989), 1-17.
14. J. Aune and O. Hallquist, "The Elkem Multipurpose Electric Smelter: A Cost Competitive Melt-Down Alternative," *Recycle and Secondary Recovery of Metals, The Metallurgical Society of AIME,* (1985), 833-855.
15. H. Haaland, "Elkem Internal Report," (1986).
16. Anonymous,"Elkem Internal Report," (1984).
17. G. Bendzsak and W. Baines, "Analysis of Heat and Mass Transfer Mechanisms in Electric Smelting Furnaces," *Proc. Int. Symp. on Non-ferrous Pyrometallurgy: Trace Metals, Furnace Practices, and Energy Efficiency. 31st Conf. Of Metallurgists, CIM, Edmonton, Canada*, (1992), 373-391.
18. A. Fallah-Mehrjardi, P. C. Hayes, and E. Jak, "Investigation of Freeze-Linings in Copper-Containing Slag Systems: Part I. Preliminary Experiments," *Metallurgical and Materials Transactions B,* vol. 44, (2013), 534-548.
19. H. Joubert, "Designing for Slag Freeze Linings on Furnace Sidewalls–an Engineering Perspective," *Molten Slags, Fluxes and Salts, Stockholm, Sweden-Helsinki,* (2000), 1-11.
20. G. Eriksson and A. D. Pelton, "Critical Evaluation and Optimization of the Thermodynamic Properties and Phase Diagrams of the $CaO-Al_2O_3$, $Al_2O_3-SiO_2$, and $CaO-Al_2O_3-SiO_2$ Systems," *Metallurgical Transactions B,* vol. 24, (1993), 807-816.
21. D. Robertson and S. Kang, "Model Studies of Heat Transfer and Flow in Slag-Cleaning Furnaces," *Fluid Flow Phenomena in Metals Processing,* (1999), 157-168.

22. S. Kang, "A Model Study of Heat Transfer and Fluid Flow in Slag-Cleaning Furnaces," Ph.D., University of Missouri-Rolla, (1992), 1-182.
23. Y. Kang and K. Morita, "Thermal Conductivity of the CaO-Al_2O_3-SiO_2 System," *ISIJ International,* vol. 46, (2006), 420-426.
24. T. Sakuraya, T. Emi, H. Ohta, and Y. Waseda, "Determination of Thermal-Conductivity of Slag Melts by Means of Modified Laser Flash Method," *Journal of the Japan Institute of Metals,* vol. 46, (1982), 1131-1138.
25. L. Zhang and S. Jahanshahi, "Review and Modeling of Viscosity of Silicate Melts: Part I. Viscosity of Binary and Ternary Silicates Containing CaO, MgO, and MnO," *Metallurgical and Materials Transactions B,* vol. 29, (1998), 177-186.
26. Y. Kang, K. Nomura, K. Tokumitsu, H. Tobo, and K. Morita, "Thermal Conductivity of the Molten CaO-SiO_2-FeO_x System," *Metallurgical and Materials Transactions B,* vol. 43, (2012), 1420-1426.
27. P. Richet, R. A. Robie, and B. S. Hemingway, "Thermodynamic Properties of Wollastonite, Pseudowollastonite and $CaSiO_3$ Glass and Liquid," *European Journal of Mineralogy,* (1991), 475-484.
28. M. Susa, M. Watanabe, S. Ozawa, and R. Endo, "Thermal Conductivity of CaO–SiO_2–Al_2O_3 Glassy Slags: Its Dependence on Molar Ratios of Al_2O_3/CaO and SiO_2/Al_2O_3," *Ironmaking & Steelmaking,* vol. 34, (2007), 124-130.
29. K. Nishioka, T. Maeda, and M. Shimizu, "Application of Square-Wave Pulse Heat Method to Thermal Properties Measurement of Cao-SiO_2-Al_2O_3 System Fluxes," *ISIJ International,* vol. 46, (2006), 427-433.
30. S. Ozawa and M. Susa, "Effect of Na_2O Additions on Thermal Conductivities of CaO–SiO_2 Slags," *Ironmaking & Steelmaking,* vol. 32, (2005), 487-493.
31. Y. Fei, "Thermal Expansion," *Mineral Physics and Crystallography: a Handbook of Physical Constants,* vol. 2, (1995), 29-44.
32. J. M. Bockris and D. Lowe, "Viscosity and the Structure of Molten Silicates," in *Proceedings of the Royal Society of London A: Mathematical, Physical and Engineering Sciences*, (1954), 423-435.
33. G. Urbain, Y. Bottinga, and P. Richet, "Viscosity of Liquid Silica, Silicates and Alumino-Silicates," *Geochimica et Cosmochimica Acta,* vol. 46, (1982), 1061-1072.
34. K. Mills. (accessed on-line, August 15, 2015). *http://www.pyrometallurgy.co.za/kenmills/.*
35. G. Urbain, "Viscosity Estimation of Slags," *Steel Research,* vol. 58, (1987), 111-116.
36. K. Mills, L. Yuan, and R. Jones, "Estimating the Physical Properties of Slags," *Journal of the Southern African Institute of Mining and Metallurgy,* vol. 111, (2011), 649-658.
37. "Handbook of Refractory Practice," Harbison-Walker Refractories Company, Pittsburgh, USA, (2005), 1-331.
38. R. Powell, C. Y. Ho, and P. E. Liley, "Thermal Conductivity of Selected Materials," NSRDS-NBS-8. National Standard Reference Data System, (1966), 1-175.
39. M. W. Kennedy, P. Nos, M. Bratt, and M. Weaver, "Alternative Coolants and Cooling System Designs for Safer Freeze Lined Furnace Operation," *Nickel-Cobalt 2013,* (2013), 299-314.
40. J. A. Aune, "Viscosity Balance in Slag Resistance Furnaces," Elkem, (2005), personal communication to M.W. Kennedy.

7th International Symposium on High-Temperature Metallurgical Processing
Edited by: Jiann-Yang Hwang, Tao Jiang, P. Chris Pistorius, Gerardo R.F. Alvear F., Onuralp Yücel, Liyuan Cai, Baojun Zhao, Dean Gregurek, and Varadarajan Seshadri
TMS (The Minerals, Metals & Materials Society), 2016

Viscosity of Partially Crystallized BOF Slag

Zhuangzhuang Liu, Bart Blanpain, Muxing Guo

Department of Materials Engineering, KU Leuven
Kasteelpark Arenberg 44, BE-3001 Leuven, Belgium

Keywords: BOF slag, viscosity, solid particles, added-value application

Abstract

In order to stabilize free lime in the basic oxygen furnace (BOF) slag to achieve high value products, quartz sand and other additives are injected into slag. However, the current hot stage slag stabilization does not work in case of slag (before treatment) with a liquid fraction lower than 90%, where the slag containing solid particles is too viscous. Therefore, it is essential to understand how solid phases affect the viscosity of the slag system. In this study, the industrial BOF slags were investigated with respect to the correlation between volume fraction of solid phases and its viscosity. Slag viscosity was measured with a rotating type viscometer at temperatures below the liquidus. The volume fraction of solid phases was quantitatively calculated with FactSage. The Einstein-Roscoe equation was employed to predict the viscosity of BOF slag.

Introduction

Basic oxygen furnace (BOF) slag is a by-product of the oxygen converter process and is a mixture of different mineral phases. The recycling of BOF slag is a significant challenge for steel industry in order to alleviate the strains of global resourcing and to achieve higher value products. On the other hand, the application of BOF slag in road construction or cement production is limited[1-3] because of the presence of free CaO or MgO in the BOF slag, which leads to a longer term volume instability due to the expansive hydration to $Ca(OH)_2$ and $Mg(OH)_2$. A hot stage slag engineering process has been developed and implemented to solve this problem[4]. The principle of this process lies in the introduction of additives in order for free CaO and MgO to react towards a stable matrix of calcium silicates and ferrites. This is achieved by the addition of SiO_2-containing materials, such as quartz sand, glass cullet and spent foundry sands. The treatment with quartz sand offers the advantage of higher SiO_2 content per mass of additives and is not introducing other components that can cause side reactions. However, the current hot stage slag stabilization process does not work in case of slag (before treatment) with a liquid fraction lower than 90%, where the slag containing solid particles is too viscous. The slag viscosity quantifies the flow properties of the slag and affects the dissolution speed of additives, the size and distribution of entrained metallic iron particles in the slag, the sedimentation of metallic Fe particles (or droplets) to the bottom of slag pot, mass and heat transfer through the slag. The high viscosity of the BOF slags hampers handling and blocks the hot stage slag processing (e.g. leading to a serious problem with the stabiliser (SiO_2) injection). Therefore, it is essential to obtain a specific knowledge of BOF slag viscosity and to understand how solid phases affect the viscosity of the slag system.

Many efforts have been attempted to study the viscosities of metallurgical slags, where distinct viscosity models for fully liquid slag have been proposed, such as Urbain model[5], Riboud model[6], Iida model[7], quasi-chemical model[8-13] and others. However, there are only limited investigations on the viscosities of the slags containing solid particles [14-17] .This is attributed to the complexity of the viscosity of heterogeneous molten slag, which is affected by many factors, such as composition and temperature of the slag matrix, size and shape of solid crystals, fraction and distribution of the crystals and the rheological properties of molten slags. The objective of this work is to obtain specific knowledge of BOF slag viscosity and to understand how the solid crystals affect the viscosity of the slag system.

In the present study, viscosity of two industrial BOF slag samples have been measured with a rotational type viscometer at temperatures below liquidus. The volume fraction of solid phases at corresponding temperatures has been determined with FactSage and the effect of solid fraction on slag viscosity has been discussed.

Experimental

Materials and sample preparation

Two different industrial BOF slag samples were taken by dipping a steel tube into molten slag in a slag pot after slag tapping and then the slags were milled into powders for further analysis. Compositions of the two samples analyzed via XRF are shown in table 1.

Table 1. Chemical composition of BOF slag samples (XRF, wt%)

Sample	FeO	Fe_2O_3	MnO	CaO	Al_2O_3	P_2O_5	MgO	SiO_2	V_2O_5	Cr_2O_3	TiO_2	SiO_2/CaO
A	8.94	22.95	3.98	51.58	2.22	2.66	2.37	7.32	0.47	0.30	0.74	7.05
B	9.12	23.61	4.33	51.28	3.50	2.58	2.41	5.77	0.45	0.34	0.57	8.88

Viscosity measurement

Viscosity of the two samples were measured with a rotational type viscometer (Ravenfield Model FG MkIV viscometer; Ravenfield Designs Ltd, Heywood, UK). The details of the set-up can be found elsewhere[18]. The crucible and spindle are made of molybdenum. Before each measurement, the viscometer was calibrated with standard oil. After calibration, 3 gram of slag powder was filled in the crucible and heated to different temperatures (1700°C, 1650°C, 1600°C, 1550°C) under Ar atmosphere(99.999%). The sample was held at the measurement temperature for 30 minutes for homogenization of the slag and then the spindle was lowering down into molten slag to measure the viscosity at a rotating speed of 200 rpm.

Results and discussion

Volume fraction of solid phases

The volume fraction of solid phases was determined by FactSage, which is expressed through the following equation[1]:

$$\Phi_S = \frac{V_s}{V_s + V_L} \quad (1)$$

Where the volume of solid phase and liquid phase (V_s and V_L) can be calculated with

$$V_S = \sum_j v_j^{(s)} \frac{m_j^{(S)}}{M_j} \quad (2)$$

$$V_L = \sum_i v_i^{(L)} \frac{m_i^{(L)}}{M_i} \quad (3)$$

where i and j indicates the liquid and solid component, respectively. m, v and M are weights, molar volumes and molar weights of i or j. In the present calculation with FactSage, the mean molar volume of each component has been taken instead of the actual molar volumes, the uncertainty introduced by this process is estimated to be within 3%[14].

Figure 1 indicates the correlation between volume fraction of solids and temperatures. Because of the high liquidus of BOF slag, the volume fraction of solids reaches to 8.5% even at 1700°C. Since viscosity models for fully liquid slag is not applicable to predict viscosity for BOF slag in the present case, modeling of partially crystallized BOF slag viscosity is necessary.

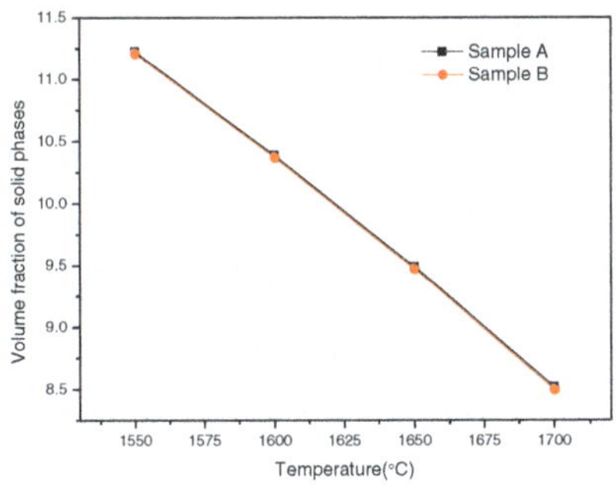

Figure 1. Volume fraction of solid phases as a function of temperature

Viscosity modelling

The first well-known theoretical work to predict viscosity of a suspension was carried out by Albert Einstein[19] as shown in the following equation:

$$\eta_R = \frac{\eta_S}{\eta_L} = 1 + 2.5\Phi_S \quad (4)$$

Where η_S and η_L are the viscosity of suspension and suspending liquid, respectively; Φ_S is the volume fraction of solids. The application of this expression is limited to a suspension with up to 5% volume fraction of rigid spherical solids with the same diameter. Another expression modified by Roscoe[20] was reported to be valid for a concentrated suspension with rigid spheres of the wide size distribution, as shown in formula (5).

$$\eta_R = (1 - 1.35\Phi_S)^{-2.5} \tag{5}$$

A more general form of equation (5) with adjustable parameters K and B is often employed and known as the Einstein-Roscoe equation for relative viscosity calculation (equation (6))[21].

$$\eta_R = (1 - K\Phi_S)^{-B} \tag{6}$$

However, there is no consensus yet for the value of parameter K and B in this equation. Kondratiev has investigated the viscosities of the partially crystallized slags in Al_2O_3-CaO-"FeO"-SiO_2 system based on the experimental dataset by Hurst[14]. The volume fraction of solid phases were also calculated with FactSage. It was found that the Einstein-Roscoe equation with K=2.04 and B=1.29 provided a very good fit for experimental results. Wright[15] studied the viscous properties of the CaO-MgO-Al_2O_3-SiO_2 slag with additions of the $MgAl_2O_4$ spinel particles with different quantity and particle size. They found that the measured viscosity was in a good agreement with the calculated one by using the Einstein-Roscoe equation with K=4.24, 3.29, and 3.56 and B=1.28, 2.36, and 2.24 for fine (0.1 to 0.21 mm), medium (0.21 to 0.44 mm), and coarse (0.44 to 0.99 mm) particles, respectively. They also tried to carry out the fitting by varying the value of K while keeping B at a constant value of 2.5. In this way, the values of K were found to be 2.6, 3.2, and 3.4 for fine, medium, and coarse particle additions, respectively. It was concluded that the two-parameter fit and the one-parameter fit had little difference with each other within experimental scatter. In another paper[16], Wright and his co-workers measured the viscosity of a CaO-FeO_x melt containing Fe_3O_4 particles with volume fraction up to 10%. It was found that the experimental data fitted well with the Einstein-Roscoe equation with K =2.5 and B=4.4. Seok[22]studied the viscosity of CaO–SiO_2–FeO–MgO–Al_2O_3 slags at 1773 K and fitted the experimental data with Einstein-Roscoe equation with B fixed at 2.5 and it was found the value of K varied from 2.74 to 3.89.

In the present work, the Einstein-Roscoe type equation was employed to predict the relative viscosity of partially crystallized BOF slag. The viscosity module in FactSage was used to determine the viscosity of the remaining liquid slag. The parameters in the Einstein-Roscoe equation were optimized to minimize the sum of squared differences between the predicted and experimental values. Table 2 shows the statistic of optimization and the concrete values of the parameters. It shows that the Einstein-Roscoe equation with empirical parameters K=7.59 and B=0.64 provides good fit to experimental data for slag A, while for slag B, the K and B are equal to 8.51 and 0.35, respectively. Apparently, the empirical parameters for two samples are very different, even though the composition and quantity of solids of the two samples at different temperatures are similar. This diversity is believed to be the result of the differences of the two samples with respect to size, shape and distribution of the solid phases present at the measurement temperature. Further work is needed in this field. The one-parameter fit by fixing parameter B at 2.5 was also conducted and the results are given in table 3. The optimized parameter K was found to be 3.20 and 6.23 for slag A and B, respectively. The adjusted coefficient of determination (Adj. R-square) was used to evaluate the goodness of fitting. Apparently the two-parameter fit could predict the viscosity better than the one-parameter fit, because of the higher adjusted R square. This is believed to be the result of the irregular shape (instead of spherical one) of solid phases in the BOF slag samples, considering that B = 2.5 is obtained for spherical particles. Figure 2 shows the measured viscosity and calculated viscosity with two-parameter fit for the two BOF slags. Considering the fact that the variance of viscosity measurement is 20%[23], it can be concluded that Einstein-Roscoe equation predicts the BOF slag viscosity well. Figure 3 demonstrates the comparison between viscosity of the solid and liquid mixture (entire slag) and that of the (remaining) liquid phase of the slag. It is found that contribution of the solid phases to the viscosity of the entire slag increases with decreasing in temperature. This solid particles' contribution is considerable even at high temperature(1700°C) with 8.5% volume fraction of solids (slag A and B).

Table 2. Empirical parameters in Einstein-Roscoe equation obtained by two-parameter fitting

Sample	K	B	Adj. R-square
A	7.59	0.64	0.966
B	8.51	0.35	0.944

Table 3. Empirical parameter K in Einstein-Roscoe equation obtained by one-parameter fitting for B=2.5

Sample	K	Adj. R-square
A	3.20	0.772
B	2.63	0.539

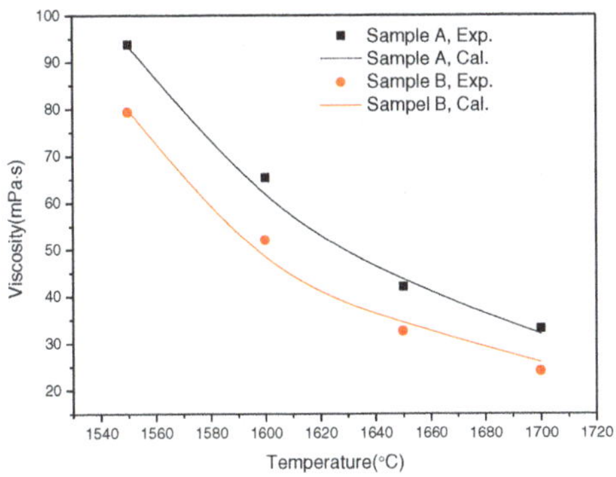

Figure 2. Experimental and calculated viscosity of BOF slag samples as a function of temperature

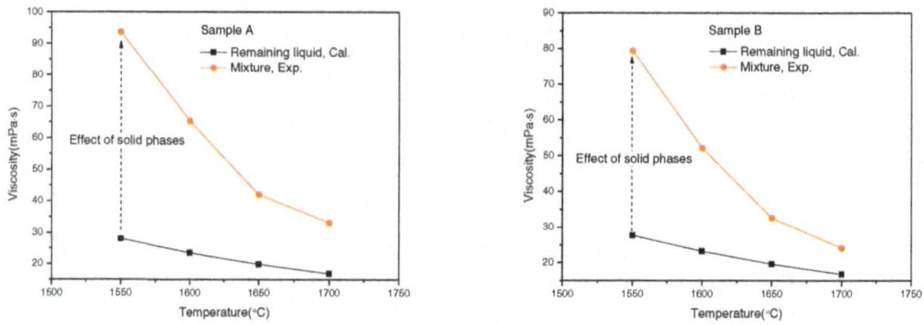

Figure 3. Effect of solid phases on the viscosity of BOF slag samples

Conclusion

The volume fractions of major solid phases of BOF slag at temperature range from 1550 to 1700°C were determined via FactSage calculation. The viscosity of the samples was measured with a rotational type viscometer at the corresponding temperatures. The effect of solid fraction on slag viscosity has been discussed based on the measured viscosity data and the Einstein-Roscoe modelling. The following conclusions were obtained

(1) The two-parameter fit of Einstein-Roscoe equation could predict the viscosity of BOF slag better than the one-parameter fit.

(2) The optimized parameters K and B in Einstein-Roscoe equation are 7.59 and 0.64, 8.51 and 0.35 for slag A and B in the present study, respectively.

(3) Contribution of solid phases on BOF slag viscosity increases with temperature decreasing. The effect of solids is considerable even at high temperature (1700°C) for BOF slag A and B in the present study with 8.5% volume fraction of solid phases.

Acknowledgements

The author would like to thank Professor Jungwook Cho (Pohang University of Science and Technology) and Dr SeungHo Shin (Pohang University of Science and Technology) for viscosity measurement and valuable discussion in this paper.

References

[1]G. Wang, Y. Wang and Z. Gao, "Use of Steel Slag as A Granular Material: Volume Expansion Prediction and Usability Criteria," J. Hazard. Mater., 184 (2010), 555–560.

[2]J.J. Emery, "Slag Utilization in Pavement Construction," Extending Aggr. Resour., 774 (1982), 95–118.

[3]S.A. Mikhail and A.M. Turcotte, "Thermal Behaviour of Basic Oxygen Furnace Waste Slag," Thermochim. Acta, 263 (1995), 87–94.

[4]M. Kühn et al., "Treatment of Liquid Steel Slags". (Paper presented at the 2nd European Slag Conference, Düsseldorf, Germany, 9 October 2000), 123-135.

[5]G. Urbain, Y. Bottinga, and P. Richet. "Viscosity of Liquid Silica, Silicates and Alumina-Silicates," Geochimica et Cosmochimica Acta, 46(1982), 1061-1072.

[6]P.V.Riboud et al., "Improvement of Continuous Casting Powders," Fachberichte Hüttenpraxis Metallweiterverarbeitung, 19(1981), 859-869.

[7]T. Iida et al., "Equation for Estimating Viscosities of Industrial Mold Fluxes," High Temperature Materials and Processes, 19(2000), 153–164.

[8]A. Kondratiev and E. Jak. "A Quasi-Chemical Viscosity Model for Fully Liquid Slags in The Al_2O_3-CaO-'FeO'-SiO_2 System," Metallurgical and Materials Transactions B. 36(2005), 623-638.

[9]A.N. Grundy et al., "A Model to Calculate the Viscosity of Silicate Melts," International Journal of Materials Research, 99(2008), 1185-1194.

[10]A.N. Grundy et al., "A Model to Calculate the Viscosity of Silicate Melts," International Journal of Materials Research, 99(2008), 1194-1209.

[11] W. Kim et al., "A Model to Calculate the Viscosity of Silicate Melts," International Journal of Materials Research, 103(2012), 313-328.

[12] E. Brosh et al., "A Model to Calculate the Viscosity of Silicate Melts," International Journal of Materials Research, 103(2013), 494-501.

[13] E. Brosh et al., "A Model to Calculate the Viscosity of Silicate Melts," International Journal of Materials Research, 103(2013), 537-550.

[14]A. Kondratiev and E. Jak. "Modelling of Viscosities of the Partly Crystallized Slags in the Al_2O_3-CaO-"FeO"-SiO_2 system," Metallurgical and Materials Transactions B, 32(2001), 1027-1032.

[15]S. Wright et al., "Viscosity of a CaO-MgO-Al_2O_3-SiO_2 Melt Containing Spinel Particles at 1646K," Metallurgical and Materials Transactions B, 31(2000), 97-104.

[16]S. Wright et al., "Viscosities of Calcium Ferrite Slags and Calcium Alumino-Silicate Slags Containing Spinel Particles," Journal of Non-Crystalline Solids, (282)2001, 15–23.

[17]R.G. Reddy and J.Y. Yen. "Effect of Solid Particles on the Viscosity of Slags," (Paper presented at the International Symposium on Extractive Metallurgy of Copper, Nickel and Cobalt, TMS, Denver, America, 21-25 February, 1993)

[18]S.H. Shin, J.W. Cho and S.H. Kim, "Shear Thinning Behavior of Calcium Silicate-Based Mold Fluxes at 1623 K," J. Am. Ceram. Soc., 97 (2014), 3263–3269.

[19]A. Einstein, "Eine neue Bestimmung der Molekuldimensionen," Ann. Phys., 19(1906), 289-306.

[20]R. Roscoe. "The Viscosity of Suspensions of Rigid Spheres," Br. J. Appl. Phys. 3(1952), 267-269.

[21]M. Duchesne et al,. "Slag Viscosity Modelling Toolbox," Fuel, 114(2013), 38–43.

[22]S. Seok et al,. "Viscosity of Highly Basic Slags," ISIJ International, 47(2007), 1090–1096.

[23]K.C. Mills et al., "'Round Robin' Project on the Estimation of Slag Viscosities," Scandinavian Journal of Metallurgy, 30(2001), 396-403.

7th International Symposium on High-Temperature Metallurgical Processing
Edited by: Jiann-Yang Hwang, Tao Jiang, P. Chris Pistorius, Gerardo R.F. Alvear F., Onuralp Yücel, Liyuan Cai, Baojun Zhao, Dean Gregurek, and Varadarajan Seshadri
TMS (The Minerals, Metals & Materials Society), 2016

ORIGIN AND EVOLUTION OF NON-METALLIC INCLUSIONS FOR Al-KILLED STEEL DURING EAF-LF-VD-CC PROCESS

TANG Hai-yan[1,2], GUO Xiao-chen[2], CHENG Peng-fei[2], LIANG Yong-cang[2], LI Jing-she[2], ZHAO Baojun[3]

[1]State Key Laboratory of Advanced Metallurgy, University of Science and Technology Beijing; No.30 Xueyuan Road; Beijing, 100083, China
[2]School of Metallurgical and Ecological Engineering, University of Science and Technology Beijing; No.30 Xueyuan Road; Beijing, 100083, China
[3]School of Chemical Engineering, The University of Queensland, Brisbane, Australia

Keywords: Al-killed steel; Inclusions; Origin; Evolution

Abstract

The origin and evolution of non-metallic inclusions for Al-killed casing steel during EAF-LF-VD-CC process were studied by industrial tracer experiments. The formations of Al_2O_3 and $MgO·Al_2O_3$ micro- and macro-inclusions from ladle slag were analyzed by theoretical calculations. The results show that micro-inclusions were formed during deoxidation and temperature decreasing in the process of casting, most of the macro-inclusions come from entrapped ladle and mold slags. In addition, most of the Al_2O_3 formed were modified into $12CaO·7Al_2O_3$ and $3CaO·Al_2O_3$ inclusions which are liquid at steelmaking temperature. MnS cannot be fully modified to CaS and it was only wrapped outside the alumina or calcium aluminate in the form of (Ca,Mn)S with the amount of Ca-Si wire fed during the test.

Introduction

Casing is one of the important materials for oil fields which can reinforce the wall of an oil well and protect the hole, and its destruction can lead to the failure of a complete well. Therefore, not only are high strength, uniform and stable qualities, and strong corrosion and wear resistances required for casings, but also high capacity to support all kinds of loads such as pulling, pressing, twisting, and bending is needed [1]. Many factors can influence the performance of casing steel. A trail in a steel factory in China showed that the non-metallic inclusions were one of the main causes. However, the origin of the inclusions is very complicated which may come from deoxidation, ladle/tundish/mold slags or furnace lining. In order to control effectively the non-metallic inclusions in production, it is necessary to understand their origin and formation mechanisms. In this paper, the origin and evolution of non-metallic inclusions for Al-killed casing steel during EAF-LF-VD-CC process were studied by industrial tracer test. The formation mechanisms were discussed base on the industrial test.

Industrial trials and study methods

The industrial trials were performed for the production of three heats casing steel with the element compositions in billets (mass/%) C: 0.39~0.40, Si: 0.21~0.23, Mn: 1.52~1.56, P:

0.014~0.015, S: 0.006~0.008, Als: 0.02~0.03, V: 0.12, T.Ca: 0.0012~0.0016, Mg: 0.0003. The production route was 150t EBT—EAF→LF→VD→Ca-Si treatment→CC.

During electric arc furnace (EAF) tapping, a certain amounts of Si-Fe, Mn-Fe alloys and Al particles were added into the molten steel in the ladle for preliminary deoxidation combined with argon stirring. Synthesis slag was also added to desulphurize and decrease the contents of FeO and MnO. When the ladle reached the refining station (LF), Al wires were fed to further deoxidize and then the graphite electrodes were switched to heat the molten steel and adjust the slag composition. After LF treatment, the ladle reached vacuum degas (VD) station to remove gases such as O_2, N_2 and H_2. Then Ca-Si wires were fed to modify the high melting point inclusions. A 30t tundish and six-strand curved type continuous caster were equipped. The cross section of the round billet is Φ350mm.

In order to define the origin of the inclusions, tracers were used in this trial. BaCO3, which was equivalent to 8 mass% ladle slag, was mixed into the ladle slag to trace if the ladle slag was entrapped into the molten steel. 7 mass% of CeO_2 was mixed into the tundish slag, and 7 mass% La_2O_3 was added into tundish paint when constructing a new tundish. They were used to trace the entrapment of tundish slag and paint. Na_2O and K_2O were used to determine the entrapment of mold slag.

Steel and slag samples were taken before and after LF refining, after Ca treatment, in tundish and billets. The compositions of the steel samples were analyzed by ICP-AES (Inductively Coupled Plasma-Atomic Emission Spectrometry) and Carbon/Sulphur analyzer. The total oxygen (T.O.) and nitrogen were analyzed by TC—436 oxygen/nitrogen analyzer from Leco, USA. The slag was analyzed by chemical method. Micro-inclusions on the cross section f each steel sample were detected and analyzed by SEM-EDS to obtain their morphologies, sizes and chemical compositions. Macro-inclusions larger than 50μm were obtained by electrolysis method with non-water solution.

Results

Total oxygen and nitrogen in steel samples

Total oxygen contents in the steels include dissolved and compounded oxygens and it can well reflect the cleanliness of the steel. The change in nitrogen content is an indicator of the air absorbed during transportation of molten steel [2,3]. The averages of total oxygen and nitrogen contents of steel samples in different steps are shown in Figure 1. It can be seen that the total oxygen was decreased from 62ppm to 37ppm during LF refining, and it was 21ppm after VD calcium treatment, showing that the LF and VD were effective for oxygen removal. The total oxygen was increased by 2ppm in the tundish, indicating reoxidation could occur during casting which was confirmed by the increase of nitrogen content from 67ppm to 80ppm during the process. This suggests that the protection measure for the long nozzle needs to be improved. The total oxygen of head billet was 29ppm, which was significantly higher than other billets such as normal, continuous casting and tail billets. This could be caused by instable casting which resulted in slag entrapment and reoxidation.

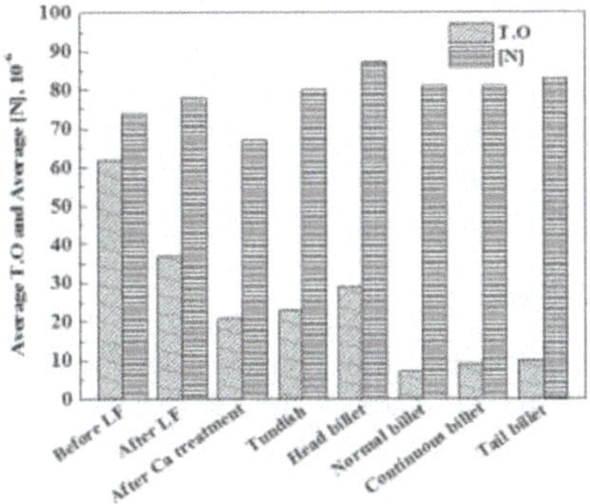

Figure1.Variation of the average T[O] and [N] in steel samples

Evolution of non-metallic inclusions in different stages

Before LF refining, the observed inclusions are mainly pure Al_2O_3 (Figure 2(a)), Al_2O_3–based inclusion (Figure 2(b)) and MnS inclusion (Figure2(c)). Al_2O_3 was formed by the reaction of Al and [O] in molten steel, Al_2O_3–based inclusion was formed by the compounded deoxidation of Al, Si and Mn, and MnS was precipitated by the solidification of molten steel due to high Mn content (1.54 mass%). Some of MnS show string shape, distributing around intergranular, which will cause anisotropy of steel.

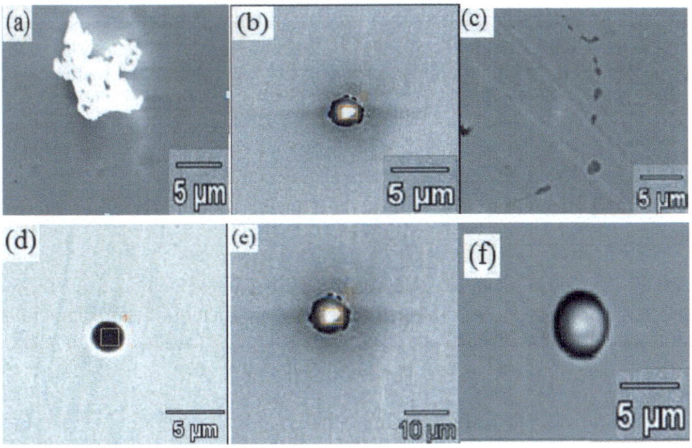

Figure 2 The typical microinclusions observed, (a) Al_2O_3, (b) Al_2O_3–based inclusion, (c) MnS, (d) MgO-Al_2O_3, (e) CaO-Al_2O_3-SiO_2-MgO, (f) $12CaO{\cdot}7Al_2O_3$

After LF, the inclusions were mainly transferred into $MgO-Al_2O_3$ (Figure 2(d)) and $CaO-Al_2O_3-SiO_2-(MgO)$ (Figure 2(e)) due to steel-slag and steel-refractory interface reactions, mostly less than 5 m in size. After VD degas, Ca-Si wire was fed into the molten steel, most of the inclusions were modified into spherical calcium aluminate $12CaO\cdot 7Al_2O_3$ (Figure3(f)) and $CaO-MgO-Al_2O_3$ type inclusions. Some of these inclusions were wrapped with (Ca,Mn)S with the size 2～20 m. The amount of MnS was significantly decreased during LF and VD refining and they were mostly coexisted with oxides in the form of (Ca,Mn)S, and hardly pure CaS was found.

Macro-inclusions in steel samples

Macro-inclusions refer to those larger than 50μm and they were extracted with electrolysis method. Figure 3 shows the typical morphologies of some macro-inclusions and their EDX patterns.

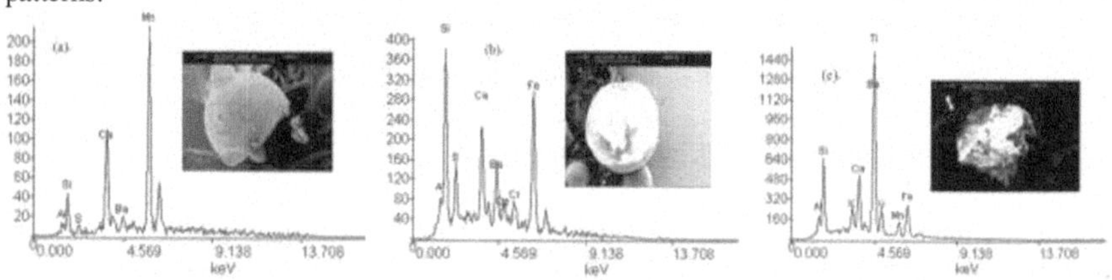

Figure 3 Morphology and energy spectrum of typical macroinclusions (a) after LF samples; (b) billet samples; (c) billet samples

It can be seen from Figure 3(a) that the macro-inclusion in the steel after LF contains Ba, indicating this type of macro-inclusions originated from ladle slag. Figures 3(b) and 3(c) show macro-inclusions in billets contain Ba, Ce and K, indicating ladle slag, tundish refractory and mold slag were entrapped into the molten steel to form complicated macroinclusions. The details will be discussed in the following sections.

Discussions

Effect of [Ca] and [Al] contents in molten steel on the compositions of the precipitated inclusions during LF refining

In order to control the inclusions in practice, the thermodynamic analysis for the formation of the inclusions during LF refining is performed. The activity coefficient of each element and activity in liquid steel is calculated by Eqs. (1) and (2).

$$\lg f_i = \sum_j e_i^j[\text{mass}\%j] \qquad (1)$$

$$a_{[i]} = f_i \cdot [\text{mass}\%i] \qquad (2)$$

Where e_i^j represents the first order interaction coefficient of elements j to i relative to diluted solution, and $a_{[i]}$ is the activity of component i relative to mass 1% diluted solution. Table 1 lists the calculated activity coefficient of each element in liquid steel at 1873K. In the calculation, the composition of the liquid steel is (mass%): C, 0.39; Si, 0.22; Mn, 0.52; S, 0.007; P, 0.015; Al, 0.02; Ca_T, 0.0016. The interaction coefficients of the elements in liquid steel are taken from the references [4,5].

Table 1 Activity coefficients of elements in liquid steel at 1873 K for casing steel [4,5]

T	f_C	f_{Si}	f_{Mn}	f_{Al}	f_S	f_O	f_{Mg}
1873K	1.139	1.259	0.938	1.062	0.953	0.492	0.767

Table 2 Deoxidation reactions in liquid steel and related equilibrium constant expressions [6]

Reaction equation	ΔG^0/(J/mol)	Eq.
$[Si]+2[O]=SiO_2(s)$	$\Delta G^0 = -581900+221.8T$ J/mol	(3)
$2[Al]+3[O]=Al_2O_3(s)$	$\Delta G^0 = -1202000+386.3T$ J/mol	(4)
$6[Al]+2[Si]+13[O]= 3Al_2O_3(s)\cdot 2SiO_2$	$\Delta G^0 = -47741543+1592T$ J/mol	(5)
$[Ca]+[Si]+3[O]=CaO\cdot SiO_2$	$\Delta G^0 = -1301477+359.73T$ J/mol	(6)
$[Ca]+2[Si]+2[Al]+8[O]= CaO\cdot Al_2O_3\cdot 2SiO_2$	$\Delta G^0 = -3088133+942.88T$ J/mol	(7)
$2[Ca]+2[Al]+[Si]+7[O]= 2CaO\cdot Al_2O_3\cdot SiO_2$	$\Delta G^0 = -3190946+861.23T$ J/mol	(8)

In order to improve the casing quality, the main task of LF refining is to decrease the amount of inclusions and control their compositions in the fully-liquid zone at the refining temperature. During LF refining, due to the steel-slag interaction, different types of inclusions may be formed. According to the ternary phase diagram $CaO\text{-}Al_2O_3\text{-}SiO_2$ [7], the possible precipitated non-metallic inclusions include Al_2O_3, SiO_2, CA_X, $3Al_2O_3\cdot 2SiO_2$ (short for A_3S_2), $CaO\cdot Al_2O_3\cdot 2SiO_2$ (CAS_2), $2CaO\cdot Al_2O_3\cdot SiO_2$ (C_2AS), $CaO\cdot SiO_2$ (CS), in which the eutectic area of CAS_2 and CS is expected with lower melting temperature. These substances could be formed by the deoxidation reactions [6].

The activities of precipitated phases in Eqs (3)~(8) are taken as unity, substituting the activities coefficient in Table 1, the stable phase zone of the above mentioned substances can be obtained as shown in Figure 4. It can be seen that Al_2O_3 inclusion is easily formed when the acidic dissolved aluminum, $[Al]_s$, is over 3ppm. To control the compositions of the inclusions in eutectic area of CS and CAS_2, the required $[Al]_s$ is between 0.1~20ppm and $a_{[Ca]} < 9.2\times 10^{-10}$. The acidic dissolved aluminum is about 0.01~0.03% during LF refining, CS and CAS_2 are not able to form and only Al_2O_3 inclusion is stable. Figure 5 shows the relationship of $[Al]_s$ and $a_{[Ca]}$ when forming CAS_2 and C_2AS. Compared with Figure 4, the concentration range to form C_2AS is in the same range to form CS and Al_2O_3, thus CS and Al_2O_3 will be formed prior to C_2AS. This means C_2AS inclusion cannot be formed in the present LF process conditions. As a result, the main inclusion during LF refining is Al_2O_3, which is consistent with the observations in the experiments.

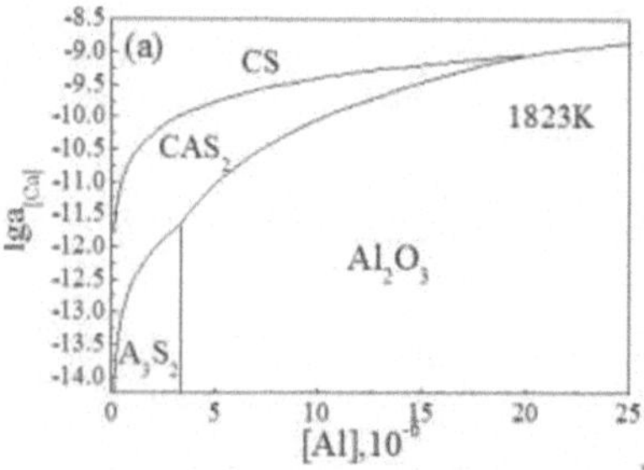

Figure 4 Effect of [Ca] and [Al] contents on deoxidation products at 1873K

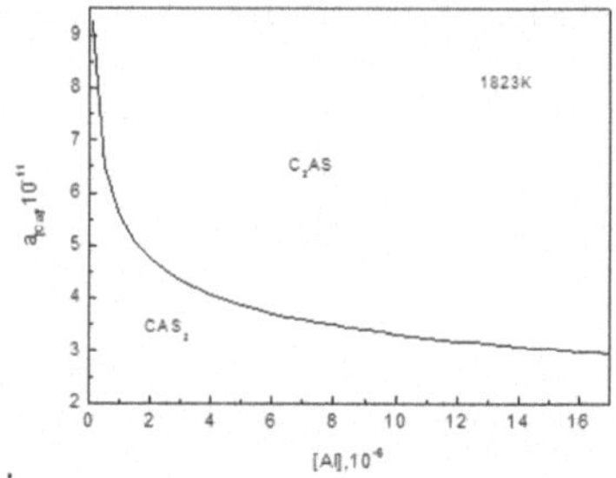

Figure 5 $a_{[Ca]}$-[Al] relationship when forming CAS_2 and C_2AS

Formation of $MgO{\cdot}Al_2O_3$: It is observed from the experiments that, $MgO{\cdot}Al_2O_3$ spinel type inclusions exist in the steel samples after LF and Ca treatment. There are three types of spinel observed. One is the isolated spherical spinel, mainly containing Mg, Al, and O elements in small quantities. The second type is the modified spinel type inclusion containing Mg, Al, Ca, O and a little Si, it is the main existence form. The third type is the MgO Al_2O_3 spinel as core wrapped with sulphides.

Comparing the steel samples taken at different steps, it was found that there was very little $MgO{\cdot}Al_2O_3$ spinel type inclusions observed before LF refining, but they began to occur after LF. This can be explained by the following equations. The acidic dissolved [Al] in molten steel reacts with MgO in slag and lining to generate [Mg], then reacting with Al_2O_3 to form MgO Al_2O_3 [8].

$$4(MgO)+2[Al]=(MgO\cdot Al_2O_3)+3[Mg] \tag{9}$$

$$\log K_1=-33.09+50880/T=\log\frac{f_{\%,Mg}^3[\%Mg]^3\cdot a_{MgO\cdot Al_2O_3}}{f_{\%,Al}^2[\%Al]^2\cdot a_{MgO}^4} \tag{10}$$

$$4(Al_2O_3)+3[Mg]=3(MgO\cdot Al_2O_3)+2[Al] \tag{11}$$

$$\log K_1=-34.37+\frac{46950}{T}=\log\frac{f_{\%,Al}^2[\%Al]^2\cdot a_{MgO\cdot Al_2O_3}^3}{f_{\%,Mg}^3[\%Mg]^3\cdot a_{Al_2O_3}^4} \tag{12}$$

The phase stability diagram of MgO / MgO Al_2O_3 / Al_2O_3 can be calculated by the Eqs. (9) ~ (12). According to Fujii et al. [9], the activity of $MgO{\cdot}Al_2O_3$ is taken as 0.8 and that of MgO as 0.99 for Eq. (2) at 1873 K due to the very small solubility of Al_2O_3 into MgO. Meantime, the activity of $MgO{\cdot}Al_2O_3$ and Al_2O_3 are taken as 0.47 and 1 respectively for Eq. (2) since there is no solubility of MgO into Al_2O_3[9]. The phase stability diagram is shown in Figure 6. It can be seen that contents of dissolved [Al] and [Mg] in molten casing steel are mainly in the formation region of $MgO{\cdot}Al_2O_3$ as signed in the Figure. The results are consistent with the experiments. The quantity of $MgO{\cdot}Al_2O_3$ spinel type inclusion was increased after VD treatment and occurred mostly in the composite oxide form. This is the modified product of MgO Al_2O_3 spinel inclusion due to Ca-Si wire feeding after VD treatment. The researches of Itoh et al.[10] and YANG et al.[11]

showed that about 1 ppm Ca in the steel could significantly decrease the stability of MgO Al_2O_3 spinel inclusions and made them modifying into MgO-Al_2O_3-CaO inclusions. The dissolved calcium is about 1.2 ~ 3.0ppm after Ca-Si wire feeding, so more MgO-Al_2O_3-CaO type inclusions were observed.

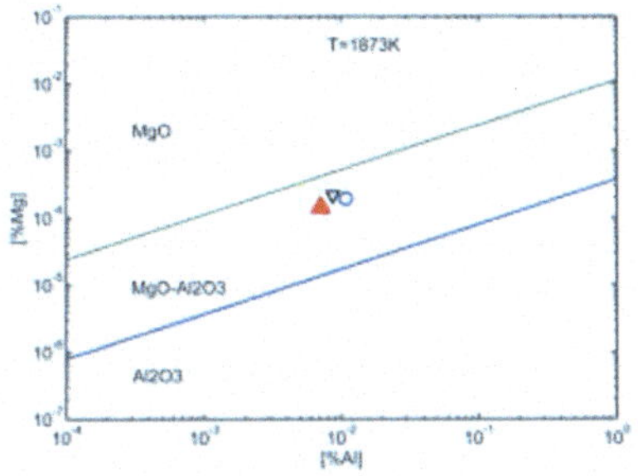

Figure 6 Phase stability diagram of MgO/MgO·Al_2O_3/Al_2O_3

Macro-inclusions from the ladle

Many observed macro-inclusions contain tracer Ba, indicating ladle slag is entrapped into the molten steel. When the argon flowrate is higher, slag entrapment will occur on the steel-slag interface. The critical Weber Number is 6.796 according to the reference [12]:

$$w_e = F_S / \sqrt{F_g F_\sigma} = \rho_s u_p^2 / [g\sigma_{s-m}(\rho_m - \rho_s)]^{1/2} = 6.796 \qquad (13)$$

$$u_p = \sqrt{6.796 \frac{[g\sigma_{s-m}(\rho_m - \rho_s)]^{1/2}}{\rho_s}} \qquad (14)$$

Where u_p is the horizontal velocity of liquid steel at slag entrapment, σ_{s-m} is steel-slag interfacial tension, ρ_m and ρ_S are densities of steel and slag respectively.

Substituting σ_{s-m} = 1.2 N/m, ρ_m = 6890 kg/m^3, ρ_S = 3000 kg/m^3 [2,12], u_p is calculated to be 0.7.

$$u_p = 4.177 \times (1-\alpha)^{1/12} \frac{Q^{1/3} H^{1/4}}{R^{1/3}} = 0.7$$

Assuming α = 8%, the critical argon flowrate Q of slag entrapment for 150t ladle is 183.3 L/min. In industrial production, the flowrate blown during LF is 510.9~608.5 L/min, much higher than that calculated, thus slag entrapment is inevitable.

Conclusions

(1) Before LF refining, the observed microinclusions are mainly pure Al_2O_3, Al_2O_3–based and MnS inclusions. Al_2O_3 is formed by the reaction of Al and [O] in molten steel, Al_2O_3–based inclusion is formed by the compounded deoxidation of Al, Si and Mn, and MnS is precipitated by the solidification of molten steel.

(2) Most Al_2O_3 inclusions are modified into $12CaO \cdot 7Al_2O_3$ 和 $3CaO \cdot Al_2O_3$ inclusions which are liquid at steelmaking temperature, while MnS cannot be fully modified into pure CaS and it can only be wrapped outside the alumina or calcium aluminate in the form of (Ca,Mn)S with the amount of Ca-Si wire fed in the trials.

(3) Macroinclusions originated from the ladle slag, tundish cover and mold slag. The gas flowrate during LF refining is much higher than that required for slag entrapment.

Acknowledgements

The authors are grateful for the support from the State Key Laboratory of Advanced Metallurgy of the USTB (No. 41603014) and the National Natural Science Foundation of China (No. 51374021).

References

[1] X.M. Dong, Q.C. Tian, Q.A. Zhang, "Corrosion behaviour of oil well casing steel in H_2S saturated NACE solution", *Corros. Eng. Sci. Techn.*, 45(2010) 181-185.
[2] H.Y. TANG, J.S. LI, "Thermodynamic analysis on the formation mechanism of magnesium-aluminum MgO Al_2O_3 spinel type inclusions in casing steel", *International Journal of Minerals, Metallurgy and Materials,* 17(1) (2010), 32-38.
[3] H.Y. TANG, J.S. LI, "Effect of flow control devices of tundish on cleanliness of billets", *Journal of Iron and Steel Research International*, 15, (2008), 499-504.
[4] G K. Sigworth and J.F. Elliot, "The thermodynamics of liquid dilute iron alloys", *Metal Science*, 18(1974), 298-310.
[5] S.W. Cho and H. Suito, "Assessment of calcium-oxygen equilibrium in liquid iron", *ISIJ Int.*, 34 (3) (1994), 265-269.
[6] Z.L. Xue, "Control on oxide inclusions composition and shape in spring steel", (PH.D. thesis, Iron and Steel Research Institute, 2001), 22.
[7] Verein Deutscher Eisenhüttenleute, Slag Atlas 2th Edition, Verlag Stahleisen GmbH, 1995.
[8] Z.Y. Deng, M.Y. Zhu., "Evolution Mechanism of Non-metallic Inclusions in Al-Killed Alloyed Steel during Secondary Refining Process", *ISIJ international,* 53(3) (2013), 450-458.
[9] K. Fujii, T. Nagasaka and M. Hino. Activities of the Constituents in Spinel Solid Solution and Free Energies of Formation of MgO, $MgO \cdot Al_2O_3$. *ISIJ Int.,* 40 (2000), 1059-1066.
[10] W. Yang, L.F. Zhang, X.H. Wang, et al., Characteristics of Inclusions in Low Carbon Al-Killed Steel during Ladle Furnace Refining and Calcium Treatment, *ISIJ Int.,*53 (8) (2013) 1401-1410.
[11] H. Itoh, M. Hino, B.Y. Shiro, Thermodynamics on the Formation of Spinel Nonmetallic Inclusion in Liquid Steel, *Metall. Mater. Trans. B* 28 (1997), 953-956.
[12] Z.Q. Xiao, L.X. Hu, "Research on the behavior of large particles and their sources of inclusions blowing steel", *Iron & Steel*, 23(1) (1988), 23~26.

7th International Symposium on High-Temperature Metallurgical Processing
Edited by: Jiann-Yang Hwang, Tao Jiang, P. Chris Pistorius, Gerardo R.F. Alvear F., Onuralp Yücel, Liyuan Cai, Baojun Zhao, Dean Gregurek, and Varadarajan Seshadri
TMS (The Minerals, Metals & Materials Society), 2016

THE DYNAMIC DISSOLUTION OF COKE WITH SLAG IN MELTING AND DROPPING ZONE

Yingli LIU[1] Qingguo XUE[1] Wentao GUO[1] Haibin ZUO[1] Xuefeng SHE[1] JingsongWANG[1]

[1]State Key Laboratory of Advanced Metallurgy, University of Science and Technology Beijing, Beijing 100083, PR China

Keywords: Oxygen blast furnace, Coke, Dissolution, Slag

Abstract

In the process of oxygen blast furnace, the lower coke ratio and the higher coke load make the skeleton coke more important. To understand the dissolution of coke in melting and dropping zone, the influence of FeO content, temperature and reaction time on the dissolution of coke with slag were studied by single-factor experiment. Meanwhile, the three factors and three levels orthogonal were conducted. The dissolution degrees were characterized by the weight loss ratio and diameter variation. The results showed that the weight loss ratio increased and the diameter decreased with the increase of FeO content, temperature and time. The three factors and three levels orthogonal test showed that time was the most significant factor on coke dissolution, FeO content of the second, and finally the temperature. The microstructure of coke after reaction indicated that the dissolution occurred on coke surface.

Introduction

The blast furnace is the conventional ironmaking process which produces molten iron directly from a charge of coke and iron ore along with the injection of the auxiliary materials such as pulverized coal. Through a hundred years' development, the carbon consumption of the traditional blast furnace is close to the theoretical minimum value of the process[1]. Under the pressure of resource scarcity and environmental degradation, researchers have developed a series of new blast furnace ironmaking technologies to reduce carbon consumption, such as carbon composite iron ore hot briquette [2], injection material (hydrogen material and reduction gas)[3-5], and oxygen blast furnace(OBF)[6].

Oxygen blast furnace is a new ironmaking process. It adopts pure oxygen at room temperature to replace the traditional hot blast operation and uses heated recycle top gas after removal CO2. Its feasibility and advantages of energy saving have been confirmed in theoretical and experimental blast furnace. Under the condition of oxygen blast furnace, coke ratio is greatly reduced and coke load is increased. Therefore, the skeleton function of coke in OBF is more important than that in BF. In order to meet the needs in new condition, the dissolution of coke with slag in melting and dropping zone were investigated.

Experimental

Raw Materials

The slag used in this experiment is composed of FeO, CaO, SiO_2, MgO and Al_2O_3. CaO, SiO_2, MgO and Al_2O_3 are high purity analytical powder. The basicity of slag is 1.3, which is the same as typical blast furnace slag. The compositions of slag are shown in Table I.

Table I. Compositions of slag with different FeO content (mass%)

	FeO (%)	CaO (%)	SiO_2 (%)	MgO (%)	Al_2O_3 (%)	R
S1	5	42.9	32.9	10.8	8.4	1.3
S2	10	40.1	30.7	10.8	8.4	1.3
S3	15	37.2	28.6	10.8	8.4	1.3
S4	20	34.3	26.5	10.8	8.4	1.3

An equilibrium technique[7] is used to prepare FeO from Fe_3O_4 powder. Experimental conditions for FeO preparation are determined based on the C-O-Fe equilibrium phase diagram. Firstly, Fe_3O_4 powder was pressed into cylinders, which was charged into a steel tube. Then the iron tube was put into the reaction tube. Secondly, the steel tube was suspended into a reaction furnace of 900℃. A gas mixture of 1:1 CO/CO_2 was injected into the steel tube from the bottom. After 6 hours' reaction time, the sample was cooled using high purity N2 to prevent reoxidation. Finally, the target production was confirmed by XRD.

Coke in this test was from Laiwu Ltd. The industrial analyses of coke were presented in Table Ⅱ. To eliminate the influence of shape, coke were made to Φ16mm×15mm. Then the coke were dried in 120℃for 3h.

Table Ⅱ Industrial analysis of coke

A_f	V_f	C_f	S	M_{40}	M_{10}
12.16	1.64	86.20	0.694	87.32	6.42

Experimental Apparatus and Conditions

The dynamic dissolution of coke with slag was carried out in a high temperature furnace, as shown in Figure1. The heating element was Silicon Molybdenum. The maximum temperature of the reaction furnace was 1873K. Under the protection atmosphere of 3L/min Ar, the furnace was heated into the set temperature. The initial weight of coke was recorded as M_0. 80g slag in Mo crucible was put into the furnace and was pre-melted for 30min. Then coke was submerged into slag and rotated at a speed of 30rpm. After the set reaction time, coke was took out and put into graphite powder to avoid oxidation. When coke was cooled to room temperature, the weight Mt and the diameter Dt were measured. The weight loss ratio was calculated as formula (1). The microstructures and element distribution of coke before and after reaction were observed by SEM and EDS.

$$\eta = \frac{M_0 - M_t}{M_0} \qquad (1)$$

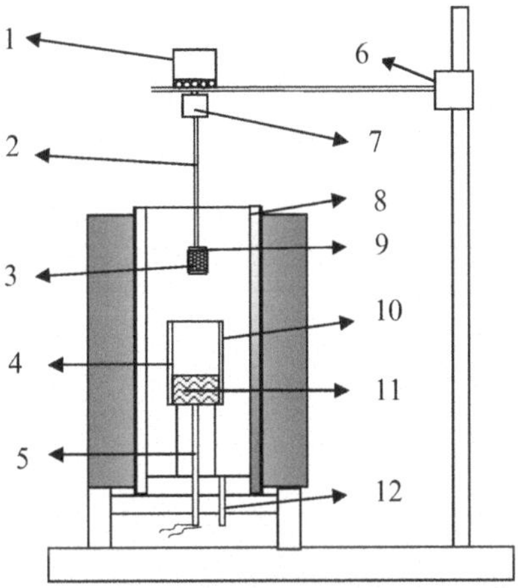

Figure 1 Experiment apparatus

1. Variable speed motor; 2. Mo wire; 3. Coke; 4 Mo crucible; 5.Thermocouple; 6.Lifting device; 7.Universal coupler; 8.Furnace tube; 9.Mo nut; 10.Grapht crucible; 11.Slag; 12.Ar inlet

Results and Discussion

Influence of FeO Content in Slag on the Dissolution of Coke

In the melting and dropping zone, the FeO content in slag is gradually decreased. In order to study the influence of FeO content in slag on the dissolution of coke, the FeO contents were set as 5%, 10%, 15% and 20%. The experimental temperature was 1450℃ and the reaction time was 30min. The specific experimental conditions were shown in table III.

Table III Experimental condition of different FeO content

	t(min)	T(℃)	FeO (%)	CaO (%)	SiO_2 (%)	MgO (%)	Al_2O_3 (%)
1	30	1450	5	42.9	32.9	10.8	8.4
2			10	40.1	30.7	10.8	8.4
3			15	37.2	28.6	10.8	8.4
4			20	34.3	26.5	10.8	8.4

The variation curves of weight loss ratio and coke diameter with slag FeO content at 1450℃ were shown in Figure 2. The results indicated that, when the FeO content was 5% and 10%, the variation of weight loss ratio was slight. But in the case of 15% and 20%, the weight loss ratio and diameter variation increased sharply. In the condition of 20%, the weight loss ratio reached to 22.68% and the diameter was 14.90mm. This showed that the dissolution of coke was serious. All above represented that the weight loss ratio was not a linear function of FeO content.

When FeO content decreased to a certain degree, the dissolution of coke with slag would be weakened obviously.

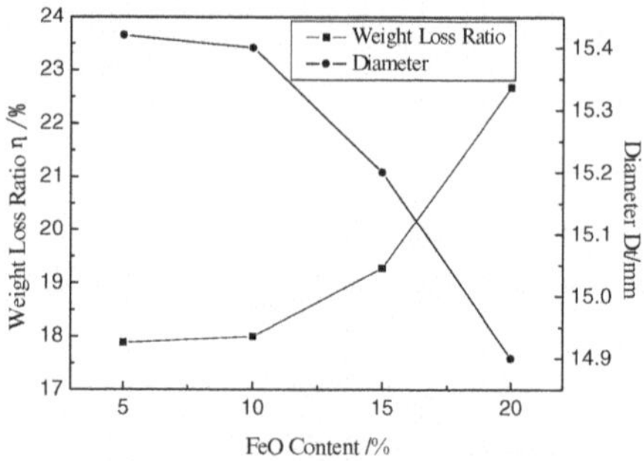

Figure 2 Weight loss ratio and diameter of coke vs. slag FeO content at 1450℃

Influence of Temperature on the Dissolution of Coke

In melting and dropping zone, the reaction temperatures were determined to be 1400℃, 1450℃ and 1500℃ according to the temperature distribution curves in BF. FeO content in slag was 10% and the reaction time was 30min.

The variation curves of weight loss ratio and coke diameter with temperature were presented in Figure 3. The results showed that the influences of temperature on the dissolution of coke were small in the range from 1400 to 1450°C, and the weight loss ratio increased from 15.32% to 18.00%. When temperature was elevated from 1450°C to 1500°C, the weight loss ratio changed from 18.00% to 29.76%. The shape of the curves indicated that the dissolution degrees of coke grown exponentially with the rising of temperature. The thermodynamic parameters show that the reaction of coke with slag is endothermic. The increase of temperature will aggravate the dissolution of coke. From a kinetic point of view, the reaction of coke with slag may be controlled by four steps[8]: 1) mass transfer of FeO to the gas/slag interface from the bulk; 2) chemical reaction at the gas/slag interface; 3)diffusion of gaseous in the gas film around the carbon, and 4) chemical reaction at the gas/carbon interface. When FeO content in slag is fixed, the mass transfer and chemical reaction will be improved with the rising of temperature.

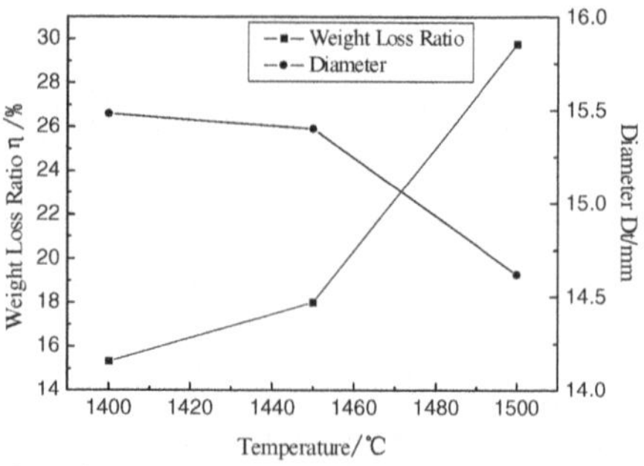

Fig.3 Weight loss ratio and diameter of coke vs. temperature

Influence of Reaction Time on the Dissolution of Coke

In the condition of oxygen blast furnace, the smelting intensity is greater than that in traditional blast furnace. The utilization coefficient increased from 2.5 to about 5. The burden went down faster in OBF than that in TBF, which made the resident time of burden shorter. Therefore, the dissolution of coke after different reaction time was investigated. With reference to the productivity of blast furnace, the reaction time was determined to 15min, 30min, 45min and 60min. FeO content in slag was 10% and temperature was 1450℃.

The variation curves of weight loss ratio and diameter of coke with reaction time were shown in Figure 4. The results turned out that the relationship between weight loss ratio and time was nearly linear. The dissolution rate was almost constant. This was estimated to be due to the fact that there was excessive FeO in slag. With the dynamic rotation of coke, the liquid iron on the surface of coke was replaced by FeO in slag. These made the reaction continuous. In the condition of oxygen blast furnace, the reaction time of coke with slag was shortened. Therefore, the dissolution of coke was weakened and the structural strength would be retained.

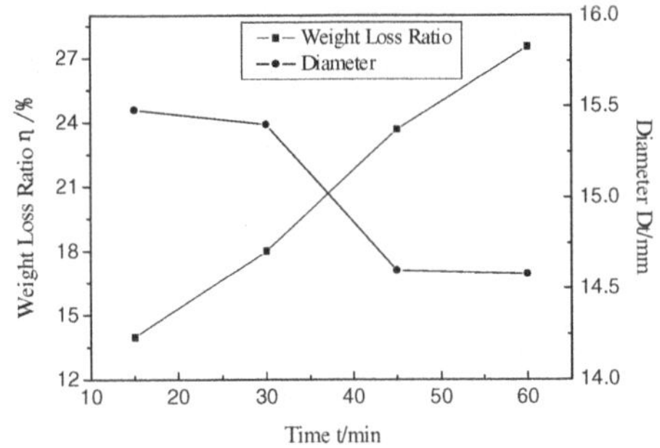

Figure 4 Weight loss ratio and diameter of coke vs. reaction time

Three Factors and Three Levels Orthogonal Experiment

To study the effect degree of the three factors of FeO content, dissolution temperature and dissolution time on the coke dissolution, the three factors and three levels orthogonal experiment were carried out. The experimental scheme and results of orthogonal experiment were shown in Table IV.

Table IV Experimental condition and results of orthogonal experiment

	A. temperature /°C	B. time/min	C. FeO content/%	Weight loss ratio/%	Diameter/mm
1	1400 (1)	15 (1)	5 (1)	9.58	15.48
2	1400 (1)	30 (2)	10 (2)	15.32	15.48
3	1400 (1)	45 (3)	15 (3)	27.94	14.72
4	1450 (2)	15 (1)	15 (3)	15.61	15.40
5	1450 (2)	30 (2)	5 (1)	18.10	15.42
6	1450 (2)	45 (3)	10 (2)	23.67	14.60
7	1500 (3)	15 (1)	10 (2)	17.33	15.30
8	1500 (3)	30 (2)	15 (3)	26.91	14.72
9	1500 (3)	45 (3)	5 (1)	22.70	14.90

The processing results of data were shown in Table V and VI by the range analysis method. The range values in Table V and VI showed that the order of significant factor was: B>C>A. Namely, time was the most significant factor on coke dissolution, FeO content of the second, and finally the temperature. The experiment with the least weight loss was $A_1B_1C_1$: temperature was 1400℃, reaction time was 15min, FeO content was 5%. These were coincident with the single-factor experiment.

Table V Orthogonal analyses of weight loss ratio

Weight loss ratio/%		A. temperature /°C	B. time/min	C. FeO content/%
	K1	52.84	42.52	50.38
Horizontal sum	K2	57.38	60.33	56.32
	K3	66.94	74.31	70.46
	k1	17.61	14.17	16.79
Average	k2	19.13	20.11	18.77
	k3	22.31	24.77	23.49
Range		4.70	10.60	6.69

Table VI Orthogonal analyses of diameter

Diameter/mm		A. temperature /°C	B. time/min	C. FeO content/%
	K1	45.68	46.18	45.80
Horizontal sum	K2	45.42	45.62	45.38
	K3	44.92	44.22	44.84
	k1	15.23	15.39	15.27
Average	k2	15.14	15.21	15.13
	k3	14.97	14.74	14.95
Range		0.25	0.65	0.32

Microstructure of Coke After Reaction

Figure 5 showed the macroscopic photos of coke before and after reaction in 1450℃ for 30min. It showed that the variation of diameter was obvious. But there was almost no change in height. To understand the microscopic change, the coke after reaction was observed by SEM and EDS. Figure 6 showed the microstructure and element distribution of coke. The granular iron adhered to the coke surface. The liquid iron film was formed in high temperature. In addition, it can be inferred from the distribution of Fe element that the reaction of coke with slag focused on the surface. This might be because that the small pore and the surface tension of slag blocked the slag's way into the interior.

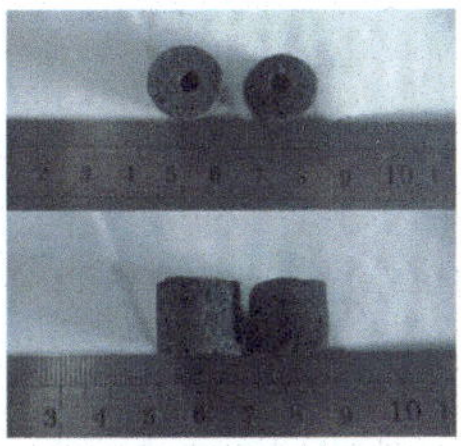

Figure 5 Coke photos before and after reaction

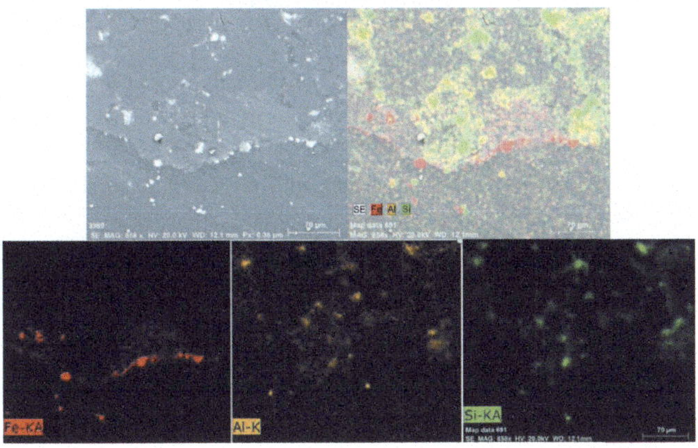

Figure 6 SEM photos and element distribution of coke

Conclusions

To understand the dissolution of coke in melting and dropping zone, the dynamic dissolution of coke with slag were carried out. The results showed that:

(1) The weight loss ratio increased and the diameter decreased with the increase of FeO content, temperature and reaction time. In the condition of oxygen blast furnace, the reaction time and FeO content in slag was decreased, which weakened the coke dissolution. These were beneficial to coke to maintain stronger high temperature property.

(2) Three factors and three levels orthogonal experiments showed that time was the most significant factor on coke dissolution, FeO content of the second, and finally the temperature.

(3) The microstructure and element distribution of coke after reaction indicated that the dissolution occurred on coke surface.

Acknowledgments

The authors gratefully acknowledge the financial support from National Natural Science Foundation of China and Baosteel (NO. 51134008) and National Basic Research Program of China (973 Program) (NO. 2012CB720401).

References

[1] GIELEN D, MORIGUCHI Y. "CO_2 in the iron and steel industry: an analysis of Japanese emission reduction potentials". *Energy Policy,* 30(10), 2002: 849-863.

[2] MATSUIY, SAWAYAMA M, KASAI A, "Reduction behavior of carbon composite iron ore hot briquette in shaft furnace and scope on blast furnace performance reinforcement," *ISIJ Int.*, 43(12), (2003):1904-1912.

[3] WAKIMOTO K.A "Feedstock Recycling System of Waste Plastics in a Blast Furnace at NKK"(paper presented at Proceedings of 60th Ironmaking Conference, Maryland)2001: 473-483.

[4] AGARWAL J C, BROWN F C, CHIN D L, et.al. "Injection of natural gas at high levels in the blast furnace with low oxygen consumption:field test results at WCI steel," Ironmaking Conference Proceedings, 1999: 143-163.

[5] Tang En et al., "Practical technical reviews in energy saving and CO2 emission reducing for the blast furnace"(paper presented in 2011 the development of the national metallurgical energy-saving emission reduction and low carbon Technology Symposium, Tangshan, China):308-311.

[6] Qin M S et al., "Blast furnace operation with full oxygen blast," *Iromaking steelmaking*, 6(15)(1988):287-292.

[7] B. Sarma et al., "Reduction of FeO in smelting slags by solid carbon:experimental results," *Meall. Trans. B*, 27B(1996):717-730.

[8] Kawang SEO and R. J.FRUEHAN, "Reduction of FeO in slag with coal char," *ISIJ Int.*, 40 (2000):7-15.

7th International Symposium on High-Temperature Metallurgical Processing
Edited by: Jiann-Yang Hwang, Tao Jiang, P. Chris Pistorius, Gerardo R.F. Alvear F., Onuralp Yücel, Liyuan Cai, Baojun Zhao, Dean Gregurek, and Varadarajan Seshadri
TMS (The Minerals, Metals & Materials Society), 2016

RESEARCH ON OXIDATION KINETICS OF SPHC STEEL AT 500~900℃

Dongchang Huang, Guangliang Wu, Xinbin Liu

(School of Minerals Processing and Bioengineering, Central South University, Changsha, Hunan, 410083, China)

Keywords: SPHC; TGA; oxidation kinetics; heat active energy; numerical simulation

Abstract

Oxidation Kinetics of SPHC was analyzed by TGA, and the curve of oxidation weight gain with different temperature was acquired. It is found that, when the oxidation temperature is between 500℃ and 600℃, all the curves are linear; when the oxidation temperature is above 700℃, the curves are close to parabolic. At the temperature of 900℃, the oxidation weight gain increases dramatically. The TGA results indicate that the activation energy of SPHC steel is 146.19 kJ/mol at the temperature of 500℃~900℃, and an oxidation kinetic model was set up for variable temperature. Based on the temperature change during rolling process, numerical simulation of oxide scales evolution was carried out for SPHC steel, and the simulation result agreed well with measured ones. It can provide a basis for estimating the scale growth and controlling the temperature on site.

Introduction

The binding target in the National Development Planning Outline of China is that when per-unit GDP energy consumption reduces by 20%, the total discharge of major pollutants will be reduced by 10%. Hence, the penalties for waste acid discharge was increased unprecedentedly, and the requirements of energy consumption for iron and steel follow-up processing was further improved, causing to prompt the cold rolling plant and automobile production enterprises to require for hot-rolled material products which would decrease pickling or even avoid pickling.

Oxide scale will be inevitably generated on the surface of hot rolled steel plate (SPHC). However, scale is one of the most serious factors which influence the quality of the strip surface [1, 2]. How to decrease the scale thickness and get the reasonable structure of that, is the key to improve the surface quality of the products.

Typical structure of scale is composed of three layers, including thinner Fe_2O_3 in the outermost scale, Fe_3O_4 in the middle and FeO near the substrate [3, 4]. Because Fe_3O_4

Corresponding author: Guangliang Wu, Professor, Doctoral supervisor, E-mail: glwu_899@sina.com.

has a lower hardness (350HB～500HB), good plasticity and superior adhesion, it will not peel and crack by bending in the subsequent deep processing. On the other hand, compared with Fe_2O_3, FeO and Fe_3O_4 are more easily removed by pickling, reducing the pickling load before cold rolling [5]. Therefore, decreasing the thickness of the scale and controlling the scale structure of hot rolled strip is an important way to reduce the acid amount and enhance the efficiency of pickling [6].

In this article, taking SPHC steel sample as the research object, oxidation kinetics was measured in laboratory by TGA. According to the TGA results, the curve of oxidation weight gain with different temperature was obtained, and an oxidation kinetics model was established with variable temperature. Based on the temperature change during rolling process, numerical simulation of oxide scales evolution was carried out for SPHC steel. Basic data can be established to estimate the scale growth. Simultaneously, it can provide a theoretical basis for designing hot rolling conditions, to control the scale structure and thickness on hot-rolled products, reaching the purpose of reducing pickling or avoiding pickling in the follow-up processing.

Experiment

Experimental material and Experimental Device

Specimen material is SPHC steel, whose composition is shown in Table I. The size of sample steel is 2mm×10mm×15mm.

Table I Chemical composition of SPHC steel /%

	C	Si	Mn	P	S
SPHC	≤0.07	≤0.05	0.20～0.45	≤0.02	≤0.02

TGA, as shown in Fig.1, produced by French company SETARAM, was used as the

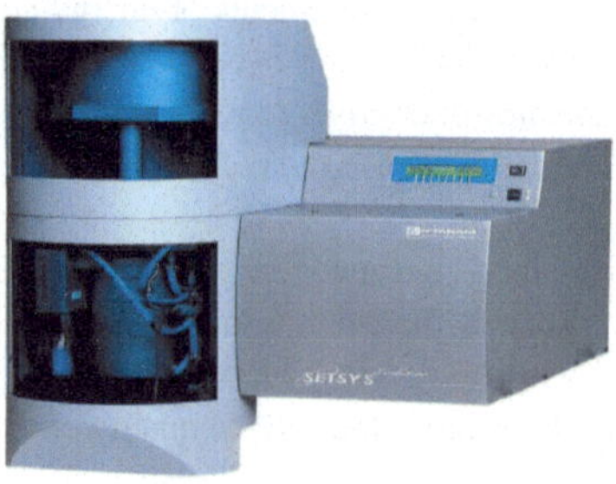

Fig.1 Image of TGA appratus

experimental device to carry out this investigation. The maximum temperature of its heating furnace is 1750℃. It can complete TG and TG/DTA measurement from room temperature to 1750℃, or TG and TG/DSC measurement from room temperature to 1000℃ in reducing atmosphere.

Experimental Method

Fig.2 demonstrates oxidation kinetics experiments. The oxidation temperature was 500℃, 700℃, 600℃, 800℃ and 900℃; The oxidation time was 0 ~ 540min, and the

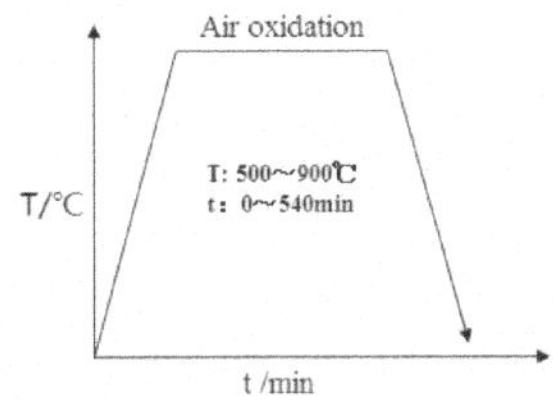

Fig.2 The project of oxidation kinetics experiment

experiments were carried out in the open air. When the furnace met the temperature requirements, the prepared sample was introduced, The sample was removed after oxidation for different times, and cooled to room temperature rapidly, then the weight gain was recorded. The gas flow rate was 5L/min during the oxidation experiments.

Results and Discussion

Oxidation kinetics curve of SPHC steel

The curves of oxidation weight gain of SPHC, at 500℃～900℃, are described in Fig.3. The results show that oxidation per unit area was small. When the oxidation temperature is at 500℃～600℃, all the curves are linear; When the oxidation temperature is above 700℃, the curves are close to parabolic. At the temperature of 900℃, the oxidation weight gain increase dramatically. It is evident that oxidation weight gain increases fast with the increase of oxidation temperature.

At the same time, it can be seen from Fig.3 that, approximately one hour later, oxidation rate gets slower at low temperature, maintaining the same level, with a plateau in the curve of oxidation weight gain. At the temperature of 900℃, oxidation rate does not get slower until five hours later. That is to say, at higher temperature, the time of plateau oxidation is longer. It is illustrated that the temperature and time will augment the steel sample oxidation. A conclusion is that the decrease of rolling temperature, would be beneficial to reduce the content of iron oxide.

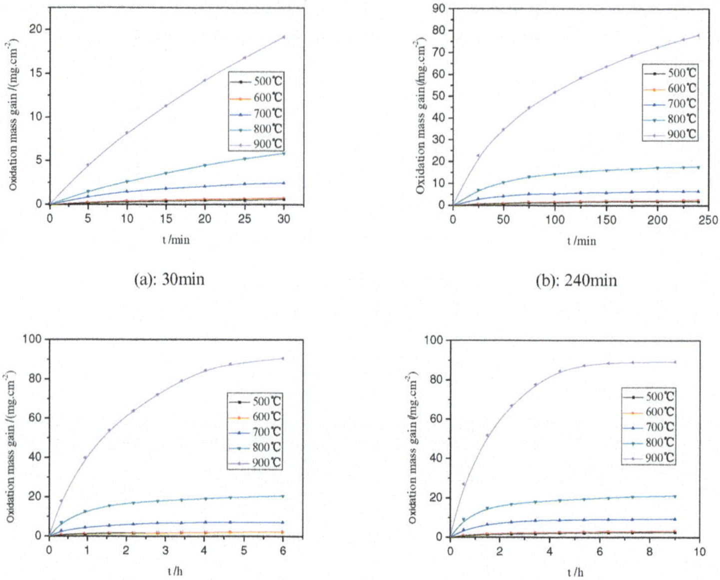

(a): 30min (b): 240min

(c): 360min (d): 540min

Fig.3 The curve of oxidation weight gain of SPHC

Calculation of activation energy

A model was set up, by combining oxidation kinetics with actual performance, and the growth of scale described with the parabolic equation [7]. According to the research of Kofstad [8], the model can be described as follows:

$$\Delta W^2 = K_P \cdot t \qquad (1)$$

where, K_p is oxidation rate constant ($mg^2 \bullet cm^{-4} \bullet s^{-1}$); and ΔW is oxidation weight gain ($mg \bullet cm^{-2}$); and t is oxidation time (s).

The rate constant can be expressed in Eq. (2):

$$K_{p,T} = A \cdot \exp(\frac{-Q}{RT}) \qquad (2)$$

where, Q is the activation energy (J/mol); T is oxidizing temperature (K); R is universal gas constant (8.314, J/mol); A is model constant.

Taking the natural logarithm on both sides of Eq. (3) yields

$$\ln K_{p,T} = \ln K_0 + (\frac{-Q}{R}) \cdot \frac{1}{T} \qquad (3)$$

The oxidation rate constants, at a certain temperature, can be calculated according to the Eq. (1) from the experimental results. After data regression of oxidation weight gain and time, the oxidation rate constants, at different temperatures, are obtained, as shown in Table II.

Table II The oxidation rate constants (Kp)

Temperature /℃	Kp /($mg^2 \cdot cm^{-4} \cdot s^{-1}$)
500	1.2E-4
600	3.7 E-4
700	2.8 E-3
800	1.4 E-2
900	2.5 E-1

Eq. (3) shows that $\ln K_{p,T}$ would have a linear relationship with 1/T, as shown in Fig.4. Q can be obtained from the slope of $\ln K_{p,T}$ as a function of 1/T, i.e. the slope is (-Q/R). Q can be obtained by the regression analysis, Q = 146.19kJ/mol (correlation coefficient r = 0.90861).

Variable temperature oxidation kinetics model

The activation energy acquired above, can be used to calculate dynamic oxidation under variable temperature conditions. Continuously temperature changing can be approximated as the effect of a number of small temperature steps, as shown in Fig.5.

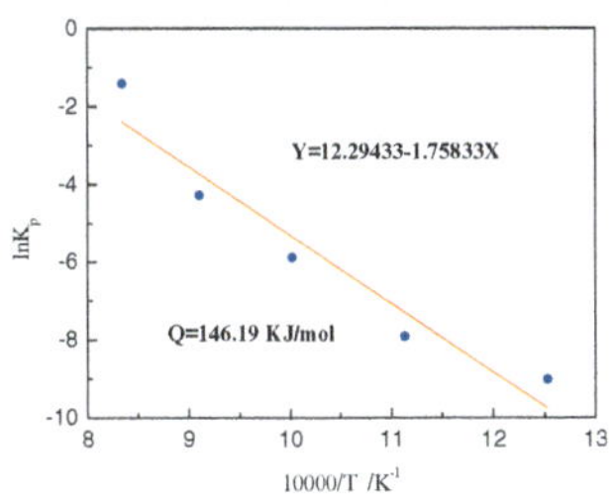

Fig.4 Relationship between lnKp and 1/T

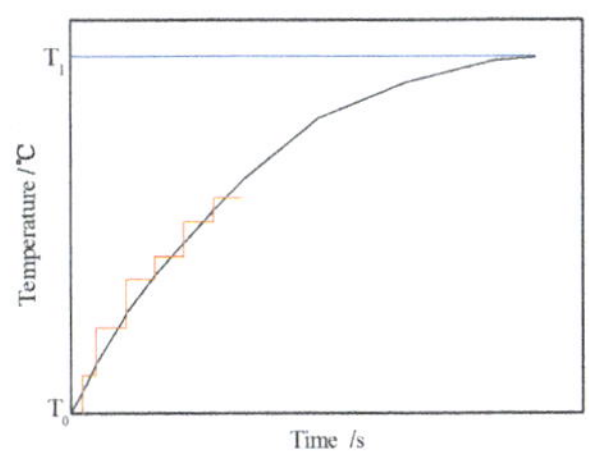

Fig.5 Calculation simplification at variable temperature

According to the previous studies, the oxidation weight gain accords with the parabola law at the specific temperature. The oxide scale weight gain, under the condition of variable temperature, can be approximated as the sum of a number of small isothermal steps. Therefore, under the condition of variable temperature, the oxidation kinetics equation can be represented as:

$$\Delta W_i^2 = \Delta W_{i-1}^2 + K_p^i \cdot \delta t_i \quad (4)$$

$$\delta t_i = \frac{\delta T}{\alpha(T_i)} \quad (5)$$

$$K_p^i = A \cdot \exp(-\frac{Q}{RT_i}) \quad (6)$$

Where, i is the step number (i=1, 2, 3…n); and δt_i is step dusation (s); and δT is temperature step (K); and $\Delta W_0=0$.

The activation energy can be obtained from the calculation. Thus, the oxidation kinetics model at variable temperature can be written as：

$$\Delta W_i^2 = \Delta W_0^2 + K_p^1 \cdot \delta t_1 + ... + K_p^i \cdot \delta t_i = \Delta W_0^2 + \sum_{i=1}^{\infty} K_p^i \cdot \delta t_i \quad (7)$$

Numerical simulation of scale thickness evolution in rolling process

Based on the oxidation kinetics model mentioned above and the actual hot rolling temperature system, the thickness growth of scale can be simulated in the whole line during the hot rolling process. Taking the center of the length and the center of the width of the strip length as the calculating point, the temperature trend of the point is calculated, as shown in Fig.6.

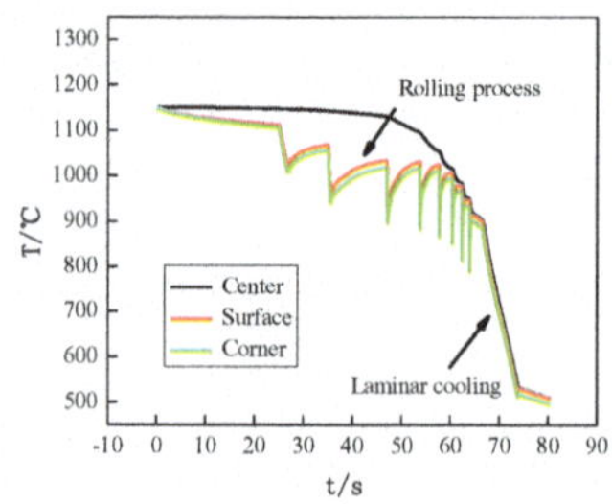

Fig.6 The temperature history of rolling process

The fraction reduction in thickness of scale during the rolling process was taken to be 50% of that of the steel substrate [9, 10]. The thickness of the scale at each stage is obtained after inputting the temperature trends to a numerical simulation program. Results of simulation for each process are shown in Fig.7, and the Cross-sectional topography of the oxide scale is shown in Fig.8.

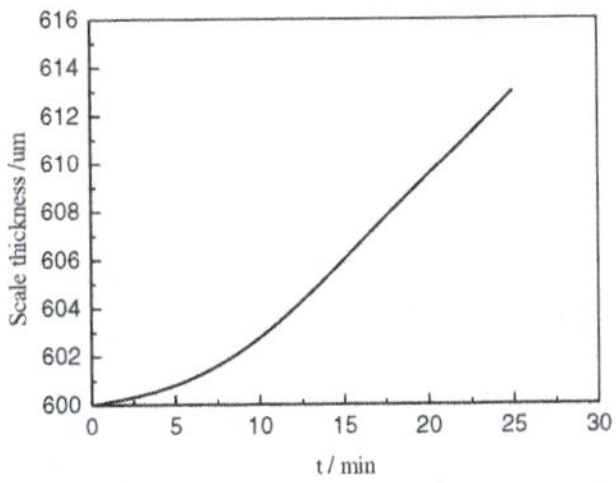

(a): in heating furnace

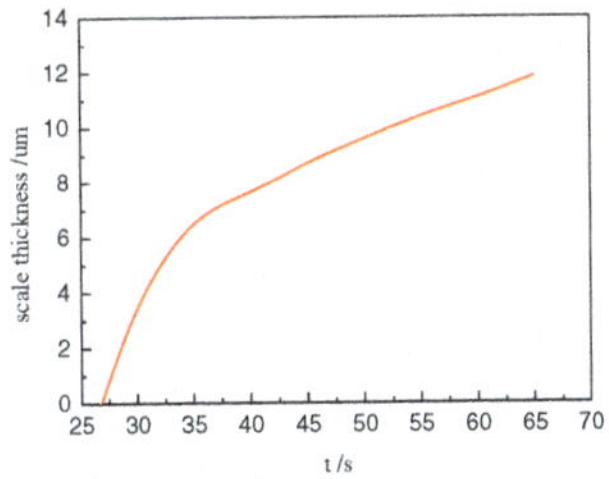

(b): in rolling process

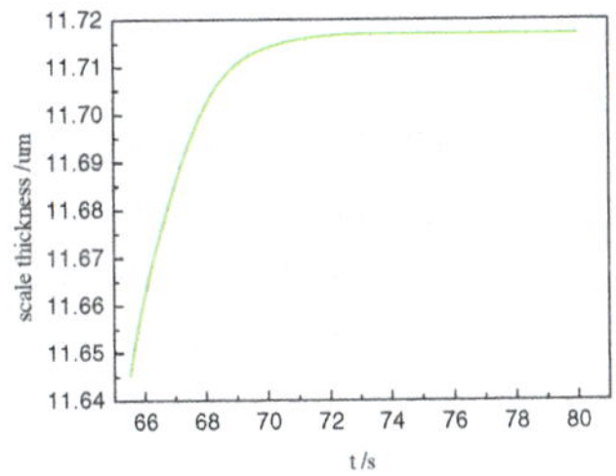

(c): in layer cooling

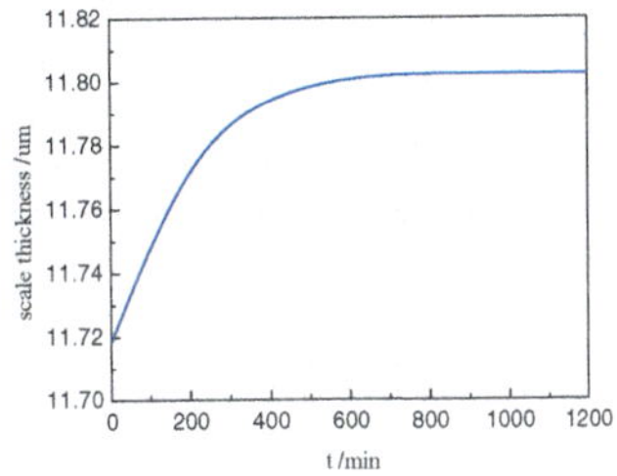

(d): in coiling process

Fig.7 Numerical simulation of scale thickness

TabIII Comparison of calculated values with measured values /um

calculated values			measured values	
center	corner		center	corner
11.98	10.81		11.12	10.08

Fig.8 Cross-sectional topography of the oxide scale

Conclusions

Based on the results of TGA, the conclusions are as follows:

According to the curves of oxidation weight gain, it was concluded that the decrease of rolling temperature, would be beneficial to reduce the content of iron oxide.

By regression analysis of the experimental data, the activation energy was obtained, Q = 146.19kJ/mol.

Making use of the activation energy, a variable temperature oxidation kinetics model was established in this article. It could be written as:

$$\Delta W_i^2 = \Delta W_0^2 + K_p^1 \cdot \delta t_1 + ... + K_p^i \cdot \delta t_i = \Delta W_0^2 + \sum_{i=1}^{\infty} K_p^i \cdot \delta t_i$$

Based on the temperature change during rolling process, the numerical simulation of oxide scales evolution was carried out for SPHC steel, and the simulation result agreed well with measured ones. It is of great help for the control of the scale thickness, to estimate the change of the rolling scale thickness at each stage, which cannot be measured under the field conditions.

References

[1] Z. M. Wu, "Formation Causes of Scale of Hot-rolled Strip and Its Countermeasures," *Steel Rolling*, 24 (2007), 56-58.

[2] C. Q. Huang, K. Zhao, M. Q. Fang, "Mechanism of Scale Formation on Strip Surface during Rolling and Its Preventing Methods," *Iron and Steel*, 32 (1997), 32-35.

[3] Torresa M, Colas R. "A Model for Heat Conduction through the Oxide Layer of Steel during Hot Rolling," *Mater Process Technology*, 105 (2000), 1483.

[4] Tomoki F, Hikaru O, Yasuhiro M. "Mechanism of Red Scale Defect Formation in Si Added Hot Rolled Steel Sheets," *ISIJ International*, 34 (1994), 906.

[5] W. Z. Cui, "Cause of Formation of Scales on the Surface of Hot Rolled Strip and the Methods of Removing Them," *Shanxi Metallurgy*, 3 (2010), 25-26.

[6] T. B. Wei, "Analysis of Scale Forming Reasons and Descaling Methods in Hot Rolling Process," *Research on Iron & Steel*, 4 (2003), 54-58.

[7] Klaus S, S. X. ZHOU, "Contribution to Scale Growth During Hot Rolling of Steel," *Metal Working*, 9 (2003), 538-542.

[8] Kofstad. P, "Prediction Model of Metal Non-Isothermal Oxidation Kinetics," *Acta Chemica Scandinavica*, 12 (1958), 701.

[9] G M. Cao et al., "Numerical Simulation of Oxide Scales Evolution During Hot Strip Rolling," *Journal of Iron and Steel Research*, 22 (2010), 13-16.

[10] B. Sun, G M. Cao, Y. Zou, "Study on Thickness Numerical Simulation and Morphology of Oxide Scale on Hot-Rolled Low Carbon Steel," *Journal of Iron and Steel Research*, 23 (2011), 34-38.

7th International Symposium on High-Temperature Metallurgical Processing

Direct Reduction and Smelting Reduction

Session Chairs:
Onuralp Yücel
Chenguang Bai

7th International Symposium on High-Temperature Metallurgical Processing
Edited by: Jiann-Yang Hwang, Tao Jiang, P. Chris Pistorius, Gerardo R.F. Alvear F., Onuralp Yücel, Liyuan Cai, Baojun Zhao, Dean Gregurek, and Varadarajan Seshadri
TMS (The Minerals, Metals & Materials Society), 2016

EXPERIMENT RESEARCH ON DIRECT REDUCTION OF CELESTINE BY ROTARY HEARTH FURNACE PROCESS

Dongping Duan[1,2], Hongliang Han[2], Siming Chen[1,2], E Zhou[2], Li Zhong[2]

[1]Qinghai Institute of Salt Lake, Chinese Academy of Sciences, Xining, Qinghai, 810008, China
[2]Key Laboratory of Green Process and Engineering, Institute of Process Engineering, Chinese Academy of Sciences, Beijing 100190, China

Keywords: Celestine, Rotary hearth furnace, Direct reduction

Abstract

In this study, physical, chemical and microscopic characteristics of Celestine were investigated firstly. Based on which, the thermodynamic analysis of Celestine reduction was discussed. And the experiments of Celestine carbon-containing pellets used in rotary hearth furnace process were studied. The results showed that it is feasible to deal with Celestine by rotary hearth furnace direct reduction process. When the temperature is 1200°C, C/O ratio is 0.8 and holding times is 20min, the conversion rate of $SrSO_4$ in Celestine is above 95%, the utilization rate of Celestine is nearly 100%. This study can provide theoretical and technical basis on economical and rational use of Celestine.

Introduction

Sr containing products which are wildly used in electronics, alloys, superconducting production are an important industrial raw materials. And Celestine is the main raw material for the production of strontium salts. The resources of Celestine proved in China accounts for 50% of the world total. The proved Celestine deposits are mainly distributed in Qinghai, Chongqing, Hubei, Jiangsu, Sichuan, Yunnan, Xingjian and other places. Among them, reserves of Celestine in Qinghai province is above 2000 million tons, which is account for 80% of total in whole China[1,2]. However, the lower ore grade of Celestine (by $SrSO_4$) in China can't meet the needs of industrial production without beneficiation[3,4].

In industrial production, strontium salts are generated from Celestine concentrate with a major method - Carbonthermal reduction, method that requiressimple process equipment and it is able to generate consistent product quality. In this process, $SrSO_4$ in Celestine ore is firstly reduced to dissoluble SrS followed by leaching, purification and carbonization, to synthetize strontium carbonate[5,6]. But up to now, most of carbonthermal reduction processes used rotary kiln or fluidized bed to react Celestine ore. A large number of raw materials are overflowed with gas, leading to lower reduction efficiency, high cost, and poor economic benefit. However, processing in rotary hearth furnace has several merits such as high reducing speed, superior product performance, flexible operation conditions and environmental eco-friendly[7,8]. All of these advantages indicate that this offer unique advantages for processing of Celestine ores.

In order to clarify the above mentioned advantages in the pyrometallurgical processing of Celestine ores, a number of experiments were conducted in order to investigated\ and report:

- Physical, chemical and microscopic characteristics of Celestine and;
- Thermodynamic analysis of Celestine reduction and;
- The experiments of Celestine carbon-containing pellets used in rotary hearth furnace process.

Materials and methods

Materials

The main components of Celestine concentrate and anthracite used in this study are listed in Table 1 and Table 2, respectively. The mineralogical composition(XRD) of Celestine concentrate is also shown in Figure 1.

Table 1. The chemical composition of Celestine (mass, %)

	$SrSO_4$	$CaSO_4$	$CaCO_3$	SiO_2	$BaSO_4$
Celestine	67.55	14.11	12.32	2.53	1.44

Table 2 The industrial analysis of the reductant (mass, %)

	Fixed carbon	Volatile	Ash	S	P
Anthracite	82.37	6.75	10.98	0.50	0.03

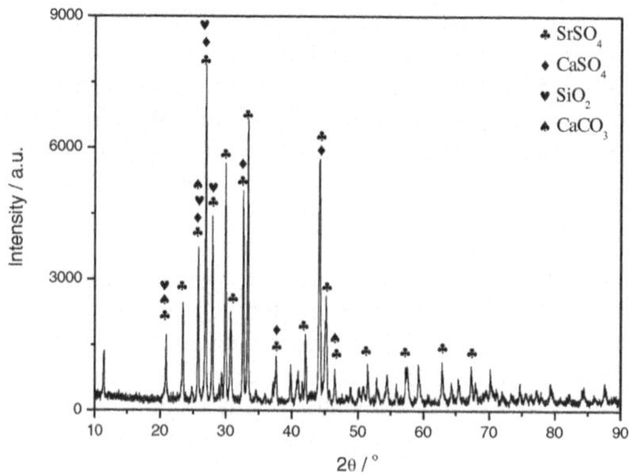

Figure 1. The mineral composition of Celestine

Experimental Method

The experiments were conducted in a high temperature electric resistance furnace (Figure 2). Celestine concentrate which were grounded fine to -0.15mm and dried anthracite powders were mixed in certain C/O proportion and 8wt% water were added into the mixture. After that, cylindrical cake samples (Celestine carbon-containing pellets) with Ø20mm×15mm were extruded by compression molding.

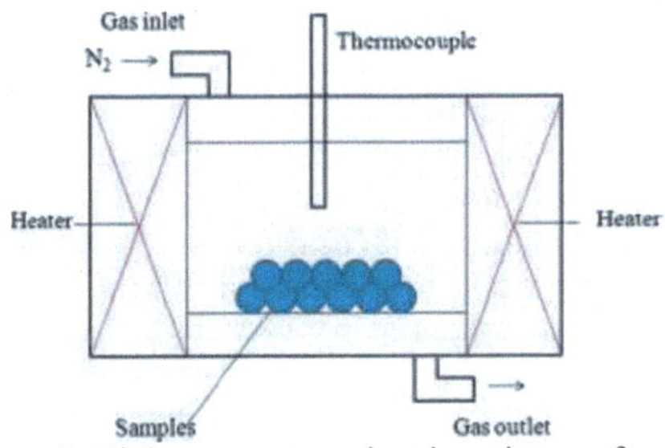

Figure 2. High temperature electric resistance furnace

The extruded samples were placed in a high temperature furnace and heated to the designated temperature in a pure nitrogen atmosphere, and held for the required time, then cooled to room temperature in the furnace. In this study, the reduction experiments with different temperatures (800, 1000, 1200 and 1300°C), different holding times (5, 10, 20 and 30min) and different C/O ratio (0.6, 0.8, 1.0 and 1.2) were conducted.

Results and discussion

Thermodynamics analysis of Celestine carbothermal reduction

Thermodynamics analysis of Celestine carbothermal reduction in this study is mainly around the reaction $SrSO_4$ reacted to SrS as follows:

$$SrSO_4 + 2C = SrS + 2CO_2(g) \quad (1)$$
$$SrSO_4 + 4C = SrS + 4CO(g) \quad (2)$$
$$SrSO_4 + 4CO(g) = SrS + 4CO_2(g) \quad (3)$$
$$SrS + 2O_2(g) = SrSO_4 \quad (4)$$
$$2SrS + 3O_2(g) = 2SrO + 2SO_2(g) \quad (5)$$
$$C + CO_2(g) = 2CO(g) \quad (6)$$

By calculating "$T - \Delta G$", reaction (1) can occur at 283°C. It described that $SrSO_4$ reacted to SrS by reaction (1) in the low temperature region. With temperature increased, reaction (2) can occur when the temperature is 492°C. Meanwhile, reducing gas CO generated could enhance, reaction (3) as CO gets in contact with $SrSO_4$. If O_2 is present in the reaction system, SrS can be oxidized to SrO or $SrSO_4$. However, when the temperature reaches700°C, carbon gasification reaction (6) occurs, and CO_2 generated in reactions (1) and (3) can be converted into CO gas in the high temperature region to participate in reaction (3), which can improve the recovery rate of strontium.

Nevertheless, as certain amount of $CaSO_4$ and a small amount of $BaSO_4$ are contained in Celestine concentrate, their thermodynamics behaviors in reduction process has to be considered. The reactions are as follows:

$$CaSO_4 + 2C = CaS + 2CO_2(g) \quad (7)$$
$$CaSO_4 + 4C = CaS + 4CO(g) \quad (8)$$
$$BaSO_4 + 2C = BaS + 2CO_2(g) \quad (9)$$
$$BaSO_4 + 4C = BaS + 4CO(g) \quad (10)$$

By calculating "$T - \Delta G$", the initial temperature of reaction (7) is 213°C. It is 70°C lower than initial temperature of reaction (1). Reaction (8) can occur at 455°C, lower than reaction (2) about 40°C. The initial temperature of reaction (9) and reaction (10) are 304°C and 502°C, respectively. Both of them are higher than the data obtained in $CaSO_4$ and $SrSO_4$.

In summary, by thermodynamic analysis of these reactions, reduction of $SrSO_4$, $CaSO_4$ and $BaSO_4$ can occur above 500°C. But protective or reducing atmosphere in reaction process is needed to suppress the SrS re-oxide. At the same time, an excess of reducing agent added into the samples (namely improve carbon amount) lead to generate excess of reducing CO gas. It has good effect on reaction.

Reduction experiment of Celestine concentrates

Following the above thermodynamic analysis, Celestine reduction experiment though direct reduction processes were conducted by simulating a rotary hearth furnace at different temperatures, carbon ratio, and holding times.

Effect of temperature: Experimental results on the reduction of Celestine at different temperatures with C/O ratios of 1.0 and a holding time of 10 min were conducted. The calculated conversion rates ($SrSO_4$ to SrS) are shown in Table 3. XRD results of a sample reduced at 1200°C are presented in Figure 3.

Table 3. Conversion rate of $SrSO_4$ at different temperature

	800°C	1000°C	1200°C	1300°C
conversion rate/%	34.52	66.75	90.52	92.24

The results shown that the $SrSO_4$ conversion rate from, Celestine is greatly affected by temperature. With the temperature increasing, the conversion rate of $SrSO_4$ significantly increased. The conversion rate of $SrSO_4$ is less than 35% at 800°C, and the conversion rate is about 70% at 1000°C, but when temperature reaches to 1200°C, the $SrSO_4$ conversion rate is up to 90%. However, less increasing is found with increase temperature, when temperature reach to 1300°C, the conversion rate only 2 percent improved. In thermodynamic analysis, $SrSO_4$ could reduce to SrS at 500°C. While the kinetic of this react is too low in dynamics analysis. Therefore, more appropriate react temperature of Celestine reduction is higher than 1200 °C.

In addition, as seen in Fig.3, all SrSO4 has been reduced to SrS and $Sr_{0.75}Ca_{0.25}S$ at 1200°C.

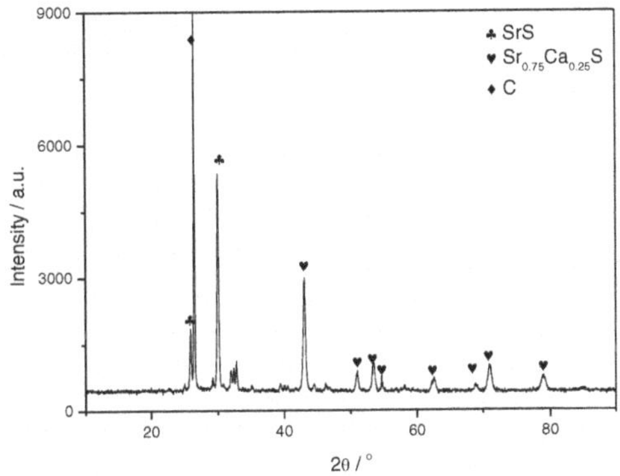

Figure 3. The mineral composition of sample

Effect of carbon ratio The effect of the C/O ratio on the conversion rate $SrSO_4$ at 1200°C and a holding time of 10 min are shown in Table 4.

Table 4. Conversion rate of SrSO4 at different C/O ratio

	0.6	0.8	1.0	1.2
conversion rate/%	68.42	90.18	90.52	91.24

As C/O ratio increase, the conversion rate of $SrSO_4$ in Celestine increase. When the C/O is 0.6, poor $SrSO_4$ reduction and lower conversion rate are found since lower carbon content. But when the C/O increases to 0.8, the $SrSO_4$ conversion rate reach up to 90%. However, less increasing is found with increase C/O ratio, the $SrSO_4$ conversion rate is 91.24% with a C/O of 1.2. As a result, more appropriate C/O of Celestine reduction is 0.8.

Effect of holding times The effect of the holding time on the conversion rate of $SrSO_4$ at 1200°C with C/O of 0.8 are shown in Table 5.

The $SrSO_4$ conversion ratio in Celestine increased gradually with increasing holding times. When the holding time is 5min, the conversion rate is below 84%. But when the holding time is 20min, the $SrSO_4$ conversion rate can reach 95.83%. However, the $SrSO_4$ conversion rate is 97% with a holding time of 30min.Therefore, more appropriate constant temperature time of Celestine pellet is 20 min.

Table 5. Conversion rate of $SrSO_4$ at different holding times

	5min	10min	20min	30min
conversion rate /%	83.34	90.18	95.83	97.06

Conclusions

(1) The composition of Celestine concentrate is mainly $SrSO_4$, $CaSO_4$, $BaSO_4$ and $CaCO_3$, and it also contains a small amount of SiO_2. The $SrSO_4$ content in Celestine concentrate can reach 67.55%.
(2) Based on thermodynamic analysis, reduction of $SrSO_4$, $CaSO_4$ and $BaSO_4$ can occur above 500°C, but protective atmosphere and excess reducing agent ratio in reduction process is needed.
(3) It is feasible to deal with Celestine by rotary hearth furnace direct reduction process. When the temperature is 1200°C, C/O is 0.8 and holding times is 20min, the conversion rate of $SrSO_4$ in Celestine is above 95%.

References

[1] Deniz Bingol, Salih Aydogan, S. Sinan Gultekin. Neural model for the leaching of celestite in sodium carbonate solution[J]. Chemical Engineering Journal, 2010,165(2): 617-624.
[2] R. Suárez-Orduña, J.C. Rendón-Angeles, K. Yanagisawa. Kinetic study of the conversion of mineral celestite to strontianite under alkaline hydrothermal conditions[J]. International Journal of Mineral Processing, 2007, 83(1–2):12-18
[3] Salih Aydoğan, Murat Erdemoğlu, Ali Aras, Gökhan Uçar, Alper Özkan. Dissolution kinetics of celestite ($SrSO_4$) in HCl solution with $BaCl_2$[J]. Hydrometallurgy, 2006, 84(3–4):239-246
[4] Murat Erdemoğlu, Salih Aydoğan, Eberhard Gock. Effects of intensive grinding on the dissolution of celestite in acidic chloride medium[J]. Minerals Engineering, 2009, 22(1):14-24
[5] Abdullah Obut, Peter Baláž, İsmail Girgin. Direct mechanochemical conversion of celestite to $SrCO_3$[J].Minerals Engineering, 2006,19(11):1185-1190
[6] N. Setoudeh, N.J. Welham, S.M. Azami. Dry mechanochemical conversion of $SrSO_4$ to $SrCO_3$[J]. Journal of Alloys and Compounds, 2010, 492(1–2):389-391
[7] Dongping Duan, Hongliang Han. Mechanisms of iron nugget forming in rotary hearth furnace process at low temperature[C]. 6th International Congress on the Science, and Technology of Ironmaking, 2012,10
[8] Hongliang Han, Dongping Duan. Dephosphorization Technology of High Phosphorus Oolitic Hematite by Rotary Hearth Furnace Iron Nugget Process[C]. 6th International Congress on the Science, and Technology of Ironmaking, 2012,10

7th International Symposium on High-Temperature Metallurgical Processing
Edited by: Jiann-Yang Hwang, Tao Jiang, P. Chris Pistorius, Gerardo R.F. Alvear F., Onuralp Yücel, Liyuan Cai, Baojun Zhao, Dean Gregurek, and Varadarajan Seshadri
TMS (The Minerals, Metals & Materials Society), 2016

INFLUENCE OF SLAG BASICITY ON THE SILICON WITHIN THE STAINLESS STEEL MASTER ALLOY PREPARED BY SMELTING REDUCTION OF Fe-Ni-Cr SINTERS

Yanhui Liu, Pingsheng Lai, Xuewei Lv, Chenguang Bai

School of Materials Science and Engineering, Chongqing University, Chongqing, 400044, China

Keywords: Stainless steel master alloy, Reduction, Thermodynamics calculation

Abstract

The purpose of this paper is to study the effect of slag basicity on the Si content in stainless steel master alloys prepared by smelting reduction of Fe-Ni-Cr sinters. Theoretical analysis focusing on the gas, slag and alloy phases of the smelting reduction process at 1600℃ was carried out by thermodynamic calculation software----FactSage, showing that the increase of CaO within the slag can result in the decrease of Si within the master alloy. Experiments were also carried out in vacuum melting furnace at 1600 ℃, and chemical analysis method was used to detect Si content in the alloys. When the MgO content is 9.00%, the experimental results were consistent with the calculated values. However, when the MgO content rises up to 14%, the content of Si in the alloy reaches the maximum when SiO_2 equals to 22.81%.

Introduction

Stainless steel [1] is characterized by high strength and ductility as well as excellent corrosion and heat resistances. Ni and Cr are important strategic alloying metal elements with wide industrial applications for the production and the excellent performance of the stainless steel. Conventional stainless steel production methods mainly depend on Fe-Cr alloy [2-9] and Fe-Ni alloy [10-14] as alloying raw materials. However, with the deterioration of high grade mineral resource like Ni-based and Cr-based deposit world widely, the production cost of Fe-Cr alloy and Ni-containing material with electric arc furnace went up continually, leading to the increase of the production cost of stainless steel. In addition, with the decrease of good quality Ni and Cr resources, more and more low-grade laterite and chromite are used into the production process of the ferroalloys. However, most of these Ni and Cr resources cannot be directly added into the electric furnace, so the concomitant agglomeration step is indispensable and the performance of sinter must be improved. Under this big background, a new and compact flow of producing austenitic stainless steel master alloys by composite sintering using the laterite and chromite and smelting reduction in a small blast furnace was proposed.

As the old adage goes, "look after the slag and the metal will look after itself," meaning that slag are very important for the production of clean metals. Slag properties have a decisive effect on both process control and on the quality of the product. However, the blast furnace slag formed in the new process is quite different from that in the traditional iron-making process, and the outstanding feature is laid in the high MgO amd Al_2O_3 content, meaning that the proper physical-chemistry properties and smelting temperature need to be further studied. Experimental study and theoretical analysis have been carried out focused on the composite sintering, showing that the quality of composite sinters obtained were improved, and the recommended ratio between laterite and chromite is 7:3, the optimal water content of the mixtures and the carbon addition are

20% and 13% respectively. [15] Both the feasibility analysis and experimental studies for direct smelting reduction to prepare the stainless steel master alloy has been carried out by us, showing that the master alloy with proper compositions and satisfying metal recovery rates can be obtained at 1 600 ℃. The Si content within the master alloy is affected by the slag compositions. Furthermore, the oxidation of Si within the master alloy can generate a lot of hot. In this study, the influence of slag basicity on the Si within the stainless steel master alloy prepared by smelting reduction of Fe-Ni-Cr sinters will be researched.

Experimental

Materials

Compositions of the Fe-Ni-Cr composite sinters are shown in **Table I**. The pure carbon fines were used as the reductant. The atomic ratio of carbon and oxygen was fixed to be 1.10. The experimental scheme to study the effect of slag basicity on Si within the alloy was listed in Table II. MgO content of samples A11 to A15 was 9%, and that of A21 to A25 was 14%. It can be clearly seen in the ternary phase diagram as shown in **Figure 1**.

Table 1 Chemical compositions of Ni-Cr composite sinter (in amss %)

Oxides	Cr_2O_3	SiO_2	Fe_2O_3	FeO	Al_2O_3	MgO	CaO	NiO
Content	14.81	4.37	30.59	25.27	14.26	4.53	3.94	0.87

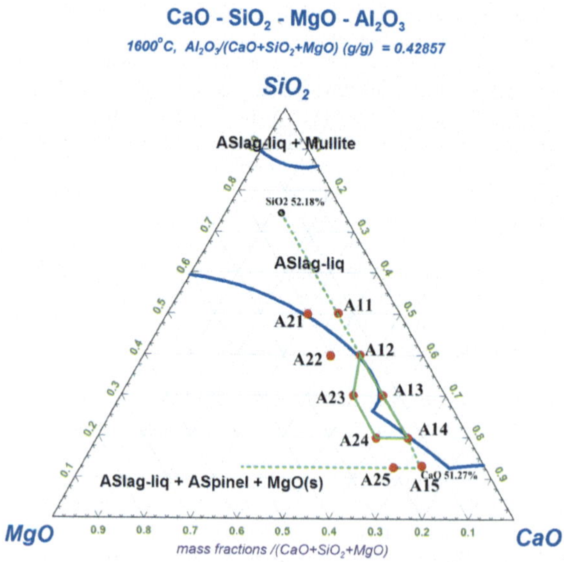

Figure 1. Experimental scheme labeled in the ternary diagram

Table II Experimental scheme of the effect of slag basicity on Si within the alloy (in mass %)

Sample No.	CaO	SiO_2	MgO	Al_2O_3
A11	26	35	9	30
A12	33	28	9	30
A13	40	21	9	30
A14	47	14	9	30
A15	52	9	9	30
A21	21	35	14	30
A22	28	28	14	30
A23	35	21	14	30
A24	42	14	14	30
A25	47	9	14	30

Experimental procedure

The general procedure for the reduction and separation behaviors, which can be divided into 8 steps: drying, grinding, mixing, briquetting, smelting reduction, water cooling, crushing, separation. The drying temperature was set to 120 ℃, and the drying time was 2 hours. The purpose of ore grinding is aimed to improve the reduction kinetics of the composite sinters because of the limited experimental conditions. Graphite heater was used in the furnace, and the maximum temperature can reach 1750 ℃. The PID temperature control program was used and the fluctuation is limited to ±2 ℃. The furnace tube was made of alumina material. Pressure for pelletizing was set to 15 MPa, and the diameter of the pellets was fixed to be 30 mm. The graphite crucible was used to study the effect of reduction temperature on the reduction and separation behaviors because the poor wettability of the graphite by the molten slag for the convenience of the separation between the slag and the crucible. The reduction time was set to be 1 hour. In this study, the reduction experiments were carried out under the N_2 (99.9%) atmosphere during the whole heating process. The Si content within the master alloy was analyzed by the chemical composition analysis.

Results and discussion

Results of the Si content within the alloy were shown in **Figure 2** under the condition that MgO within the slag was fixed to be 9%. Theoretical calculations by the FactSage were also carried out to study the change of the Si content and the calculated values were also plotted in **Figure 2**. When the MgO within the slag is 9%, the increase of slag basicity results in the dramatic decrease of the Si content within the alloy, which is consistent with the theoretical calculations. Reasons are as follows. For one thing, the SiO_2 within the slag decreased along with the increase of the CaO. For another, the viscosities of the slag are decreasing with the increase of CaO, improving the reduction kinetics condition.

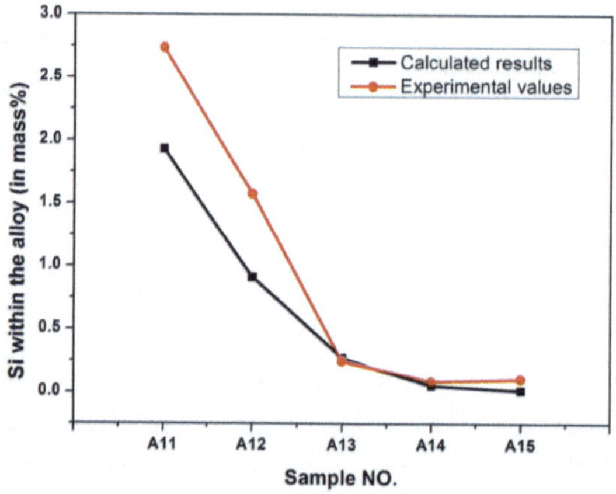

Figure 2. Variation of Si in the alloy along with different slag compositions (MgO=9%)

Variation of Si content in the alloy were shown in **Figure 3** with the MgO content within the slag is 14%, according to the calculated results, the increase of CaO results in the dramatic decrease of the Si content within the alloy, which is similar to the **Figure 2**. However, the content of Si in the alloy reaches the maximum, 1.02%, when SiO_2 equals to 22.81%, meaning that experimental values are totally different from the theoretical results. For A21 and A22, the Si content is quite low resulting from the high viscosity. For A24 and A25, the low Si within the alloy can be attributed into that the SiO_2 content in the slag is not high, only 14% and 9%. However, the Si content of A23, 1.02% is rather higher comparing with other four samples. Reasons may be attributed into the SiO_2 content within the slag (21%) and the appropriate viscosity of the slag.

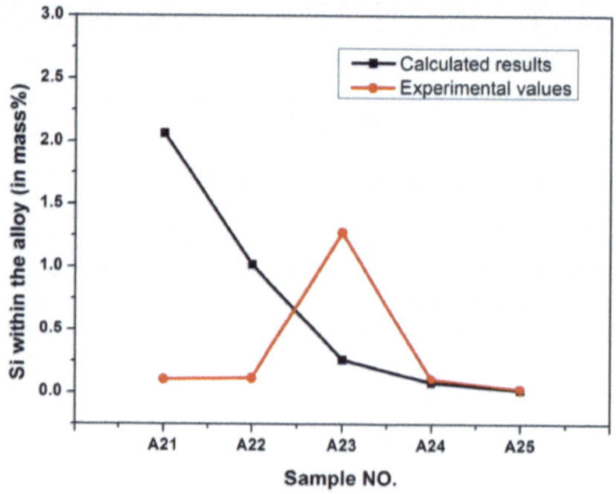

Figure 3. Variation of Si in the alloy along with different slag compositions (MgO=14%)

Conclusions

Effect of slag basicity on the Si content in stainless steel master alloys prepared by smelting reduction of Fe-Ni-Cr sinters were carried out in this work, and the main conclusions are as follows:

(1) When the MgO within the slag is 9%, the increase of slag basicity results in the dramatic decrease of the Si content within the alloy, which is consistent with the theoretical calculations. For one thing, the SiO_2 within the slag decreased. For another, the viscosities of the slag are decreasing with the increase of CaO, improving the reduction kinetics condition.

(2) When the MgO within the slag is 14%, the content of Si in the alloy reaches the maximum, 1.02%, when SiO_2 equals to 22.81%. The experimental results do not agree with the theoretical values.

Acknowledgement

The authors are especially grateful to the National Natural Science Foundation of China (NSFC) (Grant No. 51234010) and Chongqing University Graduate Students Scientific Research Innovation Project (No. CYB14026).

References

1. M.J. McNallan, and T. Debroy, "Effect of Temperature and Composition on Surface Tension in Fe-Ni-Cr Alloys Containing Sulfur", *Metallurgical and Materials Transactions B-Process Metallurgy and Materials Processing Science*, 22 (4) (1991), 557-560.
2. W.G. Pei, and O. Wijk, "A kinetic study on chromite ore smelting reduction", *Scandinavian Journal of Metallurgy*, 22 (1993), 38-44.
3. E.O. Yape, and N.M. Anacleto, "Direct Smelting of Chromite and Laterite Ores with Carbon Under Argon Atmospher", *Advanced Materials Research*, 849 (2014), 170-176.
4. Y. Wang *et al.*, "Kinetics of Carbothermic Reduction Of Synthetic Chromite", *Journal of Mining and Metallurgy Section B-Metallurgy*, 50 (1) (2014), 15-21.
5. G.U. Kapure *et al.*, "Direct Reduction of Low Grade chromite overburden for recovery of metals", *Ironmaking & Steelmaking*, 38 (8) (2011), 590-596.
6. D. Chakraborty, S. Ranganathan, and S.N. Sinha, "Carbothermic Reduction of Chromite Ore under Different Flow Rates of Inert Gas", *Metallurgical and Materials Transactions B-Process Metallurgy and Materials Processing Science*, 41 (1) (2010), 10-18.
7. D. Chakraborty, S. Ranganathan S.N. Sinha, "Investigations on the Carbothermic Reduction of Chromite Ores", *Metallurgical and Materials Transactions B-Process Metallurgy and Materials Processing Science*, 36 (4) (2005), 437-444.
8. Y.W. Li *et al.*, "Reduction mechanism of chromite ore in blast furnace", *Journal of Iron and Steel Research International*, 11 (4) (2004), 19-23.
9. M. Tsomondo, and D. Simbi, "Kinetics of chromite ore reduction from $MgO-CaO-SiO_2-FeO-Cr_2O_3-Al_2O_3$ Slag System by Carbon Dissolved in High Carbon Ferrochromium Alloy Bath", *Ironmaking Steelmaking*, 29 (1) (2002), 22-28.

10. M. Liu *et al.*, "Novel Process of Ferronickel Nugget Production from Nickel Laterite by Semi-molten State Reduction", *ISIJ International*, 54 (8) (2014), 1749-1754.
11. M.J. Rao *et al.*, "Carbothermic Reduction of Nickeliferous Laterite Ores for Nickel Pig Iron Production in China: A Review", *JOM*, 65 (11) (2013), 1573-1583.
12. W.J. Yang *et al.*, "Study on the Influence of CaO on Reduction Smelting for the Laterite Ore", *Advanced Materials Research*, 634-638 (2013), 3260-3264.
13. B. Li, H. Wang, and Y.G. Wei, "The Reduction of Nickel From Low-grade Nickel Laterite Ore Using A Solid-state Deoxidisation Method", *Mineral Engineering*, 24 (14) (2011), 1556-1562.
14. S. Mohanty, S.K. Roy, and P.K. Sen, "Thermodynamic Considerations in Reduction of Nickeliferrous Laterite by Methane", *Metallurgical and Materials Transactions B-Process Metallurgy and Materials Processing Science*, 39 (5) (2008), 639-642.
15. Y.H. Liu *et al.*, "Preparation and Consolidation Mechanism for Nickel-Chromium Composite Sinters", *Journal of Iron and Steel Research (Chinese).* (Accepted)

7th International Symposium on High-Temperature Metallurgical Processing
Edited by: Jiann-Yang Hwang, Tao Jiang, P. Chris Pistorius, Gerardo R.F. Alvear F., Onuralp Yücel, Liyuan Cai, Baojun Zhao, Dean Gregurek, and Varadarajan Seshadri
TMS (The Minerals, Metals & Materials Society), 2016

REDUCTION BEHAVIOR OF CHROMIC OXIDE IN Ti –BEARING BF SLAG

Baohua Li, Xuewei Lv*, Yun Chen, Yanhui Liu, Shengping Li
College of Materials Science and Engineering, Chongqing University, Chongqing 400044 China
*Corresponding author. E-mail: lvxuewei@163.com (Xuewei Lv)

Keywords: Hongge iron ore, Cr_2O_3, Ti-bearing BF slag containing Cr_2O_3, Reduction

Abstract

There are almost ten billion tons of vanadium-bearing titanomagnetite in Panzhihua-Xichang area. However, iron ores in Hongge, which has the most abundant reserves in the four mine areas in Panzhihua, have not been utilized sufficiently so far. The reason behind this is the relatively high Cr_2O_3 in the ores (0.49 %~0.82 %), that due to its high melting temperature will negatively influence the blast furnace reduction and slagging process. Therefore, in the blast furnace process, Cr is hope to be reduced into molten iron, which can reduce slag viscosity problems.

Slag iron reduction experiments were conducted to quantify the Cr reduction. Experimental results shows that with the increase of TiO_2, the content of Cr in the hot metal is decreased when balanced. And during the initial stage of the reduction reaction, the reaction kinetic model is a first-order reaction.

Introduction

Hundreds of millions of tons of vanadium-bearing titanomagnetite resources can be found in China's Panxi region [1]. Existed in polymetallic intergrowth composite iron ore, vanadium-bearing titanomagnetite has a high comprehensive utilization value [2]. As important metal elements which contained in vanadium-bearing titanomagnetite, the demands of iron, vanadium, and titanium are decreasing. So it has great significance to improve the technology of the comprehensive utilization of vanadium-bearing titanomagnetite [3,4]. With nearly 40 years of development, a technological process of steel and vanadium products centered on vanadium-titanium magnetite beneficiation, blast furnace process, converter vanadium extraction, converter steelmaking, continuous casting production has already established. However, for the Hongge mining area with the largest reserve, one of the four main mining areas in the Panxi region, resources utilization are still underdeveloped. The main problem is that Hongge ores contain Cr_2O_3 of 0.49 %–0.82 % in addition to iron, vanadium, and titanium [5].

As a refractory oxide, Cr_2O_3 affects the blast furnace reduction and slagging process, furthermore, chromium in molten iron significantly affects the follow-up converter vanadium extraction process. Therefore, a systematic and comprehensive investigation of these effects is relevant prior to have a large-scale development and utilization of Hongge mineral resources.

Researches on the viscous flow characteristics of Ti-bearing BF slag are extensive. Gao [6] shows that a small amount reduction of TiO_2 can decrease the slag viscosity under Ar at a fixed basicity. Zhang [7] also find that TiO_2 may function as a network modifier. Furthermore, if the reduction continues, generating the Ti(C, N), the slag viscosity will have a substantial increase. Bai [8] insists that the generating of the high temperature lyophilic colloid contributes to the increase of slag viscosity. Shankar [9], Park [10], Sohn [11,12] and Liao [13] studied quinary slag systems (CaO-SiO_2-Al_2O_3-MgO-TiO_2) and found that, TiO_2 additions can decrease the slag viscosity under Ar at a fixed basicity, and with increasing the CaO/SiO_2 ratio can also decrease the slag viscosity at a fixed TiO_2 content.

Moreover, some researches have been done on high chrome vanadium-bearing titanomagnetite. Li [14] investigates solid-state reduction properties of carbon-bearing chromite pellets, suggests that reduction temperature, carbon content in pellets and addition agents have great influences on the reduction properties. Zhao [15] investigates influential factors in the solid carbothermic reduction of chromite, shows that the influences on the solid carbothermic reduction are multifactorial.

However, few scholars have studied the reduction behavior of chromic oxide in Ti-bearing BF slag. In the present study, the influence of TiO_2 content has been carried out on the reduction of Cr_2O_3 in the slag.

Experimental

Materials

The experimental samples used in this study were based on the chemical composition of on-site slag from Panzhihua Iron and Steel Corporation, as showed in Table I. According to the similar properties of Cr_2O_3 and V_2O_5 in BF, calculated by the distribution ratio of V between slag and iron, the content of Cr_2O_3 in the slag can be obtained [16]. A pure chemical reagent was used to prepare the samples. Before preparation, the reagent was dried for 6 h at the 120 °C to remove moisture. In addition, CaO was added in the form of $CaCO_3$ with the same molar mass. The chemical components of the prepared slag sample are shown in Table II. According to the iron-carbon phase diagram, pig iron is prepared by using iron powder saturated carburizing. The slag ratio is 0.6 based on the on-site slag from Panzhihua Iron and Steel Corporation. The factors and levels of experiment are showed in TableIII.

Table I. Chemical composition of BF slag from Panzhihua Iron and Steel Corporation, wt%

CaO	SiO_2	MgO	Al_2O_3	TiO_2
27.0	24.3	8.3	14.4	22.3

Table II. Experimental compositions at present work, wt%

NO.	Compositions of slag, wt%						C/S
	CaO	SiO_2	Al_2O_3	MgO	TiO_2	Cr_2O_3	
1#	40.33	36.67	14	8	0	1	1.1
2#	37.71	34.29	14	8	5	1	1.1
3#	35.10	31.90	14	8	10	1	1.1
4#	32.48	29.52	14	8	15	1	1.1

Table III. Factors and Levels of Experiment

Factor	Level									
Time/min	10	20	30	40	50	65	80	110	140	170
TiO_2/%	0	5	10	15	-	-	-	-	-	-

Apparatus and procedure

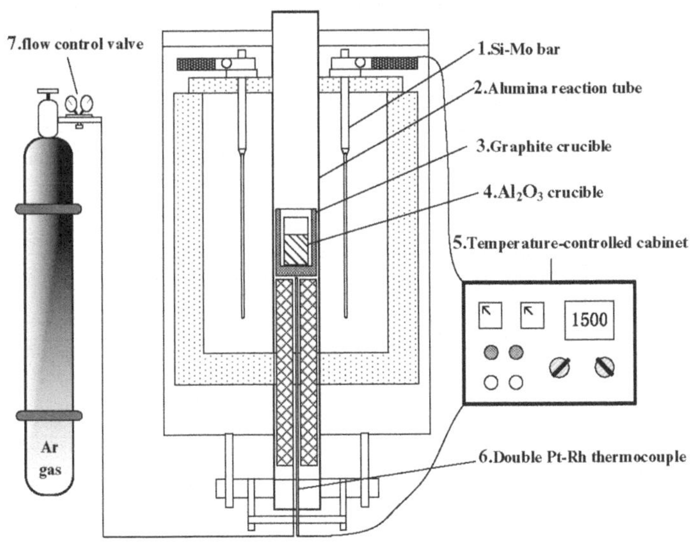

Figure 1. Experimental apparatus of $MoSi_2$ furnace

Experimental setup for the reduction reaction is shown in Figure 1. $MoSi_2$ was used as the heating element; the furnace tube is made of Aluminum oxide; the furnace bottom was sealed so that the crucible and test structure were not oxidized after the inert gas flew in. The measuring pole and measuring head in the rotary test were made of molybdenum.

In the experimental process, the furnace was protected under Ar of 99.9 % (0.1 L· min^{-1}). The prepared sample was placed in the corundum crucible (50×120 mm) for per-melting and kept at 1550 °C for 30 min. A molybdenum bar was used to stir the per-melting slag every 10 min so that the components of the slag are uniformly mixed. After finishing the experiment, samples were water quenched, dried and grinded, and slag samples were obtained; the reduction reaction was conducted in the corundum crucible (50×120 mm). 90 g pig iron was put into the corundum crucible, then adding 54 g pre-melted slag when the temperature is constant at 1450 °C. The samples, as well as the corundum crucible was put into water in case of being oxidized when reached the specified time. After crushing and grinding, the samples were sent for XRF analysis.

Temperatures were measured by thermocouples and automatically controlled under ±1 degree.

Reaction mechanism

Both Cr_2O_3 and TiO_2 can be reduced at elevated temperature. According to the oxygen potential in molten iron, TiO_2 will be superior to Cr_2O_3 to be reduced, because of its higher oxygen potential, as shown in Figure 2. However, Cr_2O_3 will be superior to be reduced in Ti-bearing BF slag, causing TiO_2 will combine with CaO to generate perovskite, which has a high melting point. This is verified through the calculation of FactSage software.

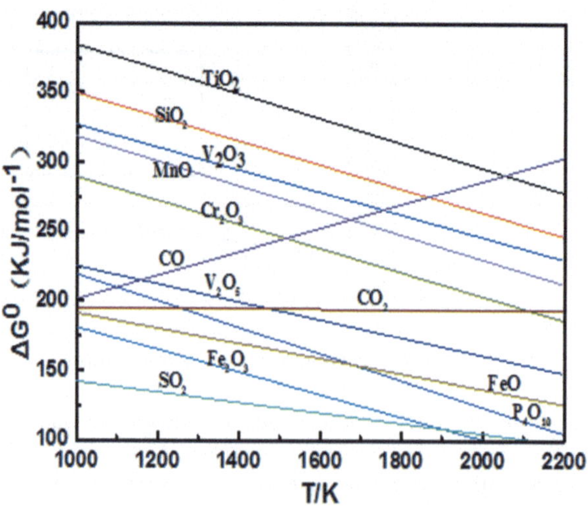

Figure 2. The relationship between $\Delta G^{\theta} - T$

Results and discussion

Effect of TiO_2 content on reduction of Cr_2O_3

Figure 3 shows the concentration of Cr in molten iron variation with time in different concentration of TiO_2. It can be seen that the concentration of Cr in molten iron increases with increasing reaction time. It is more obvious that the concentration of Cr in molten iron decreases with the increase of TiO_2 content in slag at equilibrium.

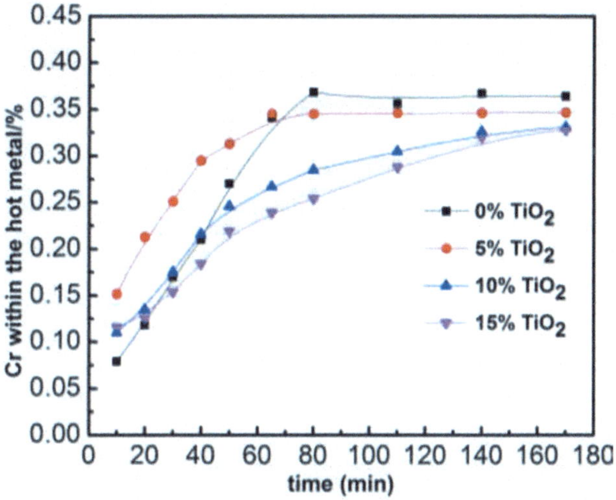

Figure 3. The concentration of Cr in molten iron variation with time in different concentration of TiO_2

Figure 4 shows the relationship between equilibrium concentration of Cr and TiO_2. It can be seen that both the theoretical calculation by FactSage software and experimental dates of the concentration of Cr in molten iron showed a decreased tendency with the increase of TiO_2 content in slag at equilibrium, and both of the corresponding dates have little different.

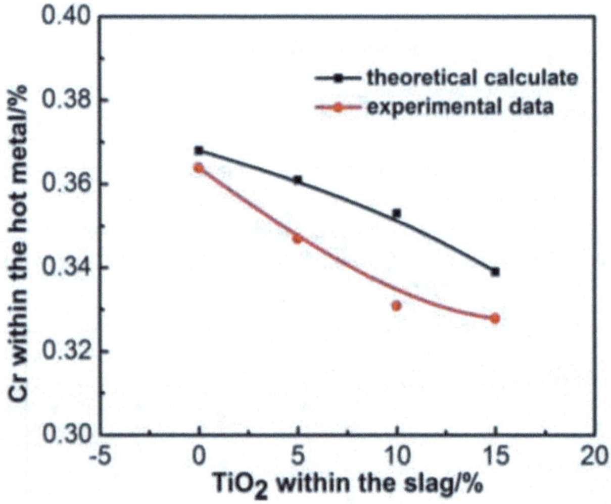

Figure 4. The relationship between equilibrium concertration of Cr and TiO_2

Kinetics model of chromic oxide reduction

Figure 5 shows the relationship between ln（C_{A0}/C_A）and *t*, where C_{A0} is the chrome content in the initial slag; C_A is the chrome content after reaction for different time; *t* is the reaction time. It is obvious that the ruduction reaction of Cr_2O_3 in Ti-bearing BF slag is the first order reaction at the initial phase.

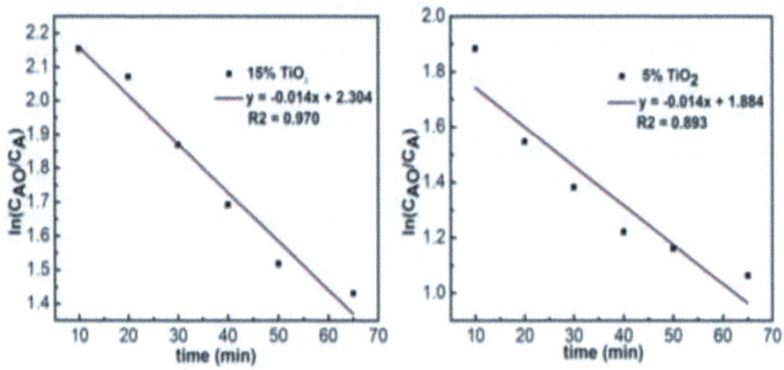

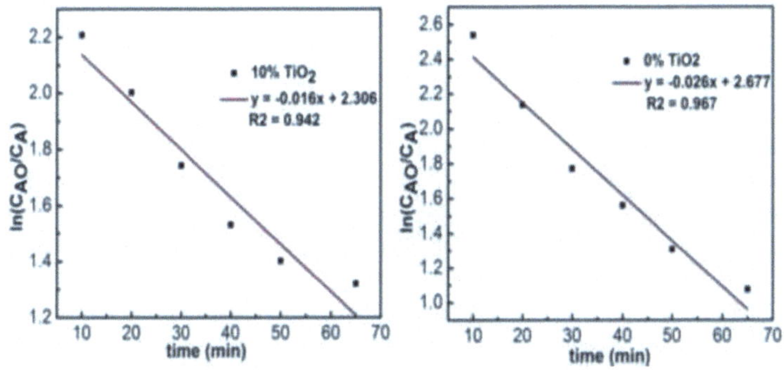

Figure 5. The relationship between ln（C_{A0}/C_A） and *t*

Conclusions

The reduction behavior of chromic oxide in Ti-bearing BF slag is experimentally demonstrated, and the following conclusions are obtained.

(1) With the increase of TiO_2 content in slag,the concentration of Cr in molten iron decreases at equilibrium.

(2) At the initial stage of the process, the reduction of the chromic oxidein Ti-bearing BF slag follows the first order reaction kinetics model.

Acknowledgement

This work was supported by "National Key Basic Research Program of China (973 Program)" (Grant No. 2013CB632603).

References

1. L. J. Xu et al., "Comprehensive Recycling Situation and Development Direction of Vanadium-titanium Magnetite in PanXi," *Sichuan Nonferrous Metals*, 1(2011), 1.
2. J. T. Zhang , "The Occurrence, Distribution and Recovery Fechnologies of Chromiun in Hongge Iron Ore," *Sichuan Nonferrous Metals*, 2005, (01): 1-4.
3. J. H. Chen et al., "Experimental Research on Improving the Recovery of Vanadium Titanomagnetite Ore in Hongge Mining Areas in Panzhihua," *Sichuan Nonferrous Metals*, 2011(2):17-20.
4. L. Li et al., "Process Mineralogy Studies on V-Ti Magnetite in Panzhihua Region," *Metal Mine*, 2010(04):89-92+109.
5. G. H. Du, "Blast Furnace Smelting Principle for Vanadium Titanium Magnetite," Science Press, Beijing, (1996), 278.
6. G. Q. Gao et al., "Titanium Slag Viscosity and Melting Point Determination," *Iron Steel Vanadium Titanium*, 1987, 1: 009.

7. L. Zhang et al., "Modelling Viscosity of TiO_2 Containing Silicate Melts," *The Seventh International Conference on Molten Slags, Fluxes and Salts*, 2004.
8. C. G. Bai, "Study on Some Physical Chemistry Problems of Blast Furnace Slag-Bearing Titania," (Doctoral degree thesis, Chongqing University, 2003), 1-66.
9. A. Shankar et al., "Sulfide capacity of high alumina blast furnace slags," *Metallurgical and Materials transactions B*, 2006, 37(6): 941-947.
10. H. Park et al., "Effect of TiO_2 on the viscosity and slag structure in blast furnace type slags," *steel research international*, 2012, 83(2): 150-156.
11. Il. Sohn et al., " A Review of the Relationship between Viscosity and the Structure of Calcium Silicate Based Slags in Ironmaking," *steel research international*, 2012, 83(7): 611-630.
12. Il. Sohn et al., "Influence of TiO_2 on the Viscous Behavior of Calcium Silicate Melts Containing 17 mass% Al_2O_3 and 10 mass% MgO," *ISIJ international*, 2012, 52(1): 158-160.
13. J. L. Liao et al., "Influence of TiO_2 and basicity on viscosity of Ti bearing slag," *Ironmaking & Steelmaking*, 2012, 39(2): 133-139.
14. J. C. Li et al., "Solid-state Reduction Properties of Carbon-bearing Chromite Pellets," The Chinese Journal of Nonferrous Metals, 2011, 21(5):1159-64.
15. Q. Zhao et al., "Influential Factors in the solid Carbothermic Reduction of Chromite," *Science & Technology Review*, 2013, 31(5-6): 40-3.
16. J. Deng et al., "Current situation and development of comprehensive utilization of vanadium-bearing titanomagnetite at PANGANG," *Journal of Materials and Metallurg*, 2007, 6 (2): 83-86.

7th International Symposium on High-Temperature Metallurgical Processing
Edited by: Jiann-Yang Hwang, Tao Jiang, P. Chris Pistorius, Gerardo R.F. Alvear F., Onuralp Yücel, Liyuan Cai, Baojun Zhao, Dean Gregurek, and Varadarajan Seshadri
TMS (The Minerals, Metals & Materials Society), 2016

EFFECTS OF MINERAL OXIDES ON THE PRECIPITATION MICRO-MORPHOLOGY OF METALLIC IRON IN THE REDUCTION OF IRON OXIDES UNDER CO ATMOSPHERE

Zhancheng GUO*, ZhiLong ZHAO, Huiqing TANG, Jingtao GAO, Lin Lin

State Key Laboratory of Advanced Metallurgy, University of Science and Technology Beijing; 30 Xueyuan Road, Haidian District; Beijing, 100083, China

Keywords: Ironmaking, Gas reduction, *In situ* observation, Iron whisker, Mineral oxides

Abstract

Grain sticking often occurs in the fluidized bed reduction of fine iron ore, which is closely related to the precipitation morphology of metallic iron on the surface of ore particles. In this work, simulating the gas-solid reduction process of iron ore, in situ observations were carried out to investigate effects of doping mineral oxides on the precipitation morphology evolution of metallic iron during reduction of iron oxides with CO. Results indicate that the precipitation morphology of metallic iron is related to the quantity of doped mineral oxides. The minimum mole fraction of doped oxide ($N_{A_yO_x}$) that can inhibit the growth of iron whiskers is related to its cation radius ($r_{A^{x+}}$), extranuclear electronic layers ($n_{A^{x+}}$) and valence electrons ($q_{A^{x+}}$). Their relation can be expressed as: $N_{A_yO_x} = 1.3\times10^{-5} \cdot \frac{r_{A^{x+}}^2 \cdot \sqrt{n_{A^{x+}}}}{(q_{A^{x+}}-1)!}$

Introduction

Gas fluidized bed reduction process of fine iron ore can greatly boost the productivity due to better reduction kinetics. However, the appearance of sticking in gas-based ironmaking process results in fluidization stagnation and could spread out over the entire fluidized bed during a very short time, making it a serious problem in the fluidized bed ironmaking process[1, 2]. Numerous studies indicated that grain sticking was closely related to the precipitation morphology of metallic iron during reduction [2-10]. A good deal of fibrous whiskers were formed on the surface of particles during gaseous reduction, and the intertextures of such whiskers would lead to sticking between particles of fine ore which could influence the normal reduction [2-9]. The differences in the precipitation morphology of reduced iron can be attributed mostly to different diffusion and nucleation conditions. In addition, the chemical compositions of ore particles, especially for some composite symbiotic mineral oxides, have a strong effect on the precipitation morphology of metallic iron [2, 3, 8, 10], the systematic study of which has still not been performed. Thus, investigating the effect of mineral oxides on the precipitation morphology of metallic iron is significant in the study of sticking mechanism and inhibition measure during gaseous reduction.

In this work, under conditions of 800 °C and volume ratio of $CO:CO_2=40:10$, simulating the gas-solid reaction system on the surface of iron ore particles, *in situ* observations were performed to investigate effects of doping mineral oxides (such as MgO, CaO, SrO, BaO, Al_2O_3, Li_2O, Na_2O and K_2O) on the precipitation morphology evolution of metallic iron after the reduction.

Experimental

<u>Experimental Apparatus</u>

In this study, experiments were performed using an in situ observation system, which mainly includes a high-temperature stage of UK Linkam TS 1500 and a stereo optical microscope of Germany Zeiss SteREO Discovery.V20. Fig.1 schematically shows this experimental apparatus. The changes of samples surface colour and morphology were used for the characterization of the reduction process and transformation of phases.

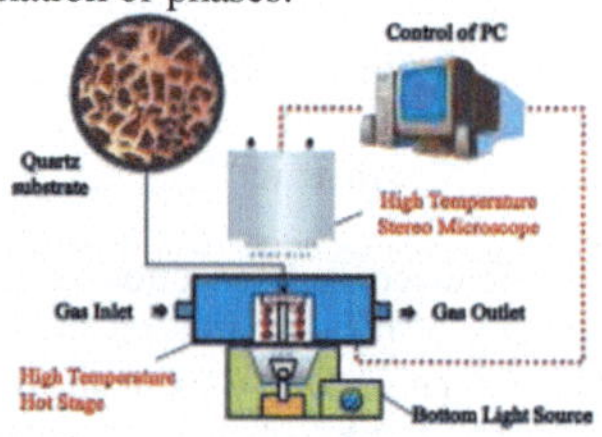

Fig.1 Experimental Apparatus of In Situ observation

<u>Samples Preparation</u>

Analytical reagent $Fe(NO_3)_3 \cdot 9H_2O$, $LiNO_3$, $NaNO_3$, KNO_3, $Mg(NO_3)_2 \cdot 6H_2O$, $Ca(NO_3)_2 \cdot 4H_2O$, $Sr(NO_3)_2$, $Ba(NO_3)_2$, $Al(NO_3)_3 \cdot 9H_2O$ and deionized water were used in this experiment. The mixture solutions of $Fe(NO_3)_3$ and $A(NO_3)_x$(A: Li, Na, K, Mg, Ca, Sr, Ba, Al, et al)were prepared in accordance with desired ratios. The polished quartz slide (Φ=6mm, h=1mm) was placed on refractory brick, and heated to 800°C, then cooled to about 500°C in air. The mixture solution was then sprayed on the quartz slide. After the solution evaporated, the sample was roasted for 10min at 1000°C to react completely and become compact. Then it was taken out and cooled to room temperature.

The principle of samples preparation was based on rapid decomposition of $Fe(NO_3)_3$ and $A(NO_3)_x$ into Fe_2O_3 and A_yO_x under high temperatures, and these oxides got deposited to the quartz slide to get the sheet sample of Fe_2O_3-A_yO_x. The reactions were:

(1) $4Fe(NO_3)_2 \cdot 9H_2O \xrightarrow{\Delta} 2Fe_2O_3 + 4NO_2\,(g) + 3O_2\,(g) + 36H_2O(g)$

(2) $nyA^{x+} + 2mFe^{3+} + (nxy+6m)NO_3^- \xrightarrow{\Delta} nA_yO_x \cdot mFe_2O_3 + (nxy+6m)NO_2(g) + \frac{nxy+3m-nx}{2}O_2(g)$

Fig.2 gives the characteristic of the sheet Fe_2O_3-A_yO_x: length diameter D is about 30~150 μm; short diameter d is about 10~60 μm; thickness h is about 10~50 μm.

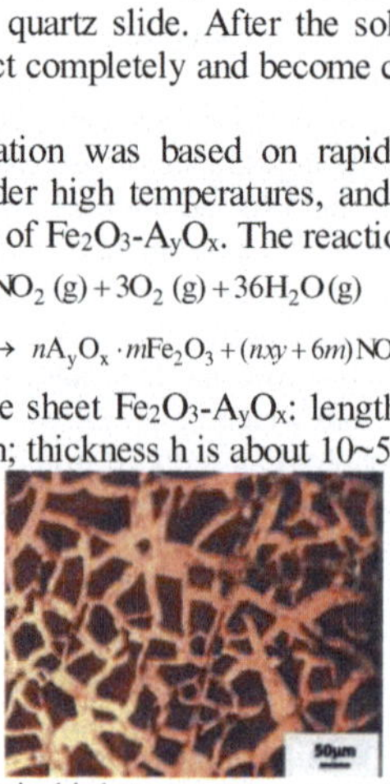

Fig.2 Images in high temperature stereo microscope

Experimental Procedure

Firstly, the quartz slide with Fe_2O_3-A_yO_x sample was placed in the high temperature hot stage, which was sealed and then put under the stereo optical microscope. The bottom of heating stage was equipped with high-brightness light source of LED. With help of these sources, the morphology variations of samples during the reduction can be observed owing to the high light transmittance for the quartz slide and sheet sample. Thus, the colour changes of sheet sample can reflect the structural evolution of iron oxides. In addition, the growth process of iron whiskers can be monitored on-line with the help of the light source from the bottom.

The heating stage was heated with the rate of 50°C/min, and kept at constant temperature after reaching the predetermined temperature (800°C), then the flow of reduction gas was kept for 30min (flow rate of CO was 40 mL/min and flow rate of CO_2 was 10 mL/min). After stopping the gas, the sample was cooled to the room temperature with the rate of 60°C/min. The whole reduction process was monitored by on-line observation under stereo optical microscope (1000x), and the images for observation were recorded at the rate of 1 sheet/s. After the above experiments, the reduced samples were subjected to SEM analysis.

Experimental Results and Discussion

In Situ Observations

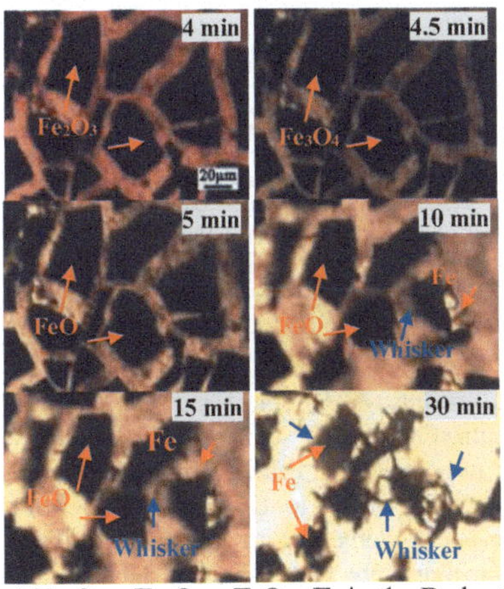

Fig.3 $Fe_2O_3 \rightarrow Fe_3O_4 \rightarrow FeO \rightarrow Fe$ in the Reduction

The precipitation morphology evolution of metallic iron after the reduction was investigated under the conditions of 800°C and the volume ratio of CO:CO=40:10. As shown in Fig.3, results reveal that iron oxide changes from red to black at about 4 min, and the area of sheet sample expands gradually to reach to the maximum after 30 s. The area increased by 8.6% as confirmed by images analysis. Afterwards, it began to shrink, and iron whiskers started to grow. The sample colour changes during reduction was owing to the conversion of $Fe_2O_3 \rightarrow Fe_3O_4$. Fe_2O_3 is trigonal system. Moreover, Fe_3O_4 is tetragonal system. The volume could expand in this conversion. These conform to the reduction rule of iron oxide. In the conversion of $Fe_3O_4 \rightarrow FeO$, the sample colour does not change, however the volume could notably shrink. Experimental results show that the rate for conversion of $Fe_2O_3 \rightarrow Fe_3O_4$ was very fast, which was followed by that of $Fe_3O_4 \rightarrow FeO$, but the rate for $FeO \rightarrow Fe$ was the slowest among these conversions. The whole reaction process completed in 30 min. Besides, during reduction, parts of sheet film of Fe_2O_3 tend to crack, because of the formation of Fe_3O_4 cracks in $(001)_H || (111)_M$ with the increase of internal heat stress of the reduced sheet sample[11]. Based on observation, iron whiskers do not grow in the conversion of $Fe_2O_3 \rightarrow Fe_3O_4 \rightarrow FeO$ in the reduction of Fe_2O_3 with CO, but do in the stage of $FeO \rightarrow Fe$.

Effects of Mineral Oxides on the Precipitation Morphology

Effects of mineral oxides on the precipitation morphology of metallic iron during reduction were investigated on the basis of the reduction of Fe_2O_3 with CO. Observations for the reduction processes of all samples are shown in Fig.4 and SEM analysis after the reduction is shown in Fig.5. Concerning doping alkaline earth oxides, iron whiskers do not grow in the case of 2 mass% MgO. However, when achieving the same effect, 8 mass% CaO is needed, and the minimum quantity of doping SrO which is the same group with CaO and MgO reaches to 25 mass%. Moreover, the quantity of doping SrO reaches up to 50 mass%. Effects of the alkali metal elements on the precipitation morphology of metallic iron were further investigated. When achieving the similar inhibition, the minimum quantity of Li_2O is 0.4 mass%, however, the quantity of Na_2O is 4 mass% and the quantity of K_2O is 12 mass%. Only 0.5 mass% of Al_2O_3 which is the same period with Na_2O and MgO can totally inhibit the growth of iron whiskers.

In addition, SEM analysis (Fig.5) after the reduction shows that cracks and pores during the reduction can decrease due to doping MgO, Al_2O_3 and Li_2O into the sample. However, when doping CaO, SrO, BaO, Na_2O and K_2O, cracks and pores can increase.

Mechanism Analysis for Mineral Oxides on the Precipitation Morphology

From observations, doping mineral oxides has an obvious effect on the precipitation of metallic iron after the reduction. The samples doped with mineral oxides were analyzed by EDS. Fig.6, such as doping MgO, shows that alkaline earth elements change the distribution of Fe, and diminish the distribution density of Fe. Thus, it is conjectured that such element distribution is one of influencing factors of the growth of iron whiskers.

On the basis of experimental results, Al_2O_3, Li_2O and MgO in small quantities can effectively and totally inhibit the nucleation and growth of iron whiskers. However, for Na_2O, CaO, K_2O, SrO and BaO it needs more to achieve that effect. Thus, the precipitation morphology variation of the reduced iron is fundamentally related to the fresh structures after doping mineral oxides.

To understand the state for every element with uniform distribution, XRD was used for the reduced samples. As shown in Fig.7, the peaks shape of all samples are similar to that of Fe_2O_3 and ferrate of corresponding oxides, but simple oxides, i.e. Fe_2O_3 and A_yO_x, are not formed. Therefore, doped elements enter the Fe_2O_3 lattice to form a series of compounds such as A_yO_x-Fe_2O_3. Taking $MgFe_2O_4$ of spinel structure as an example, oxygen is closely arranged with face-centred cubic, and Mg^{2+} and Fe^{3+} are respectively filled in the framework of O^{2-}, which contains 64 tetrahedral interstitials and 32 octahedral interstitials. It is seen that doped cations fill in these tetrahedral interstitials and octahedral interstitials to form similar spinel structure, which leads to absence of iron whiskers.

In such structures with ionic species, all the oxides have great differences on inhibition for the growth of iron whiskers, which is possibly related to the character of cations of doped oxides as seen in Table.1. Because ionic compounds are formed after doping elements, considering the coordination and electrons spinning of the cations of doped oxides and Fe, effects of all ions on the reduction were investigated by using the radius of Shannon [12] in this work.

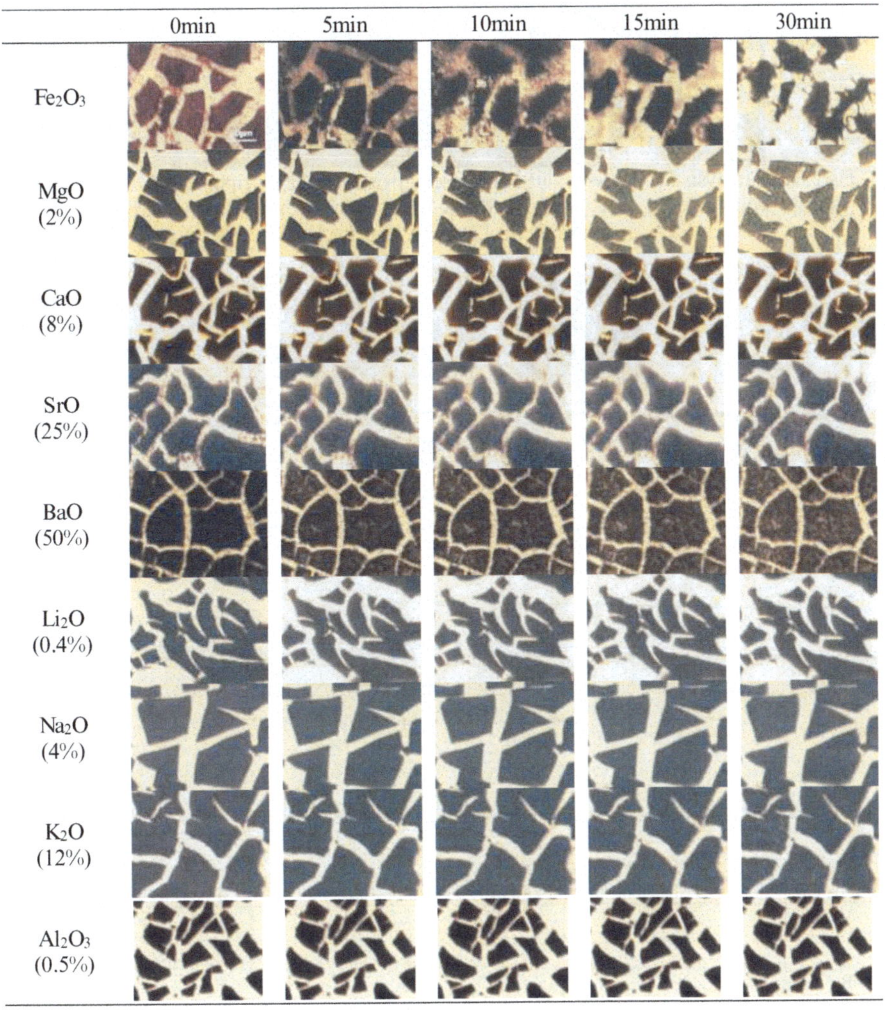

Fig.4 In situ observations for effects of mineral oxides on the reduction of iron oxides

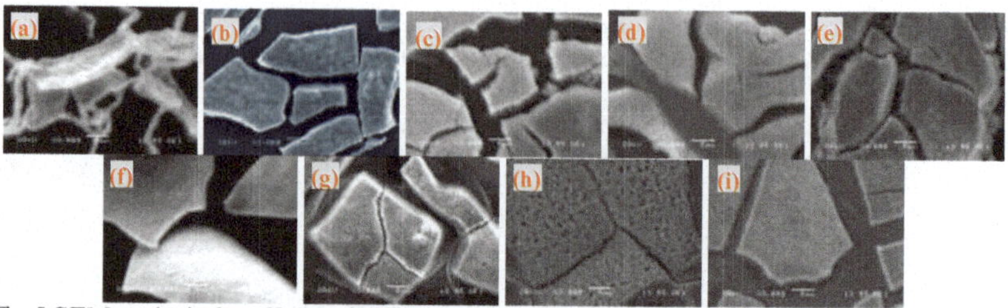

Fig.5 SEM analysis for effects of different mineral oxides on the precipitation morphology of metallic iron: (a)Fe_2O_3, (b)2%MgO, (c)8%CaO, (d)25%SrO, (e)50%BaO, (f)0.3%Li_2O, (g) 4%Na_2O, (h) 12%K_2O, (i) 0.5%Al_2O_3

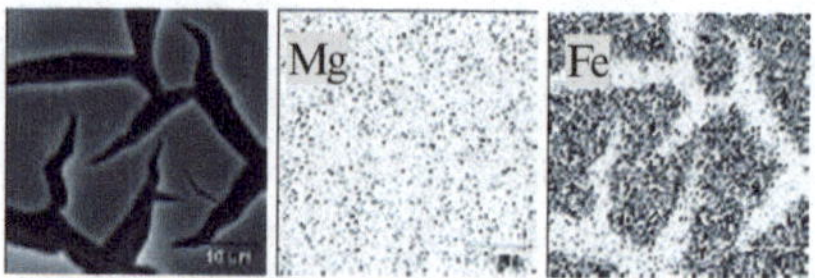

Fig.6 SEM Images for the sheet oxide iron with 2 mass% MgO and surface scanning image of elements

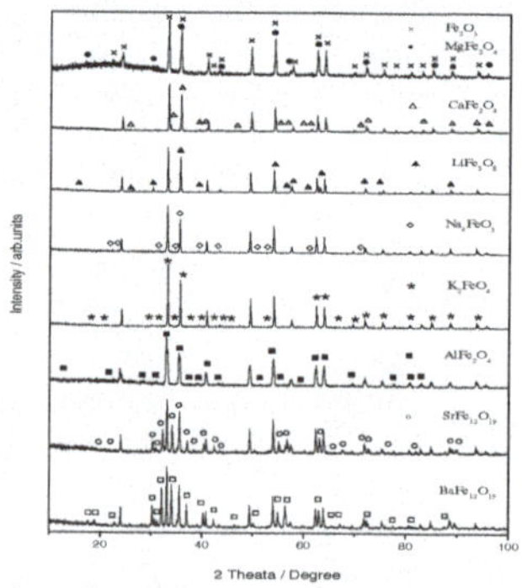

Fig.7 XRD analysis of samples after doping mineral elements with the same concentrations

Table1 Cation character of all mineral oxides

Cation (A)	Radius for Shannon(r/pm)	Extranuclear electronic layers (n/ ind)	valence electrons (q/ ind)
Mg^{2+}	65	2	2
Ca^{2+}	99	3	2
Sr^{2+}	133	4	2
Ba^{2+}	135	5	2
Li^{+}	60	1	1
Na^{+}	95	2	1
K^{+}	133	3	1
Al^{3+}	50	2	3
Fe^{2+}	76	3	2
Fe^{3+}	64	3	3

Judging from the ion characters, ions with large radius, many extranuclear electronic layers and few valence electrons have certainly few charges in the unit volume, and have a weaker adsorption for atom Fe and atom O. Therefore, only enough mineral oxides can inhibit the migration and oriented growth of Fe crystal. Furthermore, they avoid the formation and growth of iron whiskers. Thus, the minimum mole fraction of doping oxide which can inhibit the growth of iron whiskers ($N_{A_yO_x}$) is proportional to its cation radius ($r_{A^{x+}}$), and is inversely proportional to its extranuclear electronic layers ($n_{A^{x+}}$) and its valence electrons ($q_{A^{x+}}$), which can be expressed as follows:

$$F(N_{A_yO_x}) \sim \frac{f(r_{A^{x+}})\cdot f(n_{A^{x+}})}{f(q_{A^{x+}})}$$

On the basis of observations, a relation was obtained by regression analysis for the quantities of all oxides:

$$N_{A_yO_x} = 1.3\times10^{-5}\cdot\frac{r_{A^{x+}}^2\cdot\sqrt{n_{A^{x+}}}}{(q_{A^{x+}}-1)!} \quad \text{(r: pm)}$$

Table.2 shows the comparison of observations and the predictions by equation. Results indicate that the practical quantities of doping oxides are consistent with those predicted by the equation. Overall, the minimum quantities of doping mineral oxides, which can influence the precipitation morphology of metallic iron in the reduction Fe_2O_3 with CO, are related to their cation radius, their valence electrons and their extranuclear electronic layers. The larger the radius of doping cation is, the more extensive the [FeO_6] lattice structure is restructured and consequently, the reduction reaction will be accelerated, and the minimum quantity of doping oxide to inhibit the growth of iron whiskers will gradually increase. Conversely, the less the radius is, the less the [FeO_6] lattice structure is destroyed. Moreover, the cracks and pores are not formed in the reduction.

Therefore, to inhibit well the nucleation and growth of iron whiskers and furthermore to avoid the gain sticking, the particle surface can be treated by the modification of cations. The cation with small radius and many valence electrons should be first selected considering decreasing its quantity and cost.

Table 2 Comparison of the quantities of doping oxides for formula and experiment

Mineral oxide (A_xO_y)	Formula parameter			Quantity by formula		Quantity by exp.
	$r_{A^{x+}}$	$n_{A^{x+}}$	$q_{A^{x+}}$	$N_{A_yO_x}$	$W_{A_yO_x}$%	$W'_{A_yO_x}$%
MgO	65	2	2	0.078	2.06	~2
CaO	99	3	2	0.221	9.02	~8
SrO	113	4	2	0.332	24.42	~25
BaO	135	5	2	0.530	49.10	~50
Li_2O	60	1	1	0.047	0.45	~0.4
Na_2O	95	2	1	0.166	3.39	~4
K_2O	133	3	1	0.398	12.75	~12
Al_2O_3	50	2	3	0.023	0.74	~0.5

Conclusions

(1) Metallic iron is mostly formed as iron whisker in the reduction of iron oxides with CO under 800°C. The reduction rate in the conversion of $Fe_2O_3 \rightarrow Fe_3O_4$ is faster than that of $Fe_3O_4 \rightarrow FeO$. However, the conversion of $FeO \rightarrow Fe$ is slow and lasting. The changes of colour and volume can be attributed to the lattice transformation in the conversion of $Fe_2O_3 \rightarrow Fe_3O_4 \rightarrow FeO$, with consequent formation of cracks and pores. The nucleation and growth of iron whiskers also occur during the conversion of $FeO \rightarrow Fe$.

(2) From the effects of doping elements on the morphology of iron whiskers, MgO、Al_2O_3、Li_2O can inhibit the formation and growth of iron whiskers, respectively, and CaO and Na_2O can inhibit their nucleation and promote their growth with the increasing quantity of doping oxide. However, SrO and BaO promote nucleation and growth. In addition, MgO, Al_2O_3 and Li_2O can reduce the occurrence of cracks and pores during reduction, and on the other hand CaO, SrO, BaO, Na_2O and K_2O can contrary promote such cracks and pores.

(3) Doping mineral oxides can change the precipitation morphology of metallic iron after the reduction. The mole fraction of doping oxide ($N_{A_yO_x}$) is related to its cation radius ($r_{A^{x+}}$), its extranuclear electronic layers ($n_{A^{x+}}$) and its valence electrons ($q_{A^{x+}}$). The more the cation radius is, the less the quantity of doping oxide added. In contrast, the less the cation radius is, the more the quantity of doping oxide is. The minimum mole fraction of doping oxide can be expressed as:

$$N_{A_yO_x} = 1.3 \times 10^{-5} \cdot \frac{r_{A^{x+}}^2 \cdot \sqrt{n_{A^{x+}}}}{(q_{A^{x+}} - 1)!}$$

Acknowledgement

This work is supported by National Natural Science Foundation of China, Project 51234001

References

1. Istitute of Chemical Metallurgy of Academia Sinica, Institute of Scientific and Technical Information of China, *Gas ironmaking in Fluidized Bed* (Beijing: Scientific and Technology Literature Publishing House, 1977).
2. Komatina M, Gudenau H W, "The Sticking problem during direct reduction of fine Iron ore in the fluidized bed," *Metalurgija*, (2004), 309-328.
3. Degel R, "Eisenerzreduktion in der Wirbelschiht mit wasserstoffreichem Gas: Sticking und Ansätze," (Dissertation, Germany, RWTH Aachen 1996).
4. Schiller M, "Mikromorphologie der Eisenphase als Folge der Reduktion von Eisenoxiden. Dissertation," (Germany, RWTH Aachen 1987).
5. Gudenau H W, Hirsch M, Denecke H, et al, "Direct reduction of iron ore fines in a fluidized bed using hydrogen-rich gas," *Stahl und Eisen (Germany)*, 117(4) (1997), 91-99.
6. Gudenau H W, Fang J, Hirata T, et al, "Fluidized bed reduction as the prestep of smelting reduction," *Steel Res*, 60(314) (1989), 138-44.
7. Fang J, "Sticking problem in fluidized bed iron ore reduction," *China: Iron Steel*, 26 (5) (1991), 11- 14.
8. Zhilong Zhao, Huiqing Tang, Zhancheng Guo, "In situ observation and mechanism research of the influence of CaO on the growth of iron whiskers under CO atmosphere," *Journal of University of Science and Technology Beijing*, 33(7) (2011), 817-22.
9. Nicolle R, Rist A, "The Mechanism of Whisker Growth in the Reduction of Wüstite," *Metallurgical Transactions B*, 10B (1979), 429-36.
10. Bartels M, Lin W G, Nijenhuis J, "Agglomeration in fluidized beds at high temperatures: Mechanisms, detection and prevention," *Progress in Energy and Combustion Science*, 34(5) (2008), 633-66.

11. Kashiwaya Y, Yamaguchi Y, Kinoshita H, et al, "In Situ Observation of Reduction Behavior of Hematite with Solid Carbon and Crystallographic Orientation between Hematite and Magnetite," *ISIJ International*, 47(2) (2007), 226-33.
12. Shannon By R D, "Revised Effective Ionic Radii and Systematic Studies of Interatomie Distances," *Acta Cryst*, A32 (1976), 751-67.

7th International Symposium on High-Temperature Metallurgical Processing
Edited by: Jiann-Yang Hwang, Tao Jiang, P. Chris Pistorius, Gerardo R.F. Alvear F., Onuralp Yücel, Liyuan Cai, Baojun Zhao, Dean Gregurek, and Varadarajan Seshadri
TMS (The Minerals, Metals & Materials Society), 2016

INFLUENCE OF OPERATION PARAMETERS ON MASS FRACTION OF SULFUR IN THE HOT METAL IN COREX PROCESS

Laixin Wang[1,3], Shengli Wu[1,2], Minyin Kou[1], Xinliang Liu[1], Yujue Wang[1], Weidong Zhuang[3]

[1]School of Metallurgical and Ecological Engineering, University of Science and Technology Beijing, 30 Xueyuan Road, Haidian District, Beijing 100083, P. R.China
[2]School of Metallurgy and Chemical Engineering, Jiangxi University of Science and Technology; No.156, Kejia Ave.; Ganzhou, Jiangxi, 341000, P. R. China
[3]National Engineering Research Center for Rare Earth Materials, General Research Institute for Nonferrous Metals, Grirem Advanced Materials Co. Ltd., No. 2 Xinjiekouwai Street, Beijing 100088, P. R.China

Keywords: Desulfurization, Quality of hot metal, Basicity, Fuel rate, Low carbon ironmaking.

Abstract

COREX process has obvious advantages in environment protection but still has some disadvantages, for example, it has a higher sulfur content of hot metal than blast furnace. In this paper, the thermodynamics and dynamics of desulfurization have been investigated to obtain the main operation parameters related to the mass fraction of sulfur in the hot metal. In addition, statistical analyses have been carried out to study the relationship between operating parameters and sulfur content of COREX hot metal based on actual plant data. It has been found that the mass fraction of sulfur in hot metal is significantly influenced by the C content in hot metal, MgO/Al_2O_3 in slag, the concentration of MgO in slag, the temperature of hot metal and binary basicity (R_2).A linear model has been developed and subsequently used for reducing the mass fraction of sulfur in hot metal, improve the quality of hot metal and contribute to low carbon ironmaking.

Introduction

Up to now, blast furnace, a conventional ironmaking process, is still the dominant technology to produce steel due to its high efficiency. However, blast furnace inevitably uses coke as its fuel, while the coke plants have many environment problems associated with their operation. And the coke price is becoming more and more expensive due to the reducing coking coal resource. Thus, from economical and environmental points of view to produce iron without coke, numbers of smelting reduction processes have been developed. COREX is the world's first commercially established and industrially proven smelting-reduction process mainly based on noncoking coal. It is regarded as one of the most promising alternative ironmaking processes in the world[1-5]. Excepting the environment and economic advantages, the quality of hot metal of COREX process must be guaranteed in order to keep smooth and efficient operation. Sulfur content is one of the most important parameters to represent the quality of hot metal. However, the sulfur content of COREX hot metal is much higher than that of blast furnace hot metal. Therefore, desulfurization is significantly imperative for COREX to achieve high quality hot metal. There is only few literatures focusing on the desulfurization in COREX hot metal, while a lot of work have been done concerning on the desulfurization in blast furnace. Referring to the previous research and study, changing slag system is one of the best choices to achieve the aim of desulfurization. Those fundamental researches about desulfurization in slag system are suitable for not only blast furnace but also COREX and even steelmaking[6,7]. Furthermore, some

researches about gas desulfurization also have been done[8,9].
In this paper, research on desulfurization of COREX process in terms of operation parameters has been done based on actual plant data of a steel company's COREX3000 in 2010. Correlation analysis and regression analysis are used to analyze the qualitative and quantitative relationships between operation parameters and sulfur content of hot metal. The most effective operation parameters have been found for the sulfur content of COREX hot metal.

Thermodynamics and Dynamics of Desulfurization

Thermodynamics of Desulfurization

At 1773K, CaO makes a 97% contribution for desulfurization of slag and MgO for the else 3%[10]. Therefore, taking CaO as an example, the thermodynamics and dynamics of desulfurization in slag systems can be explained as follows. The ionic reaction equation of desulfurization can be represented by Eq. (1).

$$(\mathrm{O}^{2-})+[\mathrm{S}]=(\mathrm{S}^{2-})+[\mathrm{O}], \quad \Delta G^{\ominus}=124455-50.26T(\mathrm{J/mol}) \tag{1}$$

The activity coefficient at 1773K of sulfur and oxygen in hot metal can be calculated by Eq. (2) and Eq. (3)[11].

$$\lg f_O = e_O^O[\%\mathrm{O}]+e_O^C[\%\mathrm{C}]+e_O^S[\%\mathrm{S}]+e_O^{Si}[\%\mathrm{Si}]+e_O^{Mn}[\%\mathrm{Mn}]+e_O^P[\%\mathrm{P}] \tag{2}$$

$$\lg f_S = e_S^S[\%\mathrm{S}]+e_S^C[\%\mathrm{C}]+e_S^O[\%\mathrm{O}]+e_S^{Si}[\%\mathrm{Si}]+e_S^{Mn}[\%\mathrm{Mn}]+e_S^P[\%\mathrm{P}] \tag{3}$$

Thus, using the reaction given in Eq. (1) and formulas given in Eq. (2) and Eq. (3), the distribution coefficient of sulfur can be obtained from Eq. (4)[11,12].

$$L_S=\frac{(\%\mathrm{S})}{[\%\mathrm{S}]}=K_{(1)}\cdot\frac{f_S\cdot a_{O^{2-}}}{\gamma_{S^{2-}}\cdot a_{[O]}} \tag{4}$$

where $K_{(1)}$ is the equilibrium constant of Eq. (1), f_S and $\gamma_{S^{2-}}$ are, respectively, the activity coefficient of S in hot metal and free S^{2-} ions in slag, $a_{O^{2-}}$ and $a_{[O]}$ are the activity of free O^{2-} ions in slag and O in hot metal. According to Eq. (4), in order to improve the desulfurization capability of system, the following measures can be adapted.

(1) Increase $a_{O^{2-}}$: CaO is the prime choice to increase the activity of free O^{2-} in slag. So to increase the R_2 (CaO/SiO_2) in slag will be a useful method to improve the desulfurization capability of slag system.

(2) Decrease $a_{[O]}$: As we all know, the concentration of O in hot metal has a strong relationship with the oxidizability of slag on the interface, which equals to the FeO concentration in slag generally. And as the main elements (except P) increase, f_O decreases in hot metal, which is good for desulfurization.

(3) Increase *K*: Increasing the temperature is a very useful way to increase *K* since the reaction in Eq.(1) is endothermic.

(4) Increase f_S: As Eq. (3) shows, f_S increases when the contents of C, Si and P rises

appropriately.

Dynamics of Desulfurization

As the desulfurization reaction occurs only on the interface between slag and hot metal, the diffusion is a very important process. The kinetic conditions can be given as Eq. (5)[12].

$$\frac{d[\mathrm{S}]}{dt}=\frac{A}{M}\frac{[\mathrm{S}]-\frac{(\mathrm{S})}{L_S^0}}{(\frac{1}{k_S}+\frac{L_S^0}{k_m})\cdot\frac{1}{L_S^0}}=\frac{A}{M}\frac{[\mathrm{S}]-\frac{(\mathrm{S})}{L_S}}{\frac{1}{k_S L_S^0}+\frac{1}{k_m}} \tag{5}$$

where A is the interface area, M is the mass of hot metal. L_S is the distribution coefficient of sulfur in Eq. (4). k_m and k_s are, respectively, the mass transfer coefficient of S in hot metal and S^{2-} in slag. Therefore, there are some measures to improve the kinetic conditions.
(1) Increase A: A can be increased by slowing down the drop rate of hot metal through slag since the interface area mainly refers to the area that the hot metal drops fall through the slag.
(2) Decrease M: This equals to improve the slag ratio to increase the mass of desulfurization in terms of thermodynamics and improve the interface area in terms of dynamics as a result.
(3) Improve k_s and k_m: Since the mass transfer coefficient in hot metal is much larger than that in slag, it's better to decrease the viscosity of slag.

The Theoretical Influence Factors of Desulfurization

According to the thermodynamics and dynamics of desulfurization, some measures can be taken to increase the desulfurization, such as to increase the temperature, R_2, slag ratio and the content of C, Si and P, decrease viscosity of slag, the drop rate of hot metal through slag and the oxidizability of slag. However, most of these indexes are microcosmic, and they should be translated into macroscopical parameters in practical operation for better understanding. So the temperature of hot metal, slag ratio, fuel rate, coal rate, coke rate, composition of hot metal and slag, and melting rate are considered in the following analyses.

Statistical Analysis of Operation Parameters Affecting Sulfur Content in Hot Metal

SPSS (Statistical Product and Service Solutions) is used to carry out correlation analysis and regression analysis to study the relationship between the operation parameters and sulfur content in hot metal. Correlativity is a kind of incompletely determinate relationship. It means there are some relations between each other, not one determines other. And Pearson correlation coefficient (hereinafter referred to as PCC) is used to measures this strength and direction of the linear association between two variables. Its value ranges from -1 to 1, which means the relationship between two variables various from absolute negative linear correlation to absolute positive linear correlation[13-15].

Correlation Analysis

When study the correlativity between one parameter with sulfur content in hot metal, it is very important to minimize the extra effects of other parameters. Therefore, other operation parameters are limited to a narrow range in this paper. Thus, three ranges have been selected for each parameter based on the histogram of each parameter and are given in Table I. One of the ranges is selected based on the strength of a linear association between two operation parameters after pretreatment, and the screening programs are listed in Table II.

Based on the previous analysis, Table III shows the statistical results of PCC(Pearson Correlation Coefficient) and significance index. As Table III shows, the sulfur content in hot metal has positive correlations with melting rate while it has negative correlations with fuel rate, coal rate, slag ratio, temperature of hot metal, the content of Si in hot metal, the content of C in hot metal, R_2, the MgO concentration and MgO/Al_2O_3 in slag. The significance index means the probability of making mistakes, so it would be as small as possible (usually 0.05 in SPSS). And the larger the absolute value of PCC, the stronger the linear relationship between two variables. Thus, the five main operation parameters affecting sulfur content in hot metal are the content of C in hot metal, MgO/Al_2O_3 in slag, the concentration of MgO in slag, the temperature of hot metal and R_2.

Table I. The Ranges of Each Parameter

Parameters	Ranges		
	1	2	3
Fuel rate(kg/tHM)	750~1225	830~1120	880~1000
Coal rate(kg/tHM)	550~1100	700~930	740~850
Slag ratio(kg/tHM)	200~400	200~300	230~270
Temperature of hot metal(℃)	1460~1570	1480~1540	1500~1520
The content of Si in hot metal (%)	0~2.0	0.5~1.5	0.5~0.9
The content of C in hot metal (%)	3.8~4.6	4.05~4.45	4.2~4.35
R_2	1.15~1.32	1.19~1.27	1.22~1.26
The content of MgO in slag (%)	8.0~11	8.5~10.5	9~10
MgO/Al_2O_3 in slag	0.48~0.78	0.55~0.7	0.58~0.65
Melting rate(t/h)	95~165	122~158	128~150

Table II. Screening Programs of Actual Plant Data

	Fuel rate	Coal rate	Slag ratio	T	[%Si]	[%C]	R_2	MgO%	MgO/Al_2O_3	Melting rate	Coal/ coke	Coke rate
Fuel rate	0	0	2	2	1	2	2	2	2	1	5~6	0
Coal rate	0	0	2	2	1	3	2	2	2	1	0	120~150
Slag ratio	1	1	0	2	1	3	2	2	2	1	0	0
T	1	1	2	0	2	3	3	2	2	1	0	0
[%Si]	1	1	2	3	0	3	2	2	2	1	0	0
[%C]	1	1	2	3	3	0	2	2	2	1	0	0
R_2	1	1	2	3	2	3	0	2	2	1	0	0
MgO%	1	1	2	3	2	3	3	0	0	1	0	0
MgO/Al_2O_3	1	1	2	3	2	3	3	0	0	1	0	0
Melting rate	1	1	1	3	2	2	3	2	3	0	0	0

*In the table, "0" means no data filtering; "1", "2" and "3" represent the ranges of each parameter in Table I.

Multiple Regression Analysis

Based on the results of correlation analysis, multiple regression was used to study the quantitative relationship between operation parameters and sulfur content in hot metal[13,14]. This's because of the existing of collinearity between operation parameters, which makes the regression coefficients unreliable. To decrease the degree of collinearity, the "stepwise" is adopted as the input mode in SPSS[14]. The independent variables are the five main operation parameters affecting sulfur content in hot metal after the screen of collinearity. And the regression result is presented in Eq. (6).

$$[\%S] = 0.661 - 0.053[\%C] - 0.157\,R_2 - 0.050(M/A) - 1.02\times10^{-4}\,T - 0.001(\%M) \qquad (6)$$

where [%S],[%C] are, respectively, the contents of S and C in hot metal. R_2 is the binary basicity. (%M) and (M/A) are the MgO concentration and the MgO/Al_2O_3 in slag. T is the temperature of hot metal.

Table III. Correlation Analysis for Parameters and Sulfur Content in Hot Metal

Parameters	Pearson	Significance	Regression equation
Fuel rate	-0.390	0.073	$Y=0.152-0.107\times10^{-3}X_1$
Coal rate	-0.262	0.108	$Y=0.089-5.07\times10^{-5}X_2$
Slag ratio	-0.125	0.397	$Y=0.059-6.5\times10^{-5}X_3$
Temperature of hot metal	-0.457	0.013	$Y=0.483-2.922\times10^{-4}X_4$
The content of Si in hot metal	-0.292	0.148	$Y=0.051-0.011X_5$
The content of C in hot metal	-0.658	0.000	$Y=0.462-0.097X_6$
R_2	-0.386	0.024	$Y=0.176-0.108X_7$
The content of MgO in slag	-0.472	0.020	$Y=0.102-0.006X_8$
MgO/Al_2O_3 of slag	-0.610	0.002	$Y=0.105-0.101X_9$
Melting rate	0.237	0.217	$Y=0.002+0.309\times10^{-3}X_{10}$

The actual plant data in September of 2010 is used to verify the regression equation. Before this verification, the abnormal data are needed to be eliminated. The results are shown in Figure 1 and Figure 2. It can be seen from Figure 1 that the slope of the line is 0.9056 which is close to 1 and the determination coefficient is 0.655 which implies that the predicted data agrees well with plant data. The percentage of difference between plant and regressed data within±20% is 92% for sulfur content of hot metal in Figure 2. So it can be also concluded that the regression equation can be used to predict the sulfur content in hot metal in practical operation relatively.

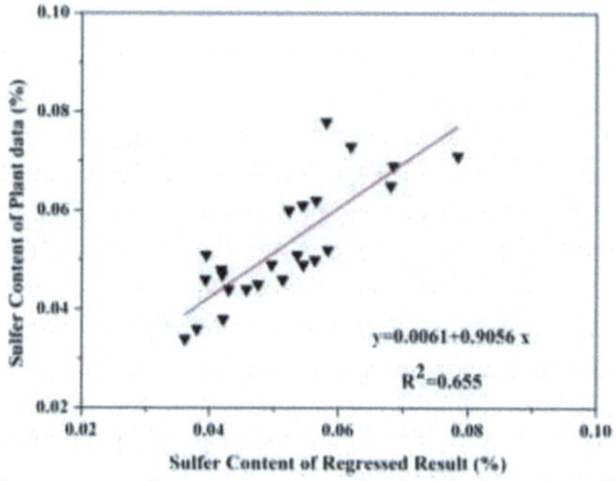

Figure 1. The contrast between regression and plant data-linear fitting

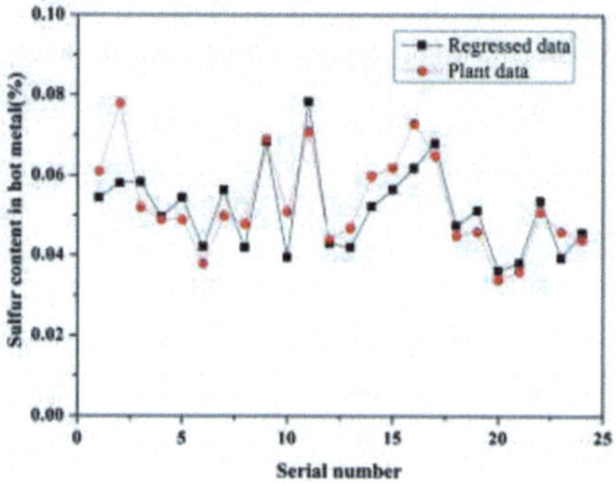

Figure 2. The comparison between regression and plant data- scatter diagram

<u>Instructions for Improving the Quality of Hot Metal</u>

Based on the results of correlation analysis and multiple regression, some main influence factors have be obtained. According to the Eq.(6), the content of C in hot metal, the binary basicity, the MgO concentration and the MgO/Al_2O_3 in slag and the temperature of hot metal are the five main factors affect the sulfur content of hot metal in COREX process. So the following instructions for decreasing the sulfur content of hot metal and improving the quality of hot metal have been given as Table Ⅳ shows.

Table Ⅳ. Methods for Decreasing the Sulfur Content of Hot Metal

Methods	Adjustment	The content of S in hot metal
Temperature of hot metal	↑ 10 K	↓ 0.001
The content of C in hot metal	↑ 1%	↓ 0.053
The content of MgO in slag	↑ 1%	↓ 0.001
MgO/Al_2O_3 of slag	↑ 1	↓ 0.050
R_2	↑ 0.1	↓ 0.016

Conclusions

(1) The theoretical analyses indicate that the following factors may have effects on mass fraction of suffer in the hot metal in COREX process: temperature, slag ratio, R_2 , viscosity of slag and the contents of C, Si and P, the drop rate of hot metal through slag and the oxidizability of slag in microcosmic scope. And for actual production, they are translated into the temperature of hot metal, slag ratio, fuel rate, coal rate, composition of hot metal and slag, and melting rate.
(2) After correlation analysis, five main operation parameters affecting sulfur content of hot metal are the content of C in hot metal, MgO/Al_2O_3 in slag, the concentration of MgO in slag, the temperature of hot metal and R_2.
(3) Based on the correlation analysis results, a multiple regression analysis was conducted to estimate the quantitative relationship between operation parameters and sulfur content in hot metal. After verification, the regression equation can be used to predict the sulfur content of hot metal in plant relatively.

Acknowledgements

The financial support of the National Natural Science Foundation of China (U1260202) is gratefully acknowledged.

References

[1] Kumar, P.P. et al., "Raw Materials for Corex and Their Influence on Furnace Performance," *Ironmaking & Steelmaking*, 36 (2) (2009), 87-90.

[2] Lee, Seung-Cheol, et al. "The Effects of Operational Parameters on the Transport Phenomena in COREX Melter-Gasifier," *ISIJ International*, 40 (11) (2000), 1073-1079.

[3] Kumar, P.P. et al., "Modelling of Corex Process for Optimisation of Operational Parameters," *Ironmaking & Steelmaking*, 33 (1) (2006), 29-33.

[4] Wu, S, et al. "Basic Characteristics of the Shaft Furnace of COREX Smelting Reduction Process Based on Iron Oxides Reduction Simulation," *ISIJ International*, 50 (7) (2010), 1032-1039.

[5] Liu, X, et al. "Mathematical Model of Lump Coal Falling in the Freeboard Zone of the COREX Melter Gasifier," *Energy & Fuels*, 25 (12) (2011), 5729-5735.

[6] Pistorius, P. C., et al. "Examples of How Fundamental Knowledge can Improve Steelmaking: Desulphurisation Kinetics Calcium and Modification." *Transactions of the Indian Institute of Metals*, 66 (5-6) (2013), 519-523.

[7] Fincham, C. J. B., and Richardson F. D. "The Behaviour of Sulphur in Silicate and Aluminate Melts," *Proceedings of the Royal Society of London A: Mathematical, Physical and Engineering Sciences*, 223 (1152) (1954), 41-62.

[8] Pelton, A. D., See, J. B, and Elliott, J. F. "Kinetics of Evolution of SO_2 from Hot Metallurgical Slags." *Metallurgical Transactions*, 5 (5) (1974), 1163-1171.

[9] Agrawal, B., Yurek, G. J., and Elliott, J. F. "Kinetics and Mechanism of the Desulfurization of Liquid Blast Furnace Slags by Ar-H2O Gas Mixtures." *Metallurgical Transactions B*, 14 (2) (1983), 221-230.

[10] Shi, C, et al. "The Thermodynamic Model of Ls Between the Carbon Saturated Hot Metal and CaO-SiO2-Al2O3-MgO Slag System," *The Chinese Journal of Process Engineering*, 10 (1) (2010), 158-164.

[11] Guo, H, *The Course of Metallurgical Physical Chemistry*(Beijing, Metallurgical Industry Press, 2004).

[12] Wang, X, et al., *Ferrous Metallurgy (The ironmaking part)* (Beijing, Metallurgical Industry Press, 2013) (in Chinese).

[13] Liu, A, et al., *Introduction to Basic Analysis Using SPSS 22*(Beijing, Peking University Press, 2014)(in Chinese).

[14] Wang, Z, et al. *SPSS Statistical Analysis & Comprehensive Application*(Shanghai, Shanghai Jiao Tong University Press, 2012).

[15] Li, H, et al. *SPSS Data Analysis*(Beijing, Posts& Telecom Press,2012).

7th International Symposium on High-Temperature Metallurgical Processing
Edited by: Jiann-Yang Hwang, Tao Jiang, P. Chris Pistorius, Gerardo R.F. Alvear F., Onuralp Yücel, Liyuan Cai, Baojun Zhao, Dean Gregurek, and Varadarajan Seshadri
TMS (The Minerals, Metals & Materials Society), 2016

INFLUENCE OF OPERATION PARAMETERS ON STICKING BEHAVIOR OF PELLETS IN COREX SHAFT FURNACE

Xinliang Liu[1], Shengli Wu[1, 2], Zhe Wang[1], Laixin Wang[1],Mingyin Kou[1]

[1]School of Metallurgical and Ecological Engineering, University of Science and Technology Beijing, 30 Xueyuan Road, Haidian District, Beijing, 100083, P. R. China.
[2]School of Metallurgy and Chemical Engineering, Jiangxi University of Science and Technology; No.156, Kejia Ave.; Ganzhou, Jiangxi, 341000, P. R. China

Keywords: COREX; Pellets; Sticking; Metallization; Operation parameters

Abstract

The sticking behavior of iron ores in the shaft furnace is one of the most important problems in the COREX melting progress, which directly influence the smooth operation and productivity of shaft furnace. Therefore, it is of great importance to study the sticking behavior of pellets in COREX shaft furnace. The results showed that porous metallic iron produced at the interface of pellets samples was the main reason for the sticking of pellets in the COREX shaft furnace. The SI (sticking index) and metallization rate increased with the reduction temperature and gas flow, and decreased with melting rate. However, the SI of pellets in the COREX shaft furnace could be decreased by the proper match of melting rate and reduction temperature, reduction temperature and gas flow, while the metallization rate of pellets were kept at a certain level. Furthermore, the matching of melting rate and reduction temperature was more effective than the matching of reduction temperature and gas flow.

Introduction

The COREX process is a well industrialized technology for economic production of hot metal on the basis of untreated coals and meeting all environmental demands. The process mainly comprises two reactors, namely: the shaft furnace, where DRI is produced; and the melter gasifier, where high quality reducing gas is generated and DRI is melted [1]. However, the sticking behaviors of iron ores in the COREX shaft furnace is one of the most important problems for COREX process [2-3]. The utilization of reducing gas, furnace operation and the productivity would be influenced by the sticking behavior [4].The sticking of iron ores in COREX shaft furnace mainly occurred between sponge irons of high metallization rate at the bottom of the shaft furnace [5]. Former studies have showed that the reduction temperature, the metallization rate and the gas compositions are the main influencing factors of sticking behaviors [6,7]. It was found that higher reduction temperature would increase the SI, but the productivity would be obviously decreased if the reducing gas temperature is reduced to avoid the sticking. For example, a decrease of reduction temperature from 850℃ to 750℃ could decrease the productivity in 30~40%. What is more, the sticking behaviors tended to increase with the increase of metallization rate and the contact area of iron ores when particles collide together [8].Former studies also showed the mechanism of sticking behaviors, the freshly precipitated iron of high surface energy, the whiskers iron precipitated on the particle surface and liquid phase stimulated by gangue and FeO [9-11].

Most of the former studies have mainly focused in understanding the sticking behaviors in the fluidized bed and the iron ore powders. However, reduction conditions such as the burden load, the gas composition, the reduction temperature and the melting rate are quite different in the COREX shaft furnace compared with those in a fluidized bed. Therefore, it is of great importance to study the sticking behaviors of pellets in COREX shaft furnace and the influence of operation parameters on sticking behaviors of pellets.

Experimental

The experiments were carried out using a static reduction furnace schematically shown in **Figure 1**. Every test were consist of 500 g pellets samples of 10~12.5 mm in diameter and 500 g pellets samples of 12.5~16 mm in diameter. The samples were heated by a rate of 5℃/min to the defined temperature under a nitrogen atmosphere. After reaching the defined temperature, reducing gas and load were introduced to the samples, and the reaction was conducted by a predefined time. Once the reduction time has been completed, samples were kept under nitrogen atmosphere for an additional time within the furnace under load, Finally, the samples were cooled down to room temperature without load, the SI and the metallization rate of pellets were measured. Experimental conditions for the defined experimental cycle are shown in **Table I**.

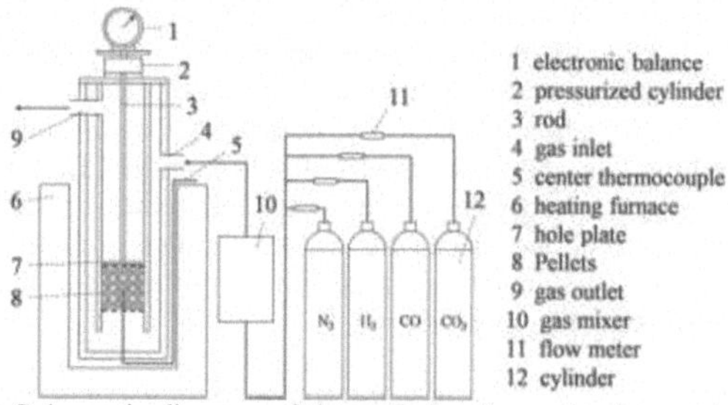

Fig.1 Schematic diagram of the static loading reduction equipment

Table I Experimental conditions

Items	Values
Defined temperature	850℃, 900℃, 950℃
Reducing gas composition	68% CO, 9% CO_2, 23% H_2
Reducing gas flow	3.4 L/min, 6.7 L/min, 10 L/min, 13.3 L/min
Reduction time under reducing gas	180 min, 210 min, 240 min
Holding time under nitrogen atmosphere	900 min, 105 min, 120 min
Gas flow of nitrogen	2 L/min
Load	14 N/cm^2

In order to measure the SI of pellets, pellets samples were fed into a drum tumble (ø 130 mm × 800 mm)at a rotation speed of 30 rpm for 10 min, Weight of the sticking pellets (pellets in diameter larger than 16 mm) was measured and the SI was calculated by the following equation.

$$SI=M_1/M_0*100\% \quad (1)$$

Where, M_1 is the weight of sticking pellets after the drum test, M_0 is the weight of pellets after reduction.

The reduction conditions of pellets under different situation are shown in **Table II**. Case 1 to case 5 were used to study the matching of melting rate and reducing gas temperature, case 6 to case 9 were used to study the matching of reducing gas temperature and gas flow.

Table II Reduction conditions for pellets

Cases	Reduction temperature /℃	Melting rate	Reduction time/min	Holding time /min	Reducing gas flow /$L \cdot min^{-1}$	SI /%
Case 1	850	Low	240	120	13.3	70.36
Case 2	900	Middle	210	105	13.3	56.78
Case 3	900	High	180	90	13.3	38.42
Case 4	950	Middle	210	105	13.3	78.50
Case 5	950	High	180	90	13.3	52.14
Case 6	875	Low	240	120	10	66.28
Case 7	875	Low	240	120	6.7	58.34
Case 8	900	Low	240	120	6.7	67.63
Case 9	900	Low	240	120	3.4	58.65

Results and Discussion

Sticking Mechanism of Pellets in COREX Shaft Furnace

The SI of pellets under Case 1 was 70.36%, and the metallization rate was 89.93%. The SEM analysis results of sticking pellets samples were shown in **Figure 2**. As shown in the figure, all the pellets samples were stuck together before the drum test, meaning the sticking level of pellets samples under Case 1 was quite high due to its high metallization rate and long reduction time. Besides, the sticking interface between pellets samples was very long and dense when the magnification is 110. However, a porous structure was observed when the magnification was increased to 2000.

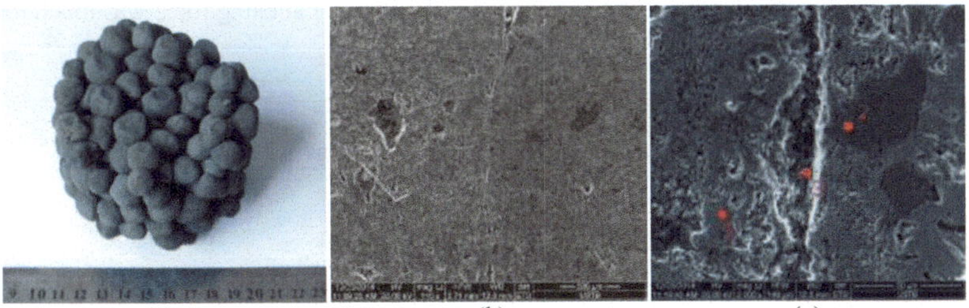

(a) (b) (c)

Fig.2 SEM photos of sticking pellets samples under Case 1.
(a), Before drum test, (b),magnification 110x, (c) magnification 2000x

Table III Chemical compositions of the sticking pellets samples (mol%)

Zone	O	Mg	Al	Si	Ca	Fe
1	33.75	1.26	0.61	2.77	1.47	60.15
2	43.78	8.10	0.76	27.51	13.89	5.96
3	8.78	0.77	1.19	0.53	0.50	88.22

The chemical compositions of the sticking pellets around the sticking interface are shown in **Table III**. Zone 1 (close to the sticking interface, grey area) mainly consists of Fe and O, meaning the main minerals were FeO and metallic iron. Zone 2 (close to the sticking interface, dark area) mainly consists of O, Si, Ca and Mg, meaning the main minerals are gangue components such as CaO, Al_2O_3, and SiO_2. Zone 3 (on the sticking interface) mainly consists of Fe and O, with Fe content been higher compared with Zone 1, meaning that the main mineral is metallic iron. Therefore, it can be concluded that the porous metallic iron produced at the interface of pellets samples is the main reason for the sticking behavior of pellets in the COREX shaft furnace.

Influence of Operating Parameters on sticking Behavior of Pellet

The SI and metallization rate of pellets samples under different reduction temperature and melting rate are shown in **Figure 3**. Under the same gas flow and melting rate, the SI and the metallization rate of pellets samples increased with the increase of reduction temperature. Under the middle melting rate, the SI of pellets raised from 56.78% to 78.50% when the reduction temperature increased from 900℃ to 950℃, while the metallization rate increased from 88.82% to 92.78%. As for the high melting rate, the SI and the metallization rate increased from 38.42% to 52.14%, 81.91% to 87.90%, respectively. The reason why the SI increased with the temperature was attributed to the stronger adhesive force at higher reduction temperature, which was determined by the surface energy of metallic iron [12]. Besides, higher reduction temperature increased the reduction rate, more metallic iron was produced at the sticking interface and the sticking force was strengthened. Therefore, the SI of pellets samples increased with the increase of reduction temperature. However, the influence of reduction temperature decreased with the increase of melting rate, the SI of pellets samples increased about 22 and 14 percent points under the middle and high melting rate when the reduction temperature increased from 900℃ to 950℃, the reason was that the reduction time was much longer under middle melting rate than that of high melting rate, so the action time of sticking force was much longer and the influence of temperature was much more obvious.

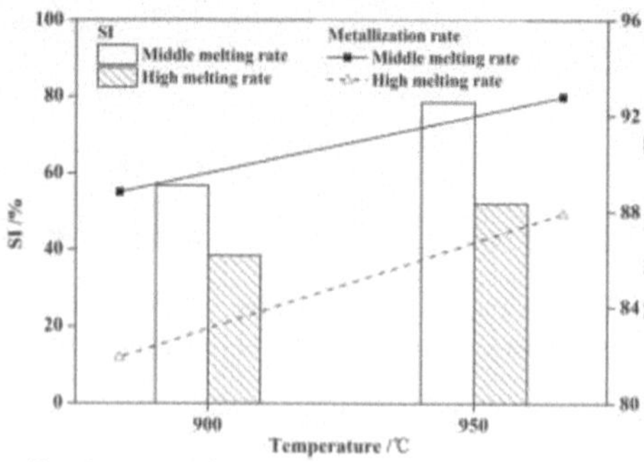

Fig.3 SI and metallization rate of pellets under different temperature and melting rate

On the other hand, the SI and metallization rate of pellets samples decreased with the increase of melting rate under the same gas flow and reduction temperature. The SI and metallization rate of pellets samples decreased about 18 and 7 percent points at 900℃, respectively. And decreasing amplitude of SI and metallization rate were 26 and 5 percent points at 950℃. The reason was that the reduction time and the holding time decreased with the increase of melting rate, so the metallization rates of pellets samples were decreased. What is more, the action times of sticking force were obviously decreased, so the SI of pellets was decreased due to the above reasons.

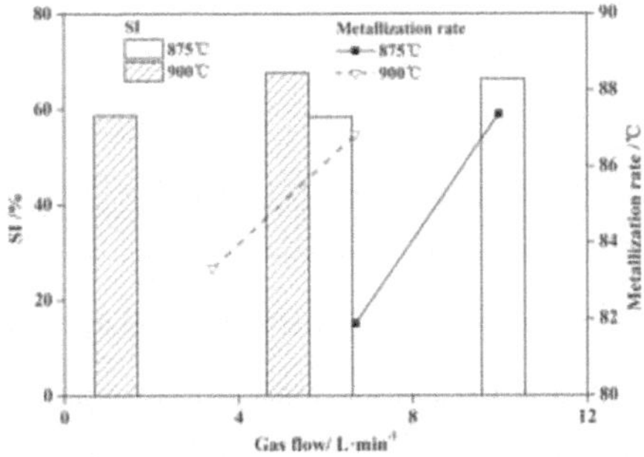

Fig.4 SI and metallization rate of pellets under different temperature and gas flow

The SI and metallization rate of pellets samples under different reduction temperature and reducing gas temperature were shown in **Figure 4**. Similarly, the SI and metallization rate of pellets samples increased with the increase of reduction temperature under the low melting rate and gas flow of 6.7 L/min. On the other hand, the SI and metallization rate of pellets samples increased with the increase of gas flow. The SI raised from 58.65% to 67.63% when the gas flow increased from 3.4 L/min to 6.7 L/min at 900℃, while the metallization rate raised from 83.33% to 86.83% under the same conditions. The SI and metallization rate of pellets samples raised from 58.34% to 66.28%, 81.89% to 87.37% respectively when the when the gas flow increased from 6.7 L/min to 10 L/min at 875℃. The reason was that higher gas flow increased the contact between pellets and reducing gas and the reducing rate of pellets was improved, so more metallic iron was produced at the sticking interface. Then the sticking force was strengthened by the higher metallization rate, resulting the increase of SI.

Influence of Matching Model of Operating Parameter

As shown in the above results, the SI and metallization rate of pellets samples increased with reduction temperature and gas flow, and decreased with the increase of melting rate. However, the influence of increasing reduction temperature was still unclear when increasing the melting rate or decreasing gas flow at the same time. Would the SI of pellets be decreased when reduction temperature properly matched the melting rate and gas flow but the metallization rate were kept at a certain level?

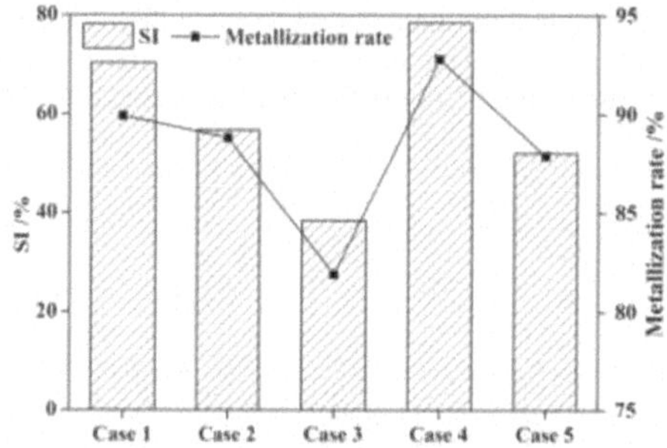

Fig.5 SI and metallization rate of pellets under different match models

The SI and metallization rate of pellets samples under different match models of melting rate and reduction temperature were shown in **Figure 5**. As shown in the figure, the higher the metallization rate was, the higher the SI was. This result indicated that high metallization rate would increase the SI of pellets samples. Compared with the reference test (Case 1), the SI and metallization under high temperature (950℃) matching middle melting rate was much higher, the increasing amplitude of SI and metallization rate were 8 and 2.9 percent points, respectively. These results showed that if the increasing amplitude of reduction temperature was very high, it was difficult to decrease the SI of pellets samples. On the other hand, the SI and metallization rate under middle temperature (900℃) matching middle melting rate, middle temperature (900℃) matching high melting rate, high temperature (950℃) matching high melting rate were all lower than that of reference test. Compared with the reference test, the SI of pellets decreased about 14, 32 and 18 percent points under the above three match models, the metallization rate decreased about 1.1, 8.0 and 2.0 percent points. The reason was that the influence of reduction time and holding time on the sticking force is much stronger than that of reduction temperature, so the SI of pellets samples were decreased.

As shown in the above results, the metallization rate decreased obviously when middle temperature (900℃) matched high melting rate, so this match model was not suitable to the COREX shaft furnace.

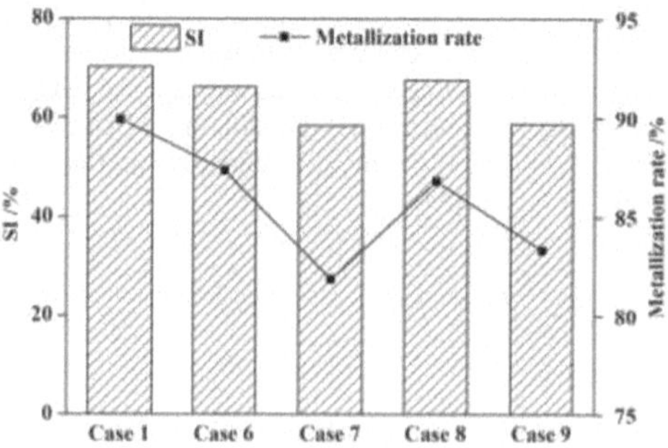

Fig.6 SI and metallization rate of pellets under different match models

The SI and metallization rate of pellets samples under different match models of reduction temperature and gas flow were shown in **Figure 6**. Similarly, the SI of pellets samples increased with its metallization rate. The SI and metallization rates under different match models were all lower than that of reference test. Compared with the reference test, the SI of pellets under case 6, 7, 8, 9 decreased about 4, 12, 3, 12 percent points, respectively. The metallization rate decreased about 2.6, 8.0, 3.1, 6.6 percent points.

As shown in the above results, the metallization rate under case 7 and case 9 decreased obviously due to the large decreasing amplitude in gas flow, so these two match models were not suitable for the COREX shaft furnace. The reason was that the influence of reduction temperature on SI was much stronger than that of gas flow, but the influence of gas flow on the metallization rate was much stronger, so the increasing amplitude must be kept at small range.

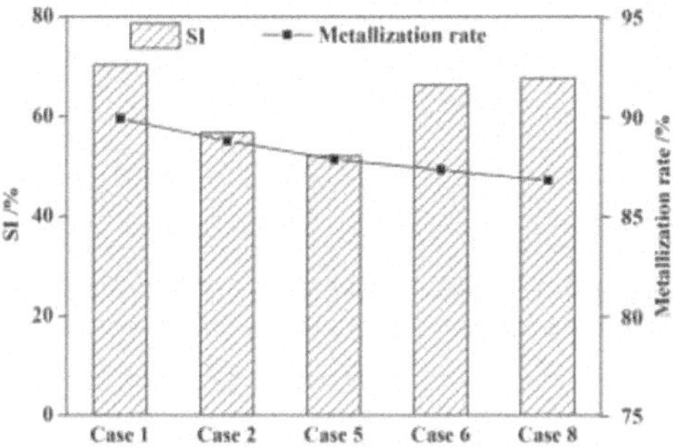

Fig. 7 SI and metallization rate of pellets under different match models

To further compare the matching models, the SI and metallization rate of pellets samples under optimized match models were shown in **Figure 7**. As shown in the figure, the SI under melting rate matching reduction temperature (case 2 and case 5) were much lower than that of reduction temperature matching gas flow (case 6 and case 8), but the metallization rate under melting rate matching reduction temperature were much higher, so the match model of melting rate and reduction temperature were much better than that of reduction temperature matching gas flow, the reason was that the influence of reduction time and holding time on the SI was much stronger than that of temperature and gas flow.

Conclusions

(1) The porous metallic iron produced at the interface of pellets samples is the main reason for the sticking of pellets in the COREX shaft furnace.
(2) The SI and metallization rate of pellets samples increased with the reduction temperature and gas flow, decreased with the melting rate.
(3) The SI and metallization rate of pellets were all decreased under the matching of increasing melting rate and increasing reduction temperature, increasing reduction temperature and decrease gas flow, but the influence the former one was more effective than the latter one.

Acknowledgements

The financial support of the National Natural Science Foundation of China (U1260202) is gratefully acknowledged.

References

1. O. Erich, *et, al*, "COREX Process: Hot Metal Production on the Basis of Coal". Metallurgical Plant and Technology, 9(6), (1986), 24, 26, 28-29.

2. P. L. M. Wong, *et al*, "Sticking Behavior in Direct Reduction of Iron Ore". *Ironmaking & Steelmaking*, 26(1), (1999), 53-57.

3. K. S. Abdel, M. I. Nasr, and A. A. EI-Geassy, "Developed Model for Reduction Mechanism of Iron Ore Pellets under Load". *Ironmaking & Steelmaking*, 38(3), (2011), 189-196.

4. H. F. Xu, *et al*, "Clustering Mechanism and Influencing Factors of the Shaft Furnace Burden in COREX Process". *Baosteel Technology*, 6, (2011), 44-47.

5. G. W. Li, "Operation Status and Technological Problems of COREX-3000" *Baosteel Technology*, 6, (2008), 11-18.

6. L. Y. Li, Z. C. Huang, and T. Jiang, "Sticking of Iron Ore Pellets during Reduction with Hydrogen and Carbon Monoxide Mixtures: Behavior and Mechanism". *Powder Technology*, 235, (2013), 1001-1007.

7. L. Y. Yi, *et al*, "Sticking of Iron Ore Pellets in Direct Reduction with Hydrogen and Carbon Monoxide: Behavior and Prevention". *Journal of Central South University*, 21, (2014), 506-510.

8. B. Zhang, et al. "Relation between Sticking and Metallic Iron Precipitation on the Surface of Fe_2O_3 Particles Reduced by CO in the Fluidized Bed". *ISIJ International*, 51(9), (2011), 1403-1409.

9. J. Fang, "Sticking Problem in Fluidized Bed Iron Ore Reduction", *Iron & Steel*, 26, (1991), 11-15.

10. B. Zhang, *et al*. "A Comparative Study of influence of fluidized conditions on Sticking Time during Reduction of Fe_2O_3 Particles with CO". *Powder Technology*, 225, (2012), 1-6.

11. Y. W. Zhong, *et al*. "Sticking Behavior Caused by Sintering in Gas Fluidization Reduction of Hematite". Ironmaking & Steelmaking, 39(1), (2012), 38-44.

12. J. H. Shao, Z. C. Guo, and H. Q. Tang, "Influence of Temperature on Sticking Behavior of Iron Powder in Fluidized Bed". *ISIJ International*, 51(8), 1290-1295.

7th International Symposium on High-Temperature Metallurgical Processing
Edited by: Jiann-Yang Hwang, Tao Jiang, P. Chris Pistorius, Gerardo R.F. Alvear F., Onuralp Yücel, Liyuan Cai, Baojun Zhao, Dean Gregurek, and Varadarajan Seshadri
TMS (The Minerals, Metals & Materials Society), 2016

RELATIONSHIP BETWEEN COKING PROPERTIES OF LUMP COAL AND ITS PULVERIZATION IN COREX PROCESS

Qihang LIU[1], Peng ZANG[2]

1. School of Metallurgical Engineering, Xi'an University of Architecture and Technology, No. 13 Yanta Road, Xi'an, Shaanxi, 710055 China

2. Xi'an Tianhong Electric Co., Ltd., No. 20 Jinye Second Road, Gaoxin District, Xi'an, Shaanxi, 710055 China

Keywords: Lump Coal, Heat Transfer, Fines Generation, Coking Rate

Abstract

Experimental work was conducted to determine the relationships of coal char conversion with temperature and heat absorption. The results show that at an earlier stage of the coke forming process, lump coal would absorb large amounts of heat but the increase of high temperature strength of lump coal was not significant. Contrarily, the lump coal's high temperature strength was greatly improved at later stage of coke forming process. A number of residual coal was found in most tuyere coke samples illustrating that the lump coals added in the melter gasifier were not completely transformed into coke when they arrived at the tuyere region. The results indicate that the improvement of lump coal's high temperature performance is limited in the melting gasifier by the low coke forming rate which could become the core factor of fines generation of lump coal in the melting gasifier.

Introduction

Coal is used in COREX smelting reduction offering a superior environmental performance compared with the coke process[1)]. The coke's function such as heat source、reducing agent and carburization could be replaced by the coal added in COREX, but the replacement of the coke's skeleton pillar is still needed more development. Since the running of COREX C3000 in BAO-Steel, the high powder ratio of lump coal was always the most important limiting factor for COREX production. Though the measures such as changing production parameters 、coal blending ratios and size of coals were taken to improve the production, the problem of high powder ratio of lump coal was unresolved. According to the operational data from BAO-Steel, the average powder ratio of lump coal reached 60%, even the dead zone's powder ratio was more than 80%. These results made the experts of BAO-Steel to reconsider the fuel structure of COREX[1-3)].

Experience on the blast furnace, shows that the heat conduction of coke could become the limiting step for iron-making when the coke's radius is more than 60-70mm[4]. In addition, main differences between lump coal and coke can be summarized as follows:

(1) The thermal conductivity of lump coal is smaller than that of coke, so the thermal ability of coal is inferior compared with coke.

(2) Coal would absorb large amount of heat during the coke forming process in melting gasifier.

(3) The caking coal used in COREX was easy to bind together during the coke forming process and thus the heat transfer was hindered by the bigger massive coal char.

It has also been observed that lump coal tends to move towards the high temperature region over 800 ℃ immediately when it is added to the top of melting gasifier enhancing a fast pyrolysis of lump coal and precipitation of tar from the lump coal. This phenomena would lead to the blockage of gas pipeline and poor reliability of equipment[5]. The reduction process would also be influenced by the precipitated tar.

Therefore, COREX coal has been selected to investigate the transformation process of the lump coal at high temperature (1000 ℃).

Experiment

Experimental method

Table 1 shows the fundamental analysis of coal used in COREX process:

Table 1 Analysis of coals (wt.%)

Proximate analysis				Elemental analysis				Caking index
FC_d	A_d	V_d	M_{ad}	C_d	H_d	N_d	O_d	$G_{R,I}$
60.35	7.71	31.94	2.84	72.89	4.87	0.96	13	55

Lump coals (20mm) weighed about 200g used in COREX were put in a self-made stainless container which was connected to a device for tar collecting. N_2 was injected into the stainless container about 30min to exclude the air in it, then sealed the container except for remaining a gas pipeline connected to the tar collecting device.

In order to simulate the rapid heating condition prevalent in the melting gasifier, an empty muffle furnace was first heated to 1000 ℃. Then a stainless steel container equipped with coal samples was quickly introduced into the furnace for coke–making.

After a period of time, the container was taken out and placed in a drying oven to cool it down. The steps above were repeated by setting the heating time respectively at 30min,

45min, 60min, 90min, 120min, 150min, 180min, 240min. The tar precipitated from lump coal and the residual coal char were weighed by an electronic balance, and the surface texture and microstructure of the residual coal char were examined by optical microscope and SEM. Finally, the coal char's reactiveness and the strength after reaction were measured as well as the density and volume of different coal chars.

Results and Discussion

Table 2 shows the analysis of different coal chars' microstructure (minerals and impurities were not listed). Here, a new index named coal char conversion ω was defined as:

$$\omega = \frac{\sum_{i=1}^{n} \mu_i}{\sum_{i=1}^{n} \mu_i + C} \quad (1)$$

Where $\sum_{i=1}^{n} \mu_i$ was the proportion of the coke's optical structure, and μ_i represents one of the coke's optical structure such as isotropic、anisotropic and so on. C was the proportion of the residual coal.

In Table 2, the coal char conversion ω was calculated by Eq. (1).

Table 2 Analysis results of coal char's microstructure /%

Heating time/min	Coke							The residual coal	ω/%
	Fusain	Isotropic	Anisotropic	Fine mosaics texture	Medium mosaics texture	Coarse mosaics texture	Residual carbon particle		
30	0.00	2.90	16.23	0.00	1.16	0.00	2.03	75.70	22.77
45	0.23	6.13	5.19	3.21	1.32	0.32	6.23	72.04	23.90
60	3.29	10.18	2.66	2.95	2.36	0.56	7.89	62.32	32.42
90	0.79	18.10	1.05	2.83	3.40	0.95	7.71	51.98	40.12
120	0.11	28.73	1.29	2.41	6.91	1.11	8.09	32.88	59.67
150	0.15	32.10	12.30	0.00	2.20	2.30	12.30	34.60	63.94
180	0.90	52.45	1.24	2.91	9.24	1.29	6.32	12.35	85.76
240	0.33	44.19	17.94	5.65	3.30	10.10	13.29	0.33	99.65

The coal char's reactivity and the strength after reaction were respectively defined as MCRI and MCSR, and the detection methods of MCRI and MCSR were the same as the coke's CRI and CSR. The relationships of the coal char conversion ω with MCRI, MCSR, weight loss G and tar yield are presented as in figure 1, figure 2 , figure 3 and figure 4, respectively.

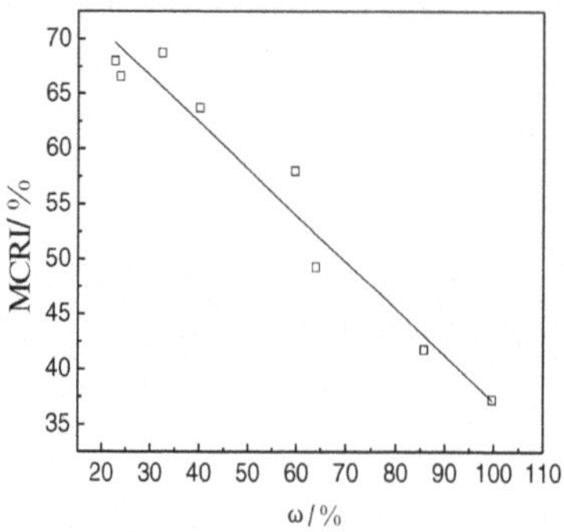

Fig.1 Relation between ω and MCRI

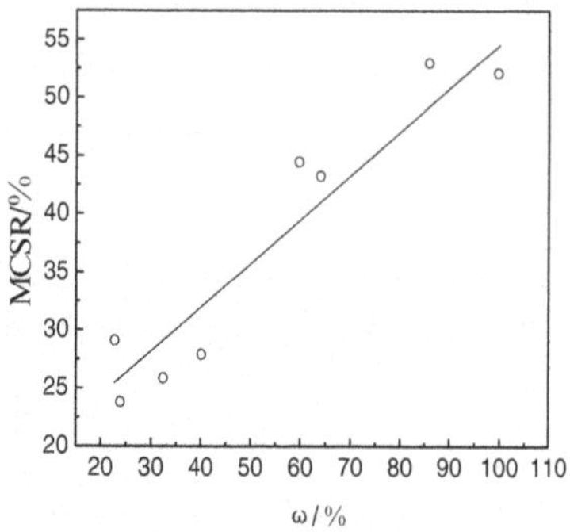

Fig.2 Relation between ω and MCSR

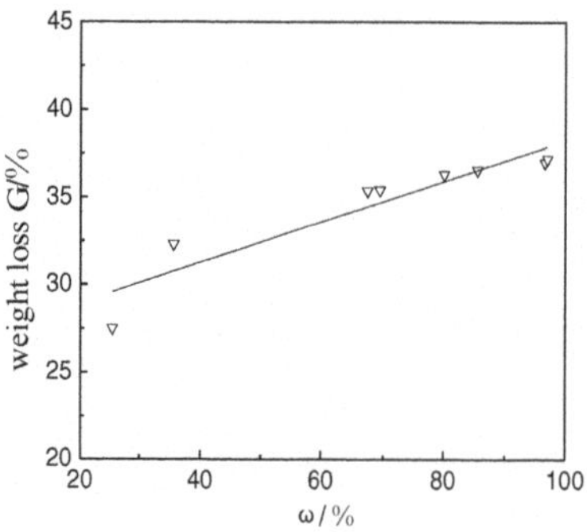

Fig.3 Relation between G and ω

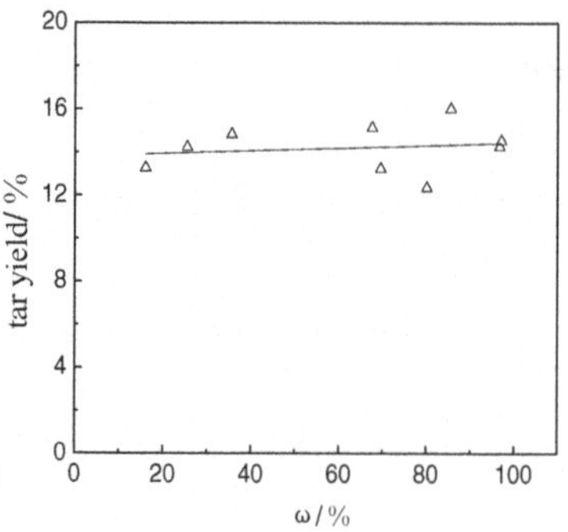

Fig.4 Relation between ω and tar yield

Relation Between Coal Char's Performance and Temperature

Calculations

A large temperature gradient would appear in the internal lump coal when suddenly added to a high temperature environment. While the coal char conversion and high temperature performance of lump coal would be affected by the variation of the internal temperature or heat absorption.

The analysis or numerical solutions of heat conduction differential equation are usually used to calculate the temperature distribution of coke oven, and the coal in coke oven is assumed to be semi-infinite plate[4]. However, the conditions in this experiment were not the same as in a coke oven, and therefore it was more appropriate to treat coal particles as spherical. The radius of lump coal is R and the initial temperature T_0 is 20℃. The lump coal was placed in a high temperature furnace whose temperature T_f is always 1000 ℃, The heat is transferred through the stainless container between the coal char and the air in the high temperature furnace. The thermal conductivity λ of stainless steel[10] is about 16 W/(m·K) which was much higher than the coal. Regardless of heat resistance of the stainless container, besides, it was assumed that the temperature of air in the high temperature furnace was kept at 1000℃.

Basic physical characteristic parameters[9-11]: when $T \le 673K$, $\lambda = 0.23$ W/(m·K), when $T > 673K$, $\lambda = 0.23 + 2.24\times10^{-5}(T-673)^{1.8}$ W/(m·K); when $T \le 623K$, $c_p = 1254$ J/(kg·K), when $T > 623K$, $c_p = 1254 - 1.75(T-623)$ J/(kg·K); the equivalent radius of the lump coal $R = \sqrt[3]{\frac{3m}{4\rho\pi}}$, the weight and density of the solid products were measured by experiment; the heat transfer coefficient α of air is about 5-10 W/(m²·K) under the condition of natural convection, the test was performed in the stationary air, then the interfacial heat transfer coefficient between the air and solid products was set as α=8 W/(m²·K).

Table 3 shows the experimentally determined densities for the obtained solid products.

Table 3 Density of different coal chars

Heating Time/min	30	45	60	90	120	150	180	240
Density ρ/kg/m³	1056	980	863	802	745	735	710	702

For one-dimension unsteady heat conduction of sphere, the following relationship is established[11]:

$$\frac{\partial(\rho c_p T)}{\partial t} = \frac{1}{r^2}\frac{\partial}{\partial r}\left(\lambda r^2 \frac{\partial T}{\partial r}\right) + q_V \qquad (2)$$

Where ρ stands is the denstiy density, r is the sphere radius, c_p the heat capacity at the constant pressure, and q_V the strength of internal heat source, W/m³.

Initial conditions : $T(r,0) = T_0$; Boundary conditions : $\frac{\partial T(r,t)}{\partial r}\Big|_{r=0} = 0$, $-\lambda\frac{\partial T(r,t)}{\partial r}\Big|_{r=0} = \alpha[T_f - T(R,t)]$

The first item in the infinite series of the analytical solution was only singled out for convenient calculation.

$$\frac{\theta(x,t)}{\theta_0} = 2\frac{\sin\mu_1 - \mu_1\cos\mu_1}{\mu_1 - \sin\mu_1}\exp(-\mu_1^2 Fo)\frac{1}{\mu_1 x}\sin(\mu_1 x) \qquad (3)$$

The curves (nomogram) of $\frac{\theta_m}{\theta_0}$(x=0) and $\frac{\theta}{\theta_m}$ change with *Fo* and *Bi* can be obtained from equation (3), and the relationship as formula (4) was adopted for each point in the sphere:

$$\frac{\theta}{\theta_0} = \frac{\theta_m}{\theta_0}\frac{\theta}{\theta_m} = \frac{T_m(t)-T_f}{T_0-T_f}\cdot\frac{T-T_f}{T_m(t)-T_f} = \frac{T-T_f}{T_0-T_f} \qquad (4)$$

Thus the temperature of different points in the sphere can be obtained by nomogram.

Figure 5 shows the distribution of internal temperature of coal char at different heating times.

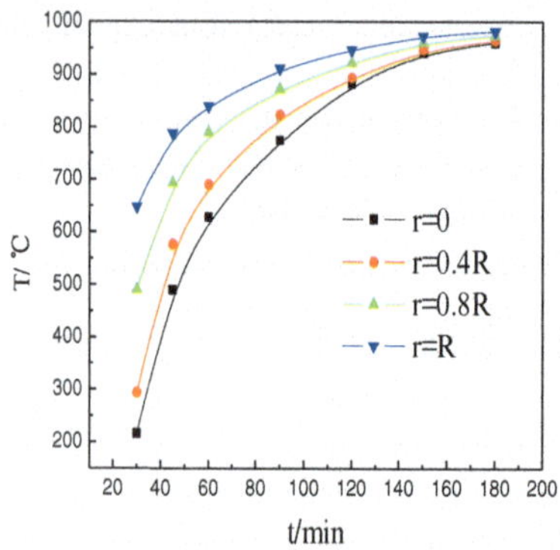

Fig.5 Distribution of internal temperature of coal

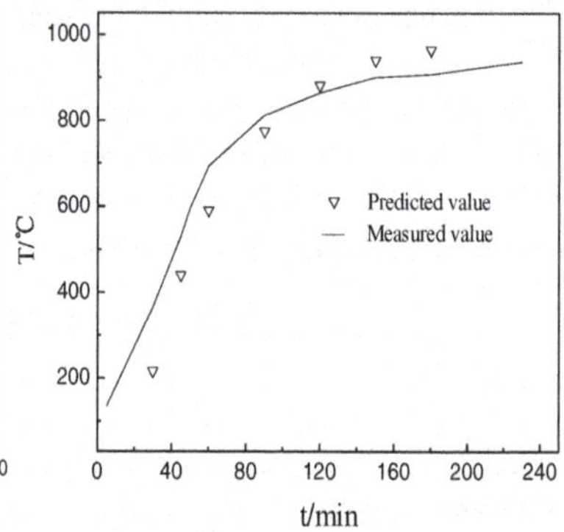

Fig.6 Measured and calculated value of coal's center temperature

Figure 6 shows the measured and estimated temperature at the centre of the coal samples. Experimental temperature was measured by a thermocouple plugged in the center of coal samples through the gas export of stainless steel container. When the empty muffle furnace was heated to 1000 ℃, the coal samples with thermocouple were quickly pushed into the furnace.

Fig 7 shows the heat absorption of coal char at different heating time.

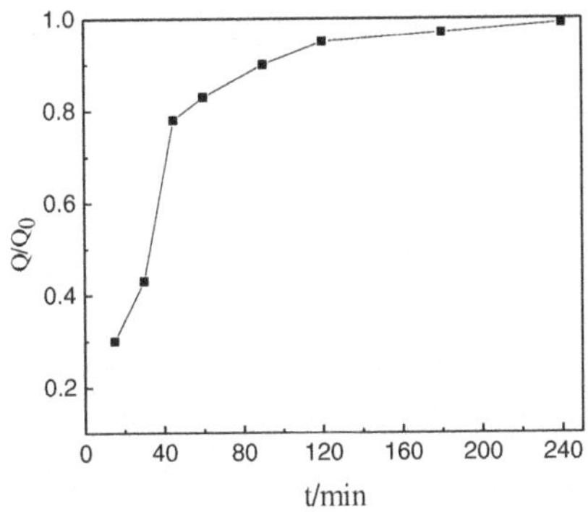

Fig.7 Heat absorption of coal char at different heating time

It can be seen that the endothermic of lump coal was concentrated on the prophase of coke forming, and the absorbed heat of the top 1/3 phase could account for 80% of total heat absorption, and the heat absorbance rate at the prophase is also higher than that at the anaphase as shown in Fig 7

Discussion

The lump coal fall from the top of COREX Melting Gasifier to the moving bed rapidly and stay in the furnace for only around three hours. The average temperature of lump coal in the furnace is about 1000 ℃[3,5]. It can be inferred from the test results that the lump coal will not transform into coke completely. Part of the lump coal may be consumed as coal in the upper-middle part of the furnace while the remaining part of lump coal will partially transform into coke and then reached front of the tuyere for combustion. In this case, the slow coke forming rate will lead to the serious chalking of the lump coal.

Conclusions

The most important restriction of COREXC3000 was the fines generation of lump coal in the melting gasifier. In this paper, the variation of high temperature performance and the cause for fines generation of lump coal were clarified through the simulation test and theoretical calculations. An important reason for the fines generation was that the low coke forming rate could not meet the requirements of coal's high temperature performance in the melting gasifier which was confirmed by detecting the tuyere coke samples. The measure of preheating the lump coal before it was added to the melting gasifier could be taken. The preheating temperature could be set to 700℃~ 800℃, with the heating rate at 20℃/min and the preheat time of 1h. Then the coal char conversion could be ensured by 30%, and the restriction of heat transfer could be dramatically reduced, even eliminated.

The internal temperature and heat changes of the lump coal were calculated by unsteady heat conduction equation. Then the relationship of the coal char conversion with average temperature, absorbed heat and high temperature performance was given by the experiment. Moreover, the effect of heat transfer process on the performance of the coke formed from lump coal was analyzed. According to the results, lump coal would absorb large amounts of heat in the early stage of coke forming process. However, the increase of high temperature strength of lump coal was not significant at early stage which leads to the fines generation of lump coal. By controlling the stage of coke forming, the coking rate could be increased and thus the high temperature performance of coal char was improved. So the fines generation of lump coal could be reduced, and the precipitation of coal tar could also be effectively controlled.

References

[1] Prachethan Kumar P, "Influence of coal size on the performance of COREX process," Steel Research International, 80 (3)(2009),179-184.

[2] Q. H LIU, K WU, "Kinetic Analysis of Tar's Separation from Lump Coal," ISIJ International, 55(5)(2015), 947-951.

[3] D Suika, "COREX and FINEX Technology Process Update," (The Second Baosteel Biennial Academic Conference. Shanghai: Baosteel Group Co.,2006).

[4] X. L WANG, "Iron and steel metallurgy," (Beijing, Metallurgical Industry Press, 2013),182-185.

[5] J. W Cumming, "Reactivity of coals Via a weighted Mean Activation Energy," Fuel, 63(10)(1984), 1436-1440.

[6] A Eberle, D Siuka, C Böhm, "Development in COREX technology," Steel World, (7)(2002), 28-32.

[7] P. Prachethan Kumar, D. Gupta, "Factors Affecting Fuel Rate in COREX Process," Ironmaking and Steelmaking, 33(4)(2006): 293-297.

[8] S Richard, "Interactions between coking coals in blends," Fuel, 82(4)(2003):439-450.

[9] Q.H LIU, K WU, "Influence of different parameters on theoretical flame temperature before tuyere in corex melter gasifier,"(PRICM8, 2013), 703-712.

[10]Y. L. Guo, W. R. Xu, "The burden structure and its consumption in the melter gasifier of the corex process," Metallurgical and Materials Transactions B: Process Metallurgy and Materials Processing Science, 44(5)(2013): 1078-1085.

[11]H. J. Zhang, "Heat conduction," (Beijing, Higher Education Press, 1992), 189.

7th International Symposium on High-Temperature Metallurgical Processing
Edited by: Jiann-Yang Hwang, Tao Jiang, P. Chris Pistorius, Gerardo R.F. Alvear F., Onuralp Yücel, Liyuan Cai, Baojun Zhao, Dean Gregurek, and Varadarajan Seshadri
TMS (The Minerals, Metals & Materials Society), 2016

THERMOGRAVIMETRIC ANALYSIS OF COAL USED IN ROTARY KILN OF IRON ORE OXIDE PELLET

Qiang Zhong,[1] Yongbin Yang,[1,2] Qian Li,[1] and Tao Jiang[1,3]

1.—School of Minerals Processing and Bioengineering, Central South University, Changsha, Hunan, 410083, China. 2.—e-mail: ybyangcsu@126.com. 3.—e-mail: jiangtao@csu.edu.cn

Keywords: Coal, Thermogravimetric analysis, Rotary kiln, Combustion, oxygen-enriched

Abstract

Thermogravimetric analysis (TGA) has been used to research the pyrolysis and combustion processes of coal used in rotary kiln of iron ore oxide pellet. Furthermore, the effects of air flow and oxygen concentration on the combustion behavior of coal were examined with TGA. The results showed that two individual stages can be distinguished for the pyrolysis of coal, the water evaporation stage and the de-volatilization stage. And the information on ignition and burnout characteristics can be obtained from the combustion process analyzed by TGA. Large air flow almost has no effect to the combustion process except that it enhances combustion intensity. However, the combustion process is further accelerated under oxygen-enriched conditions. With all experimental results, it could provide an attractive approach of analyzing coal for the rotary kiln process.

Introduction

Rotary kiln as a key equipment being employed widely in production of iron ore oxide pellet. The Fuel as a heat source is an integral part of the rotary kiln. Coal is one of the most extensively used fuels to provide the rotary kiln with heat. With the development of steel industry, the demanding of the rotary kiln for coal is huge. However, utility coal resource is becoming increasingly poor and the price of coal is consecutively rising. Furthermore, there are a low utilization efficiency of the coal and a serious waste of the coal in pellet plant. It is necessary and exigent for people to save coal and use coal efficiently. [1-5]

The rotary kiln as a complex system which is large delay, nonlinear, hyperthermal, time varying, and composition diverse and so on, the real situation of coal combustion process and the temperature distribution is difficult to be effectively regulated. [6-8] Thermogravimetric analysis (TGA) which is a simple but quite effective technique covers a wide range of applications in research, development and economic assessment of the fuels and observes the burning profile of a fuel. The combustion characteristics such as the ignition temperature, the burnout temperature and the DTG_{max} can be obtained from TGA, which is great importance for the estimation of combustion efficiency to establish the optimum operational conditions of coal combustion. [9-12] In this study, the pyrolysis and combustion processes of the coal used in the rotary kiln were investigated with thermogravimetric analysis. After that the effects of the air flow and the oxygen concentration on the combustion behavior of coals combined with the production process of rotary kiln were evaluated. With all experimental results, it could provide an attractive approach of analyzing coal for the rotary kiln process.

Experimental

Materials

YMG coal used in the experiments was provided by a pellet plant of China. Sampling was carried out carefully in order to represent the original coal properties as much as possible. The samples were dried in an oven at 363 K for 7 h, after which they were crushed and sieved to make particle size under 250 m. A comparison of the proximate and ultimate analysis results, as well as of the calorific value of the coal is made in Table 1. As can be seen, the coal belongs to the type of bituminous coal because of volatile matter (12.90%) and fixed carbon (76.47%). The contents of C and fixed carbon are high, resulting in the high heating value for the coal.

Table I. Proximate Analysis, Ultimate Analysis and Heating Value of Coal Sample

Proximate analyses (wt.%, d)			Ultimate analyses (wt.%, d)					$Q_{net.d}$ (Kcal/kg)
Volatile matter	Fixed carbon	Ash	C	H	N	O	S	
12.90	76.47	10.63	80.61	3.46	1.42	2.89	0.49	7794

Experimental Method

Proximate analyses of the coal samples were performed according to Standard Practice for the Proximate Analysis of Coal and Coke (GB/T212-2008). Ultimate analyses of the coal samples were performed according to ASTM D 5373-02 (for carbon,hydrogen and nitrogen contents) and ASTM D 4239-05 (for sulfur content) standard test methods with LECO TruspecCHN-Sultimate analysis instrument. The heating value was measured with an adiabatic oxygen bomb calorimeter.

Pyrolysis and combustion tests were performed in a NETZSCH differential thermogravimetrc analyzer STA449C (precision of temperature measurement±1 K, microbalance sensitivity 0.1 g), during which sample weight-loss value and rate as functions of temperature or time were recorded continuously under dynamic conditions. To eliminate the effects of eventual side reactions and mass and heat transfer limitations, a small amount of each material (about 5-10 mg) were thinly distributed in an Al2O3 ceramic crucible. The pyrolysis tests were conducted at atmospheric pressure and temperatures ranging from 293 to 1273 K under nitrogen atmosphere. The combustion tests were conducted at atmospheric pressure and temperatures ranging from 293 to 1273 K under artificial air atmosphere. For both tests, the flow rate was fixed at 100 ml/min and the heating rate was 20 K/min. Moreover, tests of air flow and kiln temperature were performed with isothermal thermogravimetric analysis. Firstly the temperatures rose from 293 to 1373 K with nitrogen flow rate of 100 ml/min, and then the temperature was stabilized at 1373 K with artificial air flow rate. The air flow was changed while the oxygen concentration remained constant under the experiments of different air flow. And the oxygen concentration was changed while the air flow was fixed at 100 ml/min under the experiments of different oxygen concentration.

Results and Discussion

Pyrolysis

Pyrolysis test of the coal was performed to investigate the behavior of volatiles. Pyrolysis TG and DTG profile for the coal is shown in Fig. 1.

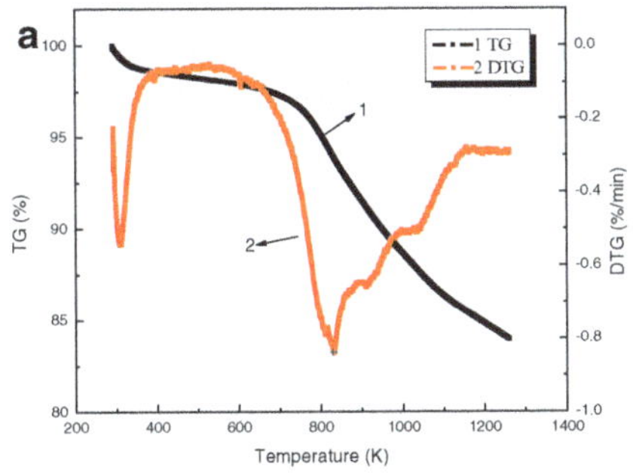

Fig. 1 Pyrolysis TG and DTG profile of coal.

From Fig. 1, two individual stages can be distinguished for the pyrolysis of the coal. The first pyrolysis stage is considered to be the water evaporation stage, during which appears from 293 K to 379 K with a peak at 307 K and mass loss 1.37%. The other stage is regarded as the de-volatilization stage, which spans from 650 K to 1150 K with a peak at 831 K and mass loss 12.16%.[13, 14]

Combustion

Combustion behavior of the coal in the rotary kiln can be estimated from combustion characteristics which obtained from TGA is a guide to select a suitable coal for the rotary kiln. Combustion TG and DTG profiles for the coal are shown in Fig. 2.

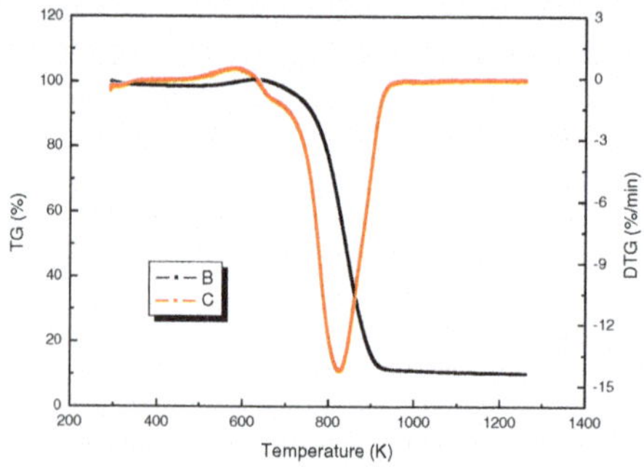

Fig. 2 Combustion TG and DTG profile of coal.

It can be seen from Fig. 2 that the combustion DTG profile shows only one peak. Combining with the proximate analysis result, it indicates that the combustion of fixed carbon makes a great contribution to the mass loss. And the combustion profile is different from that of pyrolysis, the main reaction of combustion is the combustion of fixed carbon while that of pyrolysis is vaporization and transport of volatile material.

The ignition temperature is the temperature where a sudden decrease occurs in the DTG profile, The peak temperature represents the point in the DTG profile where the rate of weight loss comes up to the maximum, and the rate at this temperature is called the maximum combustion rate (DTG_{max}), which can clearly display the combustion behavior of the coal. For the coal used in the rotary kiln of iron ore oxide pellet, it provides heat to ensure the temperature of kiln head, kiln body and kiln tail being not lower than 1523 K. And the heat of kiln tail provided by the coal combustion should preheat pellet and ensure the temperature of chain grate being not lower than 1223 K. [2, 3, 15] Based on these, Coal has the combustion characteristics of low ignition temperature and high combustion intensity is suitable for the rotary kiln. For Fig. 2, the ignition temperature of the coal is 401 K, which indicates that it is easily flammable. And the DTG_{max} rate is up to 14.24%/K, which indicates that it has a high combustion intensity and can provide enough heat for the rotary kiln and the chain grate.

Effect of Air Flow

Coal is sprayed into the kiln from the kiln head whose temperature is not lower than 1273 K, after that its temperature risen quickly and burns with air. Therefore, air flow and oxygen concentration have a great effect on coal combustion in rotary kiln. [16]

The combustion processes of the coal under different air flow are shown in Fig. 3. As the air flow increased from 100 ml/min to 300 ml/min, the burnout time is almost unchanged while the peak time shortens and the DTG_{max} rate is increased. Therefore, large air flow almost has no effect to the combustion process except that it enhances combustion intensity, which means that the combustion process of the coal is difficult to be regulated by air flow.

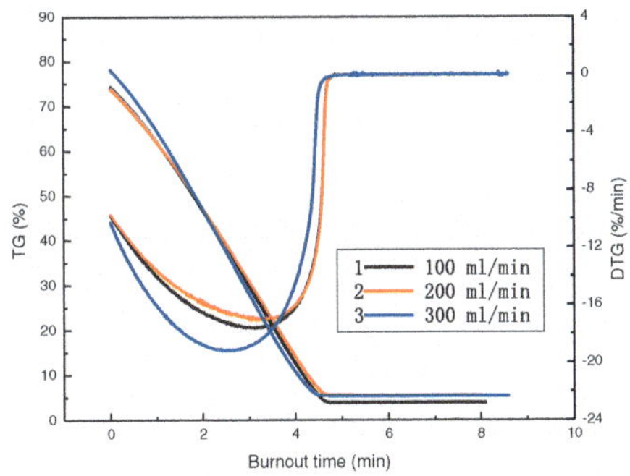

Fig. 3 Combustion TG and DTG profiles of coal under different air flow.

Effect of Oxygen Concentration

With the constant air flow of 100 ml/min, the combustion processes of the coal under different oxygen concentration (N_2:O_2=90:10, N_2:O_2=80:20 and N_2:O_2=70:30) are shown in Fig. 4. 10% O_2 (N_2:O_2=90:10), 20% O_2 (N_2:O_2=80:20) and 30% O_2 (N_2:O_2=70:30) mean the coal burned under low oxygen atmosphere, air atmosphere and oxygen-enriched atmosphere, respectively.

As the oxygen concentration increased from 10% to 30%, the combustion processes of the coal have an obvious change especially that of 30% oxygen concentration. The burnout time is reduced from 5.5 min to 4.2 min and the peak time is shorten from 3.9 min to 2.6 min, while the DTG_{max} rate is increased from 15.12 %/K to 20.79 %/K. It has been studied that the presence of oxygen would reduce the force of the diffusion layer. [17, 18] As a result, the increasing oxygen concentration results in quicker reaction of the oxygen with the volatile material and the fixed carbon to favorably enhance the combustion of coal.

Therefore, the combustion intensity of the coal is enhanced with the increase of oxygen concentration and the combustion process was further accelerated under oxygen-enriched conditions, which indicates that the combustion process of the coal is easy to be regulated by the oxygen concentration.

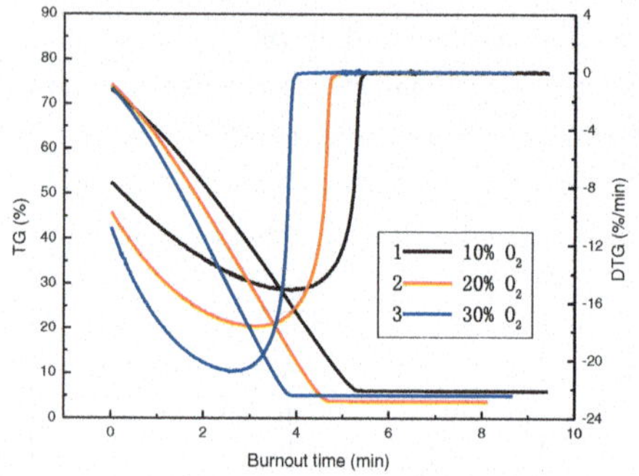

Fig. 4 Combustion TG and DTG profiles of coal under different oxygen concentration.

Conclusions

The pyrolysis and combustion processes of the coal used in the rotary kiln were investigated with thermogravimetric analysis. Furthermore, the effects of the air flow and the oxygen concentration on the combustion process of the cola were examined. A summary of the results obtained in this study is as follows:

1) There are two individual stages in the coal pyrolysis process, first the water evaporation stage and the de-volatilization stage.

2) The information on ignition and burnout characteristics can be obtained from the combustion process analyzed by TGA. Coal has the combustion characteristics of low ignition temperature and high combustion intensity maybe is suitable for the rotary kiln.

3) Large air flow almost has no effect to the combustion process except that it enhances

combustion intensity. However, the combustion process is further accelerated under oxygen-enriched conditions, which indicates that the combustion process of the coal is easy to be regulated by the oxygen concentration.

Acknowledgement

This work is supported by the National Natural Science foundation of China (No. 51074182 and No. 51234008).

References

[1]. L. Ljung et al., "Convective Drying of an Individual Iron Ore Pellet–Analysis with CFD," *Int. J. Heat Mass Tran* , 54 (17) (2011), 3882.

[2]. T. Umadevi et al., "Influence of Pellet Basicity (CaO/SiO2) on Iron Ore Pellet Properties and Microstructure," *ISIJ Int.*, 51 (1) (2011), 14-15.

[3]. S.K. Sadrmezhaad, A. Ferdowsi and H. Payab, "Mathematical Model for a Straight Grate Iron OrePpellet Induration Process of Industrial Scale," *Comp. Mater. Sci.*, 44 (2) (2008), 296-297.

[4]. X.H. Fan, Y. Wang and X.L. Chen, "Mathematical Models and Expert System for Grate-kiln Process of Iron Ore Oxide Pellet Production. Part II: Rotary Kiln Process Control," *J. Cent. South Univ.*, 19 (6) (2012), 1724-1727.

[5]. M. Georgallis et al., "Modelling the Rotary Lime Kiln," *Can. J. Chem. Eng.*, 83 (2) (2005), 212-223.

[6]. Y. Man et al., "Influence of Temperature and Time on Reduction Behavior in Iron Ore-Coal Composite Pellets," *Powder Technol.*, 256(2014), 361–366.

[7]. R. Bandyopadhyay et al., "Assessment of Ash Deposition Tendency in a Rotary Kiln Using Thermo-mechanical Analysis and Experimental Combustion Furnace," *Fuel*, 135(2014), 301-307.

[8]. D.Q. Zhu et al., "Direct Reduction Behaviors of Composite Binder Magnetite Pellets in Coal-Based Grate-Rotary Kiln Process," *ISIJ Int.*, 51 (2011), 214-219.

[9]. R. Font et al., "Analysis of the Pyrolysis and Combustion of Different Sewage Sludges by TG," *J. Anal. Appl. Pyrol.*, 58 (4) (2001), 927-941.

[10]. M.B. Folgueras et al., "Thermogravimetric Analysis of the Co-combustion of Coal and Sewage Sludge," *Fuel*, 82 (2003), 2051-2055.

[11]. R. Font, A. Fullana and J. Conesa, "Kinetic Models for the Pyrolysis and Combustion of Two Types of Sewage Sludge," *J. Anal. Appl. Pyrol.*, 74(2005), 429-438.

[12].F.X. Li, S.Z. Li and Y.Z. Chou, "Thermal Analysis Study of the Effect of Coal-burning Additives on the Combustion of Coals," *J. Therm. Anal. Calorim.*, 95 (2) (2009), 633-638.

[13].S.Y. Yorulmaz and A.T. Atimtay, "Investigation of Combustion Kinetics of Treated and Untreated Waste Wood Samples with Thermogravimetric Analysis," *Fuel Process. Technol.*, 90 (2009), 939-946.

[14].D. Vamvuka et al., "Possibility of Using Paper Sludge in Co-firing Applications," *Fuel*, 88 (4) (2009), 637-643.

[15].S. Dwarapudi et al., "Influence of Pellet Size on Quality and Microstructure of Iron Ore Pellets," *ISIJ Int.*, 48 (6) (2008), 768-769.

[16].Y.F. Liao and X.Q. Ma, "Thermogravimetric Analysis of the Co-combustion of Coal and Paper Mill Sludge," *Appl. Energ.*, 87 (11) (2010), 3529.

[17]. C.D. Blasi and R. Colomba, "Modeling and Simulation of Combustion Processes of Charring and Non-charring Solid Fuels," *Prog. Energ. Combust.*, 19 (1) (1993), 71-104.

[18].O. Marin et al., "Simulating the Impact of Oxygen Enrichment in a Cement Rotary Kiln Using Advanced Computational Methods," *Combust. Sci. Technol.*, 164 (1) (2001), 193–207.

7th International Symposium on High-Temperature Metallurgical Processing

Sintering and Pelletizing of Iron Ores

Session Chairs:
Liyuan Cai
Deqing Zhu

7th International Symposium on High-Temperature Metallurgical Processing
Edited by: Jiann-Yang Hwang, Tao Jiang, P. Chris Pistorius, Gerardo R.F. Alvear F., Onuralp Yücel, Liyuan Cai, Baojun Zhao, Dean Gregurek, and Varadarajan Seshadri
TMS (The Minerals, Metals & Materials Society), 2016

ENHANCING THE REMOVAL OF SODIUM AND POTASSIUM OF SINTER BY CO-CONTAINING FLUE GAS CIRCULATION SINTERING PROCESS

Chen Liu, Guanghui Li, Ruijun Wang, Zhengwei Yu, Qian Li, Zhao Jing, Yuanbo Zhang

(School of Minerals Processing & Bioengineering, Central South University, Changsha, Hunan 410083, China)

Keywords: Iron ore; Reductive sintering; Alkali metals; Flue gas circulation

Abstract

The high content of sodium and potassium in sinter is harmful for blast furnace ironmaking. Reductive sintering technology can remove more alkali metals than conventional sintering process. In order to improve the removal ratios of Na and K, reductive sintering with both CO-containing flue gas circulation and carbon-bearing pellet was conducted in this study. Experimental results show that all indexes except the *VSV* and *P* were improved in reductive sintering with CO circulation. Flue gas circulation technique combined with carbon-bearing pellets were beneficial to reductive sintering, because it could maintain good sintering indexes and enhance the removal of Na, K. The best removal ratios of Na, K and reduction degree of iron oxides were 20.37%, 63.73% and 23.76%. In reductive finished sinter, silicate mineral becomes the important binder phase, and main iron mineral phases are wustite (Fe_xO) and magnetite (Fe_3O_4).

Introduction

Reductive sintering technology can produce sinter based on traditional sintering process, meanwhile, it can result in the reduction of sinter. That achieves the aim of transferring a part of iron ores reduction from blast furnace iron-making to sintering process, which can reduce energy consumption and total greenhouse gas emission[1-3]. The characteristics of reductive sintering are those as follows: the using efficiency of carbon is improved because of carbon direct reduction; CO produced from direct reduction can become the indirect reduction agent; reductive sinters used for BF burden can remarkably reduce coke ratio[4].

The input raw materials for BF production contain a number of negative elements such as alkali metals, heavy metals. Especially, alkali metals concerned are two elements of Na and K. They cause a series of problems by affecting hot metallurgical properties of sinter and pellet[5, 6]. The vast majority of alkali metals are residue in sinter as traditional sintering process barely removes them[7]. Reductive sintering can largely reduce alkali metals content in sinter, but their content is still higher than the maximum content value of blast furnace requirement.

In this study, reductive sintering enhancing technology, CO-containing flue gas circulation sintering process combined with carbon-containing pellets, was adopted to further remove alkali metals. Effects on the regular indexes, the removal ratios of Na, K and reduction degree of iron oxides were investigated.

Experimental

Materials

In this study, iron-bearing materials include magnetite concentrate from China and fine iron ores from Australia. Fluxes (limestone, dolomite, burnt lime) and coke breeze are used. Chemical composition and size distribution of raw materials are shown in Table I and Table II, respectively.
Table I shows that magnetite concentrate has high content of potassium oxide (0.08 wt.%) and sodium oxide (0.22 wt.%), and fluorine (0.39 wt.%). The mass proportion of -0.074mm is 96.20%, which meets the requirement of pelleting.

Table I. Chemical Composition of Raw Materials/%

Raw Materials	TFe	FeO	SiO_2	CaO	MgO	Al_2O_3	K_2O	Na_2O	F	LOI
Iron Concentrate	65.68	26.41	2.80	0.68	0.65	0.37	0.08	0.22	0.39	1.93
Fine Iron Ore	60.86	0.22	3.78	0.13	0.12	0.99	0.01	0.02	/	5.89
Limestone	0.37	0.17	2.68	51.37	2.83	0.10	0.13	0.03	/	42.75
Dolomite	1.52	0.55	3.82	34.06	21.30	0.03	0.11	0.04	/	45.73
Burnt Lime	0.62	0.04	2.74	63.19	3.17	0.12	0.09	0.02	/	20.80
Coke Breeze	1.77	7.43	7.11	0.35	0.08	1.82	0.11	0.09	/	86.47

Table II. Size Distribution of Raw Materials/%

Size /mm	+0.074		0.074-0.045		-0.045	
iron concentrate	3.80		12.67		83.53	
Size /mm	+8	8-5	5-3	3-1	1-0.5	-0.5
Fine Iron Ore	7.02	19.42	15.56	22.55	11.87	23.58
Limestone	0.16	8.97	18.44	28.92	15.06	28.45
Dolomite	3.06	6.28	6.98	14.52	9.49	59.67
Coke Breeze	0	0.65	9.25	25.35	21.99	42.76

Methods

All experiments are conducted in a laboratory sinter pot with 500 mm in height and 100 mm in diameter. Schematic diagram of sintering apparatus is shown in Figure 1.

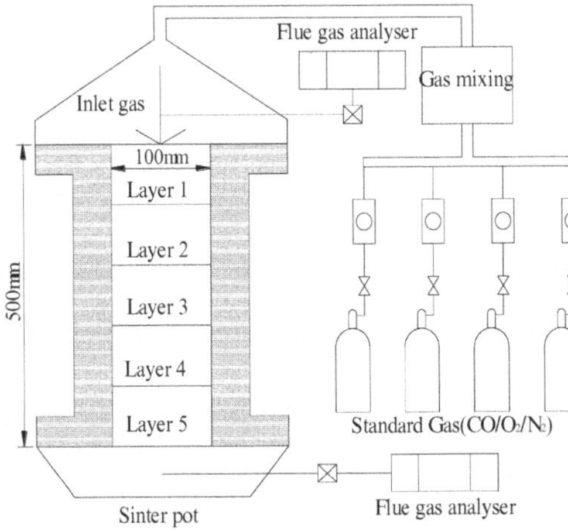

Figure 1. Schematic diagram of sinter pot in lab

Reductive sintering experiments consists of sinter mixture preparation (raw materials proportioning, the first mixing, granulation), carbon-bearing pellets preparation, the second mixing, feeding, ignition, sintering with flue gas circulation, cooling and sieving.

Sinter mixture was fixed as follows if not specified: the mass ratio between iron concentrate and fine iron ore is 4:1, return fine content is 30%, matrix materials basicity is 2.14, burnt lime dosage is 3%. Sinter mixture was granulated in a drum (Φ600×300mm) for 5 min. Ignition temperature was 1100±50 °C and the time was 2 min. At the end, sinter was cooled with protecting gas circumstance (N_2).

For reductive sintering indexes evaluation, the regular indexes, such as vertical sintering velocity (*VSV*), yield (*Y*), ISO tumbling index (*TI*) and productivity (*P*) are measured. The removal ratios of Na, K (φ, %) were calculated as following Eq. (1).

$$\varphi = (R_0 - R) / R_0 \times 100\% \tag{1}$$

where: R_0 is the calculated content of Na_2O, K_2O in sinter (%), R is the measured content of Na_2O, K_2O in sinter (%).

The reduction degree of iron oxides (δ, %) was expressed according to Eq. (2).

$$\delta = (W_0 - W) / W_0 \times 100\% \tag{2}$$

where: W_0 is oxygen content combined with iron in sinter mixture (%), W is oxygen content combined with iron in sinter (%).

Results and discussion

In our previous research[8], effects of common conditions such as CO dosage on the removal ratios of K_2O and Na_2O have been studied. In the present work, reductive sintering with CO circulation is studied first under different raw materials and sinter basicity and then reductive sintering enhancing technology is conducted.

Reductive sintering with CO circulation

Compared with traditional sintering, reductive sintering needs more fuel. In order to know the atmosphere in sinter layers, contents of sintering flue gas constitutes (CO, CO_2, O_2) were detected and shown in Table III.

Table III clearly shows that CO and CO_2 contents increase and O_2 content decreases in flue gas with coke breeze dosage increase. Flue gas with high content of CO cannot discharge directly and should be used. Flue gas circulation sintering process is the best way to solve the issue, so reductive sintering with CO circulation experiment was conducted.

Table III. CO, CO_2 and O_2 Content Ranges in Sintering Flue Gas/%

Coke Breeze Dosage	CO	CO_2	O_2
6	2-3	8-10	12-10
9	4-5	10-12	10-7
12	5-6	12-14	8-5

In the experiment, the whole coke breeze dosage was 12%, which was added into sinter mixture as follows: 40% coke breeze was directly mixed with iron ores and fluxes, and the rest was added after 2 min granulating. CO content in circulating gas was 5%. The results are shown in Table IV and the change of removal ratio of Na, K in different sinter layers are also illustrated in Figure 2.

As shown in Table IV, all indexes are improved except vertical sintering velocity and productivity. Removal ratios of Na and K raise up to 25.26% and 60.95%, indicating that reductive sintering with CO circulation evidently enhances the removal of Na and K. In Figure 2, removal ratios of Na, K increase from the 1st layer to the 4th layer, and then decrease in the 5th layer. In the fourth layer, peak values of removal ratios were 68.68% and 29.84%.

Table IV. Reductive Sintering Indexes with and without CO Gas Circulation

CO circulation	*VSV* mm/min	*Y* %	*TI* %	*P* t/m^2·h	φ_K %	φ_{Na} %
Without	18.87	88.74	68.31	1.252	50.48	14.95
With	13.57	89.55	74.67	1.074	60.95	25.26

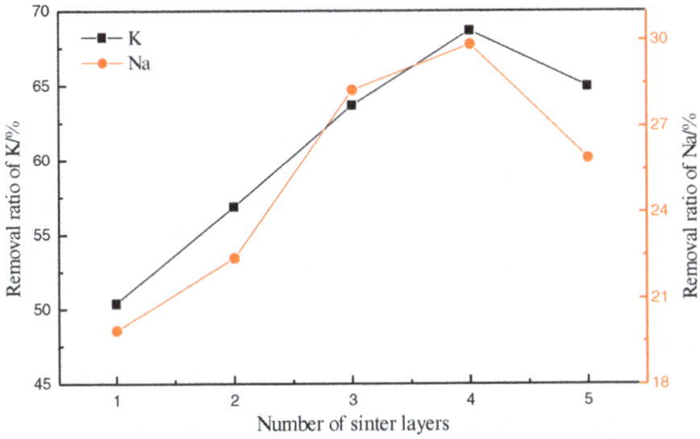

Figure 2. Removal ratios of Na and K in different sinter layers

Reductive sintering enhancing technology

The technology of CO-containing flue gas circulation sintering process combined with carbon-bearing pellets in sinter mixture is adopted to enhance reductive sintering. Clearly, through increasing CO content and decreasing O_2 content in circulating gas, it can strengthen reducing atmosphere in sinter layer and prevent finished sinter from re-oxidation. The use of carbon-bearing pellets increases the contact area of coke breeze and iron concentrate and improves the permeability of sinter layer.
Effects of circulating gas composition and coke breeze size in pellets on reductive sintering enhancing technology are studied. It is noted that coke breeze adding way also change. Here, a part of that is added into pellet, the other part is directly added into sinter mixture.

Effect of circulating gas composition.

Carbon-bearing pellets were prepared as following conditions: half of the total iron concentrate was used to pelletize by adding bentonite (2%) and coke breeze (5%, -0.074mm). Green pellet size was 8-10mm in diameter. The coke breeze dosage was 5% of the whole sinter mixture mass, which was added into mixture to granulate. O_2 content in circulating gas was 15%. The results were shown in Table V.
In Table V, it is worth noting that the *VSV* and *P* are very low while CO content is 10%. Removal ratio of K keeps at high levels with different CO content, and achieves as high as 61.90% when CO content is 8%. Removal ratio of Na increases slightly from 20.62% to 24.23%. Reduction degree of iron oxides sharply increases to 20.86%.
Sodium and potassium occur in iron concentrate in the form of silicate and aluminosilicate which are reduced more effectively by carbon direct reduction than by CO indirect reduction. More CO content and less O_2 content in circulating gas increase the proportion of coke breeze to occur direct reduction, which enhance the removal of Na and K. Meanwhile, the more CO burns, the more quantity of heat

produces, which increase sinter layer temperature contributing to the removal of Na and K.

Table V. Reductive Sintering Indexes with CO Content Variation

CO /%	*VSV* /mm·min^{-1}	*Y* /%	*TI* /%	*P* /t·m^{-2}·h^{-1}	φ_K /%	φ_{Na} /%	δ /%
6	12.39	87.64	59.53	1.302	53.33	20.62	10.93
8	10.11	87.06	60.13	1.067	61.90	23.20	11.60
10	6.93	87.38	62.44	0.604	55.71	24.23	20.86

Effect of coke breeze size in pellets.

Carbon-bearing pellets were prepared with the same proportion of iron concentrate and bentonite, but with 3% of coke breeze dosage. Coke breeze dosage in mixture for granulation decreased to 4%. O_2 and CO content was 15% and 6%, respectively. Coke breeze size in pellets is -0.074mm, 0.074-0.15mm and 0.15-0.3mm. The results were shown in Table VI.

In Table VI, with the increase of coke breeze size, the *VSV*, *TI* and *P* increase first and then decrease, while the sinter yield increases all the time. Removal ratio of K and reduction degree of iron oxides rises from 52.58% to 63.73% and from 12.62% to 23.76%, respectively. Sodium removal ratio maintains in a certain range of 21%.

Coke breeze evenly distributed in pellets contacts sufficient with iron concentrate. When coke breeze with larger particle size combusts (0.15-0.3mm), lower burning speed and longer reaction time make more sodium, potassium and iron around it be reduced smoothly. On the contrary, when coke breeze particle size is -0.074mm, removal ratio of K and reduction degree of iron oxides are the lowest because of incomplete reduction.

Table VI. Reductive Sintering Indexes with Coke Breeze Granularity Variation

Coke breeze size /mm	*VSV* /mm·min^{-1}	*Y* /%	*TI* /%	*P* /t·m^{-2}·h^{-1}	φ_K /%	φ_{Na} /%	δ /%
-0.074	12.57	84.56	62.32	1.155	52.58	21.43	12.62
0.074-0.15	13.57	84.24	66.20	1.224	60.64	21.76	17.73
0.15-0.3	13.08	86.04	64.15	1.193	63.73	20.37	23.76

Characterization of the microstructure of sinter

Representative microstructures of different binder phases of finished sinter from traditional sintering and reductive sintering were analyzed by an optical microscope, which are shown in Figure 3.

Acicular ferrite calcium as the main binder phase is abundantly generated (Figure 3(a)). Silicate mineral as the assistant binder phase bonds the magnetite mineral (Figure 3(b)). But silicate mineral become the vital binder phase (Figure 3(c)), and its crystallization degree is high during cooling stage. Study has showed that most of Na and K are concentrated in silicate glass phase[9]. Therefore, reductive sinter has little of silicate glass phase to concentrate Na and K, indicating that reductive sintering can efficiently remove more of them. Silicate mineral may have more Na and K which cannot be removed (Figure 3(d)).

It can be seen from Figure 3(c) that main iron mineral phases in reductive sinter are Fe_xO and Fe_3O_4.

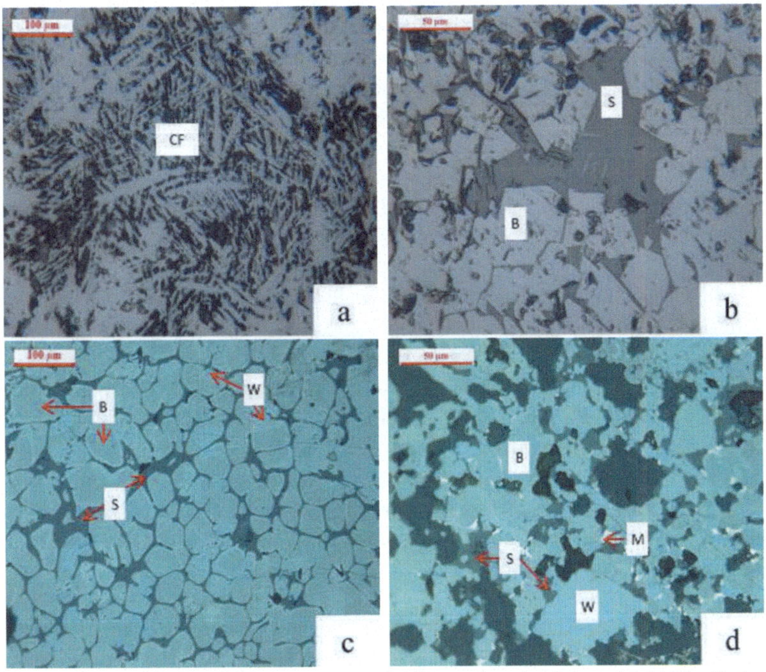

CF-Acicular ferrite calcium, W-Wustite (FexO), S-Silicate mineral, B-Magnetite, M-Metallic iron

(a, b) Traditional sinter; (c) Reductive sinter; (d) Not molten pellet

Figure 3. Different binder phases in finished sinter

Conclusions

Reductive sintering with CO circulation makes all indexes better except the *VSV* and *TI*. CO-containing flue gas circulation sintering process combined with carbon-bearing pellets as a new reductive sintering technology is adopted to enhance the removal of Na, K and reduction degree of iron oxides. Higher CO content in inlet gas and larger coke breeze particles in pellets are beneficial for improving removal ratios of Na, K. When O_2 and CO content are 15% and 6%, and coke breeze size in pellets is 0.15-0.3mm, it can achieve that removal ratios of Na and K are 20.37% and 63.73%, reduction degree of iron oxides is 23.76%. Through microstructure analysis, unlike the traditional high basicity sinter, silicate mineral is the important binder phase in reductive sinter, and high crystallization degree of them is conductive to remove more of Na and K.

Acknowledgements

The authors wish to express their thanks to the Program for New Century Excellent Talents in University (NCET-11-0515) for financial support of this research. This work was also financially supported by Co-Innovation Center for Clean and Efficient Utilization of Strategic Metal Mineral Resources.

References

1. M. Satoshi, S. Hideaki, and T. Kanji, "Development of the Process for Producing Pre-reduced Agglomerates," *JFE GIHO*, 22 (2008), 25-31.

2. H.Q. Chen, "Development of Reductive Sintering Technology," *International Metal Guides*, (6) (2009), 1-4.

3. J.G. Hu et al., "New Ironmaking Technology of Reducing CO_2 of Japanese Steel Industry," *Metallurgical Collections*, (3) (2011), 47-50.

4. X. Jiang et al., "Technological Analysis of Comprehensive Utilization of Blast Furnace Dust with Pre-reduction Sintering Technology," (Paper presented at the National Ironmaking Production Technology and Ironmaking Academic Annual Meeting, Zhengzhou, 2014), 1169-1173.

5. P. Besta, K. Janovska, and J. Haverland, "Negative Elements in Blast Furnace Process," *Metal*, (5) (2012), 23-25.

6. K. Ouyang et al., "Research Development of Alkali Metal in Blast Furnace," *Journal of Hebei Polytechnic University (Natural Science Edition)*, 33 (1) (2011), 37-41.

7. S.Z. Zhang, F. Zhang, and G.P. Luo, "Influence of Alkali Metal on the Process of Sintering," *Journal of Inner Mongolia University of Science and Technology*, 29 (2) (2010), 108-111.

8. Q. Li et al., "Research on Removal of Potassium and Sodium by Pre-reduction Sintering," (Paper presented at the 4nd International Symposium on High -Temperature Metallurgical Processing, San Antonio, Texas, March 2013), 67-74.

9. Y.Q. Yang, S.P. Lu, and X.G. Duan, "Analysis on Behavior Mechanism of K_2O and Na_2O in Sintering," *Science and Technology of Baotou Steel*, 38 (6) (2012), 33-36.

7th International Symposium on High-Temperature Metallurgical Processing
Edited by: Jiann-Yang Hwang, Tao Jiang, P. Chris Pistorius, Gerardo R.F. Alvear F., Onuralp Yücel, Liyuan Cai, Baojun Zhao, Dean Gregurek, and Varadarajan Seshadri
TMS (The Minerals, Metals & Materials Society), 2016

THE PREHEATING AND ROASTING PROPERTIES OF FLUORINE-BEARING IRON CONCENTRATE PELLETS AND MAIN INFLUENCE FACTORS

Lu Yang, Shuai Wang, Gang-hua Fu, Yu-feng Guo*, Tao Jiang

School of Mineral Processing and Bioengineering, Central South University,

Changsha, Hunan, China, 410083

Keywords: Iron Concentrates, Pellets, Fluorine

Abstract

The fluorine-bearing iron concentrate is one of the refractory raw materials for pelletization, which is mainly used in Baotou iron and steel company with lots of problems, such as narrow roasting temperature range and low compressive strength of products. According to its specific character, the effects of roasting temperature, time on the compressive strength and microstructure of pellets were researched. The results show that low melting substances formed during the roasting was the dominant reason of those problems. High melting substances were added in the roasting to investigate their effects on the roasting properties of fluorine-bearing iron concentrate pellets. Low melting substances were reduced when chromite was added. The compressive strength of products was 3100N/P when the 2% chromite was added at the roasting temperature of 1150 °C, which was increased 700 N/P compared with the pellets without chromite, and the roasting temperature range was expanded by 170 °C.

Introduction

More attention has been paid in recent years to the use of pellets in blast furnace due to their good strength and improved reducibility, swelling and softening melting characteristics. The fluorine-bearing iron concentrate is one of the refractory raw materials for pelletizing, which is mainly used in Baotou iron and steel company. Due to the development of beneficiation technology, the fluorine content is reduced to 0.5% or less [1], with more than 90% of fluorine in the form of fluorite [2]. But it's still hard to be used in production because of narrow roasting temperature range and low compressive strength of products. The quality of these finished pellets plays a vital role in decreasing the fuel rate and increasing the productivity of blast furnace. High swelling index reduces the strength of pellets after their reduction thereby causing irregularities in the furnace like high resistance to gas flow and low smelting intensity [3].The results show that low melting substances formed during the roasting, which was the dominant reason of those problems. High melting substances were added in the roasting to investigate their effects on the roasting properties of fluorine-bearing iron concentrate pellets.

Experimental

Raw Material

The following materials were used for preparing the green pellets: iron concentrates, chromite, dolomite, bentonite. The main chemical compositions of raw materials are shown in Table I. They were ground separately to get the required fineness for pelletizing. Pelletizing property and grain size distribution of fluorine-bearing iron concentrate are shown in Table II. Physical properties of bentonite are shown in Table III.

Table I. Mainly Chemical Compositions of Raw Materials

compositions	TFe	FeO	Cr_2O_3	SiO_2	Al_2O_3	CaO	MgO	F	K_2O	Na_2O
iron concentrate	61.91	26.36	-	4.93	1.74	3.04	0.51	1.56	0.290	0.156
chromite	20.55	-	40.60	4.96	14.70	0.39	11.13	-	-	-
Dolomite	0.56	-	-	0.75	0.71	33.50	22.29	-	-	-
Bentonite	2.36	0.22	-	54.58	9.90	5.18	2.72	-	0.86	1.68

Table II. Pelletizing Property and Grain Size Distribution of Fluorine-bearing Iron Concentrate

Pelletizing index	The specific surface area /($cm^2.g^{-1}$)	<0.074mm/%	<0.037mm/%
0.62	1683.37	85.19	70.52

Table III. Physical Properties of Bentonite

Moisture /%	Montmorillonite / %	<0.074mm/%
7.455	82.75	73.23

Method

The main process of this study was as follows: raw material testing-roasting experiment - mechanism analysis. Green pellets were prepared using a laboratory balling disc with a diameter of 1000 mm and a tilting angle of 45°. Before firing, pellets were dried in a hot oven at 110 °C for 5 hours to ensure that all the moisture is removed. Then pellets were preheated and roasted at various temperatures with free-access air. Preheating and roasting were carried out using tubular electric furnace. After a period of time, the roasted pellets were drawn out of the reactor and cooled down to the room temperature. The preheated and roasted pellets were examined with their compressive strength and optical microscope was used to analyze the mechanism. Properties of pellets were tested by scanning electron microscope, optical microscope and X-ray diffraction analysis methods, and so on.

Results and Discussion

Preheating and Roasting Properties

To study the effects of preheating temperature and time on compressive strength of fluorine-bearing iron concentrate pellets, several experiments were conducted. Fig.1 and Fig.2 indicated that the compressive strength of preheated pellets increased with the increasing preheating temperature from 850°C to 980°C, while the compressive strength of roasted pellets decreased from 920°C. Low melting substances formed in the pellets during the roasting with a tendency to produce liquid. Pellets cracked in the process of cooling due to different shrinkage ratio of solid and liquid, leading to a weak compressive strength of pellets. Prolonging preheating time was helpful to improve compressive strength of pellets, as enough preheating time ensures the complete oxidation of Fe_3O_4. In order to make the oxidation completely react, the preheating temperature and time were selected to be 920°C and 16min respectively.

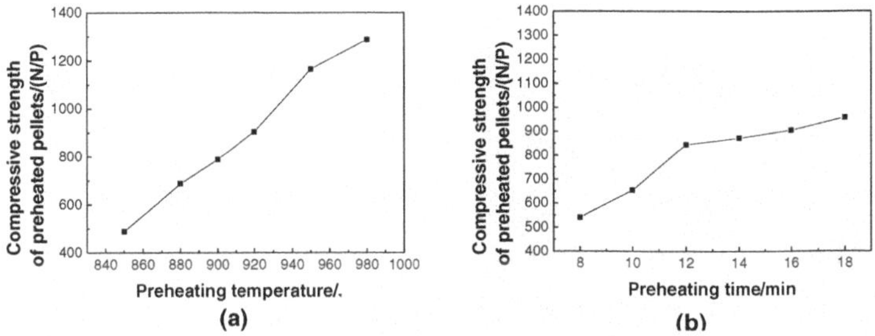

Fig.1. Effect of preheating temperature and time on compression strength of preheated pellets

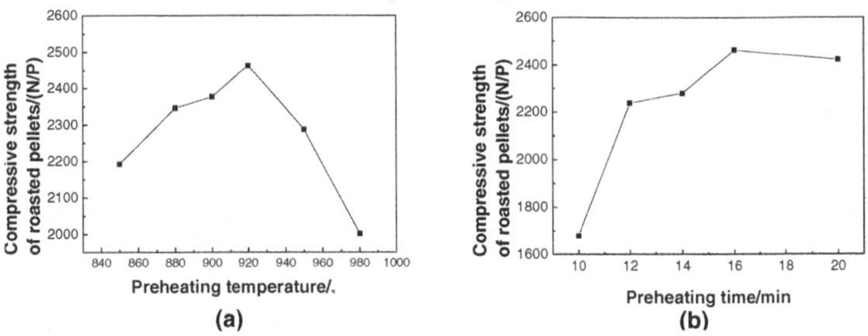

Fig.2. Effect of preheating temperature and time on compression strength of roasted pellets

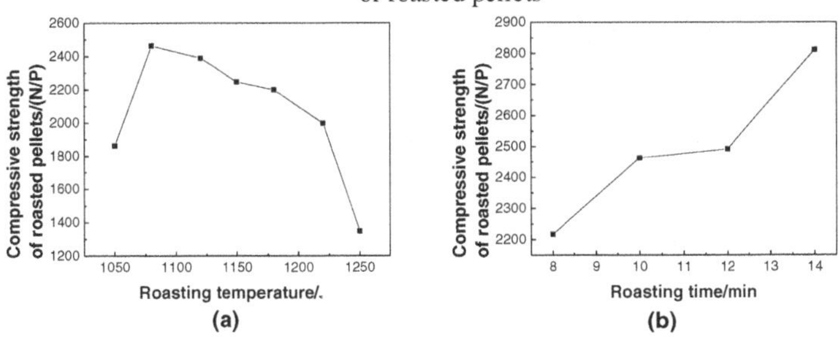

Fig.3. Effect of roasting temperature and time on compressive strength of roasted pellets

The effects of roasting temperature on fluorine-bearing iron concentrate pellets were studied in the temperature range of 1050℃ to 1250℃. Parameters were kept constant during the preheating tests as 920℃ and 16min.

Fig.3 shows that the roasting temperature and time have significant influences on compressive strength of pellets, which decreased as the roasting temperature increased from 1080°C, and increased as the roasting time extended. Extending the roasting time promoted the strength of product ball. Therefore, the low roasting temperature was recommended, and the roasting temperature range was narrow. The appropriate roasting temperature was 1080°C and appropriate roasting time was 10min.

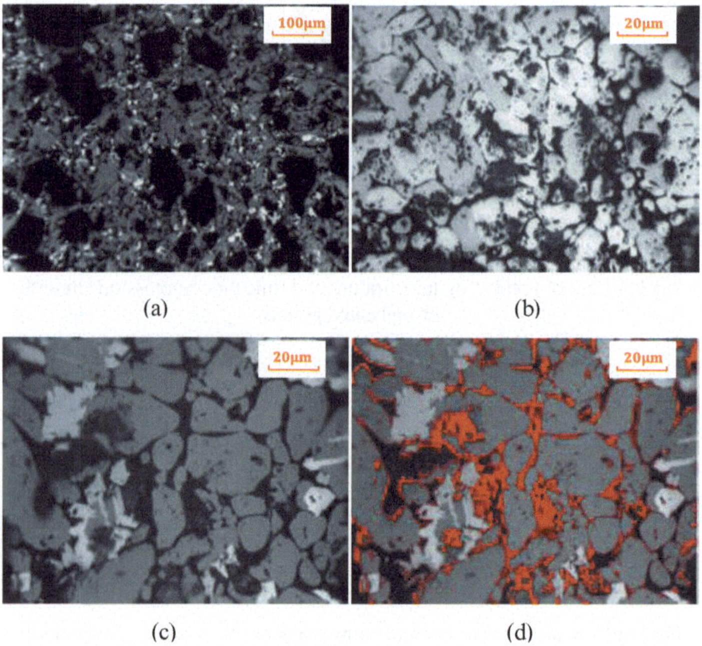

(a) (b)

(c) (d)

Fe_2O_3—white Fe_3O_4—light gray Kirschsteinite—dark gray Holes—black

Fig.4. Microstructure of fluorine-bearing iron concentrate pellets

Microstructure of fluorine-bearing iron concentrate pellets are presented in Fig.4. Fig.4 (a) shows that holes in circular or irregular shape were found in the inner pellets. Main products in the roasted pellets were hematite and magnetite, and there were certain of Fe_3O_4 in the inner layer of pellet as oxidation incompletely reacted as shown in Fig.4 (b). Amount of Kirschsteinite, which acts as binding phase, was found to be increase with increasing temperature, as was shown in Fig.4 (c) and (d). When pellets were roasted in 1100°C and 1250°C, the amounts of liquid phase were 13% and 16% respectively calculated by using the software of Image - Pro Plus. In general, the appropriate amount of liquid phase is 5%-7%. According to Fig.5, the phase diagram of SiO_2-CaO-CaF_2 system, the temperature of generating liquid phase reduced with the increasing content of CaF_2. Calcium fluoride included in iron concentrate had bad effects on compressive strength of pellets and metallurgical properties.

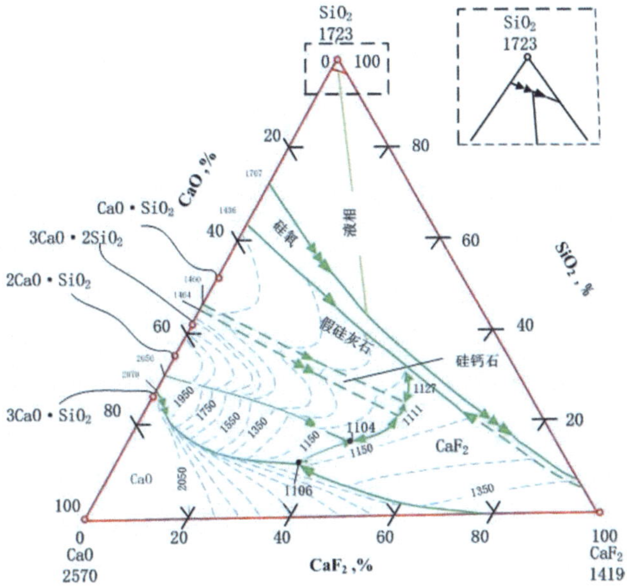

Fig.5. Phase diagram of SiO_2-CaO-CaF_2 system

The metallurgical properties were tested and Table IV shows that metallurgical properties of roasted fluorine-bearing iron concentrate pellets were not good.

Table IV. Metallurgical properties of pellets

RI/%	$RDI_{+3.15}$/%	RSI/%
56.26	96.22	22.76

Additives

High melting substances, chromite and dolomite, were added in the roasting to investigate their effects on the roasting properties of fluorine-bearing iron concentrate pellets. Fig.6 and fig.7 indicate that dolomite had little effect on compressive strength of pellets, while chromite can obviously improve roasting properties. The dolomite decomposed in 730 ~ 760 °C and 760 ~ 810 °C, the decomposition of CO_2 would affect oxygen content on the surface of the pellets and prohibited the oxidation of Fe_3O_4. Fig.7 shows that chromite can effectively broaden roasting temperature range. The compressive strength of products was 3100N/P when 2% chromite was added at the roasting temperature of 1150 °C, which increased 700 N/P compared with the pellets without chromite, and the roasting temperature range was expanded by 170 °C. Low melting substances were reduced when chromite was added.

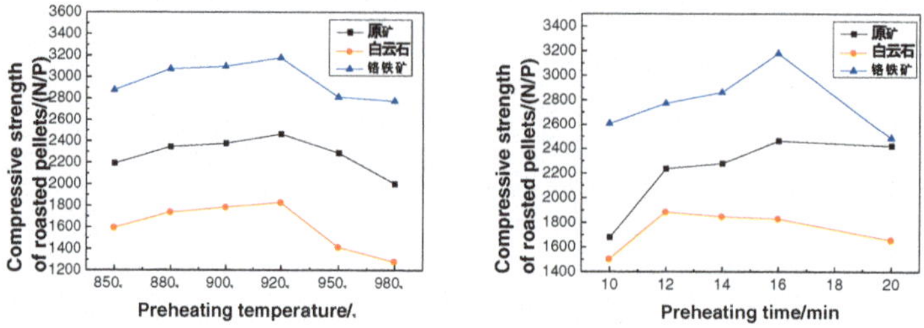

Fig.6. Effect of preheating temperature on compression strength of roasted pellets

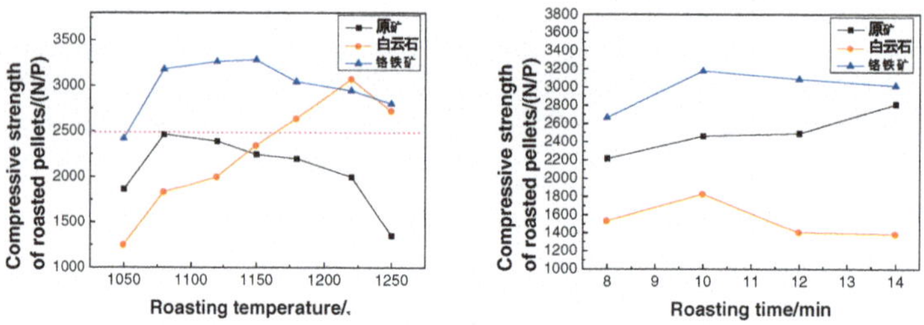

Fig.7. Effect of roasting temperature on compression strength of roasted pellets

Conclusions

The following conclusion can be drawn from this work:

There were problems of fluorine-bearing iron concentrate pellets, such as bad roasting properties at high temperature, narrow roasting temperature range and low compressive strength of products. Appropriately prolonging preheating time was helpful to improve compression strength of preheated and roasted pellets.

It's easy to generate liquid phrase because of calcium fluoride included in iron concentrate. When pellets were roasted at 1100°C and 1250°C, the amounts of liquid phase were 13% and 16% respectively. The compressive strength of pellets decreased with higher roasting temperature when there were more liquid phases.

Chromite addition considerably improved the roasting properties of the pellets. The compressive strength of products was 3100N/P when the 2% chromite was added at the roasting temperature of 1150 °C, which increased 700 N/P compared with the pellets without chromite, and the roasting temperature range was expanded by 170 °C.

References

1.Ji Junmei, "Beneficiation Test Research for Reducing Fluorine Content in Iron Concentrate of Bao Steel's Concentrator", Metal Mine, 7(2004), 35-37.

2.Wu Huling, "Study on Reasonable Allocation Technology of Iron-making Raw Materials in BaoGang Factory" (Ph.D.thesis, Northeastern University,2006),22-23.

3.Xu Bing, Xiong Ling, Yang Yong-bin, Jiang Tao, Li Qian, Wang Jie-Chao, "Study on Fluorie-Bearing Iron Concentrate Pellets Added with Dolomite", Iron and Steel, 11(2008),9-15.

7th International Symposium on High-Temperature Metallurgical Processing
Edited by: Jiann-Yang Hwang, Tao Jiang, P. Chris Pistorius, Gerardo R.F. Alvear F., Onuralp Yücel, Liyuan Cai, Baojun Zhao, Dean Gregurek, and Varadarajan Seshadri
TMS (The Minerals, Metals & Materials Society), 2016

Performance Monitoring for Grate-kiln-cooler Process Based on Quality Prediction and Statistical Analysis

Gui-ming Yang, Xiao-hui Fan, Xiao-xian Huang, Xu-ling Chen

School of Minerals Processing & Bioengineering, Central South University, No.932, South Lushan Road, Changsha, Hunan, 410083, P.R.China

Keywords: Grate-kiln-cooler; Performance monitoring; Quality prediction; Statistical analysis

Abstract

Grate-kiln-cooler is a predominant process of producing iron ore pellets in China. Since this process is of multi-variant, strong coupling and large hysteresis characteristics, a reliable performance monitoring is thus essential to ensure product quality and economic benefits. Aiming to efficiently monitor the pellet induration within this process, this study proposed an approach that combines statistical analysis and quality soft-measurement. Soft-measurement of pellet quality was accomplished based on the predicted time-temperature profile, and the soft-measured indices were on-line monitored using statistical control charts. Based on the proposed approach, a monitoring system for grate-kiln-cooler thermal process was developed for a domestic pelletizing plant. Running results demonstrate that our system can assist in the detection or diagnosis of abnormal thermal state and finally helps to stabilize the pellet production.

Introduction

Iron ore pellets are of great value to the technological & economic criteria of BF iron-making. Chinese iron ore pellet production increased to 131 million ton in 2014, which accounts for 30% of the global production. Iron ore pellets are produced mainly by three processes, since grate-kiln-cooler process has the advantages such as large handling capacity and good adaptability to various materials, and it has become a predominant process recently in China. Grate-kiln-cooler process is an integration of three devices: traveling grate for drying & preheating, rotary kiln for roasting and circular cooler for cooling. Despite the improving automatic control level, this process still encounters with control problems due to the process characteristics of multi-variant, strong coupling and large hysteresis.

A main task of operating grate-kiln-cooler process is to guarantee a good pellet consolidation. However, pellet induration within three main devices cannot be revealed straightforward, which impedes the better control of pellet quality. Based on the model established by Batterham [1], an integral of time-temperature profile to an extent represents the development of pellet quality. In this sense, pellet quality can be estimated via the easily measured process parameters, and the soft-measured quality can be monitored online. The purpose of performance monitoring is to enhance economic benefits by detecting the production abnormality. At present, monitoring and diagnosis methods for complicated industrial processes can be roughly classified as: process model based method, symptom or knowledge based method, and data based method. Data-driven method is to extract the patterns that distinguish normality and abnormality using approaches such as statistical analysis, clustering analysis, signal analysis, etc [2]. Since noise in industrial data is unavoidable, data-driven statistical process monitoring (SPM) or statistical process control (SPC) plays pivotal role in performance monitoring & diagnosis [3-5].

This paper aims to propose an effective approach to compositely monitor or diagnose the thermal state of grate-kiln-cooler process. Two indices—preheated pellet strength and roasted pellet strength were firstly predicted based on the pellet consolidation model, then control charts for these two indices were accomplished to monitor the process.

Monitoring problems in grate-kiln-cooler process

Typical grate-kiln-cooler process is depicted in Figure 1. Traveling grate is generally divided into four compartments, namely up-draught drying (UDD), down-draught drying (DDD), tempered preheating (TPH) and preheating (PH). Circular cooler is generally divided into three sections to obtain hot waste gas of different temperature, namely first section of cooling (C1), second section of cooling (C2) and third section of cooling (C3). Roasting of iron ore pellets takes place in rotary kiln with "kiln end" connected to PH and "kiln head" connected to C1. Primary air together with fuel (pulverized coal or natural gas) is injected into rotary kiln through burner-blower to supply heat for the entire process. To enhance heat exploitation, hot waste gas from the cooler is re-circulated to the grate and the kiln respectively, and hot waste gas from the kiln is re-circulated to the grate. The strong coupling feature, that is, there exists material & heat exchange among three devices and variation in one device will influence thermal states of the others, undoubtedly increases the process control difficulty.

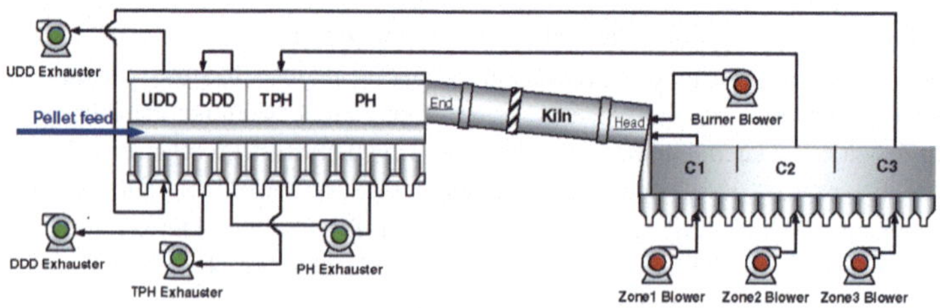

Figure 1. Schematic figure of typical grate-kiln-cooler process.

Thermal state within three devices is the most important monitored variable since it directly influences the product quality and energy consumption. However, temperature profile of iron ore pellets can be hardly online measured due to the semi-enclose and non-transparency of this process. A complete simulation model for grate-kiln-cooler process has been established in details in our previous paper [6], thus time-temperature profiles of iron ore pellets can be predicted at given measured values of process parameters. Based on the profiles, preheated & roasted pellet qualities are further proposed in this study to better monitor the process.

Soft-measurement of the monitored indices

One of the most important pellet qualities is the pellet strength. For acid iron ore pellets, strength development is associated with sintering reactions accompanied by an increase in pellet density and a decrease in volume. As heating time and heating temperature increase, pellet strength undoubtedly improves. According to this assumption, time-temperature profiles of pellets within traveling grate and rotary kiln need to be obtained firstly.

Prediction of time-temperature profiles

Hot gas in each compartment of traveling grate conducts heat exchange with pellets as it goes through the bed perpendicularly. Besides, certain physical & chemical phenomena may occur, including evaporation and condensation of moisture, oxidation of magnetite, desulphurization and bed slump caused by pellet shrinkage (for acid iron ore pellets). The entire model can be expressed as heat & mass balance equations incorporated with kinetic expressions of the above reactions. Neglecting the desulphurization and bed slump, heat balance equation for gas phase is expressed as Eq. (1) and heat balance equation for solid phase is expressed as Eq. (2).

$$GC_g \frac{\partial T_g}{\partial y} = -h_{eff} A(T_g - T_p) - 0.5 \times R_w \Delta H_w + 0.5 \times (R_m \Delta H_m + R_{cd} \Delta H_w) \tag{1}$$

$$\rho_b C_p \frac{\partial T_p}{\partial t} = h_{eff} A(T_g - T_p) - 0.5 \times R_w \Delta H_w + 0.5 \times (R_m \Delta H_m + R_{cd} \Delta H_w) \tag{2}$$

where G is superficial gas flow rate, ρ_b is bulk density of pellet bed, C_g is specific heat of gas, C_p is specific heat of pellet, T_g is gas temperature, T_p is pellet temperature, z is bed height, t is time, h is effective heat transfer coefficient, A is gas-pellet apparent contact area, $R_w(R_{cd})$ is rate of water evaporation/condensation, ΔH_w is enthalpy of evaporation/condensation, R_m is rate of magnetite oxidation, ΔH_m is enthalpy of magnetite oxidation, λ is heat distribution coefficient.

Unlike traveling grate, heat exchange within rotary kiln belongs to counter-current heat transfer. Pellets move forwards and circumferentially simultaneously, meanwhile, the injected fuel combusts and hot gas exchanges heat with the moving pellets. During modeling, one dimensional approach is used and rotary kiln is divided into a number of slices each of which has a uniform temperature. Taking the pseudo steady-state operation within each subdivided slice, temperature profiles of gas, solid and wall can be obtained along kiln axis by adopting heat-balance equations, as expressed in Eq. (3)-Eq. (6).

$$-m_g C_g \frac{dT_g}{dz} = R_{fuel}(z)\cdot \Delta H_{fuel} - (\Phi_{rgw} + \Phi_{rgs} + \Phi_{cgw} + \Phi_{cgs}) \tag{3}$$

$$m_s C_p \frac{dT_p}{dz} = \Phi_{rgs} + \Phi_{cws} + \Phi_{cgs} + \Phi_{rws} \tag{4}$$

$$\Phi_{cgw} + \Phi_{rgw} - \Phi_{cws} - \Phi_{rws} = \Phi_{rwa} + \Phi_{cwa} \tag{5}$$

$$A_{wi}\cdot (T_{wi} - T_{wo})/(\frac{b_1}{\lambda_1} + \frac{b_2}{\lambda_2}) = \Phi_{rwa} + \Phi_{cwa} \tag{6}$$

where m_g is mass flow rate of gas, m_s is mass flow rate of solid, z is position along kiln axis, R_{fuel} is rate of fuel combustion, A_i is area per unit kiln length, b_1 is thickness of refractory bricks, b_2 is thickness of steel shell, λ_1 is thermal conductivity of refractory bricks, λ_2 is thermal conductivity of steel shell, Φ_i is heat flux per unit kiln length. Subscripts *wi, wo,* represent inner wall and outer wall. Subscripts *rgw, rgs, rws, rwa* represent gas-wall, gas-solid, wall-solid, wall-atmosphere radiation respectively, and subscripts *cgw, cgs, cws, cwa* represent gas-wall, gas-solid, wall-solid, wall-atmosphere convection respectively.

In traveling grate and rotary kiln, chemical compositions of both gas and solid will change with the reaction rate. Mass balance is adopted at the same time during the calculation of the temperature profiles. Regarding to residence time, it is assumed that residence time in the grate is distributed uniformly along grate length and residence time in the kiln is distributed uniformly along kiln length. Total residence time within rotary kiln is empirically estimated by Eq. (7).

$$t_2 = \frac{1.77\sqrt{\alpha}\cdot L_2}{\beta D_i n} \tag{7}$$

where t_2 are total residence time within grate and kiln respectively, L_2 are the length of grate and kiln respectively, α is dynamic repose angle of pellets, β is the inclination angle of the kiln, D_i is the inner diameter of the kiln, n is rotational speed of the kiln.

Prediction of pellet strength

Batterham [1] proposed a formulation to describe the development of pellet strength, which is similar to the rate expression of first order kinetic reaction, as given in Eq. (8). By integrating Eq. (8), one can obtain an online model to monitor or control the grate-kiln-cooler indurator.

$$\frac{dQ}{dt} = k(Q_f - Q) \text{ with } k = \frac{A_0}{T}\exp(\frac{-E}{T}) \tag{8}$$

$$Q_p = \int_0^t \frac{A_0}{T}\exp(\frac{-E}{RT})\cdot (Q_f - Q)dt \tag{9}$$

where Q is some quality parameter, t is heating time, k is a temperature-dependent rate parameter,

Q_f is the final quality that would be attained after a long period of sintering at temperature T, A_0 is a constant for a given pellet mixture. By applying the strength model, we can predict the values of two monitored indices—preheated pellet strength and roasted pellet strength based on the time-temperature profiles of traveling grate and rotary kiln respectively.

Performance monitoring

Statistical control charts

Control charts were firstly developed and applied by W.A. Shewhart in 1924. The charts exhibit the fluctuation curve of product quality varying with time and analyze the reason that causes the fluctuation, and they assist operators in making appropriate judgments and proposing effective measures towards the abnormality. Control charts are mainly divided into variable-data control charts and attribute control charts. Preheated pellet strength and roasted pellet strength are the monitored objectives in this study, therefore the online monitoring & diagnosis is based on variable-data control charts. There are four types of variable-data control charts: mean-range charts ($\bar{x}-R$), mean-standard deviation charts ($\bar{x}-S$), median-range charts ($\tilde{x}-R$), and value-moving range charts ($x-R_s$).

A case is given in this section to illustrate the main steps of making control charts for two indices—preheated pellet strength and roasted pellet strength.

Data preparation The data were collected from field DCS of a domestic industrial grate-kiln-cooler process. Soft-measurement cycle was determined at 30 min, the values of process parameters in the past 30 min were averaged during each computation. According to the mathematical models and collected process parameters, predicted time-temperature profiles at a given time is depicted in Figure 2. As seen, total residence time of preheating was calculated at 4 min and total residence time of roasting was calculated at 20.86 min. At the beginning of PH section, the temperature of pellets (discharged from TPH section) was about 400°C, and after preheating it reached at 930°C. Pellets were further heated in rotary kiln, the highest temperature (~1250°C) occurred at around one third of kiln length away from kiln head. When the roasted pellets were discharged from kiln head, the pellet temperature retained 1200°C. Based on this profile, preheated pellet strength and roasted pellet strength were predicted.

Control chart preparation The fundamental form of control chart is the varying trend of product quality index together with the control limits calculated via statistical methods. Types of control limits include: upper control limit (UCL), lower control limit (LCL), and center line (CL). The final control charts for soft-measured pellet strength were depicted in Figure 3.

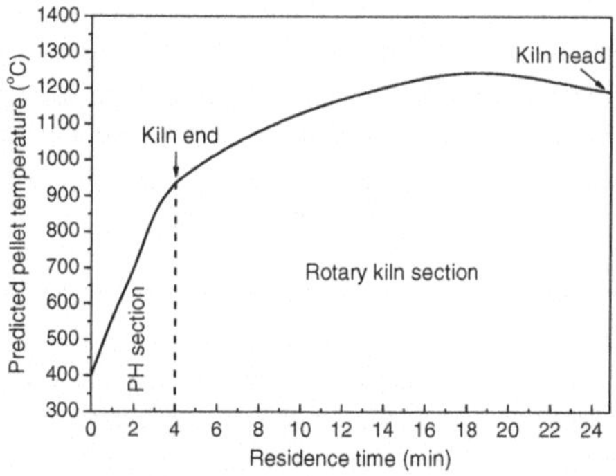

Figure 2. Time-temperature profiles of preheating and roasting at a given time.

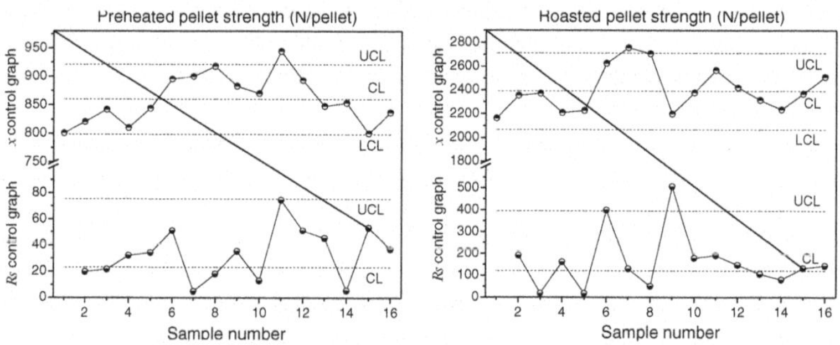

Figure 3. $x-Rs$ control charts of preheated and roasted pellet strength.

Diagnosis and remediation of abnormality

Utilization of the control charts to monitor the production state mainly refers to the analysis and judgment of production state according to the locations and varying trend of data points. The abnormal state can be represented by the following two situations: 1) the data points are out of the control limits, and 2) the data points are within the control limits, but are in anomaly permutation. When data points are out of the allowable control limits, there exists abnormal phenomenon in the production probably. Operators should look into the cause of abnormality and propose the countermeasure under these circumstances.

When predicted pellet strength is abnormal, it is important to find out the cause and deal with the abnormality. Based on our understanding of the process characteristics, the diagnostic inference procedure is proposed and shown in Figure 4. The inference system firstly reads the values of process parameters and predicts the preheated & roasted pellet strength, and uses the above abnormal rules to identify whether the process state is abnormal. If the state is not normal, the inference system will investigate the pellet temperature and residence time respectively for two devices. If pellet temperature is too high or too low, then the system looks into the parameters such as income gas temperature and gas flow rate, if these parameters are out of the allowable constraints, manipulated parameters such as fuel injection rate, primary & secondary air flow rate will be adjusted. If pellet temperature is normal but the residence time is abnormal, moving velocity of the grate and rotational speed of the kiln should be modified, but the adjustment range depends on the planned handling capacity.

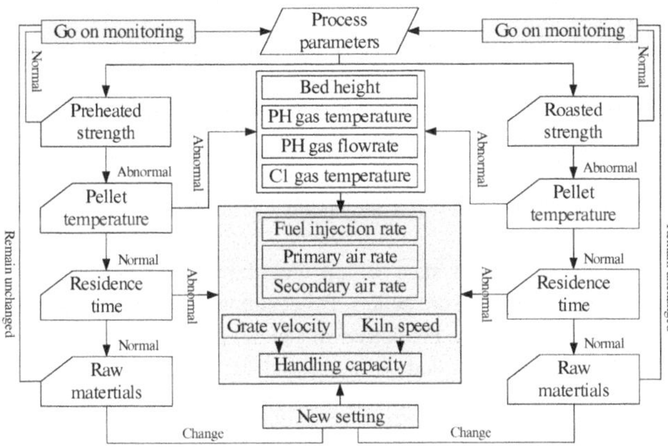

Figure 4. Diagnosis procedure (Gray regions represent manipulated parameters or actions).

Given the chart shown in Figure 3 and diagnosis procedure shown in Figure 4, preheated pellet strength at the No.11 data point is abnormally high and its moving range is also close to R_S upper control limit. Under this circumstance, the system identifies that soft-measured pellet temperature is too high while residence time of pellets within the grate has little changed. Four main variables—bed height, gas temperature in PH hood, air flow-rate of PH section, gas temperature of C1 hood are then compared with the their allowable ranges. In this case, the abnormality is attributed to high PH gas hood temperature, and the diagnostic system suggests lowering the coal injection rate from kiln head. Regarding to No.15 data point, preheated pellet strength is only about 800 N/pellet, which is close to the lower control limit. The system identifies that residence time is rational but soft-measured pellet temperature is low. By comparing the values of four variables with their normal ranges, bed height seems to be too high. At the same moving velocity of traveling grate, an increase in bed height indicates an increase of throughput per hour. When income gas temperature and air flow-rate remain unchanged, the heat received by a single pellet is reduced and thus pellet temperature decreases. In this case, the diagnostic system suggests the operators to reduce the throughput immediately or increase the coal injection rate to provide more heat.

System development & application

The monitoring contents described above were embedded into a level-2 optimized expert system developed for a large-scale pelletizing plant in China. The predicted time-temperature profiles, the control charts of preheated pellet strength and the control charts of roasted pellet strength were on-line updated with the measured process parameters at a certain time interval. The diagnostic results together with suggestions on remedying the abnormality were given in statements and were also on-line updated at a certain time interval. In the display of control charts, collecting method of the data, the recalculation of control limits and the display mode can be selected or modified by the user as well. The entire system has been applied in the plant since Oct. 2013. It aims to provide the operators with guidance to better control the process. According to the field technicians, the system helps to stabilize the thermal process by discovering the latent abnormality of thermal state in time.

Conclusion

Performance monitoring of thermal state within grate-kiln-cooler process is fundamental to steady & efficient production of iron ore pellets. Pellet strength is an important monitoring index but it cannot be on-line measured. Predictive model of preheated pellet and roasted pellet strength was proposed in this study, and the control charts of these two indices were used to monitor the process. A monitoring system containing this approach was applied in a pelletizing plant, and the running results demonstrated its reliability.

References

1. R.J. Batterham, "Modeling the development of strength in pellets," *Metallurgical Transactions B*, (17) 1986, 479-485.
2. X. Zhang, "Performance monitoring, fault diagnosis and quality prediction based on statistical theory," (Ph.D. Thesis, Shanghai Jiao Tong University, Shanghai, 2006).
3. J. Kresta, J.F. MacGregor and T.B. Marlin, "Multivariate statistical monitoring of process operation performance," *Canadian Journal of Chemical Engineering*, (69) 1991, 35-47.
4. J.F. MacGregor and T. Kourtit, "Statistical process control of multivariate proccss," *Control Engineering Piactice*, 3 (3) 1995, 403-414.
5. J. Zhang, E.B. Martin and A.J. Moris, "Process monitoring using non-linear statistical technique," *Chemical Engineering Journal*, (67) 1997, 181-189.
6. X. Fan et al., "Predictive models and operation guidance system for pellet induration in traveling grate-rotary kiln process", *Computers and Chemical Engineering*, (79) 2015, 80-90.

7th International Symposium on High-Temperature Metallurgical Processing

Utilization of Complex Ores

Session Chairs:
Varadarajan Seshadri
Guanghui Li

7th International Symposium on High-Temperature Metallurgical Processing
Edited by: Jiann-Yang Hwang, Tao Jiang, P. Chris Pistorius, Gerardo R.F. Alvear F., Onuralp Yücel, Liyuan Cai, Baojun Zhao, Dean Gregurek, and Varadarajan Seshadri
TMS (The Minerals, Metals & Materials Society), 2016

CHARACTERIZATION OF SULFIDATION ROASTING OF AN IRON-RICH MANGANESE OXIDE ORE WITH ELEMENTAL SULFUR

Tao Jiang, Li Qin, Zhixiong You, Yuanbo Zhang, Guanghui Li[1],

(School of Minerals Processing & Bioengineering, Central South University, Changsha, Hunan 410083, China)

Keywords: Sulfidation; Iron-rich manganese oxide ore; Reduction; Extraction

Abstract

Low-grade manganese oxide ore is generally characterized by high iron content. In this work, sulfidation roasting of an iron-rich manganese oxide ore with elemental sulfur was characterized. The experimental results indicated that manganese dioxide can be selectively reduced over iron oxide by using elemental sulfur. The manganese phases in roasted product included low-valence manganese oxide, manganese sulfide and manganese sulfate, while those of iron mainly depended on the sulfur addition. Iron oxide would only be reduced or sulfidized at larger sulfur additions. The effect of roasting temperature and S/Mn mole ratio on the phase transformations was also investigated; extractions of Mn/Fe by sulfuric acid were evaluated. An extraction of Mn greater than 95% was obtained while that of Fe was controlled less than 10% under the optimal conditions.

Introduction

Manganese plays an important role in steel production, chemical industry, cells and fine chemicals, dietary additives and some other applications [1]. With the depletion of high-grade manganese ores, developing economic and efficient process to recover manganese from low-grade manganese oxide ores is of great importance.

Manganese dioxide, as the major manganese mineral in manganese oxide ores, is insoluble in dilute sulfuric acid and alkaline. In order to extract manganese from the low-grade manganese oxide ores, reduction process is indispensable prior to leaching process. To date, various routes have been reported for the reduction leaching of manganese oxide ores in the existing literature. Generally speaking, the main two methods are divided into direct acid leaching in the presence of reducing agents and reduction roasting followed by acid leaching.

There are diverse reducing agents used in the direct reductive leaching process, such as pyrite, oxalic acid [2], glucose [3], aqueous sulfur dioxide [4], metallic iron and ferrous sulfate [5] and biological bacteria reduction leaching [6]. The main problems with these hydrometallurgical processes are the purification of manganese from the leaching solution, low leaching efficiency as well as co-leaching of impurities (especially Fe).

The reductants used for reduction roasting–acid leaching include coal, pyrite [7], graphite and cornstalk [8]. However, the above methods often not only require high energy consumption

[1] Corresponding author: Dr.Guanghui Li, Email: liguangh@csu.edu.cn, Tel: +86-731-88830542

(the reduction temperature is 750-900°C), but also generate a great deal of dust and poisonous gases. Therefore, those processes are not economic and environmental-friendly.

As reported, manganese dioxide can be reduced by sulfur at 300-400 ℃, which is an extremely lower temperature range [9, 11, 14]. Manganese can be leached from the roasted products by using sulfuric acid as the leaching agent. The process of manganese oxide ores by sulfur–based reduction roasting–acid leaching is a new approach for efficiently utilizing low-grade manganese oxide ores [10]. However, the sulfur based reduction process and selectivity of reducing manganese dioxide over iron oxide are not well understand.

In the present work, chemically pure sulfur is used to reduce manganese oxide ore, which is characterized by high iron content. Thermodynamics of sulfur-based reduction is previously studied. Then the sulfidation process is characterized on the basis of phase transformation and extraction of Mn/Fe. It is meaningful to improve the leaching efficiency of manganese and reduce that of iron.

Materials and Methods

Materials

Iron-rich manganese oxide ore used in this study were obtained from Jiangxi Province, China. The sample was crushed, ground and screened to 80% passing through 0.074 mm in advance. The chemical composition of the ore material is given in Table I, It shows that the total manganese grade of the sample is only 14.30% and iron grade is up to 30.35%, which is an iron-rich manganese oxide ore. The XRD pattern of the sample, presented in Figure 1, indicates that manganese mainly existed in the form of manjiroite ($(K, Na)Mn_8O_{16}\cdot xH_2O$) and that the main gangue minerals are goethite ($FeO(OH)$) and quartz (SiO_2).

Table I Main Chemical Composition of Manganese Ore/wt. %

TMn	TFe	SiO_2	CaO	MgO	Al_2O_3	P	S	K_2O	Na_2O	LOI*
14.30	30.35	18.63	0.22	0.92	2.21	0.28	0.18	2.14	0.12	10.07

*LOL: loss on ignition.

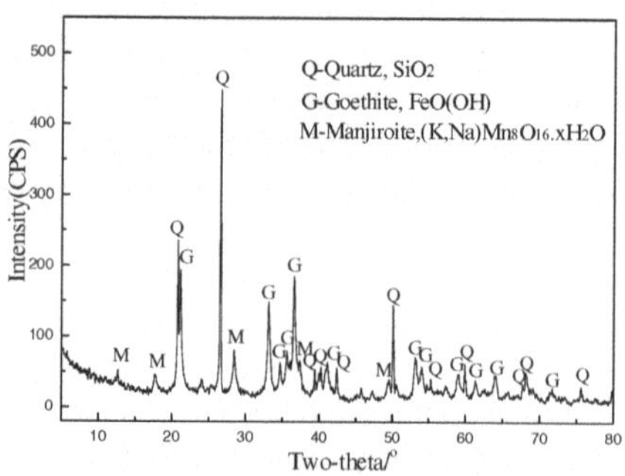

Figure. 1 XRD pattern of iron-rich manganese oxide ore

The purity of the chemically pure sulfur used as a reductant is above 99%, and the particle size of sulfur is 100% passing through 0.074 mm.

Methods

The ground manganese ores and chemically pure sulfur were put together into a sealed stainless steel reactor after being thoroughly mixed. They were then roasted at given temperature for 30 min in an electrically heated muffle furnace. After that, the roasted samples staying in the sealed reactor were cooled to room temperature in air. Finally, the cooled samples were discharged from the reactor then were prepared for analysis [11].
In each leaching test, 10 g of the ground samples was put and leached in a beaker with certain sulfuric solution, which was placed into a water bath. After the leaching process was finished, the leaching solutions were filtered. Then, the leached residues were washed three times with distilled water and dried for chemical analysis. The Mn and Fe content of reduced product and acid leach residue both were analyzed.
The leaching efficiency of Mn/Fe can be calculated as follows:

$$\gamma = (1 - \frac{m_1 \times \beta}{m_0 \times \alpha}) \times 100\%$$

Where γ is the leaching efficiency, %; m_0 is the weight of reduced product, g; α is the Mn/Fe grade of reduced product, %; m_1 is the weight of leached residue, g; and β is the Mn/Fe grade of leached residue, %.
The effects of sulfur addition on sulfidation extent were determined by the bromine–methanol leach. The phase transformation in the roasting process was measured by the using of X-Ray Diffraction.

Results and Discussion

Thermodynamics of Sulfur-based Reduction

As well known, there are many chemical reactions between iron-rich manganese oxides and sulfur during the roasting process under the anaerobic condition. With different sulfur addition, the equilibrium compositions of roasted products are different. Assumption of the manganese in manganese ore all exist in the form of manganese dioxide (22.62%) and the iron all exist in the form of ferric oxide (43.36%). The mole ratio of iron and manganese is close to 1:1. The effect of S/Mn ratio on the equilibrium composition of roasted products at different roasted temperatures is shown in Figure. 2.
As shown in the left part of Figure 2, with the increase of sulfur addition, the amount of manganese dioxide and iron oxide decrease rapidly and the manganese sulfate gradually increases at the roasted temperature of 400 °C. Some other high-valence manganese oxides increase first then reduce with the increasing sulfur ratio.
In the right part of Figure 2, it shows that SO_2 increased significantly with higher sulfur ratio at 550 °C. Compared with 400 °C, the proportion of manganese sulfate reduced and the MnS is on the increase. At the same time, the equilibrium composition of MnO is higher than in

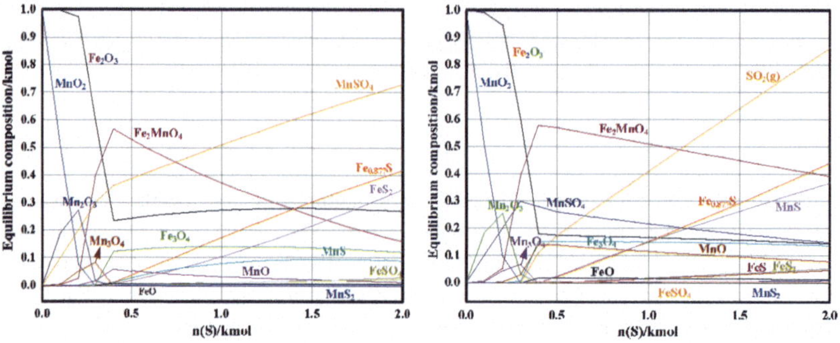

Figure. 2 Effect of S/Mn on the equilibrium composition of roasted products (1 kmol MnO_2 + 1 kmol Fe_2O_3, (left): 400 °C, (right): 550 °C)

400 °C for MnS could react with high-valence manganese oxides to form MnO. Figure 2 also shows that the manganspinel increases first then reduces with the increasing sulfur addition. From the above we could know that manganese dioxide can be selectively reduced over iron oxide by using elemental sulfur.

Effect of sulfur addition on sulfidation extent of Mn/Fe

Samples with varying amounts of sulfur were reacted at 400 °C and 550 °C in order to understand the role of sulfur additions on both the manganese and the iron recoveries. Figures 3 and 4 show the results of the sulfidation extent of Mn/Fe, which the degree of manganese is much higher than corresponding iron at all sulfur addition, either 400 °C or 550 °C.

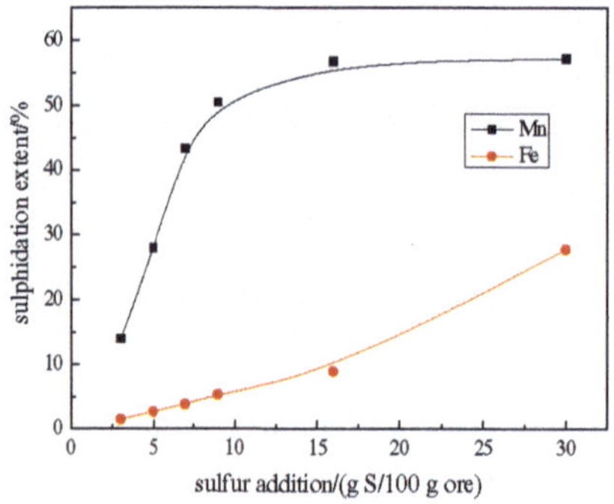

Figure. 3 Effect of sulfur addition on sulfidation extent of Mn/Fe (400 °C)

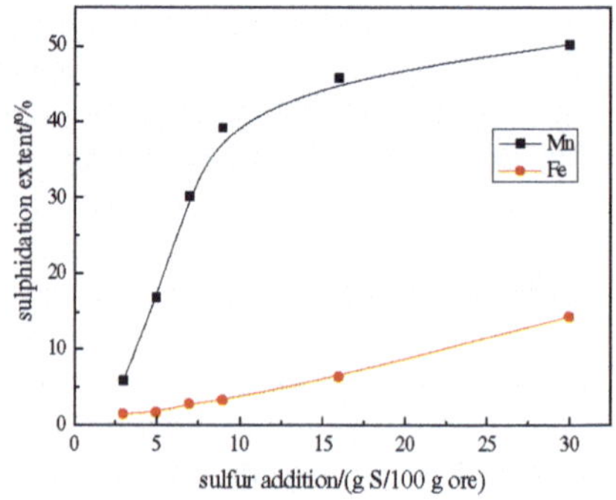

Figure. 4 Effect of sulfur addition on sulfidation extent of Mn/Fe (550 °C)

As shown in Figure 3, with the roasted temperature of 400 °C, the sulfidation extent of Mn was continues to increase at lower sulfur additions from 3% to 9%, and then kept gradually stable. One hand, part of manganese in the process of roasting translates into manganese sulfate and manganous oxide, the sulfidation extent of Mn is unable to be 100%. On the other hand, the manganese sulfide formed on the manganese oxide particles surface hinders the reaction between the internal manganese dioxide and sulfur. The manganese extraction is much higher than the corresponding iron extraction at all sulfur additions, and reaches a constant value of about 58% extraction at the sulfur addition of 16%, whereas the iron extraction continues to increase with the sulfur addition.

Figure 4 shows that with the roasted temperature of 550 °C, the sulfidation extent of Mn was continues to increase at lower sulfur additions, and then kept gradually stable. The reaction mechanism is the manganese sulfide in the roasted product could react with high-valence manganese oxide generated manganous oxide at a higher roasting temperature [12]. Then the high-valence manganese oxide used for sulfuration is less than low temperature and the highest sulfidation extent of Mn is only 40% at sulfur addition of 9%(w.t), which is less than 400 °C. The sulfidation extent of Fe also showed increasing trend with the addition of sulfur. One hand, iron oxides directly react with sulfur as pyrite at low temperature while iron oxide was gradually deacidized with sulfur as pyrrhotite at high temperature. On the other hand, the sublimation rate of sulfur accelerated significantly which result in the decrease of sulfation degree.

Increasing the sulfidation extent of Mn is advantageous to improve the leaching efficiency of manganese. From the above we could know that the process of selective sulphidation of a high iron content manganese ore is feasible, which is meaningful for us to choose suitable reaction conditions.

Effect of sulfur addition on sulfation extent of Mn/Fe

During the roasting process, part of manganese was translated into manganese sulfate and manganous oxide. As we known manganese sulfate could dissolve in the water, so we use the water leaching efficiency to evaluate the effect of sulfur addition on sulfation extent of Mn and the result is shown in Figure. 5.

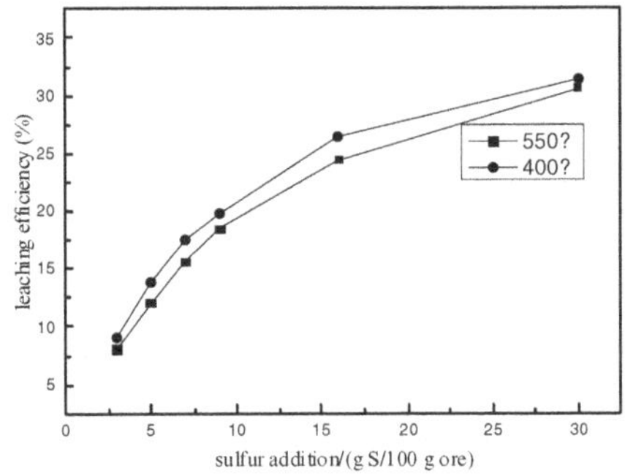

Figure. 5 Effect of sulfur addition on sulfation extent of Mn

Figure 5 shows that the water leaching efficiency of Mn with increasing sulfur addition at both 400 °C and 550 °C. With the same amount of reducing agent added, the leaching efficiency of Mn at 550 °C and 400℃ are nearly equal. The leaching efficiency remains constant when high-valence manganese oxide was completely converted to low-valence manganese oxide.

Phase transformation

Samples with varying amounts of sulfur were reacted at 500 °C, 550 °C and 600 °C to determine the impact of reaction temperature on the phase transformation of manganese ore. The analysis using X-Ray Diffraction and the diffraction patterns are shown in Figure. 6.

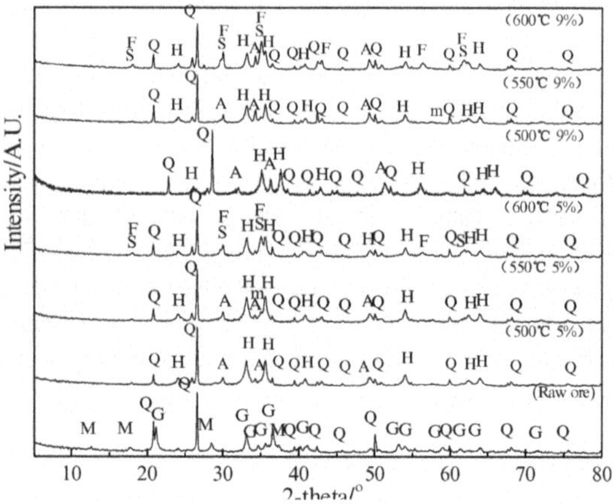

(M-Manjiroite, (K,Na)$Mn_8O_{16}\cdot H_2O$; Q-Quartz, SiO_2; G-Goethite, FeO(OH); H-Hematite, Fe_2O_3; A-Alabandite, MnS; P-Pyrite, FeS_2; m-Manganosite, MnO; S-Manganspinel, $MnFe_2O_4$; F-Magnetite, Fe_3O_4)

Figure.6 XRD patterns of the calcine roasted at different temperature with various addition of sulfur.

As displayed in Figure. 6, Manjiroite (K, Na)$Mn_8O_{16}\cdot H_2O$) and Goethite (FeO (OH)) in the raw ores disappear after the samples are roasted with sulfur, whereas many new-born compounds are generated. At the sulfur addition of 5%, MnS exists at the roasting temperature of 500 °C and 550 °C while at the temperature of 600 °C the diffraction peaks of MnS disappeared for there is not enough sulfur to react with manganese oxide, which the sublimation rate of sulfur accelerated significantly at higher temperature. MnO appears at 550 °C for MnS could react with high-valence manganese to form it [13]. The diffraction peaks of MnO and Fe_2O_3 were weaken for them formed the manganspinel ($MnFe_2O_4$) at 600 °C.

Figure 6 also shows the phase transformation on the addition of sulfur at 9%. At 500 °C and 550 °C, it is shown that there is no obvious change in the roasted product when the amount of sulfur is 5% and 9%. Comparing with the lower addition of sulfur, MnS exist at these three temperatures, including the roasting temperature of 600 °C, for the amount of sulfur is excessive, there would be superfluous manganese sulfide remained. In order to urge the low-valence generated and avoid the generation of MnS, we also need to control the mass of ratio no more than 9%.

From the above we learned that manganese dioxide can be reduced by sulfur at 400~550℃. Compared with other reductant, it could effectively reduce the energy consumption.

Leaching process

Samples with varying amounts of sulfur which roasted at different temperatures were leaching by dilute sulfuric acid. Keeping the leaching conditions as the same(i.e. leaching time, 20 min; leaching temperature, 25 °C; stirring rate, 400 r/min and liquid-to-solid ratio,10:1)[13] ,the effects of the roasting temperature and sulfur addition on leaching efficiency of manganese and iron were showed at Figures 7 and 8.
The leaching efficiencies of manganese in Figure 6 are all enhanced by increasing the sulfur ratio from 3% to 7%, no matter how the roasting temperature changed. When the sulfur addition exceeds 7%, the increase of the leaching efficiency of Mn slows and keeps about 97%. It also shows that the leaching efficiencies increase with the roasting temperatures increasing, though the range is small.

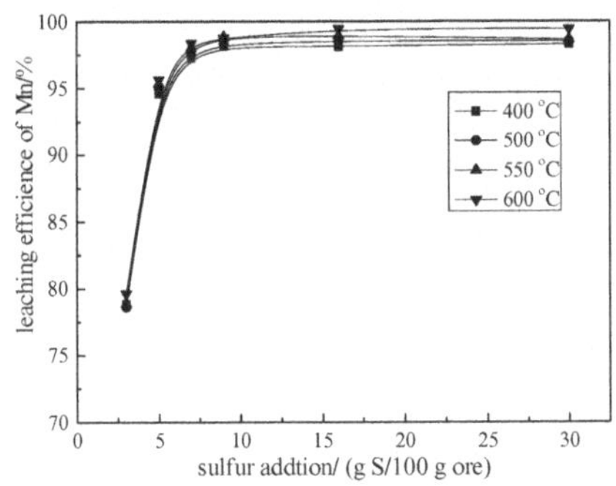

Figure. 7 Effect of sulfur addition on extraction of Mn at different roasting temperatures

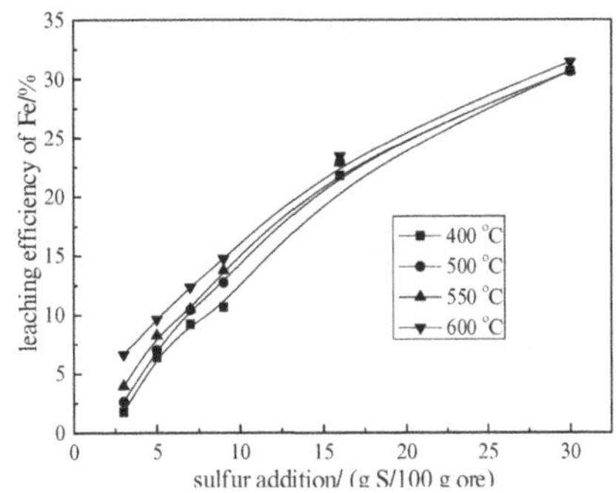

Figure. 8 Effect of sulfur addition on extraction of Fe at different roasting temperatures

The leaching efficiencies of iron in Figure 7 are all enhanced by increasing the sulfur ratio from 3% to 30%, no matter how the roasting temperature changed. Also, Figure 7 indicates that higher leaching efficiency was achieved at the samples roasted by higher temperatures, under the condition of same addition of sulfur. By Comparing with Figure 6 we can see that at the sulfur addition of 7%, the leaching efficiency of Mn have been more than 97%, which is a fairly satisfactory result. Considering on making the iron leaching efficiency as low as possible, the sulfur addition of 7% is an appropriate amount.

Conclusions

The experimental results indicated that manganese dioxide can be selectively reduced over iron oxide by using sulfur. Most of the reactions happen spontaneously at lower roasting temperatures. The manganese phases in roasted product included low-valence manganese oxide, manganese sulfide and manganese sulfate. MnS could easily reacts with Mn_3O_4 generating MnO if the roasting temperature is above 550℃. So the harmful H_2S gas was effectively avoidable in the following acid leaching process. Keeping the leaching conditions as the same, the optimal conditions for reduction roasting-acid leaching iron-rich manganese oxide ores were a rosting temperature of 550℃,sulfur addition of 7%.An extraction of Mn greater than 95% was obtained while the leaching efficiency of Fe was controlled less than 10% under the optimal conditions.

References

1. GG Mei et al., *Technology of China Manganese Industry* (Chang Sha, CS: Central South University Press, 2011), 4-9.

2. R.N. Sahoo et al, "Leaching of manganese from low-grade manganese ore using oxalic acid as reductant in sulphuric acid solution," *Hydrometallurgy*, 62 (2001) 157–163.

3. F. Pagnanellia et al, "Leaching of low-grade manganese ores by using nitric acid andglucose: optimization of the operating conditions," *Hydrometallurgy* 75 (2004) 157–167.

4. S.B. Kanungo et al, "Extraction of metals from manganese nodules of the Indian ocean by leaching in aqueous solution of sulphur dioxide." *Hydrometallurgy*, 20 (1988), 135–146.

5. M.S. Bafghi et al, "Reductive dissolution of manganese ore in sulfuric acid in the presence of iron metal." *Hydrometallurgy*, 90 (2007), 207–212.

6. T. Kai et al, "Kinetic model for simultaneous leaching of zinc sulfide and manganese dioxide in the presence of iron-oxidizing bacteria." *Chemical Engineering Science*, 55(17) (2000), 3429–3436.

7. C. LI et al. "A study on the preparation of manganese sulfate by co-calcining pyrolusite and pyrite." *Chemical World*, 4 (2000), 66–69.

8. Z. Cheng et al. "Study in reduction-roast leaching manganese from low-grade manganese dioxide ores using cornstalk as reductant." *Hydrometallurgy*, 96(1) (2009), 176–179.

9. Kh.S. Abou-El-Sherbini, "Simultaneous extraction of manganese from low grade manganese dioxide ore and beneficiation of sulfur slag," Sep. Purif. Technol. 27 (2002), 67–75.

10. T.Q. Li, "Technology of Grade Pyrolusite Ore Reduction Process and Recent Advances," *Chinas Manganese Industry*, 26(2) (2008), 5-13.

11. Y.B. Zhang et al., "Manganese extraction by sulfur-based reduction roasting–acid leaching from low-grade manganese oxide ores," *Hydrometallurgy*, 133 (2013), 126-132.

12. Z.X. You, "Fundamental Research on Novel Process of Manganese Oxide Ores by Sulfur-based Reduction Roasting-acid Leaching" (Ph.D.thesis, Central South University, 2012), 63.

13. Z.Z. Li, "Study on the Behavior of Manganese and Sulfur of Manganese Oxide Ores by Low Temperature Sulfur-based Reduction Roasting-Acid Leaching" (Ph.D.thesis, Central South University, 2012), 37-43.

14. Y.B. Zhang et al., "Researches on Technical Conditions for Sulfur Reduction Roasting of Low-grade Manganese Oxide Ores", *Mining and Metallurgical Engineering*, 32(4) (2012), 70-76.

7th International Symposium on High-Temperature Metallurgical Processing
Edited by: Jiann-Yang Hwang, Tao Jiang, P. Chris Pistorius, Gerardo R.F. Alvear F., Onuralp Yücel, Liyuan Cai, Baojun Zhao, Dean Gregurek, and Varadarajan Seshadri
TMS (The Minerals, Metals & Materials Society), 2016

RESEARCH ON RECOVERING IRON OXIDE FROM THE IRON, TIN-BEARING TAILINGS

Jun Chen, Zijian Su, Yuanbo Zhang*, Yingming Chen, Bingbing Liu

(School of Minerals Processing and Bioengineering, Central South University, Changsha 410083, China)

Keywords: Iron, tin-bearing tailings; Magnetizing roasting; Magnetic separation.

Abstract

Iron, tin-bearing tailings is a kind of hazardous solid waste generated from cassiterite dressing operation. It is difficult to utilize the tailings by present mineral processing methods due to the complex dissemination characteristics. In this study, a process of gas-based magnetizing roasting followed by magnetic separation was studied for separating and recovering iron from the tailings. The effects of roasting and magnetic separation parameters on the recovery of iron were systematically investigated. The optimum parameters were proposed as follows: $CO/(CO+CO_2)$ content of 5%, roasting temperature of 850 °C, roasting time of 60 min, grinding particle size of 90% less than 45 μm and magnetic field intensity of 800 Gs. Under these conditions, the magnetic concentrate containing 68.5% Fe and 0.06% Sn was achieved, which can be used as high-quality raw materials for direct reduction iron production.

Introduction

In China, iron, tin-bearing ores are typically refractory iron ores owing to intricate intergrowth relationship. About 500 million tons of iron, tin-bearing tailings have been discarded as waste, and it's increasing by 10 million tons annually[1]. These tailings not only occupy plenty of land, but also waste resources and pollute environment[2, 3].

In order to recover iron from the tailings, many attempts have been conducted, which can be classified as mineral processing and pyrometallurgical technologies. Gravity and magnetic separations are widely used in iron minerals beneficiation practices [4]. However, it is difficult to separate iron ores and cassiterite using gravity and magnetic separation, due to complex symbiotic relationship between iron minerals and cassiterite. Flotation is an effective method of to upgrade iron concentrate [5]. Unfortunately, the recovery rate of iron is unsatisfactory applying flotation because the tailings are characterized as fine particle size, large surface area, and slime coating. Generally, iron concentrate obtained from the tailings cannot meet the requirement for ironmaking ($Sn<0.08\%$). The tin is difficult to remove during steelmaking, and the mechanical properties of steel products will be deteriorated owing to the excessive tin content. To overcome these disadvantages, selective sulphurisation, chlorination [6] and reduction roasting processes [7] are also commonly used to obtain eligible iron concentrates from the tailings, but issues of environment pollution and equipment corrosion are difficult to avoid.

* Corresponding author: Yuanbo Zhang, email: zybcsu@126.com.

At present, many researchers have applied magnetizing roasting-magnetic separation technique to recover iron from red mud, iron ore tailings, vanadium tailings, cyanide tailings, etc. The process of magnetizing roasting followed by magnetic separation has been demonstrated to be effective for iron recovery from solid wastes [8, 9].

In this present work, a study of recovering iron oxide from the iron, tin-bearing tailings using a magnetizing roasting followed by magnetic separation process was conducted. The roasting and magnetic separation parameters were optimized, and then the phase and magnetism changes of the samples were detected by X-ray diffraction (XRD) and vibrating sample magnetometer (VSM).

Experimental

Materials

The tailings used in this study were obtained from Yunnan Province, China. Particle size distribution of the tailings and the chemical composition are showed in Table I and Table II, respectively. The total iron content is 40.15 wt% and total tin content is 0.49 wt%, X-ray diffraction (XRD) patterns of the tailings in Fig.1 show the iron oxide is goethite and the main gangue minerals are calcite and dolomite.

Table I. Particle size distribution of the tailings

Particle size(mm)	+0.074	0.045-0.074	0.037-0.045	0.030-0.037	-0.030
Mass Fraction (wt. %)	32.47	11.03	9.85	8.75	38.17

Table II. Chemical compositions of the tailings

Components	Fe_{total}	SiO_2	CaO	Al_2O_3	MgO	Sn	S	P	LOI*
Content (wt. %)	40.15	2.74	11.90	2.23	2.83	0.49	0.50	0.05	10.67

(LOI*: loss on ignition)

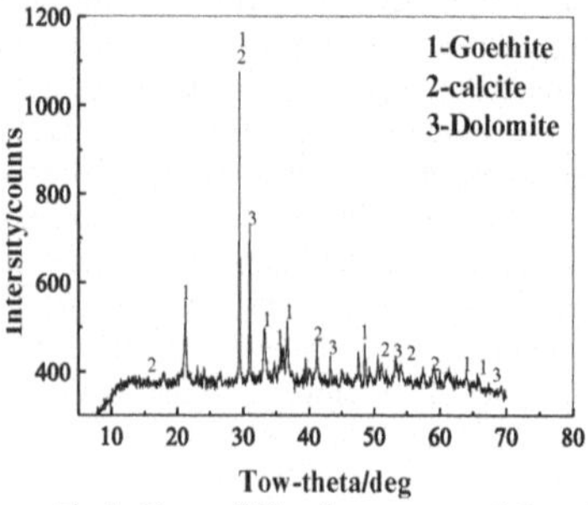

Fig.1. X-ray diffraction pattern of the tailings

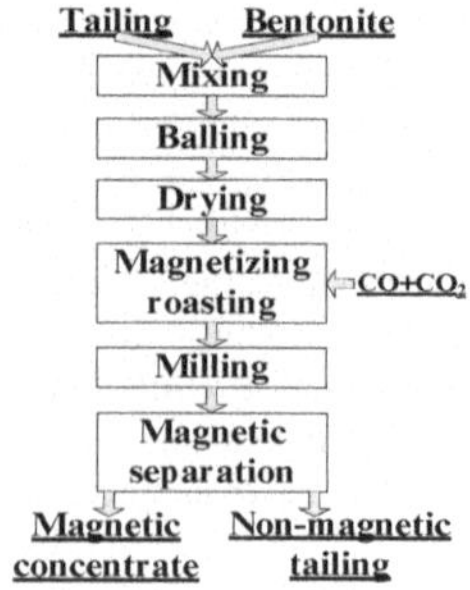

Fig.2. The experimental flow sheet.

The main chemical composition of bentonite is shown in Table III.

Table III. Main composition of bentonite

Components	TFe	Al_2O_3	SiO_2	CaO	MgO	MnO	K_2O	Na_2O
Content (wt. %)	1.63	14.15	63.17	2.24	2.01	0.05	1.25	2.29

The purity of gases (CO, CO_2, and N_2) used in this study was greater than 99.99 vol.%.

Experimental method

As depicted in Fig.2, the experiment mainly includes balling, drying, magnetization roasting, ball mill grinding followed by magnetic separation. Firstly, the iron, tin-bearing was thoroughly mixed with 0.5% bentonite for green ball production. Then the green balls were dried in an oven at 110℃ for 4 h. After drying, the dried pellets were put into a corundum crucible and roasted in the electrically heated horizontal furnace (Fig. 3). Before the dried pellets were placed into the high temperature zone of the electrically heated horizontal furnace, the furnace was preheated to the setting temperature firstly. Then, the pellets were roasted at the given temperature with setting time under weak reductive atmosphere. After that, the roasted balls were cooled to room temperature in the corundum crucible isolated from atmospheric oxygen. Prior to magnetic separation, 10g roasted balls were ground in an XMQ Φ240×90 wet ball mill. The slurry was then separated in an XCGS-73 Davis magnetic tube. The total iron content was determined by chemical titration method, as well as tin content was tested by inductively coupled plasma-atomic emission spectrometry (ICP-AES). The yield and recovery rate of magnetic concentrate were calculated according to Eqs. (a) and (b), respectively.

$$\gamma = \frac{m_1}{m_0} \times 100\% \quad \text{(a)}$$

Where: γ is the yield of magnetic concentrate, wt.%; m_0 is the feed mass of roasted sample subjected to magnetic separation, g; m_1 is the mass of obtained from magnetic concentrate, g.

$$\varepsilon = \gamma \times \frac{\beta}{\alpha} \times 100\% \quad \text{(b)}$$

Where: ε is the recovery of iron, wt.%; β is the total iron content of magnetic concentrate, wt.%; α is the total iron content of raw minerals, wt.%.

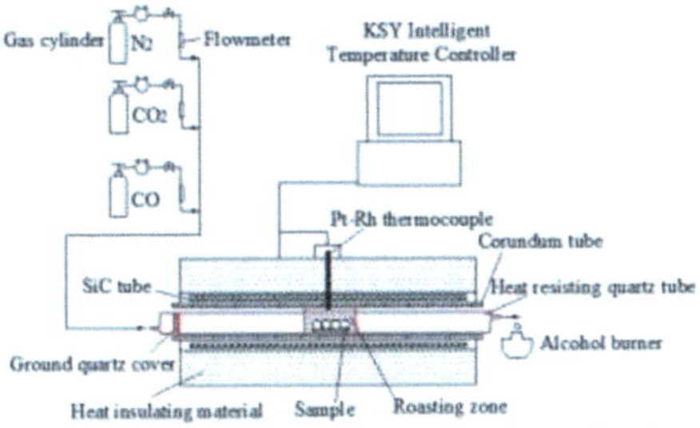

Fig. 3. Schematic diagram of the equipment for reduction roasting.

Results and discussion

Iron recovery indexes

Effect of roasting temperature Effect of roasting temperature on iron recovery indexes was studied and the results were shown in Fig. 4. It can be seen from Fig. 4, both the grade of magnetic concentrate and the recovery rate of iron increased rapidly with the rising of the temperature from 700℃ to 850℃. However, when the roasting temperature was over 850℃, both the grade of magnetic concentrate and the recovery rate began to decrease. This was probably because with the roasting temperature increasing, the reaction between carbon monoxide and goethite was sped up. While the temperature was too high, the magnetite may be over-reduced to wustite. When the temperature was 850℃, the iron grade and recovery rate were both in a high level. Hence, the optimal temperature was determined to be 850 ℃.

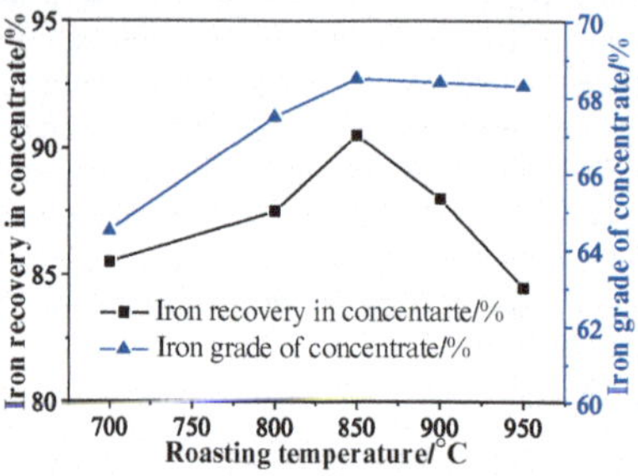

Fig.4. Effect of roasting temperature on iron recovery indexes(roasting time, 60 min; CO content, 5%; particle size, -45 m< 90%; magnetic field intensity, 800Gs).

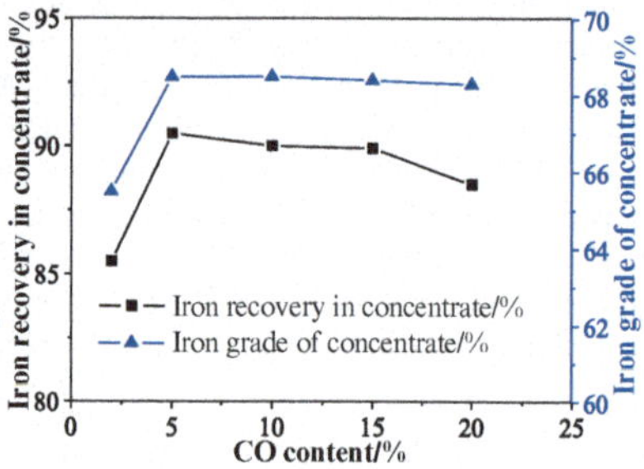

Fig.5. Effect of CO content on iron recovery indexes(roasting temperature, 850 ℃; roasting time, 60 min; particle size, -45 m< 90%; magnetic field intensity, 800Gs).

Effect of CO content. Then, the effect of CO content on iron recovery indexes was shown in Fig. 5. As shown in Fig. 5, the CO content had a significant effect on both iron recovery rate and iron grade. The results showed that while the CO content was below 5 vol.%, both the iron recovery rate and grade increase with the increasing of CO content. When the CO content further increases, the iron recovery rate and grade tends to decrease rapidly. It maybe due to that goethite was not reduced completely if the CO content was insufficient, but when the CO content was excessive, magnetite could be over reduced to wustite. Hence, the optimal CO content was chosen as be 5 vol.%.

Effect of roasting time The effect of roasting time on iron recovery rate and grade was shown in Fig. 6. The results revealed that when the roasting time was in the range of 30-60 min, both the iron recovery rate and grade in magnetic concentrate increased with the prolonging of duration time. When the roasting time was over 60 min, both the iron recovery rate and grade almost kept an invariable value. Therefore, the optimum roasting time was 60 min.

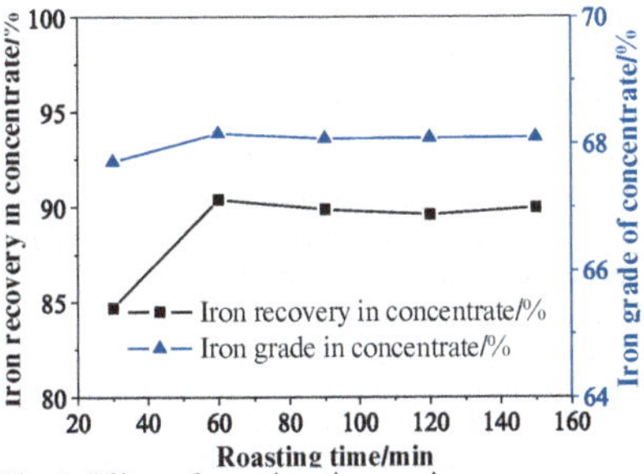

Fig.6. Effect of roasting time on iron recovery indexes (roasting temperature, 850 ℃; CO content, 5%; particle size, -45 m< 90%; magnetic field intensity, 800Gs).

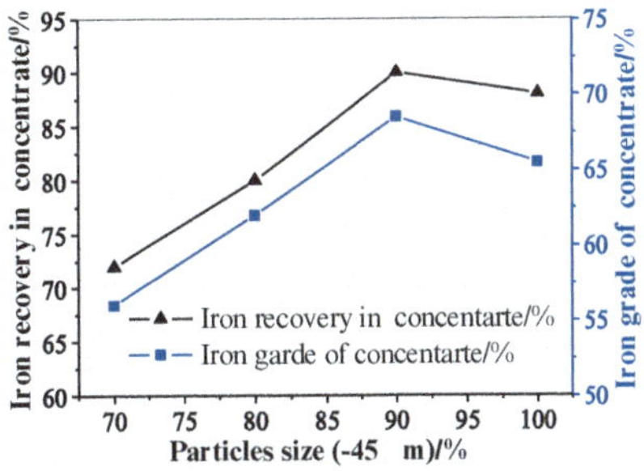

Fig.7. Effect of wet ground particle size of roasted product on iron recovery indexes (roasting temperature, 850 ℃; roasting time, 60 min CO content, 5%; magnetic field intensity, 800Gs).

Effect of grinding particle size Effect of grinding particle size on iron recovery indexes is shown in Fig. 7. From Fig. 7, the iron recovery rate and grade were dramatically affected by the particle size. With the decreasing of particle size, both iron recovery and grade of magnetic concentrate increased rapidly. At the 90% of particle size less than 45μm, the iron grade and recovery rate were both high. However, if the roasted samples were milled too fine, the particles were agglomerated easily and difficult to be separated during the magnetic separation.

Effect of magnetic field intensity Effect of magnetic field intensity on iron recovery indexes was shown in Fig. 8. It can be seen that the magnetic field intensity had a significant effect on iron recovery indexes. When magnetic field intensity was about 800 Gs, both the iron recovery rate and iron grade of iron concentrate were satisfactory.

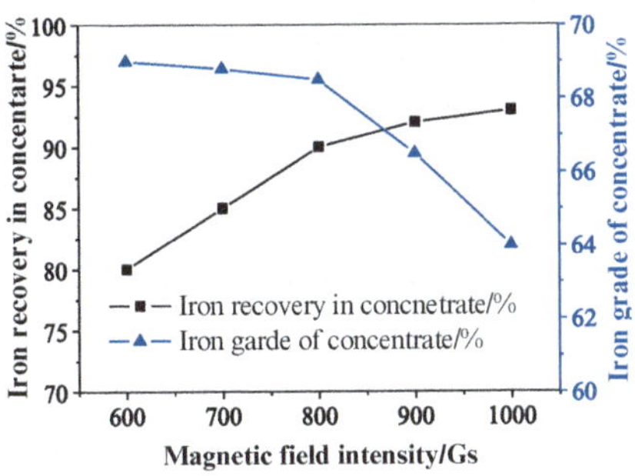

Fig. 8.Effect of magnetic field intensity on iron recovery indexes (roasting temperature, 850 ℃; roasting time, 60 min; CO content, 5%; particle size, -45 m< 90%).

Analysis of roasted products

Magnetism changes The magnetism changes (coercicity Hc and saturation magnetization) of raw material and roasted samples were shown in Fig. 9 and Table III. It can be seen that the saturation magnetization of raw material was only 0.84 emu/g Then results indicated that it is not feasible to directly separate iron oxide minerals from the gangue minerals. From Fig. 9, we can see that it was possible to solve the problem by improving the magnetism of iron oxide minerals through magnetizing roasting. Comparing to the raw material, the saturation magnetization increased from 0.84 emu/g to 53.60 emu/g at 850°C and CO content of 5%.

Table III. Magnetism changes of samples

Sample type	Hc(Oe)	Ms(emu/g)
Raw material	295.97	0.84
Roasted product (5% CO, 850℃, 60min)	194.32	53.60

(Hc: coercivity; Ms: saturation magnetization)

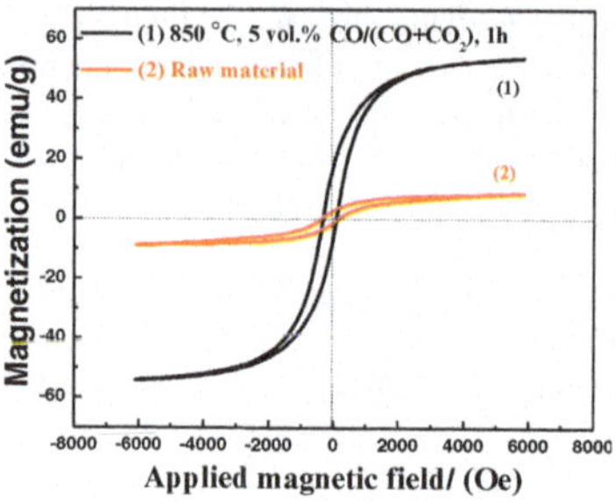

Fig.9. Magnetic hysteresis loops of raw material and roasted tailings (roasting temperature, 850 °C; roasting time, 60 min; CO content, 5%).

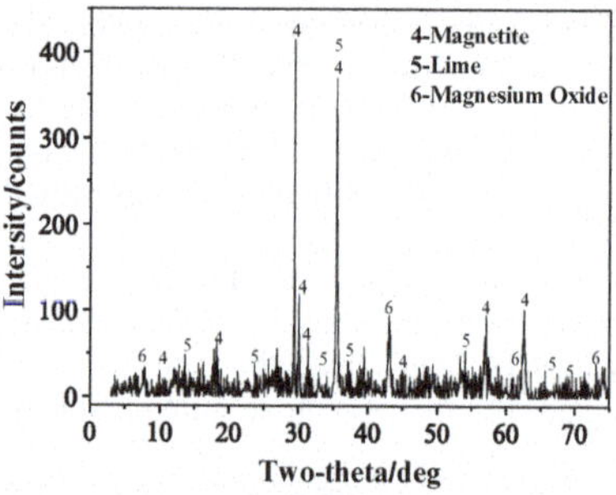

Fig. 10. XRD pattern of roasted samples (roasting temperature, 850 °C; roasting time, 60 min; CO content, 5%).

X-ray diffraction analysis The roasted product was prepared in optimized experimental conditions as following: roasting temperature of 850℃, roasting time of 60 min, CO content of 5% and particle size 90% less than 45μm was analyzed by XRD. The XRD pattern was shown in Fig.10.

From Fig. 10, it can be seen that the peak of goethite disappeared and the new peak of magnetite appears. It means that the goethite has been reduced to magnetite. Comparing to the XRD patterns of raw tailings, the peaks of calcite and dolomite were disappeared, which indicated that the calcite and dolomite in the sample were decompounded into lime and magnesium oxides, respectively.

Analysis of magnetic concentrate and tailings

X-ray diffraction analysis The magnetic concentrate and tailings were separated from roasted product under optimized process parameters as above. The XRD patterns were shown in Fig.11 (a) and Fig. 11 (b).

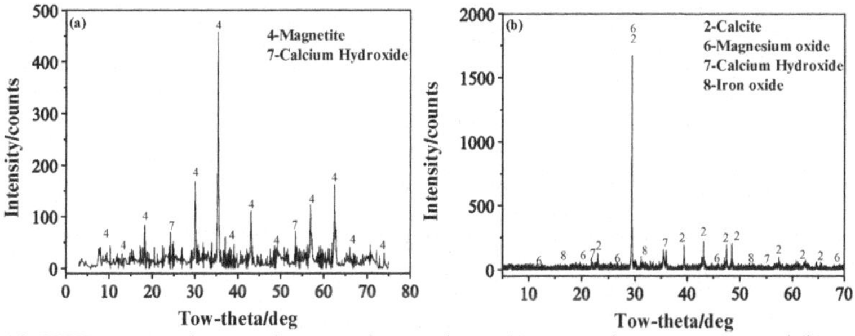

Fig. 11. XRD pattern of magnetic separation products: (a) magnetic concentrate and (b) non-magnetic tailing (roasting temperature, 850 °C; roasting time, 60 min; CO content, 5%; particle size, -45 m< 90%; magnetic field intensity, 800Gs).

From Fig.11 (a), magnetite was the predominant iron-bearing phase in the magnetic concentrate. Meanwhile, a small amount of calcium hydroxide also existed. From Fig. 11 (b) XRD pattern, it can be seen that calcite and magnesium oxide were main phase of non-magnetic tailings, indicating that most of the gangue minerals were separated by magnetic separation. The existence of calcite was that the calcium hydroxide reacted with the carbon dioxide.

The chemical analysis of products The contents of main elements in magnetic concentrate, magnetic tailing and raw material were shown in table IV. By comparing, it can be seen that the iron content increased from 40.15% in raw material to 68.50%, and the tin content decreased from 0.49% to 0.06%, where a high-quality magnetic concentrate was obtained. At the same time, a kind of tin-bearing material (1.6% Sn) was obtained, which showed that the magnetizing roasting followed by magnetic separation has prominent advantages in upgrading iron content when compared with the previous studies [10].

Table IV. The contents of main elements in the raw tailing, magnetic and non-magnetic products

	Fe/%	Sn/%	CaO/%	MgO/%	SiO_2/%	Al_2O_3/%
Raw material	40.15	0.49	11.90	2.83	2.74	2.23
Concentrate	68.50	0.06	2.83	0.66	0.42	0.35
tailing	7.85	1.60	33.56	6.53	8.82	7.57

Conclusions

The study presented a process of gas-based magnetization roasting followed by magnetic separation to recover iron oxide from the iron, tin-bearing tailings. Optimized process conditions were obtained as following: CO content of 5%, roasting temperature of 850℃, roasting time of 60 min, grinding particle size of 90% less than 45μm in the roasted product and the magnetic field intensity of 800Gs. Under the above conditions, the iron oxide was effectively separated from the iron, tin bearing tailings, and the final magnetic concentrate contained 68.5% total iron and 0.06% total tin, which was a high-quality raw material for direct reduction iron. And the recovery rate of iron was above 90%.

Acknowledgements

The authors would express their heartful thanks to National Natural Science Foundation of China (No.51574283), Co-Innovation Center for Clean and Efficient Utilization of Strategic Metal Mineral Resources, Hunan Provincial Innovation Foundation for Postgraduate (CX2015B054) and the Teachers' Research Fund of Central South University (2013JSJJ028).

References

[1] Y.P. Liu, "A Study of process mineralogy of cassiterite in old tailings of Yunnan Tin," *China Minig Magazine*, 21 (9) (2012), 84-87. (In Chinese)

[2] W. Chen, et al., "Current state of fine mineral tailings treatment: A critical review on theory and practice," *Minerals Engineering*, 58 (2014), 113-131.

[3] H. Shen, E. Forssberg, "An overview of recovery of metals from slags," *Waste Management*, 23 (2003), 933-949.

[4] A.S.S. Ahmed, E.M. Mosline, M. Abouzeid, "concentration of a Sudanese low-grade iron ore," *International Journal of Mineral Processing*, 122 (2013), 59-62.

[5] L.O. Filippov, V.V.Severov, I.V. Filippova, "An overview of the beneficiation of iron ore via reverse cationic flotation," *International Journal of Mineral Processing*, 127 (2014), 62-69.

[6] J. Zhang, "The utilization of new technology about recovery tin and iron from tailings of Yunnan Tin Corporation," *Materials Research and Application*, 6 (2) (2012), 142-145. (In Chinese)

[7] Y.B. Zhang, et al., "Tin and zinc separation from tin, zinc bearing complex iron ores by selective reduction process," *Ironmaking and Steelmaking*, 38 (8) (2011), 613-619.

[8] Y.J. Zhang, et al., "Studies on the kinetics of zinc and indium extraction from indium-bearing zinc ferrite," *Hydrometallurgy* 100 (2010), 172-176.

[9] H. F. Yang, L.L. Jing, B. G. Zhang, "Recovery of iron from vanadium tailings with coal-based direct reduction followed by magnetic separation," *Journal of Hazardous Materials*, 185 (2011), 1405-1411.

[10] C.J. Cheng, et al., "Dressing research on a goethite ores from Shanxi," *Conservation and Utilization of Mineral Resources,* 2 (2006), 33-35. (In Chinese)

7th International Symposium on High-Temperature Metallurgical Processing
Edited by: Jiann-Yang Hwang, Tao Jiang, P. Chris Pistorius, Gerardo R.F. Alvear F., Onuralp Yücel, Liyuan Cai, Baojun Zhao, Dean Gregurek, and Varadarajan Seshadri
TMS (The Minerals, Metals & Materials Society), 2016

A STUDY ON THE CHARACTERIZATION OF NICKEL LATERITES OF CENTRAL ANATOLIA

Ender Keskinkilic[1,*], Saeid Pournaderi[2], Ahmet Geveci[3], Yavuz A. Topkaya[3]

[1] Department of Metallurgical and Materials Engineering, Atilim University; Ankara, 06836, TURKEY
[2] Department of Metallurgical and Materials Engineering, Karadeniz Technical University; Trabzon, 61040, TURKEY
[3] Department of Metallurgical and Materials Engineering, Middle East Technical University; Ankara, 06531, TURKEY

Keywords: Characterization, Laterite, Ferronickel

Abstract

There are mainly three commercial nickel laterites in Turkey: Gordes, Caldag and Sivrihisar. The first two appear in Western Anatolia. The third one, Sivrihisar (Yunusemre) laterites have been found in the first decade of the millennium, in the Central Anatolia region of Turkey. Since then, mining facilities have been continued in Yunusemre. Calcination, prereduction and smelting of Sivrihisar laterites have been extensively studied and published. In the current work, some results of the studies related to characterization of Sivrihisar laterites will be presented.

Introduction

Nickel and its alloys are important to enhance some properties of materials like strength, corrosion resistance, ductility, thermal and electrical conductivity, etc. Commercially, the most important nickel and Ni-alloys are electrolytic nickel (99% Ni) and ferronickel (15-35%Ni). Among the usage areas of nickel, its use in the production of stainless steels holds approximately two third. The remaining areas can be listed as production of nonferrous alloys, steel alloying purposes, coating industry, casting, batteries and others [1].

Nickel can be found in nature as in the form of sulfide, oxide and silicate minerals. As well as the world nickel reserves are concerned, 60% of them are in the form of iron-nickel oxides called as laterites. Nickel laterite ores can be classified into two categories as limonite type and saprolite/nontronite type. Limonite ores are characterized by FeO(OH) type structure while saprolite/nontronite type ores contain significant amount of MgO together with certain amount of minerals like serpentine [$(Mg, Fe, Ni)_3Si_2O_5(OH)_4$] and smectite [$Na_{0.3}Al_2(Si_{3.7}Al_{0.3})O_{10}.(OH)_2$]. It was reported that 70% of laterites are limonite type ores [2].

One of the commercial methods of nickel extraction from laterites uses pyrometallurgical techniques. Nickel laterites are treated in a five-step process to obtain iron-nickel alloy, called as ferronickel. Drying, calcination and prereduction steps are conducted in rotary kiln furnaces

* Corresponding author: Ender Keskinkilic, Assoc. Prof., E-mail: ender.keskinkilic@atilim.edu.tr,
Tel: +90 533 3029510

while smelting is performed in electric arc furnaces. Therefore, production of crude ferronickel is based on use of two furnaces namely rotary kiln and electric arc furnace. In literature, this combined process is most of time called as rotary kiln-electric furnace (RKEF) process using the initials of the names of those two furnaces. The fifth step is refining, which is the last stage of commercial ferronickel production. Refining is commonly performed in ladle furnaces or OBM type converters and the impurities like C, P and S are eliminated.

There are three important nickel laterite reserves in Turkey. Two of them are in Western Anatolia, in the locations of Gordes and Caldag. The third one, recently found in Sivrihisar Yunusemre region at the beginning of the millennium, is in Central Anatolia. Gordes and Caldag reserves are being treated by two companies and hydrometallurgical methods are used to treat them. Related to the third one, mining facilities have been continued especially in Yunusemre location of Sivrihisar. This ore body has an average Ni content of 1.26% and it is a limonite type laterite containing relatively low MgO content and high iron content.

Possibility of pyrometallurgical nickel extraction from Sivrihisar laterites has been extensively studied and published [3-8]. In the present work, some results of the studies related to characterization of Sivrihisar laterites will be presented.

Experimental

Four barrels of representative ore samples were taken from Yunusemre Karasivritepe and Kucuksivritepe locations. Each barrel was around 100 kilograms. First, all the samples were thoroughly mixed. One representative quarter was split by the coning & quartering method. The other portions were stored in the barrels. From the quarter selected, specimens were prepared for characterization of the ore (i.e. chemical analysis, XRD and DTA-TG). During specimen preparation particle size of the ore was lowered to -74 microns using jaw crusher, roll crusher, pulverizator and disc grinder, respectively. Chemical composition of the run of mine ore is shown in Table I. The whole run of mine ore was then screened at 50 mm. The oversize lumpy materials containing chiefly gangue minerals were rejected and remaining undersize, which is indicated in this paper as "-50 mm ore", was subjected to characterization. Again, the same crushers and grinders were used for specimen preparation.

Table I. Chemical Composition for a Run of Mine Ore (%)

Ni	**SiO_2**	**Fe**	**Cr**	**Co**	**As**
1.26	30.3	32.6	1.01	0.082	0.03
Al_2O_3	**MgO**	**Fe_2O_3**	**CaO**	**MnO**	**P_2O_5**
2.68	1.03	46.6	1.69	0.62	<0.02
K	**Pb**	**S**	**TiO_2**	**Zn**	**Cu**
0.2	<0.01	0.01	0.06	0.03	0.005

The chemical analysis of the ore samples was conducted with the ICP-OES. XRD was carried out on a Rigaku SA-HF3 diffractometer with Cu Kα (λ = 1.54 Å) radiation (40 kV and 40 mA). Termogravimetric measurements were conducted in a thermobalance with DTA (differential thermal analysis) capability. DTA-TG curves were obtained by heating ~20 mg of ground sample (< 74 μm) from 25°C to 1000°C with a heating rate of 10°C/min. Dry screen analysis was performed according to ASTM E 389-03 and ASTM E 276-03 standards. Tumbler test was conducted using ASTM E 279-97 standard. For determination of clay minerals, which could not be identified by the routine sample preparation methods, the clay mineralogy technique was used [8].

Results and Discussion

Drying tests were conducted to determine the amount of physically bound water in the ore. After screening at 50 mm, oversize particles were rejected and then crushed below 20 mm size. The sample was dried at 105°C overnight. The amount of moisture was reported as 6.9% of the weight of oversize particles. -50 mm ore was subjected to the same procedure and it was reported that mechanically held water was 13.6% of the weight of this undersize, which was lower than 25-30% range reported in the literature [9]. The ore might have partly dried during the transportation and also during the sample preparation, especially coning and quartering. Calcination experiments were conducted for -50 mm fraction and the results were outlined in previous works, in detail [3,6]. Figure 1 shows the results of dry screen analysis of the run of mine ore. According to this diagram, about 50% of the ore is finer than 10 mm and a considerable fraction of the ore (about 16 wt.%) is present as dust (<600 m). A large fraction of fine particles is typical of laterites, which adversely affects operation of the rotary kiln with the high dusting rate.

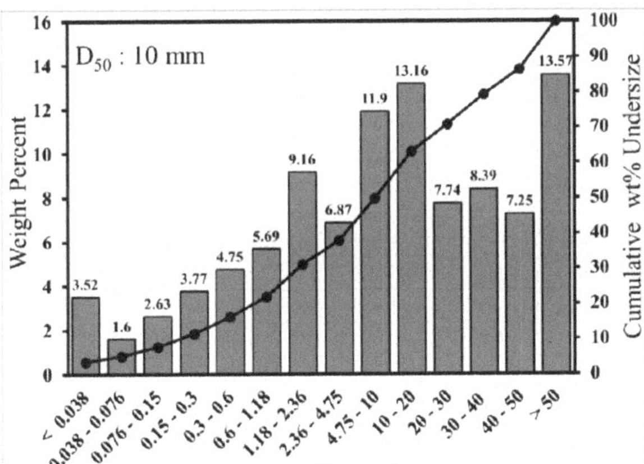

Figure 1. Particle Size Distribution of the Run of Mine Ore

Extent of upgrade in nickel content of the ore by excluding >50 mm particles was examined. Chemical composition of -50 mm ore is shown in Table II. Ni content of the ore was reported to increase from 1.26% to 1.41% by rejecting the lumps. In other words, Ni content of the ore was upgraded by 12%. Another important consequence of this rejection was the decrease in SiO_2 composition of the ore body. SiO_2 content of the limonitic laterite was found to decrease from 30.3% to 25.8%, corresponding to a decrease around 15%.

Table II. Chemical Composition of -50 mm Ore (%)

Ni	**SiO_2**	**Fe**	**Cr**	**Co**	**As**
1.41	25.8	33.7	1.26	0.093	0.04
Al_2O_3	**MgO**	**Fe_2O_3**	**CaO**	**MnO**	**P_2O_5**
3.23	1.29	48.2	1.65	0.74	0.04
K	**Pb**	**S**	**TiO_2**	**Zn**	**Cu**
0.2	0.01	0.03	0.08	0.03	0.006

In order to investigate the variation of elements/oxides with particle size, samples were prepared for chemical analysis from the size fractions obtained from dry screen analysis. Chemical compositions of size fractions were determined by ICP-OES and tabulated as Table III. Considering the weight percentages of the size fractions and Table III, it was found that ~18% of SiO_2 by weight was eliminated from the ore by rejecting >50 mm particles. During this rejection, ~10% of Ni by weight was reported as a loss of valuable element. Screening & rejection of lumps provided elimination of ~14% of the ore by weight.

Like the behavior of many valuable elements, nickel content of the ore was reported to increase as the ore particle size decreased. Ni composition of the coarsest size (i.e. >50 mm fraction) was measured as 0.923% while it was analyzed as 1.915 for the finest fraction (i.e. <0.038 mm). Similarly, iron and therefore Fe_2O_3 content of the ore were reported to rise as the particle size decreased. One of the most important gangue components, SiO_2, showed the opposite behavior: SiO_2 composition had the highest value as 46.40% for [30 mm - 40mm] fraction and it exhibited a continuous decay afterwards such that the finest fraction had SiO_2 composition of 16.00%. The behavior of nickel, iron and quartz were in accord with the expectations.

Table III. Chemical Compositions of Size Fractions of Sivrihisar Nickel Laterites (%)

Particle size (mm)	**Ni**	**SiO_2**	**Fe**	**Cr**	**Co**	**As**	**MgO**	**Fe_2O_3**	**Al_2O_3**
< 0.038	1.915	16.00	40.1	0.69	0.075	0.03	1.52	57.4	2.38
0.038 - 0.076	1.775	18.80	35.8	1.52	0.088	0.03	2.11	51.1	3.21
0.076 - 0.15	1.670	19.45	33.8	2.38	0.095	0.02	2.18	48.4	3.75
0.15 - 0.3	1.615	19.80	32.2	3.02	0.114	0.03	2.02	46.0	4.20
0.3 - 0.6	1.645	21.40	31.9	2.38	0.139	0.02	1.68	45.6	4.01
0.6 - 1.18	1.610	23.90	32.1	1.52	0.137	0.04	1.45	45.9	3.85
1.18 - 2.36	1.440	25.80	31.5	1.18	0.103	0.03	1.32	45.0	3.69
2.36 - 4.75	1.345	28.10	31.5	1.01	0.085	0.03	1.2	45.0	3.64
4.75 - 10	1.295	28.60	32.4	0.90	0.080	0.03	1.11	46.3	3.29
10 - 20	1.225	30.60	32.7	0.74	0.076	0.03	1.06	46.7	2.53
20 - 30	1.035	38.80	29.9	0.63	0.059	0.02	0.68	42.7	1.57
30 - 40	0.890	46.40	26.4	0.38	0.057	0.02	0.41	37.7	2.03
40 - 50	0.930	27.30	33.1	0.47	0.051	0.10	0.50	47.4	1.01
> 50	0.923	40.80	25.7	0.41	0.054	0.02	1.21	36.8	0.77

Table III. Chemical Compositions of Size Fractions of Sivrihisar Nickel Laterites (%) (cont'd.)

Particle size (mm)	CaO	MnO	P_2O_5	TiO_2	S	K	Pb	Zn	Cu
< 0.038	1.24	0.85	0.02	0.08	0.08	0.1	<0.01	0.04	0.030
0.038 - 0.076	2.06	0.90	0.03	0.09	0.05	0.1	<0.01	0.04	0.017
0.076 - 0.15	2.57	0.92	0.03	0.10	0.05	0.1	<0.01	0.04	0.026
0.15 - 0.3	2.78	1.04	<0.02	0.10	0.07	0.1	<0.01	0.04	0.008
0.3 - 0.6	2.54	1.26	0.02	0.09	0.04	0.1	<0.01	0.04	0.008
0.6 - 1.18	2.15	1.27	<0.02	0.08	0.05	0.1	<0.01	0.04	0.009
1.18 - 2.36	1.69	0.94	<0.02	0.08	0.05	0.1	<0.01	0.03	0.007
2.36 - 4.75	1.26	0.72	<0.02	0.08	0.04	<0.1	<0.01	0.03	0.006
4.75 - 10	1.07	0.59	<0.02	0.09	0.03	<0.1	<0.01	0.03	0.005
10 - 20	1.09	0.52	0.02	0.06	0.04	0.1	<0.01	0.03	<0.005
20 - 30	0.96	0.39	<0.02	0.03	0.03	<0.1	<0.01	0.03	0.006
30 - 40	0.12	0.35	<0.02	0.06	0.04	0.1	<0.01	0.02	<0.005
40 - 50	5.55	0.31	<0.02	0.01	0.04	<0.1	<0.01	0.03	<0.005
>50	3.00	0.39	<0.02	0.02	0.03	0.1	0.01	0.06	<0.005

Cobalt is commonly encountered in nickel laterites and its percentage in Sivrihisar ore bodies is around 0.08%. According chemical analysis of size fractions, cobalt was reported to concentrate mostly in [0.3 mm - 1.18 mm] size range with a composition of ~0.14% and was found to decrease when the particle size increased or decreased out of this range. The finest ore fraction has cobalt analysis of 0.075%, which is very near to cobalt content of the overall sample. Likewise, cobalt composition reduces to ~0.05% while approaching to coarser sizes. Similar to nickel, screening & rejection of lumps resulted in loss of ~9% of Co by weight.

Another important element in nickel laterites is chromium. Sivrihisar limonitic laterites contain chromium of around 1%. Chromium was reported to concentrate mostly in [0.076 mm - 0.6 mm] size range with a composition exceeding 2%, even reaching 3% in [0.15 mm - 0.3 mm] size fraction. The behavior of chromium is similar to that of cobalt as well as its presence in the fine and coarse fractions are concerned. Chromium content of <0.038 mm fraction was analyzed as 0.69%, while it reduced to 0.38% in coarser sizes. Screening & rejection of lumps caused a loss of ~6% of Cr by weight.

The variations of oxide components, Al_2O_3, CaO, MgO, and MnO, with particle size exhibit the similar behavior encountered in cobalt and chromium. Al_2O_3 was reported as the most abundant oxide component after iron oxide and quartz. Sivrihisar nickel laterites contain ~2.7% alumina. It is mostly concentrated in [0.076 mm - 0.6 mm] size range and its amount was reported to exceed 4% in that range. Very similar behavior was encountered for the case of CaO. Run of mine ore CaO content was analyzed as ~1.7%, while it exceeded 2.5% in the same size range. The highest CaO composition was reported in [40 mm - 50 mm] fraction having a value of 5.55%. In the same fraction, iron and iron oxide made peaks indicating that [40 mm - 50 mm] fraction is rich in iron oxides and especially calcites. MgO content of Sivrihisar ore bodies is very low like the other laterite deposits present in Turkey. Sivrihisar limonitic laterite has only around 1% MgO and only in three fractions, namely [0.038 mm - 0.076 mm], [0.076 mm - 0.15 mm] and [0.15 mm - 0.3 mm], MgO was found to exceed 2%. Sivrihisar nickel laterites contain small amounts of lead and zinc. The lead content of overall sample never exceeds 0.01% while the zinc composition of the ore was analyzed as ~0.03%. Arsenic content of Sivrihisar ore bodies is around 0.03%. This value is not regarded as a high value and low arsenic content of the ore allows pyrometallurgical nickel extraction.

Tumbler test (ASTM E279-97)[1] was used to measure the resistance of the ore to degradation by impact and abrasion. [10mm - 50 mm] size range was used for this test. Tumbler test behavior of Sivrihisar nickel laterites is shown in Figure 2. According to tumbler test results, abrasion index (i.e. +6.35 mm) was reported as 21%[2] and dust index (i.e. -600 μm) was found as 68%. Those values show that Sivrihisar ores have low resistance against abrasion and impact.

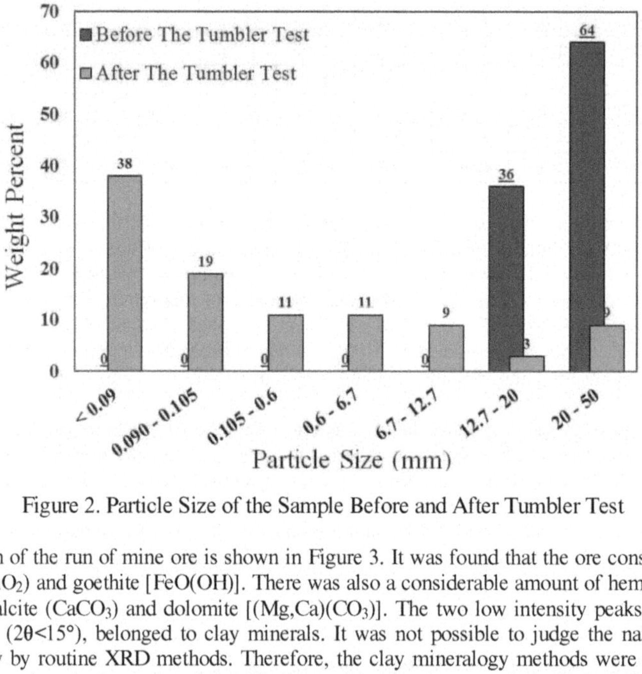

Figure 2. Particle Size of the Sample Before and After Tumbler Test

XRD pattern of the run of mine ore is shown in Figure 3. It was found that the ore consisted mainly of quartz (SiO_2) and goethite [FeO(OH)]. There was also a considerable amount of hematite (Fe_2O_3) and some calcite ($CaCO_3$) and dolomite [$(Mg,Ca)(CO_3)$]. The two low intensity peaks, recorded at low degrees ($2\theta<15°$), belonged to clay minerals. It was not possible to judge the nature of these clays simply by routine XRD methods. Therefore, the clay mineralogy methods were employed to identify these minerals. It was found that Sivrihisar nickel laterites contain kaolinite [$Al_2Si_2O_5(OH)_4$] and nontronite type smectite having general formula of $Na_{0.3}Fe_2(Si,Al)_4O_{10}(OH)_2.n(H_2O)$. According to quantitative phase analysis, it was reported that Sivrihisar ore sample was composed of goethite (53.3%), quartz (20.8%), hematite (14%), calcite (3.9%), dolomite (3.1%), nontronite (3.8%) and kaolinite (1.1%). For the details of clay mineralogy methods and quantitative phase analysis, the reader is referred to previously published work [8].

[1] This standard is actually suitable for iron ores and sinters&pellets obtained from iron ores. In the literature, a method related to tumbler test of nickel ores could not be found.
[2] Screen having 6.7 mm aperture size was used since a screen of 6.35 mm aperture size was not available.

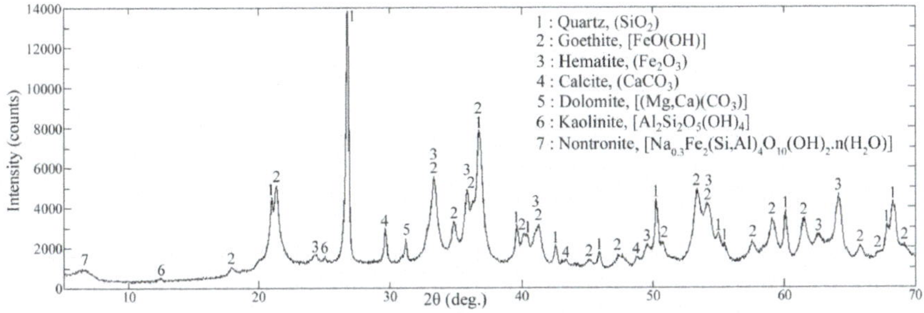

Figure 3. X-ray Diffraction Pattern of Sivrihisar Limonitic Ore

DTA and TGA curves for the run of mine ore are illustrated in Figure 4. Five main endothermic peaks and corresponding weight losses were recorded at about 60, 110, 290, 500 and 700°C. The two low-temperature endothermic peaks at 60 and 110°C were related to the evaporation of free water. A considerable amount of free water has been removed at ~60°C, while the remaining, being entrapped in pores, has evaporated at higher temperatures. The next highly endothermic peak and substantial weight loss at around 290°C were related to the dehydroxylation of goethite and formation of hematite according to reaction below:

$$2FeOOH_{(s)} \rightarrow Fe_2O_{3(s)} + H_2O_{(g)} \quad (1)$$

The other two peaks at 500 and 700°C were attributed to the dehydroxylation of clays and decomposition of calcite (and dolomite), respectively. Calcite and dolomite decompose according to the reactions (2) and (3). A more detailed information on the thermal behavior of the ore was published previously [6].

$$CaCO_{3(s)} \rightarrow CaO_{(s)} + CO_{2(g)} \quad (2)$$
$$CaMg(CO_3)_{2(s)} \rightarrow CaO_{(s)} + MgO_{(s)} + 2CO_{2(g)} \quad (3)$$

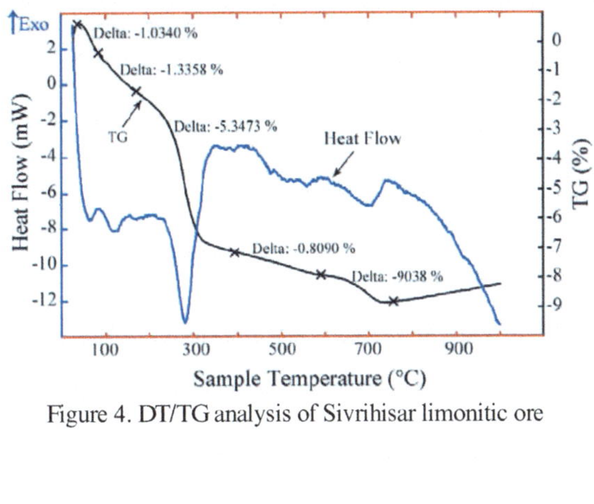

Figure 4. DT/TG analysis of Sivrihisar limonitic ore

Conclusions

- Dry screen analysis results showed that about 50% of the ore was finer than 10 mm and about 16% of the ore was present as dust (<600 m).
- Ni content of the ore was reported to increase from 1.26% to 1.41% by rejecting the lumps greater than 50 mm size.
- According to tumbler test results, abrasion index (i.e. +6.35 mm) was reported as 21%, and dust index (i.e. -600 m) was found as 68%.
- It was found that Sivrihisar nickel laterites contain goethite (53.3%), quartz (20.8%), hematite (14%), calcite (3.9%), dolomite (3.1%), nontronite (3.8%) and kaolinite (1.1%).
- Thermal analysis of the ore showed that 5 main endothermic peaks corresponds to evaporation of free water (peaks at 60°C and 110°C), dehydroxylation of goethite (peak at 290°C), dehydroxylation of clays (peak at 500°C) and decomposition of calcite and dolomite (peak at 700°C).

Acknowledgements

The authors would like to thank the Scientific and Technological Research Council of Turkey (TUBITAK) for the financial support given under Project No: 109M068 and the Scientific Research Projects (BAP) Department of the Middle East Technical University for the financial support under the project No. BAP-03-08-2012-002. The authors also wish to express their thanks to the META Nickel Cobalt Co. for supplying the lateritic ore samples from Sivrihisar.

References

1. http://www.larco.gr/nickel.php (Last accessed date: September 7, 2015).

2. M.G. King, P. Zuluani, R. Schonewille and P. Mason, "Technology Development for Processing Koniambo Laterites", ALTA Conference, (2005).

3. E. Keskinkilic, S. Pournaderi, A. Geveci, Y.A. Topkaya, "Calcination Behavior of Sivrihisar Laterite Ores of Turkey", 2011 TMS Annual Meeting, 2nd International Symposium on High Temperature Metallurgical Processing, San Diego, California.

4. E. Keskinkilic, S. Pournaderi, A. Geveci, Y.A. Topkaya, "Study of Certain Parameters in Laboratory-Scale Prereduction of Sivrihisar Laterite Ores of Turkey" (Paper presented at the ALTA 2011 Nickel-Cobalt-Copper Conference, Perth, Western Australia, May 2011).

5. E. Keskinkilic, S. Pournaderi, A. Geveci, Y.A. Topkaya, "Study of Certain Parameters in Laboratory-Scale Smelting of Sivrihisar Laterite Ores of Turkey", 2012 TMS Annual Meeting, 3rd International Symposium on High Temperature Metallurgical Processing, Orlando, Florida.

6. E. Keskinkilic, S. Pournaderi, A. Geveci and Y. A. Topkaya, "Calcination characteristics of laterite ores from the central region of Anatolia", *J. S. Afr. Inst. Min. Metall.*, 112 (2012), 877-882.

7. E. Keskinkilic, S. Pournaderi, A. Geveci, Y.A. Topkaya, "Production of Crude Ferronickel from Sivrihisar Laterite Ores of Turkey", 2013 TMS Annual Meeting, 4th International Symposium on High Temperature Metallurgical Processing, San Antonio, Texas.

8. S. Pournaderi, E. Keskinkılıç, A. Geveci, Y.A. Topkaya, "Reducibility of Nickeliferous Limonitic Laterite Ore from Central Anatolia," *Canadian Metallurgical Quarterly,* 53, no.1 (2014), 26-37.

9. H. Lagendijk and R. T. Jones, "Production of Ferronickel from Nickel Laterites in a DC-arc Furnace", Sudbury, Canada, 1997.

7th International Symposium on High-Temperature Metallurgical Processing
Edited by: Jiann-Yang Hwang, Tao Jiang, P. Chris Pistorius, Gerardo R.F. Alvear F., Onuralp Yücel, Liyuan Cai, Baojun Zhao, Dean Gregurek, and Varadarajan Seshadri
TMS (The Minerals, Metals & Materials Society), 2016

RECOVERY OF POWDERED METALLIC IRON FROM LUDWIGITE ORE VIA REDUCTIVE ROASTING WITH SODIUM SALTS - MAGNETIC SEPARATION

Guanghui Li, Huanpeng Mi, Binjun Liang, Zhiwei Peng, Yuanbo Zhang, Tao Jiang*

(School of Minerals Processing & Bioengineering, Central South University, Changsha, Hunan 410083, China)

Keywords: ludwigite ore; reductive roasting; magnetic separation; sodium salt

Abstract

Ludwigite ore has not yet been utilized on an industrial scale due to its complex mineralogy and fine mineral dissemination. In this study, sodium salts were employed to enhance the separation of iron and boron by reductive roasting-magnetic separation. It was found that the separation of iron and boron as well as the iron grade and recovery were significantly improved by the addition of Na_2SO_4 or Na_2CO_3. Under the reduction temperature of 1050 °C, reduction time of 60 min and magnetic intensity of 0.1T. Compared with the inexistence of sodium salts, the iron grade and recovery of the magnetic concentrate increased from 83.3% and 88.6% to 93.2% and 95.1%, respectively, by adding 6% Na_2CO_3. The XRD and microstructure analyses indicated that the iron oxides were transformed into metallic iron and the average metallic iron grains were grown to over 40 m which was beneficial for separation of metallic iron from gangue.

Introduction

China is rich in mineral resource of ludwigite ore, deposited mostly around its northeastern part. The total reserve of ludwigite ore is about 280×10^6 t. The composition of the ore is 30% TFe, 7.0% B_2O_3, 24% MgO, 15% SiO_2, 1.5% Al_2O_3, and 0.4% CaO[1]. The mineral composition of the ore is complex[2]. Iron and boron mainly occur in the form of szaibelyite and magnetite, and the primary gangue minerals are serpentine, mica and carbonate minerals. Due to complex mineralogy and very fine mineral dissemination, the key to exploit and utilize the ore lies in separation of iron and boron [3]. Up to now, there have been a lot of studies on the effective or comprehensive utilization of ludwigite which provide a variety of approaches to process ludwigite of different chemical compositions and phase components. Current processes for iron-boron separation of ludwigite ore include blast furnace-or pre-reduction followed by electric

* Corresponding author: Dr. Tao Jiang, Email: jiangtao@csu.edu.cn, Tel.: (+86) 731 88877656

arc furnace melting which take advantage of mature technology for iron production. However, both of them suffer undesirable iron-boron separation and insufficient alkali-dissolution activity of boron caused by rapid cooling of molten born rich slag. Evidently, these problems limit the utilization of ludwigite [4,5].

Previous studies indicate that reduction of iron oxide is accelerated even by a small quantity of alkali-metal compounds, and the growth of metallic iron grains is facilitated during the carbothermic reduction of iron ore. Besides, sodium salts react with gangues and the original structure of iron ore is destroyed [6,7]. Thus, the reduction is greatly enhanced. Aiming at recovering iron and simultaneously activating boron which was beneficial for extraction by water leaching, the process of coal-based direct reductive roasting followed by magnetic separation was developed in this article, and sodium salts were employed to enhance the reductive roasting of ludwigite ore. The effects of reducing temperature and time, type and dosage of sodium salts, grinding fineness of magnetic separation feed and magnetic field intensity were investigated. The intensification mechanisms of sodium salts on reduction and magnetic separation were revealed by XRD and optical microscopy analyses.

Experimental

Raw Materials

Ludwigite ore was taken from Liaoning province of China, and ground to 80 wt.% passing 0.074 mm. The chemical composition of the ludwigite ore is shown in Table I. XRD results (as shown in Fig. 1) indicates that the ludwigite ore is mainly composed of magnetite (Fe_3O_4) and szajbelyite ($Mg_2(OH)[B_2O_4(OH)]$), as well as ($Mg_3Si_2O_5(OH)_4$) and magnesite ($MgCO_3$). The proximate analysis of reducing coal and chemical composition of its ash are presented in Table II.

Table I Main chemical composition of the ludwigite ore sample/wt.%.

TFe	FeO	B_2O_3	SiO_2	Al_2O_3	CaO	MgO	K_2O	Na_2O	P	S	LOI
52.1	24.65	5.25	5.36	0.39	0.62	12.61	0.05	0.17	0.014	1.15	2.45

*LOI–Loss on ignition

Table II The proximate analysis of reducing coal and chemical composition of its ash.

Proximate analysis/mass%				S_{total}	Ash fusion property/°C			
M_{ad}	A_d	V_{daf}	F_{Cad}		DT	ST	HT	FT
9.66	7.03	29.25	59.42	0.28	1240	1260	1280	1300

Note: FC_d= Fixed carbon, V_d= Volatile, A_d=Ash

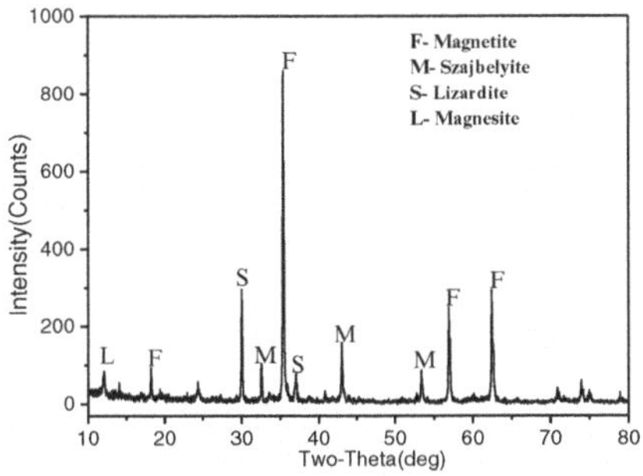

Fig. 1. XRD pattern of the ludwigite sample

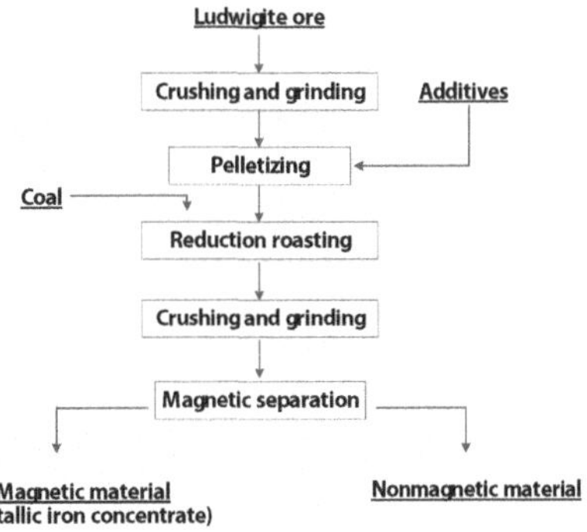

Fig. 2. Experimental flowsheet

Methods

The overall experimental flow sheet is presented in Fig. 2. The ludwigite sample was thoroughly mixed with some sodium salts for green pellets production. The green balls with diameter of 12-16 mm were dried at 120 °C for 2 h. Carbothermic reduction of the dried pellets was conducted in a vertical resistance furnace. After roasting, the reduced pellets were cooled to room temperature in a stainless steel pot isolated from atmospheric oxygen. A sub-sample of the cooled, reduced balls was ground to less than 74 m for metallic iron and total iron content determination by chemical titration method. Another sub-sample of the reduced balls was crushed to the particle size less than 1 mm, and then charged into an ball mill for grinding test at room temperature. After filtration of the slurry, filter residue was further pulped and separated in a Davis Tube. Subsequently, the magnetic material of metallic iron was obtained. The instruments and evaluation indexes of which more details were given in our previous work[3].

Results and Discussion

Effects of dosage of sodium salts

Two different sodium salts were used to intensify the reduction and magnetic separation of ludwigite ore pellet. The reducing conditions were 1050 °C of temperature, 60 min of time, and 90% passing 74 m of grinding fineness of reduced pellet, and 0.1 T of magnetic field intensity.

The effects of sodium carbonate and sodium sulfate dosages on reduction and magnetic separation are shown in Fig. 3 and Fig.4. It can be seen in Fig. 3 that, as the sodium carbonate dosage increases from 0 to 8%, the iron metallization of reduced pellet increases from 86.3% to 92.8%, and the Fe grade of magnetic material increases from 83.3% to 94.2%. The iron recovery of magnetic separation also increases from 88.6% to 94.2%. As seen in Fig. 4, with the increase of the sodium sulfate dosage, the iron recovery of magnetic separation increases initially and then decreases due to unfavorable effect of excess sodium sulfate on the iron metallization of reduced pellet.

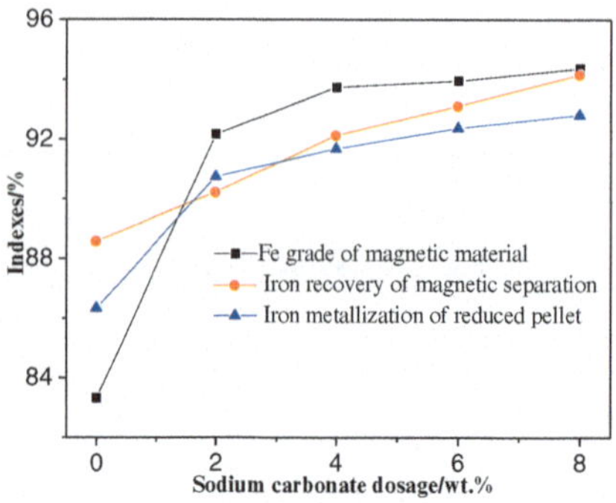

Fig. 3. Effects of sodium carbonate dosage on reduction and magnetic separation

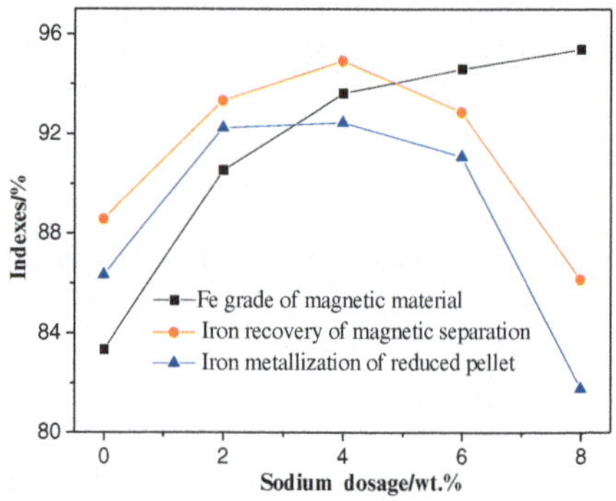

Fig. 4. Effects of sodium sulfate dosage on reduction and magnetic separation

Effects of roasting temperature

Fig.5 describes the effect of roasting temperature on the reduction and magnetic separation when the reducing time was fixed at 60 min. The sodium carbonate dosage was 6%, the grinding fineness of reduced pellet was 90% passing 74 m and the magnetic field intensity was 0.1T.

As observed from Fig5, roasting temperature has a significant impact on the reduction. It can be seen in Fig. 5 that, as the reducing temperature increases from 900 °C to 1100 °C, the iron metallization of reduced pellet increases from 89.3% to 92.8%. The Fe grade of the magnetic material and the iron recovery increase from 90.2% to 94% and from 87.9% to 94.4%, respectively. The above results show that the reducing temperature significantly influences the reduction and magnetic separation of ludwigite ore pellet. From a practical point of view, the reducing temperature was fixed at 1050 °C in the subsequent experimental series.

Effects of roasting time

The effect of roasting time on reduction and magnetic separation is shown in Fig. 6. The reducing temperature was fixed at 1050 °C. The sodium carbonate dosage was 6%, the grinding fineness of reduced pellet was 90% passing 74 m and the magnetic field intensity was 0.1T. As the reducing time increases from 30 min to 60 min, the iron metallization of reduced pellet increases from 89.9% to 92.3%, and the total iron grade of magnetic material and iron recovery of magnetic separation increase from 88.9% to 93.2% and from 91.2% to 95.1%, respectively. The iron grade and iron recovery keep unchanged when reduction time is beyond 80 min. It was inferred that the suitable reduction duration was 60 min when ferrous oxides was almost completely reduced

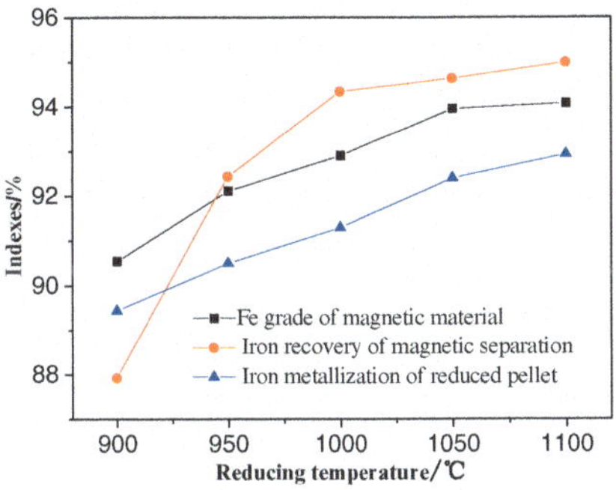

Fig. 5. Effects of roasting temperature on reduction and magnetic separation

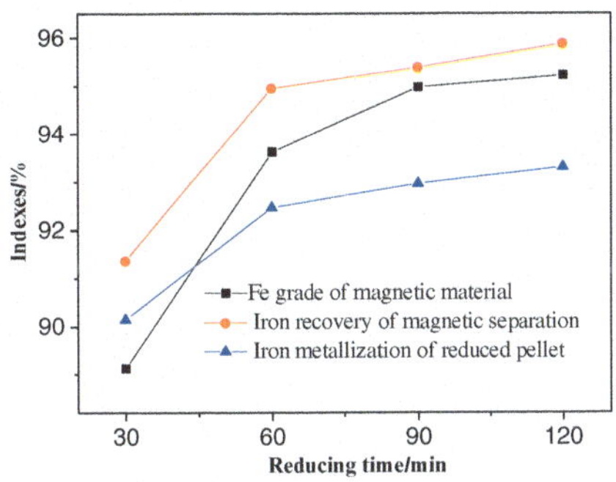

Fig. 6. Effects of roasting time on reduction and magnetic separation

Effects of grinding fineness

Fig.7 shows the effect of grinding fineness of reduced pellet on magnetic separation. The ore was reduced at 1050°C for 60 min in the presence of 6% sodium carbonate before the separation at magnetic field strength of 0.1T. The relationship between grinding fineness and grinding time of reduced pellet is shown in Table III. As seen in Fig. 7, with the increase of grinding fineness, the Fe iron grade of magnetic material increase initially, whereas too long grinding time shows unfavorable effect on the iron recovery of magnetic separation. Therefore, grinding time of 20 min is recommended.

Table III. Relationship between grinding fineness and grinding time of reduced pellet.

Grinding time/min	5	10	15	20	25	30	35
Grinding fineness (−74 m)/%	85.2	87.4	88.4	89.1	90.5	90.6	91.1

Effects of magnetic field intensity

The effect of magnetic field intensity on magnetic separation is shown in Fig. 8. The sodium carbonate dosage was 6%, and the reducing temperature was 1050 °C. The reducing time was 60 min and the grinding fineness of reduced pellet was 90% passing 74 m.

It can be seen in Fig. 8 that, as the magnetic field intensity increases from 0.05T to 2.0T, the Fe grade of magnetic material decreases by 1%, and the iron recovery of magnetic separation increases from 92.5% to 94.0%, It is mainly because more nonmagnetic material is carried into the magnetic concentrate with increasing magnetic field intensity. When the magnetic field intensity exceeds 0.1T, the iron recovery of magnetic separation increases slightly and the phosphorus content of magnetic material increases substantially. Simultaneously, the Fe grade of magnetic material decreases greatly. Thus, the proper magnetic field intensity is 0.1T.

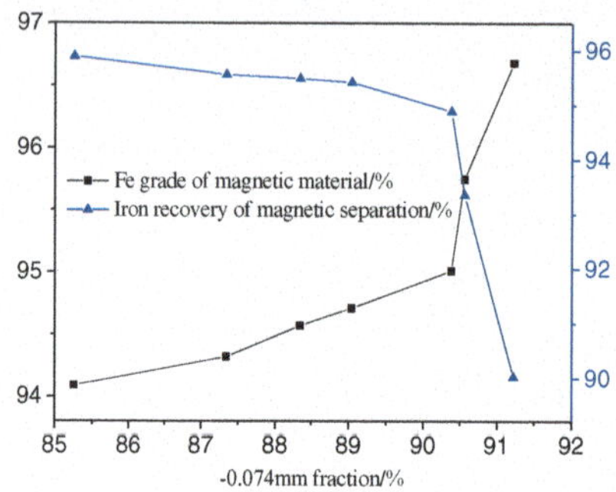

Fig. 7. Effects of grinding fineness of reduced pellet on magnetic separation

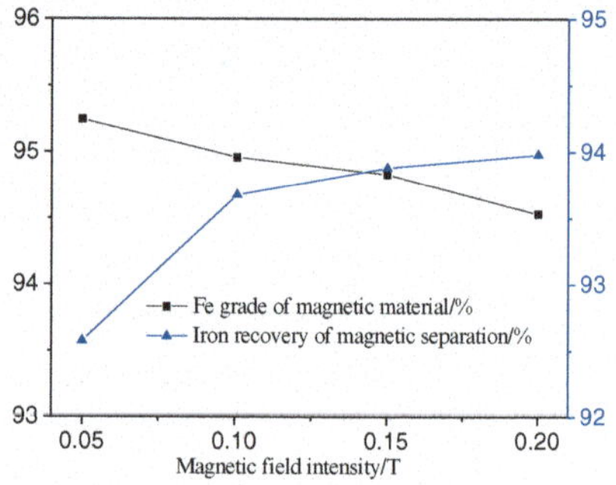

Fig. 8. Effects of magnetic field intensity on magnetic separation

Phase transformation and microstructure of reduced Ludwigite ore

Fig. 9 shows the XRD patterns of reduced ludwigite ore in the absence/presence of Na_2CO_3. When ludwigite ore was subjected to the carbothermic reduction in the absence of Na_2CO_3, magnesium borate ($Mg_3B_2O_6$), metallic iron and (forsterite, ferroan) (($Mg_{1.86}Fe_{0.14})SiO_4$) were generated resulting from decomposition of szajbelyite ($Mg_2(OH)[B_2O_4(OH)]$), lizardite ($Mg_3Si_2O_5(OH)_4$ and magnesite ($MgCO_3$) as well as the reduction of magnetite (Fe_3O_4), as shown in Fig. 9(a). In the presence of Na_2CO_3, magnesium ferrous oxide ($Mg_{1-x}Fe_xO$) instead of forsterite and ferroan is formed. This is due to the generation of sodium magnesium silicate (Na_2MgSiO_4), as demonstrated in Fig. 9(b).
The microstructures of the reduced pellet were studied by using an optical microscope in order to characterize the growing characteristics of metallic iron grains during reductive roasting. The results are shown in Fig. 10. The pellets were reduced at 1050 °C for 60 min. Fig. 10(a) shows the microstructure of the pellet reduced in the absence of sodium salts. Many brilliant white

metallic iron grains can be observed in the reduced pellet. However, the metallic iron grains with sizes of 10-20 m have not been combined with each other entirely, and are still associated with gangue minerals. The results show that the mineralogical separation of metallic iron grains and gangues is insufficient during the reduction. Thus, magnetic separation is also undesirable. Fig. 10(b) presents the microstructure of the pellet reduced in the presence of sodium salts. Compared with Fig. 10(a), most of metallic iron grains in the pellet aggregate together and grow obviously. Most of metallic iron grains exceeds 40 m. Besides, the borderline between metallic iron grains and gangues becomes clear, and thus the magnetic separation improves greatly.

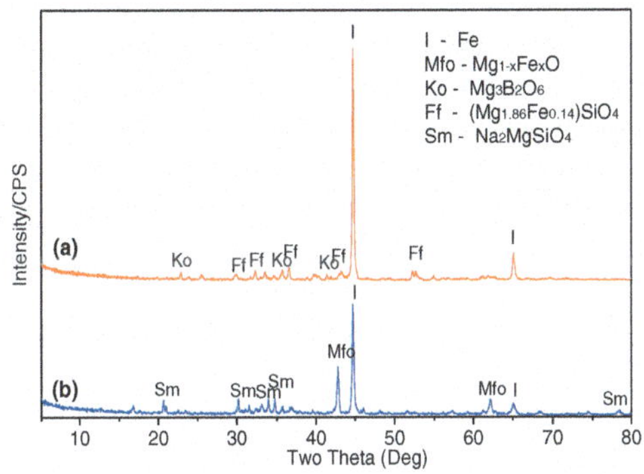

Fig. 9. XRD patterns of reduced ludwigite ore at 1100°C for 60 min, (a): in the absence of Na_2CO_3 and (b): in the presence of 6% Na_2CO_3.

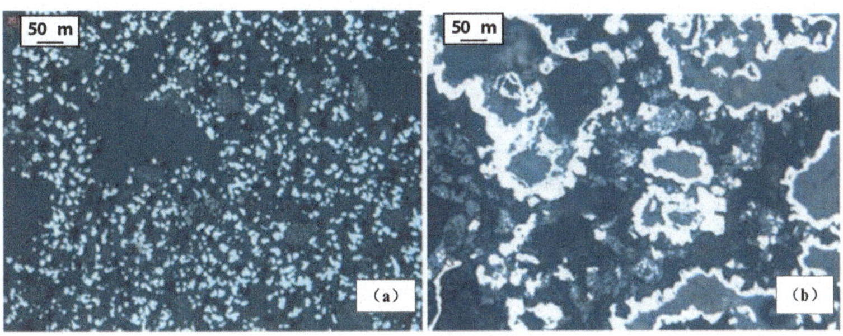

Fig. 10. Optical micrographs of polished cross-section of reduced ludwigite ore obtained at 1100 °C for 60 min; (a) in the absence of Na_2CO_3; (b) in the presence of 6% Na_2CO_3.

Conclusions

The proposed process involves carbothermic reduction of ludwigite ore in the presence of sodium salts and magnetic separation, producing metallic iron powder. The boron enriched in nonmagnetic fraction is good for subsequent boron extraction by water leaching. Ludwigite ore reduced at 1050°C for 60 min in the presence of 6% sodium carbonate yields a magnetic concentrate of 93.2% iron with a recovery of 95.1%.

The optical microscopy analysis of reduced pellet shows that metallic iron grains exist in sizes of 10-20 m and are closely associated with gangues when reduced in the absence of sodium salts. By contrast, the metallic iron grains grow markedly to the mean size of 40 m when reduced in the presence of sodium carbonate. The XRD results indicate that sodium carbonate reacts with SiO_2 and MgO in gangues to form sodium magnesium silicate (Na_2MgSiO_4) during the reductive roasting, destroying szajbelyite ($Mg_2(OH)[B_2O_4(OH)]$), lizardite ($Mg_3Si_2O_5(OH)_4$ and magnesite ($MgCO_3$) structures, which enhances the reduction of iron oxide and facilitates the growth of metallic iron grains, and eventually benefits the magnetic separation.

References

[1]Zheng, X.J., 2008. Production and Application of Boron Compounds. *Chemical Industry Press*, Beijing (in Chinese).

[2] Wang, G., Wang, J., Ding, Y., Ma, S., Xue, Q., 2012. New separation method of boron and iron from ludwigite based on carbon bearing ball reduction and melting technology. *ISIJ International* 52 (1), 45 – 51.

[3]Li,G.,Liang,B.,Rao,M.,Zhang,Y.,Jiang,T.,2014. An innovative process for extracting boron and simultaneous recovering metallic iron from ludwigite ore. *Minerals Engineering*56, 57 – 56.

[4]Weissberger, S., Zimmels, Y., 1983. Studies on concentration and direct reduction of the Ramim iron ore. *Int. J. Miner. Process*. 11, 115 – 130.

[5] .Zhan, H.R., Liu, S.R., Fan, Z.G., 2007. On the relation between cooling rate and activity of boron-rich Slag. *Journal of Northeastern University (Natural Science)* 28 (11), 1604 – 1607.

[6]Jiang,T., Jin,Y.S., Li,Q., Yang,Y.B., Li,G.H., Qiu, G., 2007. Dephosphorization technology of iron ores by Acidithiobacillus ferrooxidans. *Chin. J. Nonferrous Met*. 17, 1718 – 1722 .

[7] Li, G., Shi, T., Rao, M., Jiang, T., Zhang, Y., 2012. Beneficiation of nickeliferous laterite by reduction roasting in the presence of sodium sulfate. *Minerals Engineering*32, 19 – 26.

7th International Symposium on High-Temperature Metallurgical Processing
Edited by: Jiann-Yang Hwang, Tao Jiang, P. Chris Pistorius, Gerardo R.F. Alvear F., Onuralp Yücel, Liyuan Cai, Baojun Zhao, Dean Gregurek, and Varadarajan Seshadri
TMS (The Minerals, Metals & Materials Society), 2016

Selective Reduction of TiO_2-SiO_2 in the Carbothermal Reduction of Titanium Raw Materials for Preparation of Titanium Oxycarbide

Jiusan Xiao[1], Bo Jiang[1], Kai Huang[1], Shuqiang Jiao[2], Hongmin Zhu[1, 3]

[1]School of Metallurgical and Ecological Engineering, University of Science and Technology Beijing; 30 Xueyuan Rd.; Beijing; Beijing; 100083, China
[2]State Key Lab of Advanced Metallurgy, University of Science and Technology Beijing; 30 Xueyuan Rd.; Beijing; Beijing; 100083, China
[3]Tohoku University; 6-6-02; Aramaki-Aza-Aoba, Aoba-ku; Sendai; 980-8579; Japan

Keywords: Carbothermal selective reduction, titanium raw materials, thermodynamic analysis

Abstract

Carbothermal selective reduction of simulative titanium slag and various titanium raw materials were presented and SiO_2 was found to be partially reduced, forming a series of Fe-Si alloys during the process. Thermodynamic analysis of the selective reduction between TiO_2 and SiO_2 was performed using HSC 6.0 software, then a series of equilibrium experiments were designed and presented. The activity of Si reduced from SiO_2 was affected by the presence of Fe reduced from FeO_x and the generation of Fe-Si alloys, promoting the reduction of SiO_2 in the whole process. The results of experiment are in great accordance with the calculated data obtained from thermodynamic analysis.

Introduction

Titanium and its alloys have received great attention from researchers worldwide due to its excellent properties such as high melting point element, extraordinary corrosion-resistance. However, its wide application has been limited by high cost of Kroll process [1]. With a lot of new methods of producing titanium being proposed such as FFC Cambridge process [2], Ono Suzuki (OS) process [3, 4], molten oxide electrolysis (MOE) process [5], USTB titanium electrolysis process has been presented by Jiao and Zhu in 2006 [6, 7], indicating that high purity titanium can be collected in the cathode through the electrolysis of $TiC_{0.5}O_{0.5}$ in molten salt while the carbon and oxygen component in the solid solution evolve into the carbon oxide (CO, CO_2) gases. USTB process exhibits bright prospect and great potential for decreasing the cost of titanium production.
The raw material of USTB process ($TiC_{0.5}O_{0.5}$) can be synthesized through not only carbothermal reduction of TiO_2 but also selective carbothermal reduction of titanium raw materials such as high grade titanium slag and titanium-iron ores [8] whose utilization rate are extremely low, which could extremely contribute to reduce the cost of USTB process. According to our previous work [8], all the other elements (Mg, Al, Ca, Si) are thermodynamically supposed to maintain in oxide form during the selective reduction process when the mole ratio of graphite to TiO_2 is set as 2 : 1 for the preparation of $TiC_{0.5}O_{0.5}$. However, for the reduction of SiO_2 in titanium raw materials, the dissolution of Si into Fe will reduce the activity of Si [9], and facilitate the carbothermal reduction of SiO_2; on the other hand, the presence of other ingredients (CaO, MgO, Al_2O_3 etc.) with decrease the activity of SiO_2 [10], thus restraining the carbothermal reduction of SiO_2. Therefore, in this paper, we have presented thermodynamic analysis of selective carbothermal reduction of TiO_2-SiO_2, and then performed a series of experiments to unveil the mechanism of selective carbothermal reduction of TiO_2-SiO_2.

Experimental details

Thermodynamic analysis of carbothermal selective reduction between TiO_2 and SiO_2 was performed using HSC chemistry 6.0 software. High-purity graphite (99.9 pct, Aladdin Chemistry Co., Ltd) and oxides (99.9 pct, Sinopharm Chemical Reagent Co., Ltd) were used as raw materials in reductive experiments. Ingredients were mixed uniformly according to appropriate ratios by ball milling over 4 hours and then pressed into pellets of 20mm in diameter and 5mm in height under a pressure of 1.37GPa. The pellets were placed in a graphite crucible lined with molybdenum plate, sintering at 1873K over 4 hours under flowing argon atmosphere. The reaction rate was estimated by the mass loss in the sintering and the products were analyzed by X-ray diffraction (XRD), X-ray fluorescence (XRF), scanning electron microscopy (SEM) and energy dispersive X-ray spectroscopy (EDX).

The loss ratio of carbon is defined based on the assumption that the weight loss is completely attributed to CO produced by carbothermal reduction, including the reaction rate is also listed below:

$$\eta_C = \frac{(W_{RS} - W_{PS}) \cdot \frac{12}{28}}{W_{RS} \cdot C\%_{RS}} \tag{1}$$

$$\eta = \frac{\eta_{C,exp.}}{\eta_{C,theo.}} \times 100\% \tag{2}$$

Where W_{RS} is the total mass of raw materials, W_{PS} is the total mass of product and $C\%_{RS}$ represents the mass percentage of C in raw materials. For the carbothermal reduction of oxides, the theoretical loss ratio of carbon is 100% except for that of the formation of $TiC_{0.5}O_{0.5}$, represented by Eq. (2), is only 75%. Thus, it is convenient for us to not only decide whether the reactions are proceeding just as expected but also ensure the amount of carbon spent on converting TiO_2 to $TiC_{0.5}O_{0.5}$ using the loss ratio of carbon instead of the total mass loss.

Results and discussion

Thermodynamics of selective reduction of TiO_2-SiO_2

A lot of researches have been presented focusing the carbothermal reduction of TiO_2 [11, 12]. According to our previous work TiO_2 is carbothermally reduced in the series of:

$$TiO_2 \rightarrow Ti_nO_{2n-1} \rightarrow Ti_3O_5 \rightarrow Ti_2O_3 \rightarrow TiC_xO_{1-x} \rightarrow TiO \rightarrow TiC \tag{3}$$

According to the aim of preparing $TiC_{0.5}O_{0.5}$ from titanium raw materials, the chemical equilibrium of the selective reduction system of TiO_2-SiO_2 is:

$$Ti_2O_3 + 3C \rightleftharpoons 2TiC_{0.5}O_{0.5} + 2CO(g) \tag{4}$$

$$SiO_2 + 2C \rightleftharpoons Si + 2CO(g) \tag{5}$$

And the equilibrium constant of each reaction is:

$$K_4 = \frac{P_{CO}^2 \cdot a_{TiC_{0.5}O_{0.5}}^2}{a_{Ti_2O_3} \cdot a_C^3} \tag{6}$$

$$K_5 = \frac{P_{CO}^2 \cdot a_{Si}}{a_{SiO_2} \cdot a_C^2} \tag{7}$$

For C, $TiC_{0.5}O_{0.5}$ and Ti_2O_3, it is reasonable to set their activities approximately as 1. Since metal Si produced from the reduction of SiO_2 will dissolved in Fe, the relationship between the activity coefficient and concentration of Si at 1873K is described as below [9]:

$$\log \gamma_{Si} = -2.12(1-[Si])^2 - 0.22 \tag{8}$$

Thus the concentration of Si dissolving in Fe in 1873K is plotted as a function of the activity of SiO_2, as shown in Figure 1. Obviously the concentration of Si in Fe-Si alloys rises with the increasing of the activity of SiO_2.

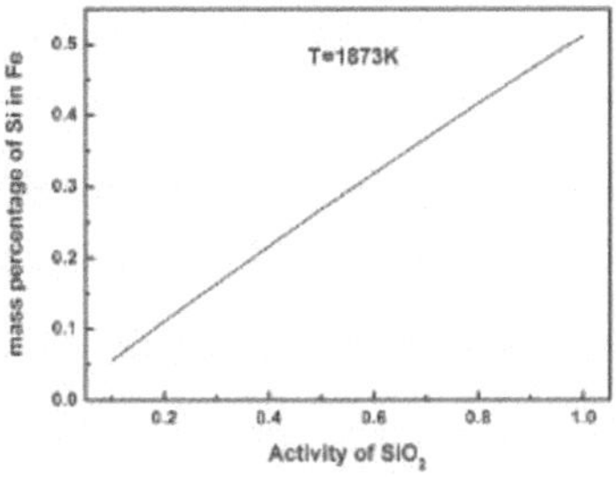

Figure 1. Concentration of Si dissolving in Fe plotted as a function of the activity of SiO_2

<u>Carbothermal selective reduction of TiO_2-SiO_2 slags</u>

In order to study the mechanism of selective reduction between TiO_2 and SiO_2, we have designed a series of experiments, as listed in Table I including the theoretical and experimental loss ratio of carbon and the reaction rate. Considering the aim of reducing TiO_2 to $TiC_{0.5}O_{0.5}$, the mole ratio of TiO_2 to C for all experiments is set as 1 : 2. Obviously all the carbon are utilized to reduce TiO_2 when Fe is not involved, according to not only the loss ratio of carbon of experiment 1 but also the XRD patterns as displayed in Figure 2. (1), where only $TiC_{0.5}O_{0.5}$ exist in the reductive product. This is in good agreement with the thermodynamic analysis. When Fe is added into the raw materials, as experiment 2 presented, loss ratio of carbon increases to 92.5%, indicating that about 70% of carbon is utilized to convert SiO_2 to Si. That is attributed to the activity-decreasing of Si which dissolves in Fe, forming the Fe_xSi phase as presented in Figure 2. (2), and TiO_2 was only reduced to Ti_2O_3 because of the loss of carbon. However, for titanium raw materials, the presence of oxides will influence the activity of SiO_2, thus we have performed the carbothermal reduction of TiO_2-SiO_2 simulative slag containing CaO as representative. Loss ratio of carbon turns out to decrease to 75% below, as experiment 3 and 4 presented, and only $TiC_{0.5}O_{0.5}$ exists in the corresponding reductive products as shown in Figure 3. (1) and (2). It turns out that SiO_2 is not reduced since the activity of SiO_2 is decreased by adding CaO, and all the carbon is reacting with TiO_2 during the whole carbothermal reduction process. Moreover, the reduction of TiO_2 is also restricted since the loss ratio of carbon is below 75%. To sum it up, even though the occurrence of Fe will facilitate the reduction of SiO_2, other oxides included in titanium raw materials will make SiO_2 less active and restrain the reduction of SiO_2, thus it is possible to selectively reduce TiO_2 of titanium raw materials to $TiC_{0.5}O_{0.5}$.

Table I. Details of designed experiments, including theoretical and experimental loss ratio of carbon and the reaction rate

	$n(TiO_2)$:$n(SiO_2)$:n(C):n(Fe):n(CaO)	η_C (theo.)	η_C (exp.)	η
1	1 : 1 : 2 : 0 : 0	75%	77.7%	104%
2	1 : 1 : 2 : 3 : 0	75%	92.5%	123%
3	1 : 1 : 2 : 3 : 0.5	75%	61.3%	82%
4	1 : 1 : 2 : 3 : 1	75%	73.7%	98%

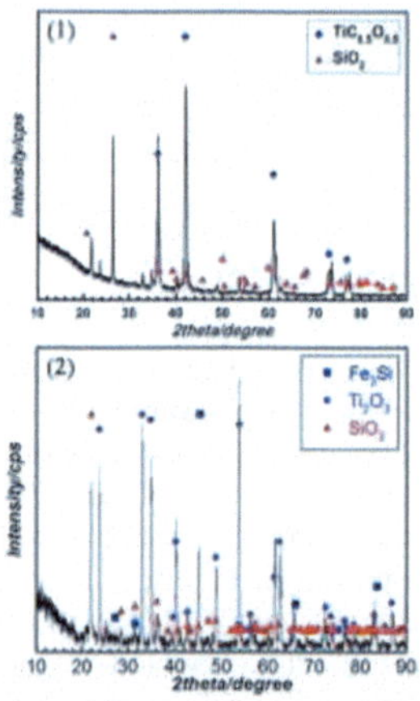

Figure 2. XRD patterns of products obtained through experiment 1 and 2 as listed in Table I

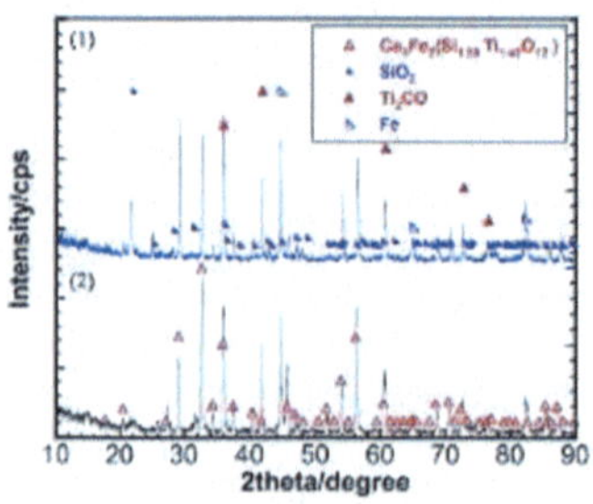

Figure 3. XRD patterns of products obtained through experiment 3 and 4 as listed in Table I

Carbothermal selective reduction of various titanium raw materials

Considering the composition of titanium materials, we have performed carbothermal selective reduction of simulative titanium slag and titanium raw materials firstly. XRD patterns of the reductive product together with that of simulative slag are displayed in Figure 4. (2) and (4), indicating that only Fe_2O_3 and TiO_2 are transformed to Fe and $TiC_{0.5}O_{0.5}$ after the selective reduction process, respectively, while other oxides stay in the slag. Then we have performed carbothermal selective reduction of various titanium raw materials. XRD patterns of products obtained through selective reduction of various titanium raw materials are displayed in Figure 4. (3). Apparently Fe_2O_3 and TiO_2 are converted into Fe and $TiC_{0.5}O_{0.5}$ after the selective reduction process, which is similar with that of simulative titanium slag mentioned above. According to the theoretical and experimental chemical composition of reductive products listed in Table II, most of SiO_2, CaO, MgO and Al_2O_3 ingredients stay in the slag instead of being reduced, and Fe_2O_3 and TiO_2 are transformed to Fe and $TiC_{0.5}O_{0.5}$, respectively, which is in great agreement with XRD results mentioned before. Therefore, the process of carbothermal selective reduction for the preparation of $TiC_{0.5}O_{0.5}$ from titanium raw materials is found to be theoretically and experimentally feasible.

Table II. Theoretical and experimental chemical composition of products obtained through carbothermal selective reduction of titanium raw materials

Composition	$TiC_{0.5}O_{0.5}$	T.Fe*	MgO	SiO_2	Al_2O_3	CaO
Theo.	72.8	9.65	7.52	5.01	2.51	2.51
Exp.	72.99	9.55	7.36	5.16	2.34	2.6

* T.Fe stands for the total amount of Fe in the sample.

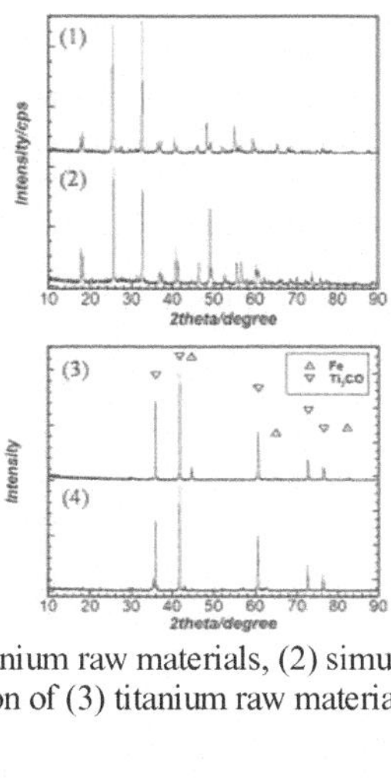

Figure 4. XRD patterns of (1) titanium raw materials, (2) simulative slag, and products obtained through selective reduction of (3) titanium raw materials and (4) simulative slag

The particle size of reductive products is about 2 m based on the SEM and EDX images of reductive products from high titanium slag, as displayed in Figure 5. Moreover, the EDX images present a uniform distribution of $TiC_{0.5}O_{0.5}$ in the reductive products, but Fe and Si seem to concentrate in almost the same area, indicating that Si reduced from SiO_2 dissolve in Fe forming Fe-Si alloys. This is consistent with the thermodynamics analysis results mentioned previously.

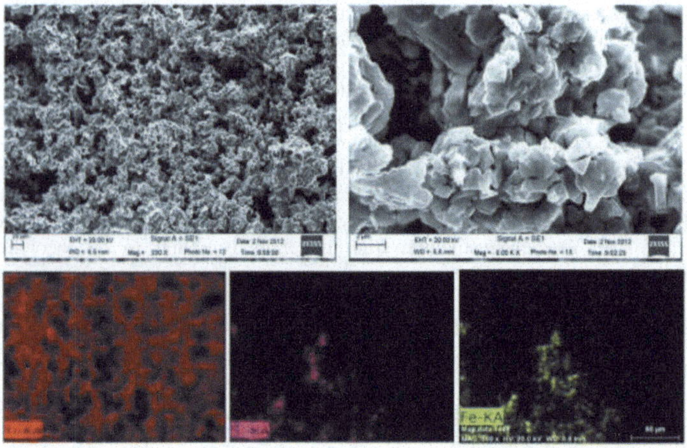

Figure 5. SEM and EDX images of products obtained through selective reduction of high titanium slag

Conclusions

The selective reduction between TiO_2 and SiO_2 have been thermodynamically and experimentally analyzed, indicating that the occurrence of Fe will facilitate the reduction of SiO_2, whereas other oxides included in titanium raw materials will make SiO_2 less active and restrain the reduction of SiO_2. Carbothermal selective reduction of simulative titanium slag and titanium raw materials have been performed and SiO_2 is found to be partially reduced, forming a series of Fe-Si alloys during the whole process, while TiO_2 has been successfully reduced to $TiC_{0.5}O_{0.5}$. The process of carbothermal selective reduction for the preparation of $TiC_{0.5}O_{0.5}$ from various titanium raw materials is found to be theoretically and experimentally feasible, which exhibits perspective potential of the recycling and utilization of titanium raw materials and reducing the cost of USTB titanium electrolysis process.

Acknowledgements

This work was supported by National Science Foundation of China (No. 50934001, 21071014, 5 1102015 and 51004008), the Fundamental Research Funds for the Central Universities (No. FRF -AS-11-002A, FRF-TP-12-023A, FRF-MP-09-
006B), Research Fund for the Doctoral Program of Higher Education of China (No. 2009000611 0005), National High Technology Research and Development Program of China (863 Program, No. 2012AA062302), Program of the Co-

Construction with Beijing Municipal Commission of Education of China (No. 00012047 and 000 12085) and the Program for New Century Excellent Talents in University (NCET-11-0577).

References

1. W.J. Kroll, "The Production of Ductile Titanium," *Transactions of the Electrochemical Society*, 78 (1940), 35-47.
2. G.Z. Chen, D.J. Fray, T.W. Farthing, "Direct Electrochemical Reduction of Titanium Dioxide to Titanium in Molten Calcium Chloride," *Nature*, 407 (2000), 361-364.
3. K. Ono, R.O. Suzuki, "A New Concept for Producing Ti Sponge: Calciothermic Reduction," *JOM*, 54 (2002), 59-61.
4. R.O. Suzuki, "Calciothermic Reduction of TiO_2 and In Situ Electrolysis of CaO in the Molten $CaCl_2$," *Journal of Physics and Chemistry of Solids*, 66 (2005), 461-465.
5. D.R. Sadoway, "The Electrochemical Processing of Refractory Metals," *JOM*, 43 (1991), 15-19.
6. S. Jiao, H. Zhu, "Novel Metallurgical Process for Titanium Production," *Journal of Materials Research*, 21 (2006), 2172-2175.
7. S. Jiao, H. Zhu, "Electrolysis of Ti_2CO Solid Solution Prepared by TiC and TiO_2," *Journal of Alloys and Compounds*, 438 (2007), 243-246.
8. C. Gao, B. Jiang, Z. Cao, K. Huang, H. Zhu, "Preparation of Titanium Oxycarbide from Various Titanium Raw Materials: Part I. Carbothermal Reduction," *Rare Metals*, 29 (2010), 547-551.
9. R.J. Fruehan, "The Thermodynamics Properties of Liquid Fe-Si alloys," *Metallurgical Transactions*, 4 (1970), 865-870.
10. H. Ohta, H. Suito, "Activities of SiO_2 and Al_2O_3 and Activity Coefficients of Fe_tO and MnO in $CaO-SiO_2-Al_2O_3-MgO$ Slags," *Metallurgical and Materials Transactions B*, 29 (1998), 119-129.
11. A. Maitre, D. Tetard, P. Lefort, "Role of Some Technological Parameters During Carburizing Titanium Dioxide," *Journal of European and Ceramic Society*, 20 (2000), 15-22.
12. P. Tristant, P. Lefort, "Approche Cinetique de la Reduction Carbothermique du Dioxide de Titane," *Journal of Alloys and Compounds*, 196 (1993), 137-144.

7th International Symposium on High-Temperature Metallurgical Processing
Edited by: Jiann-Yang Hwang, Tao Jiang, P. Chris Pistorius, Gerardo R.F. Alvear F., Onuralp Yücel, Liyuan Cai, Baojun Zhao, Dean Gregurek, and Varadarajan Seshadri
TMS (The Minerals, Metals & Materials Society), 2016

KINETICS STUDY ON THE PYROLYSIS OF LOW GRADE COALS

Ruiling DU, Keng WU*, Xiao YUAN, Daan XU, Changyao CHAO

State Key Laboratory of Advanced Metallurgy, University of Science and Technology Beijing
Beijing 100083, China
Email: 541348637@qq.com

Keywords: Kinetics, pyrolysis, coal, semi-coke

Abstract

To utilize low grade coals to produce high quality semi-coke which can be used in blast furnaces, the pyrolysis of low grade coals was studied in this paper. According to the characteristics of coal pyrolysis process, the temperature range was divided into three parts. The interfacial chemical reaction model, random pore model and the internal diffusion model with the shrinking volume of resultant were established to describe the different ranges of pyrolysis process respectively. The results showed that the experimental data was fitted well, and all the correlation coefficients (r^2) exceeded 0.97. Finally, the kinetic parameters for each temperature range have been calculated, which laid a necessary foundation for the simulation and expanded production.

Introduction

With the utilization of coals resources, high grade coals reserve is reducing sharply. Due to less carbon content, high volatile and other characteristics, low grade coals can't be directly used into blast furnaces. They should be transformed to semi-coke by retorting. The semi-coke could be widely used in iron making [1, 2].1) The semi-coke replaced part of anthracite can be injected into blast furnaces. 2) A small number of semi-cokes mixed with sinter could be added into blast furnaces. Because the reactivity of semi-coke with CO_2 is higher than coke, the consumption of coke can be reduced. 3) Semi-cokes are often used to produce ferronickel in the process of RKEF. In order to produce high quality semi-coke, the pyrolysis kinetics of coal should be further studied, which can lay a solid foundation for the process of producing semi-coke. In addition, the pyrolysis process of coals, which has obvious effects on the ignition and combustion, is an important part of coals gasification [3]. Therefore, further study on the pyrolysis process of coals can help us utilize the resources effectively [4].

According to the characteristics of pyrolysis process, the temperature range could be divided into three parts [5, 6]. First, the adsorbed gas and crystal water were removed. With a large amount of coal gas and tar vapors evaporating off, coal was transformed into semi-coke in the second phase. In the third range, the semi-coke was further coked. The reaction mechanisms were different in the diverse temperature ranges. It was unreasonable that the whole pyrolysis process was fitted with a single model [7]. Therefore, the interfacial chemical reaction model, random pore model and the internal diffusion model with the shrinking volume of resultant were established respectively to study the pyrolysis kinetics of coals. Finally, the kinetic parameters of each stage are calculated, which can be widely used in the simulation and expanded production. Meanwhile, the mechanisms and rate-controlling links can be obtained, which can be used to predict the reaction rate and degree of difficulty.

Kinetics Models

Interfacial chemical reaction model

The adsorbed gas and crystal water were removed in the low-temperature range. Gaseous product easily overflowed from the solid product. Therefore, interface chemical reaction model was used to the first phase. Assumptions: the reaction particles A is a compact sphere; the reaction type is interface chemical reaction; the chemical reaction equation is A (s) =aG (g) +bS (s) [8].

When the stage of interfacial chemical reaction is the rate-controlling link, the consumption rate of sample A is equal to the interfacial chemical reaction rate, namely $\upsilon_A = -\frac{4\pi r^2 \rho_A}{M_A}\frac{dr}{dt} = \upsilon_C = 4\pi r^2 k_{rea}$. The integral equation can be described as follows:

$$\int_{t_0}^{t_1} \delta_1 k_{rea} dt = y_1(\alpha) \tag{1}$$

Where, $\delta_1 = M_A/\rho_A R_0$; $y_1(\alpha)=1-(1-\alpha)^{1/3}$; The conversion rate α is $\alpha = (R_0^3 - r^3)/R_0^3$; M_A is the relative molecular mass of sample A; ρ_A is the density of sample A, $g \cdot m^{-3}$; r refers to the reactant A radius, m; R_0 refers to the initial radius of sample A, m.

The function $dT=\beta dt$ and the equation $k_{real} = A e^{-\frac{E_{a1}}{RT}}$ are inserted into Eq. (1). The Eq. (2) can be obtained.

$$\ln\frac{y_1(\alpha)}{T^2} = \ln\frac{\delta_1 A_1 R}{\beta E_{a1}} - \frac{E_{a1}}{RT} \tag{2}$$

Where, k_{real} is the constant of interface chemical reaction rate, $m \cdot s^{-1}$; A_1 is the former factor, $m \cdot s^{-1}$; E_{a1} is the reaction activation energy, $J \cdot mol^{-1}$.

According to the linear relationship of $\ln[y_1(\alpha)/T^2]$ and $1/T$ in equation (2), the slope and intercept of the curve can be obtained in terms of the fitting results in the low-temperature range. And then, E_{a1}, A_1, and the function of k_{real} and T can be calculated.

Random pore model

Coals were transformed into semi-coke in the middle-temperature range. The reaction rate was related to the specific surface area, which was enlarged with a large number of pores emerging. The random pore model was adopted to describe the process. Assumptions: the microporous consisted of cylindrical hole with arbitrary radius are randomly distributed in the solid reactants; pyrolysis reaction mainly occurs on the surface of micropores; in terms of ignoring the effect of diffusion, the pyrolysis rate of coal is equal to the chemical reaction rate, which has a proportional relation with the specific surface area [9, 10].

By introducing the pore structure parameters, the random pore model is successfully applied to gas-solid reaction process which has a low conversion rate (0~0.6) and exists a maximum reaction rate or gradually reduced. The relationship for fitting can be expressed as follows:

$$\ln\frac{y_2(\alpha)}{T^2} = \ln\frac{\delta_2 \psi A_2 R}{\beta E_{a2}} - \frac{E_{a2}}{RT} \tag{3}$$

Where, $y_2(\alpha)=\sqrt{1-\psi\ln(1-\alpha)}$; $\psi=\frac{4\pi L_0(1-\varepsilon_0)}{S_0^2}$; $\delta_2=\frac{3(1-\varepsilon_0)M_A}{2\rho_A R_0 S_0}$; S_0, is the initial surface area; L_0 is the total length of the initial pore; ε_0 is the initial void ratio.

Similarly, according to the relationship of $\ln[y_2(\alpha)/T^2]$ and $1/T$ in Eq.(3), the experimental data of pyrolysis at the middle-temperature range can be fitted. Finally, E_{a2}, A_2, and the fuction of k_{rea2} and T can be obtained.

Internal diffusion model with the shrinking volume of resultant

In the third range, the semi-coke was further coked with the obvious volume shrinkage. With the high temperature, the chemical reaction rate was higher than the diffusion. The internal diffusion model with the shrinking volume of resultant is adopted [11]. When the stage of gas internal diffusion is the rate-controlling link, the consumption rate of sample A is equal to gas diffusion rate. It meets the relation $\upsilon_D=a\upsilon_A$, namely $-\frac{4\pi r^2 a\rho_A}{M_A}\frac{dr}{dt}=4\pi r^2 D\frac{dc}{dr}$. The equation can be expressed as follows:

$$\ln\frac{y_3(\alpha)}{T^2}=\ln\frac{\delta_3 D_0 R}{\beta E_{a3}}-\frac{E_{a3}}{RT} \tag{4}$$

Where, $\delta_3=\frac{6M_A(c_i-c_0)}{a\rho_A R_0^2}$; $y_3(\alpha)=(1+V\alpha-\alpha)^{2/3}\left[1-3(\frac{1-\alpha}{1+V\alpha-\alpha})^{2/3}+2\left(\frac{1-\alpha}{1+V\alpha-\alpha}\right)\right]$; $D=D_0e^{-\frac{E_{a3}}{RT}}$; V is the volume ratio of solid resultants and reactants, $V=(R_X^3-r^3)/(R_0^3-r^3)$; c_i and c_0 are the gas concentration of internal and external surface the solid material respectively, $mol\cdot m^{-3}$; D is the effective diffusion coefficient, $m^2\cdot s^{-1}$; D_0 is frequency factor, $m^2\cdot s^{-1}$; a is the stoichiometric number of the gas resultants; R_X is the radius of simple A at temperature T, m.

Similarly, E_{a3} and D_0 can be acquired in terms of the relationship between $\ln[y_3(\alpha)/T^2]$ and $1/T$.

Materials and Methods

Low grade coal A was taken into the experiment. It belongs to bituminous coal and has characteristics of high volatilization. The proximate and ultimate analysis of Coal A was shown in table 1.

Table 1. Proximate and ultimate analysis of Coal A (%)

Proximate analysis				Ultimate analysis				
F_{Cd}	A_d	V_d	M_{ad}	C	H	N	S	O
53.30	8.57	31.21	6.92	65.53	6.05	1.06	0.43	8.44

The comprehensive thermogravimetric analyzer made in Germany Netzsch No. STA 409 C was adapted to the experiment. The sample was heated automatically according to the setting program. And the experimental data was collected automatically by the computer.

The size of sample was prepared to 0.074 ~0.147 mm. The mass of sample was 10~15 mg in each experiment. During the experimental process, the sample was placed at room temperature for 40 minutes to exhaust the air, heated to 105 °C at the rate of 10 K/min staying for 10 minutes to remove the adsorbed water, and heated to 900°C at the different heating rates (5, 25, and 45 K/min). In the experimental process, the gas flow 150 mL/min of high purity nitrogen (N_2≥ 99.999%) was adopted to protect the sample from oxidation.

Results

Coal is a kind of heterogeneous polymer compounds. The main decomposed products of coal are combustible gas, tar, and semi-coke. The mass loss and derivative mass loss curves of pyrolysis process at different heating rates were shown in Fig. 1 and 2.

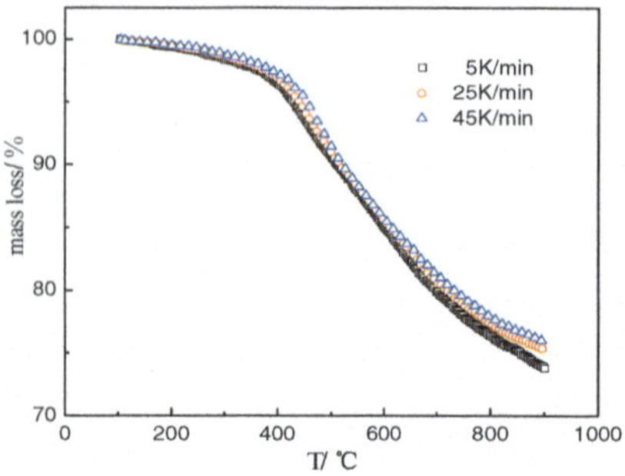

Fig. 1. TG curves of pyrolysis process at different heating rates

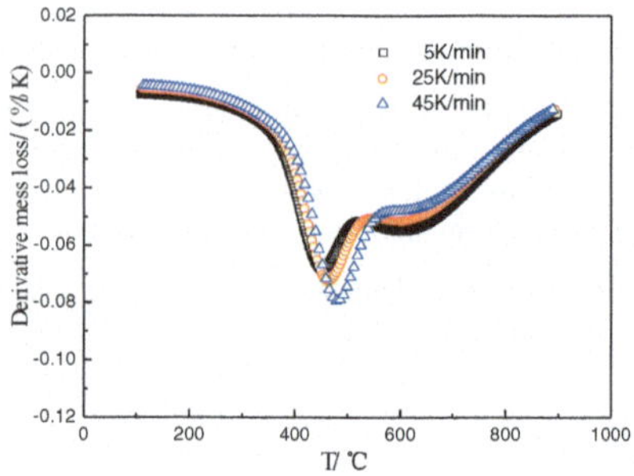

Fig. 2. DTG curves of pyrolysis process at different heating rates

According to the characteristics of curve, the pyrolysis process could be divided into three parts. First, the adsorbed gas and crystal water were removed in the low-temperature range. Meanwhile, the weak keys of coal decomposed. In this part, the heating rates had a little effect on the weight loss rates which maintained at a low level. The pyrolysis rate increased sharply at around 386 °C, which indicated the reaction mechanism and the rate-controlling link were changing.

Second, the reactions occurred in the middle-temperature range were mainly dominated by the depolymerization and decomposition reaction with a large amount of coal gas, tar vapors and etc evaporating off. With the temperature rising, the coal was transformed into semi-coke. Meanwhile, with the heating rate rising, the weight loss of coal A decreased, and the degree of pyrolysis showed a decline. The pyrolysis reaction rate reached maximum at about 470 °C. With the heating rate rising, the maximum increased, and the temperature corresponding to the peak shifted to the high temperature region.

Third, the reactions taken place in the high-temperature range was mainly dominated by Polycondensation reaction with little tar and volatile. Meanwhile, the semi-coke was further coked with the obvious volume shrinkage. At the same temperature, the faster the heating rate raised the more slowly the rate of coal a pyrolyzed. The effect of the heating rates on pyrolysis rates decreased with increasing temperature [12, 13].

Discussion

Temperature range division

According to the characteristics of pyrolysis process, the temperature range could be divided into three parts at the temperature where the variation of weight loss rate reached the maximum. The divided ranges were showed in table 2.

Table 2. Temperature ranges at different heating rates

Heating rates β (K/min)	Low-temperature range /(℃)	Middle-temperature range /(℃)	High-temperature range /(℃)
5	105~369	369~497	497~900
25	105~386	386~524	524~900
45	105~399	399~548	548~900

Based on the table 2, the temperature ranges were various at different heating rates due to the effects of heating rates in the pyrolysis process. The sectioning temperature points were higher as the heating rates enhanced.

Fitting results

According to the table 2, the experimental data of different temperature ranges at various heating rates could be fitted by Eq.(2), Eq.(3), and Eq.(4) respectively. The results were showed in Fig.3, Fig.4, Fig.5, and table 3. In the figures, "□、○、△" respectively represented experimental data at different heating rates, and the lines represented the fitting results of models.

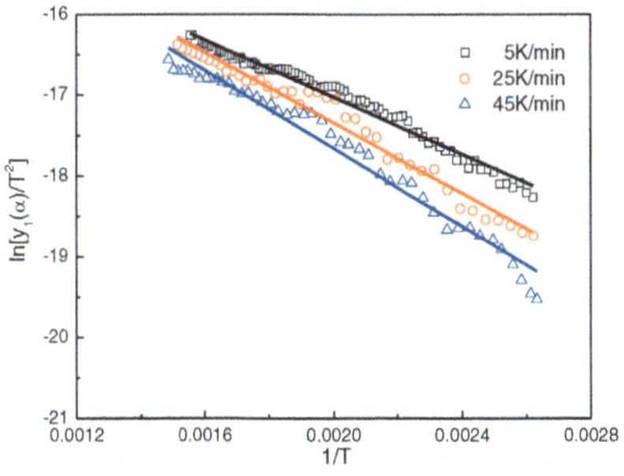

Fig. 3. Fitting Curves in low-temperature range

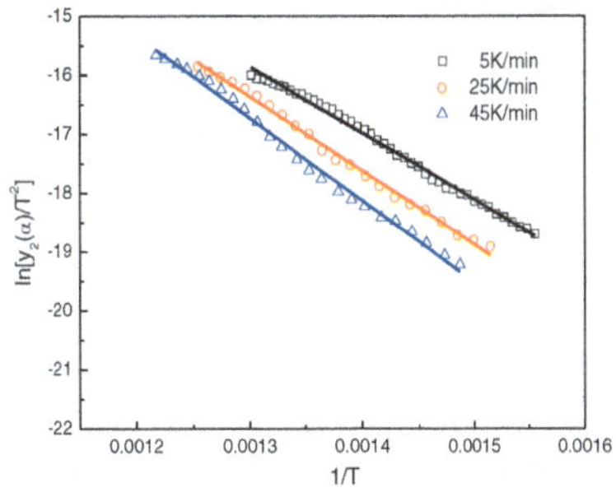

Fig. 4. Fitting Curves in middle-temperature range

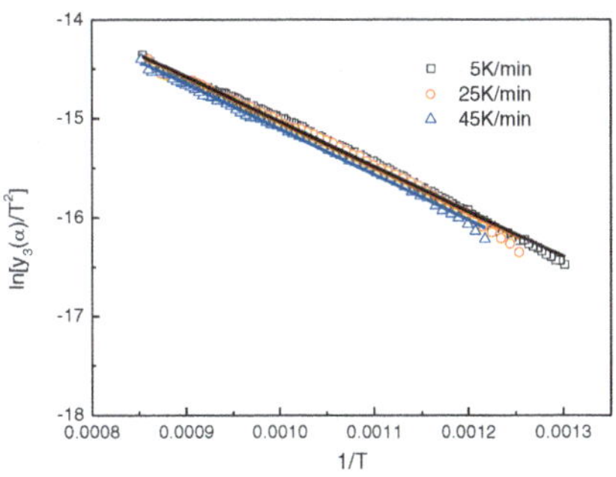

Fig. 5. Fitting Curves in high-temperature range

Table 3. Correlation coefficients (r^2) in three ranges under different heating rates

Heating rates(K/min)	r_1^2	r_2^2	r_3^2
5	0.9724	0.9928	0.9935
25	0.9716	0.9952	0.9959
45	0.9783	0.9949	0.9960

According to Fig.3, Fig.4, Fig.5, and table 3, the experimental data was fitted well, and all the correlation coefficients (r^2) exceeded 0.97. The interface chemical reaction model, random pore model and internal diffusion model with the shrinking volume of resultant were suitable to describe the three ranges of pyrolysis process respectively.

Corresponding kinetic parameters

Based on the slopes and intercepts of fitted curve and relevant parameters, the kinetic parameters in three ranges at different heating rates could be calculated. The results were listed in tables 4, 5, and 6. Where, R_0=1.11×10^{-4} m; ρ_A=1.28×10^6 g/m^3; c_i=1 mol/m^3; M_A=131; ψ=1; c_0=0 mol/m^3; a=1; V=0.733.

Table 4. Corresponding Kinetic Parameters in the low-temperature range

Heating rate β (K/min)	Temperature range T (°C)	Activation energy E_{a1} (J/mol)	Former factor A_1 (m·s^{-1})	Chemical reaction rate constant k_{rea1} (m·s^{-1})
5	105~369	1.47×10^4	2.19×10^{-4}	$\ln k_{rea1}$=-1.76×10^3/T-8.43
25	105~386	1.81×10^4	2.61×10^{-3}	$\ln k_{rea1}$=-2.17×10^3/T-5.95
45	105~399	2.00×10^4	4.49×10^{-3}	$\ln k_{rea1}$=-2.41×10^3/T-5.41

Table 5. Corresponding kinetic parameters in the middle-temperature range

Heating rates β (K/min)	Temperature range T (°C)	Activation energy E_{a2} (J/mol)	Former factor A_2 (m·s^{-1})	Chemical reaction rate constant k_{rea2} (m·s^{-1})
5	369~497	5.58×10^4	7.28×10^{-3}	$\ln k_{rea2}$=-6.71×10^3/T-4.92
25	386~524	6.04×10^4	9.45×10^{-2}	$\ln k_{rea2}$=-7.26×10^3/T-2.36
45	399~548	6.67×10^4	5.37×10^{-1}	$\ln k_{rea2}$=-8.02×10^3/T-0.62

Table 6. Corresponding kinetic parameters in the high-temperature range

Heating rates β (K/min)	Temperature range T (°C)	Diffusion activation energy E_{a3} (J/mol)	Frequency factor D_0 (m^2·s^{-1})	Effective diffusion coefficient D (m^2·s^{-1})
5	497~900	3.78×10^4	2.12×10^{-7}	$\ln D$=-4.54×10^3/T-15.37
25	524~900	3.80×10^4	4.31×10^{-7}	$\ln D$=-4.57×10^3/T-14.66
45	548~900	3.81×10^4	8.23×10^{-7}	$\ln D$=-4.58×10^3/T-14.01

In the tables 4, 5, and 6, the activation energy at the same rate-controlling link increased with the heating rates, but the magnitude of variation was less. The minimum activation energy ranged from 14kJ/mol to

20kJ/mol existed in the low-temperature range. Adsorbed gas and crystal water were removed in this process, which need less power. The maximum activation energy (55~ 67kJ/mol) existed in the second range. The main reactions were macromolecule dissociation of coal and further decomposition of macromolecules in split product, which need more power. The activation energy with minimum variation scope ranged from 37kJ/mol to 39kJ/mol existed in high-temperature range. The semi-coke coked further in this process.

Furthermore, according to the functions of chemical reaction rate constant and effective diffusion coefficient with temperature listed in the tables 4, 5, and 6, the kinetic parameters at different temperatures could be calculated. The results were shown in Fig. 6.

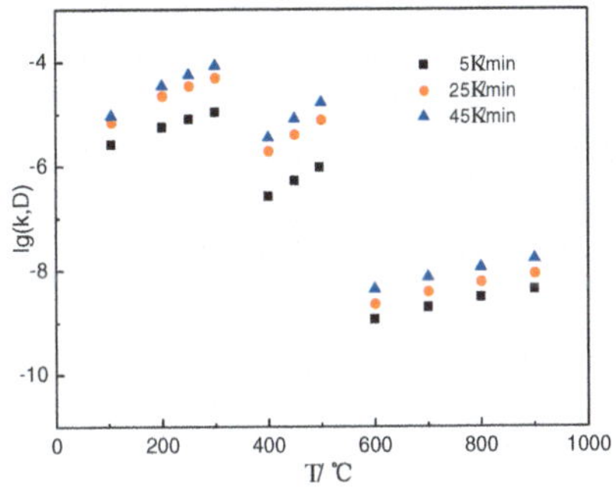

Fig. 6. Corresponding kinetic parameters (in logs) at different temperature

Magnitudes of reaction rate constant felled into 10^{-7}~10^{-5}. Magnitudes of effective diffusion coefficient ranged from 10^{-10} to 10^{-8}. The magnitudes of reaction rate constant had exceeded effective diffusion coefficient about 2~3 magnitudes, which conformed well to data scholars at home and abroad had measured [14, 15].The chemical reaction rate constant or effective diffusion coefficient increased with heating rates at the same temperature, and also increased with the temperature.

Conclusion

According to the characteristics of pyrolysis process, the temperature range was divided into three parts at the temperature where the variation of weight loss rate reached the maximum. The adsorbed gas and crystal water were removed in the low-temperature range. With a large amount of coal gas and tar vapors evaporating off, coal was transformed into semi-coke in the second phase. In the third range, the semi-coke was further coked.

The reaction mechanisms were different in the diverse temperature ranges. The interfacial chemical reaction model, random pore model, and the internal diffusion model with the shrinking volume of resultant were successfully applied to the pyrolysis process of coal. Finally, the functions of reaction rate constants and effective diffusion coefficient with temperature were obtained, which could provide necessary parameters for the utilization coal resource.

With the heating rate increasing, the whole pyrolysis process of coal revealed a thermal hysteresis phenomenon. The activation energy at the same rate-controlling link increased with the heating rates. The chemical reaction rate constant or effective diffusion coefficient increased with the heating rates at the same temperature, and also increased with the temperature.

Acknowledgements

The authors are grateful for support from the National Science Foundation China (Grant No. 51274026), the Independent Research Project of State Key Laboratory of Advanced Metallurgy (Grant No. 41603003), University of Science and Technology Beijing (USTB), China. Correspondence author: Keng WU, E-mail: 541348637@QQ.com.

References

1. Q. H. Liu, K. Wu, and H. Y. Wang, "Kinetic study of tar's separation from coals used in COREX," *Journal of China Coal society*, 37(10) (2012), 1749-1752.
2. Y. B. Zhang, G H. Li, and T. Jiang, "Reduction behavior of tin-bearing iron concentrate psellets using diverse coals as reducers," *International Journal of Mineral Processing*, 110 (2012), 109–116.
3. J. Tomeczek and H. Plaugniok, "Kinetics of mineral matter transformation during coal combustion," *Fuel*, 81 (2002), 1251-1258.
4. S. F. Zhang, F. Zhu, and C. G Bai, "High temperature pyrolysis behaviour and kinetics of lump coal in COREX melter gasifier," *Ironmaking and steelmaking*, 41(3) (2014), 219-228.
5. A. O. Aboyade, J. F. Görgens and M. Carrier, "Thermogravimetric study of the pyrolysis characteristics and kinetics of coal blends with corn and sugarcane residues," *Fuel processing Technology*, 106 (2013), 310–320.
6. H. M. Jeong, M. W. Seo, and S. M. Jeong, "Pyrolysis kinetics of coking coal mixed with biomass under non-isothermal and isothermal conditions," *Bioresource Technology*, 155 (2014), 442– 445.
7. K. Wu, Q. H. Liu, and W. L. Zhan, "Research on Tar Precipitation Kinetics Using Phasewise Analysis," *Journal of Chemical Engineering of Chinese Universities*, 28(4) (2014), 738-744.
8. H. J. Guo, *Metallurgy physical chemistry* (Beijing: Metallurgical industry press, 2006),114-126.
9. J. L. Zhang, G W. Wang, and J. G Shao, "A modified random pore model for the kinetics of char gasification," *BioResource*, 9(2) (2014), 3497-3507.
10. J. S. Gupta, and S. K. Bhatia, "A modified discrete random pore model allowing for different initial surface reactivity," *Carbon*, 38(1) (2000), 47-58.
11. C. R. Zhang, Y. G. Yang, and G. Zhang, "Kinetic model for solid state reactions controlled by diffusion," *ACTA physico-chimica sinica*, 4(5) (1988), 539-544.
12. C. J. Ping, J. H. Zhou, and J. Cheng, "Research on the pyrolysis kinetics of blended coals," *Proceedings of the CSEE*, 27(17) (2007), 6-10.
13. Z. X. Fu, Z. C. Guo, and Z. F. Yuan, "Swelling and shrinkage behavior of raw and processed coals during pyrolysis," *Fuel*, 86 (2007), 418– 425.
14. F. Ferrara, A. Orsini, and A. Plaisant, "Pyrolysis of coal, biomass and their blends: Performance assessment by thermogravimetric analysis," *Bioresource Technology*, 171 (2014), 433– 441.
15. M. Ishida, and C. Y. Wen, "Comparison of kinetic and diffusional models for solid-gas reactions," *AIChE J*, 14 (2) (1968), 311-317.

7th International Symposium on High-Temperature Metallurgical Processing
Edited by: Jiann-Yang Hwang, Tao Jiang, P. Chris Pistorius, Gerardo R.F. Alvear F., Onuralp Yücel, Liyuan Cai, Baojun Zhao, Dean Gregurek, and Varadarajan Seshadri
TMS (The Minerals, Metals & Materials Society), 2016

RESEARCH ON LEACHING OF ZINC SULFIDE ORES THROUGH SYNERGISTIC COORDINATION

Kun Yang [1,2,3], Shiwei Li [1,2,3 *], Jinhui Peng [1,2,3], Libo Zhang [1,2,3], Aiyuan Ma[1,2,3], Weiheng Chen [1,2,3], Feng Xie [1,2,3]

[1] National Local Joint Engineering Laboratory of Engineering Applications of Microwave Energy and Equipment Technology

[2] Key Laboratory of Unconventional Metallurgy, Ministry of Education, Kunming University of Science and Technology, Kunming, Yunnan 650093, China;

[3] Faculty of Metallurgical and Energy Engineering, Kunming University of Science and Technology, Kunming 650093, China;

Abstract: In this thesis, a new process of recovering zinc sulfide ores was developed. The novel synthetic coordination leaching process mainly included two steps: microwave assisted phase reconstruction and compound coordination leaching. By applying various techniques, such as chemical composition analysis, X-ray diffractometry (XRD) and Raman spectroscopy, it could be concluded that sulfide phases would directly oxidized to zinc oxides under the 500 ºC microwave roasting and adding 30% Na_2O_2. What's more, zinc leaching rate reached to 61.16% within 2 hours. Compared to the current technologies, this procedure had the advantage of avoiding SO_2 emission, high leaching efficiency, simple process, manipulate easily, and low cost, which could be expected tremendous application potentials.

Keywords: Zinc sulfide ores, Synthetic coordination, Microwave, Phase reconstruction, Compound coordination leaching

Introduction

Zinc sulfide ores is the main zinc industry resource and undertakes more than 85% of the total zinc production. The traditional zinc sulfide ores metallurgical process is consist of three steps: roasting, leaching and electrowinning (RLE) [1]. While with the development of economics and science, the shortages of RLE such as high impurity (silica, calcium, copper and iron) and SO_2 emission become more and more severer, which would set restrictions on its application [2, 3]. Thus improvement on RLE process is imperative.

Nowadays, hydrometallurgy in ammonia solutions has been considered as a prospective medium in the extraction industry of zinc for advantages of low vapor pressure, low cost, low toxicity, good complexion ability and easy regeneration [4, 5]. What's more, large part of wasteful components are neither soluble in ammonia solutions nor can chelate with ammonia, which allow selective extraction of desired metals. While application of ammonium leaching on zinc sulfide ores is rare or need huge energy cost.

* Corresponding author. Tel.:+8615025170564.
E-mail addresses: lswei11@163.com.

Microwave is one kind of nonionizing electromagnetic radiation, with wavelengths between 1 mm and 1 m (frequencies 300MHz and 300GHz). Unlike traditional electrical resistance heating, microwave energy can be dissipated in the volume of material and transferred to inner energy, thus it has the advantage of non-contact heating, energy transfer, rapid heating, selective heating, volumetric heating, quick start and stop, heating starts from interior of the material body, higher level of safety and automation [6, 7]. It can be expected that the application microwave energy on the ammonium leaching of zinc sulfide ores would have great potential.

In this review, we designed a new process associated microwave heating and ammonium coordination leaching to recover the zinc sulfide ores. By comparing the XRD patterns, Raman spectrum and chemical compositions of raw ores and leaching slag, the reaction mechanisms and microwave function were analyzed.

Experimental

Materials and Instrumentation

The zinc sulfide raw ores in this experiment was from Lanping, Yunnan. Its main chemical compositions and zinc phase distribution were shown in Table 1 and Table 2.

Table 1. Main chemical composition of zinc sulfide ores (mass fraction, %)

Zn_T	Fe	Pb	S	SiO_2	Al_2O_3	MgO	CaO
42.31	9.04	11.51	31.67	<0.5	<0.5	0.085	1.89

Note: Zn_T, total Zn content.

Table 2. Zinc phase distributions of zinc sulfide ores

Zinc phase	$ZnCO_3$	$ZnSiO_3$	zinc sulfides	Franklinite et al	Zn_T
mass fraction/%	0.22	0.35	41.29	0.45	42.31
distribution/%	0.32	0.67	98.81	0.22	100

From Table 1, it could be summarized that this ore had a high grade, with total zinc content being 42.31 %, while zinc almost existed in the form of zinc sulfides, just a little part being $ZnCO_3$, $ZnSiO_3$ and franklinite et al, thus it demanded for staff operation conditions. As the contents of Fe and Pb were 9.04% and 11.51% respectively, it could be classified as high ferrous zinc ores. Its XRD pattern was analyzed as Figure 2.

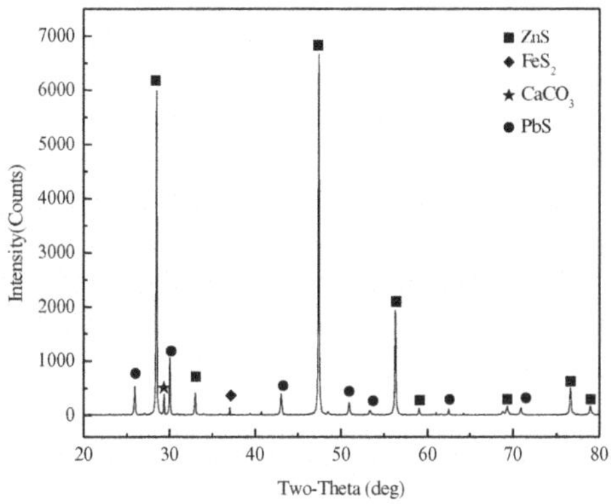

Figure 1. XRD pattern of zinc sulfide ores

As can be seen, crystalline phases in this zinc sulfide ores included ZnS (sphalerite), PbS (galena), FeS_2 (pyrite) and $CaCO_3$. Of the four phase compositions, ZnS was premiere, which was in accordance with chemical analysis and the intergrowth of galena and pyrite would improve the zinc leaching rate of sphalerite.

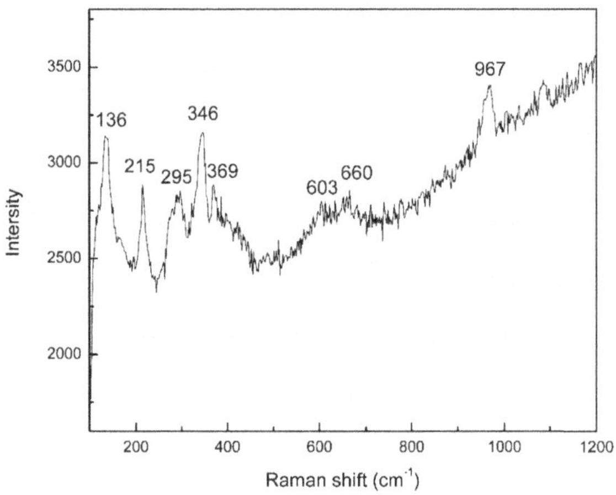

Figure 2. Raman spectrum of zinc sulfide ores

Figure 2 revealed the Raman spectrum of zinc sulfide ores. The results from Figure 2 were interpreted on the basis of published data [8-10].The polarized feature at 967 cm^{-1} was due to Si-OH stretching [11]. The sharp peaks at 600 and 490 cm^{-1} were commonly known as the "defect bands" and various structural interpretations [12, 13]. In terms of atom bonding, the LO mode (350 cm^{-1}) was assigned to the Zn-S vibrations [10-16], while the band at 295 cm^{-1} was a characteristic feature of tetrahedral Fe-S coordination [14]. The 1 LO phonon mode of PbS was at 210 cm^{-1} [17-19]. The Raman bands at 136 cm^{-1} probably originate from the lattice mode vibration of ZnS. The mode at 218 cm^{-1} was assigned to the acoustic phonon combination at W of ZnS [20]. Therefore, we deduced that there also existed hemimorphite.

The high temperature microwave reactor used in this experiment was produced by the Key Laboratory of Unconventional Metallurgy, Ministry of Education, Kunming University of Science and Technology, which used 2450 MHz magnetron tubes as microwave generator. The microwave reactor was mainly consisted of a rectangular cavity, four magnetron tubes and matched waveguide transmission tubes. Through waveguide transmission tubes, microwave produced by magnetron tubes was fed to rectangular cavity, which was designed following rules of low losses, simple structure and versatility. The microwave reactor was rated at 6 kW, and could be controlled through changing power parameters.

Analysis techniques

The zinc sulfide ores and its roast products were characterized by various techniques, including chemical composition analysis, X-ray diffractometry (XRD) and Raman spectroscopy. The Raman spectra were acquired on a Renishaw inVia Raman microscope with a rectangle semiconductor refrigeration CCD detector and coupled to a Leika microscope. The laser beam was focused on the sample by×20 lens, laser power was kept at 20 mW, and the experiments were performed at ambient conditions using a back-scattering geometry. The samples were irradiated with 514.5 nm line of an Ar+ laser. The XRD patterns were recorded on a Rigaku D/max-3B powder X-ray diffractometer using Cu Kα radiation (λ=1.5418 Å) and a graphite monochromator in the diffracted beam. A scan rate of 4° min^{-1} was applied to record a pattern in the 2θ range of 20 to 80°.

The chemicals and solvents were all analytical grade and used as without any further purification. The working solutions were prepared using analytical grade chemicals and double distilled water.

Experimental procedure

The procedure of synthetic coordination leaching of zinc sulfide ores could be divided into two steps: microwave assisted phases reconstruction and compound coordination leaching. Its diagram was shown in Figure 3.

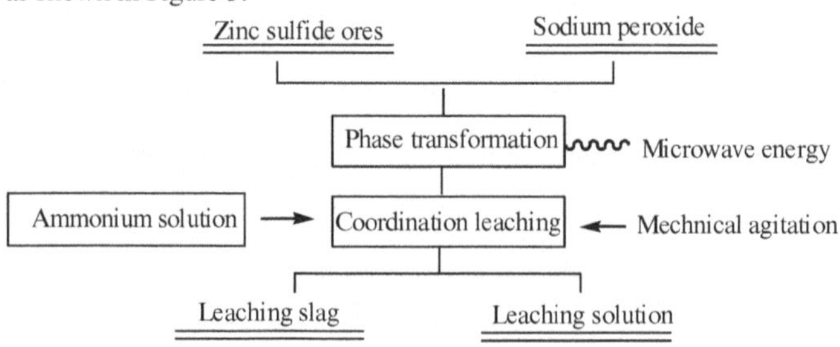

Figure 3. Leaching diagram of zinc sulfide ores

The whole process operated as adding 30% Na_2O_2 into the zinc sulfide ores, and roasting them utilizing microwave energy until 600 ℃, after that leaching it with ammonium chloride solution, which characterized as $c(NH_3)_T$=7.5 mol/L, and $c(NH_3)$: $c(NH_4^+)$=1:1. In the step of leaching, agitation speed was 300rpm, leaching temperature kept at 40℃, and added oxidizing agent- NaClO every other 0.5 h.

Resullts and Discussion

Phase transformation

Figure 4 showd the XRD pattern of zinc sulfide ores after microwave roasting, and it was obvious that part of sulfides directly changed to oxidizes, and the diffraction peaks of ZnS decreased, while PbS diffraction peak increased, meanwhile there appeared some new phases such as ZnO, $Na_2Zn_2Si_2O_7$, $Na(ZnSi_2O_4)$, $ZnSiO_3$, $NaFeO_3$, $Na_6Zn(SO_4)_4$. In the roasting process, no SO_2 emission was detected, which meant that by adding Na_2O_2, microwave energy can directly oxidize S^{2-} to SO_4^{2-}.

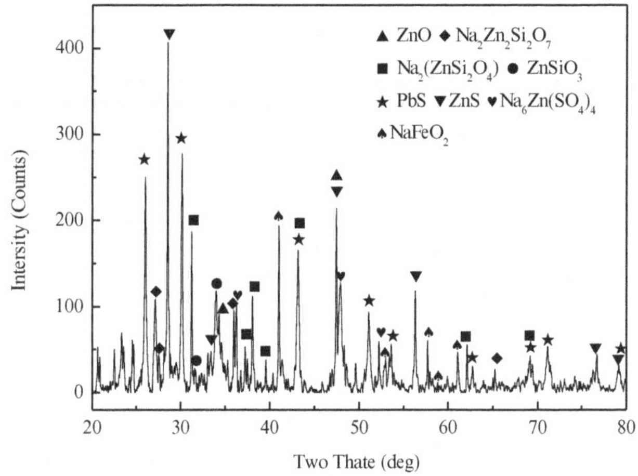

Figure 4. XRD pattern of zinc sulfide ores after microwave roasting

Zinc leaching rate

Leached the above microwave roasting sample with ammonium chloride solution. And the results showed in Figure 5.

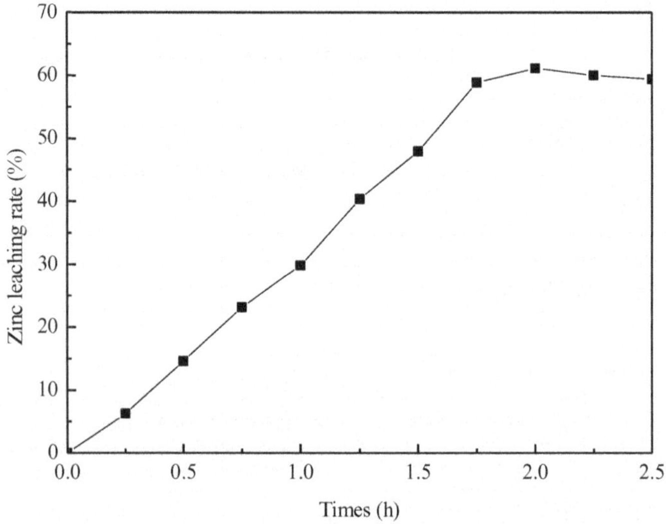

Figure 5. Leaching rate of zinc sulfide ores after microwave roasting

As can be saw from the above picture, with the adding of NaClO, zinc recovery rate was increasing until reaching 61.16%. Microwave energy truly had advantages on zinc sulfide ores leaching.

Conclusions

This thesis dealt with the recovery of zinc sulfide ores. By applying microwave assisting phase reconstruction and compound coordination leaching combined process, part of sulfide phases could directly oxidize to zinc oxides and zinc leaching rate reached to 61.16% within 2 h. Compared to the current technologies, this procedure had the advantage of avoiding SO_2 emission, high leaching efficiency, simple process, manipulate easily, and low cost, which could be expected tremendous application potentials.

Acknowledgements

This work was supported by National Program on Key Basic Research Project of China (973 Program, 2014CB643404) and the Scholarship Award for Excellent Doctoral Student granted by Ministry of Education, China (1319880206).

References

1. A. D. Souza, P. S. Pina, and V. A. Leao, "Bioleaching and chemical leaching as an integrated process in the zinc industry", *Minerals Engineering*, 20 (2007), 591-599.
2. G Deller, "World zinc supply and demand-heading for a late-decade price spike", *Lead & zinc*, 5 (2005), 17-25.
3. G Uçar. "Kinetics of sphalerite dissolution by sodium chlorate in hydrochloric acid", *Hydrometallurgy*, 95 (2009), 39-43.
4. R. Wang et al. "Leaching kinetics of low grade zinc oxide ore in NH_3-NH_4Cl-H_2O system ",

Journal of Central South University of Technology, 15 (2008), 679-683.
5. L. Feng et al. "Pelletizing and alkaline leaching of powdery low grade zinc oxide ores", *Hydrometallurgy*, 89 (2007), 305-310.
6. C. A. Pickles, "Microwaves in extractive metallurgy: Part 2–A review of applications", *Minerals Engineering*, 22 (2009), 1112-1118.
5. M. Al-Harahsheh and S.W. Kingman, "Microwave-assisted leaching-a review", *Hydrometallurgy*, 73 (2004), 189-203.
8. A. Buzatu et al. "The determination of the Fe content in natural sphalerites by means of Raman spectroscopy". *Vibrational Spectroscopy*, 68 (2013), 220-224.
9. S. Kharbish. "A Raman spectroscopic investigation of Fe-rich sphalerite: effect of fe-substitution", *Physics and Chemistry of Minerals*, 34 (2007), 551-558.
10. E. G Osadchii, Y. E. Gorbaty. "Raman spectra and unit cell parameters of sphalerite solid solutions ($Fe_xZn_{1-x}S$)", *Geochimica et Cosmochimica Acta*, 74 (2010), 568-573.
11. G Karthikeyan and A. Pandurangan, "Heteropolyacid ($H_3PW_{12}O_{40}$) supported MCM-41: an efficient solid acid catalyst for the green synthesis of xanthenedione derivatives", *Journal of Molecular Catalysis A: Chemical*, 311 (2009), 36-45.
12. D. P. Zarubin, "The two-component bands at about 4500 and 800 cm^{-1} in infrared spectra of hydroxyl-containing silicas. Interpretation in terms of Fermi resonance", *Journal of Non-Crystalline Solids*, 286 (2001), 80-88.
13. F. L.Galeener, "Planar rings in glasses", *Solid State Communications*, 44(1982), 1037-1040.
14. G A.Hope, R. Woods and C. G Munce, "Raman microprobe mineral identification", *Minerals Engineering*, 14 (2001), 1565-1577.
15. S. Kharbish, "A Raman spectroscopic investigation of Fe-rich sphalerite: effect of fe-substitution", *Physics and Chemistry of Minerals*, 34 (2007), 551-558.
16. X. Fontané et al. "Vibrational properties of stannite and kesterite type compounds: Raman scattering analysis of Cu_2(Fe, Zn)SnS_4", *Journal of Alloys and Compounds*, 539 (2012), 190-194.
17. Y. H. Zhang et al. "A highly regular hexapod structure of lead sulfide: solution synthesis and Raman spectroscopy", *Chemistry-A European Journal*, 13 (2007), 2903-2907.
18. P. G Etchegoin et al. "Temperature dependent Raman scattering of natural and isotopically substituted PbS", *Physica status solidi (b)*, 245 (2008), 1125-1132.
19. N. Wang et al. "Facile synthesis of PbS truncated octahedron crystals with high symmetry and their large-scale assembly into regular patterns by a simple solution route", *ACS nano*, 2 (2008), 184-190.
20. Y. C. Cheng et al. "Raman scattering study of zinc blende and wurtzite ZnS", *Journal of Applied Physics*, 106 (2009), 123505.

7th International Symposium on High-Temperature Metallurgical Processing
Edited by: Jiann-Yang Hwang, Tao Jiang, P. Chris Pistorius, Gerardo R.F. Alvear F., Onuralp Yücel, Liyuan Cai, Baojun Zhao, Dean Gregurek, and Varadarajan Seshadri
TMS (The Minerals, Metals & Materials Society), 2016

EFFECT OF COMPOUND ADDITIVES ON SYNTHETIC MAGNESIUM ALUMINATE SPINEL UNDER LOW TEMPERATURE

Xiaoyan Xiang, Wentang Xia*, Wenqiang Yang

School of Metallurgical and Materials Engineering, University of Science and Technology, Chongqing, 401331, China

Keywords: composite additives, high temperature solid phase sintering, magnesium aluminate spinel, temperature

Abstract

Magnesium aluminate spinel is widely used in the metallurgical industry as refractory material, however, natural magnesium aluminate spinel is rare and most of the magnesium aluminate spinel is synthetic. The traditional synthesis of magnesium aluminate spinel often needs high temperature, which leads to more energy-consumption. In this article, solid phase sintering process was employed to synthesize magnesium aluminate spinel, in which, magnesium and aluminum oxide were used as raw materials and magnesium and aluminum powder were used as composite additives. The mass ratio of magnesium and aluminum on the effect of synthesis of magnesium aluminate spinel and the characteristic of the synthesis at different temperature and time was discussed. Experiment found that the synthesis temperature was observably lowered when the addition of magnesium and aluminum was 2.0% and 1.0% respectively.

1. Introduction

Because of excellent performance in thermostability and slag corrosion resistance, magnesium aluminate ($MgAl_2O_4$) spinel has attracted much attention in the refractory industry and was widely used in the metallurgical, optical, electrical and chemical industry. For example, it used as refractory in the lining of steel-making furnaces, sidewalls and bottom of the steel ladles, melting tanks, wall lining of intermediate frequency furnace and so on [1]. However, inartificial magnesium

* Corresponding author. Tel.: +86-023-65023701.
E-mail address: wentangx@163.com.

aluminate spinel is extremely rare, and most of them are synthesized artificially. The inartificial $MgAl_2O_4$ spinel has existed for a long time, and a variety of techniques have been invented to produce $MgAl_2O_4$ spinel powders. Methods including conventional solid-state-reaction [2], hydroxide coprecipitation [3, 4], flame spray paralysis [5], spray-drying [6], freeze-drying [7], Sol-gel [8], and mechanical activation [9, 10]. Compared to other methods, conventional solid-state-reaction is the most utilized one in industrial production and more than 90% of the $MgAl_2O_4$ spinel in China is produced by solid-state-reaction method. However, because of the poor ability of nucleation, the synthesis of $MgAl_2O_4$ spinel by solid-state-reaction method often needs high temperature and complicated process, which lead to high requirement of equipment and high-energy consumption. Therefore, lower the synthesis temperature of $MgAl_2O_4$ spinel become very important in conventional solid-state-reaction method.

The nucleation and grain growth of $MgAl_2O_4$ spinel has been found to be a very strong function of additives either present in the starting materials or incorporated during the sintering process. A variety of additives were studied to enhance the sintering process, AlF_3 and CaF_2 were reported to decrease the temperature of $MgAl_2O_4$ formation by increasing the cation vacancy concentration when fluorine was incorporated in the lattice by replacing oxygen ions [11]. Better sintered properties were also reached with the addition of different rare earth oxides like Y_2O_3, Yb_2O_3 and Dy_2O_3 [12, 13]. Similarly, TiO_2 and MnO_2 were available for densification of magnesium aluminate [14, 15]. Roy et al. [16] proposed that a crystallographic seeding could stabilize a particular phase and lower the crystallization temperature to enhance the densification. According to their research, the ordered lattice of $MgAl_2O_4$ and percentage of spinel at a lower temperature could be increased by adding to 1.69wt% $MgAl_2O_4$ spinel seed during the sintering process. However, impurity ions were also introduced into the product when the above-mentioned additives (except $MgAl_2O_4$ spinel seed) were added, which adversely affects the thermostability and slag corrosion resistance of $MgAl_2O_4$ spinel. This problem was also existed when other additives were employed in the sintering process.

In our present work, magnesium and aluminum powders were used as composite additives to prepare $MgAl_2O_4$ spinel. The effect of mass ratio of Mg and Al in additives, sintering temperature and time were investigated at a different level. Finally, the sinter was analyzed using an X-ray diffractometer.

2. Experimental Procedure

The magnesium oxide, aluminum oxide, magnesium and aluminum powder used in the experiment were of analytical grade. The magnesium oxide and aluminum oxide were premixed with a molar ratio of 1.2:1, and then additives with different mass ratio of Mg and Al were added accompanying with sufficient mixing. The total mass of additives was 3.0wt% of the magnesium oxide and aluminum oxide. The sintering of the mixed powders was carried out at different temperatures with 100℃ temperature interval ranging from1000℃ to 1400℃ for 4~7h soaking at peak temperature. The powders were mixed-milled with help of high dense ZrO_2 balls in acetone media. The phases in sintered samples were analyzed using an X-ray diffractometer (Rigaku Miniflex diffractometer with Cu Kα X-ray radiation). The spinel content in sintered samples was calculated from the XRD analysis using the following equation [17]:

vol. % of $MgAl_2O_4$

$$= \left\{\frac{\text{height counts of } MgAl_2O_{4(311)}}{\sum \text{height counts}\left[MgAl_2O_{4(311)} + MgO_{(200)} + Al_2O_{3(104)}\right]}\right\} \times 100 \quad (1)$$

Finally, the optimal parameters in sintering process were determined according to the experimental results.

3. Results and discussion

3.1 Effect of additives

Fig. 1 represents the XRD pattern of sintered powders with or without additives. It can be observed that the diffraction intensity of $MgAl_2O_4$ was obviously enhanced when additives were added. This may be attributed to the increase of $MgAl_2O_4$ content and the increased crystallinity of $MgAl_2O_4$ in sintered powders. Fig. 1 also reveals the trend of unreacted magnesium oxide and aluminum oxide, which are less in the sintered powders with additives. Table 1 shows the calculated volume percent of $MgAl_2O_4$ in the sintered powders according to Fig. 1. The phase analysis demonstrates ~11 and 24 vol. % spinel phase at 1300℃ without and with additives respectively, which indicated an obvious increase of spinel formation with additives comparing to without additives. The results also proved that the compound additives containing aluminum and magnesium were conducive for the formation of $MgAl_2O_4$ spinel. The reason could be explained as follows. In the sintering process, compound additives melted first due to the low melting point of aluminum (660℃) and magnesium (648℃). Then, melted additives would induce the nucleation of $MgAl_2O_4$ spinel and accelerate the growth of the crystals.

Table I. The calculated vol. % of spinel content of sintered powders obtained from sintering with or without additives

	Sintering temperature	Sintering time	Content of spinel
With additives	1300℃	4h	24.1%
Without additives			11.4%

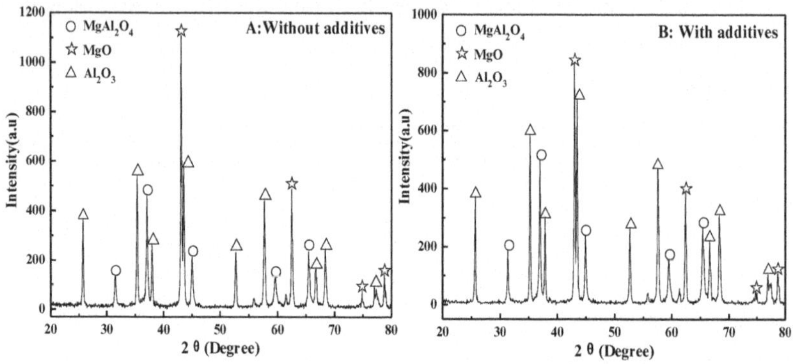

Figure 1 XRD pattern of MgO and Al_2O_3 powder sintered with or without additives (Sintering temperature 1300℃, time 4h)

3.2 Effect of Mg and Al ratio in additives

The mass ratio of Mg and Al in additives also affected the synthesis of $MgAl_2O_4$ spinel and the experimental result was shown in Fig. 2. According to Fig. 2, the initiation of spinel phase was presented with broad lines, which could be ascribed to the low crystallinity of $MgAl_2O_4$ spinel at this temperature. Similarly, the volume percent of $MgAl_2O_4$ in the sintered powders was calculated, and the results were listed in Table 2. It could be seen that the content of $MgAl_2O_4$ spinel increased slowly with the increase of the mass ratio of Mg and Al, the $MgAl_2O_4$ content reached the maximum when the mass ratio of Mg and Al was 2.0. However, further increasing in the mass ratio of Mg and Al led to the decrease of $MgAl_2O_4$ content. Therefore, the mass ratio of Mg and Al in additives was determined to 2.0 in subsequent experiment.

Table Ⅱ. The calculated vol. % of spinel content according to Fig. 2

Mass ratio of Mg and Al	Sintering temperature	Sintering time	Content of spinel
0.2(A)			3.3%
0.5(B)			3.5%
1.0(C)	1000℃	4h	3.6%
2.0(D)			3.8%
5.0(E)			2.9%

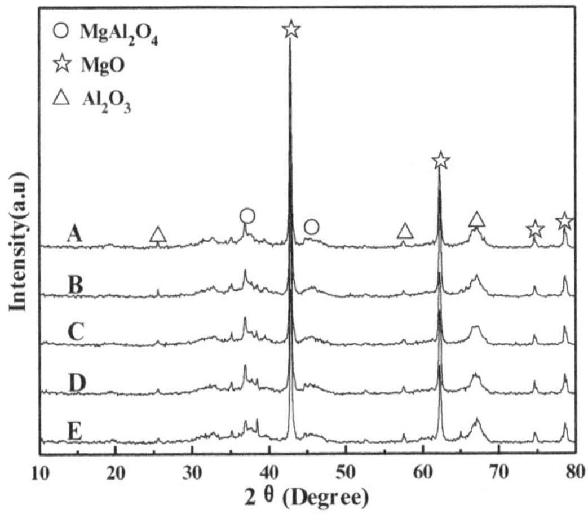

Figure 2 XRD pattern of MgO and Al_2O_3 powder sintered with different mass ratio of Mg and Al (sintering temperature 1000℃, time 4h)

3.3 Effect of sintering temperature and time

The effect of sintering temperature on the formation of $MgAl_2O_4$ spinel was shown in Fig. 3. As can be seen from Fig. 3, the diffraction intensity of MgO became to weakened as the temperature rose, while the diffraction intensity of $MgAl_2O_4$ spinel enhanced at the same time. This could be attributed to the accelerated formation of $MgAl_2O_4$ spinel caused by elevated temperature, which also reduced the MgO content in sintered powders. After a sintering at 1000℃ and 1100℃, only weak diffraction peaks of $MgAl_2O_4$ spinel phase appeared. This indicated few $MgAl_2O_4$ spinels were synthesized at 1000℃ and 1100℃. When the temperature rose to 1200℃, strong and broad diffraction peaks of $MgAl_2O_4$ spinel phase emerged, as the temperature continued to rise, the increase in the diffraction peak height accompanied by sharpening of the diffraction peaks was observed. This indicated that the crystallite size of the $MgAl_2O_4$ spinel increased with elevated sintering temperature. The volume percent of $MgAl_2O_4$ with or without additives was also displayed in Fig. 3, and the results were agreed with the XRD patterns of sintered powders: the content of $MgAl_2O_4$ spinel increased quickly with the increase of sintering temperature.

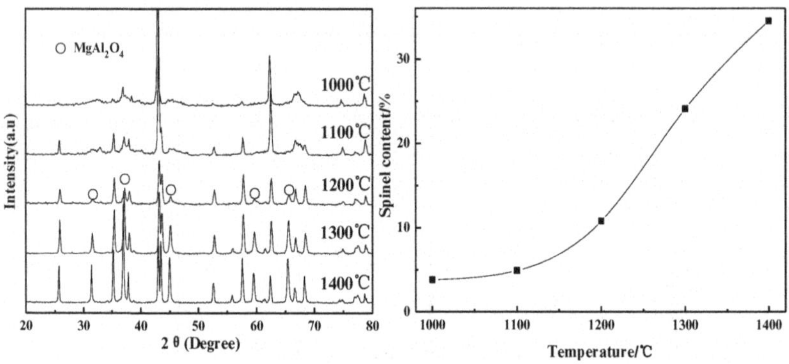

Figure 3 Effect of sintering temperature on the formation of $MgAl_2O_4$ spinel (Sintering time 4h, mass ratio of Mg and Al in additives is 2.0)

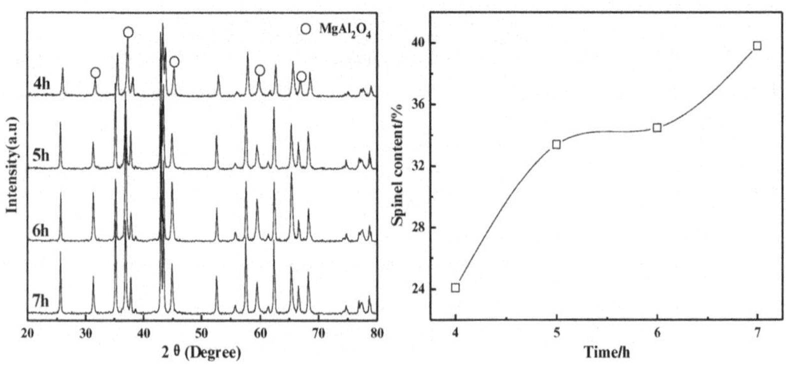

Figure 4 Effect of sintering time on the formation of $MgAl_2O_4$ spinel with additives (Sintering temperature 1300℃, mass ratio of Mg and Al in additives is 2.0)

Fig. 4 revealed the effect of sintering time on the formation of $MgAl_2O_4$ spinel with additives. It can be summarized from Fig. 4 that the diffraction intensity of $MgAl_2O_4$ spinel and the volume percent of $MgAl_2O_4$ were increased with sintering time extending. After a sintering at 1400℃ for 7h, there were still diffraction peaks of MgO and Al_2O_3 appeared. This indicated that the unreacted MgO and Al_2O_3 were existed in sintered powders, and the sintering time was deficient. However, compared to sintered powders without additives as showed in Fig. 5, the diffraction intensity of $MgAl_2O_4$ spinel and volume percent of $MgAl_2O_4$ in sintered powders with additives was increased much more. When the mixed powders were sintered at 1300℃ for 7h, the volume percent of $MgAl_2O_4$ was 39.8% with additives and 22.3% without additives. This meant a shorter time to reach same volume percent of $MgAl_2O_4$ in the sintering process. Therefore, sintering time could be shorted when additives were added in formation of $MgAl_2O_4$ spinel.

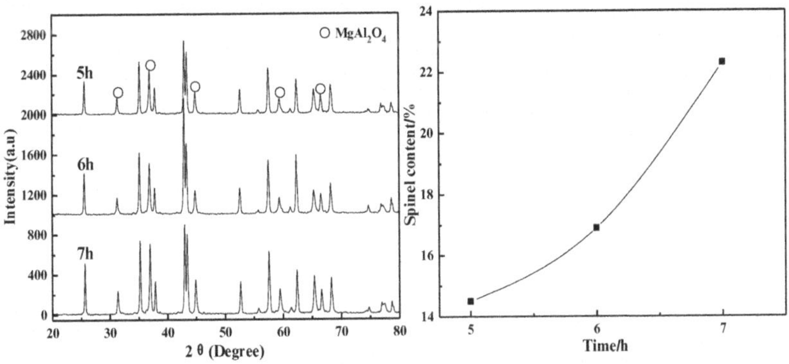

Figure 5 Effect of sintering time on the formation of $MgAl_2O_4$ spinel without additives (Sintering temperature 1300℃)

4. Conclusions

A new additive containing magnesium and aluminum powder was used to prepare $MgAl_2O_4$ spinel. The experimental results showed that the compound additives were beneficial for the formation of $MgAl_2O_4$ spinel and the optimal mass ratio of Mg and Al in additives was 2.0. The results also indicated the possibility to synthesize $MgAl_2O_4$ spinel at a lower temperature with shorter sintering time. When the mixed powders were sintered at 1300℃ for 7h, the volume percent of $MgAl_2O_4$ was 39.8% with additives, while the content was 22.3% without additives.

References

1. L.R. Ping, A.M. Azad, and T.W. Dung," Magnesium aluminate ($MgAl_2O_4$) spinel produced via self-heat-sustained (SHS) technique" , *Materials research bulletin*, 36(7) (2001), 1417-1430.

2. E. Ryskhewitch, *Oxide Ceramics* (New York, NY: Academic Press, 1960), 271.

3. J. Katanić-Popović, N. Miljević, and S. Zec." Spinal formation from coprecipitated gel" , Ceramics International, 1991, 17(1), 49-52.

4. J.G. Li, T. Ikegami, J.H. Lee, et al. " A wet-chemical process yielding reactive magnesium aluminate spinel ($MgAl_2O_4$) powder" , Ceramics international, 2001, 27(4),481-489.

5. C.R. Bickmore, K.F. Waldner, D.R. Treadwell, et al. " Ultrafine spinel powders by flame spray pyrolysis of a magnesium aluminum double alkoxide" , Journal of the American Ceramic Society, 1996, 79(5),1419-1423.

6. J.G.M. Delau. " Preparation of Ceramic Powders from Sulfate Solutions by Spray Drying and Roasting" , Am Ceram Soc Bull, 1970, 49(6), 572-574.

7. C.T. Wang, L.S. Lin, S.J. Yang. " Preparation of $MgAl_2O_4$ Spinel Powders via Freeze-Drying of Alkoxide Precursors" , Journal of the American Ceramic Society, 1992, 75(8), 2240-2243.

8. G. Ye, G. Oprea, T. Troczynski. " Synthesis of $MgAl_2O_4$ spinel powder by combination of sol-gel and precipitation processes" , Journal of the American Ceramic Society, 2005, 88(11), 3241-3244.

9. D. Domanski, G.Urretavizcaya, F. J. Castro, et al. " Mechanochemical synthesis of magnesium aluminate spinel powder at room temperature" , Journal of the American Ceramic Society, 2004, 87(11), 2020-2024.

10. W. Kim, F. Saito. " Effect of grinding on synthesis of $MgAl_2O_4$ spinel from a powder mixture of Mg $(OH)_2$ and Al $(OH)_3$" , Powder technology, 2000, 113(1), 109-113.

11. E. Kostić, S. Bošković, Š. Kiš. " Influence of fluorine ion on the spinel synthesis" , Journal of Materials Science Letters, 1982, 1(12): 507-510.

12. T.F. Baranova, I. Kurskaya, N.A. Dabizha. " Sintering of high purity fused MgO and $MgAl_2O_4$" , Ogneupory, 1981, 46(3): 54-56.

13. L.A. Skomorovskaya. " Magnesia spinel ceramics alloyed with rare-earth oxides, Glass and ceramics" , 1993, 50(4): 165-168.

14. R. Sarkar, G. Bannerjee. " Effect of addition of TiO_2 on reaction sintered MgO-Al_2O_3 spinels" , Journal of the European Ceramic Society, 2000, 20(12): 2133-2141.

15. Y.H. Baik. " Sintering of MgAl2O4 spinel and its characteristics" , Yoop Hikoechi, 1985, 22(6): 29-36.

16. J.F. Pasquier, S. Komarneni, R. Roy. " Synthesis of $MgAl_2O_4$ spinel: seeding effects on formation temperature" , Journal of materials science, 1991, 26(14): 3797-3802.
17. D. Mohapatra, D.Sarkar. " Preparation of MgO-MgAl2O4 composite for refractory application" , Journal of materials processing technology, 2007, 189(1): 279-283.

7th International Symposium on High-Temperature Metallurgical Processing
Edited by: Jiann-Yang Hwang, Tao Jiang, P. Chris Pistorius, Gerardo R.F. Alvear F., Onuralp Yücel, Liyuan Cai, Baojun Zhao, Dean Gregurek, and Varadarajan Seshadri
TMS (The Minerals, Metals & Materials Society), 2016

MICROWAVE THERMAL PREREDUCTION WITH CARBON AND LEACHING OF CHROMITE ORE FINES

Qin Guo [1, 2, 3], Linqing Dai * [1, 2, 3], Lei Li [1, 2, 3], Shenghui Guo [1, 2, 3], Jinhui Peng [1, 2, 3], Libo Zhang [1, 2, 3]

[1]Kunming University of Science and Technology; Kunming 650093, China;

[2]Key Laboratory of Unconventional Metallurgy, Ministry of Education; Kunming 650093, China.

[3]National Local Joint Laboratory of Engineering Application of Microwave Energy and Equipment Technology, Kunming, Yunnan 650093, China.

Keywords: Ferrous, Iron, Chromite fines, Microwave heating, Leaching

Abstract

The ferrous removal process after carbothermic reduction of chromite fines with microwave heating combined with leaching was investigated. Under the following conditions: Reaction temperature: 1150℃, Reaction time : 40min, Semi coke ratio: 12% , the iron content of the reduction product : 13.05% . Metallization achieved was 52.7%. Reduced chromite was leached with 8% of H_2SO_4 and $FeCl_3$ aqueous solution. The results show that: the recovery rate of chromium can reach 88.06%, the mass ratio of chromium to iron is 3.5:1 for leaching 30 minutes with 8% of H_2SO_4; the recovery rate of chromium can reach 88.91%, the mass ratio of chromium and iron is 2.56:1 for leaching 4h at room temperature with $FeCl_3$ aqueous solution (100g/L). The leaching time of latter is longer but can realize the circulation effectively and reduce the waste liquid processing problem.[1]

1. Introduction

Chromium is mainly used in three areas: metallurgy, refractory materials and chemical industries, with important strategic significance [1, 2]. Chromite resource is scarce in China, This account for 0.15% of world reserves and is scattered. Nearly half the reserves are lean ore (Cr_2O_3<35%). Therefore, China depends heavily on imported chromite concentrate, and has not sufficient capacity to resist the turbulence of international market [3]. Thus, for exploring domestic chromite resource, it is necessary to improve the utilization of chromite resources [4, 5].

Author: Qin Guo, Tel: +86-18487134538; E-mail: 303465156@qq.com

Corresponding author: Linqing Dai; Tel: +86-15911509780; Fax: 0871-65138997; E-mail:dailinqing@163.com

Chromite concentrate not only requires high-grade chromium, but the value of Cr/Fe is also an important factor. In the case of low Cr/Fe ratio chromite (Cr_2O_3/FeO <2.5), even though, the content of Cr_2O_3 is ideal, it is difficult to smelt high-grade ferrochrome alloy. How to improve Cr/Fe ratio of chromite with a cost-effective method, is also an issue of concern in the field of chrome smelting. The common agglomeration methods of Chromite fines are sintering, pelletizing and briquetting

The process, combining conventional sintering with smelting furnace, is the lowest energy consumption, in several agglomeration methods combined with smelting furnace process technology.

This paper analyses comprehensively that, at home and abroad, the process, conventional chromite ore dressing and building block, of the application current research; the exposition, combining with the basic principle and characteristics of microwave heating, of the application on microwave technology in chromite pretreatment. It is based on the feasibility and the superiority about the microwave pretreatment chromite, which develop a method, combined method of microwave heating prereduction, to choose the process of chromite with microwave heating sintering ferrochrome powder.

The experimental exploration of microwave heating is an effective ways of chromite pretreatment. The application of microwave heating in chromite provide a new view for chromium industry.

This paper studies that the chromite pretreatment process treated by the way of microwave heating, taking chromium iron ore and Semi-coke from a special ferroalloy company in Sichuan province as raw material. This experiment aims to realize the low ratio of chromium iron and the method of physical separation. The process, combining the microwave heating prereduction with leaching, is explored [6-10].

2. Experiment

2.1. Materials Characterization

In this research, chromite and semi-coke are provided by a special ferroalloy company in Sichuan province, whose main chemical components are shown in Table I and Table II, XRD pattern shown in Figure 1, SEM image shown in Figure 2.

Table I. Main Chemical Composition of Chromite (Mass Fraction, %)

Cr_2O_3	SiO_2	MgO	Al_2O_3	TFe	Cr_2O_3/FeO
42.74	5.98	8.76	11.89	19.35	1.72

Table II. Main Chemical Composition of Semi-coke (Mass Fraction, %)

FC	Ash	Volatile
82.58	9.75	7.67

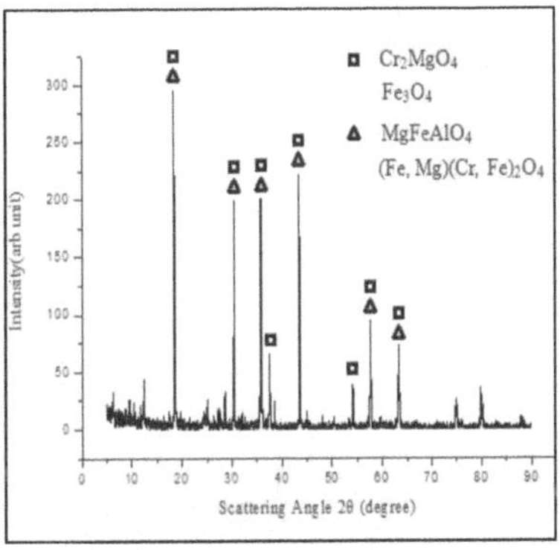

Figure 1. XRD pattern of chromite fine

Figure 2. SEM image of chromite fine

We can see from Table 1 and Figure 1, Chromium ore consists of quiet a few $MgFeAlO_4$, Cr_2MgO_4, (Fe, Mg)(Cr, Fe)$_2O_4$, Fe_3O_4. From Figure 2, It can be seen that the surface of chromite is smooth, dense; the grain is good crystallization degree, big and homogeneous. The refractory is determined by these characteristics. It is the reason of difficulty with conventional roasting.

2.2. Experimental Method

In order to investigate the process, the iron removal process, of carbothermic reduction of chromium fines with microwave heating combined with leaching.

Specific experiments are designed as follows. First, on the condition of reaction temperature is 1150℃, reaction time is 40min, semi coke ratio is 12%, the reduction products can be obtained. XRD analysis of the reduction product after it cools was conducted and then reduction products were leached using sulfuric acid and $FeCl_3$ aqueous solution.

3. Results and Discussion

XRD pattern of reduction products

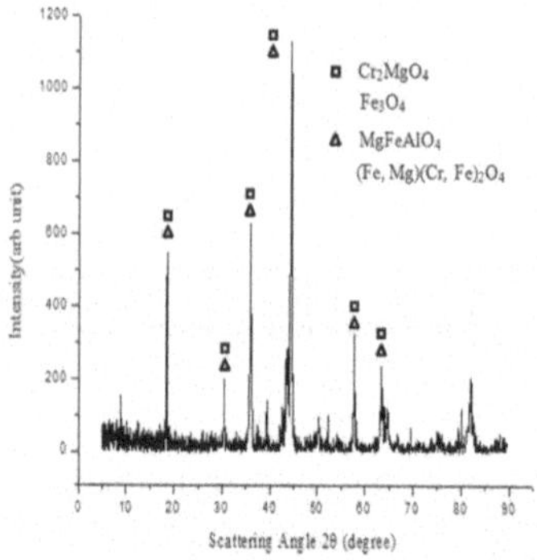

Figure3. XRD pattern of reduction products

Figure 3 shows that the microwave thermal prereduction carbon product consists of Fe, a variety of iron carbon compounds, Cr_3C_2, unreacted chromium spinel and a small amount of unreduction Fe_3O_4, which illustrates that, under the condition of selected microwave prereduction process, generate a detachable iron metal phase, create conditions for the test of ferrochrome separation of the next step.

3.1. The leaching with sulfuric acid

The acid leaching process is a mature method for the extraction of metal elements from ores and secondary resources, which is commonly used in the treatment of nickel ore[11]and iron ore[12, 13]. Cheng Li has published a thesis in regard with leaching chromite in *The Chinese Journal of Nonferrous Metals,* which indicates that using 1:3 hydrochloric acid leaching, after microwave heating chromium reduction, in order to improve the ferrochrome ratio, ideal results were obtained [14]. This experiment adopts the cheaper, more convenient transport and storage of sulfuric acid to leach. Further research of leaching comes from Qing Zhao, who states that using 10% and 30% sulphuric acid in 160℃, in order to study the leaching order of metal elements, the extraction yields of iron, magnesium and aluminum under different leaching conditions[15]. The prereduction products, as material, 8% of the excess dilute sulphuric acid, under the normal temperature table shocks and the optimum process conditions (reduction temperature 1150 ℃, reduction time 40 min, carbon ratio of 12%), will be obtained, which leached under different time, the ferrochrome ratio of leaching products and the recovery of chromium are shown in Table Ⅲ and Figure 4.

Table Ⅲ. Experiment Results of Dilute Sulphuric Acid Leaching

Leaching time/min	TFe /%	Cr_2O_3 /%	(Cr_2O_3/FeO)	Chromium recovery/%
0	25.62	45.16	1.37	--
30	13.95	54.74	3.05	88.06
60	13.99	55.09	3.06	87.49
90	13.95	54.92	3.06	87.61
120	13.87	55.13	3.09	87.25

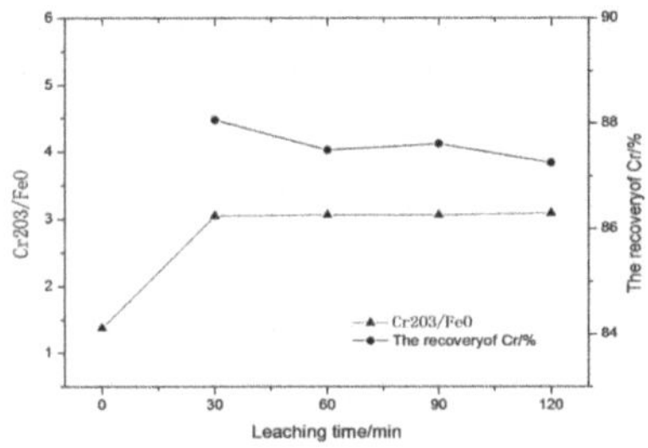

Figure 4. The influence on dilute sulphuric acid leaching of chromium recovery and Cr_2O_3 content

We can see from Figure 4, the excess of dilute sulphuric acid, on the condition of reaction time is 30mins, reaction temperature is room temperature, is leaching. The leaching product, the ferrochrome ratio of 3.05, the chromium recovery of 88.06 %, will be obtained. The leaching time becomes longer, its chromium recovery ratio decrease. The thermodynamic analysis shows that using carbon reduction chromite, at lower temperature, will be created simultaneously Cr_3C_2, elemental iron and iron carbide. Elemental iron, iron carbide and unreacted Fe_3O_4, in the dilute sulphuric acid leaching, will be dissolved in acid, Cr_3C_2 and unreacted chromium spinel, however, will be remained in the residue, so as to realize the separation of ferrochrome. Nevertheless, it will be generated Cr_7C_3 in the higher reduction temperature. This kind of carbide soluble in acid solution, which will be resulted the loss of chromium in residue. It is an effective method that the leaching, for reduction of the prereduction of products under 1150℃, of using dilute suphuric acid.

3.2. The leaching with $FeCl_3$

We try to substitute $FeCl_3$ solution for sulfuric acid solution with leaching experiment, in order to avoid the disadvantages of acid leaching. The leaching results are shown in Table Ⅳ and Figure 5

Table Ⅳ. Experiment Results of $FeCl_3$ Solution Leaching

Leaching time/h	TFe /%	Cr_2O_3 /%	(Cr_2O_3/FeO)	Chromium recovery/%
0	25.84	45.25	1.36	--
1	16.73	49.74	2.31	88.52
2	15.88	52.32	2.56	88.91
3	15.79	52.71	2.59	89.16
4	15.95	53.04	2.58	88.63

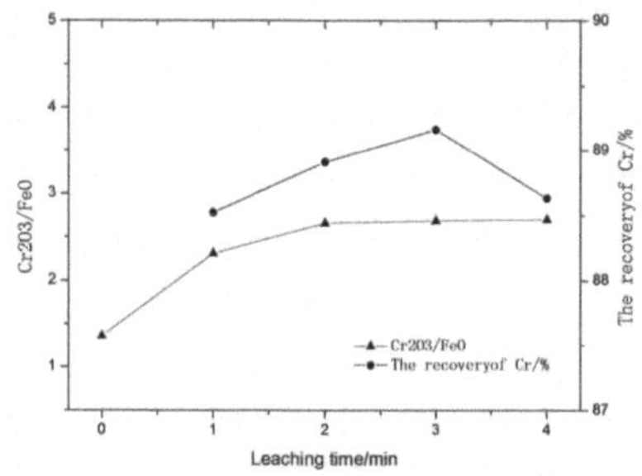

Figure 5. The influence on $FeCl_3$ solution leaching of chromium recovery and Cr_2O_3 content

The leaching liquid of $FeCl_3$ solution is made up of a large number of bivalent iron ions, a small amount of magnesium and aluminum salt. We put the air or the oxidizing agent into the waste liquid, which prompt the bivalent iron of containing oxide as trivalennt, in order to generate the circulation. $FeCl_3$ solution leaching effect slightly worse and leaching time slightly longer, comparing with the acid leaching, but the circulation can reduce effectively the cost and the treatment problem of the waste liquid.

5. Conclusions

(1) The purpose of this article shows that using microwave heating treats the ferrochrome powder ore with microwave thermal prereduction carbon, under the chrome smelting temperature, and removes the generated metallic icon by the methods of magnetic separation and leaching, to improve the ore grade of chromium and chromium iron ratio of chromium iron powder.

(2) The research of the carbon chromium iron ores in microwave thermal prereduction process shows that the reduction products, on the condition of reduction temperature is 1150℃, reduction time is 40 min, carbon ratio for 12%, the metallic iron of 13.05%, the iron metallization rate of 51.70%, will be obtained. The prereduction effect is good, which provides good conditions for ore dressing.

(3) The research of the ferrochrome powder ore prereduction product ferrochrome separation experiment shows that, using microwave thermal prereduction + chemical leaching process, can improve, relatively economic and effective, the ferrochrome ratio of chromite. The leaching products of 54.74% Cr_2O_3, ferrochrome ratio 3.05 and the chromium recovery 88.06%, on the condition of the dilute sulphuric acid of 10% is excess, leaching time is 30mins, leaching temperature is room temperature, will be obtained. The leaching products of 52.32% Cr_2O_3, ferrochrome ratio 2.56, 88.91% chromium recovery, under the condition of the $FeCl_3$ solution of 100g/L is excess, leaching time is 4 hours, leaching temperature is room temperature, can be obtained. Although $FeCl_3$ solution requires longer leaching time, comparing with the dilute sulphuric acid, it can not only realize the circulation of leaching, but also reduce effectly the problem of the waste liquid processing.

Acknowledgements

The authors are grateful for financial supported by the National Natural Scientific Foundation for Young Scholars of Behavior and Mechanism of Silicothermic Reduction Chromite Ores in Microwave field (NO. 51504114).

References

[1] Chengke Liu. University Chemistry[M]. Changsha, Central-south University of Technology Press, 1994, 165~166

[2] Wei Dai. Metallurgical Engineering of Ferrroalloy[M]. Beijing: Metallurgical Industry Press, 1999, 92:121~122

[3] Weber P, Eric R H. The reduction of chromite in the presence of silica flux [J]. Minerals Engineering, 2006, 19(3):318~324

[4] Zhenhai Jia, Techniques for Chrome Ore Fines Processing [J]. Ferroalloy, 1989, (3):39~40

[5] Katayama H G. Mechanism of Reduction of Chromite Oxide by Carbon [J]. Journal of the Japan Institute of Metals, 1976, 40(10): 993~999

[6] K. E. Haque. Microwave Energy for Mineral Treatment Processes-A Brief Review [J]. Int. J. Miner. Process, 1999, 57(1), 1~24

[7] Bryk C, W K L U. Reduction phenomena in composites of iron ore concentrate and coals [J]. Ironmaking and Steelmaking, 1986, 13(2): 70~75

[8] Kobayashi I. A New Process to Produce Iron Directly from Fine Ore and Coal [J].I&M, 2001, (9):19~22

[9] Kobayashi I. Direct Ironmaking Process Using Fine Ore and Coal [J]. Asia Steel 2000, 2000, (B):132

[10] Vazarlis H G, Lekatou A. Pelletising-sintering, prereduction, and smelting of Greek chromite ores and concentrates [J]. Ironmaking and Steelmaking, 1993, 20(1):42~53

[11] Thubakgale et al. A study of atmospheric acid leaching of a South Africn nickel laterite [J]. Minerals Engineering, 54(2013):79~81

[12] Wang et al. Characterization and atmospheric hydrochloric acid leaching of a limonitic laterite from Indonesia [J]. Hydrometallurgy, 129-130(2012):7 ~13

[13] Lekatou A, Walker R D. Solid state reduction of chromite concentrate: melting of prereduced chromite [J]. Ironmaking and Steelmaking, 1995, 22(5): 378~392

[14] Cheng Li et al. Selective Reduction of Chromite Fines by Microwave Treatment [J]. The Chinese Journal of Nonferrous Metals, 2013, 23(2):503~509

[15] Qing zhao et al. Sulfuric acid leaching of South African chromite. Part 1: Study on leaching behavior [J]. International Journal of Mineral Processing, 130(2014): 95~101

7th International Symposium on High-Temperature Metallurgical Processing
Edited by: Jiann-Yang Hwang, Tao Jiang, P. Chris Pistorius, Gerardo R.F. Alvear F., Onuralp Yücel, Liyuan Cai, Baojun Zhao, Dean Gregurek, and Varadarajan Seshadri
TMS (The Minerals, Metals & Materials Society), 2016

MECHANISMS OF STRENGTHENING THE REDUCTION OF FINE HEMATITE IN HIGH SILICON COAL-CONTAINING MINI-PELLETS BY SODIUM ADDITIVES

Zhucheng Huang, Liangming Wen, Ronghai Zhong, Tao Jiang
(School of Minerals Processing & Bioengineering, Central South University, Changsha, Hunan 410083, China)

Keywords: sodium additives; fine hematite; reducing roasting

Abstract

The reduction roasting-magnetic separation and its strengthening of high silicon coal-containing mini-pellets by sodium additives were researched. The results show that the carbon gasification reaction and reduction of fine hematite in high silicon coal-containing mini-pellets with 3% sodium additives were enhanced. Comparing with no sodium additives, the gasification rate of carbon with 3% sodium additives reacting with CO_2 for 40 minutes at 950℃ and the metallization rate of reduced pellets with 3% sodium additives after roasting for 35 minutes at 960℃ increased from 35.15% and 58.72% to 86.98% and 86.15%, respectively. The analysis of SEM equipped with EDS demonstrates that the inner pore of reduced pellets with 3% sodium additives was more developed and the iron grains size significantly enlarged, meanwhile, the iron grade of concentrate increased from 60.45% to 80.23% and the iron recovery rate decreased from 84.67% to 80.78% after grinding-magnetic separation.

Introduction

The iron ore resources of China are extremely abundant and the proven reserves is 57.872 billion tons[1], but 97% of which are poor iron ore resources with the average grade of 33% [2-3]. With the rapid development of Chinese steel industry, the consumption rate of iron ore resources is accelerated and the iron ore resources in China are unable to meet the requirement of steel production[4-5]. Therefore, developing an effective utilization technology is imminent in terms of the characteristic of iron ore resources in China. Processing the low grade iron ore by direct reduction-magnetic separation is a key direction of research. When the low grade iron ore was reduced for 90-120min in the temperature range of 1100℃ to 1200℃, the concentrate with iron grade and recovery rate both above 90% was obtained[6-7]. However, it is a process with high energy consumption and a large amount of gangue minerals (such as SiO_2) existed in the low grade iron ore produces abundant fayalite liquidoid in the temperature range of 1100℃ to 1200℃, it will cause serious influence for direct reduction production.

The gasification reaction of carbon plays an important role on the reduction process of coal-containing pellets and it is very important to research effects of additives on the reduction process of coal-containing pellets[8]. Lots of researches based on the gasification reaction of carbon show that alkali, alkaline earth metal and transition metal compounds can promote the gasification reaction of carbon effectively[9-10].

In this study, a method of reduction roasting of iron ore after granulating with coal is adapted to deal with a low-grade, micro-fine disseminated refractory iron ore and effects of sodium additive on the reduction roasting of coal-containing pellets and carbon gasification reaction were researched.

Experimental

Raw Materials

A poor hematite ore was used in this experiment and the grains less than 0.045mm accounted for 71.41%. The SEM analysis and disseminated sizes of raw ores are showed in Figure 1 and Table I, respectively. The chemical composition analysis is listed in Table II. The coal-containing was crushed to powder with grains less than 0.074mm accounting for 40% and the industrial analysis is listed in Table III.

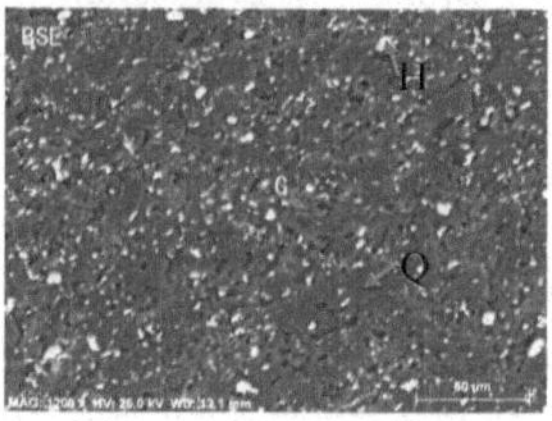

Fig 1 The microstructure image of raw ore

Table I Particle sizes of hematite in the iron ore

Size fraction(mm)	Distribution ratio/%	Cumulative distribution rate/%
-0.105+0.037	4.27	4.27
-0.037+0.026	7.53	11.80
-0.026+0.019	16.84	22.64
-0.019+0.010	31.42	50.06
-0.010	49.94	100.00

Table II Chemical analysis of iron ore %

TFe	FeO	SiO_2	Al_2O_3	CaO	MgO	P	S	K_2O	Na_2O	LOI
31.77	1.56	41.83	4.58	2.51	1.24	0.26	0.026	1.25	0.12	1.49

*LOI– Loss on ignition

Table III Industrial analysis of reductive coal %

$w(FC_d)$	$w(M_d)$	$w(A_d)$	$w(V_d)$	S
58.53	8.16	5.44	36.03	0.20

Note: FC_d= Fixed carbon, M_d= Moisture, V_d= Volatile, A_d=Ash.

Table I and Figure I show that the main ores were consisted of hematite (H) and quartz (Q) and the mass fraction of grains less than 0.01mm accounted for 49.94% of hematite. In addition, the disseminated relationship of hematite and gangue was very complex, so it is difficult to achieve effective separation of hematite and quartz by the conventional beneficiation method. Table II illustrates the grade of iron ore was 31.77% and the main gangue was SiO_2 (41.83%).

Methods

Ball Preparation and Roasting A sodium additives which dissolve easily in water was adopted in this research. Firstly, a certain amount of sodium additives was added into water and dissolved. Then the 1% bentonite and pulverized coal with the certain ratio of C_{fix}/T_{Fe} were blended with water. Lastly, hematite fines was added and made to be balls about 5mm in diameter after blending. After drying, pellets about 350 g and a special amount of coal of 1-5mm in diameter were placed into the rotary tube to reduce for desired time under predetermined temperature. At the end of the reduction roasting, the nitrogen was introduced into rotary tube to cool the reduced pellets. After cooling to room temperature, several samples were selected for milling and magnetic separation. Before milling, the pellets were crushed to be grains less than 1mm and 90% of material was milled to particles less than 0.074mm by grinding. The magnetic field strength of 183KA/m was used to magnetic separation.

Gasification of Carbon A certain quality of pulverized coal with volatile having been wipe out in protective atmosphere mixed with certain proportion of sodium additives was used as coal sample. The high pure nitrogen(400L/h) was led to horizontal tube furnace for 20 minutes to wipe the air out of tube once the temperature stabilised at the set point, then the corundum crucible loaded with coal sample was pushed into the fixed position of tube to react for desired time under predetermined temperature. When the temperature stabilised once again, the high pure nitrogen (400L/h) was replaced by CO_2 (400 L/h). At the end of carbon gasification reaction, the CO_2 (400 L/h) was replaced by purity nitrogen (400L/h) for cooling sample. After cooling to room temperature, the corundum crucible loaded with coal sample that had been weighed was put into the muffle furnace to react with air at 900℃. The residual mass was weighed until no further mass loss and the conversion rate of carbon X was computed by the formula as follows:

$$X=\frac{m_0-m_1}{m_0-m_2}$$

Where: m_0 as the original sample quality; m_1 as the remaining quality of sample after reacting with CO_2 ; m_2 as the residual mass of sample after roasting in muffle furnace.

Microscopic Analysis The phase transition and microstructures of reduced pellets were characterized by using SEM equipped with EDS.

Results and Discussion

Effects of Sodium additives on the Reduction of coal-containing Pellets

Effects of coal-containing dosage Effects of coal-containing dosage on the reduction roasting of hematite pellets with no sodium additives as shown in Fig. 2. As can be observed, the metallization rate of reduced pellets increases with the augment of coal-containing dosage and it greatly increases from 30.07% to 54.60% with the ratio of C_{fix}/T_{Fe} increasing from 0 to 0.3. When the ratio of C_{fix}/T_{Fe} is up to 0.6, the metallization rate of reduced pellets reaches to 61.23%. The results indicate that the augment of coal-containing dosage within a certain range can provide more sufficient reduction atmosphere and it can promote the reduction of hematite pellets.

Effects of Sodium Additives Dosage Effects of sodium additives dosage on the reduction roasting of hematite pellets with the ratio of C_{fix}/T_{Fe} of 0.3 as shown in Fig. 3. As can be observed, the sodium additives can promote the reduction of coal-containing pellets obviously. The metallization rate of pellets reduced for 35 minutes at 960℃ increases from 58.72% to 86.15% with the sodium additives dosage increasing from 0% to 3%. While the sodium additives dosage further increases to 4%, the metallization rate of reduction ore trends to be constant. The results indicate that a certain amount of sodium additives can accelerate the reduction of iron oxide.

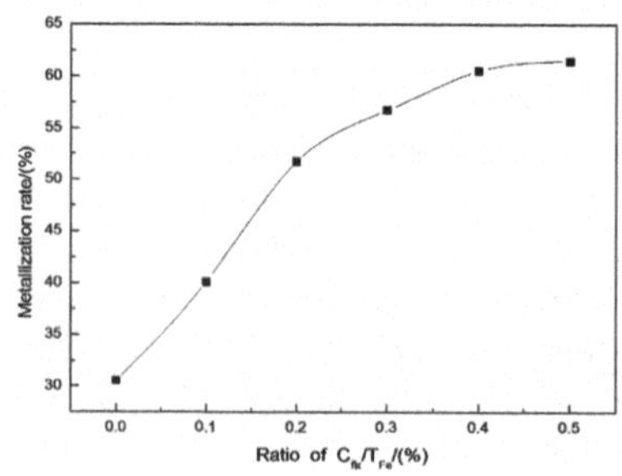

Fig. 2. Effects of the ratio of C_{fix}/T_{Fe} on the metallization rate of reduced pellets with no additive

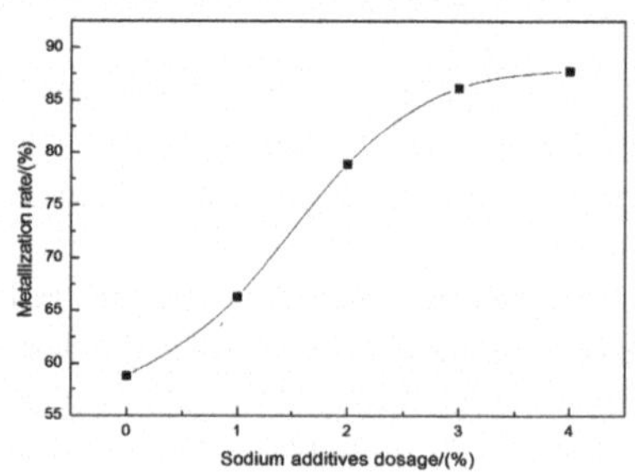

Fig. 3. Effects of sodium additives dosage on the metallization rate of reduced pellets

Effects of Sodium Additives on the Gasification Reaction of Carbon

The gasification reaction of carbon is a key process for the direct reduction of coal-containing pellets and augmenting the gasification rate of carbon can enhance reduction atmosphere, thus promotes the reduction of coal-containing pellets.

Effects of Sodium Additives Dosage Effects of the sodium additives dosage on the gasification reaction of carbon in 20 minutes at 850℃ as shown in Fig. 4. It can be seen that the gasification rate of carbon increases greatly from 10.50% to 33.10% with the sodium additives dosage increases from 0% to 0.5%. When the sodium additives dosage reaches to 3%, the gasification carbon is up to 62.49%. As the sodium additives dosage further increases to 3.5%, the gasification rate of carbon increases slightly to 63.18%. The results indicate that the sodium additives of appropriate dosage can obviously promote the gasification reaction of carbon.

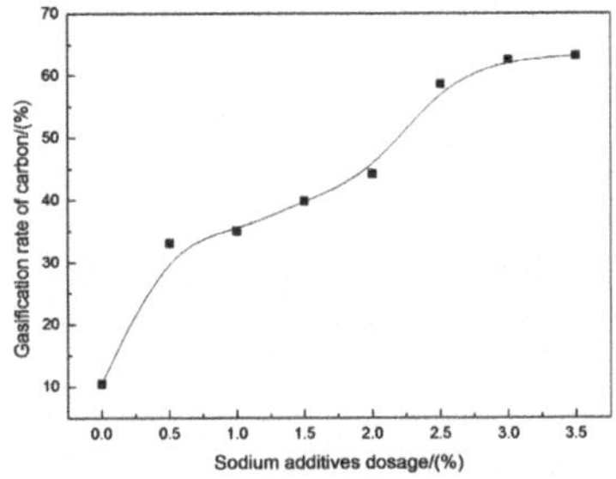

Fig. 4. Effects of sodium additives dosage on the gasification rate of carbon

Effects of Sodium Additives at Various Temperature Effects of sodium additives on the gasification reaction of carbon in the temperature range of 800℃ to 1050℃ within 20 minutes as shown in Fig. 5. As can be observed, sodium additives can significantly reduce the gasification reaction temperature of carbon. When the reaction temperature rises from 800℃ to 1050℃, the gasification rate of carbon with no additives increases from 5.90% to 65.60% within 20min. By contrast, the gasification rate of carbon with 3% sodium additives increases from 35.02% to 98.67%.

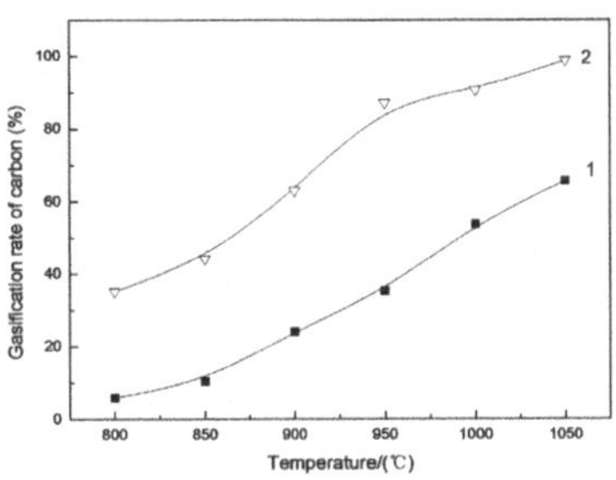

Fig.5. Effects of sodium additives on the gasification rate of carbon at various temperature
(1:carbon with no additives;
2:carbon with 3% sodium additives)

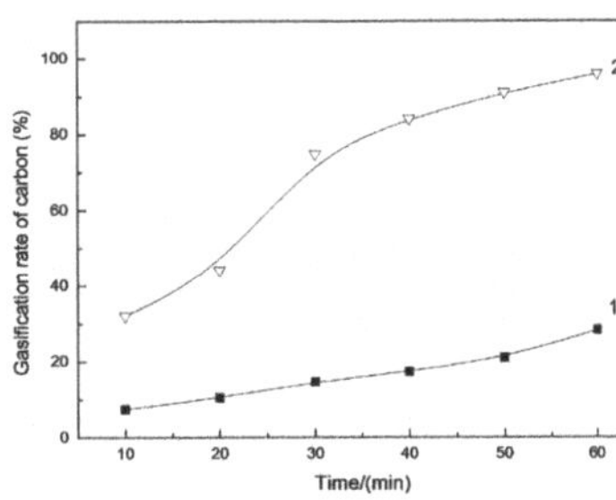

Fig.6. Effects of sodium additives on the gasification rate of carbon within various time
(1:carbon with no additives;
2:carbon with 3% sodium additives)

Effects of Sodium Additives Within Various Time Effects of sodium additives on the gasification reaction of carbon within 10min to 60min at 850℃ as shown in Fig. 6. Comparing with the carbon with no additives, the gasification rate of carbon with 3% sodium additives augments significantly within 10min to 60min and the augment extent is up to 65% within 30min to 60min. When the reaction time extends from 10min to 60min, the gasification rate of carbon with no additives just increases from

7.45% to 28.31% at 850 ℃, by contrast, the gasification rate of carbon with 3% sodium additives increases from 32.15% to 95.83%. The results indicate that sodium additives of appropriate dosage can greatly accelerate the gasification rate of carbon.

Effects of Sodium Additives on the Phase Transition and Microstructures of Reduced Pellets

Effects of sodium additives on the phase transition and microstructures of pellets reduced for 35 minutes at 960 ℃ as shown in Fig. 7 and Fig. 8. Comparing with the reduced pellets with no additives(see Figure7), the iron grains size of reduced pellets with 3% sodium additives(see Figure8) increases from about 5 m to about 10-15 m and the fayalite around silicon decreases. In addition, the inner pore of reduced pellets is more developed and the boundary between iron grains and gangue is clearer.

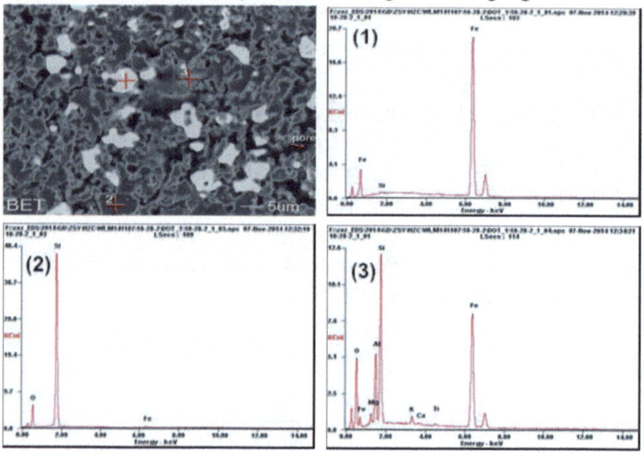

Component		Fe	SiO_2	FeO	Al_2O_3	CaO	MgO	K_2O	Matrix
1	Wt%	99.40	0.60	–	–	–	–	–	Correction
	Mol%	99.42	0.58						ZAF
2	Wt%	–	97.94	2.26	–	–	–	–	Correction
	Mol%		98.11	1.89					ZAF
3	Wt%	–	33.99	49.15	12.91	0.77	2.22	0.97	Correction
	Mol%		38.94	47.09	8.72	0.75	3.79	0.71	ZAF

Fig. 7. SEM-EDS analysis of reduced pellets without additives

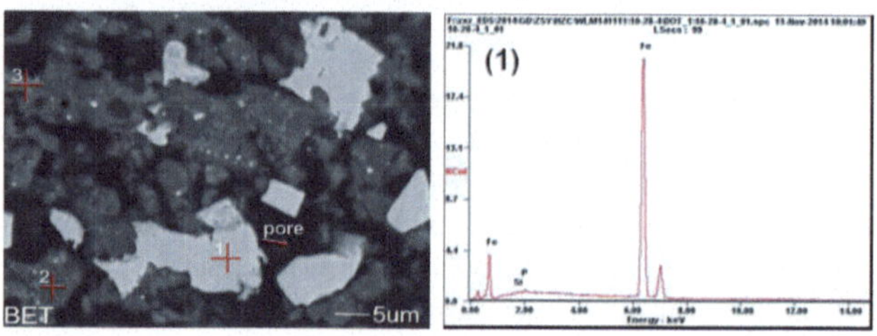

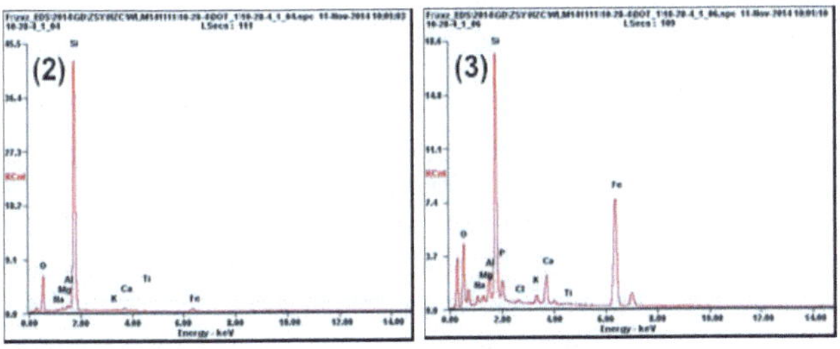

Component		Fe	SiO_2	FeO	Al_2O_3	CaO	MgO	K_2O	Na_2O	Matrix
1	Wt%	99.25	0.75							Correction
	Mol%	99.30	0.70	–	–	–	–	–	–	ZAF
2	Wt%		93.11	2.87	1.45	1.31	0.66	0.41	0.19	Correction
	Mol%	–	93.91	2.42	0.86	1.37	1.00	0.26	0.18	ZAF
3	Wt%		42.87	44.21	4.23	4.90	1.25	1.05	1.63	Correction
	Mol%	–	46.46	40.25	2.71	5.73	2.08	0.74	1.93	ZAF

Fig. 8. SEM-EDS analysis of the reduced pellets with 3% sodium additives

Effects of Sodium additives on the Magnetic-separation of reduced Pellets

Effects of sodium additives dosage on the magnetic-separation of reduced pellets as shown in Fig. 9. As can be observed, sodium additives is beneficial to the magnetic-separation of iron and gangue. When the pellets with various sodium additives dosage reduced for 35min at 960 ℃ were processed under the condition of grinding for 20min and magnetic field strength of 183KA/m, the iron grade of concentrate increases from 60.45% to 80.13% and iron recovery rate declines slightly from 84.67% to 80.78% with the sodium additives dosage increasing from 0% to 3%. While the sodium additives dosage further increases to 4%, the iron grade of concentrate essentially keep constant and the iron recovery rate obviously declines from 80.78% to 73.95%.

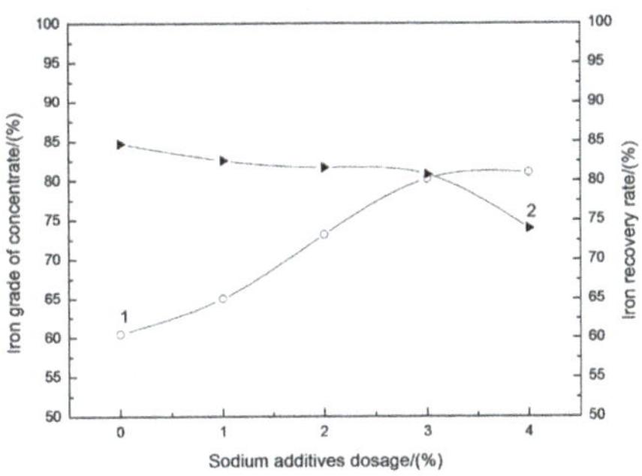

Fig. 9. Effects of sodium additives dosage on the iron grade of concentrate and iron recovery rate (1: iron grade of concentrate; 2: iron recovery rate)

Conclusions

(1) The study on direct reduction roasting of coal-containing pellets demonstrates that, sodium additives can obviously strengthen the reduction of coal-containing pellets. Comparing with reduced pellets with no additives, the metallization rate of pellets with 3% sodium additives increases from 58.72% to 80.78% reduced for 35 minutes at 960℃, and the iron concentrate with iron grade of 80.23% and iron recovery rate of 80.78% is obtained by grinding-magnetic separation.
(2) Sodium additives can significantly enhance the gasification reaction of carbon. Comparing with the carbon with no additives, the gasification rate of carbon with 3% sodium additives increases from 35.15% to 86.98% after reacting with CO_2 for 40 minutes at 950℃.
(3) The analysis of SEM equipped with EDS demonstrates that, comparing with reduced pellets with no additives, the inner pore of reduced pellets with 3% sodium additives is more developed and the boundary between iron grains and gangue is clearer and the iron grains size significantly enlarge.

References

1. J.S. ZHANG et al., "Status and trend of exploitation and utilization of iron Ore resources in China," *China Metallurgy*, 13(1) (2007):1-6.
2. Z.L. HOU et al., "Current situation and potential of iron ore resources in China," *The Exposition of Geology Prospecting*, 20(4) (2005):242-247.
3. H.J. WANG, and G.H. ZHANG, "Exploration situation and potential analysis on supply and demand of china's iron ore resources," *Natural Resource Economics of china*, 31(11) (2013):35-39.
4. J. LIU et al., "The actuality and countermeasure of the iron ore resource in China,"*China Mining Magazine*, 18(12) (2009):1-2,19.
5. Y. F. YU et al., "Processing state and technology progress of iron ore in China," *Conservation and Utilization of Mineral Resources*, 25(6) (2005):43-46.
6. B. XU et al., "study of low-grade iron ore with coal based direct reduction," *Complex utilization of valuable minerals*, 22(06) (2001):20-24.
7. D.Q. ZHU et al., "Beneficiation of super microfine low-grade hematite ore by coal-based direct reduction-magnetic concentration process," *Journal of Central South University(Science and Technology)*, 53(6) (2008):1132-1138.
8. X.M. GUO, H.F. TANG, and Y. ZHANG, "Catalytic mechanism research of alkali metal compounds on reduction of iron oxide by CO," *Journal of the Chinese rare earth society*, 20(9) (2002):440-443.
9. Sansha Coetzee, Hein W.J.P Neomagus, and John R.Bunt, "Improved Reactivity of Large Coal Particles by K_2O_3 Addition During Steam Gasification," *Fuel Processing Technology*, 114(7) (2013):75-81.
10. Z.K. ZHANG, Y.K. LIU, and X. FENG, "Catalytic Gasification Kinetics of Coal by KCl with Carbon Dioxide," *Journal of Xi'an Jiaotong University*, 37(9) (2003):955-961.

7th International Symposium on High-Temperature Metallurgical Processing

Characterization and Simulation of High Temperature Process

Session Chairs:
Baojun Zhao
Tarasankar DebRoy

7th International Symposium on High-Temperature Metallurgical Processing
Edited by: Jiann-Yang Hwang, Tao Jiang, P. Chris Pistorius, Gerardo R.F. Alvear F., Onuralp Yücel, Liyuan Cai, Baojun Zhao, Dean Gregurek, and Varadarajan Seshadri
TMS (The Minerals, Metals & Materials Society), 2016

HEAT AND FLUID FLOW MODELING TO EXAMINE 3D-PRINTABILITY OF ALLOYS

T. Mukherjee[1], J. S. Zuback[1], A. De[1] and T. DebRoy[1]

[1]Department of Materials Science and Engineering, The Pennsylvania State University, University Park, PA, 16801, USA

Keywords: Additive manufacturing, Numerical modeling, 3D-Printability

Abstract

For successful commercial applications of additive manufacturing, metallic parts need to be defect free, structurally sound and reliable. However, additively manufactured metallic parts are susceptible to distortion, lack of fusion defects and compositional changes. Methods to measure susceptibilities of various alloys to these defects are needed but not currently available. Here we show that the necessary theories can be developed based on a well-tested numerical heat transfer and fluid flow model and decades of research in the field of welding. The proposed theories are validated with experimental results reported in the literature.

Introduction

Additive manufacturing (AM) of alloys offers many advantages over the conventional techniques for producing 'near-net-shape' parts. AM allows layer-by-layer fabrication of parts with complex geometries that are used for applications in medical, aerospace, automotive and many other industries [1,2]. However, thermal distortion, lack of inter-layer bonding due to inadequate fusion and loss of alloying elements from molten pool due to vaporization are major challenges in AM.

Thermal distortion in AM originates from non-uniform heating and cooling of the deposit. The thickness of layers deposited, dimensions and mechanical properties of substrate, selection of deposition path [2], deposition strategy [3], time delay between the deposition of successive layers [4], number of layers [5] and other AM variables influence the thermal distortion in AM. Lack of fusion defects originate from inadequate penetration of the molten pool into the substrate or previously deposited layer. Important variables such as thermo-physical properties of the alloy, characteristics of the heat source and processing parameters determine the depth of penetration of the molten pool when depositing a particular layer. Adequate penetration of the molten pool into the substrate or the previously deposited layer needs to be ensured to avoid this defect. During AM of alloys, high temperatures at the surface of the molten pool can result in pronounced vaporization of alloying elements [6]. With a small melt pool size and high surface area-to-volume ratio of the molten pool that are common in AM processing, significant change of composition of the parts can occur due to selective vaporization of volatile alloying elements from the molten pool at different rates. Changes in chemical composition affect both microstructure and mechanical properties [6].

Here we show how a numerical heat transfer and fluid flow model can be used to examine the propensity of the above mentioned three common defects during AM of metallic parts. Susceptibilities of various alloys to these defects are constructed based on these theories and tested using independent experimental data. The results presented here provide a quantitative basis for 3D printing of sound metallic parts.

Heat transfer and fluid flow model

The three dimensional transient heat transfer and fluid flow model used here solves the equations of conservation of mass, energy and momentum [1]. These equations are available in standard text books [7] and published literature [8,9]. The model computes temperature and velocity fields at various locations from AM process variables, such as the laser power, power density distribution, scanning speed, chemical composition, particle size, feed rate and thermo-physical properties of the alloy powder [1, 10]. In laser-assisted AM, a fraction of the laser beam energy is transferred from the laser beam to the alloy powders and substrate. The energy is included in the energy conservation equation as a volumetric heat source, S_v as:

$$S_v = \frac{D\eta_a P}{\pi \left(r_{eff}^{z}\right)^2 h_d} \exp\left[-D\left(\frac{x^2+y^2}{\left(r_{eff}^{z}\right)^2}\right) - \beta(z)\right] \tag{1}$$

where D is the power distribution factor, η_a is the laser absorption coefficient, P is the laser beam power, $\left(r_{eff}^{z}\right)$ refers to the radius of the laser beam at any plane at a distance z from the surface, h_d is the thickness of the material layer being deposited on the substrate, x and y are the co-ordinates from the axis of the laser beam on the surface and $\beta(z)$ is laser power attenuation coefficient. [11]

The heat input from the laser beam to the substrate or the previously deposited layer is considered as a surface heat flux, S_s, which is expressed as [12]:

$$S_s = \frac{D\eta\ P_{eff}}{\pi \left(r_{eff}\right)^2} \exp\left[-D\left(\frac{x^2+y^2}{\left(r_{eff}\right)^2}\right)\right] \tag{2}$$

where P_{eff} is the available power on the surface, η is the laser absorption coefficient estimated using Bramson's equation [13] and r_{eff} is the effective radius of the laser beam at the substrate. The surface heat flux S_s is specified as a source term for the equation of conservation of energy at any surface. The laser power absorbed by the alloy powder depends on the absorptivity of the powder material. P_{eff} is the net power available that is utilized for the melting of the powder particles. The effective power P_{eff} can be expressed as $P_{eff} = P - P_L$ where, P_L is the total loss of the laser power.

The boundary conditions for the thermal analysis include heat exchange by convection and radiation with the surroundings. The boundary conditions for the velocities at the free surface are based on Marangoni convection. [1,8-9] The transient heat transfer and fluid flow calculations are performed for a solution domain representing the substrate, deposited layers, and the surrounding gas as shown in Figure 1.

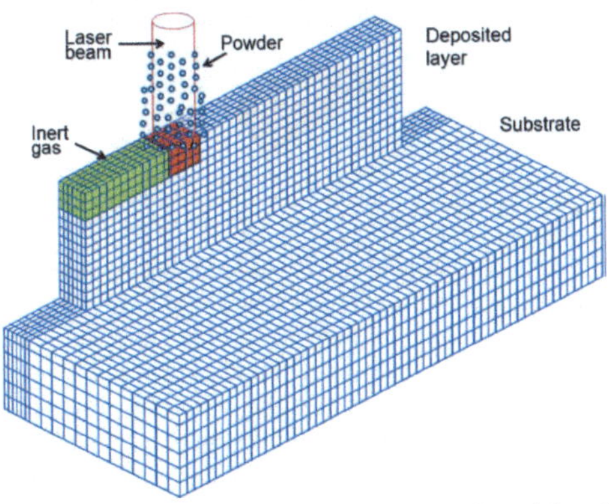

Figure 1. *Schematic representation of application of numerical model in the solution domain for AM process [1]*

The continuous movement of the laser beam is discretized using small shifts in the direction of scanning. The time-step for each shift of the laser beam is calculated from the deposition length and the laser scanning speed. An idle time is considered at the end of the simulation of each layer. The procedure is repeated until the simulation of all the layers is completed and calculations are continued until the specimen cools.

Table I. *Process conditions used in numerical calculations*

Parameter Set	Laser power (W)	Beam radius (mm)	Scanning speed (mm/s)	Layer thickness (mm)	Substrate thickness (mm)
1	2000	1.5	10.6	0.90	10
2	210	0.5	12.5	0.38	4
3	200	0.5	12.5	0.38	4

Figure 2 shows the computed melt pool geometry for the laser-based AM process of Ti-6Al-4V. Each color band in the profile represents a temperature range. The yellow colored regions in all the figures indicate that the deposited materials have reached at least the solidus temperature of the alloy. The molten pools are shown by red color. The vectors show the computed velocity fields in the molten region. A reference vector is shown by an arrow and a comparison of the length of this arrow with the vectors shown in the plots reveals the magnitudes of the computed velocities.

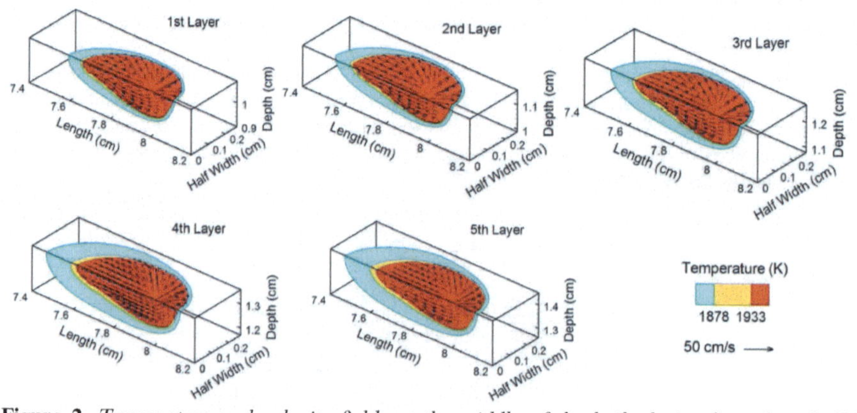

Figure 2. *Temperature and velocity fields at the middle of the built during laser based AM process of Ti-6Al-4V using process Parameter Set 1 in Table I.*

A fair agreement between the calculated build shape and size, and the corresponding measured build profile in Figure 3 indicates that the modeling results can be used to estimate thermal distortion, composition changes and lack of fusion defects with confidence.

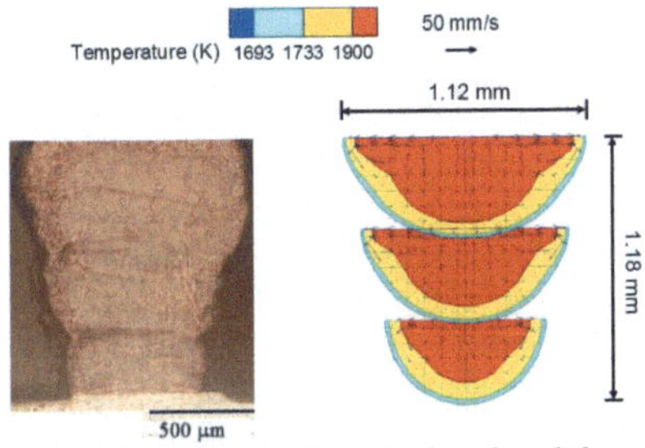

Figure 3. *Comparison between the numerically calculated and the corresponding actual build shape [1] with Parameter Set 2 of Table I.*

Thermal distortion

Thermal distortion in AM parts is caused by non-uniform expansion and contraction of different regions of the part that experience changes in temperature. Propensity for thermal distortion is calculated from the maximum thermal strain that largely depends on the molten pool dimensions. A larger pool size indicates a higher amount of contraction during solidification. Therefore, thermal strain increases with molten pool dimensions. Figure 4 shows experimentally measured maximum thermal strains (ε) reported in the literature [14-18] as a function of length of the molten pool (L). The lengths of the molten pools are calculated using the heat transfer and fluid flow model. The linear correlation between maximum thermal strain and molten pool length is given in Figure 4. Table II shows the estimated values of maximum thermal strains (ε) for various alloys using the linear correlation. Increasing layer thickness increases the thermal strain because of higher temperatures resulting from lower heat conduction from the molten pool into the substrate as shown in Table II. Thermal strain is the highest for Ti-6Al-4V, which can be attributed to its relatively low density and thermal diffusivity. The ranking of the alloys in Table II provides a relative scale of their printability considering their susceptibility to thermal distortion.

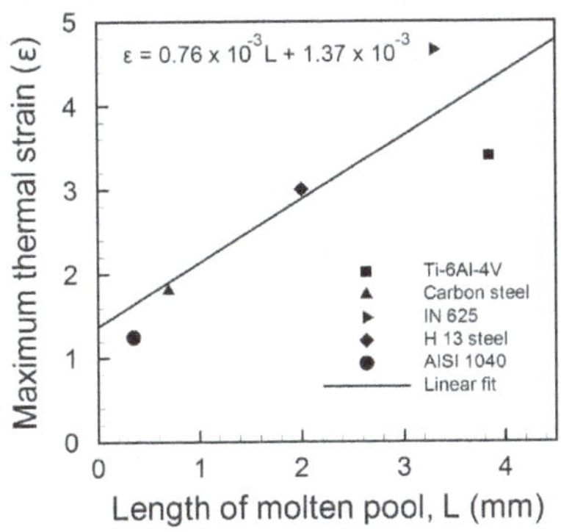

Figure 4. *Experimentally measured maximum thermal strain [14-18] as a function of length of the molten pool for five different alloys.*

Table II. *Values of maximum thermal strain (ε) in laser-based AM of a single-track three-layer deposition of SS 316, Ti6Al4V and IN 625 using Parameter Set 3 in Table I.*

Material	Layer no.	Maximum thermal strain
SS 316	1	2.4
	2	2.8
	3	3.1
Ti-6Al-4V	1	2.5
	2	2.9
	3	3.3
IN 625	1	1.8
	2	2.4
	3	2.5

Lack of fusion defects

In order to ensure appropriate inter-layer bonding, the melt pool in a deposited layer should exceed the layer thickness and adequately remelt the substrate or the previously deposited layer. Carroll et al. [19] reported a 99.999% dense part for direct energy deposition AM of Ti-6Al-4V, indicating proper inter-layer bonding. A corresponding depth was estimated about 1.02 mm for a layer thickness of 0.89 mm. This depth of penetration is about 15% higher than the layer thickness, which appears to be quite adequate for obtaining parts with near theoretical density.

The three-dimensional heat transfer and fluid flow model is used to estimate penetration depths for three alloys over a range of linear heat inputs. Table III shows that for a given heat input, Ti-6Al-4V will have the highest penetration depth (d), while SS 316 will have the lowest. Therefore, Ti-6Al-4V and SS 316 are the least and most susceptible to lack of fusion defects, respectively.

The results presented here provide a relative assessment of the susceptibilities of various alloys to lack of fusion defects. For alloys that are highly susceptible to lack of fusion defects, AM variables like laser power, scanning speed and powder feed rate should be appropriately adjusted to attain an adequate penetration depth.

Table III. *Computed penetration depth, (d), for three alloys using Parameter Set 3 in Table I.*

Heat input	SS 316	IN 625	Ti-6Al-4V
[J/mm]	d [mm]	d [mm]	d [mm]
16	0.384	0.445	0.479
20	0.426	0.479	0.517
24	0.452	0.502	0.551
28	0.467	0.521	0.574
32	0.483	0.540	0.597

Composition change

Temperatures and geometries of the molten pool are calculated using the three-dimensional heat transfer and fluid flow model. The equilibrium vapor pressures of all alloying elements are estimated from available thermodynamic data [20,21]. The vaporization fluxes of alloying elements, J_i, are estimated from the computed temperatures, T, vapor pressures, P_i, and the Langmuir equation:

$$J_i = \frac{\lambda P_i}{\sqrt{2\pi M_i T}} \tag{3}$$

where M_i is the molecular weight of element i, and λ is a positive fraction accounting for the condensation of some vaporized atoms. The surface area and volume of the molten pool are computed from the numerical heat transfer and fluid flow model. Table IV shows that Ti-6Al-4V and IN 625 will experience the most and least composition change, respectively.

Results for Ti-6Al-4V agree with results from Brice et al. [22], who reported an average composition change of approximately 0.9 wt% Al. Temperatures of the molten pool are higher for Ti-6Al-4V than the other alloys due to the relatively low thermal conductivity and density of the latter. A larger composition change for Ti-6Al-4V can be attributed to the higher temperatures and relatively high equilibrium vapor pressure of aluminum over the liquid alloy. For alloys highly susceptible to composition change, appropriate AM variables such as laser power density and scanning speed should be adjusted to reduce loss of volatile alloying elements.

Table IV. *Composition changes due to vaporization of constituting elements of three alloys during AM using the Langmuir equation (λ=0.05).*

Element	Al	Cr	Fe	Mn	Mo	Ni	Ti	V
Alloy	wt%	wt%	wt%	wt%	wt%	wt%	wt%	wt %
IN 625	−0.0008	−0.0117	−0.0002	-	0.00202	0.00984	4×10^{-5}	-
SS 316	-	0.0157	0.165	−0.220	-	0.0364	-	-
Ti-6Al-4V	−0.922	-	-	-	-	-	0.881	0.041

Conclusion

Numerical heat transfer and fluid flow calculations can provide useful information about the relative susceptibilities of various alloys to thermal distortion, loss of alloying elements and lack of fusion defects that can form during AM of metallic parts. The computed susceptibilities have been validated with independent experimental data. Results show that Ti-6Al-4V is most susceptible to thermal strain and composition change but least susceptible to lack of fusion defects. Stainless steel is more susceptible to lack of fusion defects than the other alloys considered. The results provide an understanding of the printability of various alloy powders under commonly used AM process conditions based on numerical heat transfer and fluid flow calculations.

References

[1] V. Manvatkar, A. De, and T. DebRoy, "Heat transfer and material flow during laser assisted multi-layer additive manufacturing," *Journal of Applied Physics* 116 (2014), 124905.

[2] P. Mercelis, and J. P. Kruth, "Residual stresses in selective laser sintering and selective laser melting," *Rapid Prototyping Journal* 12 (2006), 254.

[3] A. Nickel, D. Barnett, and F. Prinz, "Thermal stresses and deposition patterns in layered manufacturing," *Materials Science and Engineering: A,* 317 (2001), 59.
[4] R. Jendrzejewski, G. Sliwinski, and M. Krawczuk, "Investigation of temperature and stress fields in laser cladded coatings," *Applied Surface Science,* 254 (2007), 921.
[5] H. Pohl et al., "Thermal stresses in direct metal laser sintering," (Proceedings of the 12th Solid Freeform Fabrication Symposium, Austin, Texas, 2001).
[6] X. He, T. DebRoy, and P.W. Fuerschbach, "Alloying element vaporization during laser spot welding of stainless steel," *Journal of Physics D: Applied Physics,* 36 (2003), 3079.
[7] Patankar, S. V. *Numerical Heat Transfer and Fluid Flow* (New York, NY: McGraw-Hill 1982).
[8] T. DebRoy and S.A. David, "Physical processes in fusion welding," *Reviews of Modern Physics,* 67(1) (1995), 85-112.
[9] S.A. David and T. DebRoy, "Current issues and problems in welding science," *Science*, 257 (1992), 497-502.
[10] A. Raghavan et al. "Heat transfer and fluid flow in additive manufacturing," *Journal of Laser Applications,* 25(5) (2013), 1-8.
[11] A. Frenk et al. "Analysis of the laser-cladding process for stellite on steel," *Metallurgical and Materials Transactions B,* 28(3) (1997), 501-508.
[12] V. Neela and A. De, "Three dimensional heat transfer analysis of LENS™ process using finite element method," *The International Journal of Advanced Manufacturing Technology* 45(9-10) (2009), 935-943.
[13] W. M. Steen, *Laser Material Processing, 2nd ed* (New York, NY: Springer-Verlag, 1998) 83.
[14] J. Grum and M. Znidarsie, "Microstructure, microhardness, and residual stress analysis of laser surface cladding of low-carbon steel," *Materials and Manufacturing Processes,* 19 (2004), 2.
[15] J.C. Heigel, P. Michaleris, and T.A. Palmer, "In situ monitoring and characterization of distortion during laser cladding of Inconel® 625," *Journal Materials Processing Technolology,* 220 (2015).
[16] P. Farahmand and R. Kovacevic, "An experimental–numerical investigation of heat distribution and stress field in single- and multi-track laser cladding by a high-power direct diode laser," *Optics and Laser Technology,* 63 (2014).
[17] S. Okano et al., "Effect of welding conditions on reduction in angular distortion by welding with trailing heat sink," *Science and Technology of Welding and Joining*, 17 (2012), 4.
[18] Y.P. Yang et al. "Material strength effect on weld shrinkage and distortion," *Welding Journal,* 93 (2014).
[19] B. E. Carroll, T. A. Palmer and A. M. Beese, "Anisotropic tensile behavior of Ti-6Al-4V components fabricated with directed energy deposition additive manufacturing," *Acta Materialia,* 87 (2015), 309-320.
[20] W.F. Gale and T.C. Totemeier. *Smithells Metals Reference Book, 8th edition* (Amsterdam, Boston: Elsevier Butterworth-Heinemann 2003)
[21] C.L. Yaws. *Handbook of Vapor Pressure*, (Houston: Gulf Pub. Co, 1994).
[22] C. A. Brice, et al. 4th International light metals technology conference. Broachbeach, QLD, Australia (2009).

7th International Symposium on High-Temperature Metallurgical Processing
Edited by: Jiann-Yang Hwang, Tao Jiang, P. Chris Pistorius, Gerardo R.F. Alvear F., Onuralp Yücel, Liyuan Cai, Baojun Zhao, Dean Gregurek, and Varadarajan Seshadri
TMS (The Minerals, Metals & Materials Society), 2016

CHARACTERIZATION OF IRON-BEARING DUST PELLET IN COMPOSITE AGGLOMERATION PROCESS (CAP)

Zhuyin Chen, Bingbing Liu, Chen Liu, Xiao Kang, Yuanbo Zhang*
School of Minerals Processing & Bioengineering, Central South University, Changsha, Hunan 410083, China

Keywords: Iron-bearing dust, sintering, composite agglomeration process, behavior characterization

Abstract

Composite agglomeration process (CAP) is an effective method to utilize the iron-bearing dusts generated in the ironmaking plants, by which the dust percentage in the sinter feed can be increased up to 20 wt %. In the current study, a comparison on behavior of iron-bearing dust in the conventional sintering process (CSP) and CAP was characterized. The results showed, in the CSP products, the scattered iron-bearing dust in the mixture was disadvantageous for the sintering indexes and consolidation of the finished sinters. In the CAP products, the main phase in the iron-bearing dust pellets was well-crystallized Fe_2O_3 grain particles. In the matrix parts, there were more Fe_3O_4 particles generated, which caused to high $RDI_{+3.15}$ and low RI of the finished products. The pellets and the matrix parts were bonded by iron oxides and acicular calcium ferrite.

Introduction

Iron-bearing dusts are common solid wastes in the ferrous plants, which are generated in sintering, ironmaking, steelmaking processes, etc. The iron-bearing dusts often contain high contents of fixed carbon, iron and certain contents of calcium, magnesium and silicon. Therefore, iron-bearing dusts are able to act as partial substitutions of fluxes and fuels [1-3]. The effective utilization of iron-bearing dusts in a large scale is beneficial for the energy saving and environmental protection.

Although there are some utilizing methods of iron-bearing dusts performed, it's reported that the comprehensive utilization efficiency is only below 20% [4]. In steel plants, dusts are always sent to sintering plant directly, but the maximum utilization of the dust is limited to under 10%, because the excess dusts deteriorate the sintering indexes of the finished sinter products. Besides, the dusts are also transported out of the steel plants to produce cement, or to the dressing plant for the separation of valuable elements. So in order to reduce the transportation cost and use this kind of resource more reasonably, it's important and meaningful to realize the high proportion

* Corresponding Author: Yuanbo Zhang, E-mail: zybcsu@126.com

addition of iron dust into the iron ore sintering process.

A recent study found that the utilization efficiency of iron-bearing dust can be improved by composite agglomeration process (CAP), in which the iron-bearing dusts were used as the pellet feed. Comparing to 10% addition in conventional sintering process (CSP), the dosage of dust in CAP can be enhanced to over 20%. Moreover, the finished sintering products had good strength and metallurgical property. It demonstrated that CAP is a promising and effective process for comprehensive utilization of iron-bearing dusts in-plant.

However, the previous studies only reported the technical parameters of CAP, and seldom investigation on the mineralization mechanism was reported. This study investigated the behavior of iron-bearing dust in CSP and dust pellets in CAP and by using SEM-EDS, optical microstructure and statistics analyses in order to provide some information about mineralization mechanism of iron-bearing dusts in CAP.

Experimental

Materials

In the tests, iron-bearing dusts were used as pellet feeds. The chemical compositions, specific surface areas and size distribution of the dust samples were referred to a previous article [1]. The iron ore fines were used as matrix feeds, and the alkaline fluxes were utilized to adjust the basicity of the sintering mixture. The main chemical compositions and particle size distribution of raw materials are presented in Table I and Table II, respectively. It is worth notable in iron-bearing dusts that: 1) the fixed carbon content is 6.73%; 2) the mass fraction of the fine particle below 0.074 mm is 63 wt%.

Table I. Chemical Composition of Raw Materials (wt %)

Raw materials	TFe	SiO_2	CaO	Al_2O_3	MgO	C	LOI*
Iron ore fines	59.78	4.44	1.37	1.4	0.28	0	6.97
Iron-bearing dusts	49.09	5.45	9.58	2.12	1.37	6.73	10.79
Dolomite	0.26	1	29.66	0.03	19.95	0	46.02
Limestone	0.47	2.6	52.62	0.38	0.24	0	42
lime	0.15	1.17	75.66	0.1	0.7	0	18.26
Serpentine	4.57	38.38	1.46	0.23	36.57	0	13.93
Coke breeze	3.6	6.52	1.69	3.93	0.21	79.2	82.05

*LOI: Loss on Ignition

Table II. Size Distribution of Raw Materials (wt %)

Size(mm)	10~8	8~5	5~3	3~1	1~0.5	-0.5
Iron ore fines	16.97	14.81	19.89	25.67	8.62	14.04
Dolomite	0	0	3.85	32.29	12.87	50.99
Limestone	0	0.46	5.31	51.33	16.67	26.23
lime	0	3.7	9.1	15.7	44	31.2
Serpentine	0	0	3.18	40.5	21.25	35.07
Coke breeze	0	1.8	8.99	31.7	21.63	35.88

Methods

A sintering pot with a size of Φ 180mm×700mm was used to simulate the sinter machine. For CAP tests, iron-bearing dusts and 2.0% bentonite were firstly proportioned, mixed, and then the mixture was sent to a 1000 mm diameter disc pelletizer. In pelletizer, the moist mixture was balled into pelletized feeds of a certain diameter. Meantime, the moisture content of pellets was controlled to 11.0%. By screening, the green pellets with the size of 5-12 mm diameter were selected to be pellet feeds. The iron ore fines, return fines, fluxes and fuels were firstly granulated in a Φ 60mm×120mm drum mixer. Then, green pellets and the primary sintering mixture were mixed in the second mixer, in which the pellet feeds occupied 20 wt% of the iron ore fines. The basicity of pellet was 1.4 while the comprehensive alkalinity of the secondary mixture was 1.8. The mixture was charged into the sintering pot. The subsequent treatments including igniting, sintering, cooling and screening were carried out successively.

As a contrast, the experiment of conventional sintering process (CSP) with adding 7% iron-bearing dusts directly into the iron ore fines were conducted. The sintering conditions listing in Table III were same as the previous investigation [1]. Note that, the total fuel consumption of the two tests was equivalent to 5.0% coke breeze (calculated by the fixed carbon).

Table III. Experimental Conditions of the Sintering Test

Test	Coke breeze addition (%)	Total fuel dosage (%)	Moisture of mixture (%)	Basicity	Iron-bearing Dusts Proportion (%)	Ignition	Cooling
CAP	2.9	5.0	8.3	1.8	20	1.5min; 1100°C;5kPa	3 min; 5kPa
CSP	4.3		8.5		7		

After cooling and screening, the agglomeration indexes of finished products were measured. The polished sections of finished sinter products were prepared to observe the structure changing of the main phase and analyze the elemental content of the phases, using Optical Microscopy (LEICA MDI5000 M, Germany) and Environmental Scanning Electron Microscopy (ESEM, FEI QUANTA 200, Holland) equipped with an energy dispersive X-ray spectroscopy (EDS) detector. ESEM images were recorded in backscatter electron mode operating in a low vacuum of 0.5 Torr and 20 keV.

Optical microstructure analysis data also gave the average statistical area of each mineral phase and the total observational scope area, which was characterized by image analyzer (Image-Pro Plus 6.0 made in America). According to the statistic view, the volume fraction of mineral phase was calculated by supposing that the volume fraction of a mineral phase is equivalent to its area. Phase identification and area measurements of representative optical images were performed by Image-Pro software as illustrated in Figure 1, which have been used widely in other studies [5]. The minerals or pores with different grayscales or colors were covered by certain colors and then this software would count the area automatically. The average value of all the statistics of the same mineral phase was set as the final value.

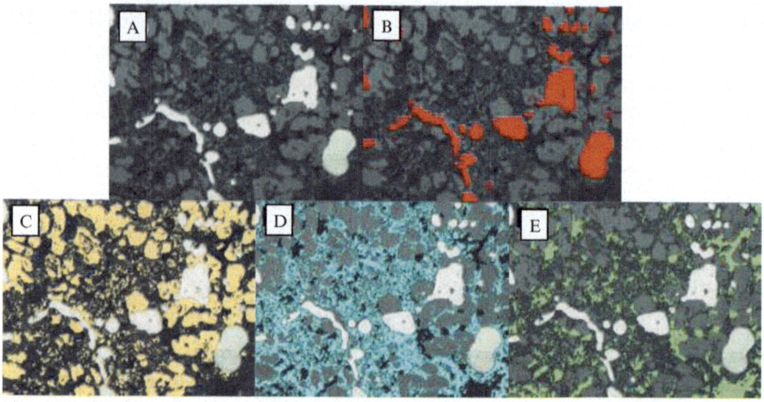

Figure 1. Area fraction statistics of different phases by image analyzer ((A): original optical image; (B), (C), (D) and (E): treated optical images for different phases) (Online version in color)

Results and Discussion

Agglomeration indexes comparison

The agglomeration indexes including yield, the International Standards Organization (ISO) tumbler index, productivity and permeability index were measured and listed in Table IV. The results show that under the condition of the solid fuel consumption decreasing by 32.6%, except for the yield, the agglomeration indexes of CAP are quite higher than those of CSP. The tumbler index, productivity and permeability index of CAP increase by 3.23%, 11.73% and 24.49%, respectively. Based on comprehensive consideration, CAP has huge advantages in the effective utilization of iron-bearing dusts in a large scale.

Table IV. Yield and Quality Indicators of Sintering Products

Test	Yield(%)	ISO tumbler index (%)	Productivity(t $(m^2 h)^{-1}$)	Permeability index (JPU)
CAP	73.69	64.83	1.476	0.305
CSP	73.63	62.80	1.321	0.245
Variation	0.06	2.03	0.155	0.060
Increment (%)	0.08	3.23	11.73	24.49

Agglomeration products properties comparison

The properties of the finished sinters were measured, and the main properties were listed in Table V. Reduction disintegration index ($RDI_{+3.15}$) of CAP products reaches to 74.62%, which is higher than that of CSP sinter by 8.79%. Reduction index (RI) of CAP product is lower than that of CSP sinter by 1.44%, which is mainly resulted from the increasing FeO content of CAP product.

Table V. Main Agglomeration Products Properties of Sintering Products

Test	$RDI_{(+3.15)}$/%	RI/%	FeO content (wt %)
CAP	74.62	83.51	7.03
CSP	68.59	84.73	6.82
Variation	6.03	-1.22	0.21
Increment (%)	8.79	-1.44	3.08

Morphology comparison

SEM-EDS analysis SEM-EDS analysis was used to confirm the chemical compositions and appearance of different phases. The representative samples including CAP and CSP (prepared conditions were listed in Table III) were selected.

The CAP products can be divided into three parts: sintering parts, pellets parts and the junction zone. The iron-bearing dust pellets were embedded in basic sinter. The liquid phase from high basicity sinter feed linked the sintering parts and pellets parts, which guaranteed the whole strength of the products. The SEM-EDS analyses of sintering matrix and connection area of CAP product were showed in Figure 2 and Table VI.

Figure 2(A) shows the elemental analyses of the main minerals of sintering matrix in CAP products. The acicular phase (Point 1) is the calcium ferrite confirmed by EDS analysis, and complex silicate containing Ca, Fe, Al and Mg is also observed (Point 3). The main phase is in the form of silico-ferrite of calcium and aluminum (SFCA), which is confirmed by EDS analysis (Point 2). As shown in Figure 2(B), in the connection area, the main phase is the melted iron oxide, which is confirmed by EDS analysis (Point 2). A small amount of silicate melt is also observed (Point 3). In pellet parts, there are hematite with smooth edges (Point 1) and large holes.

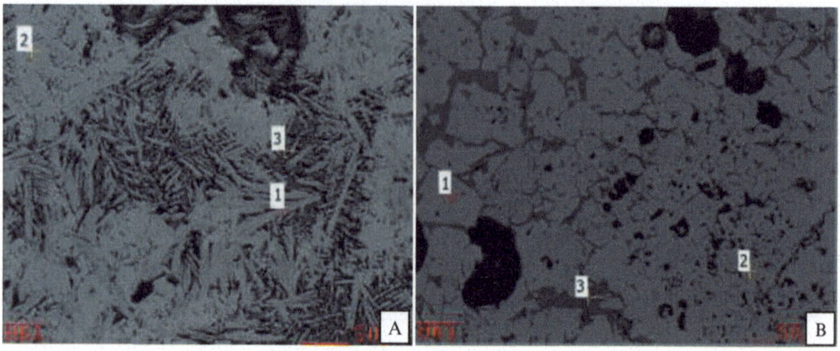

(A) Point 1: calcium ferrite; Point 2: SFCA; Point 3: silicate
(B) Point 1-2: iron oxide; Point 3: silicate

Figure 2. (A) SEM image of sintering matrix in CAP product. (B) SEM image of connection area in CAP product.

Table VI. EDS analyses results of SEM images showed in Figure 2

Element	Point					
	(A)-1 Mass (%)	(A)-2 Mass (%)	(A)-3 Mass (%)	(B)-1 Mass (%)	(B)-2 Mass (%)	(B)-3 Mass (%)
O	29.79	31.34	39.46	3.63	3.93	38.53
Fe	59.53	61.78	11.35	96.37	96.07	13.97
Si	0	0	18.2	0	0	16.21
Al	0	2.48	2.86	0	0	3.08
Ca	10.68	1.09	28.13	0	0	28.21
Mg	0	3.32	0	0	0	0

The previous investigation showed that the CSP products had uneven composition and large thin-walled holes, which caused by the scattered iron-bearing dusts. There were iron oxides, SFCA, silicate and inferior crystals existed in finished sinter.

Optical microstructure analysis Microstructure is also a vital evaluation index for the properties of agglomerate, which is shown in Figure 3. In CAP sinter, due to the high basicity, the major phase in matrix part is strip calcium ferrite with perfect crystal structure which intertwined with lump hematite closely (Figure 3(A)). In the pellets, it can be seen from Figure 3(B) that the contents of iron oxides increase, and almost all of Fe_2O_3 and Fe_3O_4 crystal grains connect each other. Silicate is filled in the interval of Fe_3O_4 crystals. There are large holes in pellets as well. The contact parts base on consolidation of hematite, magnetite and slight silicate minerals (Figure 3(C)).

As displayed in Figure 3(D)-(F), the microstructure of CSP product is heterogeneous. The crystallization size of hematite is large meanwhile calcium ferrite and kirschsteinite are also existed in CSP products (Figure 3(D)). As shown in Figure 3(E), skeleton crystals of calcium ferrite are observed in the silicates and the gaps between joined crystals of hematite or magnetite. In addition, there are so many pores in the CSP sinter owing to the scattered iron-bearing dust with high carbon content, which is blown away under 10kPa negative pressure. The lower sinter strength and poor $RDI_{+3.15}$ of CSP sinter are ascribed to the existence of skeleton crystal and pores.

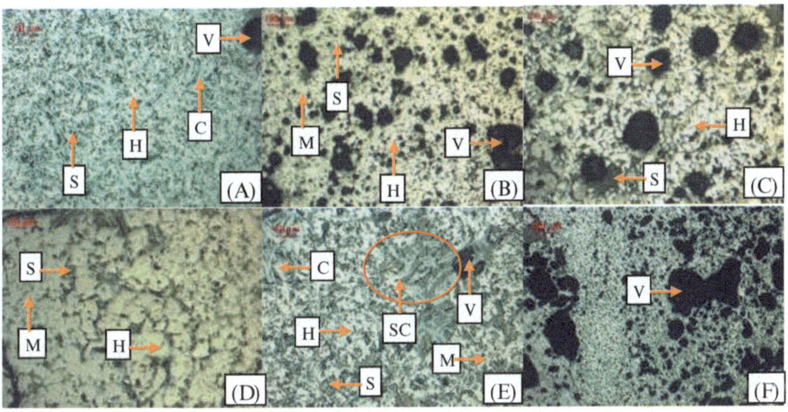

C: Calcium ferrite; H: Hematite; M: Magnetite; S: Silicate mineral;
SC: Skeleton crystal; V: Void

Figure 3. Microstructure of agglomerates. (A) SFCA in matrix part in CAP product; (B) Pelletized part in CAP product (C) Interweaved structure in transitional zone in CAP product; (D) Hematite in CSP product; (E) Skeleton crystal in CSP product; (F) Pores in CSP product

Quantitative statistics In order to further comparison, the quantitative measure of the main minerals in sinter products were calculated and listed in Table VII. As shown in Table VII: 1) The magnetite content in CAP products increases than that of CSP sinter by 31.10%. It indicates that abundant magnetite has not yet been completely oxidized in CAP product, which is the main reason for high $RDI_{+3.15}$ and low RI (Table V); 2) There is more SFCA in the CAP sinter parts, which serves as an excellent bonding phase. It can be considered as the other reason for the good strength and high $RDI_{+3.15}$; 3) The porosity and inferior crystals occupy 8.41% and 7.31% in CSP product, both are quite higher than those of CAP product, which can be considered as the other vital reasons for poor strength.

According to the statistics, in pellets of CAP product, the content of hematite is higher than that in matrix, and the Fe_2O_3 grains with rounded edge and big crystal size crystallized well. It can be concluded that the heat for Fe_2O_3 recrystallization was resulted from the combustion of carbon in dusts.

Table VII. Quantitative Measure of the Main Phases in Sintering Products

test	Main phase and its volume fraction (vol %)						
	Hematite	Magnetite	SFCA	Silicate	Holes	Inferior crystals	Others
CAP-pellet	28.30	26.17	19.21	16.87	5.85	2.65	0.95
CAP-matrix	0.47	51.29	26.04	16.23	4.85	0.40	0.72
CAP	4.64	47.52	25.02	16.33	5.00	0.74	0.75
CSP	7.67	36.25	21.52	18.53	8.41	7.31	0.31
Variation	-3.03	11.27	3.50	-2.20	-3.41	-6.57	0.44
Increment (%)	-39.45	31.10	16.24	-11.89	-40.55	-89.91	144.39

Conclusions

The main phases in the CSP and CAP products were iron oxides, SFCA, and kirschsteinite. The contents of SFCA and magnetite in CAP are higher than those of CSP, which contributed to the good strength, high $RDI_{+3.15}$ and low RI. As for the behavior of dusts in CSP product, the scattered iron-bearing dust in the mixture was bad for the consolidation of the finished sinters. Less iron oxide and SFCA as well as larger pores and inferior crystals were produced, which led to the bad consolidation performance. In CAP, the pellets embedded in matrix, and some melted phases in matrix permeated through the outer layer of pellets with forming an interweaved structure, which offset the negative effects of pores in pellets. Considering the comparative investigation, it's proved that CAP is an effective agglomeration measure treatment for iron-bearing dusts in a large scale.

References

1. B. Xu et al. "Utilization of Iron-Bearing Dusts in Iron Ore Sintering by Composite Agglomeration Process" (Paper presented at 6th International Symposium on High-Temperature Metallurgical Processing, 2015), 499-506.

2. T. Jiang et al. Composite agglomeration process (CAP) for preparing blast furnace burden. Ironmaking and Steelmaking, 2010, 37(1), 1-7.

3. X. L. Yang, H. X. Dai, X. Li. "Comprehensive Utilization and Discussion of Iron and Steel Metallurgical Slag", *Advanced Materials Research*, 807-809 (2013), 2328-2331.

4. D. Zeng et al. "Research advancement of comprehensive utilization of iron-bearing dust in iron and steel plants,"*Sintering and Pelletizing*, 36 (6) (2011), 45-49. (In Chinese)

5. S. Ren et al. "Effect of TiO_2 on equilibrium phase sinter at oxygen partial pressure of 5×10^{-3} atm". *Ironmaking & Steelmaking*, 41(2) (2014),132-137.

7th International Symposium on High-Temperature Metallurgical Processing
Edited by: Jiann-Yang Hwang, Tao Jiang, P. Chris Pistorius, Gerardo R.F. Alvear F., Onuralp Yücel, Liyuan Cai, Baojun Zhao, Dean Gregurek, and Varadarajan Seshadri
TMS (The Minerals, Metals & Materials Society), 2016

EVALUATION OF HEAT FLOW AND THERMAL STRATIFICATION IN A STEELMAKING LADLE THROUGH MATHEMATICAL MODELLING

Varadarajan Seshadri[1]; Izabela Diniz Duarte[2]; Itavahn Alves da Silva[2]; Carlos Antonio da Silva[2]
[1] Department of Metallurgical Engineering and Materials, Federal University of Minas Gerais, 6627,Av. Antonio Carlos, 31.270-901 Belo Horizonte, Brazil,
[2] Department of Metallurgical Engineering and Materials, Federal University of Ouro Preto, 35400-000, Ouro Preto, MG, Brazil

Keywords: Thermal losses, Thermal stratification, Ladle

Abstract

Mathematical modeling using computational fluid dynamics has been employed to assess the thermal losses in a steelmaking ladle. Thermal losses lead to thermal stratification and convective flow inside the ladle. The effects of preheating, mineralogical and physical characteristics of the lining as well as thickness of slag cover have been studied and the results obtained from application of the model have been discussed.

Introduction

During transit of steel making ladle between the primary refining and continuous casting station, heat losses occur. It is important to have initially liquid steel at higher temperature to compensate for the heat losses until reaching the stage of continuous casting, where the superheat in the steel melt need careful control to get a good quality end product. The thermal losses through refractory walls and the steel jacket as well as the slag cover at the top, generate convection currents and temperature stratification in the liquid steel, which requires optimization of the pouring practice. Thermal losses through slag cover and refractory linings of the ladle have been evaluated (1). For regular slag layer thicknesses the heat losses were found to be lower than that through refractory lining of the side walls and the bottom. The thermal stratification and convection phenomena resulting from thermal losses in a steel plant ladle have been investigated (2,3,4). The present study has the objective of investigating this phenomena through CFD modeling.

Materials and Methods

A typical steel making ladle with cylindrical shape with a capacity of 250 tons with refractory lining was chosen for study (Figure 1a). The slag cover in general consists of two distinct layers one liquid in contact with the liquid metal in the ladle and the other solid on the top surface with the solid to liquid fractions varying with time. This also depends on heat losses during transfer of metal in the ladle. Ansys / CFX software (5) has been used for the solution of the computational model considering that the refractory lining is monolithic with constant thickness and without the presence of the steel casing outside. The thermal conductivity and thermal diffusivity of the refractory are taken as independent of temperature. The process of heat transfer is by nature transient, inducing natural convection currents in the liquid steel. The fluid flow, spatial distribution of the velocity and pressure fields can be represented by Continuity and Navier-Stokes equations coupled to Energy Conservation equation:

Continuity equation:

$$\frac{\partial \rho}{\partial t} + \nabla . \rho u = 0 \tag{1}$$

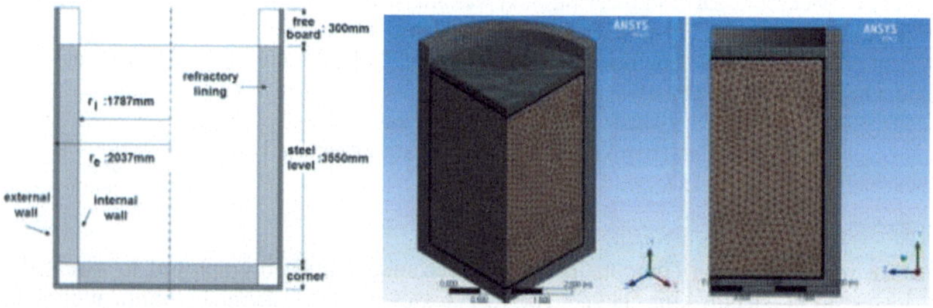

Figure 1 - a) Schematic representation of the steel ladle. b) Isometric view of the model c) Details of mesh used in the slag, steel bath and refractory

Momentum conservation equation:

$$\frac{\partial(\rho u_i)}{\partial t} + \frac{\partial(\rho u_i u_j)}{\partial x_j} = \frac{\partial}{\partial x_j}\left[(\mu_l + \mu_t)\left(\frac{\partial u_i}{\partial x_j} + \frac{\partial u_j}{\partial x_i}\right)\right] - \frac{\partial P}{\partial x_i} + \rho g \tag{2}$$

Energy equation:

$$\frac{\partial(\rho h)}{\partial t} + \frac{\partial(\rho h u_j)}{\partial x_j} = \frac{\partial}{\partial x_j}\left[\left(\frac{K}{C_p} + \frac{\mu_t}{\sigma_h}\right)\frac{\partial h}{\partial x_j}\right] \tag{3}$$

Where, ρ = density, t = time, u = average velocity, x = coordinate in a certain direction, h = enthalpy, P = pressure, g = acceleration due to gravity, K = thermal conductivity, C_p = specific heat, μ_l and μ_t = respectively, laminar viscosity and eddy viscosity, i and j = Cartesian indexes. Eddy viscosity can be estimated through $k - \varepsilon$ model of turbulence:

$$\mu_t = C_\mu \rho \frac{k^2}{\varepsilon} \tag{4}$$

Equation for kinetic energy of turbulence:

$$\frac{\partial(\rho k)}{\partial t} + \frac{\partial(\rho u_j k)}{\partial x_j} = \frac{\partial}{\partial x_j}\left(\frac{\mu_t}{\sigma_k}\frac{\partial k}{\partial x_j}\right) + \mu_t\left(\frac{\partial u_i}{\partial x_j} + \frac{\partial u_j}{\partial x_i}\right)\frac{\partial u_i}{\partial x_j} - \rho\varepsilon \tag{5}$$

Equation for rate of dissipation of kinetic energy of turbulence:

$$\frac{\partial(\rho \varepsilon)}{\partial t} + \frac{\partial(\rho u_j \varepsilon)}{\partial x_j} = \frac{\partial}{\partial x_j}\left(\frac{\mu_t}{\sigma_\varepsilon}\frac{\partial \varepsilon}{\partial x_j}\right) + C_1 \mu_t \frac{\varepsilon}{k}\left(\frac{\partial u_i}{\partial x_j} + \frac{\partial u_j}{\partial x_i}\right)\frac{\partial u_i}{\partial x_j} - C_2 \rho \frac{\varepsilon^2}{k} \tag{6}$$

Where k = kinetic energy of turbulence, ε = rate of dissipation of kinetic energy of turbulence, σ_k = Prandtl number for kinetic energy of turbulence, σ_ε = Prandtl number for the dissipation rate of turbulent energy, C_1, C_2, constants.

Considering the ladle cycle procedure usually adopted in a steel plant, after the gas rinsing and end of secondary refining, the temperature of the liquid steel (Tm) in the ladle is uniform and normally amounts about 1650 ° C. It is assumed that the liquid bath initially is quiescent, static, and it has the same temperature as the bottom refractory(Tb) and also the refractory wall (Tw). Under these conditions the initial condition can be written as:

$$\text{At, } t = 0\text{: } T_w = T_b = T_m = 1650\ °\text{C} \tag{7}$$

To determine the thermal conditions in the refractory lining, the basic assumption is that a transient thermal gradient prevails in the refractory lining that is, the temperature thereof varies with position and time and can be described by heat conduction equation for transient conditions, given by two-dimensional cylindrical coordinates:

$$\rho C_p \frac{\partial T}{\partial t} = K \frac{\partial^2 T}{\partial r^2} + \frac{1}{r}\frac{\partial T}{\partial r} + \frac{\partial^2 T}{\partial z^2} \tag{8}$$

Where, T = temperature, t = time, ρ = density, C_p = specific heat, K = thermal conductivity
At the boundary between the outer surface of the refractory and the surrounding environment there is heat loss by convection and radiation. Applying the initial condition (here T_w and T_b are regarded as the refractory soaking temperature):

$$\text{At } t=0\text{: } T_w = T_b = 700°\text{C } or\ else,\ T_e = 25°\text{C} \tag{9}$$

The heat flux can be calculated by the following equation. The first term on the right side of the equation represents heat loss by convection and the second term, heat loss by radiation:

$$-K\frac{\partial T}{\partial r} = h_{we}\,(T_w - T_e) + \varepsilon_w \sigma \left(T_w{}^4 - T_e{}^4\right) \tag{10}$$

Where, K = thermal conductivity, $T_w\ or\ T_b$ = refractory temperature, T_e= room temperature, ε_w= refractory emissivity, σ = Stefan-Boltzman constant, h_{we} = convection coefficient between the refractory and the environment, A similar equation can be written for the boundary between the bottom refractory and outer environment.

At the boundary between steel and slag is observed heat conduction, to which the initial condition applies:

$$\text{At } t = 0,\ T_s = T_m = 1650\ °\text{C},\ T_e = 25\ °\text{C} \tag{11}$$

The slag provides an insulating layer on the surface of the liquid steel by preventing direct radiation from molten steel. The equation for heat transport through the slag layer is given by (ρ= density, C_p = specific heat, T = temperature, t = time, K = thermal conductivity):

$$\rho C_p \frac{\partial T}{\partial t} = K \frac{\partial^2 T}{\partial z^2} \tag{12}$$

The steel heat flux at the top of the ladle is directly linked to the thickness of the slag layer and such that thicker slag incurs smaller losses. The heat flux transferred from the top of the slag layer to the atmosphere is given by:

$$-K\frac{\partial T}{\partial r} = h_{se}\,(T_s - T_e) + \varepsilon_s\sigma\,(T_s{}^4 - T_e{}^4) \tag{13}$$

(K = thermal conductivity, T = temperature, T_s = temperature of the slag, T_e = room temperature, σ = Stefan-Boltzmann constant, h_{se} = convection coefficient between the slag and the surrounding environment, ε_s = emissivity of the slag)

Convection as well as radiation inside the slag layer were disregarded. Although the slag is probably made by a layer of solid slag on top of a layer of liquid slag no allowance was made for describing this two phase configuration. A single value for each of the parameters K, C_p *and* ρ was taken as representative of the slag layer. Convection coefficient values can be obtained from dimensionless Nusselt number from known correlations . For vertical plates, which correspond to the inner and outer walls of the ladle, was applied a correlation cited by Incropera(6).

$$Nu = \left\{0.825 + \frac{0.387\,Ra^{1/6}}{\left[1+(0.492/Pr)^{9/16}\right]^{8/27}}\right\}^2 \tag{14}$$

Where, Nu = Nusselt number, Ra = Rayleigh number, Pr = Prandtl number.

For horizontal surface or slag layer having contact with a fluid such as air, the recommended correlations computing average Nusselt number (cited by Incropera(6)) is :

$$\mathrm{Nu} = 0.54\mathrm{Ra}^{1/4}\,(10^4 \leq \mathrm{Ra} \leq 10^7)\,;\quad \mathrm{Nu} = 0.15\mathrm{Ra}^{1/3}(10^7 \leq \mathrm{Ra} \leq 10^{11}) \tag{15}$$

In the case of refractory lining at the bottom of the ladle, for computing the heat transferred through the contact surface between the ladle and the atmosphere , heat transfer coefficient can be evaluated from the relationship cited by Incropera(6):

$$Nu = 0.27Ra^{\frac{1}{4}} \quad (10^5 \leq Ra \leq 10^{10}) \tag{16}$$

The computational model was made up of three bodies: refractory (solid physical state), steel (liquid physical state) and slag. In order to simplify the simulation due to symmetry a slice of a quarter of the ladle was taken into consideration as shown in Figure 1-b and c. The main thermo-physical properties used for this simulation are given in Table 1

Table 1 : Thermophysical properties for this simulation

Material	Steel	Dolomitic	Alumina	High Alumina	Slag
Density $kg.m^{-3}$	7080	2100	2400	2900	3807
Specific heat $J/kg.K$	754	1005	1005	1047	838
Thermal conductivity $W/m.K$	30.5	1.05	1.26	2.51	1.21
Thermal diffusivity m^2/s		4.97×10^{-7}	5.22×10^{-7}	8.27×10^{-7}	

Results and Discussion

Consistence of this work can be initially evaluated by comparing its results with some assumed analytical solution even if the same is valid only for some very particular situation. For example for an appropriate combination of high refractory thickness, low refractory thermal diffusivity and short times a semi-infinite approach to heat conduction in the lining can be assumed(1). In this case a flat temperature profile at the middle refractory region associated with a steep gradient at metal – refractory interface is expected. This is seen from this simulation in Figure 2.

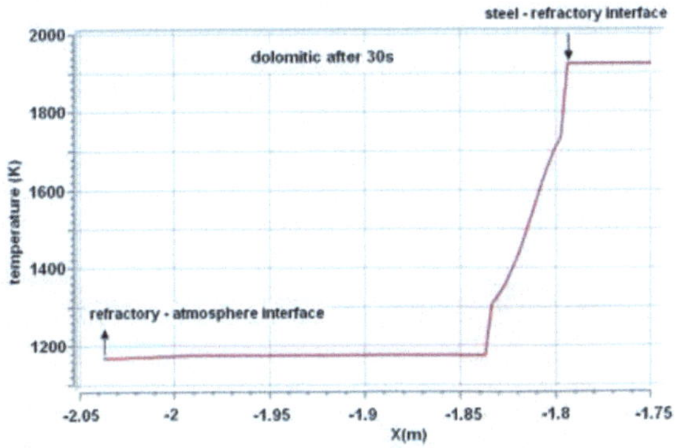

Figure 2: Temperature profile inside the refractory lining.

However the present formulation is more complete since it takes in consideration phenomena such as flow and thermal field inside the ladle (liquid steel). Based on the model described above the heat flux at the surface of the top slag was evaluated as shown in Figure 3. The initial temperature of slag was taken as 1650 °C and the temperature of atmospheric air as 25 ° C. In the early stages, the heat flux at the surface of the slag is more intense due to the steep temperature gradient between the same and the surrounding air. Gradually, as the radiation losses decrease with the decrease of the surface temperature, there is a decay in the heat flow and it tends to stabilize. Natural convection and thermal stratification are linked with each other. Heat losses from the liquid steel through both the refractory walls and also through slag , give rise to natural convection currents within the bath and as a consequence there is thermal stratification of liquid steel.

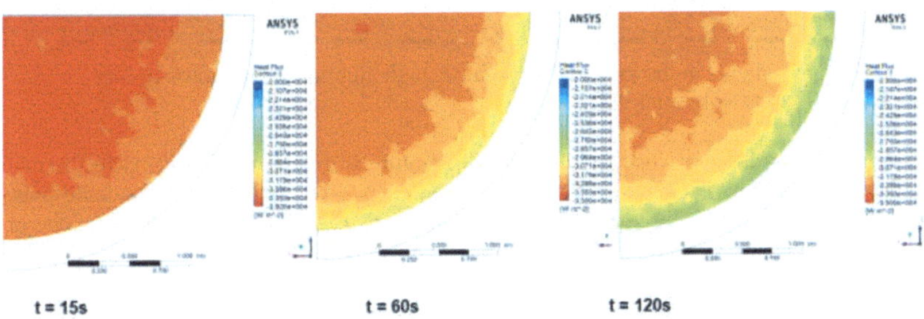

Figure 3. Heat flux at the surface of the slag after) 15s; b) and c 60s) 120s;

The convection phenomenon is represented by velocity vectors and is shown in layers with different temperatures within the bath (Figure 4). It can be noted that initially the natural convection currents are have high velocities, which stimulate thermal homogenization. However, over time the thermal stratification becomes evident, distinguished by the presence of planar isotherms except near the ladle walls, where the heat flow is through the walls is prominent. The nature of the refractory lining has a significant effect on the thermal losses, since the physical properties of the lining depend on the mineralogical composition and the mode of manufacture. In this investigation three different monolithic refractory linings were considered namely dolomite, high alumina and alumina (Figure 5). Due to its higher thermal diffusivity the high alumina refractory lining is less prone to thermal shock and wear by spalling. However, this condition makes the refractory lining exhibit a higher cooling rate. The top slag layer also serves like a refractory wall and has an insulating effect. Increasing the thickness of the slag layer decreases thermal losses (Figure 6)

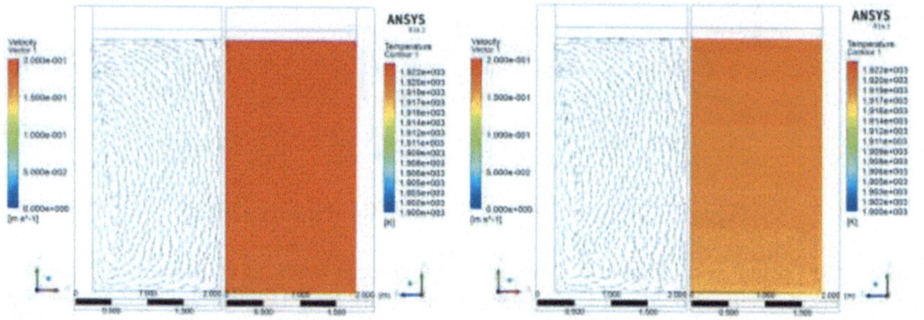

Figure 4: Velocity vectors and temperature gradient in the steel melt a) after 300s; b) after 900s.

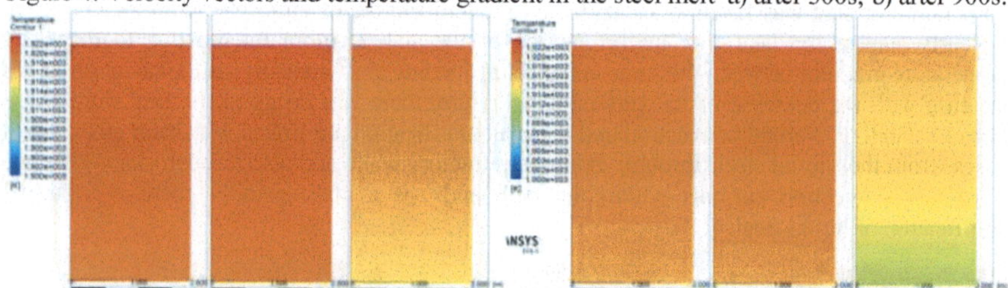

Figure 5: Thermal gradients in the ladle with dolomite, high alumina and alumina linings respectively: a) after 600s; b) after 900s.

Comparison of data from this work and data from literature is given in Figure 7. Here thermal stratification is defined as the difference of temperature at two points located at the ladle axis, one close to the surface other close to the bottom. There seems to be a good agreement in spite of differences in simulation conditions. The thermal state of the ladle is related to the soaking temperature of the refractory lining. This simulation results for two initial soaking temperatures of 700 ° C and 900 ° C are given in Figure 8. Lower soaking temperatures cause higher thermal gradients inside the refractory and higher thermal losses; as a consequence the thermal state of liquid steel can be significantly affected as seen in Figure 8.

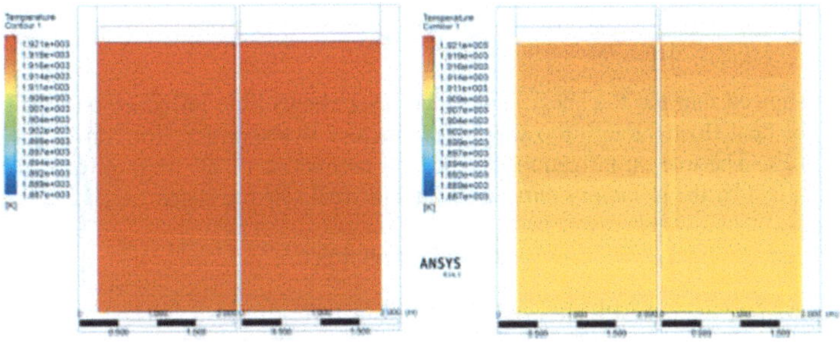

Figure 6: Liquid steel temperature gradient in a ladle with slag 200mm and 100mm slag respectively: a) after 300s; b) after 1200s

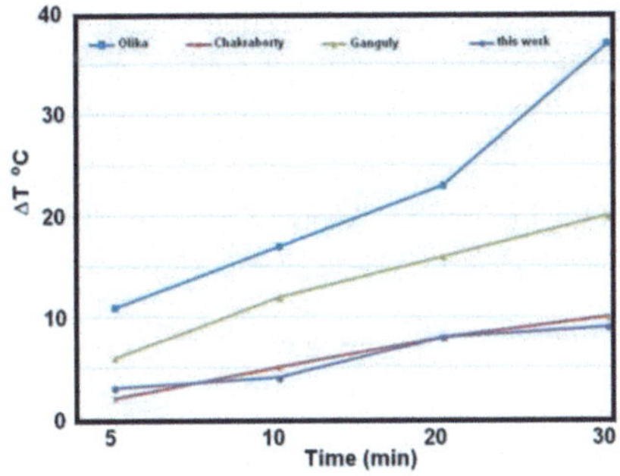

Figure 7: Thermal stratification as a function of time. Comparison of results of present work with those of Olika(7), Chakraborty(8), Ganguly(9).

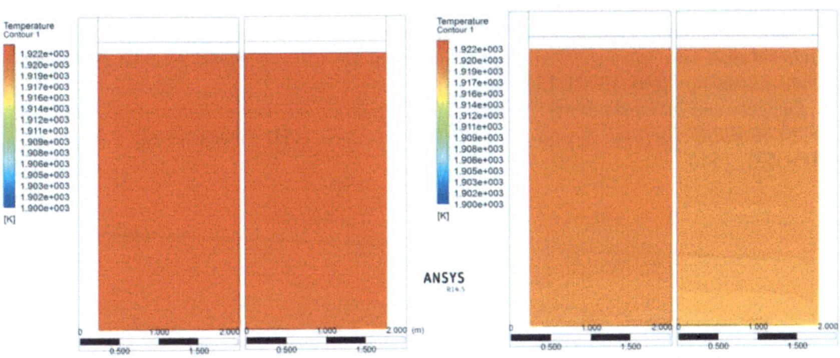

Figure 8: Thermal gradients for different soak temperatures of the refractory at 900 ° C and 700 ° C respectively: a) after 300s; b) after 900s.

Conclusions

The results showed that the thickness of the slag layer shows little influence on heat loss from the ladle. The heat flow through top slag and refractory is responsible for convection currents inside the melt. Thermal stratification can be a consequence of all these effects. The above factors also affect by the refractory mineralogy and thermal soaking state.

Acknowledgements

The authors wish to thank FAPEMIG and CNPq of Brazil for financial assistance to the project

References

1. I. D. Duarte; C. A. Silva, I. A. Silva, E. R Ferreira, A. L. Pereira and V. Seshadri, *Modelamento matemático das perdas térmicas em panelas de aciaria. (42° Seminário de Aciaria -Internacional; Porto Alegre, RS, 25 a 28 de maio/2014).*
2. B. Glaser, *A Study on the Thermal State of Steelmaking Ladles. Doctoral Thesis.* Royal Institute of Technology School of Industrial Engineering and Management (Department of Materials Science and Engineering Stockholm, September 2012)
3. K. D. Peaslee et al. *Efficiency in Steel Melting: Ladle Development.*(Proceedings of 59th SFSA T&O Conf., 2005). 42.1-42.12
4. Q. Liu et al – *Heat transfer analysis of molten steel in ladle.*(www.paper.edu.cn, 15/02/2015)
5. ANSYS Inc. PDF Documentation for Release 15.0; 148.204.81.206/Ansys/readme.html
6. F. P. Incropera et al., *Fundamentals of Heat and Mass Transfer.* (6th Edition. Wiley Asia Student Edition, John Wiley and Sons, ASIA, 2007)
7. B. Olika et al., *Numerical Simulation and Industrial Investigation on the Melt Stratification Phenomena in Ladles Holding Molten Steel.* (Scandinavian Journal of Metallurgy, v.25, 1966), 18-266
8. S. Chakraborty and Y. Sahai, *Effect of Slag Cover on Heat Loss and Liquid Steel Flow in Ladles before and during Teeming to a Continuous Casting Tundish.* (Metallurgical Transactions B, v.23B, 1992). 135-150
9. S. Ganguly and S. Chakraborty, *Numerical Investigation on Role of Bottom Gas Stirring in Controlling Thermal Stratification in Steel Ladles.* (ISIJ International, v.44, n.3, 2004)537 – 546

7th International Symposium on High-Temperature Metallurgical Processing
Edited by: Jiann-Yang Hwang, Tao Jiang, P. Chris Pistorius, Gerardo R.F. Alvear F., Onuralp Yücel, Liyuan Cai, Baojun Zhao, Dean Gregurek, and Varadarajan Seshadri
TMS (The Minerals, Metals & Materials Society), 2016

Viscous and Crystallization Characteristics of CaO-SiO_2-MgO-Al_2O_3-Fe_tO-P_2O_5-(CaF_2) Steelmaking Slags

Zhanjun Wang[1], Zuotai Zhang[2], Yongqi Sun[2], Min Guo[1] and Mei Zhang[1]

[1] State Key Laboratory of Advanced Metallurgy, School of Metallurgical and Ecological Engineering, University of Science and Technology Beijing, 100083, Beijing, P.R. China
[2] College of Engineering, Peking University, 100871, Beijing, P.R. China

Keywords: Fluorine, Viscosity, Crystallization, Structure

Abstract

The role of fluorine on the viscous and crystallization behavior of CaO-SiO_2-MgO-Al_2O_3-Fe_tO-P_2O_5-(CaF_2) slags were explored using the rotating cylinder method, non-isothermal differential scanning calorimeter (DSC), X-ray diffraction (XRD) and scanning electron microscopy (SEM) techniques. It was found that the viscosity slightly decreased with increasing CaF_2 content. It was also found that CaF_2 addition can promote the primary phase transformed from $MgFe_2O_4$ spinel to $Ca_3(PO_4)F$, which was beneficial to the enrichment of phosphorus. Additionally, the structure of the slags explained by X-ray photoelectron spectroscopy (XPS) and magic angular spinning nuclear magnetic resonance (MAS-NMR) were further analyzed, which was well in accordance with the above analysis.

Introduction

China's crude steel production in 2014 reached 822.7 million tons, an increase of 0.9% on 2013, accounting for nearly half of the world's total output. The most effective way of using the steelmaking slags is to recycle them in the iron- and steelmaking process, which will, however, result in higher phosphorus content in the slag and further burden of dephosphorization process.[1] It appears, therefore, that the selective crystallization and phase separation (SCPS) method can promote comprehensive utilization of steelmaking slags.[2]

Many researches have been carried out to study the enrichment behavior of phosphorus through adding different additives and n(2CaO SiO_2-3CaO P_2O_5) (nC_2S-C_3P) solid solution was considered as an effective phosphorous enriched phase.[1,3]

The present study aims to clarify of the thermophysical properties of CaO-SiO_2-Al_2O_3-MgO-Fe_tO-P_2O_5 slags modified through CaF_2 from the viewpoint of kinetics and XPS and NMR structural analysis, including the viscous flow and crystallization behaviors.

Experimental

The raw materials of the synthetic slags were analytical grade CaO, SiO_2, MgO, Al_2O_3, Fe_2O_3, P_2O_5 and CaF_2. The slags were prepared at 1823 K under 0.3L/min of Ar in a platinum crucible for 2 hours to obtain a homogeneous slag. The XRF results of the quenched slags were given in **Table 1**.

Table 1. Chemical compositions of slags (wt%, XRF)

Slag	C/S	$Fe^{3+}/\Sigma Fe$	CaO	SiO_2	Al_2O_3	MgO	Fe_tO	P_2O_5	CaF_2
1	1.79	0.814	38.14	21.32	5.01	6.58	19.64	9.31	0
2	1.77	0.837	35.96	20.31	4.71	6.66	20.03	9.46	2.87
3	1.82	0.825	35.04	19.21	5.12	6.84	19.86	9.37	4.56

The viscosity was measured using a calibrated Brookfield digital viscometer (*model LVDV-II+, Brookfield Engineering Laboratories, Middleboro, MA*) with a full-scale torque of 6.737×10^{-5} Nm with connecting suspension wires and spindle.

Differential scanning calorimeter (DSC) of the quenched samples was performed by a SETARAM instrument (*Setsys Evolution, S60158341, SETARAM Instrumentation, France*).

After the DSC measurement, the pre-melted slags were heated to 1823 K and then cooled to the target temperature at a cooling rate of 10 K/min. Then the precipitated crystalline phases were analyzed by X-ray diffraction (XRD) and scanning electronic microscopy (SEM).

A more quantitative analysis of structure can be obtained by using magic angular spinning nuclear magnetic resonance (MAS-NMR) and X-ray photoelectron spectroscopy (XPS).

Result and Discussion

Viscosity Measurement

Figure 1 shows the viscosity of the $CaO-SiO_2-5wt\%Al_2O_3-7wt\%MgO-20wt\%Fe_tO-10wt\%P_2O_5$ based slag with CaF_2 additions and a constant CaO/SiO_2 ratio of 1.8 at various temperatures. CaF_2 additions at 3 and 5wt% decreased the viscosity and the effect of CaF_2 was more pronounced at lower temperatures.

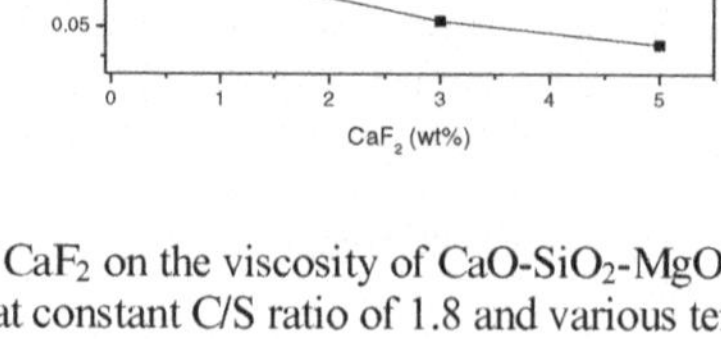

Figure 1. Effect of CaF_2 on the viscosity of $CaO-SiO_2-MgO-Al_2O_3-Fe_tO-P_2O_5$ steelmaking slags at constant C/S ratio of 1.8 and various temperatures.

DSC Measurement

The crystallization behaviors of the studied slags were investigated by employing non-isothermal DSC at various cooling rates and the crystal formation events of the slags can be determined from the exothermic peaks in the DSC curves. **Figure 2** (a) through (c) showed the

DSC curves of the slags at four different cooling rates, *i.e.*, 5, 10, 15 and 20K/min, respectively. It was observed that there were five exothermic peaks on DSC curves slags with 3 and 5wt% CaF_2, and only four exothermic peaks for slag 1 without fluorine addition. On the basis of the occurrence sequence of exothermic peaks on the DSC curves, the peaks were designated as P1, P2, P3, P4, and P5, respectively. Furthermore, with increasing cooling rate, the overall T_p (Peak temperature) shift to the lower temperature, which may be attributed to the fact that the nucleation and growth rate of crystals are functions of viscosity and undercooling. In addition, it was observed from **Figure 2** that the crystallization temperatures of the slags increase with an increase of CaF_2 content.

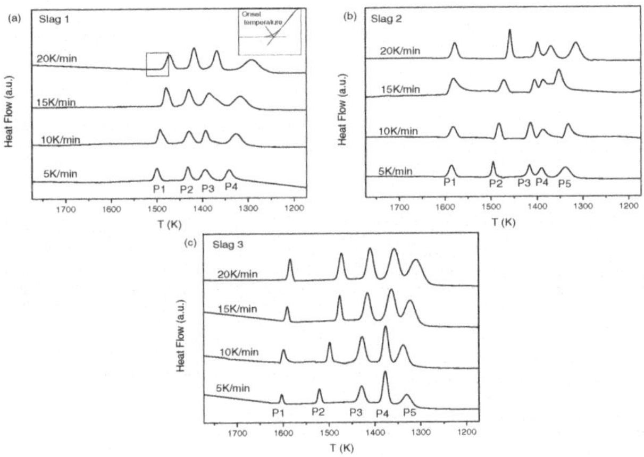

Figure 2. DSC curves of non-isothermal crystallization of slags at various cooling rates: (a) Slag 1, (b) Slag 2, (c) Slag 3

Identification of Crystalline Phases

In order to determine the sequence of crystalline phase precipitation, a series of heat treatment experiments were carried out, *i.e.*, the slags were melted at 1823K and then quenched at the target temperatures of the DSC curves. **Figure 3** and **Figure 4** showed the XRD patterns and SEM results for the water quenched slag 1 and slag 2 at target temperatures. As shown in **Figure 3**(a), $MgFe_2O_4$ spinel precipitated at 1473K first, while $MgFe_2O_4$ spinel and P-enriched phase co-precipitated at 1413K in the slag free of CaF_2. The result can also be verified from the SEM image shown in **Figure 4**(a). However, with the addition of CaF_2, it can be confirmed that the primary crystal changed from spinel ($MgFe_2O_4$) to fluorapatite ($Ca_5(PO_4)_3F$, C_5P_3F). The morphology of the crystal can also be clarified by **Figure 4**(b).

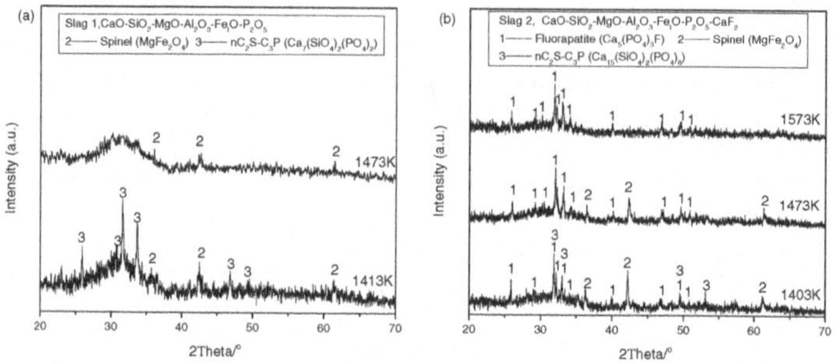

Figure 3. XRD results of slag 1 and slag 2 quenched at corresponding temperatures.

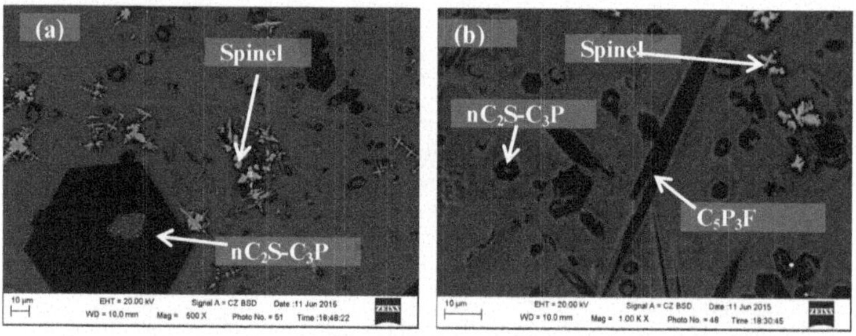

Figure 4. BSE images of slags at the cooling rate of 10 K/min and quenched at corresponding temperatures: (a) Slag 1 (1413K), (b) Slag 2 (1403K).

XPS Analysis

Generally, the structure of the slags is closely related to the viscosity and crystallization behavior. **Figure 5** shows the O^{1s} XPS result of the slags. Based on Gauss-Deconvolution method, the spectra has been fitted into three peaks at about 532 eV, 531 eV and 530.1eV, which can be attributed to bridging oxygen (O^{o}), non-bridging oxygen (O^{-}) and free oxygen (O^{2-}).[4] The two major peaks of non-bridging oxygen (O^{-}) and free oxygen (O^{2-}) increased, while the bridging oxygen (O^{o}) decreased with the addition of CaF_2. The depolymerization trend would result in the decrease of viscosity.

^{19}F MAS NMR Results

As displayed in **Figure 6**, the MAS NMR technique was also employed to analyze the slag structure. The spectra exhibit a main peak at about -105 ppm and a shoulder at about -117 ppm, which can be attributed to the bands of Ca-F and Si-F.[5-7] Because Ca^{2+} can behave as network modifier in the slags, Si-O-Ca-F band can exist in the molten slag. However, due to the greater stability of P-O-Ca than of Si-O-Ca bonds, the Ca^{2+} ions were then stripped away from the silicate network upon P inclusion. Therefore, the aforementioned F-Ca-O-Si bands would be interrupted by P-O-P and behave as F-Ca-O-P and P-O-Si, *i.e.*, the formation of F-Ca-O-P band enhanced the precipitation of $Ca_5(PO_4)_3F$.

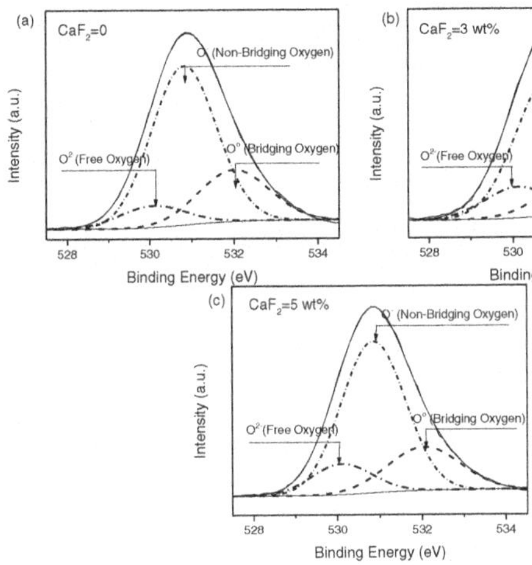

Figure 5. Fitting bands of O^{1s} binding energy of $CaO-SiO_2-MgO-Al_2O_3-Fe_tO-P_2O_5-(CaF_2)$ slags with varying CaF_2 contents.

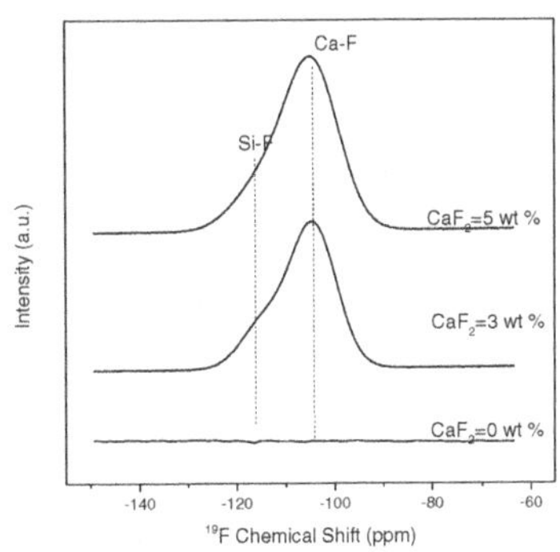

Figure 6. ^{19}F MAS NMR spectra with varying CaF_2 content of $CaO-SiO_2-MgO-Al_2O_3-Fe_tO-P_2O_5-(CaF_2)$ slags.

Conclusions

1. With increasing CaF_2 content, the viscosity decreased which can be clarified by the decreasing DOP of the structures.
2. The presence of Ca-F bonds and the greater stability of P-O-Ca than of Si-O-Ca bonds enhanced the primary crystal transformed from spinel ($MgFe_2O_4$) to fluorapatite phase ($Ca_5(PO_4)_3F$) with the addition of CaF_2.

References

1. L. Lin *et al.*, "Separation and Recovery of Phosphorus from P-bearing Steelmaking Slag," *Journal of Iron and Steel Research International*, 21(2014), 496-502.
2. Y.Q. Sun *et al.*, "The Effect of P_2O_5 on the Crystallization Behaviors of Ti-Bearing Blast Furnace Slags Using Single Hot Thermocouple Technique," *Metallurgical and Materials Transactions B-Process Metallurgy and Materials Processing Science*, 45(2014), 1446-1455.
3. X. Yang, H. Matsuura, and F. Tsukihashi, "Reaction Behavior of P_2O_5 at the Interface between Solid 2CaO SiO_2 and Liquid CaO-SiO_2-FeO_x-P_2O_5 Slags Saturated with Solid 5CaO SiO_2 P_2O_5 at 1573K," *ISIJ International*, 50(2010), 702-711.
4. H. Kim and I. Sohn, "Effect of CaF_2 and Li_2O Additives on the Viscosity of CaO-SiO_2-Na_2O Slags," *ISIJ international*, 51(2011), 1-8.
5. A. Stamboulis, R.G. Hill, and R.V. Law, "Characterization of the structure of calcium alumino-silicate and calcium fluoro-alumino-silicate glasses by magic angle spinning nuclear magnetic resonance (MAS-NMR)," *Journal of Non-Crystalline Solids*, 333(2004), 101-107.
6. J.F. Stebbins and Q. Zeng, "Cation ordering at fluoride sites in silicate glasses: a high-resolution ^{19}F NMR study," *Journal of Non-Crystalline Solids*, 262(2000), 1-5.
7. R. Hill *et al.*, " The influence of strontium substitution in fluorapatite glasses and glass-ceramics," *Journal of Non-Crystalline Solids*, 336(2004), 223-229.

7th International Symposium on High-Temperature Metallurgical Processing
Edited by: Jiann-Yang Hwang, Tao Jiang, P. Chris Pistorius, Gerardo R.F. Alvear F., Onuralp Yücel, Liyuan Cai, Baojun Zhao, Dean Gregurek, and Varadarajan Seshadri
TMS (The Minerals, Metals & Materials Society), 2016

CHEMICAL, PHYSICAL AND MORPHOLOGICAL CHANGES OF SINTERING DUST BY MECHANICAL ACTIVATION

Feng Chang[1], Shengli Wu[1, 2], Jianliang Zhang[1], Mingyin Kou[1], Hua Lu[1], Laixin Wang[1]

[1]School of Metallurgical and Ecological Engineering, University of Science and Technology Beijing; 30 Xueyuan Road; Haidian District, Beijing, 100083, P. R. China
[2]School of Metallurgy and Chemical Engineering, Jiangxi University of Science and Technology; No.156, Kejia Ave.; Ganzhou, Jiangxi, 341000, P. R. China

Keywords: Mechanical activation; Sintering dust; Characterization

Abstract

Sintering dust has been regarded as an important secondary resource in a modern steel plant. Sintering dust was mechanically treated using different milling machines and experimental conditions in air atmosphere. The changes in phase constitution, particle size, specific surface area, particle morphology, and lattice parameters of activated sintering dust were determined. It was found that the agglomeration of the particles take place during extended dry milling with accessible pores for Nitrogen gas. The maximum specific BET surface area in the stirring wet grinding and planetary wet grinding increases to 43.06 and 46.45 m^2/g after activated 360 min respectively. With the increasing of energy input, the intensity of diffraction peaks of dolomite and hematite gradually decreased. Furthermore, the previous phase almost disappeared eventually under any experimental conditions. In addition, Fourier transform infrared spectroscopy indicates that organic matter and carbonate decompose easier than other phases, especially under stirring wet milling.

Introduction

The sintering of iron ore is an agglomeration process in which blends of iron ore, fluxes, fuel in the form of coke and reverted materials from steel works are mixed together to produce a feed suitable for the blast furnace. Sintering dust is a solid waste generated during the sintering process collected by electrostatic precipitator. It is considered as a kind of dangerous dust because of its complicated components such as heavy metals and hazardous organics. According to statistics, 4 kg of sintering dust was produced per ton of steel, that is, about 3.291 million tons of sintering dust was generated in 2014 in China. The comprehensive utilization rate of sintering dust is less than 20%, and most of the sintering dust is stored at the plant. Not only does the large amount of sintering dust take up much land including farmland, it also causes environmental pollution to different degrees. The sintering dust contains a high content of SiO_2, CaO and MgO, which can cause a high incidence of silicosis to the workers, especially in the sintering of high basicity sinter. Therefore, the sintering dust must be dealt with by some kind of treatment.
The existing treatment processes for sintering dust can be roughly divided into the following four categories: sintering process, pellet process, direct reduction process and hydrometallurgical process. [1][2][3][4] With the enrichment circulation of harmful elements, it is difficult for the sinter process to totally digest the dust produced by itself. Research has shown that, the pelletability of sintering dust is much worse than the blast furnace dust or basic oxygen furnace dust. The direct reduction process is suitable for handling dust with high content of zinc. The hydrometallurgical process is mainly adopted to extract potassium in the sintering dust, however,

it needs enough water and it would cause new pollution. In addition, the recovery rate of value elements in sintering dust by pyrometallurgical or hydrometallurgical process is finite.
Mechanical activation (MA), a narrow field of the Mechanochemistry (MC) applied for the activation of chemical reaction by mechanical energy, has been used in mineral processing to produce finely ground particles, increased surface area and improved chemical reactivity of milled materials. To promote the pelletability, direct reduction kinetic conditions, and leaching kinetic conditions of sintering dust, pretreatment by mechanical activation have been investigated.
The goal of this paper is to study the chemical, physical and morphological changes of mechanically activated sintering dust by means of granulometry analysis, specific surface area analysis, X-ray diffraction (XRD), scanning electron microscopy (SEM) and Fourier transform infrared spectroscopy (FTIR).

Experimental

Raw material

The sintering dust used in this study was kindly supplied by Baosteel in China. The main chemical composition of the initial sample is given in Table I. The Total iron and calcium oxide content in sintering dust is about 45% and 11%, respectively. The total content of fluxes is around 17%. In addition, the content of potassium, sodium, zinc and lead elements are relatively low. The XRD pattern of the initial dust sample can be seen in Figure 1. It showed mainly hematite, dolomite and Potassium chloride, while small amount of limestone and Sodium chloride reflection peaks. The BET specific surface area of the starting sintering dust sample was 9.18 m^2/g. Sintering dust appear to relatively loosely agglomerated, and some of them was bright under SEM.

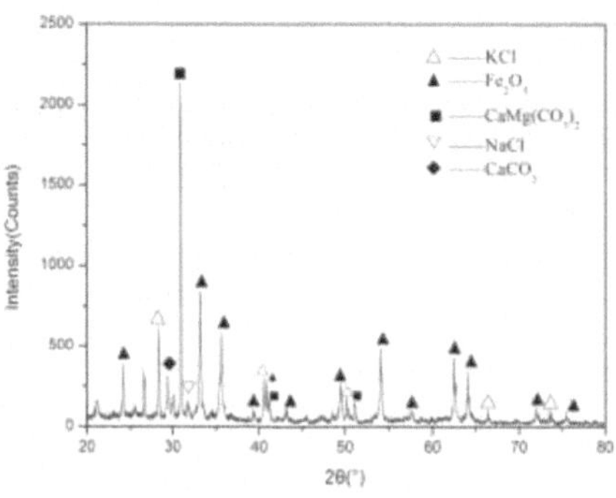

Figure 1. The XRD patterns of initial sintering dust

Figure 2. The particles morphology of initial sintering dust

Table I. The main chemical composition of the initial sample (wt. %)

TFe	FeO	CaO	SiO_2	Al_2O_3	MgO	K	Na	Zn	Pb	Cl	C
44.66	2.49	11.14	1.80	1.92	2.47	2.82	0.93	0.07	0.30	4.58	4.20

Mechanical activation and characterization

The changes brought about by different milling devices during the mechanical activation of sintering dust were investigated using two types of ball mills; planetary and stirring mills. The milling was carried out by different sizes of steel media from 3 mm to 20 mm. A statistical design for two levels of grinding methods and six levels of milling time (10, 30, 60, 120, 180, 360 min) was used. All experiments were performed in air and deionized water atmosphere. Each experiment was carried out independently and in closed condition. The experimental milling conditions are displayed in Table II. The material of mill tank and mill media is stainless steel. As milling media, the diameters of stainless steel ball used in planetary mill are 6, 10, and 20 mm. However, it is only 3 mm in stirring mill. The mass ratio of deionized water to sintering dust is 1.5 to 1 when wet grinding is conducted.

The particle size distribution of the mechanical activated samples was measured by laser diffraction (LMS-30) with anhydrous ethanol as dispersing medium. The mean particle diameter and granulometeric surface area were obtained from the particle size distribution. The specific surface area of the samples was determined by the BET method with AUTOSORB-1C (Quantachrome, USA). Samples were degassed by heating at 150 °C for 120 min immediately prior to measurements. Multi-point measurements were made to give a more accurate value. The X-ray diffraction (XRD) patterns were collected by a Rigaku TTRIII at a scanning speed of 2°/min using Cu Kα radiation (λ=0.15418). The XRD patterns of the samples were recorded in the range 2θ=20°–80°, using a step size of 0.02°. The surface structure was studied by scanning electron microscope with MLA 250 system. FTIR spectra were collected on an Infrared spectroscopy (NEXUS 670-FTIR, Nicolet, America) between 4000 and 400 cm^{-1} with a 4 cm^{-1} resolution and 32 scans. The samples were prepared as KBr pellets.

Table II. Experimental milling conditions

Milling conditions	Stirring milling	Planetary milling
Atmosphere	Air, Deionized water	Air, Deionized water
Milling time (min)	10, 30, 60, 120, 180, 360	10, 30, 60, 120, 180, 360
Ball to powder weight ratio	5:1	5:1
Speed (rpm)	400	400
Media diameter (mm)	3	20, 10, 6
L(mm) × Φ (mm)	224 × 168	100 × 117

Results and discussion

Particle size

The effects of the milling on the particle size distributions are summarized in Figure 3. The initial sintering dust has relatively wide distribution in comparison to the milled samples. The gap differences between the curves imply that the rate of breakage in the earlier stages of milling is higher than the later stages of grinding. As the milling progresses, the breakage rate decreases whatever methods and experimental condition are applied.

However, there are some differences among the particle size distributions resulted in different grinding methods and operational conditions. From the perspective of short time mechanical activation, planetary wet milling and stirring dry milling tend to produce finer particles faster. In terms of prolonged mechanical activation, the stirring grinding is better than the planetary grinding; and the wet grinding is more effective. In the case of stirring wet milling, prolonged grinding tends to produce finer particles. This indicates that further size reduction in such mills at the given circumstances is possible. In the planetary dry mill, the further size reduction is impossible if the milling time exceeds 1 h in planetary dry mill. But, intriguingly, when the dry milling time reaches to 180min, the sintering dust goes through a significant agglomeration process. For planetary wet milling, the agglomeration phenomenon lags behind the dry counterpart. In a word, dry grinding is liable to agglomerate, and planetary dry grinding tends to reunion particles earlier than planetary wet grinding. The agglomeration could be expected to decompose again with the extended milling, in particularly in dry mills. Stirring wet milling can prevent the reunion from happening. The agglomeration of particles were reported previously by many authors for various minerals in the prolonged milling and intensive milling conditions, confirming that this is a feature of extended intensive dry milling. [5] [6]

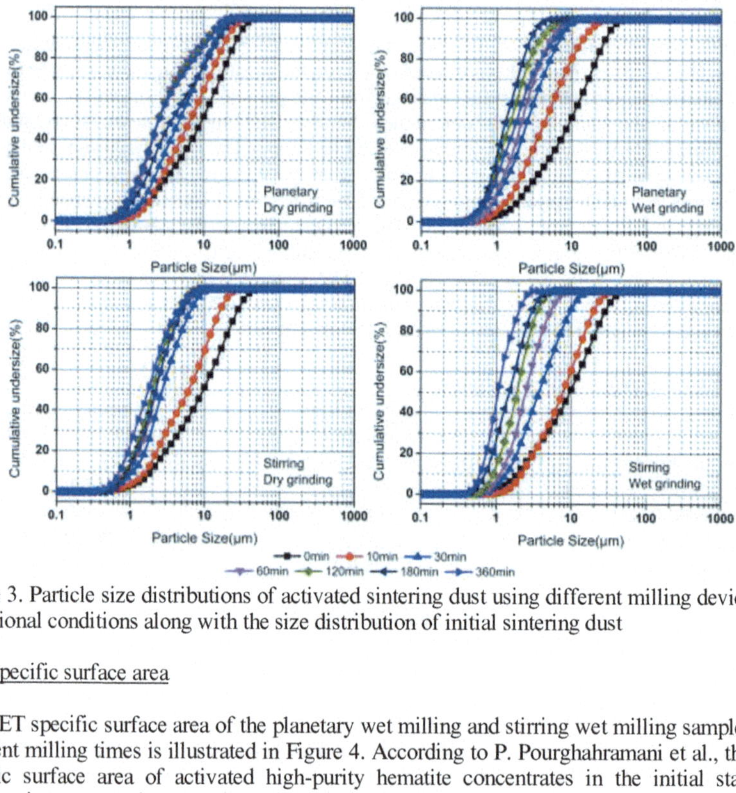

Figure 3. Particle size distributions of activated sintering dust using different milling devices and operational conditions along with the size distribution of initial sintering dust

BET specific surface area

The BET specific surface area of the planetary wet milling and stirring wet milling samples after different milling times is illustrated in Figure 4. According to P. Pourghahramani et al., the BET specific surface area of activated high-purity hematite concentrates in the initial stages of grinding increases rather sharply and continues to rise gradually. [7] Whereas, different from hematite concentrates, one most obvious feature for activated sintering dust is that the BET specific surface area decreasing a little in the initial stage whatever milling methods were applied. This is simply because that large number of fine particles of potassium chloride and sodium chloride dissolve in the water. Then, as the decomposition of hematite and dolomite particles, the BET specific surface area increased. The maximum specific surface areas in the milling with planetary wet mill and stirring wet mill are 43.06 and 46.45 m^2/g, respectively. The effect by planetary wet milling was better for short time milling, whereas, the contrary result was achieved for long time milling. Even though the agglomeration phenomenon aroused during the final stages of planetary wet milling, the specific surface areas of sintering dust continued increasing. It indicated that the agglomeration of the particles take place during extended dry milling with accessible pores for Nitrogen gas.

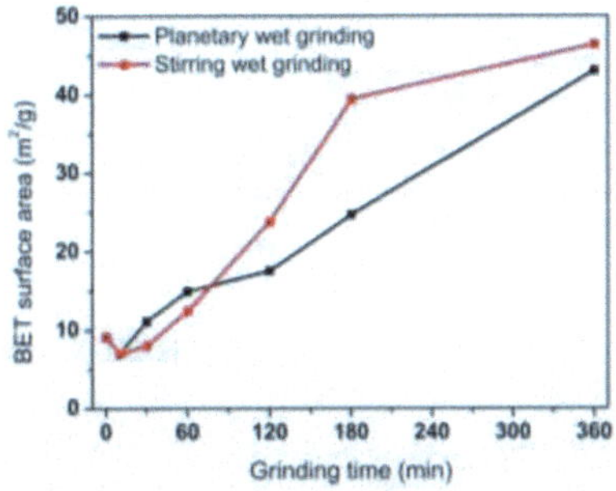

Figure 4. Changes in the BET specific surface area of sintering dust ground in planetary wet milling and stirring wet milling.

X-ray diffraction

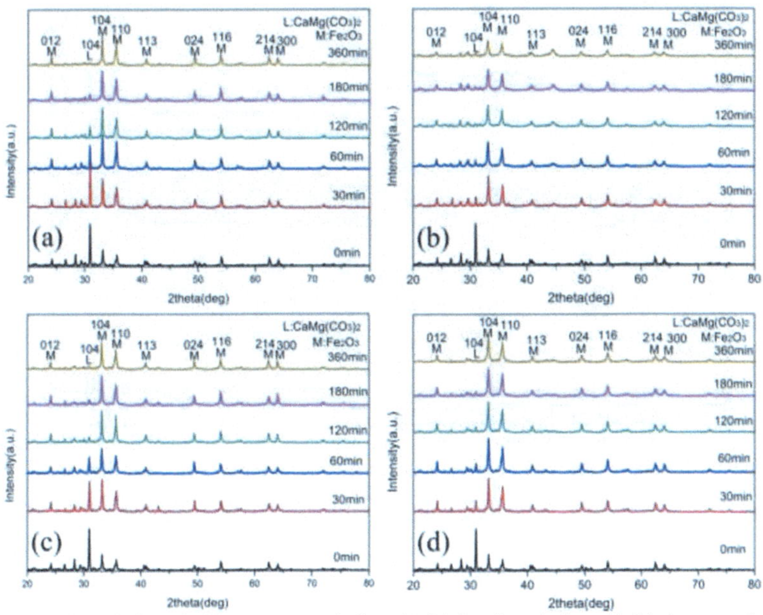

(a) Planetary dry grinding; (b) Planetary wet grinding; (c) Stirring dry grinding; (d) Stirring wet grinding.
Figure 5. The X-ray diffraction patterns of activated sintering dust as function of the grinding time using different milling devices and operational conditions.

The collected XRD patterns of the sintering dust are summarized in Figure 5 for different milling devices and operational conditions. XRD patterns show that with the grinding prolonged, the reflection peaks of potassium chloride, dolomite and hematite gradually become broad and low even disappeared, indicating that initial sintering dust did not undergo significant reactions during the milling. But the disappearance of diffraction peak at the position of 28.35 deg indicated that potassium chloride had promptly dissolved in water when stirring wet grinding was conducted. The XRD method studies do not show phases below 2 wt. %. This agrees with the observations of Kaczmarek and Ninham [8] and Stewart et al. [9]. However, our observations disagree with the findings of Kosmac and Courtney [10]. They reported a transformation of α-Fe_2O_3 to Fe_3O_4 during ball milling in a planetary mill. It is evident that our milling conditions could not change the stability of hematite.

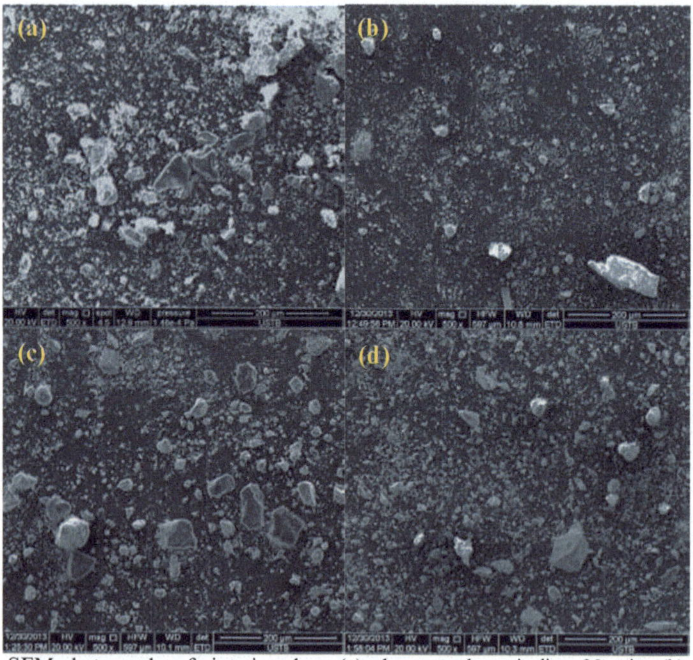

Figure 6. SEM photographs of sintering dust: (a) planetary dry grinding 30 min, (b) planetary wet grinding 30 min, (c) stirring dry grinding 30 min, (d), stirring wet grinding 30 min.

Table III. The grain size and lattice distortion of hematite under different milling conditions

Grinding type	Grinding time (min)	Air		Deionized water	
		Grain size (nm)	Lattice distortion (%)	Grain size (nm)	Lattice distortion (%)
Planetary grinding	30	107.1	0.02	65.8	0.012
	180	105.2	0.026	48.5	0.071
	360	97	0.033	25.5	0.103
Stirring grinding	30	103.6	0.014	80.7	0.035
	180	100.1	0.021	51.1	0.042
	360	87.3	0.037	50.3	0.077

In terms of the hematite phase, its grain size gradually decreased and lattice distortion gradually increased with the grinding time prolonged, as shown in Table III. The grain size and lattice distortion of hematite in the original sample were 109.5 nm and 0.009%, respectively. The effect by wet grinding was obviously better than dry grinding. The minimum grain size reached 25.5 nm after 6 hours planetary wet milling, and the corresponding maximum lattice distortion was 0.103%.

Morphological changes

The SEM micrographs of the planetary milling and stirring milling samples after milling 30 min are illustrated in Figure 6. The large agglomerated particles were basically separated and some hard bulk particles appeared as scattered distribution. As received by particle size analysis, the particle size decreases whatever methods and experimental condition are applied. Compared to dry milling, wet milling tends to produce much finer particles.

FTIR spectra

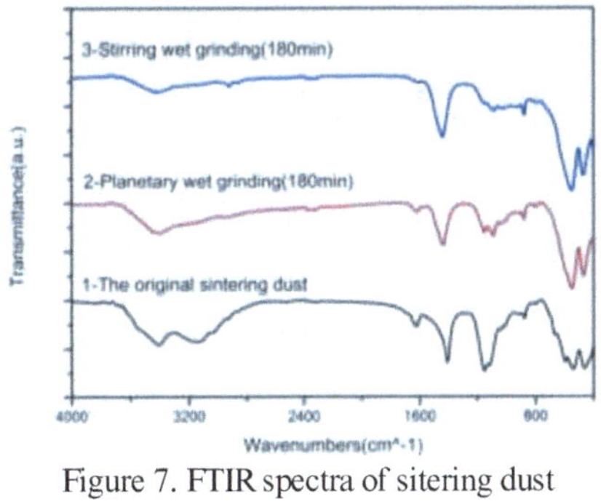

Figure 7. FTIR spectra of sitering dust

Due to high surface sensitivity, FTIR measurements were conducted to monitor the changes of the surface properties of the ground particles in both mills. The FTIR spectra of nonactivated and activated sintering dust after 180min planetary wet grinding or stirring wet grinding are presented in Figure 7. According to Figure 7, there are important changes in the relative intensities, broadening and shifting of peak positions, which confirm once again the strong structural distortion as explained by XRD analysis. The disappearance of the bands around 3138 cm^{-1} correspond to C-H stretching vibration absorption peak indicates that unsaturated polycyclic aromatic hydrocarbons decomposes easier than other phases. The disappearance of the bands around 673 cm^{-1} and 598 cm^{-1} correspond to CO_3^{2-} bending vibration absorption peak and Al-OH key respectively. The almost disappearance of the bands around 1153 cm^{-1} correspond to C-O stretching vibration absorption peak indicates that dioxin twists easier than other phases, especially under stirring wet milling. Further surface sensitive measurements will be conducted in forthcoming works to gain a better insight about the mechanochemical characteristics in high energy grinding experiments.

Conclusions

(1) Dry grinding is liable to agglomerate, and planetary dry grinding tends to reunion particles earlier than planetary wet grinding. Stirring wet milling can prevent the reunion from happening.
(2) The BET specific surface areas of sintering dust decreased first by the disappearance of fine potassium chloride, however, it increased with prolonged grinding whether the sintering dust particles reunited or not.
(3) With the grinding prolonged, the grain size of hematite gradually decreased, while the lattice distortion gradually increased. The change ratio by wet grinding was greater than dry grinding.
(4) The potassium chloride phase, unsaturated polycyclic aromatic hydrocarbons and CO_3^{2-} in sintering dust disappeared by stirring wet grinding.

Acknowledgements

The financial support of the National Natural Science Foundation of China (U1260202) is gratefully acknowledged.

References

[1] D.L. Zeng et al., "Research Advancement of Comprehensive Utilization of Iron-bearing Dust in Iron and Steel Plants," *Sintering and Pelletizing*, 36 (6) (2011), 45-49.

[2] X. Liu et al., "Removal of Potassium and Preparation of Potassium Sulfate from Sintering EAF Dust," *Metal Materials and Metallurgy Engineering*, 39 (3) (2011), 40-45.

[3] C. Peng et al., "Separation and Recovery of Potassium Chloride from Sintering Dust of Ironmaking Works," *ISIJ International*, 49 (5) (2009), 735-742.

[4] X.F. She et al., "Study on Basic Properties of Typical Industrial Dust from Iron and Steel Plant and Analysis of Its Utilization," *The Chinese Journal of Process Engineering*, 9 (S1) (2009), 7-12.

[5] P. Balaz et al., "Thiourea Leaching of Silver from Mechanically Activated Tetrahedrite," *Hydrometallurgy*, 43 (1-3) (1996), 367-377.

[6] Q.W. Zhang et al., "Mechanochemical Changes in Gypsum When Dry Ground with Hydrated Minerals," *Powder Technology*, 87 (1) (1996), 67-71.

[7] P. Pourghahramani et al., "Microstructural Characterization of Hematite During Wet and Dry Millings Using Rietveld and XRD Line Profile Analyses," *Powder Technology*, 186 (1) (2008), 9-21.

[8] W.A. Kaczmarek et al., "Preparation of Fe_3O_4 and γ-Fe_2O_3 Powders by Magnetomechanical Activation of Hematite," *IEEE Transactions on Magnetics*, 30 (2) (1994), 732-734.

[9] S.J. Stewart et al., "Effects of Milling-induced Disorder on the Lattice Parameters and Magnetic Properties of Hematite," *Journal of Magnetism and Magnetic Materials*, 260 (3) (2003), 447-454.

[10] T. Kosmaca et al., "Milling and Mechanical Alloying of Inorganic Nonmetallics," *Journal of Materials Research*, 7 (6) (1992), 1519-1525.

7th International Symposium on High-Temperature Metallurgical Processing
Edited by: Jiann-Yang Hwang, Tao Jiang, P. Chris Pistorius, Gerardo R.F. Alvear F., Onuralp Yücel, Liyuan Cai, Baojun Zhao, Dean Gregurek, and Varadarajan Seshadri
TMS (The Minerals, Metals & Materials Society), 2016

DYNAMIC THERMAL SIMULATION STUDY OF COPPER SLAG DILUTION UNDER DIRECT CURRENT FIELD

Zhang Jing[1], Sun Ying[1], Li Qiuju[1]

[1]Shanghai Key Laboratory of Modern Metallurgy & Materials processing, Shanghai University, shanghai 200072, China

Keywords: copper slag cleaning, direct-current electric field, molten copper slag, thermal simulation

Abstract

A new process of enrichment Cu in molten copper slag under electric field is presented in this article. The experiments were carried out in the laboratory on a kilogram scale. The mobility of copper slag was improved by the technique of carbon reduction. The effects of temperature, electric field strength and intermittent feeding speed were investigated on Cu migration. The results showed that the best experimental conditions were the temperature at 1523K, electric field strength was 1.5V•cm-1 and feeding rates was 15 g•$(10min)^{-1}$. The copper content of the slag decreased below 0.6wt%. The process was compared to the traditional electric furnace process of slag cleaning, there was reduced the energy consumption and heat loss, and also harmful gas emissions such as SO_2 and CO_2, accompanied by improved the efficiency of slag cleaning.

Introduction

Copper slag cleaning was an integral part in the copper metallurgical process, which was the requirement of resource re cycling and environmental stress. Now the main method of copper slag cleaning was furnace strengthening dilution method. This needs long time, low viscosity of slag and large electric energy and electrode consumption, resulting in high energy consumption and emissions[1].The development direction of copper slag cleaning was low energy consumption and high efficiency. From the special behaviors of copper or copper matte particles in the electric field, we can improve the recovery rate of copper and reduce energy consumption through direct current (dc) dilution technique, so this technique deserves further research.

Combining with previous methods and researches of slag cleaning in direct current (dc) dilution, this paper presents a new technology of processing dynamic molten copper slag in electric field. In this process, copper slag was modified with reduction by blowing industrial waste gas or natural gas, the copper particles gathered and grew up under the aid of gas stirring. Sedimentation of copper particles was accelerated in direct current field.

Compared to the traditional electric stove dilution methods, the advantages of this new process were: reduced the amount of reducing agent, such as coke and pyrite, also the energy consumption of coke production and heat loss of the reductive modification reaction, also harmful gas emissions such as SO_2 and CO_2, accompanied by Shortened the time of slag dilution, which laid a foundation for direct slag dilution.

Experimental

Dynamic thermal simulation experiment device

The dynamic thermal simulation experiment device of copper slag cleaning is the groove type reaction stove, which was specially designed and manufactured; its internal structure diagram is displayed in figure 1. The heating element is silicon molybdenum rods, the temperature control device was Shimaden SR53 temperature control program, and the temperature control precision is ±1k.The thermometric component is double platinum rhodium thermocouple. To prevent slag samples and molybdenum rods were oxidized, the experiment process was in argon atmosphere. Reaction vessel is a rectangular crucible, the size of the crucible are 26 cm (length), 3cm(width), 4 cm(height); the size of residue crucible are 35 cm (length), 8 cm (width), 4 cm(height).

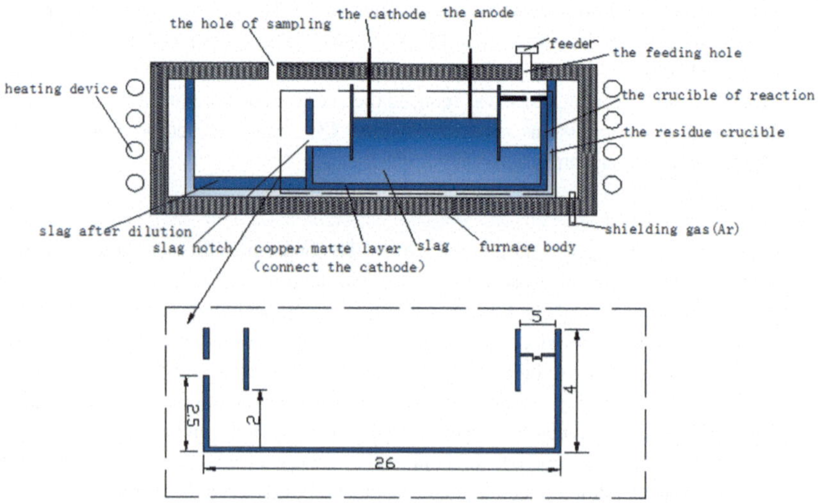

Fig. 1 Thermal simulation apparatus for molten copper slag cleaning in flow condition

Material preparation and experiment

The experimental samples are the original copper slag(S-1) and the copper slag after composition regulation (S-2).The S-2 is the mixture of original slag after reduction and CaO. The reducing conditions are: carbon slag proportions 1:67, temperature1423 k, roasting time 1 h. The main phases of samples are showed in table 1.

Table 1 Chemical Composition of Slag/ wt %

Slag number	Chemical composition/ wt %							
	FeO	Fe_3O_4	SiO_2	Al_2O_3	CaO	Cu	ZnO	others
S-1	36.56	20.70	32.52	1.19	0.39	3.85	2.55	2.24
S-2	47.38	7.37	33.54	0.10	3.00	4.36	1.90	2.35

In the crucible, add a certain amount of copper powder and slag samples, then heat melting in high temperature furnace. After molten, the thickness of liquid copper and slag at the bottom are 0.5 cm and 2 cm respectively, place thickness 1cm carbon powder upper the slag. The cathode was liquid copper at the bottom, the anode was carbon powder, molybdenum rods was a conductor between electrodes and the power, Electric field is provided by high power dc power supply. To guarantee the stability of the slag layer thickness, set the slag notch, the distance to the bottom of crucible was 2.5 cm; to make the carbon layer separate from the inlet and outlet, a baffle was arranged respectively at the both sides of crucible. To ensure the circulation of slag and copper, the distance from baffles to the crucible bottom was 2 cm. Under cold material feeding condition, to prevent slag melting speed affect dilution effect, set a corundum baffle at the bottom of the feeding mouth, the diameter of the middle holes was 0.5 cm, and the slag sample can completely through the holes into crucible in liquid form.

The experiment process is shown in figure 2, the experimental conditions were: temperature 1523 k, the electric field strength 1.0 ~ 2.0v. cm^{-1}. Every 10min add material, each feeding amount was 15~50 g, taking a slag sample every 20 minutes, the total amount of copper slag was 800 ~ 1100g at the end of reaction, which basically in the kg test scale. The copper content in the initial slag was reference standard, study the effect of various factors on the molten copper slag dilution, the experiment process was in argon atmosphere.

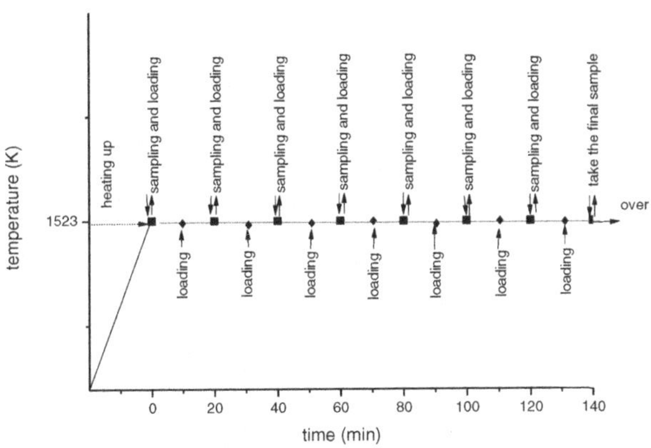

Fig.2 Operation chart of experiments

Results and Discussion

The movement of copper droplets in settling basin

Copper slag entered dilution device from the melting furnace, the copper droplets sunk into the bottom of molten pool under the action of gravity and electric field force, and formed slag and copper two layers. The movement of droplets can be seen the synthesis of horizontal and vertical direction, as shown in figure 3. In horizontal direction, copper droplets and slag moved to the exit. In the vertical direction, matte droplets through the slag layer moved to the bottom of molten pool. In order to understand the settling behavior of copper or copper matte particles in the flowing molten slag, the movement behavior of droplets in the horizontal and vertical direction was discussed respectively, and simplified as following:

(1) Regardless of the collision growth and breakup between copper droplets;
(2) Regardless of the chemical reaction between copper matte particles and slag;
(3) Assume that dilution device discharged slag continuously and steady, meaning that molten slag flowed at a constant speed in the settling basin.

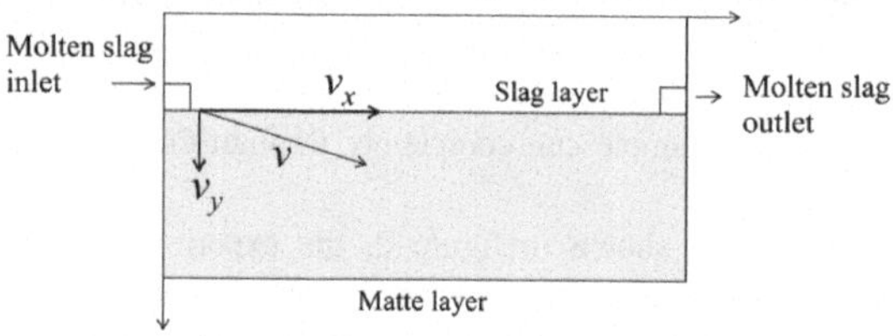

Fig. 3 Sketch of setting track of copper droplet

Horizontally, the movement speed of copper matte droplets can approximately equal to molten copper slag, can be determined by the following equation:

$$v_x = \frac{Q}{\rho_s \cdot \omega \cdot h} \qquad (1)$$

v_x----Copper droplets horizontal movement speed/m˙s^{-1}

Q----feeding speed/kg˙m^{-3}

ρ_s----the density of molten slag/kg˙m^{-3}

w----the effective width of dilution device/m

h---- the thickness of slag layer /m

When the length of the settling basin was l, the horizontal residence time of copper matte droplets in the sedimentation tank can be determined by the following equation:

$$t_x = \frac{l}{v_x} \qquad (2)$$

l----The length of the tank/m

That was,

$$t_x = \frac{\rho_s \cdot \omega \cdot h \cdot l}{Q} = \frac{\rho_s \cdot V}{Q} \quad (3)$$

V----The volume of the tank/m^3

Studies had shown that the influence of the initial velocity of copper matte droplets in the vertical direction on the droplet settling time can be neglected [2]. The instantaneous velocity of droplets on the vertical direction be marked V_{EG}, the time from the slag layer surface sunk to the matte layer was:

$$t_y = \frac{h}{V_{EG}} \quad (4)$$

The horizontal stay time determined the total time of copper matte droplets in the settling basin, the increase of t_x was advantageous to the dilution of the molten slag. Can be seen from equation 3, increasing the feeding volume ratio (V: Q) was helpful to increase the horizontal stay time of copper matte droplets. In order to make the copper matte droplets enter into the matte layer as soon as possible, the residence time need to be shortened in vertical directions, which need to speed up the settling velocity of copper matte particles or decrease the thickness of slag layer.

In fact, thin layer of slag is conducive to the sedimentation of matte, but the slag layer thickness is hard to change largely under the restrictions on the size of the tank size and the stability of the dc field. Thus the main way to shorten copper matte particles residence time in the vertical direction is increasing electric field strength, and controlling the composition of the slag. Some studies [3] had pointed out that flowing speed of slag in the longitudinal section (perpendicular to the direction of melt flow) was not uniform. The nouniform would be aggravated if increase the width of sedimentation pool. Therefore, the effective method of increasing the horizontal residence time of copper matte particle is controlling the feeding speed Q. it doesn't work by increasing width of settling basin.

The effect of main technological parameters on copper content in slag

Can be seen by the above analysis, in the process of molten slag dynamic dilution in dc field, the intensity of electric field (E) and feeding speed (Q) had important influence on copper content in final slag. In addition, modified with reduction and composition regulation can also reduce the copper content in slag.

Figure 4 shows the curve about copper contents as a function of time with different electric field intensity in the process of dynamic dilution. The main technological conditions were: temperature 1523 k, feeding speed 50 g $(10\ min)^{-1}$, the electric field intensity 1.0 ~ 2.0V. cm^{-1}, the slag was S-1. From the figure 4 can be seen that with the increase of electric field intensity, the dilution rate of molten copper slag increased, the balance time was shorten, dilution rate increased. The increasing of the electric field intensity increased the settling velocity of the copper matte particles in vertical direction, thus speeding up the dilution of copper slag. When the electric field intensity was 2.0 V.cm^{-1}, the copper content of final slag was slightly lower than 1.5 V.cm^{-1}, which basically identical with the results of static test. But too high electric field intensity can accelerate corrosion of reactor and electrode .Comprehensive was taken into consideration, the field strength should be kept in (1.5 ~ 2.0) V. cm^{-1}.

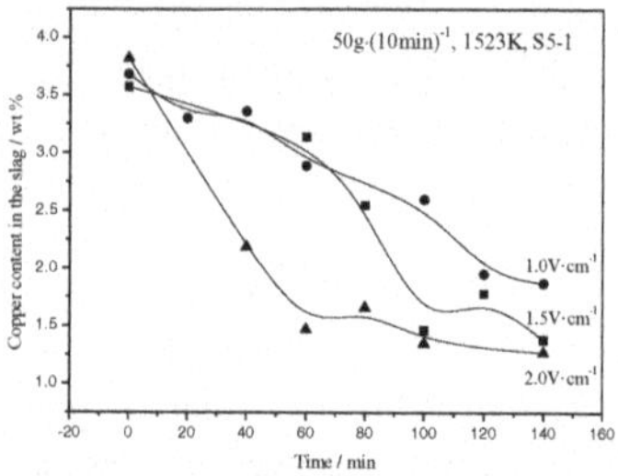

Fig.4 Copper contents as a function of time with various EPG

The influence of feeding rates on copper content in slag was showed in figure 5, the main technological conditions were: temperature 1523 k, electric field intensity 1.5 V.cm^{-1}, the slag was S-1. Fig. 5 showed that reducing feeding speed was conducive to the sedimentation of copper drops in slag. The decrease of the feeding speed reduced the flow speed of the molten slag actually, increased the actual residence time of the copper matte droplets in the sedimentation tank, and improved the recovery rate of copper in slag. When the feeding speed was less than 30 g · (10 min)$^{-1}$, the copper content in final slag basically remained unchanged. But too low feeding speed will affect the dilution efficiency, the feeding speed should be reasonable controlled according to the actual situation.

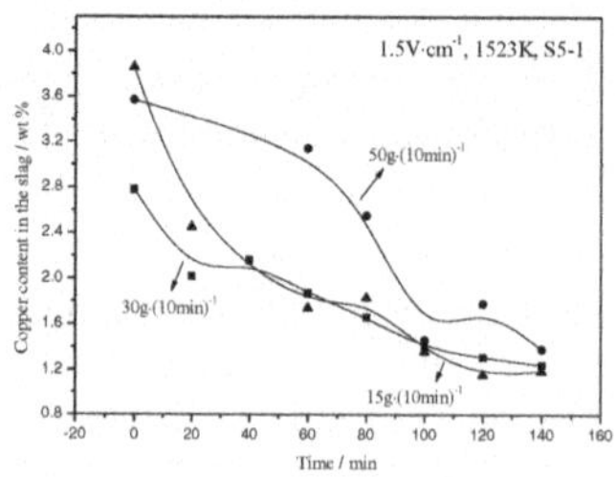

Fig.5 Copper contents as a function of time with various slag addition rates

Figure 6 shows the curve about copper content as a function of time before and after slag for component control. The main technological conditions: temperature 1523 k, feeding speed 30g ·(10min)$^{-1}$, electric field intensity 1.5V· cm^{-1}.From the figure 6 can be seen, the sedimentation rate of copper in the slag after component control was greater than that in the original slag, the balance time was also shortened, which indicated that modified with reduction and component control (moderate increase CaO content in slag, etc.) could effectively reduce the viscosity of slag and release the copper droplet, increasing the recovery rate of copper and shortened the time of copper slag dilution.

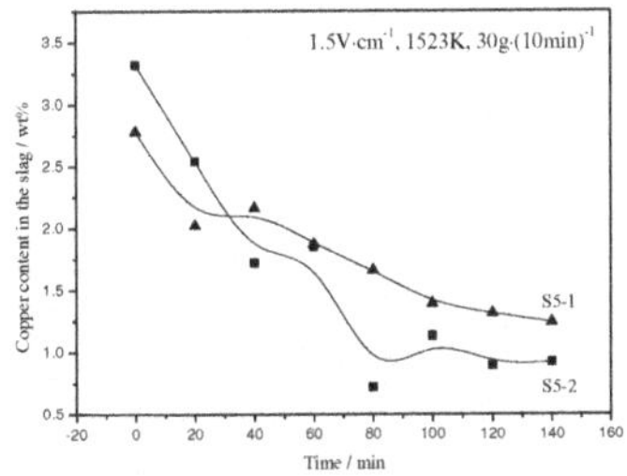

Fig. 6 Copper contents with time before and after slag component control

Parameters optimization of copper slag dilution

Adjusted the parameters of the copper slag dilution under the electric field, the optimized conditions were: electric field intensity 1.5V·cm^{-1}, feeding rate 15g· (10min)$^{-1}$, temperature 1523 k, the slag S-2. Under optimization conditions, the change trend of copper content in slag with time was showed in figure 7. In the first stage of dynamic dilution, the copper content in slag was below 0.9 wt %, the dilution time was ~80 min; After the static sedimentation in second stage the copper content in slag was below 0.6 wt%, which reached the level of copper slag dilution, the total time of two stages was 100 min.

The mainly existing method of copper slag dilution process was electric furnace depletion. The comparison about copper content in abandon slag after strengthening dilution (the world's major copper smelter furnace [4] and this new process) was showed in fig.8. From the curve , see that the optimization results were better than most of the existing electric furnace strengthen dilution process. Under the limitations of the small laboratory experiment, such as slag fluidity, thermal equilibrium and stirring condition was far from ideal condition, relatively good dilution effect still had been achieved. Which indicating the dilution process of molten copper slag in dc has a good promotion and development space.

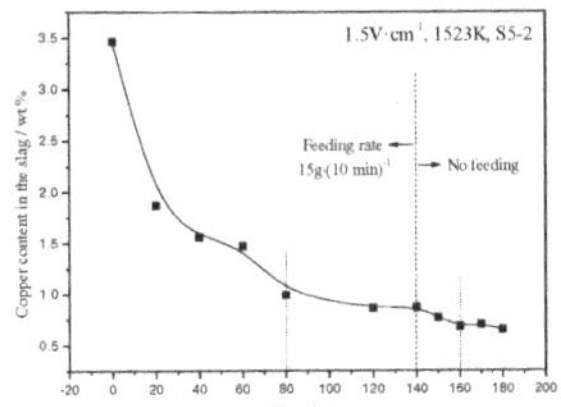

Fig. 7 Copper contents as a function of time under optimization conditions

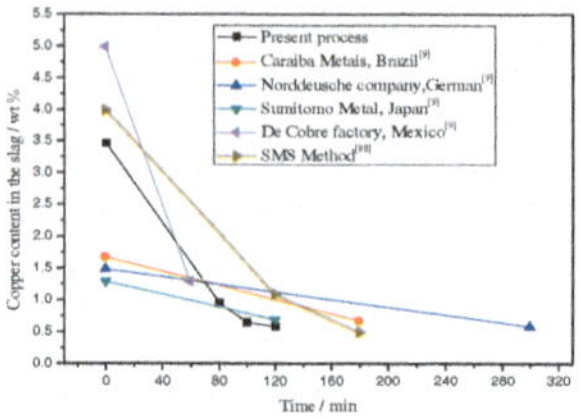

Fig. 8 Comparison of the main cleaning process

Conclusion

By studying dynamic thermal simulation of copper slag dilution under direct current field, the main conclusions were obtained as follows:

(1) In the dynamic thermal simulation of copper slag dilution, slag dilution rate increase with the increase of the electric field strength, decrease with the increase of feeding speed. In addition, a certain degree of modified with reduction and component control also conducive to reduce the copper content in slag.

(2) Through Kg scale laboratory experiment optimized the technological parameters, electric field intensity is 1.5 $V\cdot cm^{-1}$, feeding rate is 15g· $(10min)^{-1}$, temperature is 1523 k, the final copper content in slag is below 0.6 wt %.

(3)Under the limitations of the small laboratory experiment, conditions are far from ideal condition, relatively good dilution effect still have been achieved, which indicates the dilution process of molten copper slag in dc has good promotion and development space.

References

[1] Zhu Zu-ze, He Jia-qi. Modern Copper Metallurgy [M].BeiJing: Science Press, 2003:12-18.

[2] Zhou Ping, Yu Jian-ping, Chen Hong-rong, Mei Zhi. Settling mechanism and influencing factors on matte droplets in settler slag of copper flash smelting furnace[J]. Chinese journal of nonferrous metals,2006,16(12):2132-2137.

[3] Yu Jian-ping. Simulation of matte copper flash furnace settling basin subsidence mechanism analysis and vertical flow field temperature field [D]. Changsha: Central South University,2006.

[4] W.G. DaWenbote, etc. Copper Smelting Technology [M]. Translated by Yang Ji-chun, Dong Fang. The original version 4. Beijing: Chemical Industry Press, 2006:182.

7th International Symposium on High-Temperature Metallurgical Processing
Edited by: Jiann-Yang Hwang, Tao Jiang, P. Chris Pistorius, Gerardo R.F. Alvear F., Onuralp Yücel, Liyuan Cai, Baojun Zhao, Dean Gregurek, and Varadarajan Seshadri
TMS (The Minerals, Metals & Materials Society), 2016

ANALYSIS OF TURBULENCE AT THE METAL / SLAG INTERFACE IN THE MENISCUS REGION OF A CONTINUOUS CASTING MOLD THROUGH PHYSICAL AND MATHEMATICAL MODELLING

Varadarajan Seshadri[1]; Jose Dimas de Arruda [2]; Amanda Aparecida Fátima Arruda[2]; Samuel da Silva de Souza[2]; Carlos Antonio da Silva[2]; Itavahn Alves da Silva[2]
[1]Department of Metallurgical Engineering and Materials, Federal University of Minas Gerais, 6627,Av. Antonio Carlos, 31.270-901 Belo Horizonte, Brazil,
[2] Department of Metallurgical Engineering and Materials, Federal University of Ouro Preto, 35400-000, Ouro Preto, MG, Brazil

Keywords: Continuous Casting Mold, Slag, PIV

Abstract

The knowledge of the flow field inside a continuous casting mold is essential to improve operational efficiency of the process and also to achieve better inclusion control. Physical modeling, PIV technique and CFD have been employed to assess the velocity distribution in the metal- slag interface region in a slab mold. The analysis of results shows how higher casting velocity and lower immersion depth of submerged nozzle can lead to higher turbulence in the slag-metal interface. The effects of density difference between slag and metal and interfacial tension between the phases have been highlighted.

Introduction

The flow of liquid steel in a continuous casting mold is essentially turbulent giving rise to vortices and shear stresses in the steel slag interface. These are the causes of various defects like hooks and inclusions in the resulting casting and hence there have been various studies for optimizing the flow characteristics with the objective of improving the quality control to obtain an end product of acceptable quality with increased productivity(1). The metal - slag interface has been investigated by a 3-D computer model employing VOF method in conjunction with k-ϵ turbulence model; results have been compared to those obtained from experiments with a physical model (2). Similarly, the profile of the water oil interface using physical and mathematical modeling has been investigated by other researchers (3). It was shown that in the regions of higher pressure predicted by the mathematical modeling, the thickness of the oil film obtained via physical modeling was smaller, demonstrating the effect of oil flow. This work employs techniques of physical and mathematical modeling to study the instability in the interface metal - slag inside a slab mold. The effects of density difference and interfacial tension on slag entrapment have been evaluated.

Materials and Methods

The main dimensions and operational data of model and prototype are given in Table 1.

Table 1: Dimensions and operational data for model and prototype.

	Thickness (m)	Width (m)	flow rate	*SEN* immersion(mm)
Prototype	0.25	2	3.9 ~4.9ton/min	150 and 180
Model	0.15	1.2	155 ~195 lpm	90 and 110

The physical model of acrylic was built with scale λ = 0.6 (Figure 1 (a)). Dimensionless Froude $(V^2/L.g)$ and Weber $(\rho LV^2/\sigma)$ numbers were used as a criteria of similarity (v: velocity, L: characteristic length, g: acceleration due to gravity, σ: liquid/gas interfacial tension.

A submerged nozzle valve (SEN) with a square exit port with negative angle of 15° was used in this simulation. Water / silicone oil and solution of ZnCl2 / silicone oil were used for simulating slag and steel. Through the use of PIV technique vector maps of velocity profiles and kinetic energy of turbulence in the vicinity of to 1 cm from the metal / slag interface were obtained. The velocity profiles were determined by PIV using a DANTEC - 2D apparatus, which consists of dual laser system power (65mJ), wavelength of 532nm and a CCD camera Flow-Sense 2ME. Seeding was done with polyamide particles, $5\mu m$. The image analysis was performed via Dynamics software Studio - Dantec. For all the experimental conditions, about 100 images obtained with a pulse duration of 3000 s and capture frequency of 10Hz were acquired. A square region of side equal to 600 mm,(Figure 1b), in the meniscus region, was used for monitoring the steel flow between submerged valve and a narrow face. Mathematical simulations were conducted with Ansys Fluent software. The mathematical model took into consideration one fourth of the domain of the continuous casting mold assuming symmetrical flow for the remainder of the mold, as shown in Figure 1c.

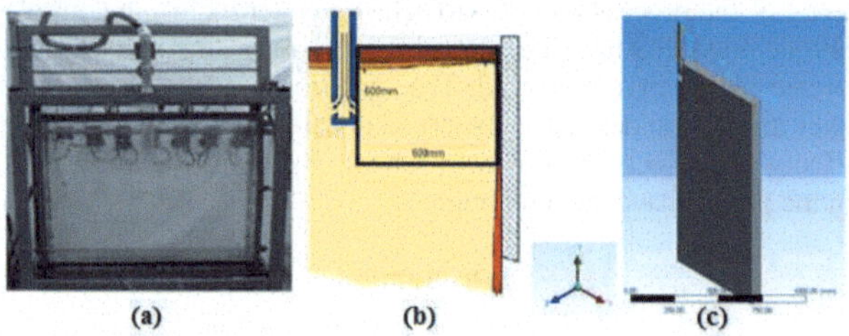

Figure 1. a) Physical model of continuous casting mold b) Image capture region by PIV technique for flow analysis in the mold; c) Geometry of the continuous casting mold chosen for solution of mathematical modeling with Ansys Fluent Software.

CFD modeling was accomplished assuming isothermal steady state conditions. The standard k-ϵ turbulence model was used. Then the following equations have been solved:

$$\frac{\partial \rho}{\partial t} + \nabla \cdot (\rho U) = 0 \quad (1)$$

$$\frac{\partial \rho U}{\partial t} + \nabla \cdot \rho U \otimes U) - \nabla \cdot \left(\mu_{eff} \nabla U\right) = -\nabla p + \nabla.(\mu_{eff} \nabla U)^T + B \quad (2)$$

$$\mu_{eff} = \mu + \mu_t \quad (3)$$

$$\frac{\partial \rho k}{\partial t} + \nabla.(\rho U k) = \nabla.\left[\left(\mu + \frac{\mu_t}{\sigma_k}\right)\nabla k\right] + P_k - \rho\varepsilon \quad (4)$$

$$\frac{\partial \rho \varepsilon}{\partial t} + \nabla.(\rho U \varepsilon) = \nabla.\left[\left(\mu + \frac{\mu_t}{\sigma_k}\right)\nabla \varepsilon\right] + \frac{\varepsilon}{k}(C_{\varepsilon 1} P_k - C_{\varepsilon 2}\rho\varepsilon) \quad (5)$$

$$\mu_t = C_\mu \rho k^2/\varepsilon \quad (6)$$

Where: ρ : fluid density (kg/m^3) ; t : time (s); ∇ : gradient operator; U : velocity (m/s) ; eff : effective viscosity (Pa.s); B : body forces, gravity; : dynamic viscosity (Pa.s); μ_t: turbulent viscosity (Pa.s);C_μ: a constant; k : turbulence kinetic energy (m^2/s^2);σ_K: a constant; Pk : rate of production of kinetic energy due to buoyancy and viscosity effects; ε : rate of dissipation of turbulent kinetic energy (m^2/s^3); C_{s1}: constant;C_{s2}: constant. Details of this procedure can be found at Ansys manual (4). Boundary conditions are: a)Non slipping condition at each liquid solid interface, nozzle and walls (then U , k and ε values are taken as zero at these locations);

b)Free surface condition at liquid atmosphere interface, with Uy equal to zero; c) An average velocity at entry nozzle can be estimated from the assumed throughput ((1.667kg/s; 2.083kg/s and 2.5kg/s for linear casting velocities of 0.78m/min; 0.98m/min e 1.2m/min, respectively); accordingly average values of k and ε have been estimated assuming: $K = 0.01\,V^2$; $\varepsilon = K^{1/3}/R$

Results and Discussion

Figure 2-a and 2-b display the images of the continuous casting mold and vector maps corresponding to the flow of 155lpm and immersion depths of 110 mm and 90mm respectively. Comparing the two immersions, there appears to be a greater thinning of the slag layer for the lower immersion depth of SEN. This is made clearer by a frame to frame analysis showing the motion of the interface. Figure 3 shows the mathematical simulation for the same conditions of experiments in the physical model, shown in Figure 2. Lower immersion depths result in greater effect on drag of the mold slag. In the case of smaller immersion one can notice a more extensive region of higher values of velocities close to the oil/water interface than in the case of higher immersion depths. The higher velocities cause higher shear stress (upper right mold, meniscus).

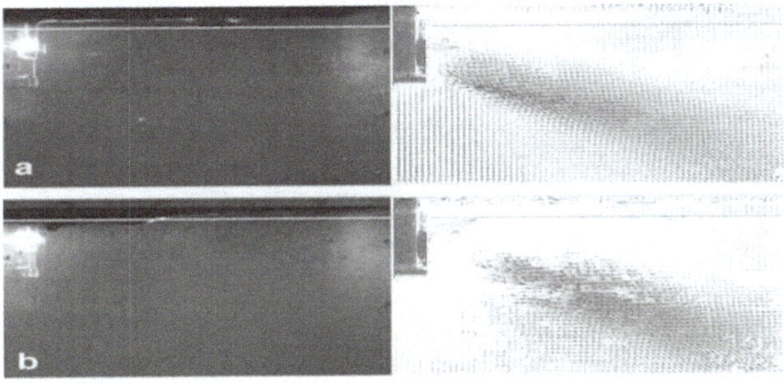

Figure 2. Images of physical model of continuous casting mold and average vetorial map for a flow rate of 155lpm for 2 depths of immersion. a) 110mm ; b) 90mm

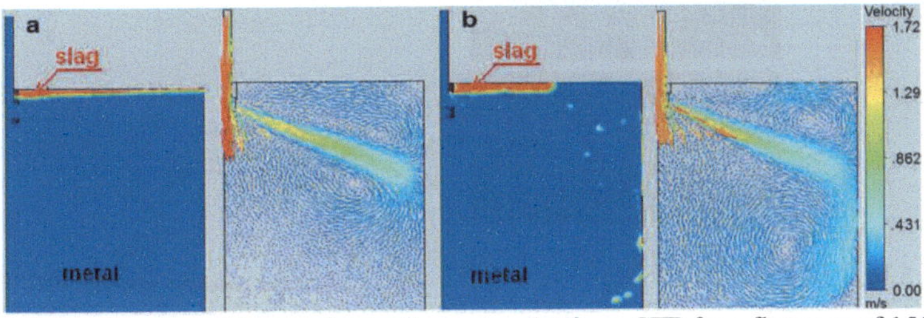

Figure 3. Meniscus conditions and velocity vector plot from CFD for a flow rate of 155lpm for 2 depths of immersion.a) 110mm ; b) 90mm

Velocity data and turbulent kinetic energy namely ½ $\overline{(V_i')^2}$ (where V_i' is the fluctuation of velocity component V_i) were acquired through a horizontal line close to metal slag interface Data from mathematical and physical simulations have been compared in the vicinity of the meniscus region defined by the straight line drawn at the top of mold, as seen in Figure 1. The velocity profile and turbulent kinetic energy versus distance from for the two immersion depths for the case of 155lpm flow rate are shown in Figure 4. One can observe high velocities along the narrow face compared with the rest of the mold. For the immersion depth of 110mm, the turbulent kinetic energy values are close to the model prediction. For the immersion depth of 90mm, the model predicts relatively larger turbulence values especially near the narrow face. Both velocity and kinetic energy of turbulence in the meniscus region are larger for smaller immersion depths of SEN.

Even in the case of flow rate of 175lpm (Figure 5), images of velocity profile and vector maps confirm that lower immersion depths lead to greater displacement of the slag layer, which can result in slag entrainment in the liquid steel in the mold.

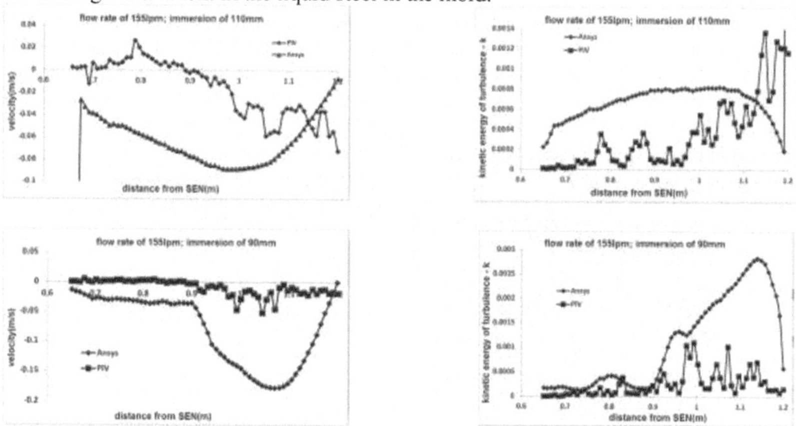

Figure 4. Velocity and kinetic energy of turbulence in the case of flow rate of 155lpm for immersion depths of 110mm e 90mm .

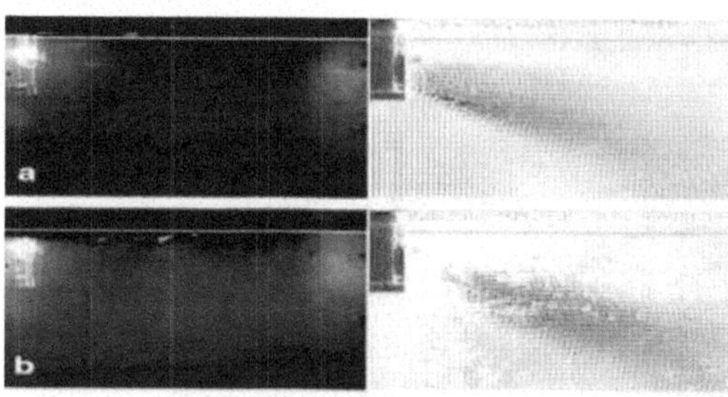

Figure 5.Images of mold and vector map of velocities in physical model experiments for flow rate of 175lpm a) 110mm immersion ; b) 90mm immersion.

Figure 6 shows the results of mathematical simulation for identical conditions of exlperiments in the physical model (Figure 5). They predict significant drag of the mold slag for smaller immersion depths of immersion. Such entrainement has not been observed in the physical model.

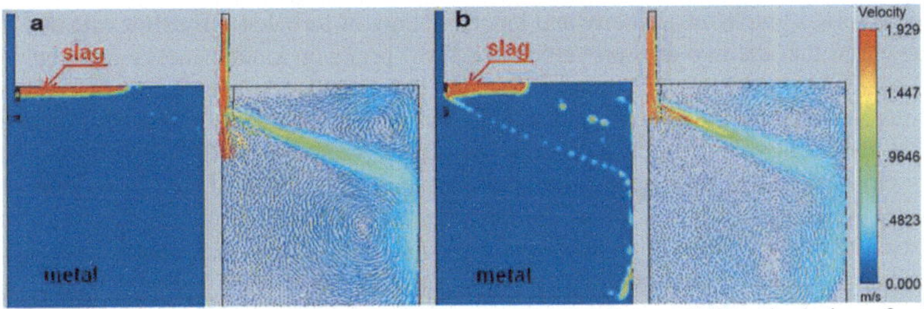

Figure 6. Conditions of the meniscus and vector map according to CFD calculations for flow rate of 175lpm a) 110mm immersion b) 90mm immersion

Figure 7 shows velocity profile and kinetic energy of turbulence for the two immersion depths for the flow rate of 175lpm. It is observed that in the case of model results and experiments the velocities are high near the narrow face compared to the rest of the mold. However these are higher in mathematical simulations and hence should lead to a greater displacement of the slag layer at the top of the mold. This higher displacement is seen in Figure 5-b and 6-b. Generally speaking the higher velocity values are associated with higher values of kinetic energy of turbulence. The curves of kinetic energy show the same tendency as velocity profile. For both the immersion depths the values of velocity are similar but the values of kinetic energy of turbulence are higher for lower immersions as expected.

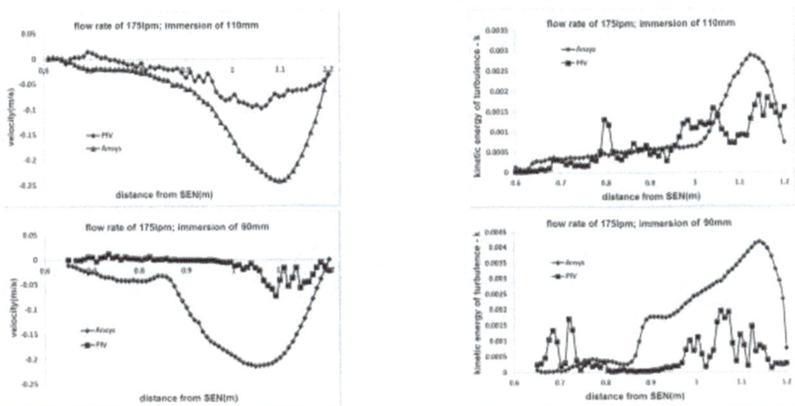

Figure 7. Velocity and and kinetic energy of turbulence for a flow rate of 175lpm and SEN immersions of 110mm and 90mm from physical model and mathematical simulations.

Similary the experimental results for the immersion depths of 195 lpm are given in Figure 8 which shows also for smaller depths of immersion greater displacement of the slag layer and also a higher deformation near the narrow face of the slag resulting in slag entrainment. Figure 9 gives the velocity profiles from mathematical simulation under conditions identical with that of physical model (Figure 8). The velocity profiles show similar trend in both the investigations in respect of velocity profiles and entrainment of slag. However in the case of the mathematical simulation the effects are more pronounced since oil entrainment was not seen often in physical modelling. CFD suggests significant entrainment for smaller immersion

Figure 10 shows graphs of velocity and kinetic energy of turbulence for flow rate of 195lpm. It is observed that the two dips on velocity and two peaks on kinetic energy of turbulence as given by CFD modeling are associated with. As expected, increased turbulence at the meniscus is shown for the lower immersion. Discrepancies identified in this work between the physical and mathematical models are due also to the fact that the PIV(velocity and kinetic energy plots) reflect an average of a hundred measurements spreaded out during some 10s of quasi steady state operation, whereas the mathematical simulation portrays the system at 10s from the beginning of the simulation which initially fluids are at rest. Hence comparison of of results of physical model and that of CFD has its limitations. The image from physical model is one out of 100 (it is a changing picture since the flow is turbulent). The vector plot is the average of 100 measurements and thus it hides the transiency. On the other hand CFD calculations were of transient nature and the results show 1shot at a given moment.

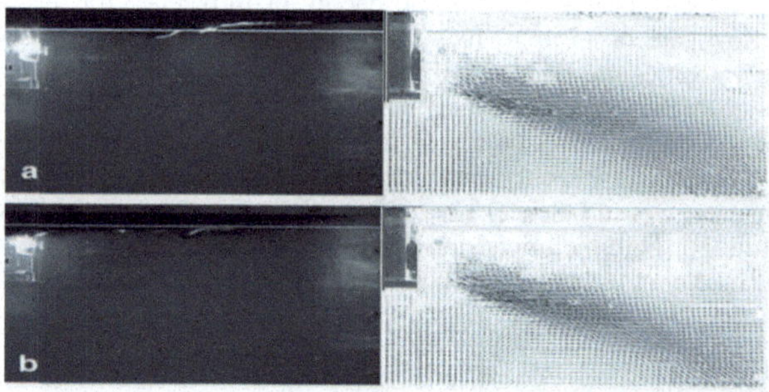

Figure 8: Images of vectorial velocity distribution in the physical model of the mold for a flow rate of 195lpm a) immersion of 110mm ; b) immersion of 90mm..

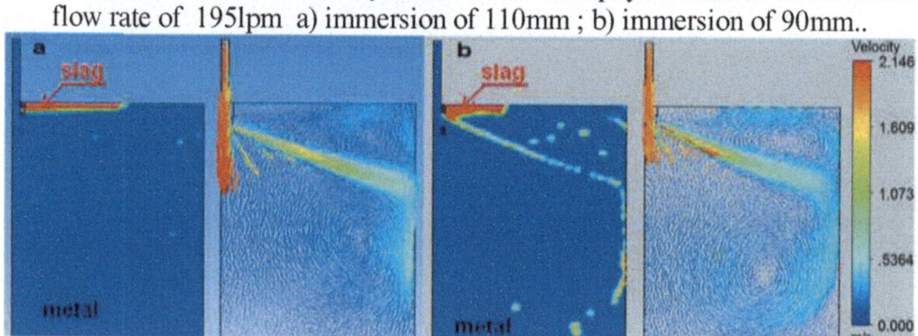

Figure 9: Meniscus conditions and vector map of velocities according to CFD calculations for flow rate of 195lpm a) immersion of 110mm b) immersion of 90mm.

Greater the density difference, greater is the stability of the interface, as highlighted below. Figure 12 shows that in the case of higher flow rate namely, 195lpm, the entrainment of the slag layer in the water/oil system is more significant compared to ZnCl2 solution/oil system. Comparing images of the interface given in the figures 11 and 12, there is a distinct difference in fluctuations of metal / slag interface in both systems analyzed. The disturbance (or deformation) caused in the oil layer in the water /oil system even for smaller flow rates can be noted and this can lead to the slag drag or slag entrainment. The velocity for slag entrainment is correlated with physical properties of steel and slag (5). It is given by (γ = interfacial tension; ρ = density; Δd = thickness of the slag layer; = viscosity):

$$V_{ent} = 3{,}065\ \gamma^{0.292} * g^{0.115} * \Delta\rho^{0.215} * \mu_{oil}{}^{0.231} / \rho_{oil}{}^{0.694} * \Delta d^{0.365} * \mu_{water}{}^{0.043} \quad (7)$$

This mathematical correlation can be used to assess the critical velocities in the two systems. By using the fluid property values shown in Table 2, one can arrive at the value of V_{ent} (oil /

water) = 0.0473 and V_{ent} (ZnCl2 / oil) = 0.0625. Therefore, the velocity ratio between these values shows that the maximum casting speed of the system $ZnCl_2$ / silicone oil is 32% more than the system water / silicone oil. According to Kasai (cited by Suzuki (6)), the entrainment of a slag drop occurs when the forces of interfacial tension and buoyancy thrust are no longer able to ensure the stability of the interface.

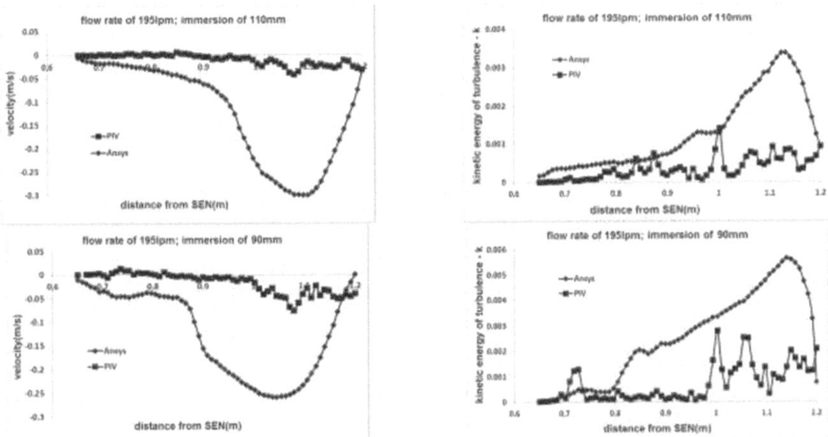

Figure 10. Velocity and kinetic energy of turbulence for a flow rate of 195lpm and immersion depths of 110mm and 90mm from physical model and mathematical simulation.

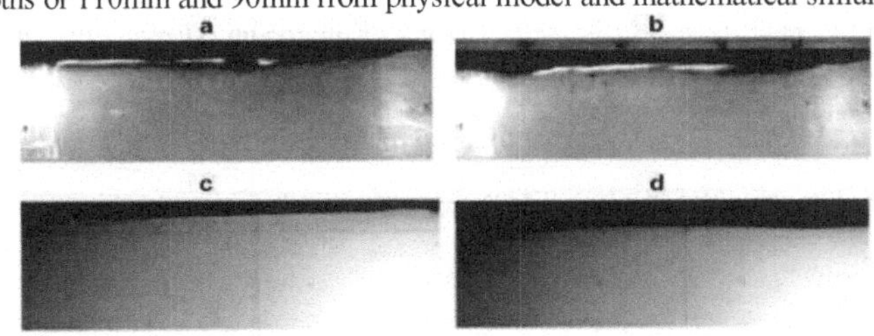

Figure 11. Interface. a) water/oil , flow rate of 155lpm, immersion 110mm; b) water / oil, flow rate of 155lpm, immersion of 90mm; c) ZnCl2 solution/oil , flow rate of 155lpm, immersion depth of 110mm; d) ZnCl2 solution/oil, flow rate of 155lpm, immersion depth of 90mm

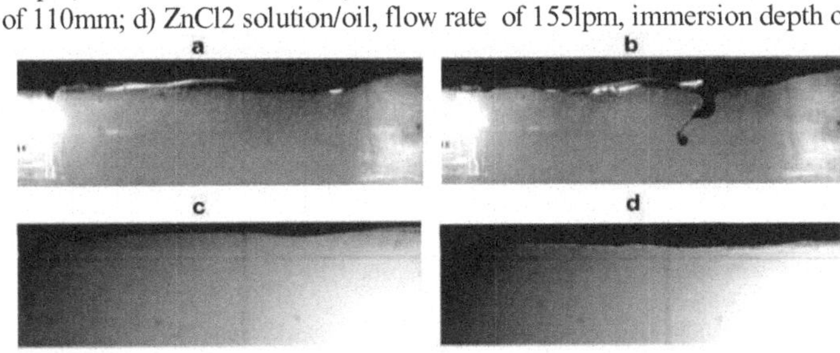

Figure 12. Interface a) water / oil, flow rate of 195lpm, immersion of 110mm; b) water / oil, flow rate of 195lpm, immersion of 90mm; c) solution ZnCl2/oil, flow rate of 195lpm, immersion of 110mm; d) solution ZnCl2/oil, flow rate of 195lpm, immersion of 90mm.

Table 2 – Properties of fluids used in the physical model

Material	Density (kg/m3)	Viscosity	Interfacial tension with oil
Water	1000	1.0 cSt ;0.001Pa.s	0.0357 (N/m)
ZnCl2 Solution	1300	2.33 cSt ;0.00303Pa.s	0.0261 (N/m)
Silicone oil	950	500 cSt; 0.526Pa.s	
Kerosene	750	3.0 cSt ; 0.00225Pa.s	

The critical velocity when the entrainment of slag occurs should be proportional to:

$$\{\sqrt{\Delta\rho\, g\, \gamma}/\rho_{oil}\}^{1/2} \qquad (7)$$

This analysis gives a greater weight to the buoyancy and suggests that in the ZnCl2 solution/oil system critical velocities are greater by about 60%. This is in support of stability of meniscus region in the case of $ZnCl_2$ / silicone oil system as against water /oil system. Hence, the possibility of emulsification of slag is amplified in the experiments involving oil and water and also explains the difference in behavior of both the fluid in the meniscus region.

Conclusions

On the basis of results obtained from the experiments and mathematical simulation the following conclusions can be reached. Higher flow and lower depths of immersion of SEN should cause greater deformation of the slag layer. Mathematical modeling and simulation predicts greater deformation of the slag layer and larger values of velocity and kinetic energy of turbulence relative to experimental results in the physical model. However, the trend was the same in both. Higher the density differences and interfacial tension values between the fluids in the system, greater is the stability of the interface.

Acknowledgements

The authors wish to thank FAPEMIG and CNPq for financial support for this research.

References

1 B. G. Thomas and L. C. Hibbeler. "*Investigation of Mold Flux Entrainment in CC Mold Due to Shear Layer Instability*", Association of . Iron Steel Technology (AIST), Warrendale, PA, CCC Report 201001, Aug. 2010 , 1-16
2 M.H. Zare et al., "*Simulation of flow field and steel/slag interface in the mold region of a thin slab steel continuous caster with tetra-furcated nozzle*", Journal of Manufacturing Processes, Volume 15, Issue 3, August 2013, 307–317.
3 C. E.P Barral, "*Efeito de parâmetros geométricos e operacionais sobre o comportamento do pó fluxante no molde de lingotamento contínuo de placa*", MSc dissertation, Fedral University of Minas Gerais (UFMG) , Brazil, 2006.
4 ANSYS Inc. PDF Documentation for Release 15.0; 148.204.81.206/Ansys/readme.html
5 M. J Harman and A. W. Cramb, "*A Study of the Effect of Fluid Physical Properties upon Droplet Emulsification*", Steelmaking Conference Proceedings, Pittsburgh, PA, March 24-27, 1996, .773-784.
6 M. Suzuki and M. Nakada, "*Perspectives of Research on High-speed Conventional Slab Continuous Casting of Carbon Steels*", ISIJ International, vol 41(2001), no 7, 670-682

7th International Symposium on High-Temperature Metallurgical Processing
Edited by: Jiann-Yang Hwang, Tao Jiang, P. Chris Pistorius, Gerardo R.F. Alvear F., Onuralp Yücel, Liyuan Cai, Baojun Zhao, Dean Gregurek, and Varadarajan Seshadri
TMS (The Minerals, Metals & Materials Society), 2016

STUDY ON THE PROPERTIES AND DAMAGE ANALYSIS ON THE LINING USED IN COOLING SECTION OF COKE DRY QUENCH FURNACES

G Xu, Y. Wang, Y. Lv, J. Sheng, H. Zhang, L. Liu
R&D Center of Wuhan Iron & Steel (Group) Co., Wuhan 430080, Hubei, China

Keywords: coke dry quenching; damage; mullite; wear; refractory

Abstract

In order to increase the service life of large scale coke dry quenching furnaces, the damage mechanism of lining was investigated, and the structure and properties of lining refractories used in cooling section were researched. The results showed that the abrasive wear volume of lining bricks at one fourth height section under skew tunnel was smaller than other sections. The wear loss of mullite brick and dense clay brick used in the lining of cooling section were respectively 8.02 cm^3 and 11.65 cm^3 in fifteen minutes. The mullite brick was laid with clay brick alternatively. By the different wear rate, the surface of lining was rough and uneven in the later service life, and short brick was easily dropped from lining. During the service, the basis wore off more easily than the aggregate. It is necessary to improve the wear resistance property, which will increase the service life of cooling section's lining for coke dry quenching furnaces.

Introduction

The technology of coke dry quenching (CDQ) can recycle energy, which saving energy and protecting environment, eliminate the pollution of the atmosphere, and also can increase strength of coke. For those advantages, it had been developed very fast. In 1985, Bao Steel built four CDQ devices of 75 $t \cdot h^{-1}$; and 140 $t \cdot h^{-1}$ CDQ device of WISCO is China's first large CDQ device, which completed and put into operation in Dec. 2003. between year 2005 to 2012, total 117 sets of CDQ devices put into production: Jing-Tang Steel company's 260 $t \cdot h^{-1}$ CDQ device is the maximum one in the world; Hubei Zhongte Xinhua Xichang Coke-making factory's 125 $t \cdot h^{-1}$ of CDQ device is used in processing ramming coke block; Sichuan Panzhihua Iron and Vanadium company's 170 $t \cdot h^{-1} \cdot$ CDQ device is used in processing the coke produced by 6.25m high coke furnace. During early 2014, there were 149 various CDQ devices in China's coke-making enterprises.

CDQ device generally consists of coke stored section, oblique wind channel, circular wind channel, cooling section, dust elimination and electric power generation system. At present, the oblique channel and circular channel of CDQ are easily damaged. From the perspective of CDQ devices operation, generally there is a common problem with refractory wear, which is primarily reflected in damaged pillars of the large circular flue of CDQ. The cooling section wears out faster, and has to repair in advance. It is seriously affected the normal operation of CDQ device [1-3]. In this article, based on the research of refractories for CDQ cooling section in WISCO, the damage causes of refractories for CDQ as well as its performance and structural factors are analyzed and discussed under the using conditions.

Damage investigation

CDQ cooling section of WISCO has been used for nearly 5 years. Because the cooling section was badly damaged in the lower part, and it was difficult to support upper lining, so it had to stop to repair. Cooling section was built by mullite bricks alternate with clay brick. Long brick was interleaved with short brick, where clay brick was used as long brick and mullite brick was used as short one. The next layer was reverse. The structure of brick is shown in Figure 1.

Figure1. Brick structure at CDQ's cooling section

Figure 2 shows different damage parts of CDQ's cooling section. The actual observation shows that the overhaul lining was damaged most in the range along clockwise 0° to 90°. There were total 55 ring bricks from lower to upper at the cooling section. The wear volume increased under the 45th ring, where some parts of the short brick wear off and sometimes whole block wear off. At 180° was the parts of dust separated wall, bricks drops badly, but the upper was better. In the range of 180° ~270° , the upper bricks wear less, compared to lower bricks. In the range of 270° ~0° , the bricks wear less than the other zone. In the process of demolish, from the 55th ring began to tear down, not regular demolish with one by one, but demolish from a fracture surface. Therefore, the research process chose the lower part near 0° separated wall, continuous measurement all the way to the next ring one by one. Thickness of height brick measurement result of cooling section is shown in Figure 3.

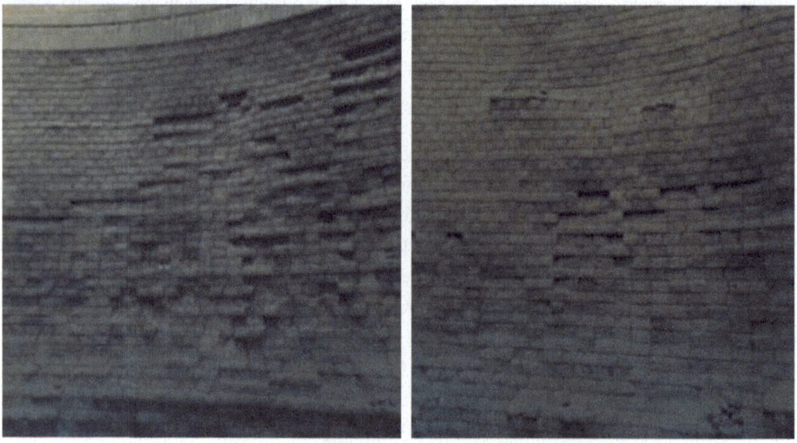

Figure 2. Breakage of different parts of CDQ cooling section

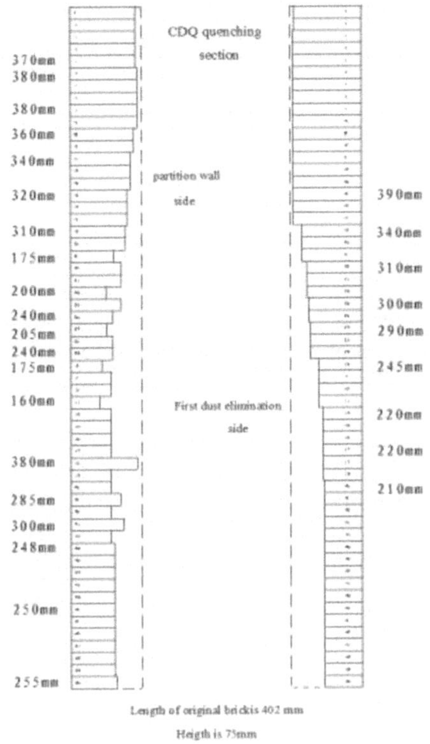

Figure 3. Size distribution of residual brick at CDQ's cooling section

Analysis from brick-size of the cooled remnants shows that bricks between the 20th to 35th wear most by 120~260 mm. The inlay-brick structure of the cooling section was long brick alternating with short brick, the long block was 230 mm and the short was 172 mm. Therefore, the short mullite bricks or clay bricks were all serious damaged, which corresponded with the practical research results. The 40th to 55th ring usually wear off 40~90 mm, while the damage was relatively small, but below the 40th ring the wear was much more serious. Changing brick's structure of CDQ's cooling section through design the lining of different thickness can achieve more balanced energy saving effect for long service life. The cooling section of CDQ will suffer from repeated impact and abrasion of the coke at high temperature and thermal destruction causing by the high flow quenching nitrogen during the time of long-term using. In the upper cooling section, the high wear capacity resulting in dark localized wear, excessive wear and tear cause the collapsing at the upper lining.

Performance of bricks used in CDQ's cooling section

NHM-3 wear-resistant device was adopted to test the abrasion resistance of mullite brick. In the test, SiC of about 80 mesh (0.178 mm) was used as wear medium, and the weight was 1Kg. The blast speed was constant and the test temperature was 450 ℃ with constant injection pressure and compressed air(0.44~0.45 MPa).Calculate wear amount of sand blasting for fifteen minutes. Wear volume = mass of the loss /brick's density. The performances of mullite brick for maintenance were analyzed and shown in table 1.

Table 1. Performances of mullite brick

Bulk Density/ ($g\cdot cm^{-3}$)	Apparent Porosity/ %	Compression strength/MPa	Linear changes on reheating/% (1150°C×3h)	Wear volume / ($cm^3\cdot 15\ min^{-1}$)
2.43	18.12	69.55	-0.04	8.02

Table 1 shows that mullite brick was worn off 8.02 cm^3 in fifteen minutes, and clay brick was worn off 11.65 cm^3 in same time, so the performance of two kinds of bricks were quite different. In the late service life of CDQ, the lining would be at the situation of the short mullite brick worn out while the long brick still remaining. In that case, the long block of clay brick would be wear fast, and the long mullite brick wear much slower. Therefore, the CDQ would be formed rugged layer.

Chemical composition of mullite brick at CDQ's cooling section is shown in table 2, the chemical composition of mortar used to brick-lay is shown in table 3. It is not in the range of mullite mineral composition, so it must be improved to meet needs of long service life.

Table 2. Chemical composition of mullite brick (w) %

Item	Al_2O_3	SiO_2	K_2O	TiO_2	Fe_2O_3
Mullite bricks at The 35th ring under the once dustingcooling section	66.15	26.65	1.06	3.39	2.75
Mullite bricks at The 35th ring under the 0°parapetcooling section	61.57	32.55	0.87	2.64	2.37

Table 3. Chemical composition of mortar used to brick-lay (w) %

Item	Al_2O_3	SiO_2	K_2O	TiO_2	Fe_2O_3	Na_2O	MgO	SO_3
Mullite fire clay at once dusting cooling section	47.55	42.1	2.12	2.65	3.66	-	-	1.92
Mullite fire clayat 0°parapetcooling section	48.23	38.67	1.7	2.5	2.89	1.38	1.08	3.56

Figure 4 shows the surface morphology of mullite brick of cooling section after wear test. As can be seen: the aggregate and powder were combined relatively dense with some porous, but failure began from the matrix, and while the matrix wear off, the slowing wear aggregate bulgy isolated and finally wear off by direct impact. Figure 5 shows microstructure of mullite brick and EDS analysis. The matrix of mullite brick had many pores, so its abrasion resistance was not good. EDS analysis that: Figure 5 (a) shows the composition of the aggregates of aggregate 1 (w%): Al_2O_3 72.94%,SiO_2 29.06% ; the composed of matrix 2 (w%): Al_2O_3 54.97%,SiO_2 35.03%. Figure 5 (b) shows the composition of the matrix 1 (w): Al_2O_3 63.70%, SiO_2 36.30%. The overall matrix composition was similar. If wear-resistant of Alumino-silicate matrix were relatively poor, the composition of the bricks should be improved to increase the hardness and strength after sintering and this leads to increase the wear resistance.

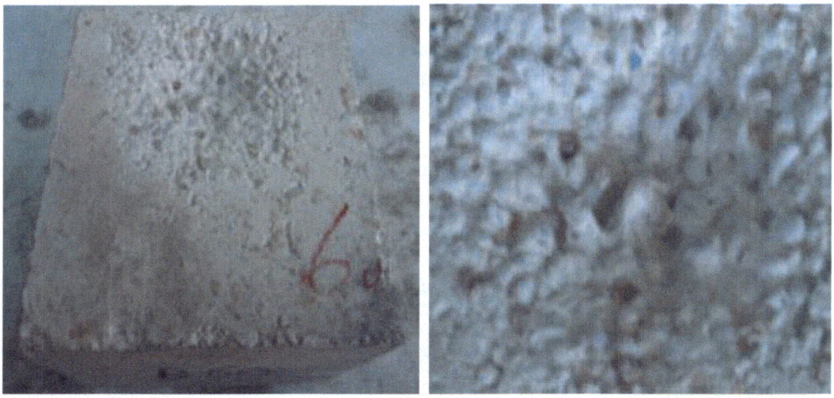

Figure 4. Surface morphology of mullite brick at cooling section after wear

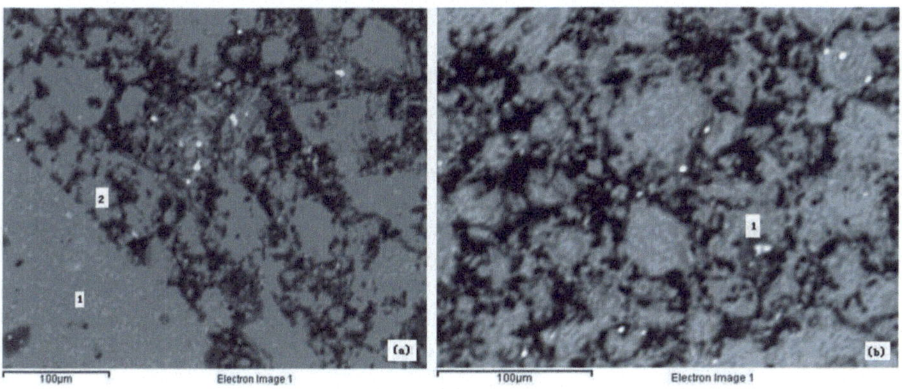

Figure 5. EDS and microstructure of mullite brick

Discussion

At early time, the refractories of CDQ's cooling section used dense clay bricks such as "QN53", which the w (Al_2O_3) =45%, cold crushing strength was 75 MPa, apparent porosity was 13% and refractoriness was 1530℃ at T_2. Such material at early time could meet the production requirements. Since the nitrogen quenching and enhanced operational conditions had changed, the destruction rate in the cooling section had increased and the reparation had become more frequently. Nippon Steel's CDQ have adopted bricks of H_{31}LL in the cooling section to improve refractories, which the w (Al_2O_3) is 62.9%, compressive strength at room temperature was equal or greater than 100 MPa, apparent porosity was less than or equal to 10%, thermal shock resistance was more than 10 times (without falling off). Domestic CDQ devices have adopted developed technology of QBM which have used mullite brick in CDQ's cooling section. It's performances showed that the compressive strength at room temperature was equal or greater than 85 MPa, apparent porosity was less than or equal to 17%, bulk density was equal or greater than 2.45 g • cm^{-3}, refractoriness was equal or greater than 1500 ℃ at T_2, heat shock resistance was greater than 22 times (water cycle, without peeling). However, from the perspective of the breakage investigation, the cooling section have badly damaged in the lower part after service for 3-5 years, which reveals that it was hard to prolong the CDQ life for one time by only rely on mullite brick. In addition, for the structure of cooling section refractories, in order to to prevent crack and stress damage, a kind of interlocked structure of mullite brick and clay brick was used by the design institutes. In the interlock structure, the long brick and short brick were interlocked alternately, where a ring used long brick of clay brick and short brick of mullite brick, another ring was reverse. Actual damage results showed that short mullite bricks were all worn out and badly damaged, while long clay bricks were maintained. In the alternating interlock structure, the wear rates of mullite bricks are higher than clay bricks, which result the partial wear darker, while the clay bricks wear more easily at the recessed part and mullite bricks wear more easily at projecting. Finally, the design of bricks should be improved to increase the wear resistance.

Analysis from the damage shows that not all cooling section was worn seriously, while the 40th to 55th ring wear minimally, under the 40th ring wear badly and the abrasion of the 20th to 35th ring was maximum. The lining of cooling section under skew wind tunnel 1/4 height wears less, while others wear heavily. Therefore, change the brick structure of the CDQ's cooling section through the rational lining design of different thickness, could also achieve the balanced longevity and energy saving.

Therefore, the property of the masonry mud needs to be cared. While the intensity of masonry mud is low, the brick joints wear quickly. At the end, while the short brick of mullite almost worn off, short bricks are easily to spall off. For the fast breakage rate of clay bricks, it is easy to protrude the long mullite bricks, forming a rugged lining and got bricks drop.

Conclusions

1. Breakage of CDQ's cooling section refractories was uneven, the cooling section lining underwind tunnel 1/4 height wear less, while the underpart wear heavily. Mullite brick has wear off 8.02 cm^3 in fifteen minutes, and clay brick wear off 11.65cm^3 in fifteen minutes and mullite brick bite masonry with the clay brick, because of different wear rate, it is easy to from rugged structure of residual lining. Therefore, the lining design should be cared to the performance match of different brick and size.

2. Mullite brick which heavily worn, porous matrix and not tight is harmful to increase one overhaul CDQ-generation time to achieve a long service life; it needs to improve properties, structure and wear resistance of refractory materials at cooling section.

3. It needs to consider refractory masonry issues of the cooling section to improve the adhesion, reduce wear of the brick joints and reduce brick spalling.

Reference

1. G.Xu, Y.Lv,J.Zhu,etc." Reason ananlysis on the abnormal damage reasons of ring wind channel for the large scale of coke dry quenching furnaces", Research on Iron & Steel, 2015, 43(4):39-41,46.
2. G.Xu, Y.Lv,J.Zhu,etc." Study on the properties and damage analysis of refractories used in the cooling section of large scale of coke dry quenching furnaces", Refractories, 2015, 49 (2):128-130.
3. X.Wang, G.Cai, J.Qing,etc. " Damage mechanism and selection of refractories used in the skew tunnel and circular tunnel section of large scale of coke dry quenching furnaces", Refractories, 2014, 48 (2):135-138.

7th International Symposium on High-Temperature Metallurgical Processing

Treatment and Recycling of Solid Slag/Wastes

Session Chairs:
Tao Jiang
Matthew Andriese

7th International Symposium on High-Temperature Metallurgical Processing
Edited by: Jiann-Yang Hwang, Tao Jiang, P. Chris Pistorius, Gerardo R.F. Alvear F., Onuralp Yücel, Liyuan Cai, Baojun Zhao, Dean Gregurek, and Varadarajan Seshadri
TMS (The Minerals, Metals & Materials Society), 2016

DEVELOPMENT OF RELIABLE VISCOSITY MODEL FOR IRON SILICATE SLAGS

Mao Chen[1], Zhixiang Cui[2], Leonel Contreras[3], Baojun Zhao[1]

[1]The University of Queensland, Brisbane, Australia
[2]Dongying Fangyuan Nonferrous Metals, Dongying, China
[3]National Copper Corporation of Chile, Santiago, Chile

Keywords: Iron Silicate Slag, Copper smelting, Viscosity, Modified QCV Model

Abstract

Viscosity is one of the important physiochemical properties of metallurgical slags which is important for slag tapping, metal loss in slag, freeze lining and fluid dynamics. High temperature viscosity measurements are time- and cost-consuming and the number of experiments is limited. Viscosity model is an effective way to predict the viscosities of the complex slags at high temperature. However, a reliable viscosity model can only be developed on the base of accurate experimental data. A unique high-temperature viscosity measurement apparatus developed at The University of Queensland enabled accurate viscosity data to be measured for the "FeO"-SiO_2-CaO-MgO-Al_2O_3 system in equilibrium with Fe. An accurate viscosity model has been developed based on Eyring's theory and extensive viscosity measurements. The viscosities of copper smelting, converting and cleaning slags over a wide range of compositions and temperatures can be predicted and applied in the process control.

Introduction

Viscosity is always an important property of slag in high-temperature extractive metallurgical operations and other industrial processes. Optimum slag viscosity should be obtained to maintain the fluent operations. The viscosity cannot be too high as a slag with low viscosity is easy for tapping and good metal/slag separation. Fayalite slag "FeO"-SiO_2 is the base of the copper smelting slags. In addition, Al_2O_3, CaO and MgO are also present in the copper slags in relatively low concentrations. Understanding the viscosity of fayalite slag and the effects of Al_2O_3, CaO and MgO to the viscosity is crucial to optimise the smelting process. In particular, the viscosity appears to be more important when the copper smelting tends to be operated at lower temperature.

Various researches have been done in the viscosity measurements of copper slags, however, the viscosity data on fayalite slag are not consistent due to the difficulties associated with the high temperature viscosity measurements. A reliable viscosity model is an effective way to predict the viscosities of the complex slags at high temperature for optimisation of the industrial operations. However, comprehensive viscosity models can only be developed on the base of accurate experimental data on low order systems. A large number of viscosities have been accurately measured by the authors for the systems "FeO"-SiO_2 [1], "FeO"-SiO_2-Al_2O_3 [2], "FeO"-SiO_2-MgO [3] and "FeO"-SiO_2-CaO [4] in equilibrium with metallic iron. These systematic and accurate viscosity data have formed solid base for development of viscosity model. The aim of the present study is to develop a reliable viscosity model for the "FeO"-SiO_2-CaO-MgO-Al_2O_3 system in equilibrium with Fe. With an accurate model, the viscosities of Cu smelting, converting and cleaning slags over a wide range of compositions and temperatures can be predicted and later can be applied in the process control and modelling.

Experimental

A high-temperature viscosity measurement apparatus was developed at The University of Queensland, Australia [1-4]. A Brookfield digital rotational rheometer (model LVDV III Ultra) controlled by a PC was used in the present study. A Pyrox furnace with lanthanum chromite heating elements (maximum temperature 1650 °C) was employed. A Pt-6Rh/Pt-30Rh (B type) thermocouple was bound to the suspension tube platform, the tip of which remained adjacent to the crucible at the level where the spindle rotated. This enabled accurate temperature of the slag to be measured during the viscosity experiments.

The main feature of this experimental setup with top suspension provides a possibility of quenching the slag sample immediately after the viscosity measurement. After the viscosity measurement, the Pt suspension wires were sheared by alumina tube placed inside the suspension tube to enable the crucible to drop directly into a water bucket beneath the furnace. The ability to quench the slag sample is important since this enables the microstructure of the slag and compositions of the phases present in the slag at the temperature to be retained and characterised directly after the viscosity measurement. In this way the uncertainty associated with the changes of the slag composition during the measurements can be minimised. In addition, presence of crystals, if any, during the experiment can be identified.

Approximately 35 ml of melted slag is required for each viscosity measurement. In order to maintain the Fe saturation condition, approximately 2 grams of Fe plate, cut from high purity iron rod, was placed at the bottom of the Mo crucible. The measurements of viscosity were usually carried out in the order from high temperature to low temperature in 50 degree interval. After decreasing to a lower temperature, the sample was kept for enough time to ensure that new equilibration has been achieved which was confirmed by the stable torque readings. The lowest measuring temperature was limited by the liquidus temperature of the slag which was predicted by FactSage 6.2 [5].

When the viscosity measurements have been completed, the sample was dropped into water so that phase assemblages at temperature can be retained. The quenched samples were sectioned, mounted and polished for EPMA measurement. A JXA 8200 Electron Probe Micro analyser with Wavelength Dispersive Detectors was used for microstructure and composition analyses. The analysis was conducted at an accelerating voltage of 15 kV and a probe current of 15 nA. The standards used for analysis were supplied by Charles M Taylor Co., Stanford, California, USA: $CaSiO_3$ for Si and Ca, Al_2O_3 for Al, Fe_2O_3 for Fe, MgO for Mg and $CaMoO_4$ for Mo. The average accuracy of the EPMA measurements was within 1 wt pct. Under reducing conditions, most of iron in the slag is present as Fe^{2+} and the Fe^{2+}/Fe^{3+} ratio is very high. However, only the metal cation concentrations can be measured by EPMA. The phase compositions are recalculated to oxides on the assumption that all iron is present as Fe^{2+} for presentation purpose only.

Possible uncertainties associated with high temperature viscosity measurements have been systematically discussed in a previous study [1]. More accurate viscosity data will be obtained by carefully following these steps. The same set of rheometer, crucible and spindle used for room temperature calibration and high temperature measurement will minimise the error associated with the equipment. A thermocouple placed next to the crucible enables the temperature to be accurately measured. EPMA measurement of the quenched slag minimises the errors associated with composition and presence of solid phase. The possible errors associated with the relative position of spindle in the slag and crucible, thermal expansion of the spindle and crucible, and the weight of the spindle have been estimated and minimised [1]. One of the comparison of present viscosity measurement is shown in Figure 1 with other data from literatures [6-9] to demonstrate the advantage of present experimental techniques.

Discrepancies of viscosities in the "FeO"-SiO_2 system from different researchers [6-9] can be seen in Figure 1, which indicates the large uncertainties exist. Whether the local maximum is exist at around 33 wt% SiO_2 has been controversial for a long time. The present measurements clearly demonstrate that the viscosity increases continuously with the increasing SiO_2 concentration and no local maximum exists.

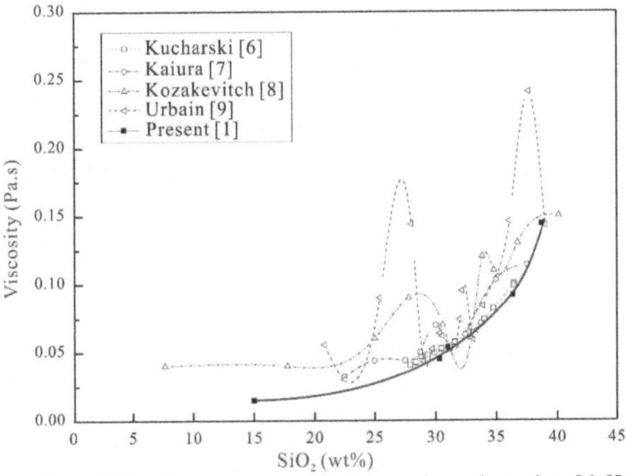

Figure 1. Comparison of data between present study and previous data [6-9] in "FeO"-SiO_2 system at iron saturation at 1573 K

In the present study, the viscosity data in the slag systems "FeO"-SiO_2 [1], "FeO"-SiO_2-Al_2O_3 [2], "FeO"-SiO_2-MgO [3], "FeO"-SiO_2-CaO [4], "FeO"-SiO_2-CaO-MgO, "FeO"-SiO_2-Al_2O_3-CaO, "FeO"-SiO_2-Al_2O_3-MgO and "FeO"-SiO_2-Al_2O_3-CaO-MgO have been systematically investigated in equilibrium with metallic iron for the development of viscosity model.

Model developments

The modified Quasi-chemical Viscosity (QCV) Model [10] is used in the present study. The modified QCV model was based on the Erying equation [11]:

$$\eta = \frac{2R\sqrt{2\pi k}}{\Delta E_v} \frac{m_{SU}^{1/2}}{v_{SU}^{2/3}} T^{\frac{3}{2}} \exp\left(\frac{E_a}{RT}\right) \tag{1}$$

where R is the gas constant (8.314 J/mol*K), k is the Boltzman constant ($1.38*10^{-23}$ m^2*kg*s^{-2}), T is the absolute temperature in Kelvin, ΔE_v is the energy of vaporisation, E_a is the activation energy, m_{SU} and v_{SU} are the weight and volume of a structure unit (SU, the calculation of *mij* and v*ij* of each SU can be seen in [12]), respectively. The parameters E_a, ΔE_v, m_{SU} and v_{SU} are the function of the pair fractions extracted from FactSage [5] with the partial properties [10].

The expressions of partial properties $\overline{E}_{a,ij}$ and $\varepsilon_{V,ij}$ in current system were shown as follows:

$$P_{Si-Si} = P_{Si-Si}^{0} + \sum_{i=Al,Fe,Ca,Mg} [P_{Si-Si}^{Si-i} \frac{X_{Si-Fe}}{(1-X_{Si-Si}-X_{Fe-Fe})^{1-\gamma_{Si/Fe}}}] + \sum_{j=Fe,Ca,Mg} [\Delta P_{Si-Si}^{Si-Al(T-O,j)} \left(X_{AlO_4}^{ch.j} X_{Si-Al}\right)^{\gamma_{Si/Al}}] \tag{2}$$

$$P_{Si-i} = P_{Si-i}^{0} \tag{3}$$

$$P_{i-i} = P^0_{i-i} \quad (4)$$

$$P_{Al-j} = P^0_{Al-j} + \Delta P^{T-O,j}_{Al-j} X^{ch,j}_{AlO_4} \quad (5)$$

$$X^{ch,j}_{AlO_4} = \frac{\left(\sum_n X_{Al-n}\right)^{\alpha_{Al/j}} \left(\sum_n X_{Fe-n}\right)^{4-\alpha_{Al/j}}}{\left(\sum_n X_{Al-n} + \sum_n X_{Fe-n}\right)^4} \left(\sum_n X_{Al-n} + \sum_n X_{j-n}\right) \quad (6)$$

where P represents $\overline{E}_{a,ij}$ and $\varepsilon_{V,ij}$; i = Al, Fe, Ca and Mg; j = Fe, Ca and Mg; n = Si, Al, Fe, Ca and Mg; $X^{ch,j}_{AlO_4}$ is related to the probability to have a tetrahedral-coordinated Al^{3+} involved in the Al-O-Fe structural units [10], the power $\alpha_{Al/Fe}$ is related to the proportion of the tetrahedral-coordinated Al-containing structural units formed as a result of charge compensation by basic oxides cations (specifically Fe, Ca and Mg in the present study), $\Delta P^{Si-Al(T-O,j)}_{Si-Si}$ and $\Delta P^{T-O,j}_{Si-Si}$ describe additional contribution to the partial molar activation and vaporisation energies of corresponding structural unit due to the presence of tetrahedral-coordinated Al^{3+} [10].

The parameters of the model are derived directly from the present viscosity measurements. The model has the ability to predict the viscosity in "FeO"-SiO_2-Al_2O_3-CaO-MgO system related to the copper process. The comparisons of the model predictions and the experimental data show good agreements as seen in Figures 2 and 3. Figure 2 shows the effects of adding Al_2O_3, CaO and MgO to replace "FeO" in the "FeO"-SiO_2 slags at fixed 33 mol % SiO_2. It can be seen in the figure that the model predictions can well reproduce the measured viscosities in three systems. At SiO_2 = 33 mol%, in all three systems, the replacements of FeO by Al_2O_3, CaO, and MgO increase the viscosity. The viscosity increases more rapidly with increasing Al_2O_3 compared to MgO and CaO.

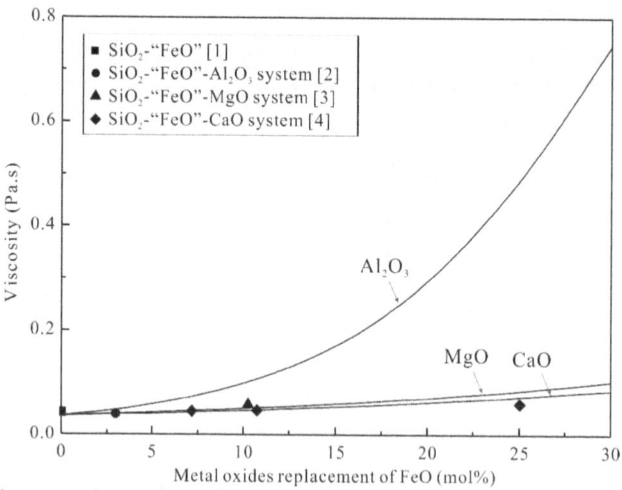

Figure 2. The comparisons of experimental data [1-4] and models predictions (lines) for the effects of replacements of "FeO" by Al_2O_3, CaO and MgO for "FeO"-SiO_2 system at around 33 mol% SiO_2 at 1623 K

Similar comparisons were carried out at 45 mol pct SiO_2 in all three systems "FeO"-SiO_2-Al_2O_3, "FeO"-SiO_2-MgO, "FeO"-SiO_2-CaO in Figure 3. As it can be seen in Figure 3, the predictions of viscosity by the modified QCV model clearly represent the trend of the experimental data in three systems. When Al_2O_3 and MgO are substituting FeO in the systems, the viscosities show increasing trends in Figure 3. However, at SiO_2 = 45 mol%, the replacement of "FeO" by CaO decreases the viscosity firstly, followed by the increasing of viscosity when CaO is higher than 13 mol pct. The modified QCV model well models the special effects of CaO placing FeO in the SiO_2-"FeO" system.

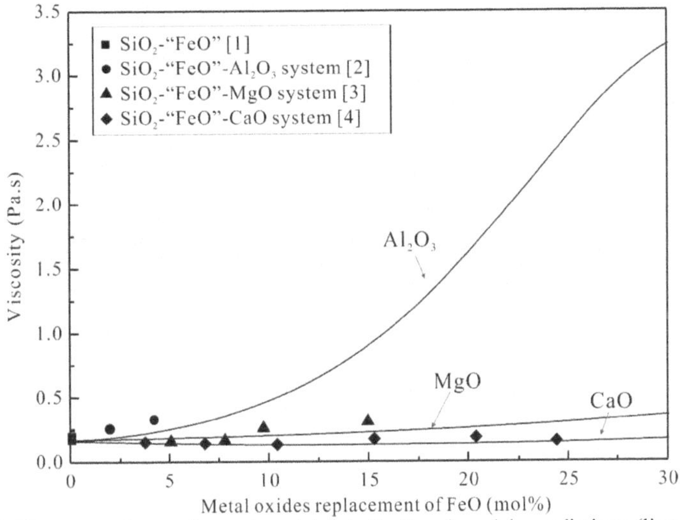

Figure 3. The comparisons of experimental data [1-4] and models predictions (lines) for the effects of replacements of "FeO" by Al_2O_3, CaO and MgO for "FeO"-SiO_2 system at around 45 mol% SiO_2 at 1623 K

The advantage of the present model is that it integrates all the systems and their subsystems using one set of parameters. The parameters for the higher order systems were derived and then combined from the lower order systems. The modified QCV model with its new parameters has also been used to predict the viscosities in higher order systems "FeO"-SiO_2-CaO-MgO, "FeO"-SiO_2-Al_2O_3-CaO, "FeO"-SiO_2-Al_2O_3-MgO and "FeO"-SiO_2-Al_2O_3-CaO-MgO and shows good agreements.

The addition of Al_2O_3 into the slag systems significantly increases the viscosity. As it can be seen in Figure 4, the replacement of FeO by Al_2O_3 in multicomponent system "FeO"-SiO_2-Al_2O_3-CaO-MgO also increase the viscosity. The predictions at two different levels of MgO well agree with the experimental data and clearly show the increasing trends. It can be also indicated from Figure 4 that more substitution of FeO by MgO increases the viscosity. The addition of small amount of Al_2O_3 into the "FeO'-SiO_2 based systems increase the viscosities, which could be explained by the "charge compensation effect" [2]. All basic cations including Fe^{2+}, Ca^{2+} and Mg^{2+} can charge balancing with Al^{3+} cations which is in tetrahedral coordination and form a network structure as SiO_2 network. The formation of network structure increases the viscosity dramatically.

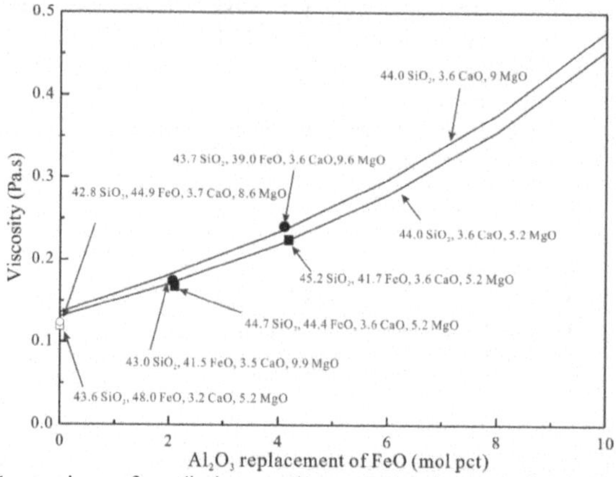

Figure 4. Comparison of predictions and experimental data at fixed SiO_2 and MgO concentrations at 1673 K in the SiO_2-"FeO"-Al_2O_3-MgO system (symbols: experimental data; lines: modified QCV model predictions)

Same tendency was predicted in the replacement of FeO by MgO in multi-component systems as the SiO_2-"FeO"-MgO system. In Figure 5, the viscosity increase with the addition of MgO into SiO_2-"FeO"-Al_2O_3 system. The experimental data in close compositions were in vicinity of the predicted values.

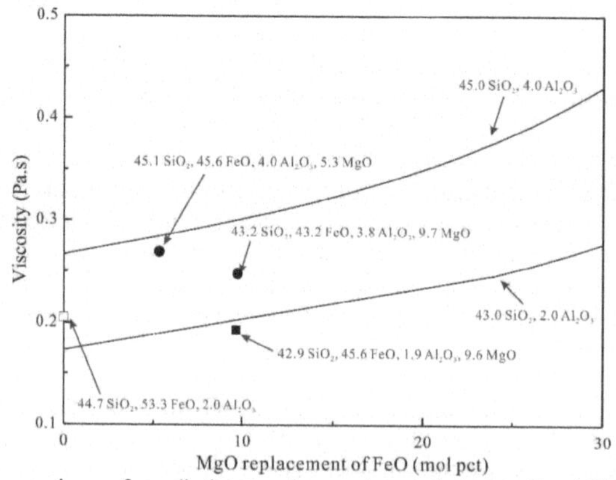

Figure 5. Comparison of predictions and experimental data at fixed SiO_2 and Al_2O_3 concentrations at 1673 K in the SiO_2-"FeO"-Al_2O_3-MgO system (symbols: experimental data; lines: modified QCV model predictions)

The predictions of the viscosities in system SiO_2-"FeO"-Al_2O_3-CaO-MgO system for the replacement of FeO by CaO are mainly based on the experimental data and the general tendency in ternary system SiO_2-"FeO"- CaO due to the lack of experimental data in higher CaO concentrations. The present predictions show clear decreasing tendencies at CaO addition lower than 8 mol pct.

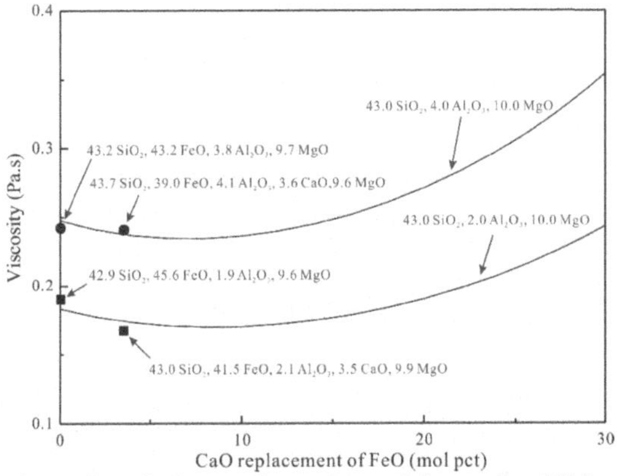

Figure 6. Comparison of predictions and experimental data at fixed SiO_2, Al_2O_3 and MgO concentrations at 1673 K in the SiO_2-"FeO"-Al_2O_3-CaO-MgO system (symbols: experimental data; lines: modified QCV model predictions)

MgO and Al_2O_3 are commonly present in the copper slags and CaO is often introduced into the slags as flux. In copper making, the Fe/SiO_2 ratios vary with different smelting process. However, in the same process, the productions are normally performed at a relatively constant Fe/SiO_2 ratios. The concentration of Al_2O_3, MgO and CaO will affect the viscosity of liquid slags. The Modified QCV model is used to predict the effects of additions of Al_2O_3, MgO and CaO to the viscosity of "FeO"-SiO_2 system and shown in Figure 7. As it has been clearly shown from Figure 7 that at fixed Fe/SiO_2=1.6, the viscosity of Al_2O_3-contianing system keeps increasing with increase of Al_2O_3 concentration. However, the viscosities in both of MgO- and CaO-containing systems decreases firstly then increase slightly. CaO is commonly used as flux to decrease the viscosity of slag. In the present calculations, it shows that little additions of CaO can helps to decrease the viscosity, but when the CaO concentration exceed 9 wt%, the viscosity of the slags increases.

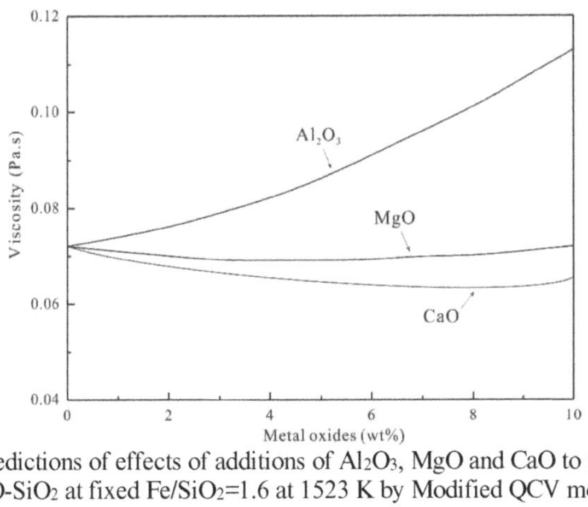

Figure 7. Predictions of effects of additions of Al_2O_3, MgO and CaO to the viscosity of FeO-SiO_2 at fixed Fe/SiO_2=1.6 at 1523 K by Modified QCV model

Conclusions

Viscosities of fayalite slags with replacements of "FeO" by MgO, CaO or Al_2O_3 in equilibrium with metallic iron in multi component systems have been systematically investigated. The viscosities in "FeO"-SiO_2-CaO-MgO-Al_2O_3 system and its subsystems were experimentally studied. The Modified Quasi-chemical Viscosity (QCV) model for copper smelting process were developed based on the newly measured viscosity data in "FeO"-SiO_2-CaO-MgO-Al_2O_3 system. The predictions shows good agreements with the experimental data. The effects of replacements of FeO by Al_2O_3, MgO and CaO were discussed based on the models predictions. The Modified Quasi-chemical Viscosity shows potentials for the viscosity optimizations for various smelting process with a wide ranges of compositions and temperatures.

Acknowledgement

The authors would like to thank Ms Yu for lab assistance and National Copper Corporation of Chile (Codelco), Dongying Fangyuan Nonferrous Metals (Fangyuan) and The University of Queensland for financial support through the "Codelco-Fangyuan Professorial Funding".

References

1. M. Chen, S. Raghunath, and B. Zhao, "Viscosity Measurements of "FeO"-SiO_2 Slag in Equilibrium with Metallic Fe," *Metallurgical and Materials Transactions B*, 44(2013), 506-515.
2. M. Chen, S. Raghunath, and B. Zhao, "Viscosity of SiO_2-"FeO"-Al_2O_3 System in Equilibrium with Metallic Fe," *Metallurgical and Materials Transactions B*, 44(2013), 820-827.
3. M. Chen, S. Raghunath, and B. Zhao, "Viscosity of SiO_2-"FeO"-MgO System in Equilibrium with Metallic Fe," *Metallurgical and Materials Transactions B*, 45(2014), 58-65.
4. M. Chen and B. Zhao, "Viscosity Measurements of SiO_2-"FeO"-CaO System in Equilibrium with Metallic Fe," *Metallurgical and Materials Transactions B*, 46(2015), 577-584.
5. C. Bale, E. Bélisle, P. Chartrand, S. Degterov, G. Eriksson, K. Hack, I. H. Jung, Y. B. Kang, J. Melançon, and A. Pelton, "FactSage thermochemical software and databases--recent developments," *Calphad*, 33(2009), 295-311.
6. M. Kucharski, N. Stubina, and J. Toguri, "Viscosity Measurements of Molten Fe-O-SiO_2, Fe-O-CaO-SiO_2, and Fe-O-MgO-SiO_2 Slags," *Canadian Metallurgical Quarterly*, 28(1989), 7-11.
7. G. Kaiura, J. Toguri, and G. Marchant, "Viscosity of Fayalite-Based Slags, " *Metallurgical Society of CIM Annual*, 1977, 156-160.
8. P. Kozakevitch, "Tension superficielle et viscosité des scoriessynthétiques, " *Rev. Metall.*, 46(1949), 505-516,572-582.
9. G. Urbain, Y. Bottinga, and P. Richet, " Viscosity of liquid silica, silicates and alumino-silicates," *Geochimica et Cosmochimica Acta*, 46(1982), 1061-1072
10. M. Suzuki and E. Jak, "Revision of Quasi-chemical Viscosity Model for Viscosity Estimation of Molten Multi-component Oxide Slag," (Paper presented at the VIIII International Conference on Molten Slages, Fluxes and Salt, Beijing, China, 2012).
11. H. Eyring and J. Hirschfelder, "The theory of the liquid state," *Journal of physical chemistry*, 41(1937), 249-257.
12. A. Kondratiev, P. Hayes, and E. Jak, "Development of a Quasi-chemical Viscosity Model for Fully Liquid Slags in the Al_2O_3-CaO-'FeO'-MgO-SiO_2 System. Part 1. Description of the Model and Its Application to the MgO, MgO–SiO_2, Al_2O_3–MgO and CaO–MgO Sub-systems," *ISIJ international*, 46(2006), 359-367.

7th International Symposium on High-Temperature Metallurgical Processing
Edited by: Jiann-Yang Hwang, Tao Jiang, P. Chris Pistorius, Gerardo R.F. Alvear F., Onuralp Yücel, Liyuan Cai, Baojun Zhao, Dean Gregurek, and Varadarajan Seshadri
TMS (The Minerals, Metals & Materials Society), 2016

REMOVAL OF IRON IMPURITY FROM ZINC CALCINE AFTER MAGNETIZATION ROASTING

Junwei Han, Wei Liu*, Wenqing Qin, Fen Jiao, Dawei Wang

School of Minerals Processing & Bioengineering, Central South University, Changsha, Hunan 410083, China

Keywords: Iron, Zinc ferrite, Calcine, Magnetization roasting, Magnetic separation

Abstract

The removal of iron impurity from zinc calcine was studied by magnetization roasting and magnetic separation. The effects of roasting temperature, magnetic intensity, ore grinding, and leaching pretreatment on iron grade and recovery were investigated. The results indicated that iron recovery increased with the increase of roasting temperature or magnetic intensity, while iron grade was always less than 22%, depending on the fact that the magnetite were not liberated completely. The good liberation was difficult to be achieved by ore grinding but was easier by acid leaching, since the magnetite grains were very fine. After the leaching pretreatment, the recovery of 76.31% with iron grade of 53.24% was obtained by one roughing and cleaning. The SEM/EDS analysis showed that a part of sphalerite, zinc ferrite and other gangues were mixed into the iron concentrate, while some of fine magnetite was reported to the tailing.

Introduction

Zinc is primarily produced from zinc sulfide concentrates by hydrometallurgical process, which consists of oxidative roasting, neutral and low acid leaching, purification, and electrolysis steps [1, 2]. In the oxidative step, iron impurity contained in the concentrates tends to react with zinc oxide into zinc ferrite ($ZnFe_2O_4$), which is structurally stable and hardly soluble in the mild conditions [3, 4]. The formation of zinc ferrite therefore causes considerable amount of zinc leaching residue generated in the subsequent leaching step. On the other hand, with the depletion of low iron bearing ores, the problem brought by iron impurity is becoming more and more serious due to the increase of iron content in the feedstock, which is based on the fact that high iron bearing ores are abundant in the world, especially in China [5].

Both hydrometallurgical and pyrometallurgical processes for zinc recovery from zinc leaching residue have been used industrially. In pyrometallurgical methods, the most common is Waelz process [6, 7], in which zinc leaching residue is subjected to a reduction roasting using large quantity of coke or coal at high temperature for the volatilization of metallic zinc. Although 90-95% of zinc can be recovered from the leaching residue, some fatal drawbacks, such as high energy consumption, high economic cost and strict requirement on dust collection system, make this technology being eliminated gradually. For hydrometallurgical processes [8, 9], the most widely used technology is hot acid leaching process, in which the low acid leaching step is

* Corresponding author: Liu Wei; Tel: +8613787007421, E-mail address: ase.6520@163.com

replaced by the leaching with hot and concentrated sulphuric acid to destroy the ferrite structure. By this technology, zinc extraction can be increased to 95%, but more than 80% of iron is also dissolved into the leaching solution at the same time. The dissolved iron must be removed prior to the purification and electrolysis steps for obtaining high quality zinc. Therefore, some methods of iron precipitation including jarosite, goethite or hematite were developed commercially [10-12]. At present, although iron precipitation is a well established technology, the generated secondary residues have never found any application and require further disposal.

In this study, a magnetization roasting was developed to convert the zinc ferrite contained in zinc calcine into zinc oxide (ZnO) and magnetite (Fe_3O_4) under weak reduction atmosphere. The generated magnetite was then separated by low intensity magnetic separation. Hence, the iron impurity can be not only removed but also recycled by this way. This paper focused on the effects of roasting temperature, magnetic intensity, ore grinding and leaching pretreatment on the removal efficiency of iron impurity. The mineralogical changes during this process were detected by XRD, SEM/EDS and optical microscope.

Experimental

Materials

The zinc calcine used in this study was kindly supplied by a hydrometallurgical plant in China. Its main chemical composition was analyzed by chemical methods for zinc and iron and X-ray fluorescence spectroscopy (Rigaku, model ZSX Primus II) for other elements, and the results are shown in table 1. The phase composition characterized by X-ray diffraction analysis (Rigaku, TTR-III) is given in Fig. 1. It can be seen from Table 1 that the zinc calcine contains more than 13% of iron, which indicates that it is a high iron-bearing zinc calcine. Fig.1 shows that the zinc calcine is mainly composed of ZnO and $ZnFe_2O_4$. The presence of ZnS indicates that the desulphurising roasting was insufficient in the previous process.

Table I. Main chemical composition of the zinc calcine (wt. %)

Zn	O	Fe	S	Pb	Cu	Si	Ca	Mn	Al	As	Ag
56.15	22.9	13.31	2.44	1.27	0.91	0.89	0.57	0.53	0.24	0.04	0.01

Methods

The magnetization roasting was carried out in the rotary kiln designed by the authors [13]. Before the roasting, the zinc calcine was ground and screened to -75μm, then was dried at 105°C for 8h. For each test, about 150g of the prepared zinc calcine was taken into the rotary kiln and heated under N_2 atmosphere to a desired temperature. A weak reductive gas containing 8% CO, 32% CO_2 and 60% N_2 was introduced into the rotary kiln for magnetization roasting. After 1h, the gas mixture was replaced by N_2 and the obtained product was cooled to room temperature. Then it was removed, analyzed and used for the following experiments. Magnetic separation was performed using a magnetic separator (model: XCGS-Φ50). An appropriate proportion of sample and water was mixed to obtain slurry with 10% of pulp density. The slurry was dispersed fully by stirring and then was used for magnetic separation at 65kA/m. After 0.5h, the iron concentrate and tailings were filtered, dried, weighed and subjected to analysis for iron grade and recovery.

Reaction mechanism

The objective of this study is to selectively convert the zinc ferrite contained in zinc calcine into zinc oxide and magnetite for facilitating zinc and iron separation. The reactions occurred possibly during the roasting process are listed as follows:

$$3ZnFe_2O_4 + CO = 3ZnO + 2Fe_3O_4 + CO_2 \quad (1)$$

$$ZnFe_2O_4 + CO = ZnO + 2FeO + CO_2 \quad (2)$$

$$ZnFe_2O_4 + 3CO = ZnO + 2Fe + 3CO_2 \quad (3)$$

$$Fe_3O_4 + CO = 3FeO + CO_2 \quad (4)$$

$$Fe_3O_4 + 4CO = 3Fe + 4CO_2 \quad (5)$$

$$FeO + CO = Fe + CO_2 \quad (6)$$

$$ZnO + CO = Zn + CO_2 \quad (7)$$

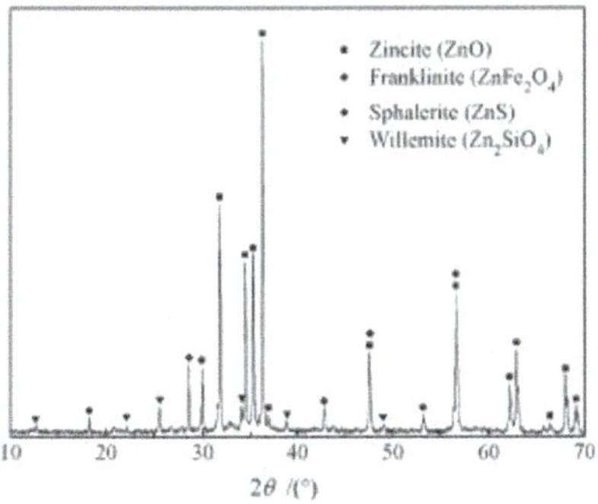

Fig. 1. XRD pattern of high iron-bearing zinc calcine

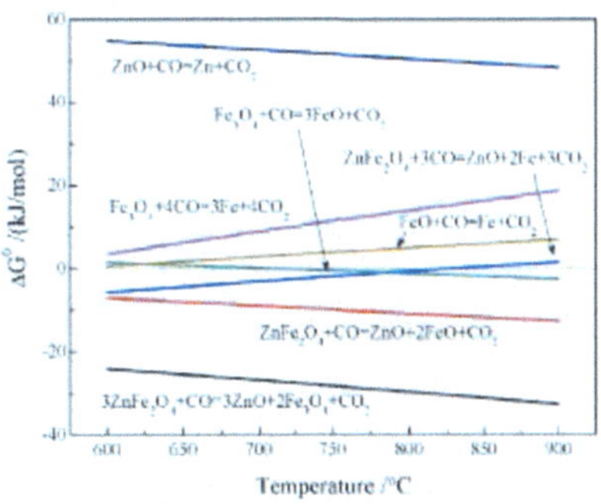

Fig. 2. Standard Gibbs free energy changes of the reactions occurred possibly

Fig. 2 shows the standard free energy changes (ΔG^0) of these reactions in the temperature range of 600-900°C. For the selective decomposition of zinc ferrite, Eq. (1) exhibits large negative ΔG^0 at all temperatures in the range investigated and the values decrease gradually with the increase in temperature, indicating that the desired reaction is feasible. The ΔG^0 for Eq. (2) and Eq. (3) show that zinc ferrite can also be converted to FeO at 600-900°C and Fe at below 780°C, respectively, and the formation of FeO is easier than Fe in thermodynamics. Additionally, the reduction of Fe_3O_4 to FeO, i.e. Eq. (4), is favourable at above 730°C, but the reaction of Fe_3O_4 or FeO to Fe, i.e. Eq. (5) or (6), is unfavourable in the whole temperature range. It is expected that ZnO will not be over-reduced to Zn, which is conducive to the experimental goal.

Results and discussion

Effect of roasting temperature

The key of this process is to selectively convert the zinc ferrite into magnetite and zinc oxide by magnetization roasting, in which reaction temperature is one of the most important factors. Therefore, the effect of temperature on iron recovery was investigated firstly at 65kA/m, and the results are shown in Fig. 3. It is seen from Fig. 3 that the iron recovery rate increased significantly with the increase of roasting temperature in the range of 650-750°C, indicating that the decomposition of zinc ferrite could be promoted by raising temperature. Thereafter, further increasing the temperature the recovery rate had no significant variation and even began to decrease above 800°C. This is attributed to the over-reduction of zinc ferrite according to Eqs. (2) and (4), which resulted in the increase of nonmagnetic FeO amount. On the other hand, it is found that the iron grade of concentrate decreased gradually, especially beyond 750°C, as the temperature was increased in the experimental range. Therefore, the optimal temperature was determined as 750°C.

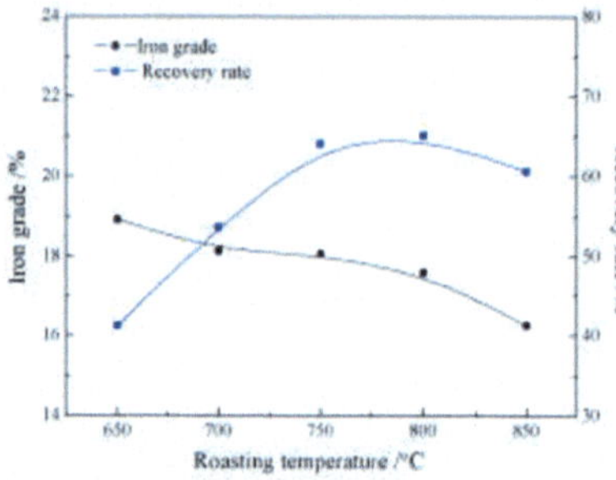

Fig. 3. Effect of roasting temperature

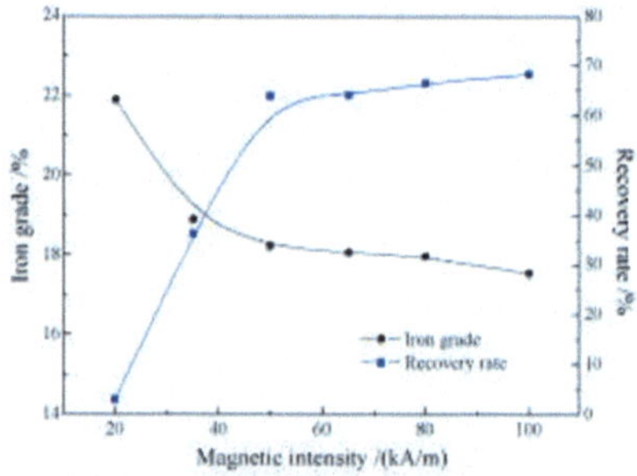

Fig. 4. Effect of magnetic intensity

Effect of magnetic intensity

Magnetic intensity, in general, has an important influence on magnetic separation process. The effect of magnetic intensity on iron recovery was investigated, and the results of which are presented in Fig. 4. The iron recovery rate increased from 3.04% to 63.91% when the magnetic intensity was increased from 20 to 50kA/m. Further increasing the intensity, the recovery rate had no significant variation. For the iron grade, it decreased significantly when the magnetic intensity was increased from 20 to 50kA/m, and thereafter, the decrease trend became very negligible. In addition, it is found that the iron grade was low in the whole experimental range and its maximum was less than 22%. This is accounted by the phenomenon that the grains of magnetite had not been liberated completely.

Effect of ore grinding

Fig. 5 shows the variations of iron grade and recovery rate with respect to different particle sizes of feedstock. The corresponding experiments were performed at 50kA/m. As shown in Fig. 5, the iron grade increased from 18.22% to 20.89% when the particle size was decreased from -150 to -38μm, while the recovery rate decreased gradually from 63.91% to 53.92%. This indicates that the recovery efficiency of iron was not ideal although the iron grade could be improved by ore grinding. This is because the magnetite grains were very fine, most of which were smaller than 10μm, which resulted in that good liberation of the magnetite grains was hardly achieved by mechanical grinding. Therefore, the roasted zinc calcine would be subjected to pretreatment by low acid leaching prior to magnetic separation process.

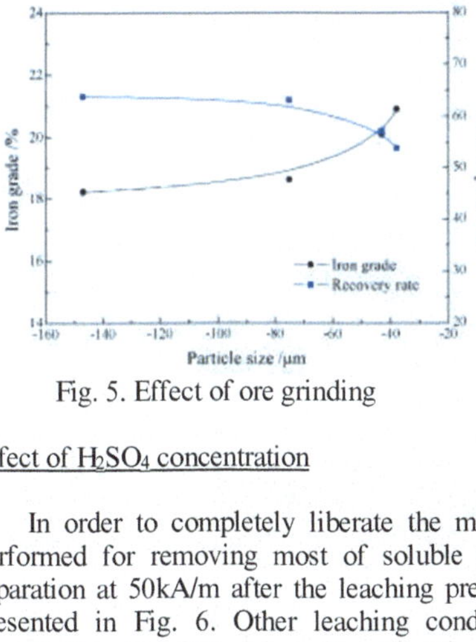

Fig. 5. Effect of ore grinding

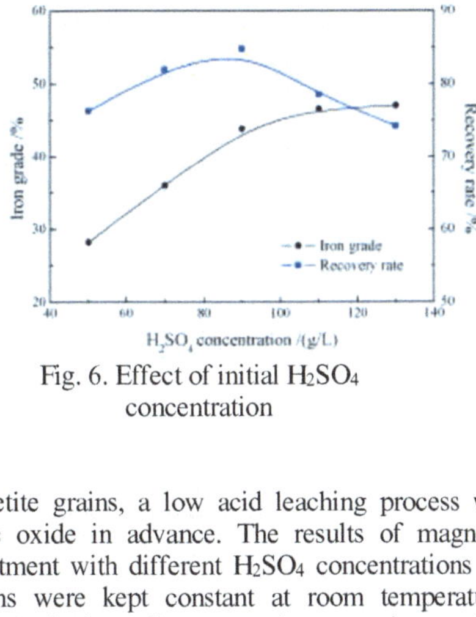

Fig. 6. Effect of initial H_2SO_4 concentration

Effect of H_2SO_4 concentration

In order to completely liberate the magnetite grains, a low acid leaching process was performed for removing most of soluble zinc oxide in advance. The results of magnetic separation at 50kA/m after the leaching pretreatment with different H_2SO_4 concentrations are presented in Fig. 6. Other leaching conditions were kept constant at room temperature, liquid/solid ratio of 10mL/g and leaching time of 10min depending on previous experiments. It is seen from Fig. 6 that the iron grade of concentrate increased significantly with the increase of H_2SO_4 concentration in the whole experimental range, while its recovery rate also increased in the range of 50-90g/L. Thereafter, the recovery rate began to decrease gradually when the acid concentration was beyond 90g/L. This is attributed to the iron dissolution was accelerated by concentrated H_2SO_4 and thus considerable amount of iron was lost in the solution. Therefore, the optimal H_2SO_4 concentration was determined as 90g/L. After the pretreatment, a concentrate with 43.88% of iron was obtained, while the iron recovery could reach up to 84.81%. This result was much better than that of without the pretreatment.

Open circuit test

In order to further increase the iron grade, the iron concentrate obtained after one roughing was subjected to a cleaning under lower magnetic intensity. The Flow sheet and the results are given in Fig. 7 and Table 2, respectively. It is seen from Table 2 that the grade of iron concentrate reached up to 53.24% after one roughing at 50kA/m and one cleaning at 35kA/m, while about 76.31% of iron could be recovered from the zinc calcine. The results suggest that this recommended process for iron recovery from iron-bearing zinc calcine is promising and further study is necessary. However, the tailing contained high iron, and the reason needed to be investigated further.

Table II. Results of open circuit test (wt. %)

Name	Iron grade	Recovery
Leachate	1.27g/L	9.52
Iron concentrate	53.24	76.31
Middling	36.55	8.77
Tailing	14.10	5.40
Zinc calcine	13.31	100.00

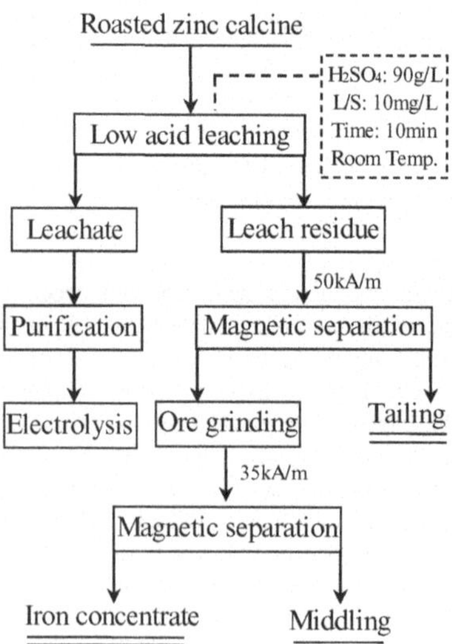

Fig. 7. Flow sheet of open circuit test

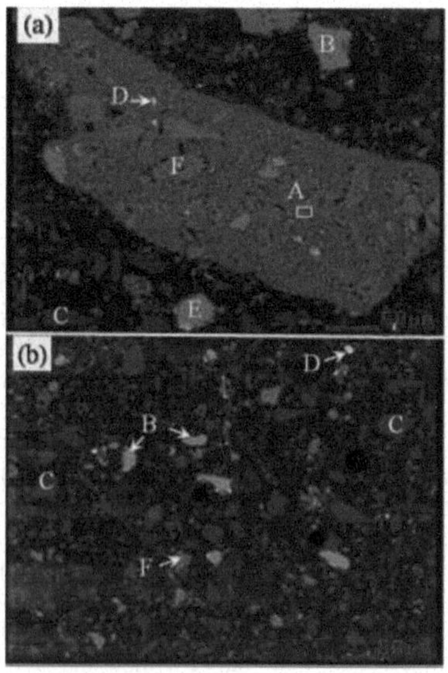

Fig. 8. SEM images of the concentrate (a) and the tailing (b)

Scanning Electron Microscopy associated with Energy Dispersive Spectroscopy techniques were used for the morphology study of the iron concentrate and tailing. The SEM images of the concentrate and the tailing are shown in Fig. 8. According to EDS analysis, as shown in Fig. 8(a), a large quantity of fine particles aggregation (A), rich in iron and oxygen, which indicates that the iron concentrate was mainly composed of magnetite, was observed. The particles (F) embedded in particles aggregation (A), containing mainly iron, zinc and oxygen, were probably unreacted zinc ferrite. The presence of large blocky particles (B) and (E), high in zinc and sulfur, was also observed, suggesting that some nonmagnetic sphalerite be mixed in the concentrate. The presence of black grey particles (C), rich in calcium, magnesium, oxygen and silicon, indicate that some of other gangues were also mixed in the concentrate. The presence of sphalerite, zinc ferrite and other gangues in the concentrate was the main reason for low iron grade. As shown in Fig. 8(b), the tailing was mainly composed of sphalerite, zinc ferrite and other gangues. And it is worth noting that some very fine magnetite was reported into the tailing due to the fact the smaller mineral grains the weaker their magnetism, which is the main reason for high iron content of the tailing. Besides, it is found that some point-like bright white phase (D) dispersed in the concentrate and tailing, high in lead, was probably metallic lead. It is obvious that the content of metallic lead in the tailing is much more than that in the concentrate, indicating that the lead, to some extent, was concentrated in the tailing, which is conducive to lead recovery in the subsequent study. Based on above analysis, in order to achieve the removal of iron from zinc calcine, further studies, including multistage magnetic separation process and grain size growth, will be carried out in the future.

Conclusions

The results of this study reveal that the removal of iron impurity from zinc calcine by the developed process, including magnetization roasting, pretreatment of acid leaching and magnetic separation steps, is promising. The iron recovery rate increased with the increase of magnetic field intensity or roasting temperature, while the iron grade of concentrate was always less than 22%, due to the fact that the grains of magnetite was severely enclosed by ZnO. Ore grinding was ineffective to the liberation of magnetite, for which was very fine. The pretreatment using low acid leaching was effective to selectively dissolve the zinc oxide. After the pretreatment, about 76.31% of iron was recovered from zinc calcine after one roughing at 50kA/m and one cleaning at 35kA/m, while the iron grade of concentrate reached up to 53.24%. The SEM/EDS analysis indicates that a part of sphalerite, zinc ferrite and other gangues were mixed into the iron concentrate, while some of fine magnetite was reported to the tailing, which are the main reasons for affecting the iron removal efficiency. Therefore, further studies, including multistage magnetic separation and increasing grain size of magnetite during the roasting process, is supposed to be carried out in the future.

Acknowledgements

The authors would like to thank the Innovation Project for Postgraduates of Central South University (2015zzts090), Co-Innovation Center for Clean and Efficient Utilization of Strategic Metal Mineral Resources and National Natural Science Foundation of China (51204210) for funding this research.

References

[1] J.C. Balarini, L.d.O. Polli, T.L.S. Miranda, et al., "Importance of roasted sulphide concentrates characterization in the hydrometallurgical extraction of zinc," *Minerals Engineering*, 21 (1) (2008), 100-110.

[2] M.K. Jha, V. Kumar, and R. Singh, "Review of hydrometallurgical recovery of zinc from industrial wastes," *Resources, Conservation and Recycling*, 33 (1) (2001), 1-22.

[3] T. Chen and J. Dutrizac, "Mineralogical changes occurring during the fluid-bed roasting of zinc sulfide concentrates, *JOM*, 56 (12) (2004), 46-51.

[4] P.C. Holloway, T.H. Etsell, A.L. Murland, "Roasting of La Oroya zinc ferrite with Na_2CO_3," *Metallurgical and Materials Transactions B*, 38 (5) (2007), 781-791.

[5] J. Han, W. Liu, W. Qin, et al., "Recovery of zinc and iron from high iron-bearing zinc calcine by selective reduction roasting," *Journal of Industrial and Engineering Chemistry*, 22 (2015), 272-279.

[6] M. Copur, T. Pekdemir, S. Colak, et al., "Industrial symbiosis: High purity recovery of metals from Waelz sintering waste by aqueous SO_2 solution," *Journal of Hazardous Materials*, 149 (2) (2007) 303-309.

[7] H. Shen and E. Forssberg, "An overview of recovery of metals from slags, *Waste Management*, 23 (10) (2003) 933-949.

[8] N. Leclerc, E. Meux, and J.M. Lecuire, "Hydrometallurgical extraction of zinc from zinc ferrites," *Hydrometallurgy*, 70 (1) (2003) 175-183.

[9] M.D. Turan, H.S. Altundoğan, and F. Tümen, "Recovery of zinc and lead from zinc plant residue," *Hydrometallurgy*, 75 (1) (2004) 169-176.

[10] J. Claassen, E. Meyer, J. Rennie, et al., "Iron precipitation from zinc-rich solutions: defining the Zincor Process," *Hydrometallurgy*, 67 (1) (2002) 87-108.

[11] M.R. Kendall, A.S. Madden, M.E. Elwood Madden, et al., "Effects of arsenic incorporation on jarosite dissolution rates and reaction products," *Geochimica et Cosmochimica Acta*, 112 (2013) 192-207.

[12] H. Yan, L. Chai, B. Peng, et al., "Reduction Roasting of High Iron-Bearing Zinc Calcine Under a $CO\text{-}CO_2$ Gas: An Investigation of the Chemical and Mineralogical Transformations," *JOM*, 65 (11) (2013) 1589-1596.

[13] W. Liu, J. Han, W. Qin, et al., "Reduction roasting of high iron bearing zinc calcine for recovery of zinc and iron," *Canadian Metallurgical Quarterly*, 53 (2) (2014) 176-182.

7th International Symposium on High-Temperature Metallurgical Processing
Edited by: Jiann-Yang Hwang, Tao Jiang, P. Chris Pistorius, Gerardo R.F. Alvear F., Onuralp Yücel, Liyuan Cai, Baojun Zhao, Dean Gregurek, and Varadarajan Seshadri
TMS (The Minerals, Metals & Materials Society), 2016

MICROWAVE HEATING OF WASTE TIRES

Yuzhe Zhang[1], Jiann-Yang Hwang[1,3], Zhiwei Peng[1,2], Matthew Andriese[1], Bowen Li[1,3], Xiaodi Huang[1,3], Xinli Wang[1,4], Likun Li[3]

[1]Department of Materials Science and Engineering, Michigan Technological University, Houghton, Michigan 49931, USA
[2]School of Minerals Processing and Bioengineering, Central South University, Changsha, Hunan 410083, China
[3]Advanced Materials R&D Center of WISCO, Beijing 102211, China
[4]School of Technology, Michigan Technological University, Houghton, Michigan 49931, USA

Keywords: Waste Tires, Rubber, Heating, Microwave

Abstract

With the development of the world, the waste tires have been a serious environmental problem. There are about 2 billion waste tires produced all over the world every year and about 500 million waste tires are produced in the United States. Most of the waste tires are land filled, which waste a lot of space. It takes hundreds of years to resolve the macromolecules of waste tire into the standard which does not pollute the environment. Microwave irradiation is a thermal processing way for waste tires decomposition. In this study, the relative dielectric loss factor of the tire at various temperatures was determined to understand the behavior of tires under microwave heating.

Introduction

Waste tire is a serious environmental concern. The reason is that with the increasing demand for tires and their short lifetime, more waste tires are produced. So it is necessary to find methods for waste tire recycling [1]. For most countries of the world, the policy includes that recycling and, particularly, the use of waste tires as building material are encouraged for preventing environmental problem.

There are various methods for waste tire recycling, such as landfill, stockpiles, combustion and pyrolysis. There are various waste tire recycling in different regions.

Table 1. Recycling of discarded tire in some countries and regions [2]

Place	Year	Amount	Disposing						
		(10000 Tons)	Thermal-utilizing	Making powder	Regenerated rubber	Renew	Export		Burying
US	1992	280	23	6	4		3		63
Japan	1992	84	43		12	9	25		8
Germany	1993	55	38	14	1	18	18		2
British	1992	45	9	6		18			67
EEC	1990	197.5	30		20				50

The landfill and stockpiles are the most common way for tire recycling. However, landfill and stockpiles will consume a lot of space and they can cause health problems. In this study, the heating approach is investigated. Pyrolysis is a technique that heats whole or shredded waste tires in a reactor vessel under an oxygen-free atmosphere. In the reactor the waste tire will be softened and broken down into oils, solid residue, fuel gas and carbon black [2]. Microwave heating is considered in this study since it is able to provide heating without combustion. Hence it can be carried out easily under an oxygen free condition.

Microwave is a kind of electromagnetic wave having the frequency from 300MHz to 300GHz. Microwave also has corresponding energies from 1.24×10^{-6} to 1.24×10^{-3}eV [4]. Compared to the normal ionization energies, the corresponding energies of microwave is much lower. The law of optics is the one which microwave obeys. Microwaves can be transmitted, absorbed and reflected. There are three kinds of materials according to various interactions between materials and microwave: opaque materials, absorbers and transparent materials. The opaque materials are able to not allow wave spread and reflect microwave (e.g., most metals). The absorbers can absorb various waves. The transparent materials only have little reciprocity with microwave. Be based on the materials properties, absorbers can be heated by microwave under different temperature [4].

The tire sample for this study was characterized in our previous paper [5]. In that study, the relative dielectric constant of the sample was determined. The goal of this study is to investigate the relative dielectric loss factor for further understanding of the tire behavior under microwave heating.

Principles

Depending on the frequency and temperature range the equation for describing the dielectric response of a substance is related to the permittivity (ε) for non-magnetic dielectrics, which can be shown as

$$\varepsilon = \varepsilon_0 \varepsilon_r = \varepsilon_0 (\varepsilon_r' - j\varepsilon_r'') \tag{1}$$

Where ε_0 is the permittivity of free space (8.8854 × 10-12F/m), and j is imaginary unit where $j^2 = -1$.

The symbol ε_r is complex relative permittivity and it is used to describe the constitutive relation between the electric field intensity in loss dielectrics and the electric flux density [6].

The symbol ε_r is consisted of two parts: one is ε_r'' which is the relative dielectric loss factor or imaginary part of complex relative permittivity. The other is ε_r' which is the relative dielectric constant or real part of complex relative permittivity. Although the imaginary part of complex relative permittivity represents the loss of electrical energy in dielectrics, the real part of complex relative permittivity is a measure of the ability of the dielectrics to stock electrical energy. The energy lost from the electric area to the dielectric is finally transformed to heat or thermal energy. As a result, the imaginary part of complex relative permittivity determines the rate of heating when applying microwave energy for dielectrics with no magnetic response.

It was assumed that the free electron conductivity mainly caused the dielectric loss, and the final-state dielectric loss factor was assumed at room temperature by measuring the dc resistance of the heated sample. The theoretical relationship between the dielectric loss factor and the dc resistance is:

$$\varepsilon'' = \frac{1}{2\pi f \varepsilon_0 R\left(\frac{\pi D^2}{4\,L}\right)} \tag{2}$$

In this equation R is the resistance measured in the experiment, the f is the frequency of microwave, the L is the sample length and the D is the diameter of the sample. The value of dc resistance was used to calculate a theoretical dielectric loss factor. Then the dielectric loss factor was compared with the measured dielectric loss factor for testing of the microwave loss mechanism [7].

The electric susceptibility ($\chi_e = \chi_e' - j\,\chi_e''$) with the data of the quality factor Q was calculated by the equation:

$$\frac{\Delta f}{f_e} + j\left[\frac{1}{2Q_{L,S}} - \frac{1}{2Q_{L,S}}\right] = \frac{-\chi_e}{1+F_{sh}\chi_e}\left[\frac{V_S}{V_C}\right]A \qquad (3)$$

In this equation the Δf is the frequency shift which is included by the sample, the f_e is the special cavity mode frequency which is 915 or 2466 MHz, $Q_{L,S}$ is the loaded cavity quality factor with the empty holder, V_C and V_S are the volumes of respective cavity and sample, F_{sh} is a real number which is based only on the shape of sample and A is a real calibration constant which is based only on the shape of electric areas in the absence of sample. The dielectric loss factor (ε_r'') and relative dielectric constant (ε_r') could be calculated after the susceptibility was determined [8].

Experiment

In this research, the cavity perturbation technique was used to measure the dielectric properties (ε_r' and ε_r'') of the tire rubber during pyrolysis at frequency of 915 and 2466MHz. This technique measures the differences which are changes in quality factors and shifting frequency caused by the microwave cavity response between the cavity with a sample –holder plus the sample and the same cavity with an empty sample-holder. Then calculation of the permittivity needs to use these differences. The experiment was done in stagnant air. The experiment was done in stagnant air with a short sample holder which is 80mm high.

During the experiments, the waste tire rubber sample was run consisted of measurements at room temperature (RT). The temperature was starting at 25°C then increased in 25°C steps to 725°C. Then the temperature was cooled 100°C steps to room temperature which is 25°C. After the measurement, the waste tire rubber sample mass was measured immediately. Then the waste tire rubber sample was removed and the temperature of holder was measured up to 725°C. There is no contamination in the holder.

All of the measurement sequences were programmed with temperature calibration in the furnace, movement of the linear actuator (Figure 1). A personal computer running lab view control software controlled all of the network analyzer analysis.

As shown in Figure 1, a cylindrical TM_{0n0} resonant mode cavity (ø 580mm × 50mm) and a resistive heating furnace are the main parts of the measurement system. The top part of the holder involving sample was raised in the furnace and held at least 5 min to make sure that the sample reach the equilibrium furnace temperature for measuring the permittivity at a specific temperature. Next, the holder with no sample lowered into the central maximum electric area region of the thick-walled, well-cooled copper TM0n0 cavity quickly with the time small than 1.5s for each frequency, The frequency shift Δf and the quality factor Q because of the sample insertion were recorded in a Hewlett-Packard 8753B vector network analyzer.

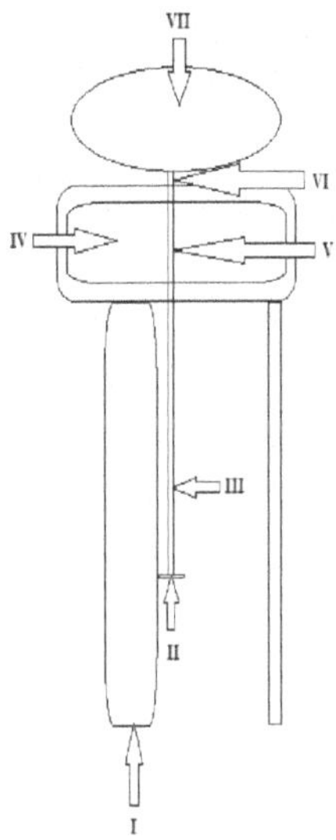

Figure 1. Diagram of the TM_{0n0} cavity system. (I) support stand, (II) sample stage on linear actuator, (III) quartz sample holder, (IV) TM_{0n0} right cylindrical cavity, (V) sample in holder, (VI) entry hole and (VIII) high temperature resistance furnace.

Microwave Absorption Properties

Figure 2 shows that the relative dielectric loss factor of tire rubber varies with temperature at 915MHz. The variation can be separated into five distinct stages. It keeps increasing in the range from room temperature to 366°C. In the range from 366°C to 650°C, it decreases first and then keeps increasing. Afterwards, the relative dielectric loss factor of tire rubber at 915MHz decreases in the range from 650°C to 700°C and then it is increasing in the range from 700 to 800°C. The relative dielectric loss factor of tire rubber at 915MHz behaves similarly to the relative dielectric loss factor of tire rubber at 915MHz. The relative dielectric loss factor of tire rubber at 915MHz is higher than that at 2466MHz in the range from room temperature to 400°C. However, the relative dielectric loss factor of tire rubber at 915MHz is lower than the relative dielectric loss factor of tire rubber at 2466MHz in the range from 400 to 800°C.

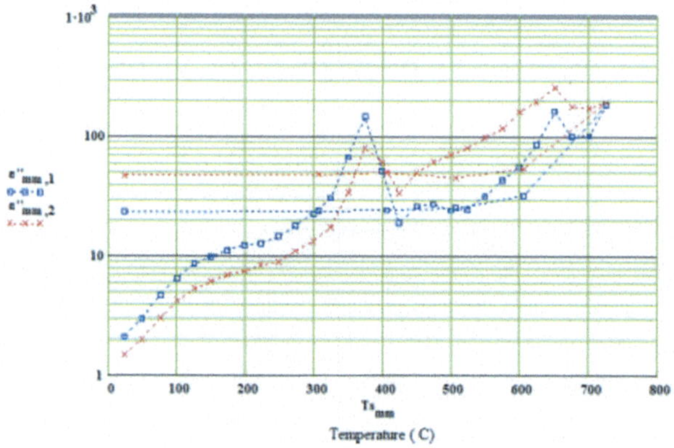

Figure 2. Temperature dependences of log(relative dielectric loss factor) under stagnant air. The ε''mm,1 is the log(relative dielectric loss factor) at 915MHz, and ε''mm,2 the log(relative dielectric loss factor) at 2466MHz.

Figure 2 shows that the relative dielectric loss factor of tire rubber at 915MHz is smaller than the relative dielectric loss factor of tire rubber at 2466MHz in the temperature range. The relative dielectric loss factor does not fit the equation $\varepsilon'' = \frac{1}{2\pi f \varepsilon_0 R(\frac{\pi D^2}{4\,L})}$, so that the microwave loss mechanism for pyrolyzed tire rubber is not pure free electron conduction. With temperature increasing, the increasing content of carbon black causes an increase in the relative dielectric loss factor. However, little air in container can react with carbon black. There is a dielectric loss peak in the range of 300 to 400°C. This is a typical interfacial/relaxation polarization phenomena behavior, usually indicating a change in the material associated with the loss of an insulating barrier. So the relative dielectric loss factor decreases in this the range.

For most materials, the relative dielectric loss factor rises with temperature. For the waste tire rubber, the major part which absorb microwave is probably carbon black. The carbon black has strong polarity so that it has great microwave absorb ability. It is not surprising to be in sight of common effects on the sample material. With the temperature increasing, the rubber also keeps producing carbon black so that the ratio of carbon black increases. As a result, the relative loss factor increases. The relative dielectric loss factor indicates dissipation of microwave power. The relative dielectric loss factor is in direct proportion to ability of microwave absorption. As a result of this, the microwave absorption ability of the sample increase with the temperature.

Conclusions

Waste tire has become a serious environmental problem. Efforts on recycling are needed. In this study, to determine the favorable conditions for the microwave heating of tire rubbers, microwave absorption properties of waste tire rubber at different temperature are characterized. It shows that the relative dielectric loss factor rises with the increasing of temperature in the range from room temperature to 800°C. This demonstrates that microwave heating of tires are favored in all of the temperature range and the higher the temperature, the better will be the absorption.

References

[1] "Tire recycling", July 20, (2015), http://en.wikipedia.org/wiki/Tire_recycling

[2] "pyrolysis", July 20, (2015), http://en.wikipedia.org/wiki/Pyrolysis

[3] Yi Fang, Maosheng Zhan*, Ying Wang, *Materials & Design*, 22 (2001) 123-127, "The status of recycling of waste rubber".

[4] Xiang Sun, Master thesis, Michigan Technology University, (2006), "Treatment of electric arc furnace dust by microwave heating".

[5] Y. Zhang, J. Hwang, Z. Peng, M. Andriese, B. Li, X. Huang, and X. Wang, "Microwave absorption characteristics of tire", Characterization of Minerals, Metals, and Materials 2015, John Wiley & Sons, Inc., 2015, pp.235-244.

[6] Zhiwei Peng, Jiann-Yang Hwang*, Matthew Andriese, *Ceramics International*, 39 (2013) 6721–6725, "Design of double-layer ceramic absorbers for microwave heating".

[7] Zhiwei Peng, Jann-Yang Hwang, Joe Mouris, Ron Hutcheon, Xiang Sun, *Metals & Materials Society and ASM International*, (2011) 2259-2263, "Microwave Absorption Characteristics of Conventionally Heated Nonstoichiometric Ferrous Oxide".

[8] Zhiwei Peng, Jiann-Yang Hwang*, Byoung-Gon Kim, Joe Mouris, Ron Hutcheon, *Energy Fuels*, 26 (2012)5146−5151, "Microwave Absorption Capability of High Volatile Bituminous Coal during Pyrolysis".

7th International Symposium on High-Temperature Metallurgical Processing
Edited by: Jiann-Yang Hwang, Tao Jiang, P. Chris Pistorius, Gerardo R.F. Alvear F., Onuralp Yücel, Liyuan Cai, Baojun Zhao, Dean Gregurek, and Varadarajan Seshadri
TMS (The Minerals, Metals & Materials Society), 2016

PREPARATION OF HIGH-QUALITY TITANIUM-RICH MATERIAL FROM TITANIUM SLAG WITH HIGH Ca AND Mg CONTENT BY ACTIVATION ROASTING PROCESS

Wenting Duan, Feng Chen, Fuqiang Zheng, Tao Jiang, Yufeng Guo*

School of Minerals Processing and Bioengineering, Central South University, Changsha, Hunan, PR.China

Keywords: Titanium slag; Activation roasting; Acid leaching; Titanium-rich material

Abstract

High-quality titanium-rich material is of significance in the development of titanium industry as it is the main raw material to produce sponge titanium and titanium white using chloride process. Activation roasting of titanium slag with high Ca and Mg contents as adding H_3PO_4 followed by acid leaching was investigated to prepare titanium-rich material. The effects of H_3PO_4 dosage, roasting temperature and time, hydrochloric acid quantity, acid leaching temperature and time on TiO_2 grade and removal of impurities were studied. The results show that after the finely ground titanium slag roasted with 8% H_3PO_4 at 1000℃for 2h, most Mg impurities in anosovite solid solution structure were converted into simple acid–soluble compounds. The impurities such as Ca and Mg obviously decreased when the hydrochloric acid concentration was 20%, leaching at 100 ℃ for 2h, from which a titanium-rich material containing 82.56% TiO_2 and 1.15% (CaO+MgO) was obtained.

Introduction

Titanium dioxide (TiO_2) is the top performing white pigment in the world, which is widely used in the manufacture of paint, papermaking, plastics, printing ink, chemical fiber, rubber and other areas of chemical industry [1]. Currently, titanium dioxide is commercially manufactured by two main processes, the Sulfate Process and the Chloride Process [2, 3]. The Sulfate Process is the traditional technology used to produce titanium dioxide, in which low-titanium-content feedstock are leached with sulfuric acid with a lot of environmental problems, and the quality of the production is low. On the other hand, in the chloride process, natural or synthetic rutile is used as the raw material to produce high-quality titanium dioxide pigment via a $TiCl_4$ stage at high temperatures which is environmental friendly and economic saving [4]. Due to the advantages above, the chloride process is considered more favorable for TiO_2 pigment production than the sulfate process. However, the chloride process has strict requirement about the feedstock that the content of Ca and Mg must be lower than 1.5%.

Titanomagnetite resources are riched in Panxi region of Sichuan Province, China [5]. The titanium slag produced by electric smelting of Panzhihua ilmenite concentrate has high contents of impurities such as Ca and Mg, which cannot meet the demand of the chloride process [6], thus upgrading the titanium slag is of great importance to promote the development of titanium industry.

Previous studies showed that titanium slag can be upgraded for the chloride process by several treatments, including oxidation roasting, oxidation-reduction roasting and alkali salt roasting, in which impurities were removed by the following water or diluted acid leaching[7-9]. However, most of these developed processes have not been commercially applied or were not suitable to treat Panzhihua titanium slag because of the high Ca and Mg content.

In the present work, a new process was proposed to achieve high efficiency transformation of titanium slag, which can be used to upgrade Panzhihua titanium slag. Impurities such as Ca and Mg obviously decreased and the obtained product can be used as the raw material of the chloride process.

Experimental

Raw materials

The raw material used in the work was titanium slag provided by Panzhihua Iron & Steel Company, China. The main chemical compositions analyzed by X-ray fluorescence are shown in Table 1 and morphology characterized by X-ray diffraction is presented in Figure 1, respectively. It can be seen from Table 1 that the titanium slag has a low TiO_2 grade and the impurities, especially CaO and MgO, are almost up to 4%. The main mineral phases of the titanium slag shown in Figure 1 are anosovite and silicate. All the chemicals and reagents used in the work were of analytical grade, and deionized water used throughout the experimental process.

Table 1. Chemical Composition of Titanium Slag/ wt%

TiO_2	TFe	Al_2O_3	CaO	MgO	SiO_2
73.88	9.61	2.71	0.86	3.12	6.52
MnO	Cr_2O_3	V_2O_5	P	S	
0.86	0.13	0.27	0.01	0.12	

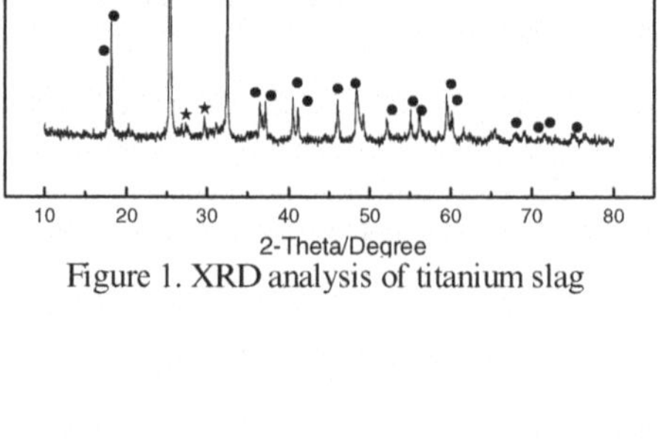

Figure 1. XRD analysis of titanium slag

Methods

The slag samples were finely ground to less than 74 m (>80%) and mixed with phosphoric acid to be pelletized. The prepared pellets were dried in an oven at 105℃for 24 h and then roasted in a tube furnace under the designed conditions. After the reaction finished, the roasted products were drawn out of the reactor and naturally cooled to room temperature. The roasted samples were then ground to the total size less than 74 m for preparation of the following acid leaching process. A 500 ml three-necked flask with a mechanical Teflon-coated agitator and a reflux condenser was the reactor, into which the required concentration of hydrochloric acid was put. A thermostatically controlled water bath was used to heat the reactor and keep it at the setting temperature. The prepared roasted samples mixed with the hydrochloric acid at a Solid /Liquid ratio of 1:5 was added into the reactor and the leaching kept for 2h. After the leaching experiment, the leaching product was filtered and washed, and the residual solid was then calcining at 900℃for 30 min to remove residual carbon and sulfur impurities to get titanium-rich material. The TiO_2 content of the obtained products was chemically analyzed by the ammonium ferric sulfate method [10].

Results and Discussion

In this work, the effects of H_3PO_4 dosage, roasting temperature and time, hydrochloric acid concentration, acid leaching temperature and time on TiO_2 grade and removal of impurities were studied.

Effects of roasting temperature

In order to research the effects of roasting temperature on the TiO_2 grade of the titanium slag, several experiments were conducted in roasting temperature ranges from 800℃to 1300℃. Figure 2 shows the phases of slag at different temperatures. The XRD pattern of the obtained roasted product indicates that raising temperature can efficiently transform titanium slag partly into rutile phase. The rutile phase of the diffraction peak intensity increased gradually when the temperature increased to 1000℃while the diffraction peak of the anosovite phase continues to weaken and without any new phase formed. However, when the roasting temperature raised above 1000℃, the slag phase was with no obvious change. The roasted products were subjected to leach in boiling hydrochloric acid, with a fixed acid leaching condition, to research the effects of roasting temperature on TiO_2 grade and recycle. It can be seen from Figure 3 that the TiO_2 grade was improved with the increasing roasting temperature from 800℃ to 1300℃, the TiO_2 grade can reach to 76.4% at 1000℃, but the TiO_2 grade started to decrease above 1000 °C while the recycle of TiO_2 little changed. Thus, the roasting temperature in this work was fixed at 1000℃.

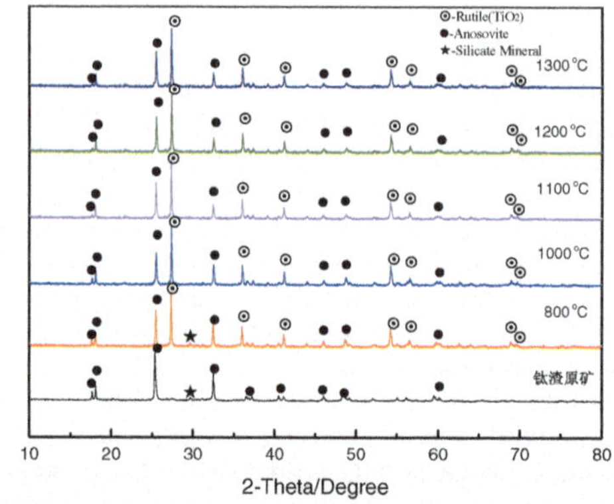

Figure 2. Effects of roasting temperature on XRD patterns of products

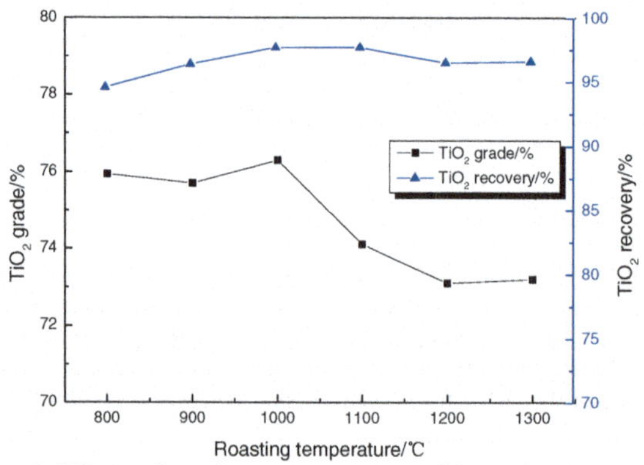

Figure 3. Effects of roasting temperature on TiO_2 grade and recovery

Effects of roasting time

To study the effect of roasting time on the TiO_2 grade, experiments were conducted with different roasting duration periods (1-4h) under same other conditions. The results are shown in Figure 4, the TiO_2 grade of the acid leaching product increased from 74.25% to 76.4% as the roasting time increased from 1 h to 2h. However, it had little effect on the TiO_2 leaching degree when the roasting time continually extended. Therefore, the roasting time was selected as 2h.

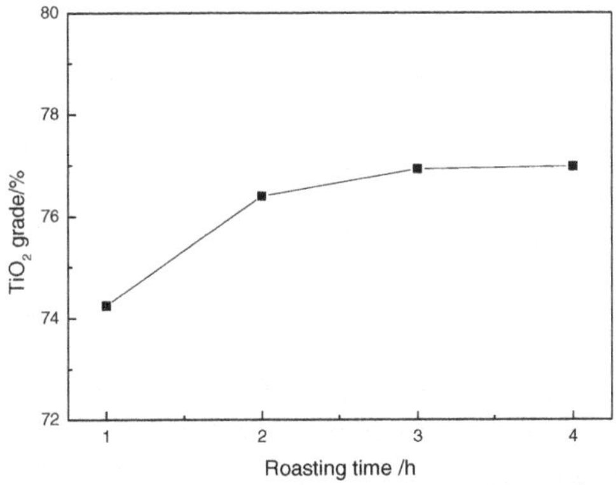

Figure 4. Effects of roasting time on TiO_2 grade

Effects of phosphoric acid dosage

The results shown in Figure 2, Figure 3 and 4 indicate that the titanium slag cannot efficiently transform to obtain Ti–rich material for the chloride process after treated with direct oxidation roasting. Therefore, the phosphoric acid was added during the roasting process to enhance the impurity removal process and improve the TiO_2 grade of the final product. In order to study the effects of phosphoric acid dosage on TiO_2 grade, several experiments were conducted. The oxidative conditions such as roasting temperature and time were fixed at 1000℃, 2h .The results of the TiO_2 grade and the impurity removal of the leaching product are shown in Figure 5 and 6, respectively.

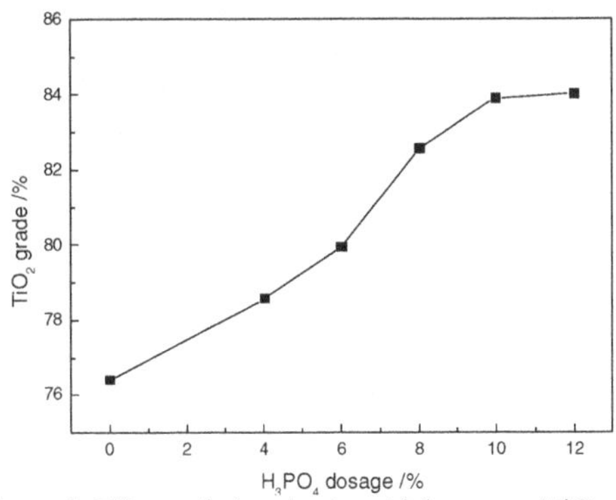

Figure 5. Effects of phosphoric acid dosage on TiO_2 grade

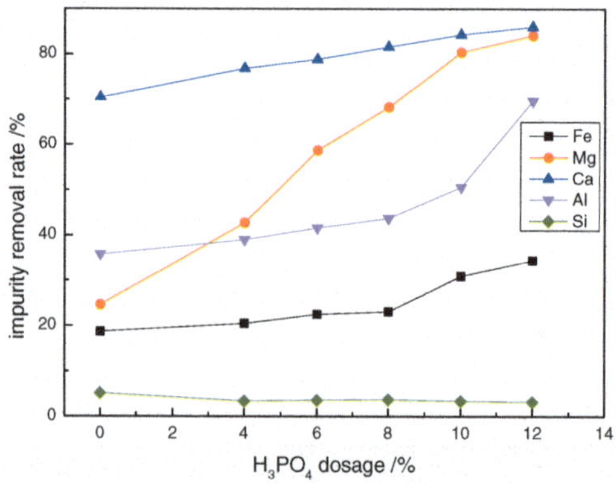

Figure 6. Effects of phosphoric acid dosage on impurity removal

The Figure 5 and 6 indicate that the phosphoric acid dosage had a significant influence on the TiO_2 grade, which increased rapidly as the phosphoric acid dosage increased from 0 to 12%, which could damage the stable structure of anosovite in titanium slag [8]. As the phosphoric acid content increased to 12%, the contents of Fe, Mg, Ca, Al decreased obviously and the TiO_2 grade reached 84.01%. These results indicate that the transformation of titanium slag into rutile was promoted by roasting with H_3PO_4.Thus,in the subsequent acid leaching experiments , the roasted products were activation roasted with 8% H_3PO_4 at 1000℃for 2h.

In order to improve the TiO_2 grade of the activation roasting products efficiently, the effects of acid concentration, leaching temperature and time on the TiO_2 grade and recovery were studied. The parameters such as the liquid-to-solid mass ratio and stirring speed were kept constant in the following experiments.

Effects of acid concentration

The effects of acid concentration on TiO_2 grade and recovery were investigated in the range of 5% to 25% when the leaching temperature and time were fixed at 100℃ and 2h. The results were shown in Figure 7 that the TiO_2 grade increased from 79.57% to 82.56% when the HCl concentration increased from 5% to 20%. However, the TiO_2 grade and recovery decreased obviously at a higher acid concentration for TiO_2 can partly dissolved in HCl with high concentration. Thus, the proper acid concentration used in the following experiments was 20%.

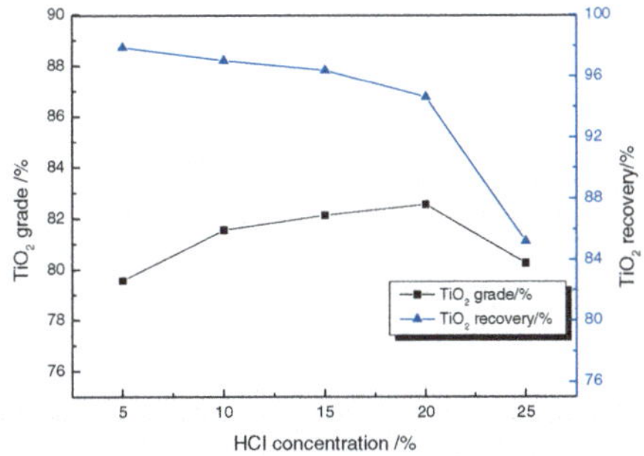

Figure 7. Effects of acid concentration on TiO_2 grade and recovery

Effects of leaching temperature and time

The effects of leaching temperature and time on the TiO_2 grade and recovery were studied. Parameters were kept constant during the tests. The results were shown in Figure 8, which shows that temperature had a certain effect on the leaching products. The TiO_2 grade increased from 72.57% to 82.56% as the temperature increased from 30℃ to 100℃. The removal rates of impurities can be improved when the leaching time increased from 1h to 2h. Extending the leaching time from 2 to 4 h, however, the TiO_2 grade of the product was not changed much. Therefore, the proper acid leaching condition selected was leaching with 20% HCl at 100℃for 2h. The chemical composition of the final product was shown in Table 2.

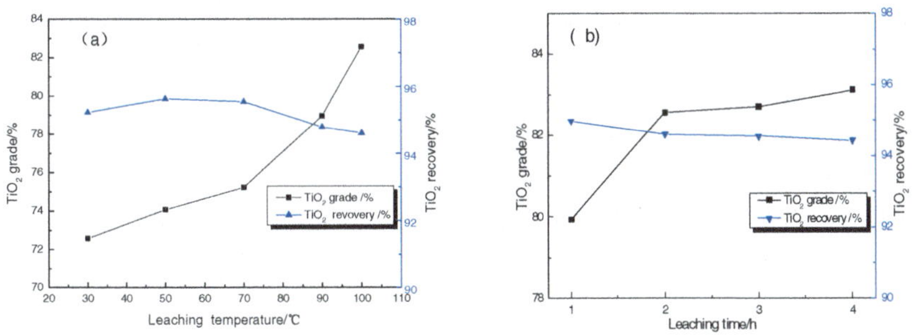

Figure 8. Effects of leaching temperature and time on TiO_2 grade and recovery ((a)temperature; (b) time)

Table 2. The impurity removal rate and chemical composition of the final product (8% H_3PO_4, 20% HCl, 100℃, 2h)

Impurity element		Fe	Mg	Ca	Al	Si
Removal rate / %		23.09	68.33	81.65	43.77	3.76
Chemical composition	TiO_2	TFe	MgO	CaO	Al_2O_3	SiO_2
Content / wt%	82.56	7.39	0.99	0.16	1.52	6.27

Conclusions

1) Titanium slag activation roasting with H_3PO_4 can efficiently enhance the removal of Ca and Mg impurities in the following acid leaching process and improve the TiO_2 grade.

2) The titanium-rich material containing 82.56%TiO_2 and 1.15% (CaO+MgO) was obtained under the conditions of activation roasting with 8% H_3PO_4 at 1000 °C for 2 h and leaching with 20% HCl at 100℃ for 2h.

References

1. T. A. Lasheen, "Sulfate digestion process for high purity TiO_2 from titanium slag," *Frontiers of Chemical Engineering in China*, 3 (2) (2009), 155-160.

2. Middlemas Scott, Fang Z. Zak, Fan Peng, "A new method for production of titanium dioxide pigment," *Hydrometallurgy*, 131 (1) (2013), 107–113.

3. Yongjie Zhang, Tao Qi, Yi Zhang, "A novel preparation of titanium dioxide from titanium slag," *Hydrometallurgy*, 96 (1-2) (2009), 52-56.

4. Wenqian Tang, Jinbao Zhang, "Energy consumption analysis and comments of manufacturing titanium dioxide by sulfuric acid process and chloride process," *Inorganic Chemicals Industry,* 43 (6) (2011), 7-9.

5.Yufeng Guo et al., "Enrichment of moderate and low grade titanium feedstock by activated roasting-acidic leaching," *The Chinese Journal of Nonferrous Metals*, 15 (9) (2005), 1446-1451.

6. Hui Lu et al., "Preparation of Synthetic Rutile Using Activation Modification-Selective Leaching Process from Preoxidation Titanium-containing Slag," *Chinese Journal of Process Engineering,* 11 (6) (2011), 994-1001.

7. Shuishi Liu, "Preparation of high-quality titanium-rich material from electro-titanium slag by phase deconstruction process" (Ph.D. thesis, Central South University, 2014), 1-16.

8. Sahu K. K., "An overview on the production of pigment grade titanium from titanium-rich slag," *Waste Management & Research*, 24 (1) (2006), 74-79.

9. Xuehua Mao, Daijun Liu, Peng Zhang, "Research progress of preparation of chlorination-grade feedstock from electro-titanium slag at home and abroad," *Modern Chemical Industry,* 31 (1) (2011), 72-75.

10. *Methods for chemical analysis of high-titanium slag and rutile—Part 1: Determination of titanium dioxide content--Ferric ammonium sulfate titration* (Beijing, Standards press of China, 2009).

7th International Symposium on High-Temperature Metallurgical Processing
Edited by: Jiann-Yang Hwang, Tao Jiang, P. Chris Pistorius, Gerardo R.F. Alvear F., Onuralp Yücel, Liyuan Cai, Baojun Zhao, Dean Gregurek, and Varadarajan Seshadri
TMS (The Minerals, Metals & Materials Society), 2016

PREPARATION OF TiC FROM TITANIUM BEARING BLAST FURNACE SLAG BY CARBOTHERMAL REDUCTION IN VACUUM

Fangqing Yin[1], Zhengfeng Qu[1], Mengjun Hu[2],Qingyu Deng[1], Meilong Hu[1*]

1 College of Materials Science and Engineering, Chongqing University, Chongqing 400030, P.R.China

2 Chongqing Iron & Steel (group) Co., Ltd. Chongqing 400080, P.R.China

Keywords: Vacuum, Carbothermal, Reduction, Titanium-bearing, Blast furnace slag, TiC

Abstract

Titanium bearing blast furnace slag, which contains 20%-30% TiO_2, is a valuable secondary titanium resource. Based on the thermodynamics calculation effect of the temperature on the reduction of the slag in vacuum were studied at certain vacuum degrees. The results show that magnesium vapor and silicon monoxide vapor can be obtained at the condition of high temperature and low pressure. The purity of TiC increases with temperature increasing. When the temperature arrives at 1623K TiC can be obtained besides the reductant carbon. XRD result of the powder collected from the furnace cover show that magnesium and silicon in the titanium bearing BF slag leave the slag system during carbothermal reduction in vacuum. From the above it can be concluded that TiC can be obtained from the titanium bearing BF slag and silicon and titanium in the slag can be separated by carbothermal reduction in vacuum and acid leaching.

Introduction

Panzhihua, located in the southwest of china, owns more than 10 thousands million tons vanadium-titanium magnetite, ranked the first in the world[1]. Fig.1. shows utilization of the vanadium-titanium magnetite ore in China.

In the process of Pan steel of the blast furnace smelting vanadium titanium magnetite, about 54 wt % of titanium resource in ore has entered into the blast furnace slag. It not only pollutes the environment but also waste the valuable resource. Therefore, many researches pay attention to the use of titanium-bearing blast furnace slag. There are many research methods that had been put forward by them such as raw materials of photocatalytic degradation, pyroceram and stone-cast-making raw materials, materials for building using and enriching by chemical method[2-4].

*Corresponding author. E-mail: hml@cqu.edu.cn; (Meilong Hu)

However these methods do not fundamentally solve the problem to use efficiently titanium-bearing blast furnace slag due to separation of titanium and silicon, the second pollution and effective utilization of the titanium resources. Aiming at the problem of difficult to use titanium-bearing blast furnace slag, preparation of TiC using the titanium-bearing blast furnace slag by the union process of carbothermal reduction in vacuum and acid leaching is put forward in this research. In this process SiO_2 and MgO can be reduced to SiO and Mg which can be leave from the slag system.

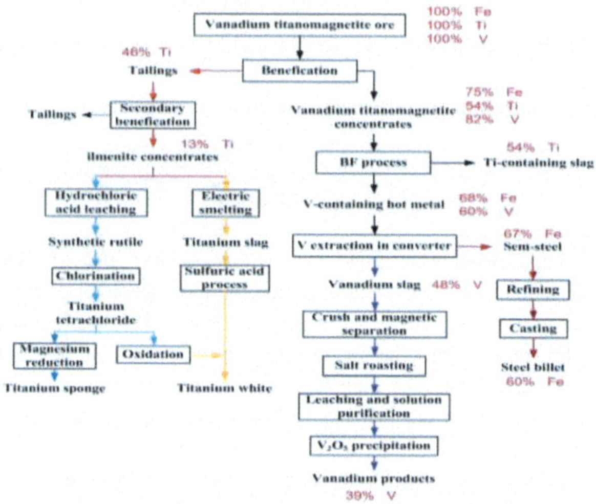

Fig.1. Diagram of the utilization of the vanadium-titanium magnetite ore-in China

Titanium carbide, with cubic crystal structure of NaCl type, is extensively used in various fields because of its outstanding properties including high strength, high hardness, high melting temperature, high electrical and high solvency with other carbides[5-6]. In particular it is used for the manufacturing of cutting tools or coated cutting tips.

The characteristics of titanium are more valence and strong affinity with oxygen [7], therefore it is more difficult to deoxygenation for TiO_2. There are two opinions about process of the carbothermal reduction of TiO_2. In the process of carbothermal reduction, TiO is observed and the last step TiC is obtained [8-10]. The formation sequence of products are $TiO_2 \rightarrow Ti_3O_5 \rightarrow Ti_2O_3 \rightarrow TiO \rightarrow TiC$. However, others think TiO is not obtained but solid solution of TiC_xO_y and the last step TiC is obtained [11-12]. The process of sequence is $TiO_2 \rightarrow Ti_3O_5 \rightarrow Ti_2O_3 \rightarrow TiC_xO_y \rightarrow TiC$. Therefore in the condition of high temperature, deoxygenation of titanium oxide is undefined. In fact, TiO_2, Al_2O_3, MgO, CaO and SiO_2 are contained in titanium-bearing blast furnace slag[13], which make process of deoxygenation for TiO_2 more difficult[14]. So the problem of deoxygenation of TiO_2 in titanium-bearing blast furnace slag has not been solved and it greatly limit recycle of titanium-bearing blast furnace slag.

Experimental

Titanium-bearing blast furnace slag was used as raw material shown as table 1 and carbon(C≥ 99.99%,mass fraction) was used as reductant in experiments. The average particle size of titanium-bearing blast furnace slag was about 75 m.

Table I. main composition of the BF slag in Pan steel, mass%

TiO_2	Al_2O_3	MgO	CaO	SiO_2	Fe_2O_3	FeO	MnO
20.78	14.61	8.57	26.17	26.48	<0.5	>1.61	0.73

For preparing TiC powders, the mixture of the titanium-bearing blast furnace slag and carbon with certain mass ratio were milled 1h in mortar. The mixed powders were compressed into blocks by cylinder proof press under 30 MPa. Then put the crucible held the blocks into the vacuum tube furnace (25kw,6.67×10^{-3}Pa, 1 7 2 7 K).After arriving to target vacuum degree, the feed was heated to different reaction temperature at a heating rate of 30 °C /min and keep to 2h. After 2 hours ,decreased the temperature and switched off the vacuum tube furnace power when the temperature gauge were cooled down to 80°C. The reduced products were obtained and crushed into powder. After reduction in vacuum the slag was acid leaching in HCl to obtain TiC.

Phase composition of the prepared production were investigated by X-ray diffraction instrument(D/max-3B) using Cu, K radiation in the range of 20-80°(2θ) with a step of 5°/min.

Results and discussion

Thermal analysis

To obtain the suitable experimental condition that TiO_2 in the slag can be completely used to product TiC and others oxides and chemical compounds detach by the carbothermal reduction in vacuum and acid leaching, thermodynamics was firstly calculated using FactSage 6.2. The possible reactions of the titanium bearing slag during carbothermal reduction in vacuum are showed as flowing:

$$2TiO_2+C\text{-}Ti_2O_3+CO \quad (1)$$
$$3TiO_2+C\text{-}Ti_3O_5+CO \quad (2)$$
$$TiO_2+3C\text{-}TiC+2CO \quad (3)$$
$$SiO_2+C\text{-}SiO+CO \quad (4)$$
$$SiO+2C\text{-}SiC+CO \quad (5)$$
$$SiO_2+3C\text{-}SiC+2CO \quad (6)$$
$$FeO+C\text{-}Fe+CO \quad (7)$$
$$CaO+C\text{-}Ca+CO \quad (8)$$

$2Al_2O_3+9C\text{-}Al_4C_3+6CO$ (9)

$Al_2O_3+3C\text{-}2Al+3CO$ (10)

$MgO+C\text{-}Mg+CO$ (11)

$CO_2+C\text{-}2CO$ (12)

Fig.2 shows the relationship between Gibbs free energy and temperature from reaction(1) to reaction(12) at 5Pa. Here, the condition of the carbothermal reduction of the oxides contained in the slag and the temperature zone of TiC formed at 5Pa are considered.

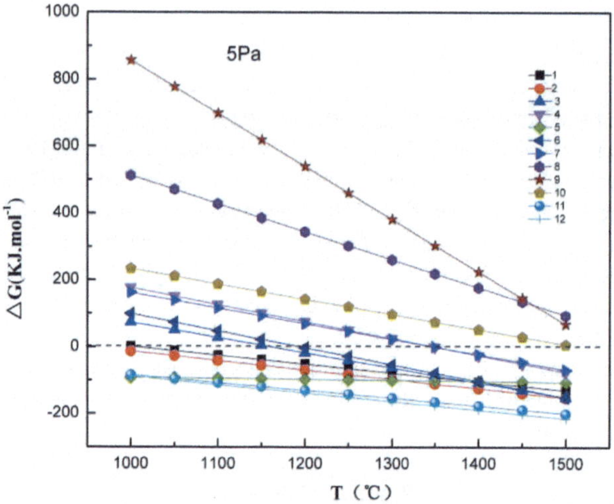

Fig.2. Changes of the Gibbs free energy with the temperature increasing at of reaction (1)-(12) as function of temperature at 5 Pa

Further, from Fig.2 it can be seen that with the temperature increasing the reaction more easily occur respectively. TiC, Ti_2O_3 and TiO can be produced between 1150 °C and 1500 °C at 5Pa, while Al_2O_3 and CaO almost can not be reduced. At the same time Fe also appear and SiO_2 is reduced into SiC and SiO. Then Mg is produced behind 1350 °C at 5 Pa.

Carbon reduction and acid leaching experiments

Fig.3 shows the XRD patterns with slag/carbon mass ratio of 100:40 at 5Pa for 2h at different temperature (1373K, 1473K, 1573K, 1623K and 1673K). The results are not completely consistent with the result of the thermodynamic calculation. At 1373K, there is not TiC produced. While the temperature reaches 1473K, TiC begins to appear at the same reduction condition. Carbon residue is observed at 1623K and 1673K because of high carbon content in experiment. Before 1623K with the carbothermal reduction temperature increasing, more and more TiC is produced.

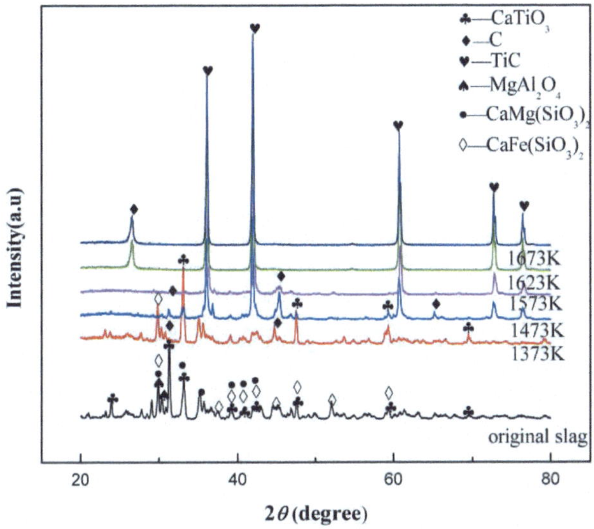

Fig.3. effect of the temperature on carbothermal reduction in vacuum

However, the result of 1623K consistent with 1673K, so at 5Pa the optimum temperature of carbothermal reduction is 1623K. Then titanium oxide is not observed in Fig.3.

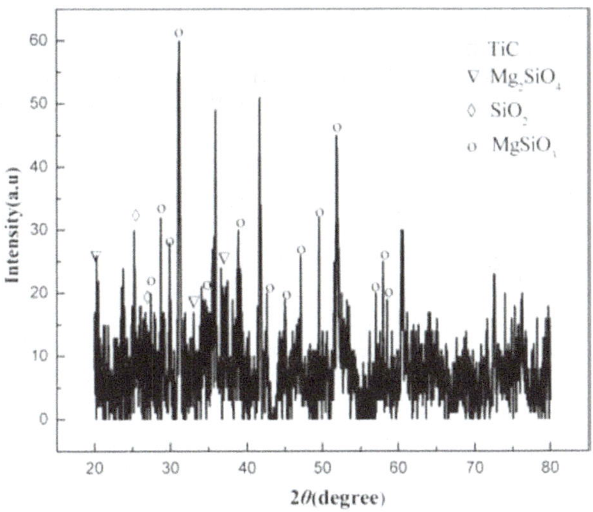

Fig.4. XRD patterns of the powder collected from the furnace cover

Fig.4 shows XRD result of the powder collected from the furnace cover at 1623K with 5Pa. It can be seen that SiO_2, $MgSiO_3$ and Mg_2SiO_4 appear due to magnesium and silicon monoxide which have properties of higher steam pressure oxidized by air. Some TiC mixed in the mixture of the furnace cover. So the high temperature and vacuum are help for separation of silicon and titanium in slag and obtaining TiC.

$$2SiO+O_2\text{-}2SiO_2 \tag{13}$$
$$Mg+O_2+SiO\text{-}MgSiO_3 \tag{14}$$
$$4Mg+3O_2+2SiO\text{-}2Mg_2SiO_4 \tag{15}$$

Conclusions

(1) Base on the thermodynamics analysis, with the temperature increasing of the carbothermal reduction in vacuum, TiC , SiC, SiO and Mg can be obtained. When the vacuum degree of 5Pa and the slag particle size of 75 m, the optimal temperature to prepare TiC is 1623K.
(2) Magnesium and silicon in the titanium bearing BF slag leave the slag system during carbothermal reduction in vacuum due to evaporating of SiO and Mg which have properties of high steam pressure. Meanwhile titanium and silicon in the slag can be separated completely.
(3) The process of preparation TiC from the titanium bearing BF slag is from $CaTiO_3$ to TiC and no intermediate founded.

Acknowledgements

The authors are especially grateful to the project 51404044 supported by National Natural Science Foundation of China and project No. 106112015CDJZR135502 supported by the Fundamental Research Funds for the Central Universities.

References

1. Kwauk, Mooson, *Legacy and growth-chemical engineering in China*(Chemical engineering science 44.11,1989),2421-2434.
2. Ooka, Chihiro et al., *Highly hydrophobic TiO_2 pillared clay for photocatalytic degradation of organic compounds in water*(Microporous and Mesoporous Materials 67.2 ,2004),143-150.
3. Robin, James E., and Paul Nordin, *Spectral absorptance at 3.8- wavelength for aluminum and pyroceram at elevated temperatures*(Applied Physics Letters 27.9 ,1975),493-495.
4. Portinari, L., C. Chiosi, and A. Bressan, *Galactic chemical enrichment with new metallicity dependent stellar yields*(Astronomy and Astrophysics 334,1998), 505-539.
5. Ahuja, R., et al, *Structural, elastic, and high-pressure properties of cubic TiC, TiN, and TiO* (Physical review B 53.6 ,1996), 3072.

6. Vallauri, Dario, IC Atias Adrian, and A. Chrysanthou, *TiC–TiB$_2$ composites: a review of phase relationships, processing and properties*(Journal of the European Ceramic Society 28.8 ,2008),1697-1713.
7. Lefebvre, Louis-Philippe, and Eric Baril, *Effect of oxygen concentration and distribution on the compression properties on titanium foams*(Advanced Engineering Materials 10.9,2008), 868-876.
8. Ono, L. K., D. Sudfeld, and B. Roldan Cuenya, *In situ gas-phase catalytic properties of TiC-supported size-selected gold nanoparticles synthesized by diblock copolymer encapsulation*(Surface science 600.23,2006), 5041-5050.
9. Bellucci, A., D. Gozzi, and A. Latini, *Overview of the TiC/TiO$_2$ (rutile) interface*(Solid state ionics 172.1,2004), 369-375.
10. Lefort, Pierre, Alexandre Maitre, and Pascal Tristant, *Influence of the grain size on the reactivity of TiO 2/C mixtures*(Journal of alloys and compounds 302.1 ,2000),287-298.
11. Lee, Gil-Geun, and Gook-Hyun Ha, *Carbothermic reduction of oxide powder prepared from titanium dioxide and cobalt nitrate*(Materials transactions 47.12 ,2006), 3007-3011.
12. Zhang, Haijun, et al, *Preparation of titanium carbide powders by sol–gel and microwave carbothermal reduction methods at low temperature*(Journal of Sol-Gel Science and Technology 46.2 ,2008), 217-222.
13. Bai, C., G. Qiu, and H. Pei, *Research on the foaming property of blast furnace slag bearing TiO2*(Teoksessa: VII International Conference on Molten Slags, Fluxes and Salts, the South African Institute of Mining and Metallurgy,2004).
14. Dickerson, Theodore, and Juan Soria, *Catalytic fast pyrolysis*(a review Energies 6.1 ,2013), 514-538.

7th International Symposium on High-Temperature Metallurgical Processing
Edited by: Jiann-Yang Hwang, Tao Jiang, P. Chris Pistorius, Gerardo R.F. Alvear F., Onuralp Yücel, Liyuan Cai, Baojun Zhao, Dean Gregurek, and Varadarajan Seshadri
TMS (The Minerals, Metals & Materials Society), 2016

STUDY ON PREPARATION OF ACTIVATED CARBON FROM HAWAII NUT SHELL VIA STEAM PHYSICAL ACTIVATION

Jianbo Lan[1,2,3,4], Shenghui Guo[1,2,3,4], Hongying Xia[1,2,3,4], Lihua Zhang[1,2,3,4], Libo Zhang[1,2,3,4], Jinhui Peng[1,2,3,4]

[1]State Key Laboratory of Complex Nonferrous Metal Resources Clean Utilization, Kunming University of Science and Technology, Kunming, Yunnan 650093, China;
[2]Yunnan Provincial Key Laboratory of Intensification Metallurgy, Kunming 650093, Yunnan, China;
[3]National Local Joint Laboratory of Engineering Application of Microwave Energy and Equipment Technology, Kunming, Yunnan 650093, China;
[4]Faculty of Metallurgical and Energy Engineering, Kunming University of Science and Technology, Kunming 650093, China

Keywords: Hawaii nut shell; Steam activation; Activated carbon

Abstract

Spent Hawaii nut shell was converted into a high-quality activated carbon via steam physical activation. Analysis of the experiment including activation temperature, activation duration, steam flow rate on the adsorption capability and yield of activated carbon were identified. Response surface methodology showed that the activation temperature was identified as 866℃, an activation duration of 36 min and a steam flow rate of 2.5mL/min with the methylene blue adsorption number of 195 mg/g and yield of 77.75%. The key parameters that characterize quality of the activated carbon such as the BET surface area, total pore volume and average pore diameter were estimated to be 1125m2 /g, 0.94mL/g and 5.21 nm, respectively. Additionally, by the SEM analyses of char and activated carbon, the surface of activated carbon has large number of pores of irregular and heterogeneous morphology.

Introduction

Hawaii nut also known macadamia nut which with high economic value, nutritional value and medicinal value. Since the nut processing leads to a lot of husk litter, a serious waste of agricultural and forestry resource unless processed [1]. Hard texture [2] is one excellent material characteristics of high quality activated carbon. Macadamia nutshell physical activation processing for preparing high quality activated carbon waste recycling will be achieved, but also to further increase its economic benefits [3].

Activated carbons (AC) are one of the most important carbon materials in the modern society, which possess high specific surface area and well developed pore structure. They have a very broad range of application in different fields [4-7], including medicine, chemicals, metallurgy, food, national defense and environmental protection as adsorbent and catalyst owing to their superior absorption and thermal stability.

Most of preparation of activated carbon materials were hard texture and high carbon content, such as bamboo, coconut shell, walnut shell sunflower seed cases, etc. [8-10]. Economic development and the importance of environmental protection, the activated carbon plays an important role. In this study, the Hawaii nut shells become high-quality activated carbon by steam physical activation [11].

Experimental Materials and Equipment

Hawaii nut shells, ceramic crucibles, tube resistance furnace, ultrasonic cleaning tank, oven, crushers, electronic scales. Laboratory equipment showed in Figure1.

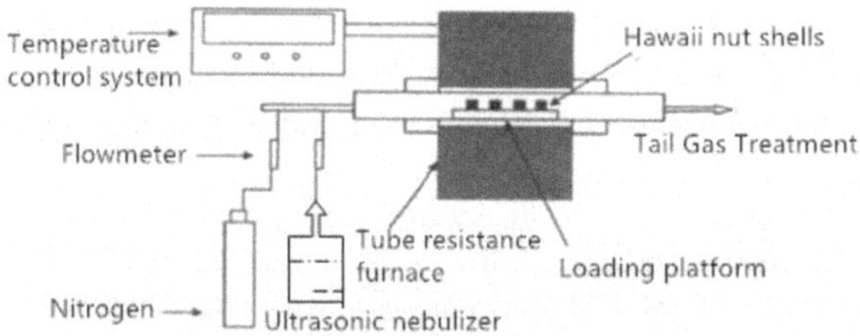

Figure1. Schematic of tube resistance furnace heating

Experiment

First, Hawaii nut shells crushed in the crusher to about 5mm, then ultrasonic cleaning for 10 minutes, and then put into the oven drying to constant weight, next into the ceramic crucible at 850 ~ 950 ℃ nitrogen blanket carbonation 30 ~ 40min. Hereafter weigh the right amount of carbonized material into tubular resistance furnace to explore the activated carbon adsorption performance effects, including activation temperature, activation time, steam flow. After activation is completed with distilled water ultrasonic clean the activated carbon, oven drying, detection the performance adsorption of activated carbon, and then calculate the yield.

Experimental Design

We used response surface methodology (RSM), which is a statistical technique for the modeling and analysis of problems in which an experimental response is dependent on several factors and central composite design (CCD) was used to design the activation experiments [12]. In this study, the effects of three independent variables were measured: X1 (Activation temperature), X2 (Activation time), and X3 (Steam Flow). The upper and lower limits of the process variables are shown in table I methylene blue value (Y1) and yield (Y2) of the activated carbon were chosen as the response values. Total of 20 sets of experiments consisted of eight factor points, six pivot points, and six center points. The responses to each experimental condition corresponded to the methylene blue adsorption value and yield, and they are listed in table I experimental error was determined by repeating the center points (15–20runs). As shown in table I, the methylene blue adsorption value varied from 66 to 193 mg/g, whereas the activated carbon yield ranged from 60.02 to 86.91 %. The experimental data were analyzed using the Design Expert software version 7.1.5(Stat-Ease Inc., Minneapolis, USA).

Table I Experimental Design Matrix and Results

Run	X1(℃)	X2(min)	X3(ml/min)	X3(ml/min)	Y2(%)
1	850.00	30.00	2.00	105	85.62
2	950.00	30.00	2.00	82	75.14
3	850.00	40.00	2.00	108	79.91
4	950.00	40.00	2.00	72	63.01
5	850.00	30.00	3.00	121	86.91
6	950.00	30.00	3.00	108	75.61
7	850.00	40.00	3.00	135	81.26
8	950.00	40.00	3.00	112	60.02
9	815.91	35.00	2.50	93	83.09
10	984.09	35.00	2.50	66	62.61
11	900.00	26.59	2.50	109	81.06
12	900.00	43.41	2.50	86	68.11
13	900.00	35.00	1.66	132	80.91
14	900.00	35.00	3.34	192	78.25
15	900.00	35.00	2.50	188	77.63
16	900.00	35.00	2.50	183	77.69
17	900.00	35.00	2.50	191	77.85
18	900.00	35.00	2.50	185	77.72
19	900.00	35.00	2.50	182	77.75
20	900.00	35.00	2.50	190	77.32

The relationship between the responses (Y1, Y2) and three independent variables (X1, X2, X3) were established using an appropriate regression model, utilizing the software. For iodine number and yield, the two-factor interaction model was selected as suggested by the software. Experimental runs performed at the center point of all the variables (15–20 runs) were utilized to determine the experimental error. The final empirical models in terms of coded factors (excluding the insignificant terms) for iodine number (Y1) and yield (Y2) are shown in Equations (1) and (2), respectively:

$$Y1=-13966.45548+27.62824X1+96.796X2+70.09547X3-0.0115X1X2+0.115X1X3+1.25X2X3-0.015399X1^2-1.28538X2^2-37.32104X3^2 \quad (1)$$

$$Y2=-656.8909+1.38974X1+9.75591X2+13.64222X3-8.18E-0.03X1X2-0.0258X1X3-0.17X2X3-6.53962E-004X1^2-0.040860X2^2+2.97804X3^2 \quad (2)$$

The quality of the model equation is always assessed using the correlation coefficient value R^2, which was 0.9887 for Eq. (1) and 0.9861 for Eq. (2). Both the R^2 values for the methylene blue adsorption value and the activated carbon yield are high, demonstrating that the predicted data and the experimental data are in close agreement using the model. Therefore, the credibility of the experimental data produced using this model is high.

The analysis of variance (ANOVA) is carried out to verify the accuracy of the model. Verifying the model accuracy is an import part of data analysis because it can help people decide if the model is applicable. The ANOVA results for the methylene blue adsorption value are shown in table II, it can be seen that the chosen model produced significant data [(Pb>F<0.0001]. Table II indicates that the X3 and along with the interaction effects of X2

2 and X32 on methylene blue adsorption value are significant, but the interaction effects of X2, X1X2, X1X3 and X2X3 are not, based on the low "Pb>F" values (B<0.05).

Table II Analysis of Variance For The Methylene Blue Adsorption Value

Source	Sum of squares	Degree of freedom	Mean	F-value	Pb>F
model	38089.95	9	4232.22	97.51	<0.0001
X1	1443.56	1	1443.56	33.26	0.0002
X2	56.11	1	56.11	1.29	0.2821
X3	3226.31	1	3226.31	74.33	<0.0001
X1X2	66.13	1	66.13	1.52	0.2453
X1X3	66.13	1	66.13	1.52	0.2453
X2X3	78.13	1	78.13	1.80	0.2094
X12	21359.38	1	21359.38	492.10	<0.0001
X22	14881.43	1	14881.43	342.85	<0.0001
X32	1254.56	1	1254.56	28.90	0.0003

Figure 2a shows the comparison of the predicted methylene blue adsorption value the experimental methylene blue adsorption value of the activated carbon. It can be seen that the predicted data match well with the experimental data, and the experimental data are evenly distributed on both sides of the model prediction.

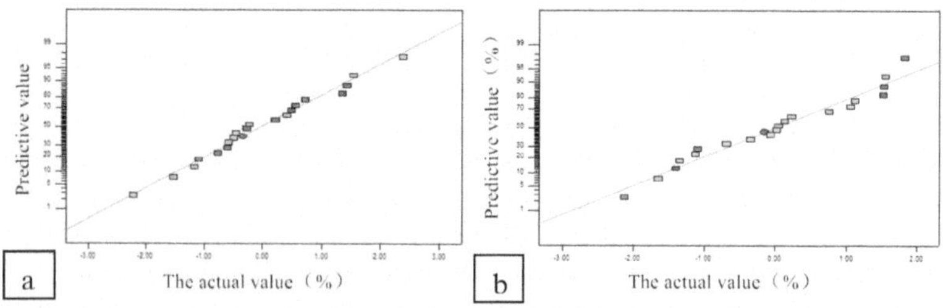

Figure 2 a. Predicted versus experimental methylene blue adsorption value of activated carbon b. Predicted versus experimental yield of activated carbon

Table III Analysis of Variance For The Yield

Source	Sum of squares	Degree of freedom	Mean	F-value	Pb>F
model	1026.32	9	114.04	78.77	<0.0001
X1	652.01	1	652.01	450.40	<0.0001
X2	271.21	1	271.21	187.35	<0.0001
X3	1.39	1	1.39	0.96	0.3506
X1X2	33.46	1	33.46	23.11	0.0007
X1X3	3.33	1	3.33	2.30	0.1604
X2X3	1.44	1	1.44	1.00	0.3413
X12	38.52	1	38.52	26.61	0.0004
X22	15.04	1	15.04	10.39	0.0091
X32	7.99	1	7.99	5.52	0.0407

The ANOVA for the response surface quadratic mode of the activated carbon yield is shown in table III. It can be seen that the chosen model produced significant data (Pb>F<0.0001). Table III indicates that the X1, X2, and along with interaction effects of X12 , X1X2 on yield are significant, but the interaction effects of X3, X1X3 and X2X3 are not. The significance of the parameter can be determined by "Pb>F" values; if the "Pb>F" value is greater than 0.05, the parameter is not significant. Figure 2b shows the comparison of the predicted yield versus the experimental yield of the activated carbon. It can be seen that the experimental data are distributed on both sides of the model prediction, indicating that the predicted data are close to the experimental data.

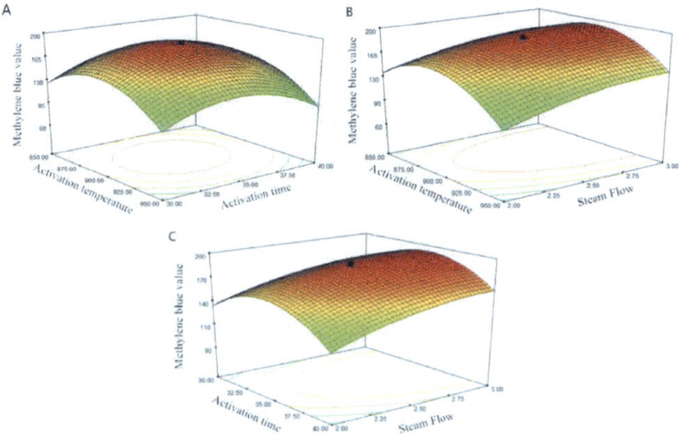

Figure 3. Three-dimensional response surface plot of methylene blue adsorption value: effect of activation duration and activation temperature (A), effect of steam flow rate and activation temperature (B), effect of steam flow rate and activation duration (C).

<u>Response Surface Analysis of Activated Carbon Methylene Blue Value</u>

Three-dimensional response surfaces were created to show the relationships between the independent variables on methylene blue adsorption value as shown in Fig. 3A–C. Figure 3A, with the activation temperature and activation time increased, activated carbon adsorption value of methylene blue gradually increased, activation reaction of water vapor and carbonization material to get through the internal channels of carbon. Later, activation temperature and activation time at 866 ℃ and 36min respectively, activated carbon adsorption value of methylene blue begins to decrease. Because of the formation of activated carbon channels have been destroyed when the increase of activation temperature and activation time. B and C, the increase of the steam flowrate, methylene blue adsorption changes little, almost unchanged after 2.5ml/min, instructions steam at 2.5ml / min is already saturated.

<u>Activated Carbon Yield from Response Surface Analysis</u>

Figure 4A, B, C, respectively activation temperature, activation time, steam flow corresponding to the response surface plot of activated carbon yield. From figure A, B, C can be found obviously with the increase of activation temperature, activation time, the yield of activated carbon has been declining, steam flow has little effect. This is mainly because of the activation of carbonized material is to open up more channels form more pores, which itself is the process of burning charcoal body, so the yield is has been declining.

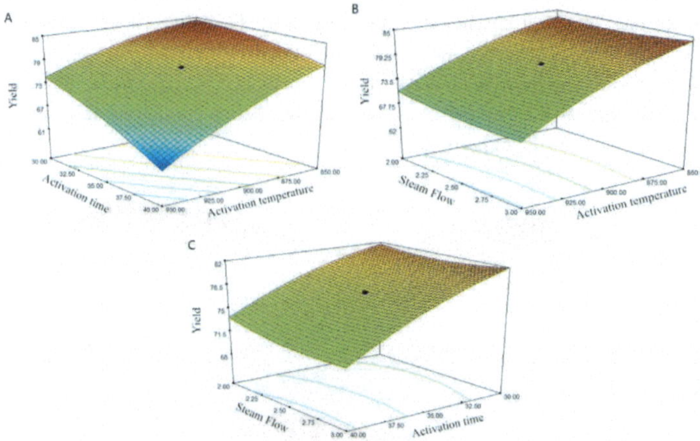

Figure 4 Three-dimensional response surface plot of yield: effect of activation duration and activation temperature (A), effect of steam flow rate and activation temperature (B), effect of steam flow rate and activation duration (C).

Pore Structure and Surface Area Analysis

Surface analysis one of the most important characteristics of activated carbon is its adsorption capacity. The adsorption capacity can be characterized by pore volume, pore size, and surface. It is well known that the adsorption depends on the pore structure. It is reported that the pore diameter should be at least 1.7 times of the molecular widest dimension in order to be good adsorption site to capture a molecule [13, 14].

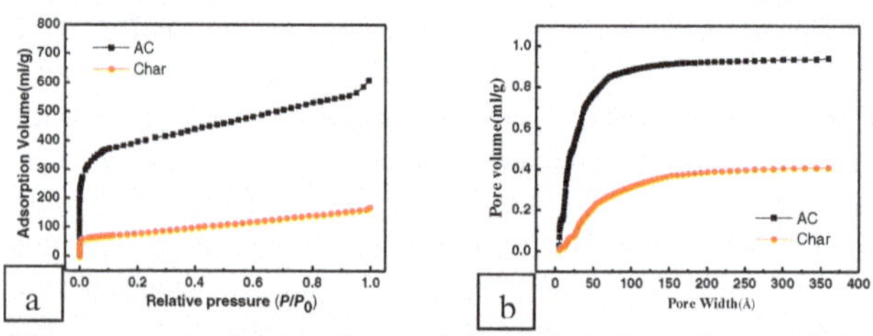

Figure 5 a. Nitrogen adsorption isvotherm of the activated car, b. cumulative pore volume distribution chart for the activated carbon and char.

Structure nitrogen adsorption is a standard procedure for determination of porosity of the carbonaceous adsorbents. Which were measured using the Autosorb-1-C instrument at 77K. It can be seen from Fig. 5a that the adsorption isotherms are type I isotherms according to IUPAC classification. Comparing the adsorption isotherms of activated carbon and char, we can see that the nitrogen adsorption isotherm of AC is exceedingly higher than that of char, clearly indicating the higher amount of pores in AC. The adsorption capacity continued to increase beyond of the relative pressure value of 0.1 that is typical for mesoporous materials. Microporous materials do not exhibit an increase in adsorption capacity when the relative pressure exceeds 0.1. Figure 5b shows the cumulative pore volume of the activated carbon and char, substantiate the amount of pores in the mesoporous range, with the average pore diameter estimated to be 5.21 nm. In general steam activation process is known to produce highly microporous material contradicting to the results of the present work. Details of pore structure of char vs activated carbon in table Ⅳ.

Table Ⅳ. Details of Pore Structure of Char Vs Activated Carbon

Project	char	activated carbon
The specific surface area (m2/g)	332	1125
Total pore volume (ml/g)	0.41	0.94
The average pore size（nm）	6.09	5.21

SEM Characterization of Activated Carbon Analysis

SEM analysis of micro structure the microscopic structure of the char (before activation) and the AC are shown in Fig. 6. Fig. 6a shows the SEM microstructure of char while Fig. 6b shows the SEM microstructure of AC. It can be seen that the surface of precursor was devoid of any tangible pores, while the surface of AC (Fig. 6b) has large number of pores of irregular and heterogeneous morphology, which at tests a significant development of pore structure. A comparison of the microstructure of AC with the char indicates that the activation process plays an important role contributing to pore-formation.

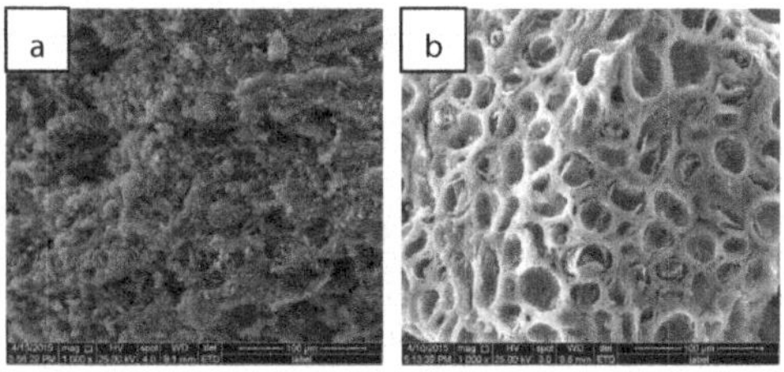

Figure 6 SEM images of char (a) and AC (b).

Conclusions

Hawaii nut shell being nullified incidental product of agriculture and forestry[25,26]，preparation of activated carbon by steam Physical activation. When the activation temperature at 886℃, activation time at 36 min and steam flow at 2.5ml/min, methylene blue value and yield were averaged to 195.00mg/g and 77.75%. The key parameters that characterize quality of the porous carbon such as the BET surface area and average pore diameter were estimated to be 1125 m2/g and 5.21 nm, respectively. The results indicate that using Hawaii nut shells for the preparation high specific surface area activated carbon is feasible by steam activation. High surface area mesoporous carbons are highly favored for liquid phase application. As a material of removed the contaminants which in the gas or liquid phase will bring huge economic benefits.

Reference

1. Baccar, R., et al, "Preparation of activated carbon from Tunisian olive-waste cakes and its application for adsorption of heavy metal ions," Journal of Hazardous Materials 162 (2) (2009): 1522-1529.

2. G. M. WALKER, and L. R. WEATHERLEY. "Textile wastewater treatment using granular activated carbon adsorption in fixed beds," Separation Science and Technology, 35 (9) (2000), 1329-1341.

3. Kaifang Fu. et al, "Preparation, characterization and application of lignin-based activated carbon from black liquor lignin by steam activation," Chemical Engineering Journal, 228 (2013), 1074-1082.

4. Wei S U, Li Z, and Yaping Z, "Preparation of microporous activated carbon from raw coconut shell by two-step procedure," Chinese Journal of Chemical Engineering, 14(2) (2006), 266-269.

5. Huang, Yuxiang, and Erni Ma,. "Thermal and structure analysis on reaction mechanisms during the preparation of activated carbon fibers by KOH activation from liquefied wood-based fibers," Industrial Crops and Products, 69 (2015): 447-455.

6. Yan-Juan Zhang, Zhen-Jiao Xing, Zheng-Kang Duan, Meng Li, and Yin Wang, "Effects of steam activation on the pore structure and surface chemistry of activated carbon derived from bamboo waste," Applied Surface Science, (315)(2014),279-286.

7. J.A. Muthanna, S.K. Theydan, "Microporous activated carbon from Siris seed pods by microwave-induced KOH activation for metronidazole adsorption", Anal. Appl. Pyrolysis, 99(2013), 101-109.

8. E. Kacen, et al, "Adsorption of strontium from aqueous solution using activated carbon produced from textile sewage sludges," Journal of Analytical and Applied Pyrolysis, 97 (2012), 149-157.

9. RAO Yi-fei, LUO Han-jin, WEI Chao-hai, and LUO Ling-feng, "Catalytic ozonation of phenol and oxalic acid with copper-loaded activated carbon," Journal of Central South University of Technology, 17 (2) (2010), 300−306.

10. X.H. Duan, C. Srinivasakannan, J.H. Peng, L.B. Zhang, and Z.Y. Zhang, "Comparison of activated carbon prepared from Jatropha hull by conventional heating and microwave heating," Biomass Bioenergy, 35 (2011), 3920–3926.

11. Bouchelta C, Medjram M S, Bertrand O, et al, "Preparation and characterization of activated carbon from date stones by physical activation with steam," Journal of Analytical and Applied Pyrolysis, 82 (1) (2008), 70-77.

12. Duan, Xinhui, et al, "Regeneration of spent catalyst from vinyl acetate synthesis as porous carbon: Process optimization using RSM," Chemical Engineering Research and Design, 92 (7) (2014): 1249-1256.

13. Körbahti, et al, "Response surface methodology (RSM) analysis of photoinduced decoloration of toludine blue," Chemical Engineering Journal, 136 (1) (2008): 25-30.

14. Liu, Qing-Song, et al, "Preparation and characterization of activated carbon from bamboo by microwave-induced phosphoric acid activation," Industrial Crops and Products, 31(2) (2010): 233-238.

15. Klijanienko, et al, "Development of mesoporosity during phosphoric acid activation of wood in steam atmosphere," Bioresource technology, 99 (15) (2008): 7208-7214.

7th International Symposium on High-Temperature Metallurgical Processing
Edited by: Jiann-Yang Hwang, Tao Jiang, P. Chris Pistorius, Gerardo R.F. Alvear F., Onuralp Yücel, Liyuan Cai, Baojun Zhao, Dean Gregurek, and Varadarajan Seshadri
TMS (The Minerals, Metals & Materials Society), 2016

COMPOSITION MODIFICATION OF ZnO CONTAINING FAYALITE SLAG FROM SECONDARY SOURCE COPPER SMELTING

Huayue Shi[1], Liugang Chen[1], Tom Peter Jones[1], Bart Blanpain[1], Muxing Guo[1]

[1]Department of Materials Engineering, KU Leuven
Kasteelpark Arenberg 44 bus 2450, 3001 Leuven, Belgium
Keywords: copper smelting, fayalite slag, zinc oxide

Abstract

Fayalite based slag is formed during the smelting process of secondary copper resources. It was found that with the presence of high ZnO, the slag became viscous, causing a slag tapping problem. In the present paper, the phase relations of the slag system ZnO-"FeO"-SiO_2-Al_2O_3-CaO (ZnO = 0-12 wt%, Al_2O_3% = 0-15 wt%, CaO = 0-1.5 wt%, with iron saturation) at 1150 °C is studied by using quenching technique followed by phase identification and compositional analysis of equilibrated samples. The dependence of FeO and Al_2O_3 solubility in the liquid slag on the chemical composition of slag is characterized. Based on the experimental data, the solid fraction of the slag, which may considerably, influences the slag viscosity, at the tapping temperature can be reasonably estimated and, therefore an appropriate modification of the slag composition can be made.

Introduction

A ZnO containing iron silicate-based (fayalite) slag is formed during secondary copper smelting process. Of particular importance is the slag viscosity which quantifies the flow properties of the slag and affects the reaction kinetics, the smelter's refractory corrosion and smooth operational practice. Experimental measurements and modelling of the slag viscosity suggested that ZnO behaves as a silicate network breaker, therefore decreasing the viscosity. Industrial copper smelting practice, however, identified that at elevated ZnO contents the slag becomes very viscous, which leads to a high copper loss and blocks the smelter. This latter effect is probably related with the precipitation of solid phases at the process temperature (1150°C for present case). Thus, more elaborated studies are necessary to understand the phase relations and solid phase precipitations of ZnO containing fayalite slags with respect to copper smelting conditions.

Literature provides experimental data for the ternary FeO-SiO_2-ZnO system [1]. It is also found in the system ZnO-FeO-Fe_2O_3-CaO- SiO_2 with CaO/ SiO_2 = 0.33, 0.71, 0.93 and 1.2 [2, 3] that increasing CaO/SiO_2 ratio significantly decreases the range of compositions at which the system is fully liquid. Although Al_2O_3 is often considered as raw material contamination and may increases the slag viscosity due to its amphoteric behavior [4], it deceases the liquidus temperature of fayalite slag [5]. On the other hand, Al_2O_3 also expands the fayalite primary phase field towards lower FeO concentrations [6].

However, this is not exactly the case for the present industrial copper smelter slag. The solubility of the major components and the extent of liquid region in the smelting temperature range (1100-1200°C) for the specific smelter slag compositions (see Table 1), i.e. high FeO and SiO_2 and minor CaO and Al_2O_3 contents, are not available. In addition, there are only limited studies on the solid precipitation behaviour of the slag during the copper smelting process, which is closely related to the slag viscosity. No information on the on-set of solid precipitation of the slag has been found.

Table1: Chemical composition of ZnO-containing fayalite slag, all Fe considered as FeO, measured with XRF

Compound	"FeO"	SiO_2	ZnO	Al_2O_3	Cr_2O_3	CaO	MgO
Wt%	53.3	28.8	8.8	5.0	1.2	1.2	0.9

The present work, therefore will experimentally investigate the behavior of the industrial ZnO-"FeO"-SiO_2-Al_2O_3-CaO (ZnO = 0-12 wt%, Al_2O_3 = 1-15 wt%, CaO = 0-1.5 wt%, with iron saturation) slag system in the secondary copper smelting process conditions. The solubility of the major components including FeO and Al_2O_3 in liquid slag will be measured. This is conducted by a quenching technique followed by EPMA micro-chemical analysis of equilibrated samples at 1150 °C. In accordance with the common industrial situation, the equilibrium test is carried out in a reducing atmosphere, ensured by iron saturation.

Experimental Procedure

To measure the FeO solubility, a ZnO containing industrial slag with composition listed in Table 1 was used as master slag. Additionally, an amorphous synthetic fayalite slag (70% FeO - 30% SiO_2) was employed as master slag for ZnO-free and low- ZnO samples. This synthetic slag was also used as master slag for the Al_2O_3 solubility measurement.

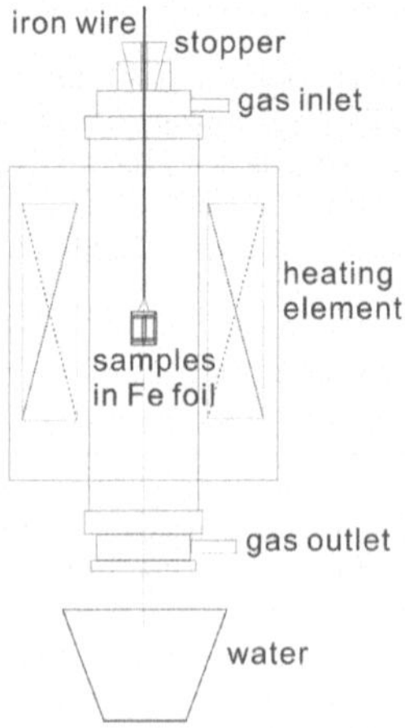

Figure 1. Quenching experiment set-up in the vertical tube furnace

The master slag was dried and ground into fine powders with particle size smaller than 80 m. Pure ZnO and SiO_2 were added into the master slag to modify the overall composition. The powder was then treated in a mixer with milling balls for 12 hours to ensure homogeneity.

3 gram of the well-mixed slag sample was then wrapped in a capsule made from 0.15 mm thick iron sheet to prevent volatilization of Zn at reducing atmosphere. The tube furnace (GERO HTRV 100-250/18, equipped with $MoSi_2$ heating elements) was used for the equilibration experiment and the furnace temperature at the constant temperature zone was controlled within +/-2 °C. The heating rate of the furnace was 5 °C / min, thus it took around three and half hours to reach 1150 °C. The capsules were suspended by an iron wire through a silicone stopper. In order to avoid excess zinc loss during the heating process, the capsule was placed at the top of the furnace tube before the desired temperature was reached, and then by inserting the wire, it moved into the constant temperature zone and rapidly reached the desired temperature. The capsule was placed in the hot zone of the furnace for 1-2 hours to ensure the equilibrium. Purified Ar was blown into the furnace with a flow rate of approximately 0.5 L / min during the entire process for atmosphere control. The experiment setup is shown in Figure 1.

To enable quenching, the bottom lid of the tube was removed. By pushing the iron wire down, samples then dropped into water underneath the furnace tube. The liquid phase was retained as glass and the crystalline solid phase was also preserved. The quenched sample was then mounted in resin, grounded and polished before further analysis.

The micro analysis of the quenched samples was carried out with an electro probe micro-analyzer coupled with wavelength dispersive spectroscopy (EPMA-WDS, JEOL JXA-8530F). The analysis was performed using an acceleration voltage of 15 kV and a probe current of 15 nA. Hematite (Fe_2O_3), willemite (Zn_2SiO_4), fluorite (CaF_2) and obsidian ((Na, K, Al, Fe) silicate glass) were used as standards, respectively for Fe, Zn, Ca, Al and Si. In order to increase the electrical conductivity, the polished sample was coated with a layer of carbon.

Results And Discussion

The industrial fayalite based slag is partially crystalized at the tapping temperature (1150 °C) as can be seen from Figure 2, which is captured from EPMA. Fayalite ($Zn_x Fe_{1-x}SiO_4$) and Spinel (Fe, Zn, Mg)O (Al, Fe, $Cr)_2O_3$ are the two major solid phases that can be found in the present slag system, the presence of metallic iron ensured the Fe saturation

As the most important component in the present slag system, FeO, suggested by the experiment data, has direct relation with the precipitation of solid phase. In the present slag system, the saturation of FeO in the liquid slag will lead to the precipitation of fayalite.

The solubility measurement of FeO is done by EPMA-WDS micro-chemical analysis of the amorphous phase which was in equilibrium with fayalite. The FeO solubility in the liquid slag at 1150 °C has been measured in the slag system ZnO-"FeO"-SiO_2-Al_2O_3-CaO (ZnO = 0-12 wt%, Al_2O_3 = 4.5 wt%, CaO = 1.5 wt%, iron saturation). As shown in Figure 3, the solubility of FeO is 56 wt% in ZnO free slag. As ZnO increases to 12 wt%, the FeO solubility is gradually decreased to 48 wt%, leading to solid fayalite precipitation. The latter may cause a considerable increase in the slag viscosity. This may partially explains the phenomenon observed at industrial practice that at elevated ZnO contents the slag becomes very viscous, resulting in a slag tapping problem. Therefore, in order to protect the slag from solid fayalite precipitation, it is quite essential to control the FeO composition of the slag.

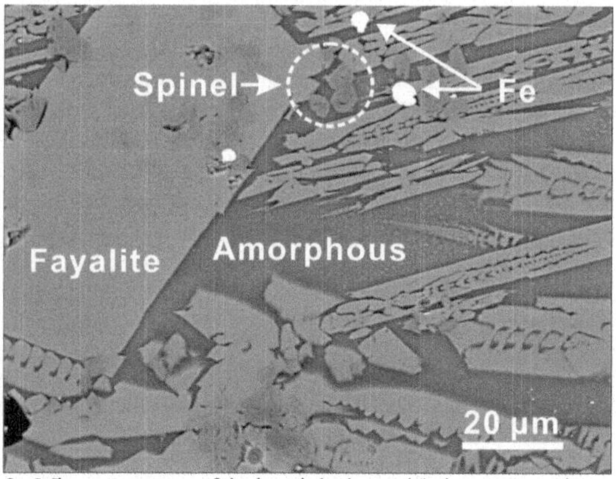

Figure 2. Microstructure of industrial slag with iron saturation, sample quenched from 1150 °C, analyzed by EPMA-WDS

The present slag system can be characterized by composition tetrahedron FeO-ZnO-SiO_2-(CaO+3Al_2O_3). A common approach is to fix the ratio of CaO/SiO_2 and Al_2O_3/SiO_2 (respectively 0.5 and 0.15 as in the present case) in the liquid slag as shown in figure.4. Thus the liquid composition lies on the pseudo-ternary plane FeO-ZnO-(SiO_2+0.05CaO+0.15Al_2O_3), while the fayalite composition lies on the tie line Fe_2SiO_4-Zn_2SiO_4 which is on the section ZnO-FeO-SiO_2 due to the low solubility of CaO and Al_2O_3 in fayalite.

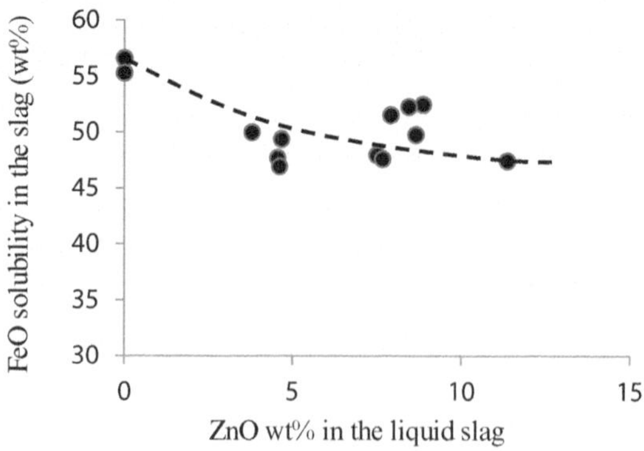

Figure3. FeO solubility in liquid slag as function of ZnO content in liquid slag at 1150°C

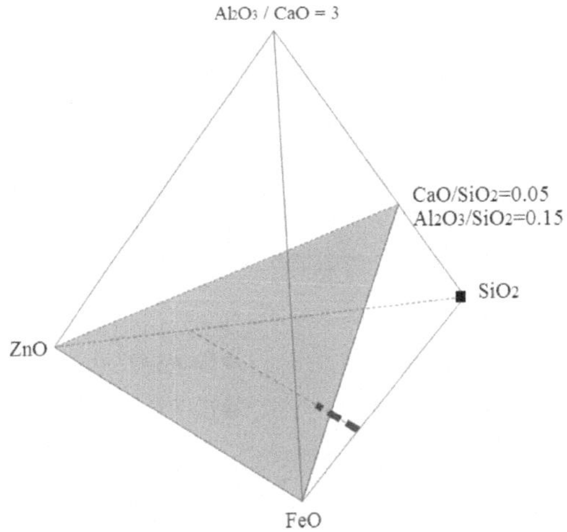

Figure 4. Composition tetrahedron $FeO-ZnO-SiO_2-(CaO+3Al_2O_3)$, with liquid composition on section $FeO-ZnO-(SiO_2+0.05CaO+0.15Al_2O_3)$

The liquid surface is projected on the $FeO-ZnO-SiO_2$ plain, as shown in Figure 5. For slag with high FeO/SiO_2 ratio, at elevating ZnO content, the liquid line draws closer to the fayalite composition, which leads to a large amount of fayalite precipitation. For slag with a lower ratio of FeO/SiO_2, increasing ZnO may suppress the precipitation of fayalite, as the composition of slag approaches the liquid surface.

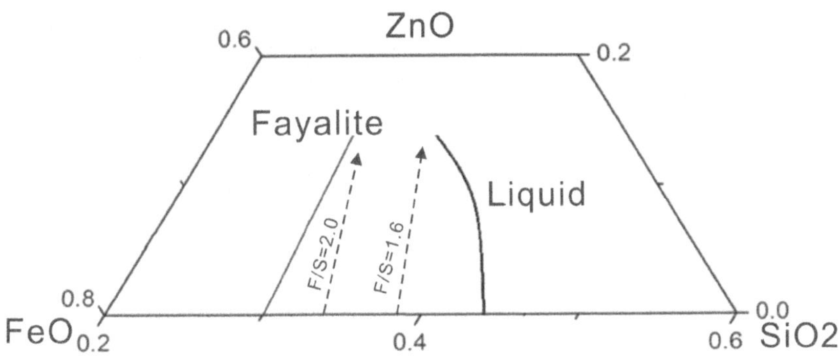

Figure 5. Pseudo-ternary phase diagram based on experimental data projected on the ZnO-FeO-SiO2 section, with CaO/SiO_2=0.05, Al_2O_3/SiO_2=0.15 at 1150°C under metallic iron saturation condition. The effect of increasing ZnO in slag with FeO/SiO_2=2.0 and 1.6 is indicated.

The presence of spinel primary phase in the fayalite slag is brought by Al_2O_3 [7]. In order to eliminate the spinel phase, the solubility of Al_2O_3 in the liquid slag is also important information. The Al_2O_3 solubility was measured in the quaternary system FeO-ZnO-SiO_2-Al_2O_3 (FeO/SiO_2=0.9, 1.4, ZnO%=0-12 wt%) at 1150 °C with iron saturation. The solubility of Al_2O_3, as similar as in the case of FeO, gradually decreases with elevating ZnO content in the liquid slag. At the FeO/SiO_2 ratio of 1.4, the solubility of Al_2O_3 in the liquid slag decreased from 13 wt% to 7 wt% as ZnO increased from 0 wt% to 9 wt%. Similarly, at the FeO/SiO_2 ratio of 0.9, the solubility of Al_2O_3 decreased from 16 wt% to 9 wt% as ZnO increased from 0 wt% to 12 wt%.

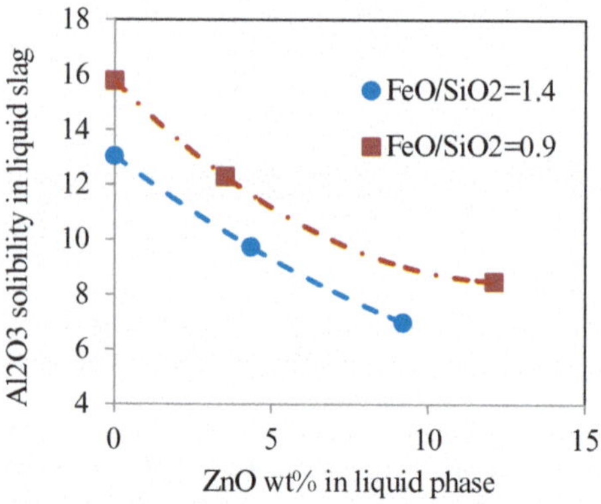

Figure 6. Al_2O_3 solubility in liquid slag as a function of ZnO content in the liquid slag at 1150 °C under metallic iron saturation

In fact, by comparing those two groups of data, FeO plays a similar role as ZnO in decreasing the Al_2O_3 solubility. Although Al_2O_3 can effectively lower the liquidus temperature of fayalite slag, the system suffers a risk of increasing spinel precipitation when excess Al_2O_3 was introduced, especially in a ZnO rich fayalite slag.

Conclusion

The phase relations of the slag system ZnO-"FeO"-SiO_2-Al_2O_3-CaO (ZnO=0-12 wt%, Al_2O_3%=0-15 wt%, CaO=0-1.5%, with iron saturation) at 1150 °C was studied. The solubility of FeO in the liquid slag is measured to be 56 wt% in ZnO free fayalite slag. As ZnO content increases from 0 to 12 wt%, the FeO solubility in the liquid slag gradually decreases to 48 wt%. A similar trend is also found in the Al_2O_3 solubility. At FeO/SiO_2 ratio of 1.4, the solubility of Al_2O_3 in the liquid slag decreased from 13 wt% to 7 wt% as ZnO increased from 0 wt% to 9 wt%, while at FeO/SiO_2 ratio of 0.9, the solubility of Al_2O_3 decreased from 16 wt% to 9 wt% as ZnO increased from 0 wt% to 12 wt%. Thus, in order to avoid tapping problems due to the solid precipitation, it is essential to control the FeO and Al_2O_3 contents in the slag when operating with ZnO-containing fayalite slag in the secondary copper smelting process.

References

1. E.Jak, B.Zhao and P. C. Hayes, "Experimental Study of Phase Equilibria in the Systems Fe-Zn-O and Fe-Zn-Si-O at Metallic Iron Saturation" *Metallurgical and Materials Transactions B*, 31 (6) (2000), 1195-1201.

2. E. Jak, B.Zhao and P.C. Hayes, "Experimental study of phase equilibria in the "FeO"-ZnO-(CaO + SiO_2) system with the CaO/SiO_2 weight ratio of 0.71 at metallic iron saturation", *Metallurgical and Materials Transactions B*, 33 (6) (2002), 865-876.

3. E. Jak, B.Zhao and P.C. Hayes, "Experimental study of phase equilibria in the "FeO"-ZnO-(CaO + SiO2) system with CaO/SiO2 weight ratios of 0.33, 0.93, and 1.2 in equilibrium with metallic iron", *Metallurgical and Materials Transactions B*, 33 (6) (2002), 877-890.

4. H. Park, S. Park, I.Sohn, "The Viscous Behavior of FeOt-Al_2O_3-SiO_2 Copper Smelting Slags", *Metallurgical and Materials Transactions B*, 42 (4) (2011), 692-699.

5. S. Chen, E. JaK, P.C. Hayes, "Effects of Al_2O_3 on Phase Equilibria in the Olivine Primary Phase Field of the MgO-"FeO"-SiO_2-Al2O_3 System in Equilibrium with Metallic Iron", *ISIJ International*, 45 (8) (2005), 1101-1105.

6. B.Zhao, and Peter C. Hayes, "The effect of Al_2O_3 on liquidus temperatures of fayalite slags" *Metallurgical and Materials Transactions B*, 30 (4) (1999), 598-605.

7. B. Zhao, E. Jak, and P.C. Hayes, "Experimental investigation of liquidus temperatures of ISP slags" (Paper presented at 7th International Conference on Molten Slags Fluxes and Salts, The South African Institute of Mining and Metallurgy, 2004), 243-248.

7th International Symposium on High-Temperature Metallurgical Processing
Edited by: Jiann-Yang Hwang, Tao Jiang, P. Chris Pistorius, Gerardo R.F. Alvear F., Onuralp Yücel, Liyuan Cai, Baojun Zhao, Dean Gregurek, and Varadarajan Seshadri
TMS (The Minerals, Metals & Materials Society), 2016

RESEARCH ON THE MELTING CHARACTERISTICS DURING THE RECYCLING PROCESS OF Q345B STEEL LF REFINING WASTE SLAG

De-qing Geng, Guang-liang Wu*

(School of Minerals Processing and Bioengineering, Central South University, Changsha, Hunan, 410083, China)

Keywords: Q345B; LF refining slag; melting Characteristics

Abstract

The waste slag generated by the LF refining process of Q345B low alloy hot-rolled steel, still has some refining capabilities. If recycled, it can reduce environmental pollution, and save the economic costs. A research on melting Characteristics of Q345B steel LF refining waste slag was investigated by using orthogonal experimental method, and the effect laws of composition on that were analyzed. Through variance analysis and significance criterion, it is concluded that the optimum ingredient of the regenerated LF refining slag is 25%Al_2O_3, 12%CaF_2, 7%MgO and R=ω (CaO)/ω (SiO_2) =2.5. And the regression analysis is carried out by using SPSS software, to get the mathematical relationship between ingredient with high reliability and melting temperature of refining slag: $T=1451.3-6.25CaF_2+2.36\ R$。

Introduction

As the user demand for low alloy Q345B cold-rolled steel increases gradually, the more and more corresponding waste of that is generated. General speaking, Ladle Furnace (LF) is adopted for Q345B refining. With the development of LF refining equipment, an increasing amount of waste slag occur, bringing about more and more serious environmental pollution problems. As a result, it is very urgent to take effective ways to recycle the waste slag to reuse. Large amounts of ingredient, such as CaO, SiO_2, Al_2O_3 and MgO, are exist in LF refining slag, which are able to take the place of the raw material s used in the metallurgical production process. At the same time, the LF refining slag has the characteristic of the clinker after high temperature melting, that can promote the rapid melting of making slag material. So the LF refining slag resource recycling in the metallurgical industry, not only can greatly improve the utilization rate of waste residue and solve the environmental problems, but also can save metallurgical resources, shorten time of smelting. Therefore, waste

Corresponding author: Guangliang Wu, Professor, Doctoral supervisor, E-mail: glwu_899@sina.com.

recycling is the most convenient and economic way to deal.

The low-alloy steel Q345B LF refining slag was taken as the research object in this research, to investigate the melting characteristics during the recycling process. The slag is pentatomic slag system, concluding CaO - A1203 - SiO2 - MgO2 - CaF2 five components.

Experiment

The experimental scheme

The melting Characteristics of Al_2O_3-CaF_2-MgO-CaO-SiO_2 pentatomic slag system, needed to be investigated, which was set as Y. Refer to the present research status of refining slag[1-3] , four factors were designated. Respective of that were dualistic basicity (R, R= CaO / SiO_2), set as for X1, $\omega(Al_2O_3)$, $\omega(CaF_2)$, $\omega(MgO)$, set as for X2, X3 and X4.

Taking the actual production situation into consideration, four factors were tested, and each influence factor has four levels, as shown in Table I. It was obtained the ratio of 16 kinds of slag composition by making the most of orthogonal experimental design method to arrange the experiment, as shown in Table II.

Table I Factor level list

factor	level			
X1	2.5	4.0	5.5	7.0
X2	15.0	25.0	35.0	45.0
X3	3.0	6.0	9.0	12.0
X4	3.0	5.0	7.0	9.0

Experimental facilities and raw material

Experimental facilities Ash melting furnace multifunctional properties were used as experiment facilities, as shown in Fig.1. Screen specimens were observed the height of the change after electrify heats up, taking the temperature at which the sample height down to 1/2 for melting temperature.

Raw material The Q345B refining waste slag was used as raw material, whose composition was shown in Table III. First, separately broken and screen the refining slag and analytical chemistry (CaO, Al2O3, SiO2, MgO and CaF2, etc.) required for experiment, to making it less than 0.074 mm (200 mesh). Through the composition fine-tuning, make it meet the requirements of composition of orthogonal table. Slag powder was put in evaporating dish, added a little water and anhydrous alcohol to mix, then put in the sample preparation to made it to 3 mm * 3 mm cylindrical specimens.

Table II Ingredient and melting temperature (T_m) of orthogonal experimental

	Ingredient /%					T/℃		
	Ca	SiO_2	Al_2O_3	CaF_2	MgO	T_1	T_2	T_3
1	56.43	22.57	15	3	3	1540	1540	1540
2	45.71	18.29	25	6	5	1361	1362	1368
3	35.00	14.00	35	9	7	1378	1370	1374
4	24.29	9.71	45	12	9	1394	1395	1400
5	57.60	14.40	15	6	7	1400	1395	1405
6	50.40	12.60	25	3	9	1363	1357	1357
7	40.00	10.00	35	12	3	1341	1336	1341
8	32.80	8.20	45	9	5	1363	1362	1360
9	56.69	10.31	15	9	9	1325	1325	1326
10	47.38	8.62	25	12	7	1283	1281	1280
11	48.23	8.77	35	3	5	1388	1384	1389
12	38.92	7.08	45	6	3	1404	1402	1407
13	59.50	8.50	15	12	5	1380	1383	1380
14	55.13	7.88	25	9	3	1321	1322	1326
15	43.75	6.25	35	6	9	1361	1363	1364
16	39.38	5.63	45	3	7	1400	1396	1395

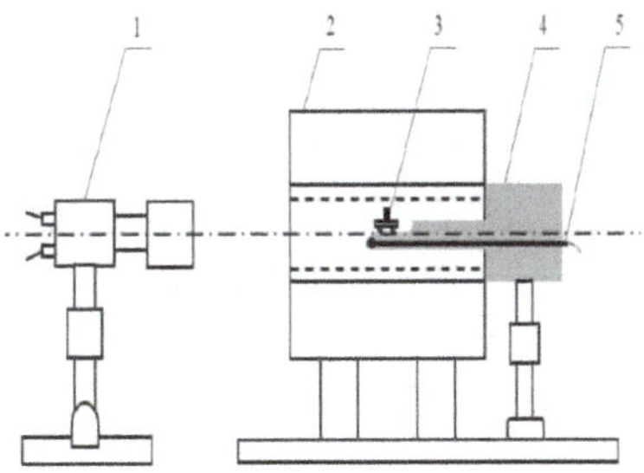

1——camcorder; 2——high-temperature furnace; 3——slag;
4——prototype sending device; 5——thermocouple

Fig.1 Melting point device

Table III The main composition of Q345B refining waste slag

CaO	Al_2O_3	SiO_2	MgO	MnO	TiO_2	Fe_2O_3	P_2O_5	S
48.841	25.602	6.088	7.869	0.157	0.445	1.247	0.0352	0.478

Results and Discussion

The test results of the melting temperature of each refining slag were listed in Table II. Through the analysis of variance, the results of four kinds of influence factor on the influence of melting temperature were listed in Table IV.

Table IV Results of variance analysis

factor	quadratic sum	degree of freedom	Significance test
R	32576.4631	3	*
Al_2O_3	41190.3651	3	* *
CaF_2	42583.5745	3	* *
MgO	13275.4335	3	-
error	6308.4926	35	-

We can know from the analysis of variance that the factor of CaF2 and A12O3 are the most significant contribution to the melting point.

For the two factors, a relational graph about CaF_2-melting point and $A1_2O_3$-melting point were drawn, as shown in Fig. 2. It is indicative of that the melting point of the refining slag reduce with the increase of CaF_2, but a turning point occurs with the content of $A1_2O_3$ changing: As the content of $A1_2O_3$ under 25%, the melting point of the slag system reduced with the increase of $A1_2O_3$, while that above 25,it increase with the increase of the content of $A1_2O_3$.

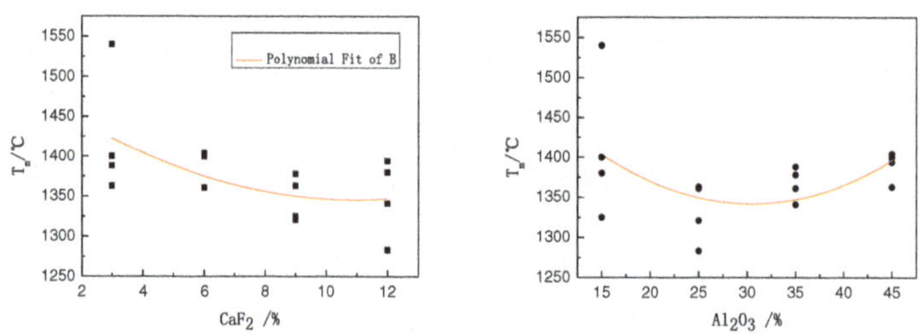

Fig.2 Effect of CaF2 and Al2O3 on melting temperature of LF refining waste slag

According to the above research, it is concluded that optimum proportion of CaF_2 and $A1_2O_3$ is 12% and 25%, respectively. The dispersion degree of melting point of the refining slag which with the low content of CaF_2 and $A1_2O_3$ is bigger, that indicates the performance of the slag is unstable to use. Making the use of SPSS software to analyze the experiment result, a correlation analysis was shown in Table V .

According to Table V, we find that R and CaF_2 are negatively relatived to melting temperature. It means that the melting temperature shows a trend of decrease occurring with the both content increase, and this was consistent with the experiment results.

Table V List of correction coefficient and significant coefficient

factor		R_2	Al_2O_3	CaF_2	MgO	T_m
R_2	r	1	-0.085	-0.085	-0.085	-0.247
	P		0.575	0.575	0.575	0.097
Al_2O_3	r	-0.085	1	-0.085	-0.085	0.156
	P	0.575		0.575	0.575	0.300
CaF_2	r	-0.085	-0.085	1	-0.085	-0.467
	P	0.575	0.575		0.575	0.001
MgO	r	-0.085	-0.085	-0.085	1	-0.132
	P	0.575	0.575	0.575		0.381
T_m	R	-0.247	0.156	-0.467	-0.132	1
	P	0.097	0.300	0.001	0.381	

The relationship between the melting temperature and the two significant factors: T_m=1451.3-6.25 CaF_2 -7.36R. According to the optimal proportion : CaF_2=12%, R_2=2.5, Al_2O_3=25%, estimating the expectations for the melting point of this slag system is T=1451.3-6.25CaF_2+2.36 R=1381.9℃.

Conclusion

The relationship between the melting temperature and the two significant factors: T_m=1451.3-6.25 CaF_2 -7.36R

The optimum ingredient of the regenerated LF refining slag is 25%Al_2O_3, 12%CaF_2, 7%MgO and R=ω (CaO)/ω (SiO_2) =2.5.

References

[1] J. Zhang, W. X. Yuan, "Calculation Model of Mass Action Concentrations for CaO-Al_2O_3-SiO_2 Melts", *Journal of University of Science and Technology Beijing*, 17(5) (1995), 418-423.

[2] Turkdogan E T. "Slag and Fluxes for Ferrous Ladle Metallurgy", *Iron and Steelmaking*, 12(2) (1985), 64.

[3] Hideaki Suito, Hajime Inoue, Ryo Inoue. "Aluminum-Oxygen Equilibrium Between CaO-Al_2O_3 Melts and Liquid Iron", *Iron & Steelmaking*, 31(12) (1991), 1389

7th International Symposium on High-Temperature Metallurgical Processing
Edited by: Jiann-Yang Hwang, Tao Jiang, P. Chris Pistorius, Gerardo R.F. Alvear F., Onuralp Yücel, Liyuan Cai, Baojun Zhao, Dean Gregurek, and Varadarajan Seshadri
TMS (The Minerals, Metals & Materials Society), 2016

SEPARATION OF HAZARDOUS IMPURITIES FROM BLAST FURNACE DUST BY WATER VAPOR ENHANCED MICROWAVE ROASTING

Aiyuan Ma [1,2,3], Chenyu Sun [3,4], Guojiang Li [4], Yongguang Luo [1,2,3,4],

Xuemei Zheng [1,2,3], Jinhui Peng [1,2,3], Libo Zhang * [1,2,3], Chao Liu [1,2,3]

[1]Yunnan Provincial Key Laboratory of Intensification Metallurgy, Kunming 650093, Yunnan, China;
[2]Key Laboratory of Unconventional Metallurgy, Ministry of Education, Kunming 650093, China;
[3]Faculty of Metallurgy and Energy Engineering, Kunming University of Science and Technology, Kunming 650093, China.
[4]Yunnan Chihong Zn & Ge Co., Ltd, Qujing 655011, China

Keywords: blast furnace dust, fluorides, chlorides, microwave roasting, water vapor

Abstract

In this paper, the removal of fluorides and chlorides from blast furnace dusts by water vapor enhanced microwave roasting was investigated, combined with the advantages of microwave selective heating and based on a thermodynamic analysis of zinc and lead halides. According to proposed reactions in the process, detailed experiments were systematically conducted to study the effect of roasting temperature, holding time, air flow, water vapor flow rates, and stirring speed on the efficiency of the removal of F and Cl. The results show that 97.26% of F and 64.23% of Cl could be separated when the condition of the roasting temperature of 650°C, stirring speed of 80 rpm, holding time at 60 min, air flow of 350 L/h and water vapor flow rate of 8 mL/min was optimized.

1. Introduction

The blast furnace dusts are a mixture of oxides expelled from the top of the blast furnace, and its major components are iron oxides [1]. The blast furnace dusts consist of zinc (Zn), lead (Pb), cadmium (Cd), chromium (Cr), nickel (Ni), other non-ferrous metals, and a small amount of rare indium (In) metals [2-7]. The direct recycling of the dusts is not usually possible since it contains some undesirable elements (fluorine and chlorine), which strongly affected the utilization of the dust in the zinc based electrical products [8]. F and Cl ion concentrations in electrolytes must meet electrolysis requirements (F < 80 mg/L, Cl < 100 mg/L) in the zinc electrolysis

Author: Aiyuan Ma, Tel: +86-15887855033; E-mail: may_kmust11@163.com

Foundation item: Project supported by the National Basic Research Program of China(2014CB643404), the Yunnan Province Young Academic Technology Leader Reserve Talents (2012HB008), the 2014 PhD Newcomer Award in Yunnan Province (1319880207).

Corresponding author: Libo Zhang; Tel: +86-13888310177; Fax: +86-5138997; E-mail: libozhang77@163.com process [9,10].

The microwave metallurgy is a highly efficient, clean, and green metallurgy technology that has developed rapidly [11]. The unique microwave heating characteristics include the low processing time, the selective and volumetric heating, and the controllable heating process. Currently, many pyrometallurgical processes for the removal of fluorides and chlorides from zinc oxide fumes [12] and CuCl residue [13] by microwave roasting were developed, and the results showed that the zinc dross is a good microwave absorbing mineral and can be heated quickly in the microwave field, confirming the feasibility of removing fluorine and chlorine from zinc oxide dust by microwave roasting [14].

The objective of the present work is to strengthen the dechlorination and defluorination process of the blast furnace dusts by microwave roasting under the atmosphere of water vapor. The influences of the roasting temperature, the holding time, the air flow, the water vapor flow rates, and the stirring speed on the dechlorination rates and defluorination rates were investigated.

2. Materials and methods

2.1. Materials

The blast furnace dust used in the experiments was received from a steel plant in Hunan province of China. The particle size distribution of blast furnace dust was analyzed, and the average particle size of D_{av} = 1.01 m, 99.87% of dust is below 3 m in particle size. The sample composition was characterized by X-ray diffraction and fluorescence (XRF) measurements. The main chemical composition of the blast furnace dust is listed in Table 1, which reveals that blast furnace dust has a high content of lead, zinc and chlorine, and the low content of fluorine. The XRD spectrum of the blast furnace dust samples is shown in Figure 1, which shows that the main mineral phases are ZnO, PbS, PbFCl, and KCl.

Table 1 The chemical compositions of blast furnace dust (mass fraction, %)

Composition	Zn	Pb	Fe	Cl	K	S	Bi	Sn	Si	Ca	F
Content (%)	67.22	15.94	3.96	3.08	2.93	2.46	1.32	0.77	0.72	0.61	0.43

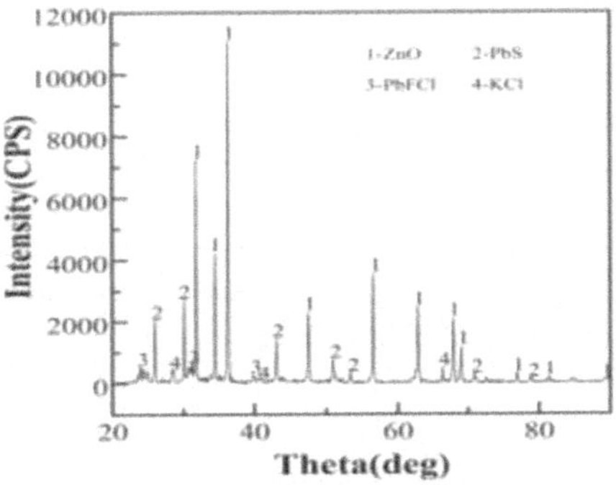

Figure 1. XRD pattern of blast furnace dust

2.2. Apparatus

A 3-kW box-type microwave reactor developed by the Key Laboratory of Unconventional Metallurgy in the Ministry of Education of Kunming University of Science and Technology was utilized for experimentation. The schematic of the experimental set up is shown in Figure. 2.

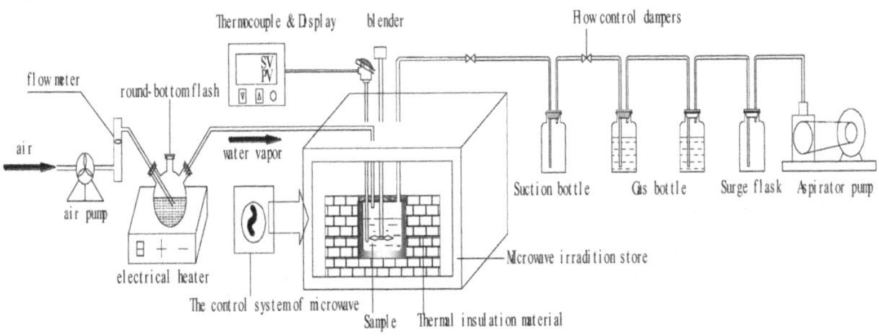

Figure 2. Connection diagram of microwave-roasting experiment equipment

The microwave heating frequency is 2450 MHz, while the power can be varied from 0 kW to 3 kW. A thermocouple was used to measure the temperature, in the range of 0 °C to 1300 °C. A mullite crucible with an inner diameter of 90 mm, height of 120 mm, having good wave-transparent and heat shock properties was used as the material support. Water vapor along with air was passed into the reactor after the material was heated to the set roasting temperature. The flow rate of steam and air into the reactor was controlled by adjusting the power to the electric heater. A dust absorption system was developed to collect the flue dust generated in the experimental process. This system comprises of a dust collection bottle, a water bottle, an alkali-absorption bottle, a buffer bottle, and a miniature pump.

2.3. Experimental Method

300 g of samples was dried, ground and placed into the mullite crucible, which was surrounded with heat preservation material. The microwave output power was 900W, and the average heating rate of the blast furnace dust were 100℃/min. Water vapor along with air was passed into the reactor after the material was heated to the set roasting temperature. The effect of the different roasting temperatures, the holding times, and the stirring speeds on the dechlorination rates and the defluorination rates were investigated under different air and steam flow rates.

2.4. Calculation of Removal Rate in Blast furnace dust

After the microwave roasting of blast furnace dusts, the samples were cooled to room temperature, and silver chloride turbidimetric method [15] was used for determination of the chlorine while fluorine was determined by fluoride ion selective electrode method. The % Cl (F) removal was expressed as,

$$\eta_{(Cl,F)} = \frac{C_{(Cl,F)} - C'_{(Cl,F)}}{C_{(Cl,F)}} \times 100\% \quad (1)$$

Where $C_{(Cl, F)}$ represents initial Cl (F) content in blast furnace dust; $C'_{(Cl, F)}$ denotes Cl (F) content in blast furnace dust after roasting; and $\eta_{(Cl, F)}$ represents % Cl(F) removal.

3. Results and discussion

3.1. Effect of Air Flow Rate

The roasting temperature and time were constant at 650 °C and 60 min, respectively. The samples were fully stirred at a speed of 60 rpm. The effects of different air flow rates (200 L/h to 400 L/h) on the F and Cl removal rate from the blast furnace dust are displayed in Fig. 3, which shows that the Cl and F removal rate increased with the rise of the air flow rate. Obviously, with air flow rate increasing from 200 L/h to 400 L/h, it was observed that the F removal rate increased significantly from 44.98 % to 82.65 %. The Cl removal rate increased obviously from 29.99% to 52.49% with the increase of air flow rate from 200 L/h to 300 L/h. While the air flow rate was more than 300 L/h, the Cl removal rate changed slowly.

However, the air flow rate should not be excessively high because the increases in air volume cause strong system heat dissipation, and the required temperature must be maintained. The optimum air flow velocity was 350 L/h, as observed in the following experiment. The purpose of this experiment is to reduce energy consumption.

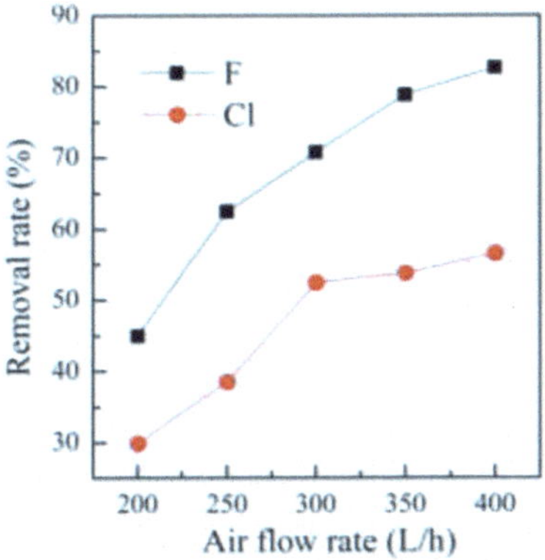

Figure 3. Effect of air flow rate on % Cl and % F removed

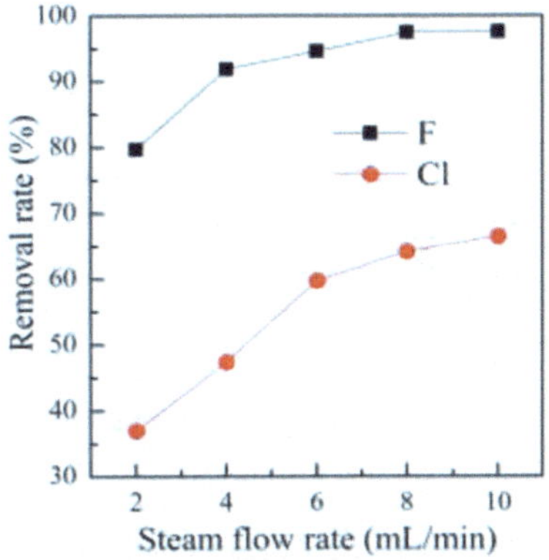

Figure 4. Effect of steam flow rate on % Cl and % F removed

3.2. Effect of Water Vapor Flow Rate

Keeping other roasting conditions as shown in section 3.1, and the water steam be carried by the 350 L/h air flow. F and Cl removal efficiency at different water vapor flow rates (2 mL/min to 10 mL/min) are plotted in Fig.4. Water vapor is observed to have a significant positive effect on the F and Cl removal rate. F and Cl removal rate increased from 79.69 % and 36.99 % at 2 mL/min to 97.59 % and 66.54 % at 10 mL/min respectively. When steam flow passes 6 mL/min, the removal rate increase slowly. Besides, controlling the steam flow rate will reduce energy consumption. Hence an optimal steam flow rate of 6 ml/min was utilized to assess the effect of other parameter.

3.3. Effect of Holding Time

The removal of F and Cl from blast furnace dust was examined with respect to time when an 6 ml/min water steam was carried by air flow of 350 L/h, temperature was 650 °C, and the samples were fully stirred at a speed of 80 rpm. The effects of holding time on the F and Cl removal rate from the blast furnace dust are shown in Fig. 5, which shows that the removal efficiency of F and Cl significantly increases from 79.51 % and 23.90 % to 97.34 % and 57.69 % when the holding time ranges from 20 min to 60 min. The reactions are carried out more thoroughly with time. The removal rate increases slowly when holding time more than 60 min. Therefore, the holding time should be set to 60 min.

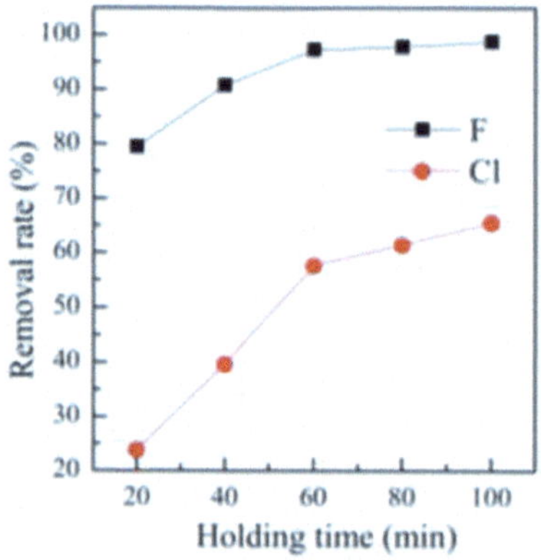

Figure 5. Effect of holding time on % Cl and % F removed

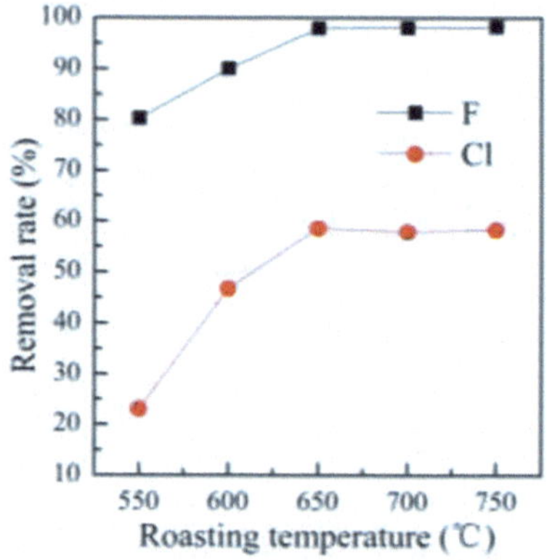

Figure 6. Effect of roasting temperature on % and % F removed

3.4. Effect of Roasting Temperature

300 g of Blast furnace dust was fully stirred quickly for 1 h at 60 rpm under different temperatures. Meantime, the water steam of 6 ml/min was blown into the microwave reactor which was carried by an air flow of 350 L/h. The results are shown in Fig. 6. Fig. 6 shows that the roasting temperature owns a significant effect on F and Cl removal efficiency, which increases rapidly from 80.26 % and 23.14 % at 550 °C to 97.94 % and 58.65 % at 650 °C. Further increase in temperature beyond 650 C was not found to have any effect on the Cl and F removal rate. Therefore, the roasting temperature of 650 °C may be optimal.

3.5. Effect of Stirring Speed

The removal efficiency of F and Cl was examined by changing the stirring speed. The conditions employed were as follows: the roasting temperature of 650°C, holding time at 60 min, air flow of 350 L/h and water vapor flow rate of 6 mL/min. The results are shown in Fig. 7.

The F removal efficiency was found to increase significantly until 80 rpm, while no significant improvement was observed at rpm in excess of 80. At the same time, the Cl removal rate increases significantly from 36.97 % to 59.69 % with the increase stirring speed from 20 rpm to 60 rpm, and Cl removal efficiency was decreased when the stirring speed exceeds 60 rpm.

Therefore, the appropriate mixing speed facilitates the microwave irradiating onto the surface of blast furnace dust more even, increasing the contact area between microwave and blast furnace dust, and enhanced heat transfer. As a result, the stirring speed should be set to 80 rpm in the follow-up study.

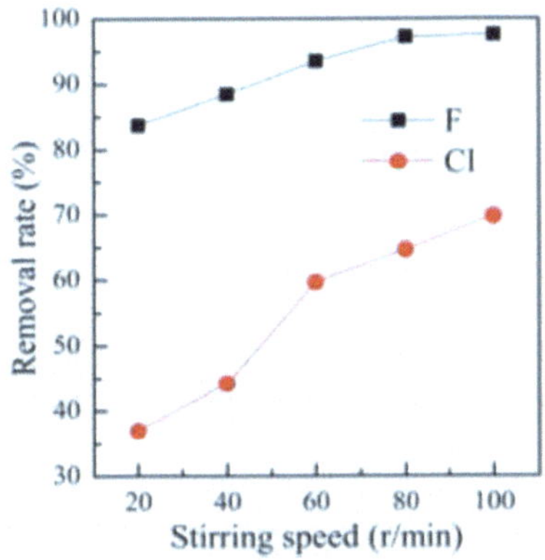

Figure 7. Effect of stirring speed on % Cl and % F removed

Based on the above research, for the water vapor enhanced separating the impurity from blast furnace dust by microwave roasting, the roasting temperature of 650°C, stirring speed of 80 rpm, holding time of 60 min, air flow of 350 L/h and water vapor flow rate of 8 mL/min might be optimal. 97.26% of F and 64.23% of Cl could be separated.

4. Conclusions

(1) The removal of F and Cl from the blast furnace dust by water vapor was enhanced microwave roasting. The results show that 97.26% of F and 64.23% of Cl could be separated under the conditions as follows: the roasting temperature of 650°C, stirring speed of 80 rpm, holding time at 60 min, air flow of 350 L/h and water vapor flow rate of 8 mL/min.

(2) The F removal efficiency was significantly from the blast furnace dust by water vapor enhanced microwave roasting, and the F could be completely removed. Nevertheless, the Cl removal efficiency was relatively poor.

Acknowledgements

The authors are grateful for financial supported by the National Basic Research Program of China (2014CB643404), the Yunnan Province Young Academic Technology Leader Reserve Talents (2012HB008), the 2014 PhD Newcomer Award in Yunnan Province (1319880207).

References

[1] B. Asadi Zeydabadi, D. Mowla, M.H. Shariat, J. Fathi Kalajahi, "Zinc recovery from blast furnace flue dust," Hydrometallurgy. 47(1997), 113–125.

[2] Y.C. Zhao, Q. Li, C.L. Zhang, and J.C. Jiang, "Production of ultrafine zinc

powder from wastes containing zinc by electrowinning in alkaline solution," Brazilian Journal of Chemical Engineering. 30(2013), 857-864.
[3] R.J.E. Martins, V.J.P. Vilar, and R.A.R. Boaventura, "Kinetic modelling of cadmium and leadremoval by aquatic mosses," Brazilian Journal of Chemical Engineering. 31(2014), 229 -242.
[4] A.D. Souza, P.S. Pina, and V.A. Leao, "Bioleaching and chemical leaching as an integrated process in the zinc industry," Minerals Engineering. 20(2007), 591-599.
[5] H. S. Altundoğan, M. Erdem, R. Orhan, A. Özer, and F. Tümen, "Heavy metal pollution potential of zinc leach residues discarded in Çinkur plant," Turkish Journal Engineering Environmental Sciences. 22(1998), 167-177.
[6] M.K. Jha, V. Kumar, and R.J. Singh, "Review of hydrometallurgical recovery of zinc from industrial wastes," Resources, conservation and Recycling. 33(2001), 1-22.
[7] M.Q. Xie, "Study on technique for extracting zinc and indium from reduction and volatilization zinc oxide dust," Mining and Metallurgical Engineering(in Chinese). 28(2008), 63-65.
[8] N. Güresin, and, Y.A. Topkaya, "Dechlorination of a zinc dross," Hydrometallurgy. 49(1998), 179-187.
[9] F. Cinar Sahin, B. Derin, and O. Yücel, "Chloride removal from zinc ash," Scandinavian Journal of Metallurgy. 29(2000), 224-230.
[10] Y.Z. Lan, Q.R. Zhao, and R.W. Smith, "Recovery of zinc from high fluorine bearing zinc oxide ore," Mineral Processing and Extractive Metallurgy. 115(2006), 117-119.
[11] K. Onol, and M.N. Saridede, "Investigation on microwave heating for direct leaching of chalcopyrite ores and concentrates," International Journal of Minerals, Metallurgy and Materials. 20(2013), 228-233.
[12] Z.Q. Li, L.B. Zhang, A.Y. Ma, J.H. Peng, J. Li, C.H. Liu, "Dechlorination of zinc oxide dust from waelz kiln by microwave roasting," High Temperature Materials and Processes. 34(2015), 291-297.
[13] Z.Y. Guo, T. Lei, W. Li, H.L. Luo, S.H. Ju, J.H. Peng, L.B. Zhang, "Clean utilization of CuCl residue by microwave roasting under the atmosphere of steam and oxygen," Chemical Engineering and Processing. 92(2015), 67-73.
[14] A.Y. Ma, L.B. Zhang, J.H. Peng, G. Chen, C.H. Liu, H.Y. Xia, and Y.G. Zuo, "Dielectric properties and temperature increase of zinc oxide dust derived from volatilization in rotary kilns," Journal of Microwave Power and Electromagnetic Energy. 48(2014), 25-34.
[15] M. Zenki, and Y. Iwadou, "Repetitive determination of chloride using the circulation of the reagent solution in closed flow-through system," Talanta, 58(2002), 1055-1061.

7th International Symposium on High-Temperature Metallurgical Processing
Edited by: Jiann-Yang Hwang, Tao Jiang, P. Chris Pistorius, Gerardo R.F. Alvear F., Onuralp Yücel, Liyuan Cai, Baojun Zhao, Dean Gregurek, and Varadarajan Seshadri
TMS (The Minerals, Metals & Materials Society), 2016

EXPERIMENTAL RESEARCH OF CONVERTER DUST FOR MOLTEN IRON DEPHOSPHORIZATION

Jian-hua Ren, Guang-liang Wu

(School of Minerals Processing and Bioengineering, Central South University, Changsha, Hunan, 410083, China)

Keywords: converter dust; dephosphorization agent; hot metal; phosphorus distribution ratio

Abstract

A lot of dust is produced in iron and steel enterprises, giving rise to serious environment pollution, with which all countries are trying to find methods to deal. The influence of relevant technological factors on dephosphorization was investigated in laboratory, and the effect laws of that were analyzed. The results demonstrate that, in the given conditions of the best slag-metal proportion for hot metal dephosphorization is 1:10, and the optimum temperature is 1350℃, The phosphorus distribution ratio of dephosphorization agent containing converter dust of 30% has the maximal value, with dephosphorization rate reaching 85%. Under the optimal conditions, premelting dephosphorization agent can significantly improve the ability of dephosphorization, dephosphorization efficiency increasing to 92%. Therefore, the dust can substitute the oxidant and fixative of dephosphorization agent, be used in hot metal dephosphorization, to enhance the efficiency of utilization.

Introduction

Metallurgical dust (hereinafter referred to as dust) is the iron-containing waste, produced by steel enterprises. The amount of dust produced in Chinese steel enterprises annually is about 100 million tons, which generate 8% ~ 15% of steel output [1]. After entering the 1980s, along with the importance of environmental protection and comprehensive utilization of resources is increasingly being recognized, the study of dust processing technology also becomes a hot topic in metallurgical industry. But for more than 20 years, processing techniques with good economic benefit and universal significance have not yet seen. So it becomes a tough problem in metallurgy industry.

Nowadays, there are mainly 3 ways to handle the dust [2-5]: distributing into sintering, selling and landfill. In order to achieve effective use of dust resources and environmental requirement of zero emission during steel production, all efforts have been making to explore new ways to handle dust, including direct reduction, smelting

Corresponding author: Guang-liang Wu, Professor, Doctoral supervisor, E-mail: glwu_899@sina.com reduction process, cold-bond pelletization and extracting iron powder etc[6-9]. But the utilization ratio is still less than 20 percent.

Components of dust are complex, about 50% of iron included [6]. It was found by optical microscopy and XRD study [1] that iron in dust mainly exists as Fe_2O_3 and FeO, as well as a small amount of Fe_3O_4 and elementary iron. Meanwhile, the dust contains some CaO, MgO, MnO and some more of K_2O, Na_2O. So the dust presents oxidizability and strong basicity. In view of this, the components of dust above can replace part of oxidant and fixation of dephosphorization agent.

In order to increase utilization of dust, converter dust is applied to experimental study of hot metal dephosphorization pretreatment in this paper.

Calculation of the formula model for dephosphorization agent

High basicity and oxidizablity are needed for dephosphorization, but the content of oxidant (Fe_2O_3) and fixative (CaO) cannot be increased at the same time when their total quantity is fixed. Therefore, it is important to determine the ratio of them. Research shows that [10,11], when CaO/Fe_2O_3 is 0.95~1.05, the content of flux is 7%~10%,and other flux is less than 10%, it is helpful to achieve good dephosphorization effect. Based on the above principle and on the basis of 100 g dephosphorization agent, the needed amount of converter dust used to substitute a certain quality score of dephosphorization agent(ωi, ωi=10%, 20%, 30%, 40%, 50%)is calculated.

The effective activity of iron oxide (FeO_n) is obtained through multiple regression by origin [12] as follow

$$\gamma(FeO_n) = 100\begin{bmatrix} -0.182\omega(CaO) - 0.037\omega(SiO_2) - 0.038\omega(MgO) \\ +0.258\omega(FeO_n) - 0.278\omega(MnO) + 0.282\omega(P_2O_5) \end{bmatrix} + 8.346 \quad (1)$$

The effective activity of iron oxide is as follow,

$$\alpha(FeO_n) = \chi(FeO_n) \times \gamma(FeO_n) \quad (2)$$

According to the equations of various substances react with calcium oxide and the activities of various substances after reaction, the effective activity of calcium oxide in dust is

$$\alpha(CaO) = \omega(CaO) - 2.716\omega(SiO_2) - 0.33\omega(Al_2O_3) - 1.18\omega(P_2O_5) \quad (3)$$

The needed amount of converter dust to substitute fixative is

$$\omega_{dust1} = \omega_i \times \omega_f \times \omega(CaO, in \cdot fixation) / \alpha(CaO) \quad (4)$$

Where, ω_f is the mass fraction of fixative in dephosphorization agent; ω(*CaO, in fixation*) is the mass fraction of calcium oxide in fixative we used. The needed amount of converter dust to substitute calcium oxide is

$$\omega_{dust2} = \omega_i \times \omega_o \times \omega(Fe_2O_3, in \cdot oxidant) \times 3 / \alpha(FeO) \quad (5)$$

Where, ω_o is the mass fraction of oxidant in dephosphorization agent; $\omega(Fe_2O_3$, *in oxidant*) is the mass fraction of iron oxide in fixative we used, among which Fe_2O_3 has been converted into FeO by whole oxygen method.

Taking $\omega dust = \min(\omega dust1, \omega dust2)$.

If $\omega dust1 \geq \omega dust2$, then the amount of fixative needed to be supplemented is

$$\omega_{f-plus} = \omega_i \times \omega_f \times \omega(CaO, in \cdot fixation) - \omega_{dust2} \times \alpha(CaO) / \omega(CaO, in \cdot fixation) \quad (6)$$

If $\omega dust2 \geq \omega dust1$, then the amount of oxidant needed to be supplemented is

$$\omega_{o-plus} = \omega_i \times \omega_o \times \omega(Fe_2O_3, in \cdot oxidant) - \omega_{dust1} \times \alpha(FeO) / [3 \times \omega(Fe_2O_3, in \cdot oxidant)] \quad (7)$$

Then the remaining content of fixation and oxidant in the original dephosphorization agent can be known.

$$\omega_{f-left} = (100 - \omega_i) \times \omega_f \quad (8)$$

$$\omega_{o-left} = (100 - \omega_i) \times \omega_0 \quad (9)$$

So, the formula composition of dephosphorization agent is

$$\omega_{dust} = \min(\omega_{dust1}, \omega_{dust2})$$

$$\omega_{fixation} = \omega_{f-plus} + \omega_{f-left}$$

$$\omega_{oxidant} = \omega_{o-plus} + \omega_{o-left}$$

$$\omega(CaF_2) = 10\%$$

Experiment

Experimental device and experimental material

Dephosphorization experimental device is a molybdenum silicide electric resistance furnace with nitrogen protection (maximum temperature is 1600 ℃). Experimental pig iron and converter dust are taken from a steel plant, the chemical composition of pig iron and dust are shown in Table I, and Table II, respectively. Additional chemicals required in dephosphorization agent are chemical pure reagents. According to the formula model of dephosphorization agent calculated above, the composition (mass fraction) of dephosphorization agent used in the experiment is shown in Table

III. Screening all the raw materials to below 0.15 mm after being grinded, then drying them and at last prepare them into dephosphorization agent according to a certain proportion.

Table I Chemical composition of experimental pig iron /%

C	Si	Mn	S	P
3.95	0.03	0.15	0.02	0.147

Table II Chemical composition of converter dust /%

TFe	FeO	CaO	SiO_2	MgO	Al_2O_3	MnO_2	S	P_2O_5
56.16	34.38	13.01	1.61	1.35	0.27	0.01	0.24	1.31

Table III Chemical composition of dephosphorization agent /%

Fe_2O_3	CaO	CaF_2	Dust
10.7~29.1	29.3~61.9	10	0~50

Experimental method

Hot metal dephosphorization experiments were conducted in molybdenum silicide furnace. In the experiment, corundum crucible which fitted with 300g pig iron was heated in the furnace, and the corundum crucible was covered by high pure graphite crucible as a protection. The preset temperature(1300~1400℃) would be preserved for 30min after the furnace temperature rose to it, then the initial molten iron was sampled by quartz tube of Φ6 mm and the prepared dephosphorization agent was quickly added to the liquid iron, timing was began at the same time. Then the liquid iron was sampled for every 5min, while a little slag was sampled by stainless-steel rod. The phosphorus contents of liquid iron and slag were analysed at last. During the experimental process nitrogen was inlet as protective gas, whose flow is 5 L/min.

Results and discussion

Influence of slag-metal ratio on dephosphorization effect

Influence of various slag-metal ratios(1:7.5, 1:10, 1:15, 1:20)on dephosphorization effect is researched when experimental temperature and dephosphorization agent components are fixed. The result is shown in Fig.1. It is observed that dephosphorization rate increases with the increase of slag quantity, it reaches the maximum when slag-metal ratio is 1:10, then slag quantity continue to be increased, the dephosphorization rate otherwise declines. It is mainly because effective components of dephosphorization agent increase with the increase of slag quantity, while the amount of liquid iron decreases, so the dephosphorization rate increases. However, if the slag quantity continues to increase, the dephosphorization rate is hard to increase furthermore, otherwise, splash is easily caused.

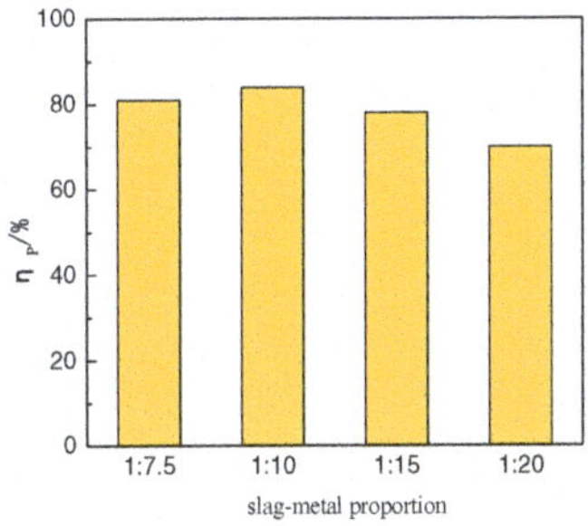

Fig.1 Dephosphorization efficiency of hot metal with different slag-metal proportion

Influence of experimental temperature on dephosphorization effect

In this experiment, slag-metal ratio (1:10) and dephosphorization agent components are fixed. The result is shown in Fig. 2.

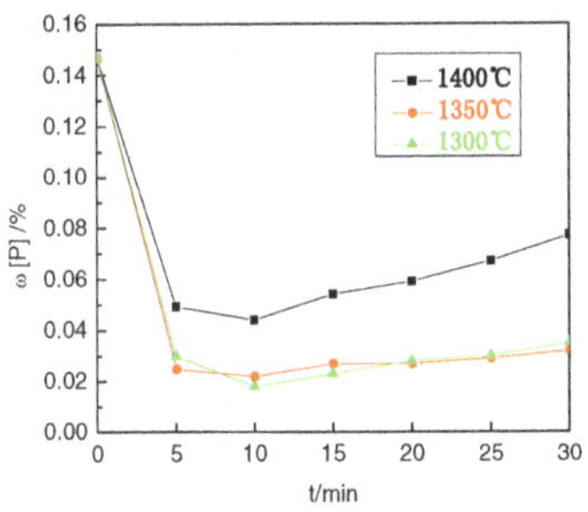

Fig.2 Variation of ω[P] in hot metal with time at different temperature

It is presented in Fig.2 that phosphorus content in hot metal reaches the minimum after 10 min of dephosphorization. When temperature is reduced from 1400℃ to 1350℃, dephosphorization rate increases from 70.06% to 70.06% while phosphorus content of hot metal reduces from 0.044% to 0.022%.When the temperature is further reduced to 1300℃, the change of dephosphorization rate is not obvious. The main reason is that, dephosphorization reaction releases heat, so low temperature is beneficial to dephosphorization. But if the temperature is too low, the fluidity of slag will be too bad which will worsen the dynamics condition of dephosphorization reaction and eventually lead to the decrease of the dephosphorization rate.

Influence of dust quantity in dephosphorization agent on dephosphorization effect

In this experiment, slag-metal ratio (1:10) and experimental temperature (1350℃) are fixed. Variation of ω[P] in hot metal with different dust proportion in dephosphorization agent at different time is shown in figure 3,while the relationship between distribution ratio of phosphorus and dust proportion in dephosphorization agent is shown in figure 4.

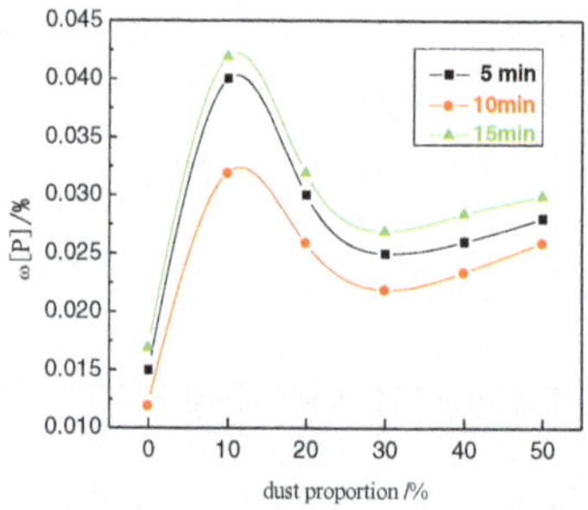

Fig.3 Variation of ω[P] in hot metal with different dust proportion in dephosphorization agent at different time

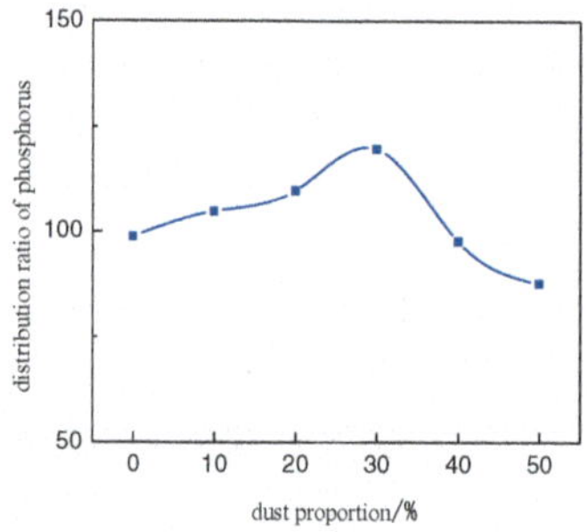

Fig.4 Relationship between distribution ratio of phosphorus and dust proportion in dephosphorization agent

It is demonstrated in figure 3 that when the dephosphorization agent is made from pure chemical reagents, its dephosphorization effect is the best. In addition, it can be observed from figure 4 that the dephosphorization rate increases firstly and then decreases with the increase of the amount of dust, the distribution ratio of phosphorus reaches the peak when dust proportion in dephosphorization agent is 30%.It is mainly because the increase of dust quantity will increase the phosphorus content of dephosphorization agent, but the change of dephosphorization rate is small, so the distribution ratio of phosphorus increases. When dust proportion in dephosphorization agent is over 30%, although phosphorus content still increases, the dephosphorization rate reduces rapidly what resulting in the decrease of distribution ratio of phosphorus. Therefore, it is feasible to substitute part of oxidant and fixative in dephosphorization agent with converter dust. The distribution ratio of phosphorus reaches the maximum when dephosphorization agent contains 30% of converter dust.

Influence of pre-melting treatment on dephosphorization effect

In this experiment ,slag-metal ratio (1:10), experimental temperature (1350℃) and dust proportion in dephosphorization agent (30%) are fixed. The dephosphorization agent has been pre-melted before dephosphorization experiment is carried out. The result is shown in figure 5. The dephosphorization rate of pre-melted dephosphorization agent (92%) is significantly greater than the dephosphorization agent without pre-melting (85%). This is mainly because each component of dephosphorization agent has been fully melted and they have reacted fully after pre-melting, besides, some low melting compounds are formed, like calcium-ferrite oxide [13], it not only accelerate the speed of melting slag, but also reduce heat loss during melting slag. Thus, dephosphorization reaction in the metal-slag interface turns from solid-liquid reaction to liquid -liquid reaction in a shorter time, the dynamics condition of dephosphorization reaction is great improved.

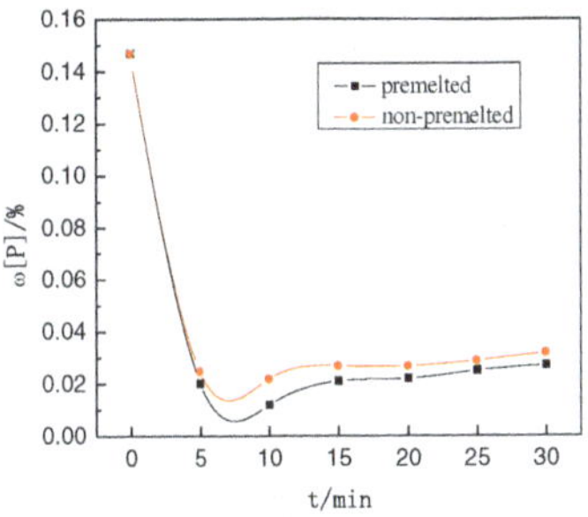

Fig.5 Variation of ω[P] in hot metal with time for premelted and non-premelted dephosphorization agent

It can be observed from the above studies that converter dust can completely be used to replace part of oxidants and fixation of dephosphorization agent for hot metal dephosphorization. Conventional dephosphorization agent is composed of sinter and lime, in this study converter dust is added into dephosphorization agent which can not only reduce cost by reducing use of sinter and lime, but also can reduce environmental pollution by improving the utilization rate of converter dust. It is presented in figure 2, figure 3 and figure 5 that the phosphorus content in hot metal decreases with time at the beginning of the reaction and rephosphorization phenomenon occurs significantly after it reaches the minimum. It is mainly due to the reduction of oxygen potential of slag. But the reaction conditions in industry is different from in the laboratory, the rephosphorization phenomenon is not significant, so in this study dephosphorization ability of dephosphorization agent is only characterized by the minimum of phosphorus content in hot metal, without considering the rephosphorization problem.

Dephosphorization effect will be influenced by P_2O_5 brought into by converter dust, and how to reduce the activity of P_2O_5 and increase the dephosphorization rate will be introduced in the next literature.

Conclusions

Based on the studies above, conclusions are listed as follows:

The optimum dephosphorization temperature for dephosphorization agent containing converter dust is 1350 ℃, the optimum slag-metal ratio is 1:10, phosphorus content in hot metal reduces to the minimum after 10 minutes of dephosphorization reaction and then rephosphorization phenomenon occurs.

With the increase of the amount of dust, dephosphorization rate increases firstly and then decreases, the distribution ratio of phosphorus reaches the peak when dust proportion in dephosphorization agent is 30%. Besides, dephosphorization rate will be improved by pre-melting the dephosphorization agent.

Converter dust can completely be used to replace part of oxidants and fixative of dephosphorization agent for hot metal dephosphorization, thus, comprehensive utilization of dust will be improved, environmental pollution will be reduced, and the enterprise's economic benefit will be raised.

References

1. X.F. She et al., "Study on Basic Properties of Typical Industrial Dust from Iron and Steel Plant and Analysis of Its Utilization," *The Chinese Journal of Process Engineering*, 9 (1) (2009), 7-12.

2. Das B et al., "Effective utilization of blast furnace flue dust of integrated steel plants," *European Journal of Mineral Processing and Environmental Protection*, 2(2) (2002), 61-68.

3. Ökvist L S et al., "The effect of BOF slag and BF flue dust on coal combustion Efficiency," *ISIJ International*, 44(9) (2004), 1501-1510.

4. L.F. Wang, "The New Development of Steel-making Dust Disposal Technics," *Energy for Metallurgical Industry*, 25 (4) (2006), 46-49.

5. H.M. Zhu, "Development and Practice of Utilizing Wet Steelmaking Dust from Converter for Sintering," *Energy Conservation Technology*, 24(2) (2006), 173-175.

6. J.C. Gao et al., "Research of Extracting Iron Powder from BOF Dust," *Powder Metallurgy Technology*, 20(1) (2002), 24-27.

7. Z.B. Shen and Y.Z. Sha, "Study on Cold Solidification Pelletizing Technology of Dust from Steel Industry," *Iron and Steel*, 38(12) (2003), 1-5.

8. Z.X. Zhang, "Resource Utilization of Iron-Bearing Dust and the Carbonation Pellet Process," *Sintering and Pelletizing*, 36(2) (2011), 44-49.

9. C.Q. Hu, S.M. Liu, and Y.Z. Zhang, "Experimental Research on Carbonation Solidification of Converter LT Dust," *Sintering and Pelletizing*, 38(4) (2013), 34-38.

10. K.X. Le et al., "Study on Technology and Influencing Factors on Hot Metal Pretreatment Dephosphorization by CaO-Fe_2O_3-CaF_2 Flux," *Iron and Steel*, 37(7) (2002), 20-22.

11. Kitamura S et al., "Effect of Stirring Energy, Temperature and Flux Composition on Hot Metal Dephosphorization Kinetics," *ISIJ international*, 31(11) (1991), 1322-1388.

12. C.X. Hu et al., "Thermodynamic Study of Utilizing Converter Slag for Hot Metal Pretreatment Dephosphorization," *Journal of Baotou University of Iron and Steel Technology*, 20(3) (2001), 223-226.

13. Y. Zhou, et al., "Experimental Study on Hot Metal Dephosphorization with Pre-melted Slag," *Steel Making*, 21(5) (2005), 33-36.

7th International Symposium on High-Temperature Metallurgical Processing

Poster Session

7th International Symposium on High-Temperature Metallurgical Processing
Edited by: Jiann-Yang Hwang, Tao Jiang, P. Chris Pistorius, Gerardo R.F. Alvear F., Onuralp Yücel, Liyuan Cai, Baojun Zhao, Dean Gregurek, and Varadarajan Seshadri
TMS (The Minerals, Metals & Materials Society), 2016

A REVIEW OF MICROWAVE TREATMENT ON COAL

Haibin Zuo[1], Siyang Long[1], Cong Wang[1], Pengcheng Zhang[1]

[1]State Key Laboratory of Advanced Metallurgy, University of Science and Technology Beijing; 30 Xueyuan Rd; Haidian District, Beijing, 100083, China

Keywords: Microwave, Coal, Thermal effect, Non-thermal effect, Potential application

Abstract

This paper introduced the research progress of microwave treatment on coal. The mechanics of microwave treatment mainly include thermal effects and non-thermal effects. Thermal effects, which lead to cracks and changes of pore structure, attribute to the selective heating of microwave. Non-thermal effects involve the removal of oxygen functional groups, improvement of coal rank and decomposition of minerals. Therefore, microwave treatment of coal has a lot of potential applications, mainly including drying, desulfurization, improvement of grindability, enhancement of coal-water slurry property and change the combustion and gasification properties.

Introduction

Microwave is a kind of electromagnetic wave whose frequency ranges from 300 GHz to 300 MHz and its wavelength ranges from 1 mm to 1 m, including decimeter wave, centimeter wave and millimeter wave. Thus, the theory of electromagnetic wave and electromagnetic field can be employed to solve some microwave problem, especially in the microwave propagation in space and its interaction with matters. At present, microwave has been widely used in many research fields, such as chemistry, biology, medicine et al. In these research fields, it can be used for heating and thawing samples, promoting the chemical synthesis, drying the raw material, sterilization and insecticide [1]. Microwave has also been widely applied in metallurgical industry. Many kinds of Ores may be decomposed (such as manganese dioxide, pyrolusite and basic nickel carbonate) or reduced (such as iron ore, ilmenite and pyrolusite) by Microwave irradiation. In some situation, microwave is used to pretreat the refractory ores, like refractory gold concentrate, in which gold can be separated more easily through leaching after microwave pretreatment [2].

Coal, as very important fuel and chemical materials in the word, is being consumed in a huge amount and its consumption is increasing constantly. However, coal is non-renewable energy and the unreasonable utilization may lead to serious environment problems. Therefore, more and more people are concentrating on how to utilize the coal efficiently and reasonably. A lot of methods are put forward, such as Integrated Gasification Combined Cycle, liquefaction, gasification and so on [3]. Pretreatment of coal by microwave is taken into account naturally by researchers due to the wide use of microwave in daily life and industries. At present, the research of microwave pretreatment on coal mainly involves dewatering [4], desulfurization [5], the improvement of grindability [6], enhancement of rheological behavior [7] and changes of combustion and gasification properties [8].

Mechanism of Microwave Pretreatment

Thermal effects

Microwave can penetrate into the object at the speed of light, absorbed by movement of electronic, ion and polarization of dipole. By this way, microwave energy can transform into heat, heating the whole object in the same time, greatly decreasing the heat loss, reducing the heating time, achieving the purpose of fast heating and energy saving [9].

The microwave absorption ability of material depends on many factors, including the frequency, the relative dielectric constant, the specific heat capacity, the water content and so on. The loss of microwave energy during penetrate into different materials in unite time and volume, which is regarded as absorption power, can be calculated according to the follow formula [2].

$$P=2\pi f \varepsilon_0 \varepsilon'' E^2 \tag{1}$$

Where P is absorption power density of material (W/cm^3), f is the frequency of microwave (1/s), ε_0 is vacuum dielectric constant (8.85×10^{-14} F/cm), ε'' is dielectric loss factor in complex permittivity ($\varepsilon=\varepsilon'-i\varepsilon''$), and E is electric field intensity (V/cm).

In the fixed electric intensity and frequency of microwave, the absorption power depends on the dielectric loss factor. The absorption power increases with a rise in dielectric loss factor value, leading to a higher heating speed. For different mineral components in coals, the dielectric properties are usually different. In the same microwave field, different mineral components are heated at different rates for their diverse dielectric properties, and the thermal expansion coefficients of different mineral components are not equal, which leads to the formation of thermal strain between different components of coal and then cause cracks and fissures [10].

In addition, the dielectric loss factor of water is much greater than that of coal [11]. Therefore water is heated rapidly and evaporated out of the coal, resulting in the change of pore structure, including pore surface area, average pore diameter and pore volume. On one hand, the gas evaporation breaks the closed pores, increasing the specific surface area and pore volume ratio. But on the other hand, most of inherent moisture in coals vaporizes rapidly, resulting in macropores broken and collapsed, leading the pore structure developed to the micropore region [8]. The final pore structure depends on these two reasons, contributing to different results by different researches [12-14].

Non-thermal effects

Microwave treatment removes the oxygen functional groups, including carboxyl, hydroxyl, and carbonyl and so on, which is observed by chemical titration [12] and infrared spectrum [15]. According to present researches, the more oxygen functional groups will be removed with microwave power improvement. The different parts of coal have an important influence on the removal of oxygen functional groups. The remained oxygen functional groups in the edge are much more than that in the center after microwave treatment. In addition, the particle size is also a vital influence factor. The bigger the coal particle is, the larger the removal of oxygen functional groups is [16]. The researchers also find that the aromatic/aliphatic ratio increases after microwave treatment, and the aromatic/aliphatic ratio is a parameter related with aromaticity and coal rank, indicating the improvement of coal rank [8].

Microwave treatment may make the minerals in coal converted and decomposed. The H_2S and SO_2 release from the coal during microwave irradiation, while the element sulfur appears in the surface. The reason is that extremely active molecules promote the old bond broken and the new bond formed, which means the Fe-S bond cleavage, and then free S^{2-} reacts with the surrounding O_2 and H_2 to generate H_2S and SO_2 [5]. Besides, the pyrite (FeS_2) can convert into pyrrhotite ($Fe_{1-x}S$) and troilite (FeS), attributing to the reaction $FeS_2 \rightarrow Fe_{1-x}S \rightarrow FeS$ proceeding continuously with the microwave irradiation time. Therefore, the sulfur can be separated from the coal by magnetic separation due to the $Fe_{1-x}S$ is in paramagnetism [17].

Potential Application of Microwave Treatment

Drying

Water in coals can be heated rapidly by microwave. Compared with conventional drying method, the water transfer condition is improved by microwave drying, and the reason is that the temperature gradient, pressure gradient and moisture transfer in coals are in the same direction, and the direction of heat transfer and mass transfer are consistent. The microwave drying process can be divided into three stages, which are preheating stage, quasi-steady drying stage and the reducing drying rate stage respectively [12]. In the preheating stage, the moisture and coals are mainly preheated and the removal of moisture is very limited. The quasi-steady drying stage is the mainly stage to remove the water in coals. In this stage, water can turn into vapor, resulting in a pressure gradient to promote the mixture of gas and water to spread out of coals. In the final stage, the moisture content of coals becomes very small, and the volatile of coals may volatilize easily, even worse, the coals may combust spontaneously. In addition, the microwave can improve the interfacial stability of coals and prevent the reabsorption of coals according to Yang's research [18].

Wu et al [19, 20] studied the dewatering of fine pulverized coals by microwave. The results show the whole drying process also can be divided into three stages and the microwave power and treatment time is the most important influence factor for the drying of coals while the effect of particle size is not obviously.

The energy consumption of microwave drying is much lower compared to other drying methods. Seehra [21] designed a microwave drying machine in the laboratory, in which the coals was fed on a conveyor belt and exposed to microwave with 2.45 GHz frequency and 800 W power. The efficiency of this machine for dewatering of fine coal slurries is estimated to be over 80% and a cost estimate of about \$3/ton for the power consumed for 10% reduction of moisture is made. Motevali [22] studied various drying systems including hot-air convection, use of microwave pretreatment with convection dryer, microwave drying, vacuum drying and infrared drying. The experimental results show that minimum consumption is microwave drying and microwave drying in hot dryer can save time and energy compared with conventional drying method.

Desulfurization

Zavitsanos [5] put forward the method of desulfurization by microwave treatment as early as 1978, by which 50% sulfur could be removed from coals and released by SO_2 and H_2S. It is the first time to study the desulfurization by microwave. Sui [23] indicated that the pyrite could be converted to pyrrhotite, separated easily by magnetic. Furthermore, the conversion rate of pyrite is promoted along with the pulverized coals particle size turning to small.

A lot of researchers are trying to remove the sulfur by adding some leaching reagent accompanied with microwave treatment. Wang [24] investigated desulfurization behavior of two kinds of coals by microwave treatment, in which the desulfurization rate could reach 20% on average. If the sample was extracted by 5% dilute hydrochloric acid, the desulfurization rate could reach 68%. Sheng [25] studied desulfurization coefficients by taking sodium hydroxide solution (NaOH) as the leachant under microwave irradiation. The results show that the sulfur can be removed efficiently and the heat value have no obvious difference. Jorjani [26] studied desulfurization by microwave treatment with peroxyacetic acid washing. The results show that the pyrite, organic sulfur and total sulfur content are reduced, and the removal rate of pyrite is even reached 91%. Zhao [27] studied desulfurization of coals by microwave combined with acetic acid and hydrogen peroxide, and the maximum desulfurization coefficient can reach up to 60.2%. The reason for this is the acetic acid and hydrogen peroxide react to get peroxide acetic acid. Furthermore the peroxide acetic acid decomposes to produce hydroxyl ions (OH^+), and the OH^+ can compose with sulfur atoms easily, resulting in the sulfur removal.

Improvement of grindability

The grindability characterizes the difficult degree and the energy consumption when crushing coals. Many researches show their interest in improving the grindability to save the crushing energy, and microwave treatment is regarded as a promising method.
Sahoo[6] studied grindability of the Indian coal with high ash under the condition of irradiation by microwave, and the results show that coals treated by microwave are ground much more rapidly than the untreated coals, with the specific rate of breakage increasing by an average of 15%. Lester [28] studied the property variety of England coal when treated by microwave in a short time, the results show that the grindability improved about 20%, and this change is related to the particle size. Ruisánchez [29] pretreated the metallurgical cokes by short pulses microwave. There is an obvious enhancement of the coke grindability compared to continuously microwave irradiation. Moreover, the energy consumption saved about 40% compared to the conventional grinding method. The mechanism for improving grindability is related to the micro cracks and fissures and the change of structure of pores, which are discussed in the second part.

Enhancement of coal-water slurry property

Coal-water slurry (CWS) is a mixture of coal, water and a few additives, and it is a promising method to utilize the coal. Wang [13] indicated that the slurry concentration increased continuously with a rise in the microwave treatment time and power, and the maximum slurry concentrate reached to 51.65wt% compared to the raw coal (46.67wt%). Cheng [30] treated the modified Indonesian lignite by microwave in a tunnel type microwave oven. The results show that the maximum solid concentration of coal–water slurry prepared from modified lignite increases from 41.6wt% to 54.0wt% with an enhanced rheological behavior. In addition, Cheng [31] indicated that the unit energy consumptions for CWS concentration promotion and inherent moisture removal by thermal heat were 214 and 22.5 times higher than those by microwave irradiation, respectively. Sahoo [7] investigated the rheological characteristics of coal-water slurries which consisted of two kinds of India coals pretreated by microwaves. The results indicate that all slurries exhibit pseudoplastic characteristic and the rheological properties of microwave-treated coal are better than untreated coal.

Change of the combustion and gasification properties

The combustion and gasification properties are very important for coal. Therefore, a lot of researches are devoted to improve the combustion and gasification properties by many different methods. With the microwave technology development, some researchers are trying to improve the combustion and gasification properties by microwave pretreatment.

Lester [32] studied the effect of microwave preheating on five different kinds of coals from around the world. The result shows that a few volatiles has emitted during treatment, according to the proximate analysis. In addition, the reactivity is affected by microwave treatment, but the severity of these changes appears to depend on coal type. Three coals have minimal changes in peak and burnout temperatures while others show more marked changes. Ge [8] studied the effect of microwave irradiation treatment on physicochemical characteristics of Chinese low rank coal. He uses the index S, which is a comprehensive combustion parameter related to burnout, ignition, peak temperature, the maximum and average combustion rates, to indicate the combustion property. The result shows that the index S decreases with the ignition and burnout temperate increasing after microwave treatment. Pang [33, 34] studied the combustion characteristics of microwave irradiated pulverized coals and the kinetics of coals combustion using the Flynn-Wall-Ozawa (FWO) model. It proves that the activation energy of irradiated pulverized coals is lower than that of untreated coals when the conversion is lower than 50%. However, the difference in activation energy between raw pulverized coals and irradiated pulverized coals decreases with the accretion of conversion. Moreover, when the conversion exceeds a particular value, the activation energy of irradiated pulverized coal is slightly higher than that of raw pulverized coal.

Ge [35] also studied the CO_2 gasification properties of three kinds of typical Chinese lignite before and after irradiated by microwave. The sample was placed into a microwave oven to irradiate for several seconds, and then transferred into a tube furnace to coke. The thermo-gravimetric analyzer was used to analyze the gasification properties. The results show that the treated coal has a lower temperature when the conversion reached 50%, and the gasification rate increases as well, indicating the increase of gasification activity. Cheng [14] studied the coal-water slurry gasification properties. The result shows that the activation energy and the start gasification temperature of treated coal are both lower than untreated coal.

Corresponding to the drying and desulfurization, the study of gasification property is far from enough. We studied a bituminous coal from Changzhibei, China, the conversion and conversion rate curves are showed in figure.1. The conversion curves do not show an obvious difference. The possible reason for this result is the competition of two effects discussed in part two. On one hand, the microwave treatment will change the pore structure, causing a positive effect of gasification, but on the other hand, the microwave treatment may improve the coal rank, causing a negative effect in gasification. Therefore, it shows different results to different coals when being treated in difference power, frequency and time.

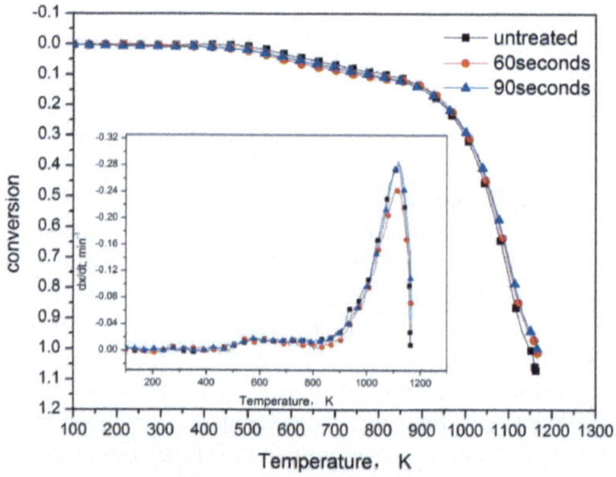

Fig.1. Experimental conversion and conversion rate curves of changzhibei coal

Conclusions and Prospect

Nowadays, microwave technology is widely used in our daily life and industries. It is also widely used in the coal treatment. Drying, desulfurization and improvement of grindability by using microwave treatment are studied by many researchers for a long time, while improvement of coal-water slurry property and change of the combustion and gasification properties are rarely studied compared to the former three applications. The mechanic of these applications are complex, it mainly includes the selective heating which can lead to the formation of micro fissures, the change of pore structure, the removal of oxygen functional groups and the improvement of the coal rank.

However, there are still many problems which are not known very clearly. For example, the change of coal structures during the microwave treatment is not very clear at present. Besides, the microwave desulfurization combined with other methods will be the hotspot.

In a word, microwave treatment is a processing method to improve the properties of coals, especially the low rank coals. The low rank coals can be irradiated by microwave to dry and desulfurize, and then the low rank coals can be used for combustion, gasification or preparing for water-slurry. The other applications of microwave pretreated coals also need be studied further.

References

1. Q.Y. Mou, and X.J. Li, "Applications of microwave heating technology." PHYSICS-BEIJING, 6 (33) (2004): 438-442.
2. Q.H. Jing, S.H. Dai, and K.M. Hua, "Microwave Chemistry." Beijing: Science Press, 2001, 282-318.
3. S.E. Hui, et al., "Clean utilization of coal and pollution control." Beijing: China Power Press, 2008.
4. Tahmasebi A, Yu J, Han Y, et al., "Study of Chemical Structure Changes of Chinese Lignite upon Drying in Superheated Steam, Microwave, and Hot Air." Energy Fuels, 6(26) (2012), 3651-3660.

5. J.H. Xiong, X.Q. Duan and Y.J. Sun, “Research progress of microwave technology in the coal desulfurization.” Coal Preparation Technology, 5(2013).
6. Sahoo B K, De S, Meikap B C. “Improvement of grinding characteristics of Indian coal by microwave pre-treatment.” Fuel Processing Technology, 10(92) (2011), 1920-1928.
7. Sahoo, B. K., et al. “Enhancement of Rheological Behavior of Indian High Ash Coal– Water Suspension by Using Microwave Pretreatment.” Industrial & Engineering Chemistry Research, 6(49) (2010), 3015-3021.
8. Ge, Lichao, et al., “Effects of microwave irradiation treatment on physicochemical characteristics of Chinese low-rank coals.” Energy Conversion & Management, 3(71) (2013), 84–91.
9. Ma, Shuangchen, et al. “Progress for Thermal and Non-thermal Effects of Microwave Chemistry.” Chemistry 1(74) (2011), 41-46.
10. A.W. Zhao, “Test and research on auxiliary coal desulfurization with microwave.” Coal Science and Technology, 30(2002), 45-46.
11. H. Xia, “Study on Microwave Pyrolysis of Low Rank Coal.” Master thesis, Henan Polytechnic University, 2012.
12. Z.Q. Xu, F.W. Xin, Y.N. Tu, “Moisture migration and interfacial modification of lignite in microwave dehydration process.” Journal of China Coal Society, 1(39) (2014), 147-153.
13. A.Y. Wang, “Promoting mechanism of microwave irradiation on slurry ability of lignite.” Master thesis, Zhejiang University, 2012.
14. X.G. Cheng, “The Effect on CWS’s Slurring Property and gasification by microwave and lightwave.” Master thesis, Zhejiang University, 2008.
15. Tahmasebi A, Yu J, Han Y, et al., “Study of Chemical Structure Changes of Chinese Lignite upon Drying in Superheated Steam, Microwave, and Hot Air.” Energy Fuels, 6(26) (2012),3651-3660
16. F.W. Xin, “Moisture migration and interfacial modification of lignite in microwave dehydration process.” Doctor thesis, China University of Mining and Technology, 2014.
17. S. Weng, J. Wang, “Mossbauer Spectroscopy Study of Enhanced Magnetic Separation of Pyrite from Coal by Microwave Irradiation.” Journal of Fuel Chemistry and Technology.” 4(20) (1992).
18. X. Yang, et al. “Characteristics of Lignite Dehydration in Microwave Field.” Coal Engineering, 2(47) (2015), 109-112.
19. H.Y. Wu, J.Q. Zeng and W.U. Wei, “Characteristic and Mechanism Study on Fine Coal Dewatering by Microwave Heating.” China Metallurgy, 9(20) (2010).
20. H.Y. Wu, J.Q. Zeng and W.U. Wei, “Orthogonal experiment and SEM research on coal fines dewatering by microwave heating.” Research on Iron & Steel, 1(38) (2010).
21. Seehra M S, Kalra A, Manivannan A, “Dewatering of fine coal slurries by selective heating with microwaves.” Fuel, 5(86) (2007), 829–834.
22. Motevali A, Minaei S, Khoshtagaza M H, “Evaluation of energy consumption in different drying methods.” Energy Conversion & Management, 2(52) (2011), 1192–1199.
23. J.C. Sui, et al., “Experimental study of the influence of desulfurization by microwave radiation.” Journal of Huazhong University of Science & Technology, 6(32) (2004), 93-95.
24. J. Wang, J.K. Yang and S.H. W, “Changes of Iron-Sulfur Compounds in the Process of Coal Microwave Desulfurization.” Journal of East China University of Science & Technology 1(16) (1990), 45-49.

25. Y.H. Sheng, X.X. Tao and N. Xu, "Experimental study on influence factors of coal desulfurization under microwave irradiation." China Coal (2012).
26. Jorjani E, Rezai B, Vossoughi M, et al. "Desulfurization of Tabas coal with microwave irradiation/peroxyacetic acid washing at 25, 55 and 85 C." Fuel, 7(83) (2004), 943-949.
27. J.L. Zhao, et al., "The Desulfurization of High Organic Sulfur Coal by Oxidation under Microwave Radiation." Journal of Microwares 2(18) (2002), 80-84.
28. Lester E, Kingman S, "Effect of microwave heating on the physical and petrographic characteristics of a UK coal." Energy & fuels, 1(18) (2004), 140-147.
29. Ruisánchez E, Arenillas A, Juárez-Pérez E J, et al., "Pulses of microwave radiation to improve coke grindability[J]." Fuel, 102(2012), 65-71.
30. Cheng J, Zhou F, Wang X, et al., "Physicochemical properties of Indonesian lignite continuously modified in a tunnel-type microwave oven for slurribility improvement." Fuel, 150(2015), 493-500.
31. Cheng J, Zhou J, Li Y, et al., "Improvement of coal water slurry property through coal physicochemical modifications by microwave irradiation and thermal heat." Energy & Fuels, 4(22) (2008), 2422-2428.
32. Lester E, Kingman S. "The effect of microwave pre-heating on five different coals." Fuel, 14(83) (2004), 1941-1947.
33. Q.H. Pang, J.L. Zhang, R. Mao, et al., "Kinetic Analysis on Combustion Characteristics of Microwave Irradiated Pulverized Coal." Journal of Iron and Steel Research, 11(25) (2013), 11-18.
34. Q.H. Pang, J.L. Zhang, R. Mao, et al.," Mechanism of Effect of Microwave Modification on Pulverized Coal Combustion Properties." Journal of Iron and Steel Research, International, 3(21) (2014), 312-320.
35. C.L. GE, et al., "Influence of microwave irradiation treatment on gasification characteristics of typical Chinese lignite." Journal of Zhejiang University, Engineering Science 4(48) (2014), 653-659.

7th International Symposium on High-Temperature Metallurgical Processing
Edited by: Jiann-Yang Hwang, Tao Jiang, P. Chris Pistorius, Gerardo R.F. Alvear F., Onuralp Yücel, Liyuan Cai, Baojun Zhao, Dean Gregurek, and Varadarajan Seshadri
TMS (The Minerals, Metals & Materials Society), 2016

Central Segregation of High-carbon Steel Billet and Its Heredity to the Hot-rolled Wire Rods

Yuan Ji[1], Yujun Li[1], Shaoxiang Li[1], Xiaofeng Zhang[2], Jiaquan Zhang[1]

[1]School of Metallurgical and Ecological Engineering, University of Science and Technology Beijing, Beijing 100083, P.R. China
[2]Beijing Metallurgical Technology Research Institute, CCTEC, Beijing 100028, P.R. China

Keywords: High-carbon steel, Heredity, Central segregation, Solidification structure, Mechanical property

Abstract

The present study describes central segregation of the High-carbon steel billets of 150mm square and its heredity to the following hot-rolled wire rods, whose sizes are 10mm and 6.5mm in diameter separately. The relationship between center segregation of the billets and solute element distribution of rods, the mechanical properties and microstructure of rods were studied experimentally. The results show that segregation index of C is higher than that of Mn. With the increase of central segregation, the stability of the mechanical property of the rods becomes worse. The formation of carbide network and central martensite in the rolled products can be attributed to the heredity of the as-cast central segregation. The 6.5mm-diameter-rod has higher tensile strength than that of 10mm diameter which might be due to bigger reduction ratio. Thermo Mechanical Control Process (TMCP) should be used together with Billet Center-segregation-controlling (BCSC) to improve the overall properties of rolled products effectively.

Introduction

High carbon steels are widely used to manufacture bearings, hard wire rods, spring steels and other high value-added products which are applied to different aspects of our daily life. Because of their wide solidification temperature range (T_L-T_S), solidification time and the mushy zone will be prolonged, and the center macro-segregation will be more severe due to the redistribution of solute elements and the flow of the residual melts. These will deteriorate the homogeneity of not only as-cast structure but also the subsequent hot rolling products, and further, can lead to serious quality problems during the downstream operation of the wire rods, such as breakage of drawing [1-2]. The severe center segregation of billets can be one important factor to induce breakage problems of the subsequent wire rod products during drawing and twisting operation.
In this paper, the 70 steel billets of 150mm square under different casting conditions (such as casting speed and specific water flow, etc.) were adopted for the experimental investigation. For each billet, it was rolled into two kind wire rods with 10mm and 6.5mm in diameter separately. The as-cast structure of billets and the quality of rolled wire rods (such as mechanical properties, element distribution and sorbite rate, etc.) were analyzed comparatively. The heredity of central segregation of high-carbon billets to their hot-rolled wire rods was also discussed.

Experimental Schemes and Plant Trials

The 70 steel production process can be briefly presented as follows:
170t BOF→180t LF→ CC→ Hot Rolling→ Drawing.
The fundamental information of the caster is shown in **Table Ⅰ**. The steel compositions are shown in **Table II**. In order to achieve different experimental effects of casting parameters on the quality of as-cast billet structure, the plant trials are designed as demonstrated in **Table Ⅲ**.

Table Ⅰ. Billet caster details

Casting section (mm ×mm)	150×150
Number of strands	12
Curved radius (m)	10
Casting speed (m/min)	1.6~2.2
EMS	M-EMS+F-EMS
Superheat (℃)	35
Mold length (mm)	800
Spray mode	1/0.34, 2/1.94, 3/2.39

Table II. Chemical compositions of the carbon steel 70 (wt %)

C	Si	Mn	P	S	Cr	Ni	Cu
0.70	0.27	0.65	0.02	0.02	0.1	0.1	0.1

Table Ⅲ. Operating conditions and experimental schemes

Casting Scheme	Speed (m/min)	Total Soft Reduction amount	Specific water flow (l/kg)	Wire rod number Φ10mm	Wire rod number Φ6.5mm
1	2.0	6	0.5	1-1	1-2
2	2.0	6	0.9	2-1	2-2
3	2.1	6	0.9	3-1	3-2
4	2.1	10	0.5	4-1	4-2
5	2.1	6	0.5	5-1	5-2
6	2.0	10	0.5	6-1	6-2
7	1.9	0	0.5	7-1	7-2

Laboratory experiments of billets

Figure 1 shows the experimental billet samples (15×150×150 in mm) taken along the casting direction for the following experiments. Drilling methods were used to analyze chemical elements of C and Mn with the apparatus of ϕ5 mm-alloyed-drill functioning in the cross section on both sides. In order to accurately analyze the center segregation, segregation indexes in both solidification center (As) and geometric center (Ag) of the billet were tested by the following formula. The Ci in both formula indicates the point from sample's center to surface respectively, and Cs and Cg refer to the contents of elements which located in solidification center and geometric center separately.

$$A_s = C_s / \sum_{i=1}^{n} \frac{C_i}{n} \qquad A_g = C_g / \sum_{i=1}^{n} \frac{C_i}{n}$$

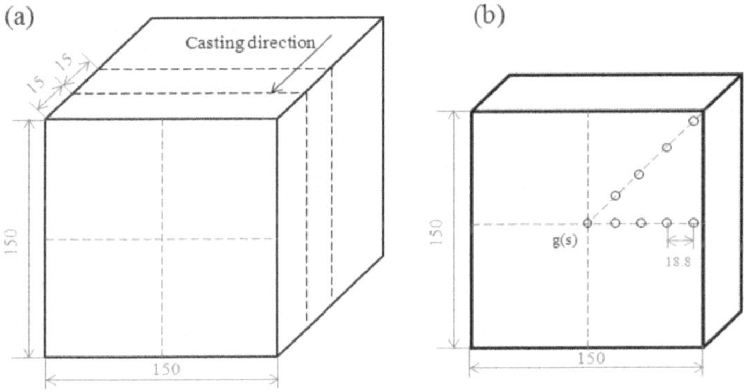

Figure 1. Schematic diagram of experimental methods

Laboratory experiments of wire rods

Following the procedural required apparatus of the first rough mill, steckel mill and down coiler, the temperature of billets after heating respectively demonstrated as 965~983℃, 904~927℃ and 839~852℃. Samples were taken from the wire rods of different diameters for the following experiments.

A. Mechanical properties testing. In order to achieve comparatively more accurate results, three cylinder samples were taken from the same wire rod to do the tensile test. The average tensile strength and elongation percentage of different samples were measured and compared separately.

B. Chemical composition analysis. The element distributions of C and Mn along the diameter of the wire rod were analyzed by the electron probe micro-analyzer (EPMA). The relationship between the element distributions of billet and its rolled wire rods was also discussed.

C. Microstructure observation experiments. After surface grinding and buffing, the wire rod slices of 20×ϕ10(in mm) or 20×ϕ6.5(in mm) were etched by 4% nitric acid alcohol and later, were observed by the optical microscope under the magnification of 500. Took the center of the wire rod as the center of a circle whose semi-diameter was one fourth the diameter of the wire rod; the sorbite rate was observed on four points on the circle, equally ranged by one fourth of the circle's perimeter. The point with the lowest sorbite rate in four observational points needed further experiment. Three more observational points were established near the aforementioned point to calculate their sorbite rates and their average value was regarded as the final sorbite rate of the whole wire rod.

Results and discussion

Center segregation of billets

The segregation indexes of C and Mn are shown in Figure 2. As can be seen, the result of scheme "5" and "6" should be the better option to choose. In addition, the results of As and Ag are different in most schemes, which should be noted that the two segregation index evaluations should both be used in order to get a more precise as-cast structure assessment.

The segregation index of C(1.02-1.26) is higher than that of Mn (0.98-1.06). In other words, compared to manganese, carbon is more likely to segregate in the billet center. In addition, carbon segregation and manganese segregation have the same variation trend from billet surface to center. This fact may be caused by the effect of Mn on the activity of carbon. Kirkaldy[3] proposed that manganese lowers the activity of carbon, and therefore Mn-rich regions were likely to attract C. Furthermore, the center segregation of carbon may attributed to the formation of martensite or secondary cementite in the following hot rolling process [4-6]. These hard brittle phases will decrease the toughness of steel drastically. As demonstrated by Strecken et al [7], in the view of heredity, the central segregation formed during the continuous casting process would remain to the following rolled products. It is extremely difficult to eliminate the chemical segregation by the following procedures which depend strongly on the time and temperature of reheating and hot rolling operations [8-9]. Center macro-segregation remains in the downstream products, having chances of causing quality issues. If observed on a microscopic scale, segregation which occurs between dendrites is problematic as well. For example, the micro-segregation of billets is a prerequisite for the ferrite/pearlite band formation after hot rolling, which will deteriorate the mechanical properties of steels. It is shown that both micro- and macro-segregation are closely related to the quality of the finished steel products, therefore, the performance of the rolled products is dominated by the solidification structure of billets, and this is the manifestation of the heredity.

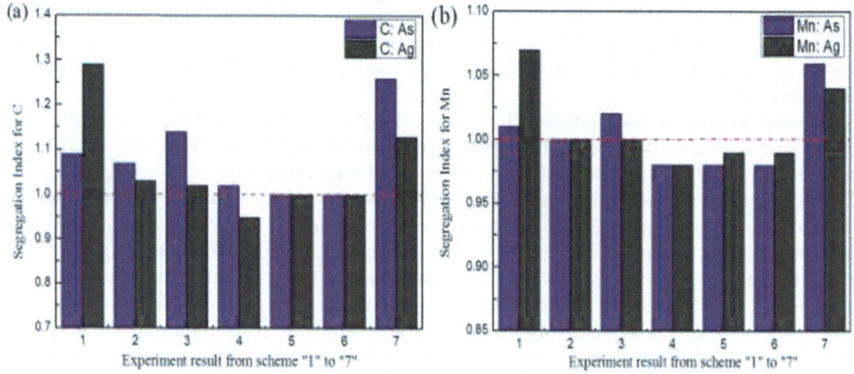

Figure 2. Segregation index of C and Mn in all experimental schemes

Mechanical property tests

Figure 3(a) and Figure 3(b) demonstrate the effect of average carbon content of billets on the mechanical property of the hot-rolled wire rods. The results show that with the increase of the average carbon content of billets, the average tensile strength and average reduction of area of the hot-rolled wire rods increase accordingly. In this article, the average range of elongation is chosen to represent the stability of mechanical property of wire rods. The relationship between the average range of wire rods and the centre carbon segregation index of billets is shown in Figure 4. The figure shows that the average range increases with the increment of central segregation, which means the stability of the mechanical property of the rods is downgraded.

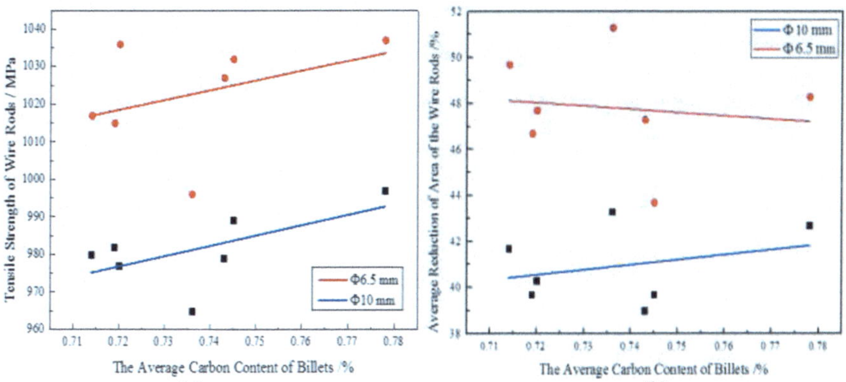

(a) (b)

Figure 3. (a) Tensile strength (b) Reduction of area of wire rods against its average carbon content of billets

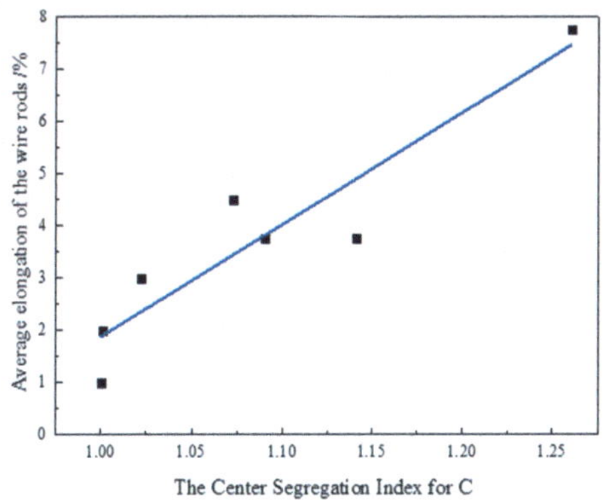

Figure 4. Average elongation of wire rods against their billet centre carbon segregation index

In conclusion, the central segregation of billets does manifest the heredity function of the mechanical property of its hot-rolled wire rods. During the continuous casting process, solidification produces nonuniform partitioning of chemical elements between parent liquid and growing solid crystals. This partitioning or segregation of chemical elements inherits in the downstream products after hot rolling which can affect the processing property of products indirectly [10-11]. The technology of Thermo Mechanical Control Process (TMCP) is widely used in improving the homogeneity of organization of wire rods in many practical productions. Considerable literatures demonstrate that it needs a very long period of time of heat preservation process for billets to make chemical elements distribute more evenly. But the austenite grain would grow fast if the temperature was too high, and the big austenite grain would also deteriorate the processing property of wire rods. The reason why the thin wire rods perform better strength and plasticity than the thick ones may be that as the rolling reduction increases, the homogeneity of chemical elements and organization of the thin wire rods is improved.

Chemical composition analysis of wire rods

The relationship between the element distributions of C and Mn of billets and those of its rolled wire rods were studied by the electron probe micro-analyzer (EPMA). Taking the No.7 billet sample (casting conditions: casting speed is 1.9m/min, specific water flow is 0.5L/kg, without SR) and its 6.5mm-diameter-rod as an example, Figure 5(a) shows the element distributions of C and Mn of billet from surface to center, while Figure 5(b) displays the element distributions of its wire rod along the diameter. As can be seen, for the corresponding positions of billet and its wire rod, the element distributions of C and Mn from surface to the center of billet and its wire rod varies according to the same changing trend. Though after hot rolling, there is still chemical segregation remaining in the center of the wire rod which is formed in the center of billet during solidification process.

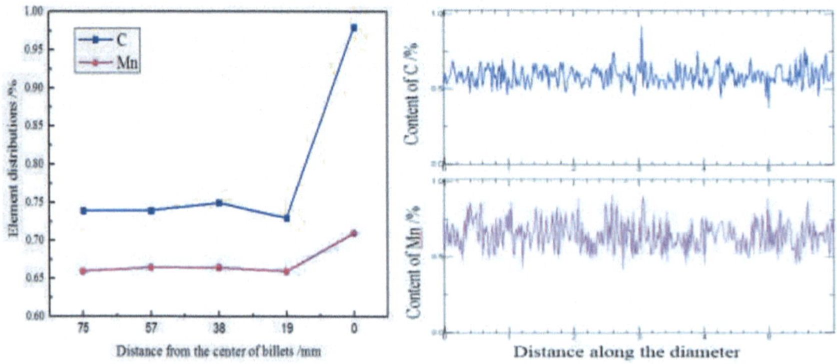

(a) (b)

Figure 5. Carbon and manganese distributions of (a) billets (b) wire rods

Microstructure observation experiments

Sorbite is a kind of pearlite whose interlamellar spacing is 0.1~0.3 m, which performs fine plasticity and high strength. The most economic and effective way to improve the strength of the wire rod is to increase its sorbite rate. After surface grinding and buffing, the wire rod slices were etched by 4% nitric acid alcohol. The light microphotograph of organizations in the wire rod samples is shown in Figure 6. Figure 7(a) indicates that sorbite rate of wire rods changes with the central segregation of billets in a consistent relation. It shows that the sorbite rates of both 10mm-diameter-rods and 6.5mm-diameter-rods decline with the increase of the central segregation of billets, and the 10mm-diameter-rods declines in a faster manner. Taking two kinds of wire rods into consideration, the comparison of sorbite rates is shown is Figure 7(b). As can be seen, for wire rods of different diameters rolled by the same billet, the sorbite rate of the 6.5mm-diameter-rod is higher than that of the 10mm-diameter-rod. The reason may be that the thin rod has a higher cooling rate in the inside of the wire rod than the thick one, which promotes the formation of the sorbite.

Figure 6. Schematic diagram of pearlite and sorbite in wire rods

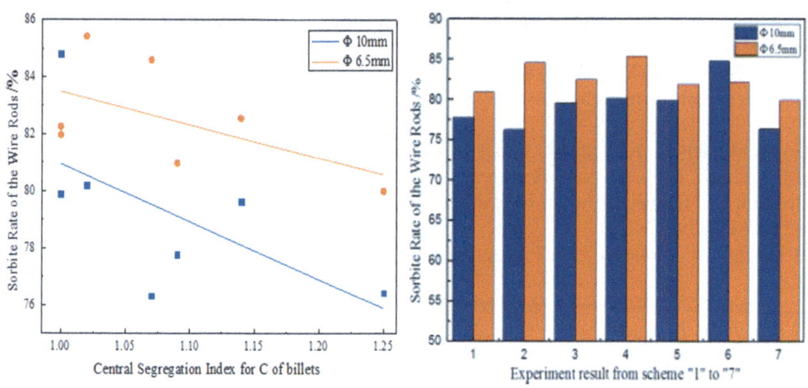

(a) (b)

Figure 7. The sorbite rate changes with (a) segregation index (b) billet number

Conclusions

(1) For the 70 steel billets studied, the center segregation index of carbon (1.02~1.26) is significantly higher than that of manganese (0.98~1.06). The segregation of carbon and manganese varies in an according trend, which shows the definite heredity function on the component homogeneity of the following hot rolled wire rods.
(2) The average carbon content of billets is closely related to the tensile strength and the reduction of area of wire rods. When the average carbon content of billets increases 0.06%, the tensile strength of wire rods increases by 33Mpa, and the reduction of area of wire rods increases by 4.2%.
(3) The average range of elongation of wire rods increases with the increment of the center segregation of billets. When the central segregation index of carbon increases 0.24, the average range of elongation increases by 6.65%. It is shown that the more carbon segregates into the center of billets, the worse the stability of the mechanical property of the rods becomes.
(4) Under the same production conditions, the sorbite rate of the thin rods is obviously higher than that of thick ones.

References

1. Yamashita T, Torizuka S, Nagai K. "Effect of manganese segregation on fine-grained ferrite structure in low-carbon steel slabs," *ISIJ international, 2003, Vol. 43, pp. 1833-1841.*
2. Majka T F, Matlock D K, Krauss G. "Development of microstructural banding in low-alloy steel with simulated Mn segregation" *Metallurgical and Materials Transactions A, 2002, Vol. 33, pp. 1627-1637.*
3. Kirkaldy J S, von Destinon-Forstmann J, Brigham R J. Simulation of banding in steels[J]. *Canadian Metallurgical Quarterly, 1962, 1(1): 59-81.*
4. Carsi M, Ruano O A, Peñalba F, et al. High strain rate torsional behavior of an ultrahigh carbon steel (1.8%C-1.6%Al) at elevated temperature [J]. *Metallurgical and Materials Transactions A, 1997, 28(9): 1913-1920.*
5. Han K, Smith G D W, Edmonds D V. Developments in ultra-high-carbon steels for wire rod production achieved through microalloying additions [J]. *Materials & Design, 1993, 14(1): 79-82.*
6. Grossterlinden R, Kawalla R, Lotter U, et al. Formation of Pearlitic Banded Structures in Ferritic-pearlitic steels [J]. *Steel Research, 1992, 63(8): 331-336.*
7. Strecken K, Pfennig M, Anelli E, et al. Optimising the structure of high-carbon steel wire [J]. *EUR, 1998(18434): 1-169.*
8. Flemings M C. Solidification Processing [J]. *Metallurgical Transactions, 1974, 5(10): 2121-2134.*
9. Carpenter H C H, Robertson J M. The Formation of Ferrite from Austenite [J]. *Jour. Iron and Steel Inst, 1931, 123: 345-394.*
10. Won Y M, Thomas B G. "Simple model of microsegregation during solidification of steels," *Metallurgical and Materials Transactions A, 2001, Vol. 32, pp. 1755-1767.*

11. Ganguly S, Choudhary S K. "Quantification of the Solidification Microstructure in Continuously-Cast High-Carbon Steel Billets," *Metallurgical and Materials Transactions B, 2009, Vol. 40, pp. 397-404.*

7th International Symposium on High-Temperature Metallurgical Processing
Edited by: Jiann-Yang Hwang, Tao Jiang, P. Chris Pistorius, Gerardo R.F. Alvear F., Onuralp Yücel, Liyuan Cai, Baojun Zhao, Dean Gregurek, and Varadarajan Seshadri
TMS (The Minerals, Metals & Materials Society), 2016

TRANSFORMATION OF SODIUM JAROSITE TO HEMATITE IN HYDROTHERMAL IRON PRECIPITATION PROCESS

Yizhao WANG, Cunxiong LI*, Zhigan DENG, Xingbin LI, Chang Wei, Gang FAN
Kunming University of Science and Technology, Faculty of Metallurgical and Energy Engineering, Yunnan Kunming, 650093, P. R. China
Keywords: zinc hydrometallurgy, iron precipitation, hematite; sodium jarosite, phase transformation

Abstract

During the process of hematite precipitation, the transformation of Na jarosite convert into hematite has been studied in a batch reactor at different temperatures, ranging from 170℃ to 200℃ with 3h of the retention time in the presence of 30g/L Fe (II) and 4g/L Na^+. The composition of precipitation was seriously depended on the temperature, at 170℃ hematite and sodium jarosite are the reaction products of iron precipitate, with an increasing in reaction temperature the sodium, sulfur content in precipitated decreased and sulfur mainly exist in forms of SO_4^{2-}, and when temperature over 200℃ hematite was the only phase in product. An increasing temperature was not only beneficial to iron precipitation, also conducive to jarosite conversion to form high quality hematite precipitate.

Introduction

Up to two thirds of the zinc is produced by the roast-leach-electrolysis process in the world [1]. In this process, iron is one of the major problems that a mass of iron is disposed as waste residues in a number of zinc plants. Krauss [2] reported that there are over 11 million tons iron needed to disposed in the electrolytic zinc plants around the world. Meanwhile, the presence of dissolved iron will result in ferrite formation which disables the recovery of valuable metals from leach residues [3]. In view of above reasons, the dissolved iron in the zinc sulfate solution must be removed.

The zinc industry has developed three main types of iron eliminated processes namely: jarosite, goethite and hematite process. Although these iron removal technologies have been available for many years, the jarosite process is the most widely used owing to its simplicity and the lowest cost. Jarosite process can easily be carried out at atmospheric pressure [4,5]. The type compound of jarosite precipitated is $MFe_3(SO_4)_2(OH)_6$ where M can be K, Na, NH_4, Ag, 1/2Pb, etc, and it is readily filterable and washed. Although Goethite process also have these advantages, it is relatively expensive than jarosite process in some way. The jarosite process has been widely used in the hydrometallurgical process of zinc, however, it also has some inevitable problems. These problems become more prominent, especially the implementation of strict environmental regulations on residue disposal. For instance, jarosite incorporate some divalent base metals and contain other pollutants such as compounds of sulfur and arsenic, and the alkali metals [5, 6]. High volumes of iron residues, produced in these two processes, which has no potential use is an also a major issue.

The hematite process is the most costly of the three processes since the hematite operates at elevated temperature (>180℃) in autoclaves to effect the hematite precipitation [7], but this process have some distinct advantages. For example, among the three iron residues, the iron content of hematite residues is the highest, as well as the density. Compared with the other two methods the hematite process would reduce a large volume of iron residues. Hematite is also the most stable iron-containing solid at room temperature, and this stability is in favour of the stockpiled in storage ponds without cause environmental pollution.

Akita Zinc Company [8] which is the only industrial operator of hematite process in the hydrometallurgical process of zinc has sold its hematite precipitate to cement manufacture. Moreover, hematite precipitation has some potential commercial value for iron production.

Although hematite is the main output during hematite process it also contain trace amount of jarosite-type compounds [3], and this would cause either decreasing the quantity of iron residues or increasing sulfur content in solid product. A series of experiments were undertaken to determine the transformation behavior of sodium jarosite into hematite during the process of hematite precipitation.

The conversion of sodium jarosite into hematite from a ferrous sulfate solution involves the following sequence of reaction:

Oxidation of ferrous sulfate to ferric sulfate:

$$2FeSO_4(aq)+1/2O_2(g)+H_2SO_4(aq) = 2Fe_2(SO_4)_3(aq) + H_2O(I) \quad (1)$$

Hydrolysis of ferric sulfate sulfate to produce hematite or react with Na^+ to form sodium jarosite:

$$Fe_2(SO_4)_3(aq)+H_2O(I)= Fe_2O_3(s) +3H_2SO_4 (I) \quad (2)$$

$$3Fe_2(SO_4)_3(aq) + Na_2SO_4 (aq) +12H_2O (I) = 2NaFe_3(SO_4)_2(OH)_6 (s) + 6H_2SO_4 (I) \quad (3)$$

The total reaction of jarosite covert to hematite precipitation:

$$2NaFe_3(SO_4)_2(OH)_6 (aq)=3Fe_2O_3(s)+ Na_2SO_4 (aq)+ 3H_2O(I)+3H_2SO_4 (aq) \quad (4)$$

The hydrothermal conversion of sodium jarosite to hematite involve two stages [9] :

$$2NaFe_3(SO_4)_2(OH)_6 (aq)+ 6H_2SO_4 (aq)= 3Fe_2(SO_4)_3(aq) + Na_2SO_4 (aq)+12H_2O(I) \quad (5)$$

$$3Fe_2(SO_4)_3(aq) + 9H_2O(I)=3Fe_2O_3(s)+9H_2SO_4(aq) \quad (6)$$

This paper discusses the results of an investigation on sodium jarosite to hematite precipitation which was focused on the conditions of temperature and reaction time.

Experimental Section

Analytical grade ferrous sulfate and sodium sulphate were used to prepare initial solutions containing 30g/L Fe(Ⅱ) and 5g/L Na^+, The testes were carried out in a 2L PARR titanium autoclave which was of the completely sealed type with electric heating.

To start an experiment, after the ferrous sulfate completely dissolved, 1.5 L prepared solution and desired amounts of sodium sulphate were transferred into the reactor and then purged using nitrogen exclude oxygen from the system until the target temperature was reached. The autoclave was heated to the set temperature with agitated at 500rpm and the oxygen partial pressure was then fixed to the desired level which was maintained constant throughout the experiment. During the process of each experiment, samples were withdrawn at regular time and immediately filtered to separate the liquor from the residue. At the end of the test, agitation was stopped, the content in the autoclave was immediately cooled by circulating cold water. The product slurry was filtered by using a filter press and a fine pore (2.0m) filter paper. The collected solids were washed with distilled water for two times and dried in an oven at 50℃ for 24 hours.

All the solution samples were analyzed for iron by titration with potassium dichromate. Some of the solid products were analyzed for iron and sulfur, and submitted X-ray diffraction to identify the phases in it. Additionally, Fourier transform infrared was applied to analysis SO_4^{2-} group.

Results and discussions

The effect of temperature on precipitate phase transformation

The effect of temperature on transformation of sodium jarosite into hematite at temperatures varying from 170℃ to 200℃ was studied. The efficiency of iron removal and the iron contents in hematite precipitate obtained at different temperatures were shown in Fig.1 and 2, and the precipitate was detected by XRD shown in Fig. 3.

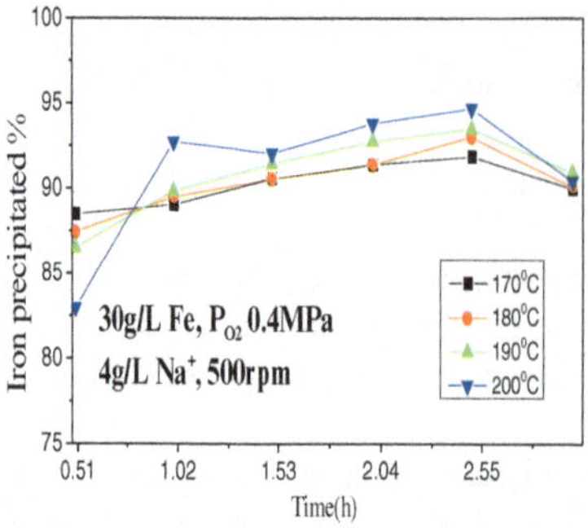

Fig1. Effect of the temperature on precipitation rate of Fe

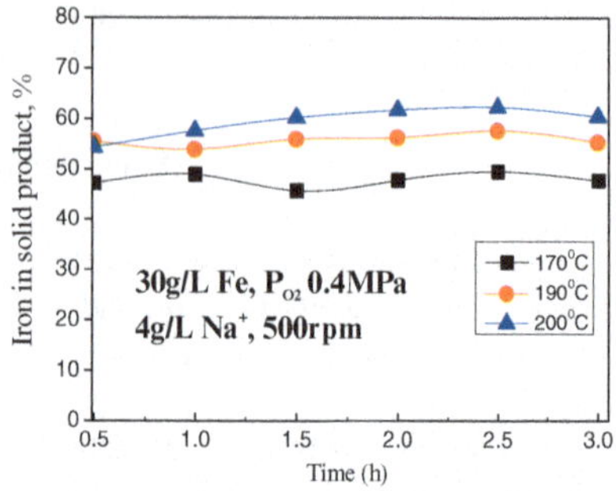

Fig2. Effect of the temperature on the contents of iron in hematite residues

We could observe in Fig. 1 that after 60 min of reaction period, more than 80% of iron had precipitated as hematite or jarosite at various temperatures. With the extended response time, both the iron-removal efficiency and the iron concentration of precipitate were found to increase slightly. Probably because of the presence of acid in the solution seemed to suppress the reaction 2,3,4 and a large proportion of sodium jarosite did not translate into hematite precipitate, especially at 170℃. This suggested that reaction 2, 3 contributed to iron precipitation but with the rise of heating temperature and the extension of response time, the reaction of hematite (2) appeared to be dominant. According to these data, after 3h in response, both efficiency of iron-removal and the iron content was decreased (about 1% ~ 2% lower). The reason for this phenomenon was the solubility of Fe_2O_3 increased in the process of cooling which would lead to the smaller particles of hematite crystals dissolve into the solution, this result was in good agreement with Umetzu et al [10].

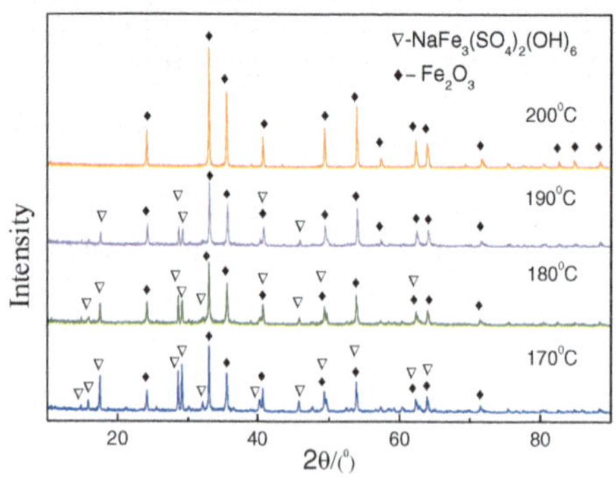

Fig.3 the effect of temperature on precipitate phase transformation obtained at 170, 180, 190 and 200℃. Conditions: 30g/L Fe^{2+}, 4g/L Na^+, 3h, P_{O2} 0.4MPa, 500r/min and 3h

It can be seen from Fig.2 that precipitation of hematite and sodium jarosite or their mixture depended on the different temperatures. We could detect a large amount of Na jarosite and hematite in the products at the temperature range from 170 to 180℃. The products were consisted of dominant hematite and minor Na jarosite at 190℃. The hematite phase was the only phase at 200℃, this precipitation contained 62% Fe, 1.7% S and 0.033% Na. these results were also verified our results are the same with the trends that Na jarosite converts to hematite at the temperature of over 190℃[11]

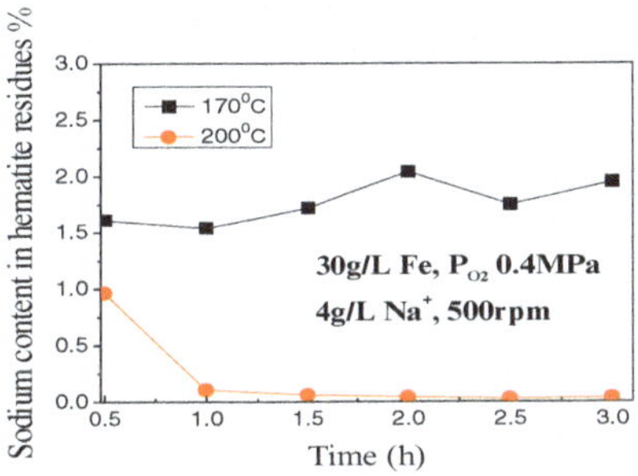

Fig.4 Effect of retention time and temperature on content of sodium in hematite residues

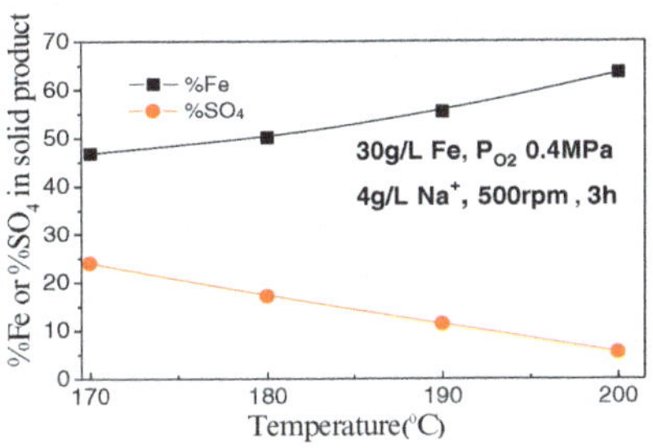

Fig.5 Effect of temperatures on the content of iron and sulphur in precipitate

Characterization of Solid Products

Sodium content in hematite residues obtained at temperatures of 170℃ and 200℃at 60min and 120min were analyzed, and the result was shown as Figure 4.

The Figure 4 showed that the sodium contents in the products were significantly different at 170℃and 200℃. At 200℃, the sodium was less than 0.066% in the precipitates at the first 1h,which indicated that the jarosite had completely translated into hematite. Nevertheless, the amount of sodium in hematite was as high as 1.5~2.04% in the whole reaction of 170℃. We found that the precipitate mainly contained Fe_2O_3 and a trace amount of Na jarosite at 1h, 2h, 3h detected respectively by the X-ray diffraction analysis. It may be attributed to the sodium jarosite remained essentially constant or only small amounts of sodium jarosite converting to hematite within 3h.

The content of iron and sulphur in iron precipitates at the different temperatures were determined, and the respective composition data was shown in Figure 5.we could observe from Figure 5 that the iron content in the precipitates increase with increasing temperature; whereas the S content in precipitates were significantly reduced with an increase in temperature. The precipitates had a composition of 47% Fe and 24% SO_4 at 170℃ after 3 hours reaction with a mixture of hematite and sodium jarosite. As the temperature increased, the Fe content increased to 62% and the %SO_4 rapidly dropped to 5% at 200℃ after 3 hours reaction, at the meantime, there was only hematite existed in the precipitates. It was apparent that temperatures above 190℃ had an advantage for sodium jarosite converting into hematite and obtaining the pure hematite products.

The previous date showed that all the sulfur was present as SO_4^{2-} in the hematite precipitates [3,12]. The incorporation of SO_4^{2-} in the hematite precipitates was further confirmed by FTIR analysis. The FTIR spectroscopy was made at 200℃ after 3 hours reaction , and the hematite precipitate contained 1.7% S (or 5% SO_4^{2-})(Figure 6).Three peaks emerged in the v SO_4^{2-} region (S-O stretch) between 900 and 1300cm^{-1}. According to the previous literature [11], the existence of 3 bands indicated the hematite adsorbed with SO_4^{2-}. Furthermore, it also made clear that SO_4^{2-} was uniformly distributed on the surface of hematite precipitates. Our result was supported by A.Pratt and Dutrizac [13,14]

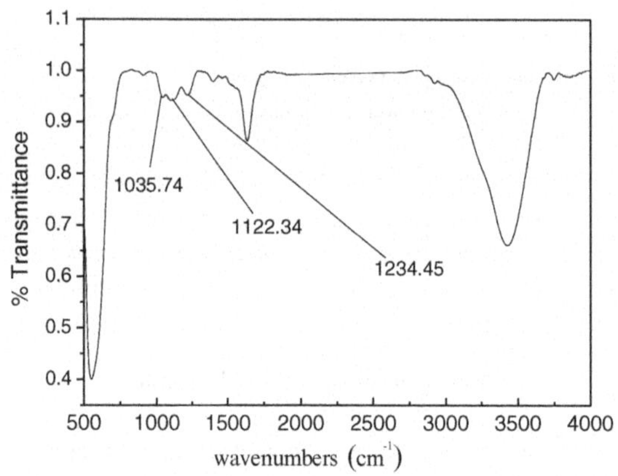

Figure6. FT-IR spectra of hematite precipitate produced at 200^{0}C for retention time of 180 min.

Conclusions

We can put forward the following conclusion from our experimental results that the hydrothermal conversion of sodium jarosite to hematite in the process of hematite.

1.The residues were detected by X-ray diffraction analysis after 3h showed that Fe_2O_3 was the only iron compound formed, and the hydrothermal conversion of Na jarotite was rapidly complete within 1h at 200℃. At lower temperatures (<190℃), however, major amounts of both Na jarosite and hematite, and that longer retention times had fewer impact on the transformation response.

2.In the process hematite, the presence of sodium sulfate in the solution contributed to speed up the rate of iron precipitate and enhance the purity of hematite at 200℃.

3.There were two forms of sulphur in hematite at lower temperatures, denoted as $\%S_{adsorb}$ and $\%S_{jarosite}$. The remaining sulphur at higher temperature (200℃) can be attributed to SO42- chemisorbed on hematite.

4. An increasing temperature was not only beneficial to jarosite conversion, also conducive to iron enrichment and reduce contents of sulphur in residues.

Acknowledgement

Financial support from the National Natural Science Foundation of China (Grant No.51364022, 51474117) and the Major Projects in the Scientific Research Foundation of the Education Department of Yunnan Province of China (Grant No. ZD2014003) is gratefully acknowledged.

References

1 D.A.Temple, "Zinc—From Mine to Market, Some Views on the Industry," Zinc'85, ed. K.Tozawa, MMIJ, 1985, pp.3-29

2 C.J.Krauss, "Effect of minor elements on the production of electrolytic zinc from zinc sulphide concentrates" in Zinc' 85, eds. K. Tozawa, MMIJ, 1985, pp. 467-481.

3 John E. Dutrizac and Tzong T. Chen "Behaviour of various impurities during the precipitation of hematite from ferric sulphate solutions at 225℃", TMS,eds, J.E. Dutrizac,et,al, Ottawa,Canada, 2012, 489-499.

4 V. Arregui, A. R. Gordon and G Steintveit: "The jarosite process–past, present and future", In 'Lead-Zinc-Tin'80', eds. J. M. Ciganet et al, The Metallurgical Society of AIME, Warrendale, PA, 1979, 97–124.

5 J. E. Dutrizac and J. L. Jambor: "Jarosites and their application in hydrometallurgy", Rev. Mineral. Geochem, 40(2000), 405–451.

6 D. Paktunc and J.E.Dutrizac: "Characterization of arsenate-for sulphate substitution in synthetic jarosite using X-ray diffraction and X-ray absorption spectroscopy", Can.Mineral, 41(2003), 905–919.

7 J.E Dutrizac, "Converting jarosite residues into compact hematite product", JOM, 1990, 36-39.

8 T. Yamada, S. Kuramochi, S. Sato, and Y. Shibachi, "The recent operation of the hematite process at the Iijima Refinery" in Zinc and Lead Processing,eds. J.E. Dutrizac et al,CIM, Montreal, Canada, 1998, 627-638.

9 J. E. Dutrizac and A. Sunyer. "Hematite formation from jarosite type compounds by hydrothermal conversion", Canadian Metallurgical Quarterly, 51(1) (2012), 11-23.

10 Umetz. Y, Tozawa. K, Sasaki. K, "The hydrolysis of ferric sulfate solutions at elevated temperatures", Metall. Soc. of CIM, Montreal, 1977, 111-117.

11 J.E. Dutrizac, "The hydrothermal conversion of jarosite-type compounds", in "productivity and technology in the metallurgical industries", eds. M. Koch and J. C.Taylor, TMS, 1989, 587-612.

12 Terry C. Cheng and George P. Demopoulos, "Hydrolysis of Ferric Sulfate in the Presence of Zinc Sulfate at 200 ℃ : Precipitation Kinetics and Product Characterization", American Chemical Society, 43(2004), 6299-6038.

13 A. Pratt, "XPS study of synthetic hematite power", CANMET-MMSL Mineralogy Report M-5438, Ottawa, Canada, 2000, 14pp.

14 J.E. Dutrizac and J.L Jambor, "Characterization of the iron arsenate-sulphate compounds precipitated at elevated temperatures", Hydrometalluirgy, 86(2007), 147-163.

7th International Symposium on High-Temperature Metallurgical Processing
Edited by: Jiann-Yang Hwang, Tao Jiang, P. Chris Pistorius, Gerardo R.F. Alvear F., Onuralp Yücel, Liyuan Cai, Baojun Zhao, Dean Gregurek, and Varadarajan Seshadri
TMS (The Minerals, Metals & Materials Society), 2016

EFFECT OF DIFFERENT COOLING SYSTEM ON THE SOLIDIFICATION OF THE SINTERS

Haibin Zuo[1], Jiangwei Shen[1], Cong Wang[1]

[1]State Key Laboratory of Advanced Metallurgy, University of Science and Technology Beijing; 30 Xueyuan Rd; Haidian District, Beijing, 100083, China

Keywords: cooling system, SFCA, sinters, microstructure, solidification

Abstract

The research on the formation of Silico-Ferrite of Calcium and Aluminium (SFCA) in the process of cooling and solidification is rarely reported. The crystallization process was observed by using scanning electron microscopy (SEM) and optical microscope in this paper. The effect of different cooling system on the microstructure and phase composition of the sinters was investigated to obtain the appropriate cooling rate to improve the metallurgical properties of sinters. The experimental results show that the crystallization of calcium ferrite is getting better when the cooling rate is decreased from water quenching to cooling in the air. There are a large number of hematite phases with interconnected crystal structure in the sample cooled in the air. In the case of the compressive strength, it was determined by the mineral composition and pore structure, higher SFCA and hematite contents in sinter cooling in the air lead to a higher strength. Glass phase in the water quenching sinter was the major reason for the lowest strength. High porosity contributes to a decrease in compressive strength for the sinter cooling in the furnace. Therefore the air cooling was the appropriate cooling rate to achieve high quality sinter under this experimental condition.

Introduction

The steel industry in 21st century continues to develop and progress. Iron and steel materials are still the most important structural materials and the largest amount of engineering materials. Sinter is the main raw materials of the blast furnace. With the increasing steel production, the quality requirements become more and more important for the sinter. The properties of different sinter are determined by their mineral composition and microstructure of sinter ores, which are derived from the different sinter materials, production conditions such as moisture of blending ores, alkali, carbon content, sintering temperature and cooling rate [1].

In sintering process, the main source of heat energy is the combustion of coke breeze, and the mixture is heated in a weak reducing atmosphere and cooled in an oxidizing atmosphere [2]. Firstly, during the heating stage, calcium ferrite phases are formed from the reaction of iron oxides which contained in ores, lime calcined from calcite and gangue in solid state. In the following liquid step, gangue materials start to dissolve into the melt. Finally, in the cooling stage, the development of sinter microstructure is completed through solidification and precipitation [3] [4].

The process of cooling and solidification of calcium ferrite formation remains to be clarified. The forming mechanism of the acicular calcium ferrite during the cooling process will be an important breakthrough to improve the quality of sinter. This paper states the effects of different cooling systems on the sinter microstructure and physical phase composition from the microscopic point of view to find a suitable cooling rate to improve metallurgical properties of the sinter.

Experiment procedure

Sample Preparation

The chemical composition of the sample in the experiment is as follows: CaO=15 wt. %; Fe_2O_3=75 wt. %; SiO_2=5 wt. %; Al_2O_3= 5 wt. %. Particle size was less than 50 m. The four reagents were added into the agate crucible with the anhydrous ethanol as solvent, and then dried at 70℃. After drying, the reagents were mixed in the agate crucible for 10 min. Then the mixed sample was divided evenly into three parts, every part was pressed into a cylindrical shape with 5 MPa. The samples loaded in three aluminum oxide crucibles were placed into muffle furnace to heat up to the experimental temperature at a rate of 20 ℃/min.

Experimental Method

The samples were heated to 1300℃ in a muffle furnace and the holding time was 8 h. After the experiment, the three samples were respectively cooled in water, in air and in the furnace. The different cooling ways have different cooling rates, in which the cooling rate in water is the fastest while the cooling rate in the furnace is the slowest.

Phase Determination (XRD, SEM, Optical and the Tensile Strength Measurement Device)

A part of annealed samples was ground into fine powder by using an agate mortar and pestle, and then the grinding powder was analyzed by XRD to determine the phase composition [3] [5]. The XRD data were obtained using a Rigaku D/MAX-RB X-ray diffractometer (Rigaku Corporation, Tokyo), using Cu-Kα as a radiation source (40 kV, 300 mA) and a graphite monochromator in the diffraction beam path. Continuous scanning mode was adopted to obtain data, where the scanning speed was 10 deg/min and the diffraction angle 2θ range was from 10 to 90 deg.

The remaining part of the annealed sample apart from the alumina crucible was polished and prepared for SEM and optical microscope detection. AZEISS ULTRA 55 field-emission scanning electron microscope (CARLZEISS Optical Instrument Co. Ltd., Jena) was employed. Energy dispersive spectroscopy (EDS), an Oxford X-Max 50 EDS detector (Oxford Instruments, Oxford), was coupled to the SEM.

The schematic diagrams of tensile strength measurement apparatus were shown in Figure 1. The device was composed of two parts, which are the tensile strength at room temperature and the high temperature tensile strength. The range of the pressure sensor with an accuracy of 1 N is 0~6000 N. The measuring range of the displacement meter was 25 mm, and the precision was 0.01 mm. When the tensile strength was tested at room temperature, the pressure sensor was moved to the left of the operating table, and connected with steel pressure bar. The sinter sample was put on the upper part of the steel base. The tensile strength experiment was made by using the control system of the upper part of the device which falls down at a certain rate. The system automatically record the pressure (P) and displacement (U) value. The pressure rapidly decreased when the sinter was crushed.

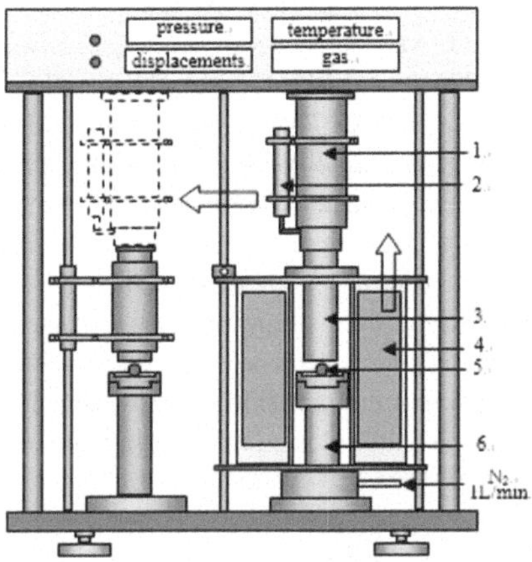

Figure 1. Schematic diagrams of tensile strength measurement apparatus
1-pressure sensor; 2-displacement meter; 3-Al_2O_3 compression bar; 4-heating furnace; 5-sinter; 6-Al_2O_3 base

The measurement principle of tensile strength was shown in figure 2. The relationship tensile strength of sinter and pressure were shown in the following formula (1), in which d refers to the diameter of the sample and l refers to the length of the sample. The experiment was repeated for 5 times under the same conditions. The average value refers to the tensile strength of sinter.

$$\sigma = \frac{2P}{\pi d l} \quad (1)$$

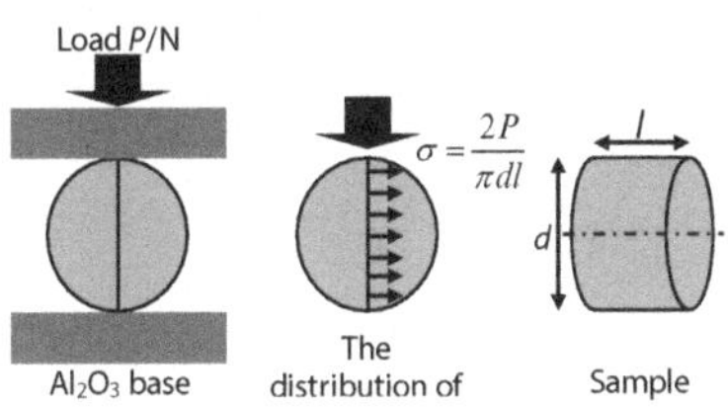

Figure 2. Measurement principle of tensile strength

Results and Discussion

The Crystallization Phenomenon

As the temperature decreases, the various minerals begin to crystallize from the liquid phase. The solidification of water quenching is taken as an illustration to investigate the solidification process, which is shown in the figure 3. Point A in figure 3 refers to complete crystallization phase, which composes of 59.65%Fe, 12.19%Ca, 6.17%Si, 1.4%Al and 20.69%O. Point B indicates the incomplete crystallization phase, in which the components are 47.81%Fe, 23.77%Ca, 10.61%Si, 1.88%Al and 15.94%O. Point C is liquid phase, which includes 42.74%Fe, 26.29%Ca, 11.37%Si, 2.34%Al and 20.69%O. Point D refers to Fe_2O_3. The melting point of A, B and C were calculated by Factsage software. The melting points of A, B and C were 1500 ℃, 1372 ℃and 1333 ℃ respectively. The melting point of Point A is the highest, and the melting point of Point C is the lowest. It can be deduced that the minerals with the high melting point first begin to crystallize, and the melting point of the residual liquid phase decreases gradually. Decreasing the temperature continuously the minerals with the low melting point were crystallized gradually. Therefore, the mineral structure is characterized with minerals with high melting point surrounded by the low melting point minerals. Comparing the melting points which were calculated to the actual sample melting points, it has a little difference. The possible reason is the deviation of components measured by EDS and the omission of Al when calculating melting points by Factsage software, both of which will lead to a higher melting point in the calculation.

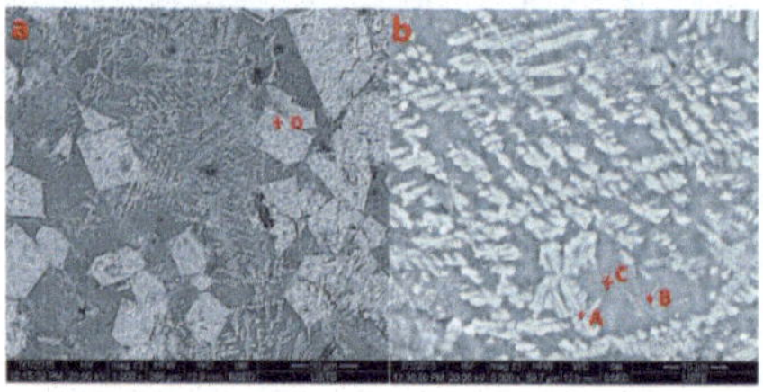

Figure 3. The SEM images of water quenching. a. 1000x b. 5000x

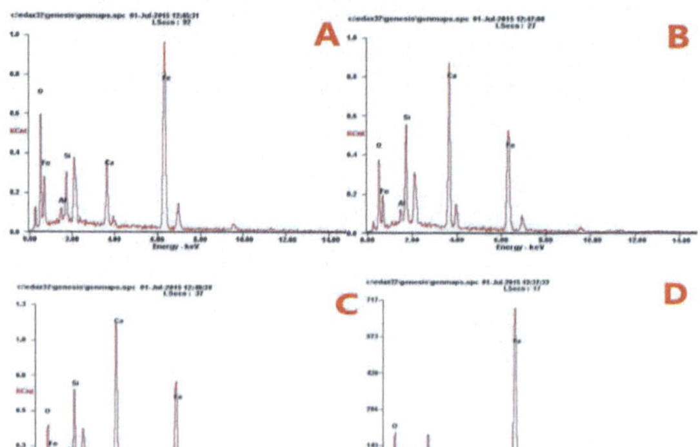

Figure 4. EDS analysis results

The Analysis Results of X-Ray Diffraction

Based on the X-ray diffraction in the figure 5, hematite, SFCA and floating body as major minerals were detected in each sample. The glass phase could only be found in the sample with water cooling. Magnetite could hardly be detected.

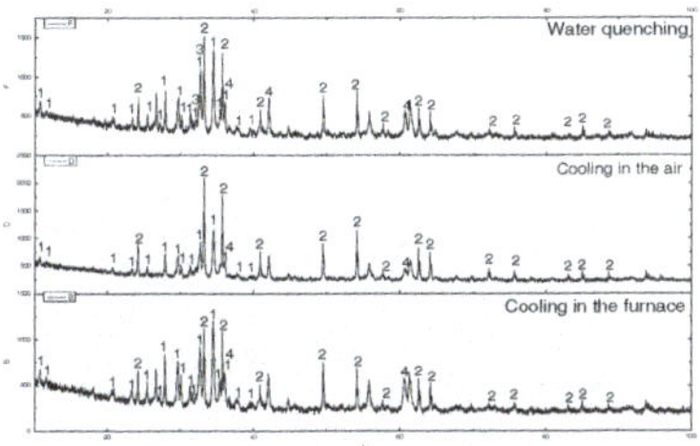

Figure 5. X-ray diffraction analysis of different cooling systems
1-SFCA 2-Hematite 3-Silicate 4-Wustite

Mineralogical Investigation

The microstructures of sintered samples with different cooling rates were compared as follows.

When the cooling rate was very rapid, such as water quenching, the hematite was distributed around the pore in the sintered sample and its shape was the skeletal crystal. This structure leads to the serious low-temperature reduction degradation. Therefore, some methods have to be employed to improve the cementation between the liquid phase and solid phase through reducing amount of the skeleton crystal hematite. Figure 6 presents most of the platy and columnar SFCA and a little acicular SFCA in the sample with water quenching. The possible reason was that the high alumina content promotes the SFCA growing from the needle to the columnar [6]. Another reason is that the crystallite formation is too late to grow into the acicular SFCA due to the rapid solidification rate of liquid phase [7]. At meantime, the liquid phase forms a lot of glass phase surrounding the SFCA leading to a poor reducibility of sintered sample.

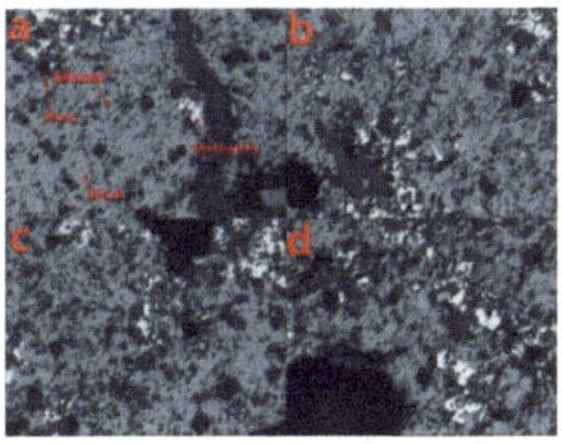

Figure 6. The microstructure of sample with water quenching, 200x

When the cooling rate was slow, such as cooling in the air, the hematite presented interconnected crystal and euhedral crystal structure in the sintered samples. It can be seen that the head and the tail of the hematite were connected in the figure 7. The more interconnected crystal was contained, the better the structure strength was. The SFCA was columnar and molten corrosion and the interlaced columnar structure was well developed. The reason is that when the cooling rate of the liquid phase was slow, the SFCA develops well from the acicular SFCA to the columnar SFCA. Compared to the water quenching, the glass phase decreases dramatically under the condition of air cooling. The increase of porosity, smaller pore size and pore distribution uniformity make the strength and reducibility better.

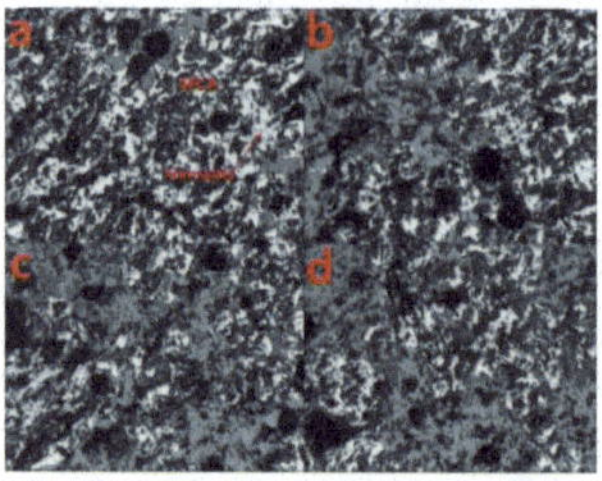

Figure 7. The microstructure of sample with air cooling, 200x

For the sample cooling in the furnace, the hematite shows skeleton crystal and granular, distributing evenly in the sample in the figure 8. Compared to the other cooling regimes, the porosity of the sample increases. Although it is beneficial to the reduction of sinter, it is not conducive to its structure strength.

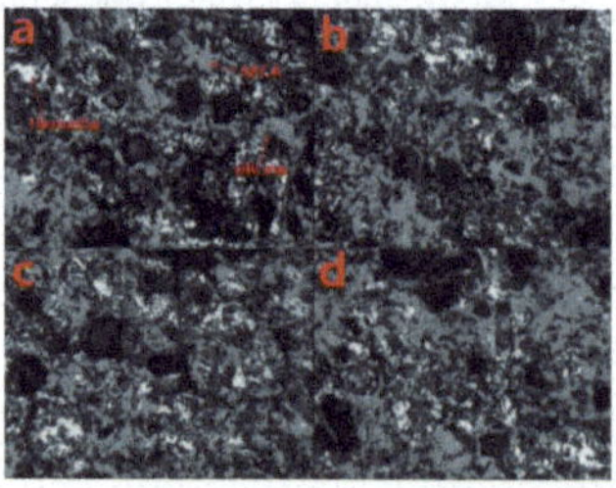

Figure 8. The microstructure of sample with cooling in the furnace, 200x

The Compressive Strength of the Sinters

The compressive strength at high temperature was measured using the measurement device of high-temperature compressive strength, as shown in the table I. According to table I, the compressive strength of the sample with water cooling was 7.5 MPa. When decreasing cooling rate, using air cooling, the compressive strength was improved, reaching up to the maximum 53.1 MPa. Decreasing cooling rate furthermore, cooling in the furnace, the compressive strength of the sample reduced to 20.9MPa, higher than that of cooling in the water. In the process of crystallization, the cooling rate of the water quenching was too fast, the liquid phase formed the glass phase directly, leading to the non-releasing of the internal energy which was an important reason for the reduction of sintering strength. A large number of grains were generated and conflicted with each other at the same time. The expansion coefficients of various minerals are different, leading to an internal stress appearance in the process of crystallization which is not easy to be eliminated. Even the fine cracks in the sinter can be found to reduce the strength of sinter.

In the process of cooling, the thermal stress occurs due to the temperature difference between the surface and the center of the sinter. The stress is mainly determined by the cooling condition and can be eliminated by slow cooling or heat treatment. But it cannot get agreeable effect by reducing the cooling rate when the temperature is below 800 ℃. Because the low temperature phase transition of $2CaO \bullet SiO_2$ in this temperature range played a decisive role in reducing the sinter strength.

The big internal stress results in a lower mechanical force correspondingly, and the low temperature phase transition caused by $2CaO \bullet SiO_2$ aggravates the degradation of mechanical strength. Through the counting method, the mineral phases composition was analyzed. The maximal production of calcium ferrite and hematite was obtained in the sinter cooling in air, while the most glass phase is in the sinter cooling in water. The compressive strength of the SFCA was higher than that of the hematite, and the compressive strength of the hematite was higher than that of the glass phase. It means that the sinter composed of high strength minerals such as SFCA and hematite was prone to have a high strength. Moreover pore structure is another important factor to the strength. The strength of sinter cooling in the furnace turns to decrease even though the cooling rate was still reducing compared to the air cooling, which can attribute to the highest porosity of sinter when cooling in the furnace.

Table I. The compressive strength of the sinters

Cooling systems	water quenching	air cooling	Cooling in the furnace
Compressive strength(MPa)	7.5	53.1	20.9

Conclusions

The crystallization of calcium ferrite is getting better when the cooling rate is decreased from water quenching to cooling in the air. There are a large number of hematite phases with interconnected crystal structure in the sample cooled in the air. In the process of cooling, the thermal stress occurs due to the temperature difference between the surface and the center of the sinter. The stress was mainly determined by the cooling condition and it can be eliminated by slow cooling or heat treatment. We can not get a consistent effect by reducing the cooling rate when the temperature was below 800 ℃. The low temperature phase transition of $2CaO \bullet SiO_2$ in this temperature range plays a dominant role in degradation of the strength of the sinter. It is the reason that the strength reduce from cooling in the air to cooling in the furnace. Through the counting method, the mineral phase content was measured. The most quantity of calcium ferrite and hematite were obtained in sinter in the cooling air and the most quantity of glass phase was obtained in the water quenching sinter. In the case of the compressive strength, it was determined by the mineral composition and pore structure, higher SFCA and hematite contents in sinter lead to a higher strength. Glass phase in the water quenching sinter was the major reason for the lowest strength. High porosity contributes to a decrease in compressive strength for the sinter cooling in the furnace. Therefore the air cooling was the appropriate cooling rate to achieve high quality sinter under this experimental condition.

References

1. Ji-Won JEON, Sung-Mo JUNG, "Formation of Calcium Ferrites under Controlled Oxygen Potentials at 1273 K". ISIJ International, Vol. 50 (2010), No. 8.
2. X. D, "The Sintering Characteristics of Mixing SiO2 with Calcium Ferrite at 1473 K (1200℃)". METALLURGICAL AND MATERIALS TRANSACTIONS B.2010.
3. L.F. Li, S.H. Z, Q. L, N.H. Z, "Present situation and Prospect of research on sinter matching". Journal of Iron & Steel Research, 2013.25(9):2-5.
4. H.M. LONG et al, "Sintering Principle and Technology of Iron Powder".Metallurgical Industry Press, 2010.8:61-64.
5. X. D. X.M. Guo, "Effect of solid state SiO2 on the formation process of binary calcium ferrite". Iron & Steel, 2015.50(3):33-38.
6. S.J. Zhang, S.T. Wang. Experimental study on the formation mechanism of calcium ferrite like needle. Iron & Steel, 1992.27(7):7.
7. X.M. Guo, L. Z, Q. Li, H.B. Shen, M.S. Zhou, "Mineral composition and structural characteristics of high basicity sinter". Iron & Steel, 2007.42(1):17-19.

7th International Symposium on High-Temperature Metallurgical Processing
Edited by: Jiann-Yang Hwang, Tao Jiang, P. Chris Pistorius, Gerardo R.F. Alvear F., Onuralp Yücel, Liyuan Cai, Baojun Zhao, Dean Gregurek, and Varadarajan Seshadri
TMS (The Minerals, Metals & Materials Society), 2016

Filtration Property of Pure Willemite Acid Leaching Sludge Under Pressure

Hailong Yang, Cunxiong Li*, Chang Wei, Zhigan Deng, Xingbin Li, Gang Fan and Minting Li[9]

Faculty of Metallurgical and Energy Engineering, Kunming University of Science and Technology, 650093, Kunming,China

Keywords: willemite, leaching liquor, filtration property, viscosity.

Abstract

The filtration problems on acid leaching sludge of high silicon zinc ore have been paid extensive attention in zinc hydrometallurgy. It was found that soluble silica can be effectively converted into silica precipitate with good settling, filtration and washing properties, and the filtration rate can be in excess of 800 $L/m^2 \cdot h$ under certain leaching conditions. According to investigating on the filtration rate, viscosity and concentration of silica of leaching sludge or liquor, the filtering temperature can tremendously improve the filtration property of pulp. At the same time, the experimental data indicated that the viscosity of leaching liquor was the key factor influencing filtration property of leaching sludge.

Introduction

With high grade zinc sulfide ore exhausted, the abundant zinc oxide ores of silicon-bearing attract tremendous attention by metallurgical industry. Although traditional technology of pyrometallurgy is mature, the restriction of energy consumption and increasingly strict environmental protection make this method impractical. From this point, the conventional wet zinc treatment process gets its position, but the soluble silica is easy to generate gel in the leaching process, which seriously affects the filtration property of leaching pulp and reduces the recovery rate of zinc, even brings difficulties in purification, electrolysis and subsequent treatment procedure. In order to control the harm of silica sol in leaching process, researches are mainly focused in three aspects: one is to control the dissolution of silica, avoiding the aggregation of silica to generate colloidal silica through this pathway. Xu[1] has investigated that reducing the initial acid concentration can decrease the dissolution ratio of silica in leaching solution. Castro[2] has found that bioleaching can inhibit the dissolution of silica. And Dufresne[3] has researched on quick leaching method to control the dissolution of silica through restricting the volume of solution strictly. On the other hand, effective solid-liquid separation is got through strengthening polymerization of silica generate precipitation by adding flocculant [4-7], it is the most mature method in industry application, such as VM, EZ and Radina technology. The third aspect, enhancing process reaction by introducing techniques such as

*Correspondence author: Cunxiong Li; professor; Tel./fax: +86 871 5188819. E-mail address: licunxiong@126.com.

pressure [1,8-9] and microwave[10] make the dissolution of silica transform into silica precipitation. In this way, the filtering property of pulp is improved.

Considering the complex behaviors of silica in solution, many researchers have dedicated to investigating the structures of silicate, the behaviors of decomposition and polymerization of silica, and the behaviors of colloidal silica. However, in view of the interactive reactions of decomposition and polymerization of silica, the passing of extensively polymerized silica directly into solution cannot be envisaged. Murata[11] concluded that silicate ions of up to $Si_6O_{18}^{12-}$ can and will pass into solution intact. From this point, the polymerization molecule of $Si_6O_{18}^{12-}$ maybe exist in leaching solution. During the synthetic route for the preparation of outstanding silica gel materials, investigations of the sol-gel process are of considerable interest due to the behaviors of colloidal silica. The transformation of sol to gel was studied by many methods such as NMR, vibration spectroscopy, small angle X-ray scattering, methods of viscometry, acoustic resonance technique, electrospray-scanning mobility particle sizer (ES-SMPS), atomic force microscopy (AFM) etc[12-16]. Along with so many problems working out, many questions are published concomitantly and the problem of filtration property of leaching pulp is still patient to solve. In this paper, the crucial factor of affecting filtration property of pulp is verified by the viscosity test on pulp and leaching liquor, what's more, a series of tests about viscosity and concentration of silica are launched.

Experimental

Apparatus

A UV-1801 ultraviolet spectrophotometer is used to determinate the concentration of silica by spectrophotometric. A glass calomel electrode pH meter is used to test the pH of solution. A NDJ-5S rotational viscometer is used to measure the viscosity of pulp and leaching liquor. And a 2XZ-4 sliding vane rotary vacuum pump is used to measure the filtration rate of pulp.

Procedures

The pure willemite is leached in a autoclave under the condition of total gas pressure 0.8 MPa, liquid-solid ratio10, reacting temperature 393 K, average particle size 55 m, mole ratio of H_2SO_4:Zn 1, stirring rate 600 rpm, leaching time 30 min(From Fig.1 to Fig.3, all of the experiments are single factor researches), then the leaching pulp is taken out from autoclave for other tests, such as measure the filtration rate of pulp, measure the viscosity of pulp and leaching liquor and determinate the concentration of silica.

Results and discussion

Influence of leaching conditions on filtration property of pulp

The effects of leaching conditions on filtration property of pulp have investigated, such as pressure, temperature, stirring rate, leaching time, liquid-solid ratio and mole ratio of H_2SO_4:Zn. The pressure can improve the filtration property evidently with the increasing of pressure, as shown in Fig.1,when the pressure increases from 0.8 MPa to 2 MPa, the filtration rate of pulp increases by more than 100 L/m²·h. From Fig.2 to Fig.3, with the liquid-solid ratio or mole ratio of H_2SO_4:Zn increasing, they have a distinctly negative influence on filtration property of pulp. It is noteworthy that the pulp cannot be filtered approximately when the mole ratio of H_2SO_4:Zn accesses to 1.2. Then the temperature, stirring rate and leaching time have hardly any influence on filtration property of pulp, but the filtration rate of pulp keeps stable relatively and has a minor fluctuation around 820 $L/m^2 \cdot h$. Comprehensively, the pure willemite acid leaching sludge under pressure can obtain an excellent filtration property and the filtration rate can be in excess of 800 $L/m^2 \cdot h$ under certain leaching conditions.

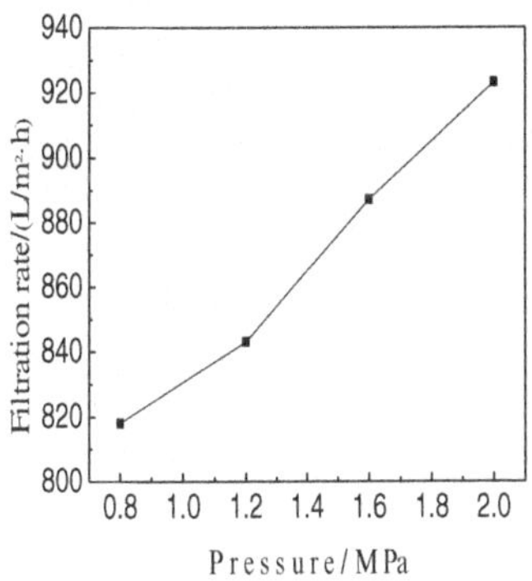

Fig.1 Effect of pressure on filtration rate of pulp

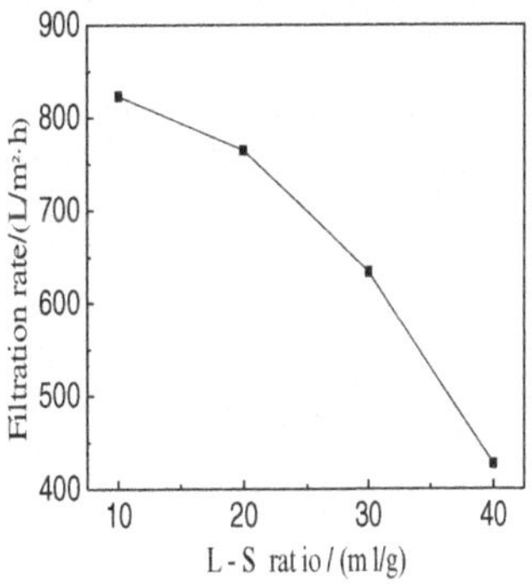

Fig.2 Effect of liquid-solid ratio on filtration rate of pulp

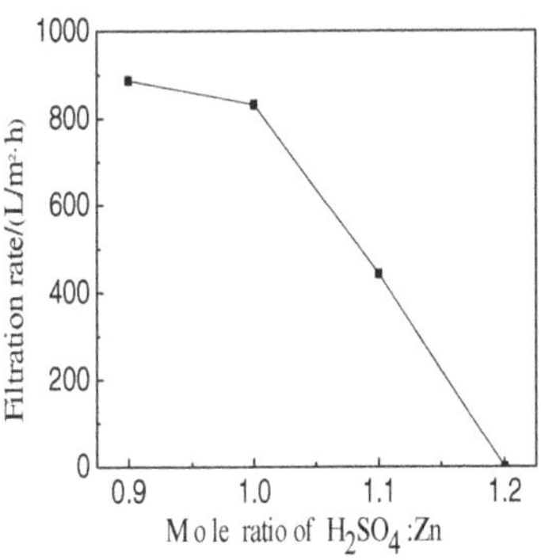

Fig.3 Effect of mole ratio of H_2SO_4:Zn on Filtration rate of pulp

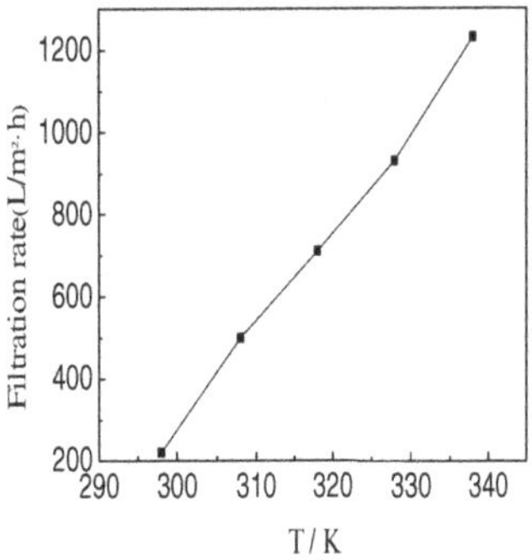

Fig.4 Effect of filtering temperature on Filtration rate of stored pulp

Factor of determination the filtration property of pulp

During practical industrial production, the pulp will be remained for a certain period of time before filtering. However, it is found that the pulp has a serious problem in filtering after storing few days. Then an interesting phenomenon draws public attention, the pulp filtration property of storage improved tremendously after heating, as shown in Fig.4. It indicates that the filtering temperature has a crucial influence on pulp filtration property of storage. In order to explore the influence further, the viscosity of stored pulp has been tested and the results are showed in Fig.5. It can be seen that the viscosity of pulp increases with the increasing of filtering temperature, what a contradiction, that increasing viscosity contributing to improve filtration property of pulp gets confused. Then the viscosity of leaching liquor storing for several months has been investigated by accident and an important phenomenon that the leaching liquor cannot be filtered, the results of viscosity testing of the leaching liquor as shown in Fig.6. It can be seen that the viscosity of the leaching liquor decreases with the heating temperature increasing. And when the leaching liquor is heated to 313K, it can be filtered swimmingly. The foregoing confusion can make sense, as the viscosity of leaching liquor relative to viscosity of the whole pulp can be neglected, but the viscosity of leaching liquor is the key factor of filtration property of pulp. On the basis of Mark-Houwink equation:$[\eta]= kM^a$, where $[\eta]$ is intrinsic viscosity, k is proportional constant, M is average molecular weight, and a is empirical parameter correlation with molecular shape, the average molecular weight of polymer molecule can be predicted by the viscosity of solution[17-19]. From this point, when the size of polymer molecule in leaching liquor reaches definite magnitude, the pore of filter paper will be blinding and the filtration of pulp can't proceed smoothly. In conclusion, the conclusive factor of the filtration property of zinc oxide ore silicon-bearing acid leaching pulp is the size of silica polymer molecule in leaching liquor.

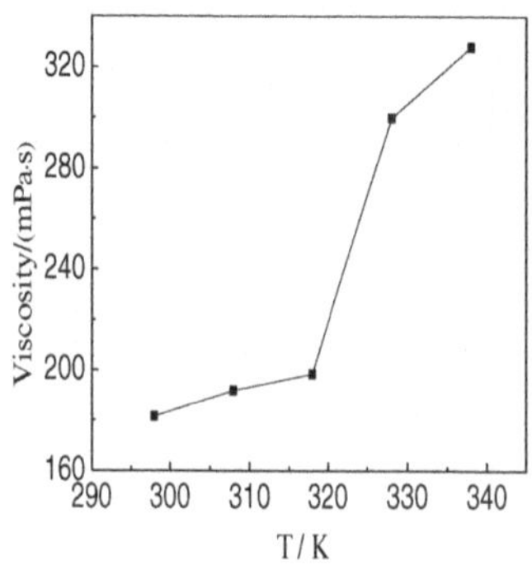

Fig.5 Effect of heating temperature on viscosity of stored pulp

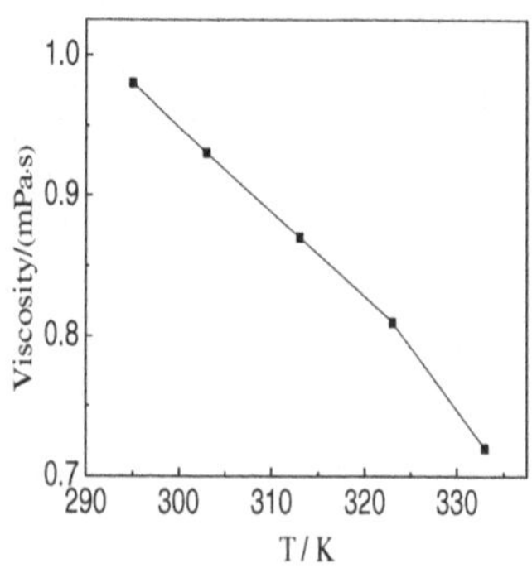

Fig.6 Effect of heating temperature on viscosity of leaching liquor storing for several months

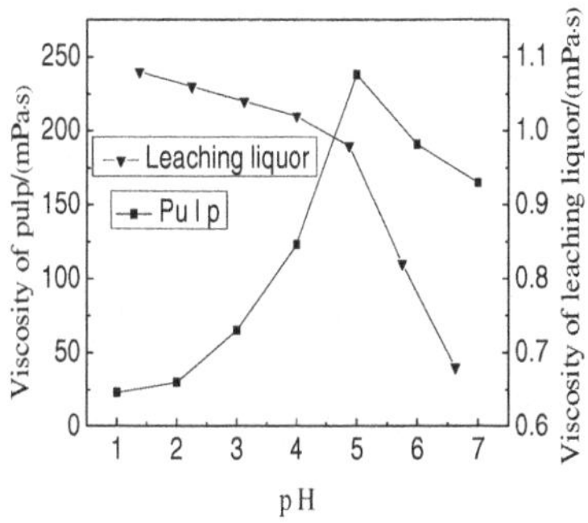

Fig.7 Effect of pH on viscosity of pulp and leaching liquor

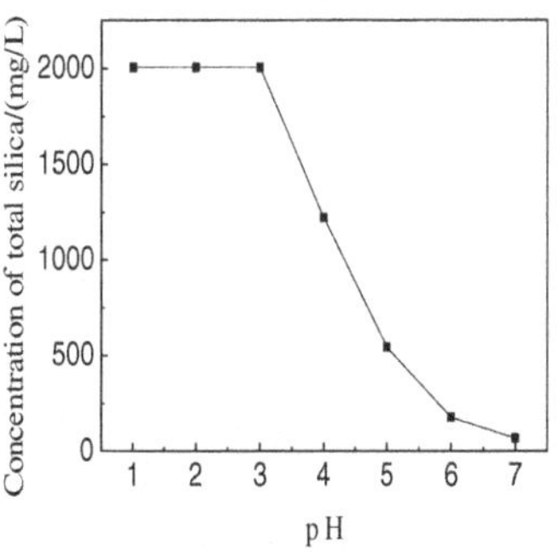

Fig.8 Effect of pH on concentration of total silica in leaching liquor

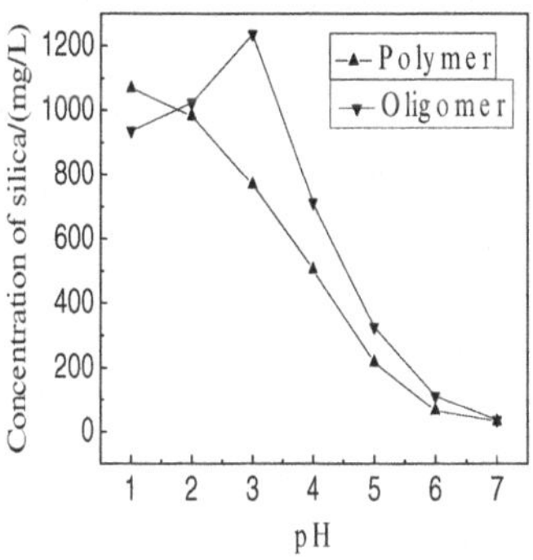

Fig.9 Effect of pH on silica concentration of oligomer and polymer in leaching liquor

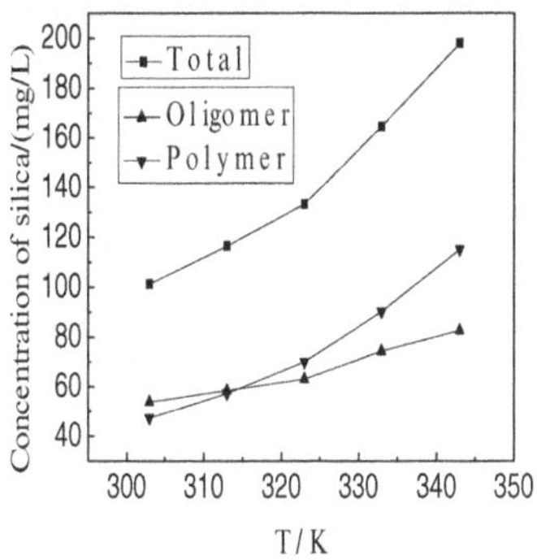

Fig.10 Effect of temperature on concentration of total silica,silica concentration of oligomer and polymer in leaching liquor

Determination the viscosity of leaching liquor

The pH of pulp and leaching liquor is adjusted by sulfuric acid and ammonia water. When the pH is 4, the leaching liquor begins to generate flocculent white precipitate. With the addition of ammonia water, the quantity of white precipitate also increases largely, thus the viscosity of leaching liquor must be tested after filtering. As shown in Fig.7, the viscosity of leaching liquor decreases gradually with pH increasing. Combined with Fig.8 and Fig.9, at the pH range of 1 to 3, the concentration of total silica is still stable and the silica concentration of polymer is decreases gradually, thus the viscosity of leaching liquor have a close-knit relationship with the silica concentration of polymer. At the pH range of 4 to 7, the concentration of total silica decreases vastly and the percentage of silica concentration of polymer is more and more close to 50%, so this phenomenon can indicates that the viscosity of leaching liquor is affected by the concentration of total silica. Comprehensively, the concentration of total silica and the silica concentration of polymer are crucial factors for viscosity of leaching liquor(the method of determination the concentration of silica refers to reference [20]).

At the pH range of 1 to 5, as shown in Fig.7, the viscosity of pulp increases gradually, it can demonstrate that the polymerization reaction of silica occupies dominant situation gradually with pH increasing. The viscosity of pulp begins to decrease when pH is above 5, the reason is that the molecular size of white precipitate is less than the silica polymer. Along with the white precipitate generated continually, viscosity of the whole pulp displays a tendency to decrease.

The discussion of heating temperature

In Fig.4, the temperature of filtering increased is conducive to improve filtration property of pulp, but the leaching temperature has hardly any influence on filtration property of pulp, the reason for this is that the pulp taken out from autoclave needs a cooling treatment, which leads the filtration temperature of pulp to be around 313K. Combined Fig.6, Fig.7 and Fig.8, the viscosity of the leaching liquor decreases with the heating temperature increasing, meanwhile, the viscosity of the leaching liquor decreases with the concentration of total silica decreases, however, the concentration of total silica has a opposite relationship with the heating temperature from Fig.10, the leaching liquors of Fig.10 are obtained from pulp filtered under the heating temperature and the detection of concentration of total silica under ambient temperature, so the actual statuses of leaching liquors under the heating temperature can't be presented. Anyway, what can be seen from Fig.10 is that the solubility of silica will increase with the temperature increasing. From this point, although the filtration temperature can tremendously improve the filtration property of pulp, the concentration of total silica also will increase with filtration temperature increasing, so a prediction that the oligomer of silica dominates in the leaching liquors of silica species with heating temperature increasing can be concluded.

Conclusions

The viscosity of the leaching liquor has an important influence on filtration property of pure zinc silicate acid leaching sludge. From molecular level, the size of silica polymer molecule in leaching liquor does determine filtration property of zinc oxide ore silicon-bearing acid leaching pulp. Further studies conclusively demonstrate that the size of silica polymer molecule in leaching liquor is dependent on the concentration of total silica and the percentage of silica concentration of polymer in leaching liquor. It can sufficiently account for that increasing of liquid-solid ratio and mole ratio of H_2SO_4:Zn will lead bigger trouble in filtration of pulp. Most of all, the increasing filtration temperature can tremendously improve the filtration property of pulp and the formation silica species in leaching liquors is affected by the heating temperature vastly.

Acknowledgement

Financial support from the National Natural Science Foundation of China (Grant No.51364022, 51474117) and the Major Projects in the Scientific Research Foundation of the Education Department of Yunnan Province of China (Grant No. ZD2014003) is gratefully acknowledged.

References

[1] H.S. Xu et al., "Sulfuric acid leaching of zinc silicate ore under pressure," Hydrometallurgy, 105 (1) (2010), 186-190.

[2] I.M. Castro et al., "Bioleaching of zinc and nickel from silicates using Aspergillus niger cultures," Hydrometallurgy, 57 (1) (2000), 39-49.

[3] R.E. Dufresne, "Quick leach of siliceous zinc ores," JOM-Journal of Metals, 28 (2) (1976), 8-12.

[4] I.G. Matthew and D. Elsner, "The hydrometallurgical treatment of zinc silicate ores," Metallurgical Transactions B, 8B (1977), 73-83.

[5] B.Terry, "The acid decomposition of silicate minerals part II. Hydrometallurgical applications," Hydrometallurgy, 10 (1983), 151-171..

[6] J. Baldyga et al., "Precipitation of amorphour colloidal sillica from aqueous solutions-Aggregation problem," Chemical Engineering Science, 77 (2012), 207-216.

[7] D. Solberg and L. Wagberg, "Adsorption and flocculation behavior of cationic polyacrylamide and colloidal silica," Colloids and Surfaces A: Physicochemical and Engineering Aspects, 219 (1-3) (2003), 161.

[8] C.X. Li et al., "Pressure leaching of zinc silicate ore in sulfuric acid medium," Transactions of Nonferrous Metals Society of China, 20 (5) (2010), 918-923.

[9] C.X. Li et al., "Pressure acid leaching of high silicon zinc oxide ore," Chinese Journal of Nonferrous Metals, 19 (9) (2009), 1678-1683.

[10] Y. Hua, Z. Lin and Z. Yan, "Application of microwave irradiation to quick leach of zinc silicate ore," Minerals Engineering, 15 (6) (2002), 451-456.

[11] K.J. Murata, "Internal structure of silicate minerals that gelatinize with acid," The American Mineralogist, 28 (1943), 545-562.

[12] M. A. Tenan, D. M. Soares and C. A. Bertran, "Aqueous sol-gel process in the silica-metasilicate system," A Microrheological Study Langmuir, 16 (2000), 9970-9976.

[13] AC.J. H. Johnsson, M. C. Camerani and Z. Abbas, "Combined

electrospray-scanning mobility particle sizer (ES-SMPS) and time-resolved synchrotron radiation-small-angle X-ray scattering (SR-SAXS) investigation of colloidal silica aggregation. part II. Influence of aggregation initiator on gel stability," The Journal of Physical Chemistry B, 115 (2011), 9547-9555.

[14] M. Jokinen, JB. Rosenholm and E. Gyorvary, "Viscoelastic characterization of three different sol–gel derived silica gels," Colloids and Surfaces A, 141 (2) (1998), 205-216.

[15] R.K. Das and M. Das, "Catalytic activity of acid and base with different concentration on sol–gel kinetics of silica by ultrasonic method," Ultrasonics Sonochemistry, 26 (2015), 210-217.

[16] E.I. Mayo, D.D. Pooré and A.E. Stiegman, "Catalysis of the silica sol−gel process by divalent transition metal bis(acetylacetonate) complexes," Inorganic Chemistry, 39 (5) (2000), 889-905.

[17] Z. Burkus and F. Temelli, "Determination of the molecular weight of barley β-glucan using intrinsic viscosity measurements," Carbohydrate Polymers, 54 (2003), 51-57.

[18] M. Kawahigashi, H. Sumida and K. Yamamoto, "Size and shape of soil humic acids estimated by viscosity and molecular weight," Journal of Colloid and Interface Science, 284 (2) (2005), 463-469.

[29] S. Ishida et al., "Average molecular weight and viscosity of acetone solutions of phenolic resins," Polymer Bulletin, 10 (11-12) (1983), 533-537.

[20] H.L. Yang et al., "Molybdenum blue photometry method for the determination of colloidal silica and soluble silica in leaching solution," Analytical Methods, 7 (13) (2015), 5462-5467.

7th International Symposium on High-Temperature Metallurgical Processing
Edited by: Jiann-Yang Hwang, Tao Jiang, P. Chris Pistorius, Gerardo R.F. Alvear F., Onuralp Yücel, Liyuan Cai, Baojun Zhao, Dean Gregurek, and Varadarajan Seshadri
TMS (The Minerals, Metals & Materials Society), 2016

HYDROTHERMAL SULFIDATION OF CERUSSITE WITH ELEMENTAL SULFUR

LI Cun-xiong*, Yang Hai-long, WEI Chang, LI Xing-bin, DENG Zhi-gan, FAN Gang, LI Min-ting
Faculty of Metallurgical and Energy Engineering, Kunming University of Science and Technology, Yunnan, Kunming 650093 China

Abstract

Hydrothermal sulfidation of pure cerussite with elemental sulfur has been reported in the present paper. The variables considered in the study were temperature, time, particle size and elemental sulfur fraction in the reacting mixture. Temperature was the most important factor, at temperatures above 180°C, PbS and $PbSO_4$ were the only stable reaction products in the sulfidation of cerussite. The experimental data indicated that under the hydrothermal conditions with a particle size of -74+58 m and sulfur fraction in reacting mixture of 15% at 200 °C for 120 min, and 92% of lead sulfidation rate was achieved.

Keywords: Cerussite, hydrothermal sulfidation, elemental sulfur, lead sulfide, lead sulfate

1. Introduction

Normally, the lead-bearing deposits contain, besides lead sulfides, relatively large amounts of oxidized minerals of this metal such as cerussite, lead oxide and basic lead carbonate and so on. It is generally known that the oxide-type minerals of lead are more difficult to float than their sulfide counterparts. This fact is due to the higher solubilities of oxide minerals and also to the extensive hydration of oxide surfaces (Rey, 1953; Fuerstenau et al., 1985; Herrera et al., 1998). Oxidized lead sulfide minerals also respond poorly to flotation. Therefore, oxidized lead minerals and oxidized lead sulfide are frequently treated with sulfidizing agents prior to flotation for converting the surface of the minerals into a more insoluble lead sulfide compound that will respond to flotation (Fuerstenau et al., 1987).

Cerussite was chosen for this investigation because it is the most commercially important lead oxide minerals. Considerable work has been carried out in the past to increase the recovery of cerussite by sulfidation applied prior to flotation (Caproni et al., 1979; Guy and Trahar, 1985; Marabini et al., 1985; Malghan, 1986; Fuerstenau and Raghavan, 1986; Sotillo and Fuerstenau, 1988; Barbaro, 2000; Önal et al., 2005). Based on these and other studies, sodium sulfide is one of the most widely used alkali metal sulfides in the sulfidization-flotation process, its effectiveness is not entirely satisfactory, possibly due to the fact that the sulfidized layer detaches readily. When an alkali sulfide is used as the reagent, a careful control of the sulfidization stage is critical to maximizing the flotation recovery (Malghan, 1986). Insufficient amount of sulfide gives poor recoveries because of inadequate sulfidization, while an excess sulfide causes poor flotation due to the depressant action of sulfide ions. A major difficulty in sulfidization process is to establish and maintain an optimum sulfide concentration.

In the present study cerussite was sulfidized with elemental sulfur under hydrothermal conditions. In such a sufidizing operation, the oxidized ore constituents may be merely filmed with the stable metal sulfide compounds or they may be converted to a greater or less rate into the stable metal sulfide compounds. The main variables expected to have significant influence on the sulfidation of cerussite are the temperature, time, particle size and composition of the reacting mixture. Determination of the effects of these variables should provide the optimum conditions for a high conversion of cerussite. Subsequent flotation of the sulfidized product should provide the optimum conditions for high lead recovery from the cerussite concentrate. This article is concerned solely with the study of the variables affecting the conversion of cerussite by sulfidation under hydrothermal conditions.

2. Experiment

2.1 Materials

In the laboratory scale tests, the cerussite used in this study was analytical grade >97 pct. Elemental sulfur (>99% pure, powder) was used as the sulfidizing agent. For the sulfidation experiments, both materials were mixed thoroughly with various cerussite/sulfur ratios.

2.2 Experimental Apparatus

The experimental apparatus used for sulfidation of the cerussite consisted mainly of 0.5 L, autoclave which was equipped with a heating mantle, a PID temperature controller, a variable speed stirrer and an internally mounted cooling coil a horizontal tube furnace and ancillary equipment.

2.3 Operational Procedure

The procedure followed in each run consisted of heating the sealed autoclave containing about 200 mL of water to the set temperature under slow N_2 slow. Once it reached the set temperature, the nitrogen flow was shut and the stirrer was turned on, and about 20 g of the mixture sample was mixed with water. At the end of the predetermined time of reaction, the autoclave was rapidly water-cooled to about 30°C under nitrogen atmosphere, and the solution was filtered for collecting the reacted sample.

The reacted sample was divided in two fractions. The first fraction was characterized by X-ray diffraction (XRD) to determine the compounds formed during the sulfidation process.

For the calculation of the sulfidation rate of lead in the sample, the second fraction first was extracted with carbon disulfide (CS_2) in a Soxhlet apparatus to remove the excess unreacted elemental sulfur in the sulfidized material, because an excess of sulfur was usually present in the reacted sample and this would be confirmed in our present study. And then the second fraction was chemically analyzed to determine the content of lead sulfide in the reacted sample.

3. Results and discussion

For all hydrothermal sulfidation of cerussite experiments, the ratio of solid mass to liquid volume (S/L) was 1:10, the slurry was stirred at a speed of 500 rpm, and the slurry was controlled in mild alkaline solutions of pH 9.

The appearance of the sulfidized product for different experimental conditions changed only in color. Samples treated at temperatures lower than 180 ℃ had a dark gray color, while samples treated at temperatures above 200 ℃ showed a slight gray color.

3.1 Effect of temperature

The effect of temperature on sulfidation of cerussite was studied using the cerussite of size fraction -74+58 m, with elemental sulfur fraction of 15%, at temperatures varying from 120°C to 200 ° for 60 and 120 minutes, and the results are shown in Fig. 1. It can be seen from Fig.1 that the temperature has a significant effect on the sulfidation rate of lead in cerussite. The sulfidation rate of lead increased from 3.3% to 90.9% when the temperature increased from 120 to 220 °C for 120 minutes.

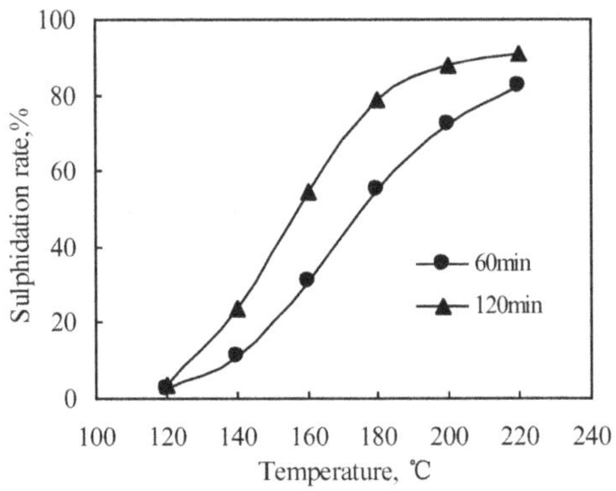

Fig.1 Effect of temperature on hydrothermal sulphidation of cerussite with elemental sulfur.

As will be discussed next in detail, XRD analysis of the reaction products showed that at temperatures below 160°C, the sulfidation reaction products were unsulfidized cerussite and PbS while at temperatures above 180°C, the diffraction lines for cerusstie are no longer observed, but strong lines for PbS and weak lines for $PbSO_4$ were identified instead. Therefor the temperature of hydrothermal sulfidation of cerussite should be over 180°C for more desirable products of lead sulfide.

3.2 Effect of the Fraction of Sulfur in the Mixture

Experiments were carried out using cerussite mixed with different weight fractions of sulfur to determine the amount of sulfur necessary to sustain the sulfidation reaction. In the experimental runs, additional sulfur was necessary to compensate for sulfur losses due to condensation since, during the course of the reaction; some condensed sulfur was observed in the lower-temperature regions of the autoclave wall. The results of experiments carried out under different weight fractions of sulfur with sample of size -74+58 m reacted at 200°C for 60 and 120 minutes are presented in Figure 2.

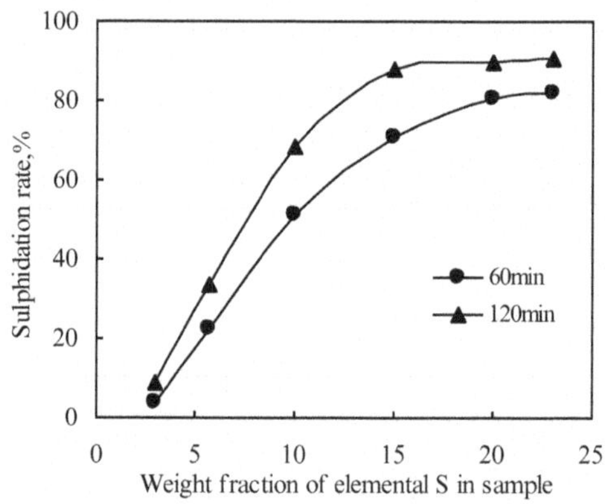

Fig.2 Effect of weight fraction of elemental S in sample on hydrothermal sulphidation of cerussite with elemental sulfur.

It can be observed in Fig. 2 that there was a transparent increase in sulfidation extern of lead with weight fraction of elemental sulfur in sample increased from 3 to 15 percent. A further increase in weight fraction of elemental sulfur in sample didn't produce a substantial increase on sulfidation rate. The subsequent experiments were carried out using an elemental sulfur fraction of 15 to guarantee the presence of enough sulfur for the reaction to reach completion in the various experimental conditions of temperature and times studied.

3.3 Effect of Time of Sulfidation

The effect of holding time on the sulfidation of cerussite was studied using the sample size fraction -74+58 m , elemental sulfur fraction 15%, for time prolonged from 30 min to 300 min at temperature of 160 and 200°C。 The results are shown in Figure 3.

We can observe in this figure that the sulfidation rate of cerussite increased greatly when the reacting time was prolonged from 30 min to 90 min at 220°C. Beyond 180 min, there was no significant increase in the sulfidation rate of lead both at 160°C and 220°C. We can also observe that the temperature indeed has an important effect on the rate of sulfidation of cerussite. At 160°C, the sulfidation is slow and maximum sulfidation is reached in more than 180 minutes, while at 220°C, the maximum sulfidation is attained in less than 120 minutes.

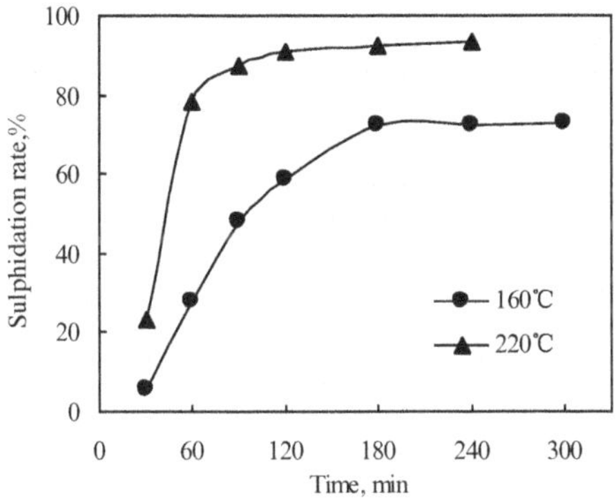

Fig.3 Effect of reacting time on hydrothermal sulphidation of cerussite with elemental sulfur.

3.4 Effect of Particle Size

Experiments were carried out at 180°C with cerussite of various particle sizes (-106+74 m, -74+58 m, -58+48 m, -48+42 m), and the results are shown in Fig.4. These results in Fig.4 indicate that the sulfidation rate is slow for the larger particle sizes, and there is a transparent increase with a decrease in particle size. We can also observe that the maximum conversion attained is nearly the same in all cases.

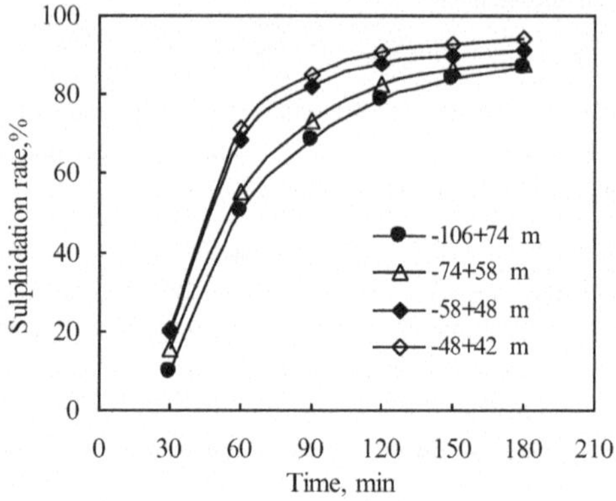

Fig.4 Effect of particle size on hydrothermal sulphidation of cerussite with elemental sulfur.

3.5 Characterization of the sulfidized products

The identification of the product compounds was performed by the conventional X-ray diffraction method. Sulfidation products obtained under various experimental conditions were analyzed accordingly, and the results are shown in Figures 5 and 6.

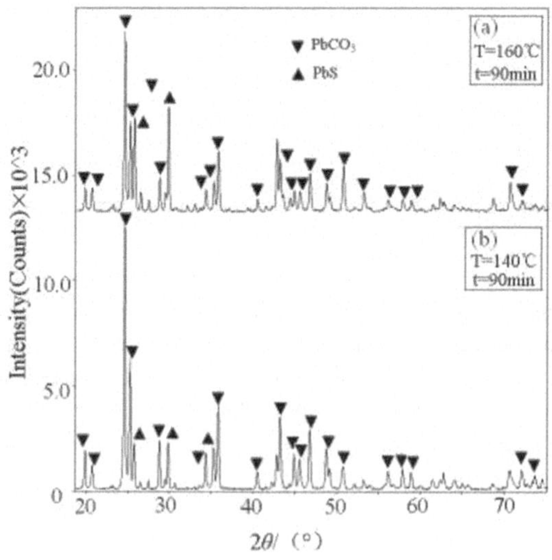

Fig.5-(a) and (b) The XRD patterns of sulfidation products

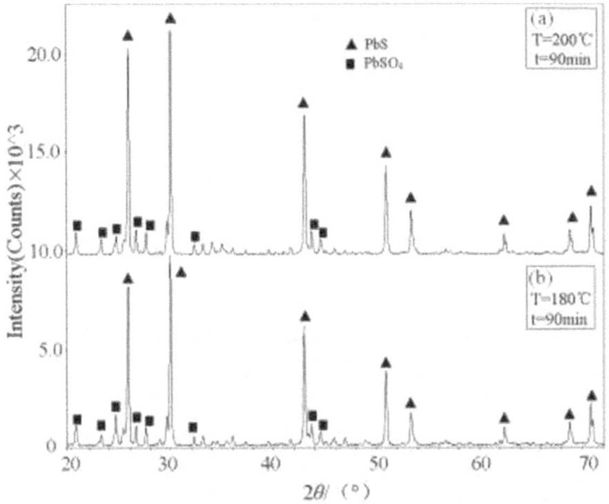

Fig.6- (a) and (b) The XRD patterns of sulfidation products

In Figure 5 we can see XRD patterns for reacted samples which were obtained by with cerussite particle sizes of -74+58 m, which was sulfidized for 90minutes at temperature of 140 and 160°C. Pattern (a) corresponds to sulfidation at 140°C, and pattern (b) corresponds to sulfidation at 160°C. Since at 140°C and 90 minutes of reaction the sample was only partially reacted, we can observe diffraction lines for the reactants cerussite and as well as for the products PbS. Pattern (b) is similar to the pattern obtained at 140°C; again, here, we can only see diffraction lines for cerusstie and PbS, and the characteristic PbS peak became higher with an increase in sulfidation temperature.

Figure 6 shows XRD patterns for samples reacted at 180°C and 200°C for 90 minutes. For the sample sulfidized at 180°C (pattern (a)), the diffraction lines for cerusstie are no longer observed, but strong lines for PbS and weak lines for $PbSO_4$ were identified instead. Therefore, at 180°C, PbS and $PbSO_4$ are the stable reaction products of cerussite sulfidation. As was shown in pattern (b) the characteristic peaks of PbS become higher with an reaction temperature increased to 200°C. The absence of the cerussite peaks in this case may indicate that at these conditions, the sulfidation reaction was completed or near completion, and PbS and $PbSO_4$ were formed as stable reaction products in the sulfidation of cerussite.

Conclusions

In the present study cerussite was sulfidized with elemental sulfur under hydrothermal conditions and the following results can be concluded.

1. The variables considered in the study were temperature, time, particle size and elemental sulfur fraction in the reacting mixture. Temperature was the most important factor on sulfidation of cerussite, with an increase in temperature the sulfidation rate of lead improved greatly and the constituent of products changed a lot.

2. At temperatures above 180°C, PbS and $PbSO_4$ were the only stable reaction products in the sulfidation of cerussite.

3. The cerussite was sulfidized with elemental sulfur under hydrothermal conditions with a particle size of -74+58 m and sulfur fraction in reacting mixture of 15% at 200 °C for 120 min, and 92% lead sulfidation were achieved.

Acknowledgements

The authors gratefully acknowledge the NSFC (Natural Science Fund Council, China, 50904030, 51364022, 51474117) and Yunnan Province Applied Basic Research Item (project no. ZD2014003) for the financial support of this work.

References

Barbaro, M.,2000. *Lead and zinc ores-flotation* (Rome: Academic Press), 15-34.

Caproni, G, et al., "The processing of oxidized lead and zinc ores in the Campo Pisano and San Giovanni Plants," Minerals Engineering, 3(13)(1979), 71-91.

Fuerstenau, D.W., Raghavan, S., "Surface chemical properties of oxide copper minerals" (paper presented at the Advances in Mineral Processing meeting, Littleton 1986)

Fuerstenau, D.W., Sotillo, F., and Valdivieso, A. "Sulfidization and Flotation Behavior of Anglesite, Cerussite and Galena" (paper presented at the 15th International Mineral Processing Congress , France, 1985), 74-86.

Fuerstenau, M.C. et al, "The Surface Characteristics and Flotation Behavior of Anglesite and Cerussite," International Journal of Mineral Processing, 20(1986.), 73-85

Guy, P.J., Trahar, W.J, "The effects of oxidation and mineral interaction on sulphide flotation" (paper presented at the Flotation of Sulphide Minerals meeting , Amsterdam, 1986), 91–109.

Herrera-Urbina, R., Sotillo, F.J., and Fuerstenau, D.W., "Amyl xanthate uptake by natural and sulfide-treated cerussite and galena," International Journal of Mineral Processing, 55(1986), 113–128.

Malghan, S.G., "Role of sodium sulfide in the flotation of oxidized copper, lead, and zinc ores," Miner. Metall. Process., 3(1986.), 158-163.

Marabini, A.M., Alesse, V., and Garbassi, F., "Role of sodium sulphide, xanthate and amine in flotation of lead–zinc oxidized ores" (paper presented at the IMM Reagents in the Minerals Industry, London), 125–136.

Önal, G, et al., "Flotation of Aladag oxide lead–zinc ores. Minerals Engineering, 18(2)(1986), 279–282.

Rey, M., "The flotation of oxidized ores of lead copper and zinc" (paper presented at the IMM Recent Developments in Mineral Dressing, London, 1953) 541–548.

Sotillo, F.J., Fuerstenau, D.W., 1988. The sulfidization and flotation of cerussite and galena. In: Castro Flores, S.H., Alvarez Moisan, J. (Eds.), Froth Flotation. Elsevier, Amsterdam, pp. 271–288.

7th International Symposium on High-Temperature Metallurgical Processing
Edited by: Jiann-Yang Hwang, Tao Jiang, P. Chris Pistorius, Gerardo R.F. Alvear F., Onuralp Yücel, Liyuan Cai, Baojun Zhao, Dean Gregurek, and Varadarajan Seshadri
TMS (The Minerals, Metals & Materials Society), 2016

INDIRECT EXPERIMENTAL STUDY ON THE OXIDATION OF HOT METAL BEARING VANADIUM AND CHROMIUM

Xuan Liu[1], Jiang Diao[1*], Yong Qiao[1], Tao Zhang[2], Bing Xie[1]

[1]College of Materials Science and Engineering, Chongqing University, Chongqing 400044, P. R. China
[2]Department of Scientific Research, Chongqing University of Education, Chongqing 400065, P. R. China

Keywords: Oxidation; Vanadium; Chromium; Distribution ratio; Equilibrium experiment

Abstract

The equilibrium experiments between the $FeO-SiO_2-MnO-TiO_2-V_2O_3-Cr_2O_3-MgO$ slag system and solid pure iron were performed to investigate the vanadium and chromium distribution ratio under Ar atmosphere. The influence of w(FeO), w(V_2O_3) and w(Cr_2O_3) in slag on vanadium and chromium distribution ratios were studied at 1653 K. The experimental results were compared with calculated distribution ratios based on molecule and ions coexistence theory. Both of them showed that higher FeO content in slag would improve the distribution ratios of V and Cr between slag and hot metal. The distribution ratio of Cr or V increased with the increase of Cr_2O_3 or V_2O_3 content, respectively. To optimize the oxidation of V and Cr from hot metal, a high oxidizing slag and suitable content of Cr_2O_3 and V_2O_3 in the slag were recommended. The results provided reference for the further study on the extraction of V and Cr in V-Ti magnetite.

Introduction

The most important V-Ti magnetite ore bodies which are located in Panzhihua, China, are rich in multiple valuable elements e. g. titanium, vanadium and chromium. Vanadium and chromium would be dissolved into hot metal, which gained from V-Ti magnetite ore, in BOF. In order to make full use of valuable resources in iron ores, valuable elements in hot metal would be oxidized in BF. The current technique route for recovery of valuable elements in V-Ti magnetite of Panzhihua plant is shown in figure 1. Titanium is extracted from titanium concentrate or from blast furnace slag. V and Cr would dissolve in hot metal since the oxidation reactions of V and Cr could not proceed in complete reduction atmosphere in BF. [1] To meet the different thermodynamics demand for the oxidation of V, Cr and C, the hot metal would be blew twice.

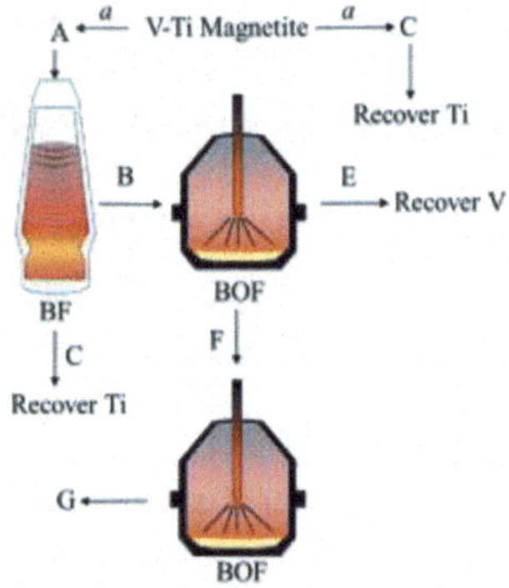

Fig. 1. Current technique route of recovery of valuable elements in V-Ti magnetite. A: Iron concentrate; B: Hot metal; C: BF Slag; D: Titanium concentrate; E: Vanadium Slag; F: Semi-steel; G:Molten steel; *a*: Beneficiation;

Choosing an efficient oxidizing slag with proper slag composition is the key to extract as much as possible amount of V and Cr from hot metal. Many fundamental researches have investigated the oxidation behavior of V from hot metal or molten steel. Dong et al. and Zhang et al. [2,3] studied the oxidation of V from hot metal with the FeO-SiO_2-MnO(-TiO_2) slag system under argon atmosphere and analyzed the effects of influential factors on the oxidization ratio of V such as slag composition, temperature, Si, Ti content in hot metal . Zeng et al. [4] studied the oxidation of V from hot metal under atmosphere at 1623 K, 1653 K and 1669 K. It was found that FeO has obvious effect on the oxidation ratio of V. The molecule and ions coexistence theory (abbreviation as IMCT) also was widely used in metallurgical thermodynamics researches such as distribution ratios, capacity, etc.[3,4]

However, the oxidation of Cr is extensively studied in the refining process of molten steel for the content of Cr should be controlled strictly for many grades of steel. The distribution ratio of Cr between slag and molten steel has been studied with the variation of temperature, quantity of slag, slag composition.[5-7] Kunikat et al.[8] investigated the distribution ratio of Cr between BaO-MnO binary slag and Fe-Cr-Mn-C_{satd} alloys. It can be found from these researches that higher oxidation ability of slag is benefit for the oxidation of Cr.

Single element systems are widely studied in previous researches. However, the single element system may not fit the present hot metal which contains both V and Cr elements. Thus, the distribution ratios of V and Cr between hot metal and slag need to be studied deeply. In the current work, the distribution ratio of V and Cr between FeO-SiO_2-MnO-TiO_2-V_2O_3-Cr_2O_3-MgO slag and hot metal is investigated through laboratory indirect experiments, where the influences of the contents of FeO, Cr_2O_3 and V_2O_3 in slag have been examined. On another hand, the distribution ratios calculated by IMCT was used to compare with experimental results. The aim of this paper are thus to obtain a basic fundamental understanding of the oxidation of V and Cr with oxidizing slag system and thus provide some data to Panzhihua plant.

Principles

As the limit of the thermodynamics conditions of V, Cr and C oxidation, the oxidation temperature V and Cr should be controlled around 1653 K. [4] In this case, the carbon in hot metal cannot be oxidized completely, or the temperature of hot metal is so high that the oxidation of V and Cr would be restrained. Since there is nearly 4% carbon existing in the semi-steel, the oxidation reaction between the interface of slag and metal at the end of first blowing cannot reach equilibrium state. Therefore, it is difficult to study the oxidation of V and Cr from semi-steel directly since the existent FeO in the slag would prior react with the carbon dissolved in the semi-steel, which causes the change of the composition of the slag during oxidizing reaction period. Therefore, we adopt the indirect method in present work.

When a solid iron (phase I) is in equilibrium with a slag, the semi-steel (phase II) which has the same chemical potential for V and Cr as that of solid iron should be also in equilibrium with the slag regarding V and Cr. [12,13] When the solid iron has the same chemical potential for V and Cr as that of semi-steel, the relation of the mass fractions of V and Cr in phase I and phase II can be deduced. Equation (1) shows the relation which was transformed from related thermodynamics data.

$$\ln(A) = \ln\left(\frac{w[\mathrm{V/Cr}]^{\mathrm{II}}}{w[\mathrm{V/Cr}]^{\mathrm{I}}}\right) = \frac{\left(\mu_{\mathrm{V/Cr}}^{\theta_{\mathrm{I}}} - \mu_{\mathrm{V/Cr}}^{\theta_{\mathrm{II}}} + \mathrm{R}T\ln\gamma_{\mathrm{V/Cr}}^{\mathrm{I}}\right)}{\mathrm{R}T} - \ln\gamma_{\mathrm{V/Cr}}^{\mathrm{II}} \tag{1}$$

Where, $w[\mathrm{V/Cr}]^{\mathrm{I}}$ and $w[\mathrm{V/Cr}]^{\mathrm{II}}$ represent the mass fraction of V or Cr in phase I and II, respectively; $\mu_{\mathrm{V/Cr}}^{\theta_{\mathrm{I}}}$ and $\mu_{\mathrm{V/Cr}}^{\theta_{\mathrm{II}}}$ represent the standard chemical potential of V or Cr in phase I and II; T is kelvin temperature; R is gas constant the value equal 8.314; $\gamma_{\mathrm{V/Cr}}^{\mathrm{I}}$ and $\gamma_{\mathrm{V/Cr}}^{\mathrm{II}}$ are the activity coefficients of V or Cr in phase I and II with reference to 1 Mass% of V and Cr at infinite dilute solution.

The values of $\mu_{\mathrm{V/Cr}}^{\theta_{\mathrm{I}}}$ and $\mu_{\mathrm{V/Cr}}^{\theta_{\mathrm{II}}}$ are same, if the slag and hot metal are in equilibrium condition system with $\mu_{\mathrm{V/Cr}}^{\theta_{\mathrm{I}}}$ and $\mu_{\mathrm{V/Cr}}^{\theta_{\mathrm{II}}}$ reference to same standard. $\mu_{\mathrm{V/Cr}}^{\theta_{\mathrm{I}}}$, $\mu_{\mathrm{V/Cr}}^{\theta_{\mathrm{II}}}$, $\gamma_{\mathrm{V/Cr}}^{\mathrm{I}}$ and $\gamma_{\mathrm{V/Cr}}^{\mathrm{II}}$ are calculated with references are listed in table I. The calculated $\ln(A)$ values of Cr and V with equation (1) are 0.09 and 2.00, respectively.

The contents of V and Cr in semi-steel are calculated with equation (2).Thus, the V and Cr distribution ratio between the slag and a solid iron can be converted to the distribution ratios between the slag and a semi-steel by multiplying the values A of V and Cr between solid iron and semi-steel, and then equation (3) can be derived. The equilibrium distribution ratio of V and Cr between semi-steel and oxidizing slag can be calculated with equation (3),

$$w[V/Cr]^{II} = w[V/Cr]^{I} A \tag{2}$$

$$L_{V/Cr} = \frac{w(V/Cr)_s}{w[V/Cr]^{I} A} \tag{3}$$

Table I. The activity coefficients data of V and Cr in solid iron (I) and hot metal (II)

T/K	γ_V^I (Ref.11)	γ_{Cr}^I (Ref.12)	γ_V^{II} (Ref.13)	γ_{Cr}^{II} (Ref.14)
1653	0.97	0.20	0.13	0.18

where, $w(V/Cr)_s$ is the mass fraction of V and Cr in slag determined by XRF, $w[V/Cr]^{I}$ is the mass fraction of V and Cr in solid iron, which is determined by Optical Emission Spectrometer. In the present calculation, the activity of Fe was assumed as 1 according to the references related to thermodynamic application of metallurgical slag and metal system. [15]

Experiments

Experiments were carried out in a $MoSi_2$ furnace which equipped with an automatic temperature control system with temperature errors range of ±2 K. Analytical grade reagents of SiO_2, $MnCO_3$, TiO_2, V_2O_3, Cr_2O_3, Fe_2O_3, Fe and MgO powers were uniformly mixed to prepare the slags which used in experiments. FeO was obtained through the reaction between Fe_2O_3 and Fe powders. An iron crucible has been put into an Al_2O_3 crucible which is used to protect the iron crucible reactor. In the experiments, 80g oxidative FeO-SiO_2-MnO-TiO_2-V_2O_3-Cr_2O_3-MgO slag was melt in a pure iron crucible for 1 hour under argon atmosphere. After an hour, a solid iron piece was added into the slag with constant temperature and argon atmosphere was kept during experimentation. Every solid iron piece has the same weight (40g) and same cylinder shape. The contents of C, Si, Mn, P, S, Cr, V and Ti in the solid iron are listed in table II. It can be see that the content of these elements are extremely small. The experiment time was determined with the pre-experiments. The pre-experiment results show the content of V and Cr do not change with time greatly after 6 hours. Thus, the experimental time was chosen as 7 hours.

After experiments, the slag along with the crucible was quenched in water. Then, the slag was ground and solid iron sample was cleaned carefully before analysis. The composition of the slag and the content of V and Cr in solid iron pieces are detected by X-ray fluorescence (XRF) and Optical Emission Spectrometer, respectively. Based on experimental data and the known thermodynamic data, the contents of V and Cr in semi-steel equilibrate with the same composition slag at same temperature can be calculated with equation (2). Thus, the distribution ratios of V and Cr between semi-steel and the oxidative slag could be determined.

Table II. The mass content of C, Si, Mn, P, S, Cr, V and Ti in pure iron samples used in experiments.

Sample No.	wt-%							
	C	Si	Mn	P	S	Cr	V	Ti
1	0.0026	0.0006	0.0174	0.0053	0.0124	0.0120	<0.0001	<0.0001
2	0.0030	0.0006	0.0250	0.0050	0.0100	0.0200	<0.0001	<0.0001

Results and Discussion

The content of FeO, Cr_2O_3 and V_2O_3 are the influential factors in experiments. The contents of C, Cr and V in solid iron samples and FeO, V_2O_3 and Cr_2O_3 in the slag after experiments are listed in table III. The calculated distribution ratios based on IMCT were used to compare with experimental results in discussion.

The experimental and calculated distribution ratios of V and Cr vary with w(FeO) are shown in figure. 2. V and Cr distribution ratios are increasing with the FeO increasing from 41.67% to 54.30%, which is in agreement with the study of Zeng et al. [4] FeO provides large amount of oxygen ions and increases the oxygen potential which benefits the oxidation of V and Cr. Generally. With the increase of FeO in slag, the activity of FeO and the oxidation property in slag increase. [16]

Table III. The content of C, Cr and V in solid iron samples and FeO, V_2O_3 and Cr_2O_3 of final vanadium slag in this experiments.

Variable	No.	solid iron wt-%			slag composition wt-%			
		C	Cr	V	FeO	V_2O_3	Cr_2O_3	T(K)
w(FeO)	1	0.004	0.019	0.0018	41.67	7.49	5.96	1653
	2	0.0078	0.0105	0.0011	44.07	8.38	7	1653
	3	0	0.0061	0.0003	45.2	7.11	5.57	1653
	4	0.001	0.0062	0.0003	48.41	6.81	5.51	1653
	5	0	0.012	0.0003	52.12	5.88	4.72	1653
	6	0	0.004	0.0002	54.3	5.83	4.64	1653
w(Cr_2O_3)	7	0	0.0168	0.0016	51.42	7.44	2.54	1653
	8	0.0001	0.017	0.0026	52.78	10.61	5.16	1653
	9	0.0001	0.009	0.0014	46.53	9.79	6.59	1653
	10	0.0078	0.0105	0.0011	44.07	8.38	7	1653
	11	0	0.018	0.0019	49.75	6.48	8.24	1653
w(V_2O_3)	12	0	0.0013	0.0004	51.88	6.83	7.72	1653
	13	0.001	0.0011	0.0016	44.07	8.38	7	1653
	14	0.001	0.0013	0.0007	51.42	10.35	7.37	1653

	15	0.001	0.0005	0.0006	52.76	10.32	6.38	1653
	16	0	0.0013	0.0007	46.71	11.19	5.64	1653
	17	0.001	0.001	0.0007	46.41	19.6	5.07	1653

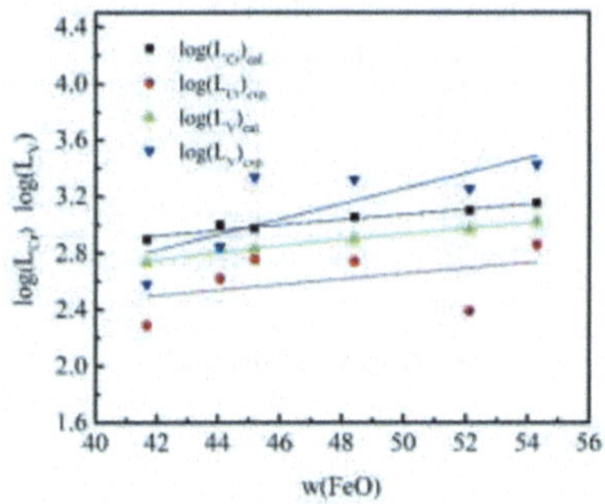

Fig. 2. The variations of the distribution ratio of V and Cr with the increase of w（FeO）in slag.

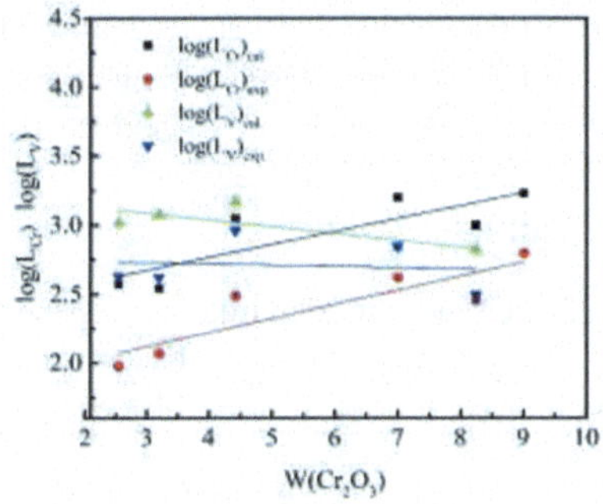

(a)

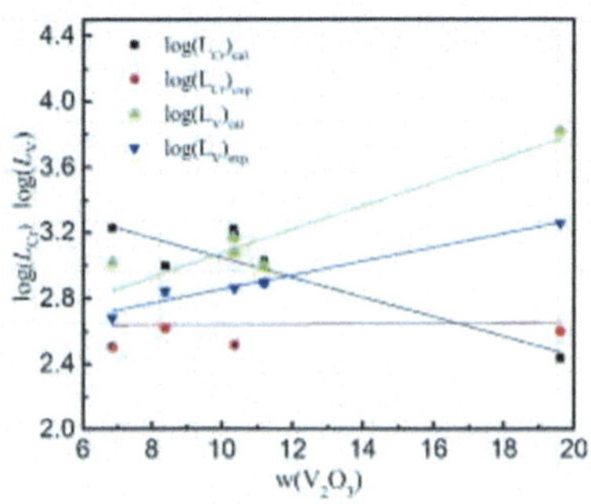

(b)

Fig. 3. The variations of the distribution ratio of V and Cr in slag with the increase of (a) Cr_2O_3 and (b) V_2O_3.

The effect of Cr_2O_3 and V_2O_3 content in slag on the V and Cr distribution ratios between FeO-SiO_2-MnO-TiO_2-V_2O_3-Cr_2O_3-MgO and hot metal are illustrated in figure. 3. Cr distribution ratio increases with the increase of Cr_2O_3 content in slag in figure. 3a. At the same time, it can be seen from figure. 3b that the distribution ratio of V increases with the increase of V_2O_3 in slag. As Cr_2O_3 content increases from 2.52 to 8.24 or V_2O_3 content increases from 6.83 to 19.60, the value of V and Cr distribution ratio does not show a significant change. The mainly distribution ratios calculated with IMCT shows same tendency as experimental results present. It reveals that the indirect experimental method is reasonable in the study of the oxidation behavior of hot metal.

Conclusion

1. FeO can improve the oxidation of V and Cr. The distribution ratios of V and Cr increase with the increase of FeO. Increasing Cr_2O_3 or V_2O_3 content only improve the distribution ratio of Cr or V, respectively. The increase of SiO_2 content decreases the distribution ratio of V and Cr.

2. To optimize the oxidation of V and Cr from hot metal, a high oxidative slag and suitable content of Cr_2O_3 and V_2O_3 in the slag is recommended, with temperature control around 1653 K.

3. Most magnitudes of V and Cr distribution ratios are in the range from 2.0 to 3.5. Cr has higher values compares with V. This indicates that Cr can be oxidized easier than V from hot metal.

4. The calculated distribution ratios were compared with experimental results and mainly of them verified the experimental results. The indirect experimental method and model is reasonable in this study.

Acknowledgements

This work was partially supported by the Fundamental Research Funds for the Central Universities (Project CDJZR 14130001), National Natural Science Foundation of China (No. 51090382) and National Basic Research Program of China (No. 2013CB632604). We express our gratitude to all those who helped us during the preparation of this paper.

References

1. H.J. Du (eds), Z. X. Yang, Y. Z. Li, W. Z. Wang and X. Y. Shi: *The mechanism of the smelting of V-Ti magnetite in Blast furnace* (Beijing, Science Press, 1996),44.

2. Y. L. Zhang, F. Zhao, and Y. G. Wang, "Effects of Influencing Factors on Distribution Behaviors of Vanadium between Hot Metal and FeO-SiO_2-MnO (-TiO_2) Slag System," *Steel Research International*, 82 (8) (2011), 940-950.

3. J. M. Dong, F. Zhao, Y. L. Zhang, S. T. Qiu, and Y. Gan, "Distribution Behaviors of Vanadium between Hot Metal and FeO − SiO2 − MnO(− TiO2) Slag System and Influential Factors," *The Chinese Journal of Process Engineering*, 10 (6) (2005), 1076-

1083.
4. X.P. Zeng, B Xie, C.Y. Zhao, and Q.Y. Huang, "Vanadium distribution behavior at the end point of vanadium extraction by converter process," *Journal of University of Science and Technology Beijing*, 12 (2012), 1371-1378.
5. S. Kitamura, H. Aoki, and K. Okohira, "Dephosphorization by CaO-based Flux Reaction of ChromiumContaining Molten Iron," *ISIJ International*, 34 (5) (1994), 401-407.
6. T. Nakasuga, K. Nakashima, and K. Mori, "Recovery Rate of Chromium from Stainless Slag by Iron Melts," *ISIJ International*, 44 (4) (2004), 666-672.
7. G. J. Albertsson, F. Engström, and L. Teng, "Effect of the Heat Treatment on the Chromium Partition in Cr- Containing Industrial and Synthetic Slags," *Steel Research International*, 85 (10) (2014),: 1418-1431.
8. Y. Kunikate, K. Morita, F. Sukihashi, and N. Sano, "Equilibrium BaO-MnO Distribution Ratios of Phosphorusand Chromiumbetween Melts and Carbon Saturated Fe-Cr-Mn Alloys at 1573K,". *ISIJ International*, , 34 (10) (1994), 810-814.
9. K. Ito, and N. Sano, "Phosphorus Distribution between Slag and Metal at Hot Metal Temperatures,". *Tetsu-to-Hagané*, 67 (3) (1981), 536-540.
10. G. Li, C. Zhu, Y. Li, X. Huang, and M. Chen, "The Effect of Na_2O and K_2O on the Partition Ratio of Phosphorus between $CaO–SiO_2–Fe_tO–P_2O_5$ Slag and Carbon-Saturated Iron," *steel research international*, 84 (7) (2013), 687-694.
11. O. Kubaschewski, H. Probst, and K. H. Gkiger, "The Thermochemical Properties of Solid and Liquid Iron-Vanadium Alloys," *Zeitschrift für Physikalische Chemei Neue Folge*, 104(1977), 23-30.
12. R. Benz, J. F. Elliott, and J. Chipman, "Thermodynamics of the Carbides in the System Fe-Cr-C,"*Metallurgical Transactions*, 5 (1974), 2235-2240
13. G. Qi, C. Ji, and Y. Che," Thermodynamics Activity Interaction Coefficients in Fe-C-V Melts," *Transaction of the Japan Institute of Metals*, 28 (5) (1987), 412-423.
14. H. G. Hadrys, M. G. Frohberg, J. F. Elliott, and C. H. P. Lupis, "Activities in the Liquid Fe-Cr-C(sat), Fe-P-C(sat) and Fe-Cr-P Systems at 1600℃," *Metallurgical Transactions*, 1 (1970), 1867-1874.
15. X.-m. Yang, M. Zhang, J.-l. Zhang, P.-c. Li, J.-y. Li, and J. Zhang. "Representation of Oxidation Ability for Metallurgical Slags Based on the Ion and Molecule Coexistence Theory," Steel Research International, , 85 (3) (2014), 347-375.
16. S. Basu, A. K. Lahiri, and S. Seetharaman, "Activity of Iron Oxide in Steelmaking Slag," *Metallurgical and Materials Transactions B*, 39 (3) (2008), 447-456.

7th International Symposium on High-Temperature Metallurgical Processing
Edited by: Jiann-Yang Hwang, Tao Jiang, P. Chris Pistorius, Gerardo R.F. Alvear F., Onuralp Yücel, Liyuan Cai, Baojun Zhao, Dean Gregurek, and Varadarajan Seshadri
TMS (The Minerals, Metals & Materials Society), 2016

Influence of CaO on Non-isothermal Crystallization Kinetics of Spinels in Vanadium Slag

Wang Zhou[1], Bing Xie[1*], Wen-Feng Tan[1], Jiang Diao[1], Hong-Yi Li[1], Zhang Tao[2]

1 College of Materials Science and Engineering, Chongqing University, Chongqing 400044, China
2 Department of Science and Technology, Chongqing University of Education, Chongqing 400065, China

*Corresponding author: bingxie@cqu.edu.cn

Abstract

Adding CaO into the vanadium extraction converter is an effective way to decrease the dephosphorization burden in subsequent steelmaking process. The crystallisation process of spinel is one of the most decisive factors in the subsequent extraction. The theoretical calculation results show that the temperature of the peak total crystallization rate will decrease 5K with basicity increasing 0.1. The relation between cooling conditions and grain size of spinels is also discussed. It is found that low cooling rate and high basicity benefit spinel crystal growth. And the non-isothermal crystallization kinetics results indicated that the restriction factors of spinel growth is that solid crystals accept new components from liquid-solid interfacial. While the sample was cooled down to 1250℃(1523K) at cooling rate of 5K/min, the crystal growth is consistent with Ostwald coarsening grew up law, and controlled by diffusion of particle.

Keywords: vanadium slag, spinel crystals, crystallization, kinetics

Introduction

Due to the excellent physical properties such as high tensile strength, hardness and fatigue resistance of vanadium alloys, vanadium plays an extraordinarily important role in steel making, chemical industry and aviation field [1]. There is no single recoverable vanadium mineral, and the distribution of vanadium is fragmented, it is mainly associated to patronite, vanadinite, carnotite and vanadium-titanium magnetite [2]. In BF or other iron making process, vanadium-titanium magnetite is the major raw material to produce vanadium-containing hot metal which is then oxidized in LD converter to obtain vanadium slag [2, 3]. The obtained vanadium slag is traditionally sodium roasted followed by water leaching to extract vanadium. The sodium roasting-water leaching process requires low CaO content to avoid the formation of water-insoluble calcium vanadate [4]. However, low CaO content vanadium slag lacks the ability of dephosphorization, then a heavy dephosphorization burden follows in the subsequent steelmaking process. At the same time there are some environmental issues during the sodium roasting-water leaching process [4], and consequently a great deal of studies focus on calcium roasting extracting vanadium process which is an environment-friendly technique [5-7]. In contrast with the prior technic, the calcium roasting technique allows high CaO contents in the vanadium slag, and consequently inspires the idea of removing phosphorus contained in the hot metal by adding CaO during the production of vanadium slag. Not only does it decrease the stress of phosphorus removal during steelmaking, but also avoids the problems of sodium roasting-water leaching process for that the high calcium vanadium slag can be roasted without salt addition followed by alkaline leaching.

Its ability of dephosphrization will be improved by increasing CaO content during slagging, but the relationship between CaO content and crystallization behavior of vanadium spinel is

not clear and definite. The non-isothermal crystallization process of vanadium spinel in synthesized high calcium and high phosphorus slag based on the industrial slag (the composition was shown in Table 2) was studied by high temperature quenching samples-image analysis method. For describing the influence on spinel crystal, the basicity of vanadium slag, which is usually represented by R(CaO/SiO2), substitutes the CaO content.
This work was financially supported by National Natural Science Foundation of China (grant no.21105134 and no.51474041), National Basic Research Program of China (973 Program, grant no.2013CB632604) and Fundamental Research Funds for the Central Universities of China (grant no.CDJRC10130010, no.CDJZR12135502 and no. CDJZR 14130001).

Theoretical Calculation

It is widely accepted that the crystallisation of spinels in vanadium slag can be divided into two stages: nucleation and crystal growth. According to the classic crystallization kinetic model reported by Turnbull [8,9], Zhang et al.[10] calculated the nucleation rate and growth rate of spinel in vanadium slag. The nucleation rate and growth rate can be expressed as

$$I = \frac{N_V \kappa T}{3\pi a^3 \eta} \exp\left[\frac{-16\pi\alpha^3\beta}{3T_r(\Delta T_r)^2}\right] \tag{1}$$

$$U = \frac{f\kappa T}{3\pi a^2 \eta}\left[1 - \exp\left(\frac{-\beta\cdot\Delta T_r}{T_r}\right)\right] \tag{2}$$

where I is the nucleation rate per unit volume, U is growth rate, N_v is the number of molecules (or atoms) per unit volume, κ is Boltzmann's constant, T is the absolute temperature, a is the lattice parameter of the crystal, η is viscosity, f is the fraction of acceptor sites in the crystal surface, α is the reduced crystal/liquid interfacial tension, β is the reduced molar heat of fusion, $T_r = T/T_m$ is the reduced temperature, $\Delta T_r = 1 - Tr$ and T_m is the melting point of the crystal and expressed in K.
The main spinel phases of vanadium slag in Pan-steel are MgV_2O_4, $Fe_xV_{3-x}O_4$, $Mn_xV_{3-x}O_4$ ($1 \leq x \leq 3$) and Fe_2TiO_4 spinels. Diao et al. [11] studies showed that V and Ti are concentrated in the spinel phases. Nevertheless, since the concentrations of Mg and Mn are fairly low to be compared with Fe, and the similarity of Fe spinels and (Mg, Mn) spinels is relatively high, only the Fe spinels are taken into consideration in the present study, namely FeV_2O_4 and Fe_2TiO_4. The parameters of structure and melting point of these spinel crystals are given in Table 1 [12].

Table 1 Values of parameter used in calculation

Spinel type	a (Å)	N_v	T_m (℃)	α	β	f	κ
FeV_2O_4	8.543	1.60387×10^{27}	1750	1/3	1	$0.2\Delta T_r$	1.38×10^{-23}
Fe_2TiO_4	8.378	1.70051×10^{27}	1395				

Experimental

Preparation of Slag Samples

Initial materials were chemicals with purities of $FeC_2O_4\cdot 2H_2O$ (≥98.0%), SiO_2 (≥99.0%), V_2O_3 (≥99.0%), TiO_2 (≥99.0%), $MnCO_3$ (≥99.0%), MgO (≥98.0%), CaO (≥98.0%), $3CaO\cdot P_2O_5$ (≥98.0%) supplied by Chengdu Kelong Chemical Co. Ltd. Reagent of CaO was calcined at 1373 K for 12 h before weighing and mixing. The initial materials were mixed to synthesize vanadium slag samples containing 2 mass% P_2O_5 according to the chemical compositions in Table 2. The mixture of initial materials was placed in 10 kW box-type electric resistance furnace with proportional-integral-differential (PID) controller (±2 K) for heating.

Table 2 Chemical composition of slag samples, wt%

	FeO	SiO_2	Al_2O_3	CaO	V_2O_3	TiO_2	MgO	MnO	P_2O_5	R(CaO/SiO_2)
industrial slag	39.38	16.07	3.08	2.40	13.01	11.11	4.12	8.32	0.00	0.15
Ca1	30.00	13.48	3.00	17.52	13.00	10.00	3.00	8.00	2.00	1.30
Ca2	30.00	12.40	3.00	18.60	13.00	10.00	3.00	8.00	2.00	1.50
Ca3	30.00	11.92	3.00	19.08	13.00	10.00	3.00	8.00	2.00	1.60

Experimental method

200 g mixture was charged into a corundum crucible (inner diameter, 46 mm; height, 120 mm) which was then placed in another graphite crucible (inner diameter, 53 mm; height,140 mm). They were placed in the furnace at 1073 K and heated to 1373 K at the rate of 3 K/min to allow the reagents of $FeC_2O_4 \cdot 2H_2O$ and $MnCO_3$ to decompose to FeO (totally assumed as FeO) and MnO. Then the mixture was heated to 1773 K (1500℃) and held at 1773 K for 30 min to ensure homogeneity. And subsequently cooling down the mixture to 1723K(1450 ℃), 1673K(1400℃), 1623K(1350℃), 1573K(1300℃), 1523K(1250℃) at the cooling rate of 5K/min and 15K/min respectively to get the corresponding samples. The samples were quickly cooled by water, then slag samples are polished by silicon carbide. Morphology of the vanadium slag is studied by scanning electron microscopy (SEM). Mean diameter of V-containing spinel phases are measured in five different view fields by image analysis software.

Principle

In this paper, the non-isothermal crystallization kinetics of vanadium spinel is analyzed by Crystal Size Distribution (CSD) which is put forward to describe the nucleation and growth of crystal. And in order to investigate the distribution of crystal size, the method of Wager [14] was applied to transform quantity of crystal unit area (N_A) into unit volume (N_V), expressed as Eqs. (3).

$$N_V = N_A^{1.5} \tag{3}$$

The lognormal distribution function is used to describe distribution of crystals, which is expressed as Eqs. (4). The α and β^2 represent the mean and the variance of the lognormal distribution function respectively. The relationship of them in divers cooling rate is expressed with corresponding scale drawing. The crystal growth mechanism could be deduced from the relationship of α and β^2. When the curve of CSD appears the lognormal distribution function, the crystal growth is controlled by the surface as β^2 have linear increase with α increasing. Namely, the kinetic restrictive step of crystal growth is that solid crystals accept new components from liquid-solid interfacial. However, the crystal growth is controlled by the diffusion as α increase while β^2 remains a constant. It means that the diffusion of particle from liquid to interface becomes the kinetic restrictive step.

$$f(D) = \frac{1}{D\beta\sqrt{2\pi}}\exp[-\frac{(\ln D-\alpha)^2}{2\beta^2}] \tag{4}$$

Results and discussion

Influence of CaO on mean diameter of spinels

The mean diameter of vanadium spinel in different basicity was obtained by image statistics as shown in Fig.1. It can be seen that the mean diameter of spinel increased significantly with the basicity increasing from 1.3 to 1.6. There is a common phenomenon that the grain size of spinel increased slowly from 6 m to 20 m approximately with the temperature decreasing from 1450℃ to 1300℃. However the accelerated growth of spinel occurs while the samples were cooled down to 1250℃(1523K).

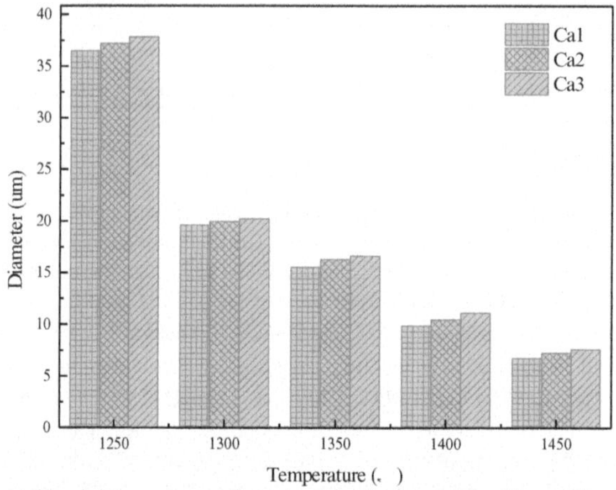

Fig.1 The effect of basicity on mean diameter of spinel crystal with cooling rate of 5K/min

Nucleation rate and growth rate of spinels

Effect of temperature on nucleation rate and growth rate is calculated by MATLAB7.0 and shown in the Fig.2 and Fig.3. It can be seen that there is an optimum temperature range for the spinel nucleation and growth. Fig.2 and Fig.3 (a~c) respectively show that the nucleation rate and growth rate of spinel increased with the basicity increasing from 1.3 to 1.6. And the optimum temperature range for the spinel nucleation and growth is basically the same in different basicity. But there is an obvious difference between the optimum temperature range of nucleation and growth in the same basicity. The optimum growth temperature range of FeV_2O_4 is 1400K~1700K and the optimum nucleation temperature range is 900K~1200K.
It is still difficult to give an accurate optimum temperature range for the spinel crystals. Therefore, it is necessary to combine the nucleation and growth rates. Zhang et al. [10] defined the crystallisation rate r as

$$r = \frac{\pi}{3} I U^3 \tag{5}$$

And the total crystallization rate r_{total} is given by

$$r_{\text{total}} = \sum_i \omega_i r_i \tag{6}$$

The calculation results is shown in Fig.4 (a~c). These figures indicate that the optimum temperature range of total crystallization rate in accordance with that of FeV_2O_4. It can be seen from Fig.4 (a~c) that the temperature of the peak total crystallization rate for spinel decreased 5K with the basicity increasing 0.1, and the total crystallization rate increased significantly with the basicity increasing from 1.3 to 1.6. Therefore, the increase of basicity benefits growth of spinel crystals.

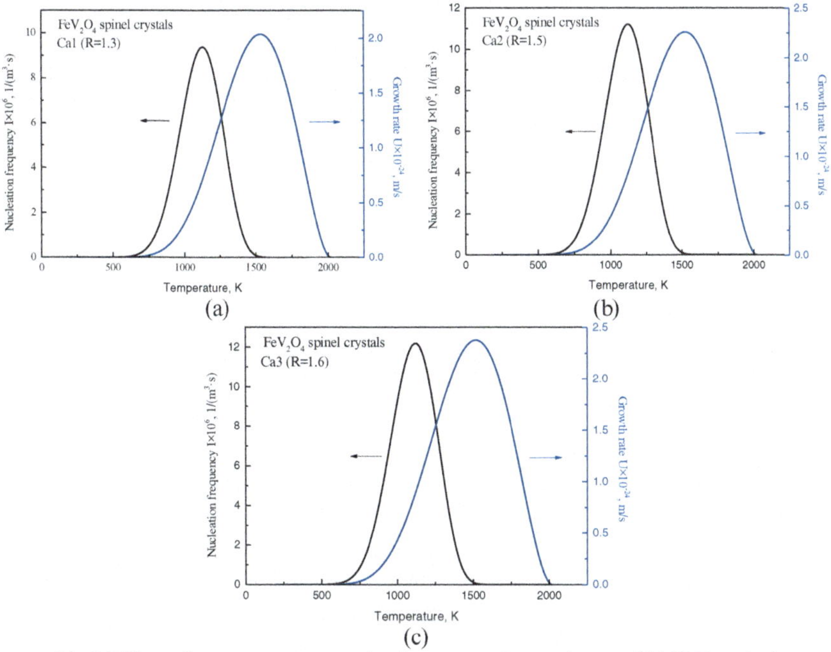

Fig.2 Effect of temperature on nucleation rate and growth rate of FeV_2O_4 spinel

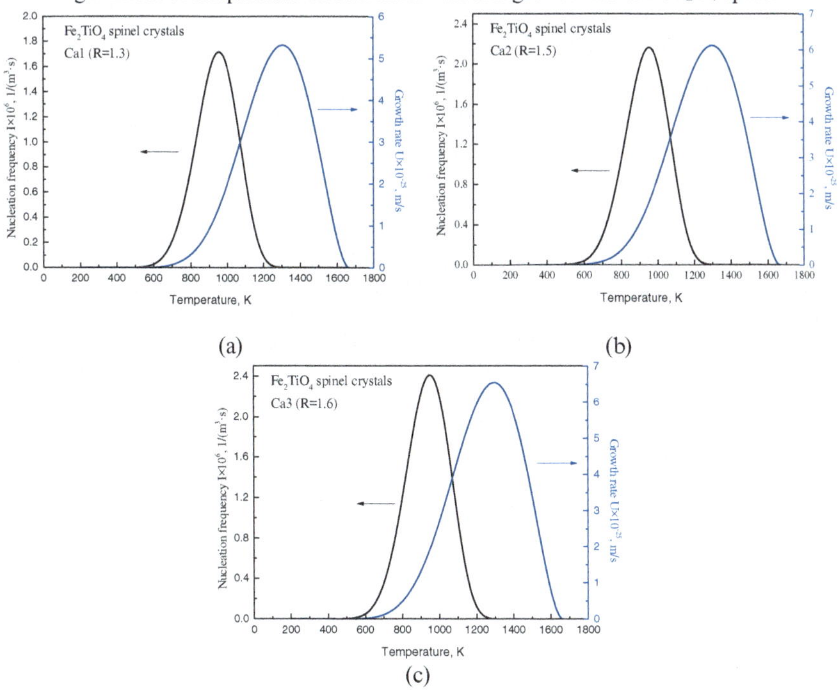

Fig.3 Effect of temperature on nucleation rate and growth rate of Fe_2TiO_4 spinel

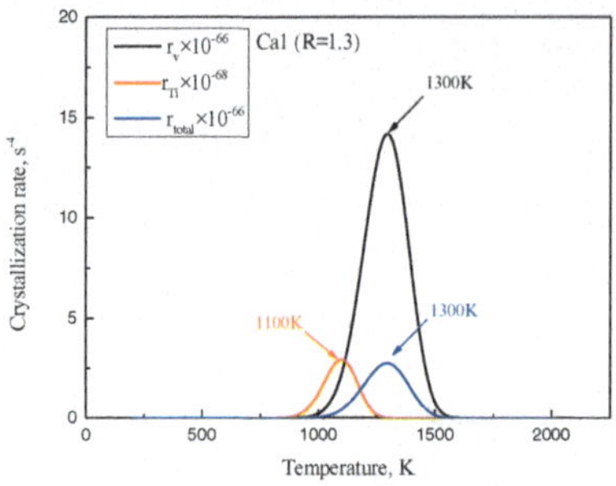

(a)

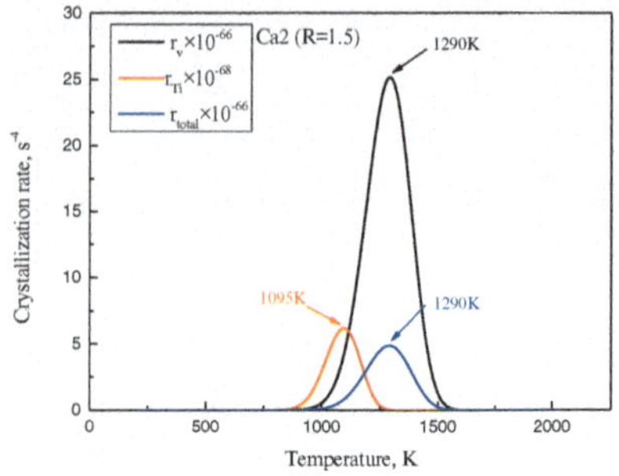

(b)

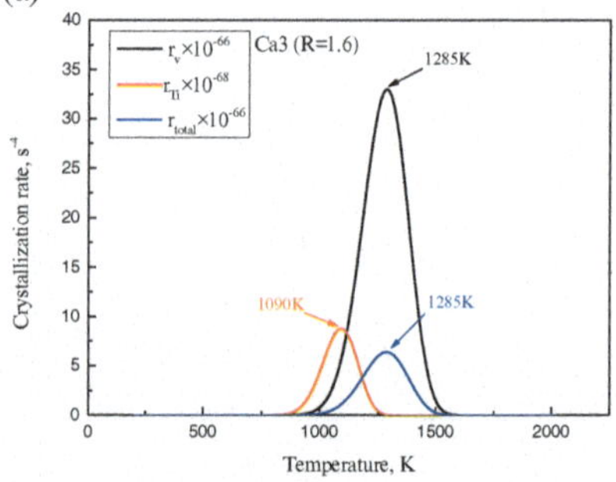

(c)

Fig.4 The relationship of crystallization rate as a function of temperature

The non-isothermal crystallization kinetics of spinels

The curves of CSD involving diverse cooling rate in Ca1 slag whose basicity is 1.3 are plotted in Fig.5. While the cooling rate is 5K/min in Fig.5 (a), it can be seen that the curves peak range is 5~15 m in high temperature, with the temperature decreased, the curves peak is moved to the direction of grain size increment, which reach to 15~25μm. While the cooling rate increased to 15K/min, as Fig.5 (b) shown that the curves peak range is 4~7 m in high temperature, with the temperature decreased to 1250℃, the curve peak range only reaches to 10~20 m. It means that low cooling rate benefits the growth of spinel crystal.

With regard to Ca1, the solution of α and β^2 of the curve of CSD at divers temperature is plotted in the top right corner of Fig.5 (a~b). The corresponding increasing between β^2 *and* α indicates that the crystal growth controlled by interface in the temperature range above. It must be pointed out that the curve of CSD deviates from the lognormal distribution when the cooling rate is 5K/min at the temperature of 1250℃. According to the study of Eberl [15] and Kile [16], it can be deduced that the crystal growth is consistent with Ostwald coarsening grew up law, and controlled by diffusion of particle, while the sample was cooled down to 1250℃ (1523K) at cooling rate of 5K/min. The means of grain growth is that bigger grains swallow smaller grains during the later period of cooling crystallization.

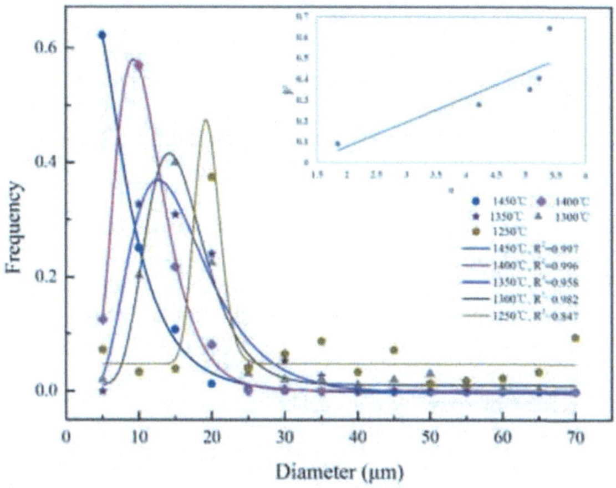

(a) The cooling rate is 5K/min

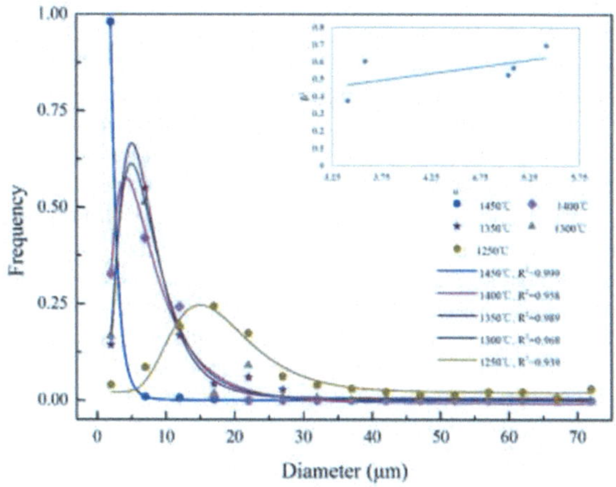

(b) The cooling rate is 15K/min

Fig.5 The crystal size distribution of spinels in Ca1 slag sample at different cooling rate

Conclusions

(1) The temperature of the peak total crystallization rate for spinel decreased 5K with the basicity increasing 0.1, and the total crystallization rate increased significantly with the basicity increasing from 1.3 to 1.6.

(2) The mean diameter of spinel increased significantly with basicity increasing from 1.3 to 1.6. The grain size of spinel increased slowly from 6 m to 20 m approximately with the temperature decreasing from 1450℃ to 1300℃, however, the accelerated growth of spinel occurs while the samples were cooled down to 1250℃(1523K).

(3) Low cooling rate benefits the growth of spinel crystal. The kinetic restrictive step of crystal growth is solid crystals receive new components from liquid-solid interfacial. While the sample was cooled down to 1250℃(1523K) at cooling rate of 5K/min, the crystal growth is consistent with Ostwald coarsening grew up law, and controlled by diffusion of particle.

References

1. R.R. Moskalyk and A.M. Alfantazi, "Processing of vanadium: a review," *Minerals Engineering*, 16(2003), 793-805.
2. S.Z. Yang, *Vanadium metallurgy* (Beijing, BJ: Metallurgical Industry Press, 2010).
3. D.X. Huang, *Vanadium Extraction and Steelmaking* (Beijing, BJ: Metallurgical Industry Press, 2000).
4. H. A. Barolin, tran. by C. L. Wang, *Oxidation of vanadium.* (Beijing: Metallurgical Industry Press, 1982).
5. S.X.Li, "Mechanism research on oxidation roasting and leaching process of high calcium lowgrade vanadium slag," (PH.D. thesis, Chongqing university, 2011)
6. H. Y. Li, N. Wang, B. Xie, Y. Wang, C. Li and H. X. Fang, China Patent, No.2012100667355, (2012).
7. X.J. Li, "The effect of roasting parameters in calcium roasting on the leaching rate of vanadium," *Journal of process engineering*, 12(2012), 56-57.
8. D. Turnbull, "Formation of crystal nuclei in liquid metals," *J. Appl. Phys.*, 1950, 21, (10), 1022–1028.
9. D. Turnbull, "Under what conditions can a glass be formed," *Contemp. Phys.*, 1969, 10, (5), 473–488.
10. X. Zhang X, B. Xie B, J. Diao, X.J. Li. "Nucleation and growth kinetics of spinel crystals in vanadium slag," *Ironmaking & Steelmaking*, 2012, 39(2): 147-154.
11. J. Diao, B. Xie, C. Ji, X. Guo, Y. Wang and X. Li: "Growth ofspinel crystals in vanadium slag and their characterization," *Cryst. Res. Technol.*, 2009, 44, (7), 707–712.
12. X.J. Li, "Research on spinels growth law and calcification roasting mechanism of vanadium slag," (M.D. thesis, Chongqing University, 2011).
13. K.C.Mills, S.Sridhar.Viscosities of Ironmaking and steelmaking slags[J]. Ironmaking and Steelmaking. 1999, 26(4):262-268.
14. L.R. Wager. "A note on the origin of ophitic texture in the chilled olivine gabbro of the Skaergaard intrusion," *Geological Magazine*, 1961, 98(05): 353-364.
15. D.D. Eberl, V.A. Drits, J. Srodon. Deducing growth mechanisms for minerals from the shapes of crystal size distributions[J]. American journal of Science, 1998, 298: 499-533.
16. D.E. Kile, D.D. Eberl, A.R. Hoch A R and M.M. Reddy. An assessment of calcite crystal growth mechanisms based on crystal size distributions[J]. Geochimica et Cosmochimica Acta, 2000, 64(17): 2937-2950.

7th International Symposium on High-Temperature Metallurgical Processing
Edited by: Jiann-Yang Hwang, Tao Jiang, P. Chris Pistorius, Gerardo R.F. Alvear F., Onuralp Yücel, Liyuan Cai, Baojun Zhao, Dean Gregurek, and Varadarajan Seshadri
TMS (The Minerals, Metals & Materials Society), 2016

RECOVERY OF NICKEL AND COPPER FROM POLYMETALLIC SULFIDE CONCENTRATE THROUGH SALT ROASTING USING NH_4Cl

Cong Xu[2], Hongwei Cheng[1,2*], Guangshi Li[1], Changyuan Lu[1], Xingli Zou[1,2], Xionggang Lu[1,2*], Qian Xu[1,2]

[1] School of Materials Science and Engineering, Shanghai University, Shanghai 200072, People's Republic of China
[2] State Key Laboratory of Advanced Special Steel, Shanghai University, Shanghai 200072, People's Republic of China
*Corresponding author: Ph.D., Prof.; Tel.: +86 21 56334042; Fax: +86 21 56335768;
E-mail: hwcheng@shu.edu.cn (H.W. Cheng); luxg@shu.edu.cn (X.G. Lu).

Keywords: sulfide concentrate; ammonium chloride; chlorination; roasting

Abstract

The roasting of nickel-bearing sulfides is a complex but important process which has not been adequately characterized. Combination of X-ray diffraction and scanning electron microscopy has been proved to be useful for such experiments. The leaching results of nickel, copper, iron and cobalt was analyzed by ICP methods. Process possibility of polymetallic sulfide concentrates through a three-step procedure (oxidation roasting - chlorination - water leaching) was suggested. Ammonium chloride could react with nickel/copper oxides to produce intermediate complexes. At the temperature range of 280~340 °C, these complexes further decomposed to the corresponding simple chlorides. Roasting sulfide concentrate using NH_4Cl is effective for complex ores.

Introduction

Nickel is an important transition element and ranks twenty-fourth in abundance among the elements in earth's crust. It is widely used in the production of stainless steel, electroplating, and other miscellaneous purposes such as catalysts. Nickel deposits in the world are mainly found in two forms: sulfide (27%) or laterite (73%). In China, by contrast, the majority of nickel ores (about 91%) is existed in sulfide deposits. As raw materials, however, nickel sulfides generally coexist with other sulfides such as chalcopyrite, pyrrhotite in basic igneous rock intrusions. This makes it difficult to separate them completely from the raw ores. And with the high-grade deposits depleting, the complex multi-metallic nickel deposits are being growingly exploited.

The conventional pyrometallurgical processes involving smelting, converting and electro-refining are applied commercially to recover valuable metals. But these techniques are poor in extracting valuable metals comprehensively, causing serious metal loss in the slag such as cobalt or PGE. Taking account that several metals existed in the ore, it is unsuitable to employ conventional pyrometallurgy directly. So alternative methods are pressing to deal with the complexity of raw ores than before[1-4].

Several studies have been devoted to the chlorination process as an alternative to recovering the metals from sulfides. Typical chlorination reagents are Cl_2, HCl and other alkali or alkaline metal chlorides. However, there is a general preference for sodium chloride or ammonium chloride as it provides a cheaper source of chlorine than Cl_2 or HCl gas. And it was shown that solid ammonium chloride could react with metal oxides to generate soluble metal chlorides at relatively low temperatures compared with the chlorination with sodium chloride[5, 6]. Separation of rare earth through roasting with ammonium chloride has been extensively investigated by Chi et al[7, 8]. And authors developed a process for the chlorination of copper-cobalt oxide ores with ammonium chloride. They observed that the chlorination of metal oxide with ammonia chloride was feasible, and it converted oxides to metal chlorides at about 300 °C. Also, this process shows great preference to transition elements such as iron, cobalt, nickel and copper but cannot react with silicon, aluminium or magnesium[9].

Ammonium chloride, is a by-product of producing sodium carbonate, sodium hydrogen carbonate and glycin in chemical industries. There are 90% NH_4Cl products from the Hou's Process in China[10], but they are largely limited in agriculture as a result of the competitive edge of urea. Moreover, the chlorination of oxide or sulfide ores with ammonium chloride can provide a way to dispose of the superfluous chemical by-product. And ammonium chloride begins to decompose into ammonia and hydrogen chloride at 200 °C and decomposes completely at 338 °C[5].

$$NH_4Cl=NH_3+HCl \quad (1)$$

In this paper, nickel sulfide concentrate was used to recover the valuable metals such as nickel, copper, iron and cobalt. At first, sulfide concentrate was transformed into metal oxides by oxidation roasting. Then the chlorination of roasted ore with ammonium chloride was conducted at 300 °C to avoid the complete decomposition as well as ensure a better reaction kinetics[11]. Also, water leaching was carried out to determine the recovery of desired metals. And the reaction mechanism of nickel/cooper oxide with ammonium chloride was established on the base of previous studies and this experiment.

Table 1. The main chemical composition of nickel sulfide concentrate ore

Element	Ni	Cu	Fe	Co	Mg	Si
Content/%	9.41	6.66	34.31	0.23	2.83	3.70

Materials and methods

A. Materials

Sulfide nickel concentrate originating from Jinchuan (Gansu Province, China) was used in this study. The main chemical composition of the ore is given in Table 1. According to the X-Ray diffraction pattern of nickel-copper sulfide ore shown in Fig. 1, the main mineral phases are pentlandite, chalcopyrite, pyrite, pyrrhotite and iron oxide, while serpentine is the major gangue mineral.

Mineral compositions of the complex multimetallic sulfide nickel concentrate, the chlorination roasting products/calcinate and the leaching residue were characterized by X-ray diffraction on a Rigaku model 3 kW D/MAX2200V PC and scanning electron microscopy (SU70, Hitachi, Japan) equipped with an energy-dispersive X-ray spectrometer and operated at a voltage of 15 kV.

The chemical element analysis of the raw ore and the leaching solution were performed by inductively coupled plasma-atomic emission spectroscopy (PERKINE 7300DV). And all reagents used in this study were of analytical grade.

B. Experimental

Oxidation roasting

The purpose of oxidation roasting was to transform the sulfides into oxide. And quite a few work have been done to investigate the oxidation process of nickel-copper sulfides [12-14]. In their studies, the oxidation of pentlandite and chalcopyrite could take place in the temperature range of 350-700 °C, including the breakdown of $(Ni,Fe)_9S_8$ forming MSS, the decomposition of $CuFeS_2$ into Cu_2S and FeS, as well as the sequential oxidation of metal sulfides. Therefore, the sulfide concentrate was roasted at 400 °C, 500 °C and 600 °C respectively in a muffle furnace for two hours.

Chlorination of roasted concentrate by NH_4Cl

The roasted nickel concentrate samples under different roasting temperature were ground and sieved below 74 m, and then thoroughly mixed with NH_4Cl (particle size below 300 m) in a alumina crucible and heated in a muffle furnace at 250 °C and 300 °C for two hours respectively. And the roasted slag was leached by hot water (80 °C, 100 mL) for two hours. Then the supernatant was filtered and the leach residue cake was washed with pure water. The content of nickel, copper, iron and cobalt in the solution was analyzed by ICP—AES. And the metal extraction was calculated based on the mass balance of the main elements.

Reaction of pure metal oxides with NH_4Cl

In order to investigate the chlorination mechanism, the reaction process between nickel/copper oxide and ammonium chloride (particle size below 300 m) was studied with regard to the effect of roasting temperature and NH_4Cl addition on the reaction process. The roasting temperature was varied from 250 °C to 350 °C with intervals of 50 °C, whilst keeping the mass ratio (ammonium chloride to the roasted ore) and holding time constant (2:1 and 2 h respectively). Then the influence of the mass ratio on the products was conducted at roasting temperature of 300 °C and holding time of 2 h.

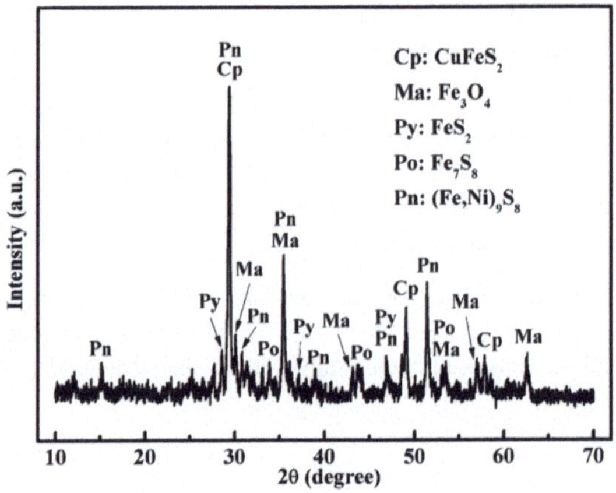

Fig. 1. The X-ray diffraction of nickel sulfide concentrate ore

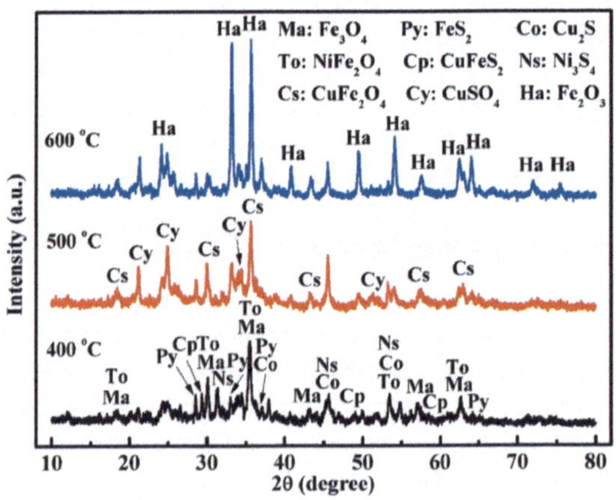

Fig. 2. The X-ray diffractions of the sulfide concentrate after oxidation roasting at 400 °C, 500 °C and 600 °C for 2 h

Results and discussion

A. The oxidation process of complex multi-metallic sulfide ore

To identify the oxidation roasting process of sulfide concentrate, the X-ray diffraction patterns of sulfide concentrate after oxidation roasting at different temperature are shown in Fig. 2. It could be observed that pentlandite decomposed into simple sulfides such as FeS, FeS_2, Ni_3S_2 and partly oxidised at 400 °C. When it heated to 500 °C, chalcopyrite transformed into $CuFe_2O_4$ and $CuSO_4$. And higher roasting temperature led to the formation of iron oxide[12].

And from SEM images of the raw ore and the sulfide concentrate after oxidation roasting shown in Fig. 3, there were huge changes in the surface of the ore. Especially particles broke into different product layers remarkably at 500 °C. And the formation of iron oxide at 600 °C resulted in grain refinement.

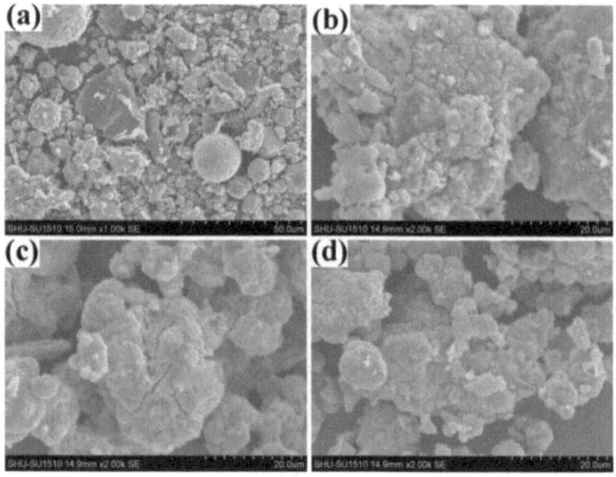

Fig. 3. Micromorphology of (a) raw ore, (b) – (d) sulfide concentrate after oxidation roasting at 400 °C, 500 °C and 600 °C for 2 h

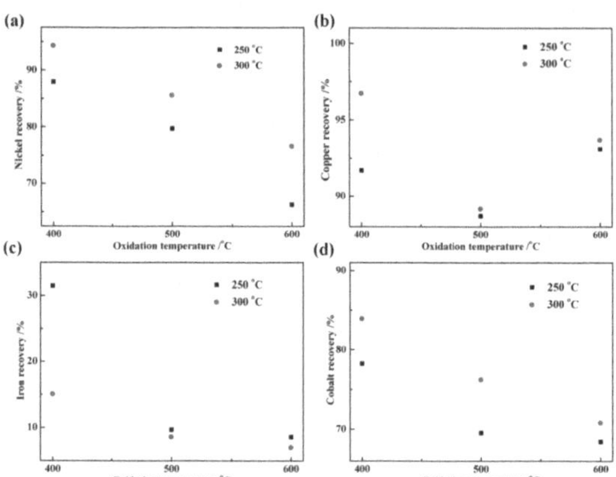

Fig. 4. Effect of oxidation temperature on the recovery of (a) Nickel, (b) Copper, (c) Iron, (d) Cobalt

B.Effect of oxidative roasting temperature on the metal recovery

Fig. 4 shows the effect of oxidation temperature on the metal recovery. It can be seen that an increase in oxidation temperature led to a decrease in the metal recovery. It is interesting to note abnormal leaching behaviour of copper for a roasting temperature of 600 °C. This might be caused by the decomposition of copper sulfate and accelerated the chlorination reaction[12]. However, iron had low recovery owing to the formation of ammoniojarosites and iron oxides in water solution. This was confirmed by our previous experiments. And higher chlorination temperature could favor the increase in metal recovery except iron.

C.The reaction mechaism between NiO/CuO and NH_4Cl

In order to have a better understanding of reaction mechanism about chlorination roasting, the chlorination of pure metal oxides with ammonium chloride was conducted. The X-ray diffraction patterns of roasting products of pure oxides are shown in Fig. 5. Ammonium chloride could react with nickel oxide or copper oxide to generate complex intermediate products—ammonium chlorometallates[15]. And excess salts were essential for the reaction. It could be concluded that the formation of ammonium chlorometallates (NH_4NiCl_3, $(NH_4)_2NiCl_4$ and $(NH_4)_2CuCl_4$) occurred at low roasting temperature, which decomposed to simple chlorides at over 350 °C. This was also confirmed by other previous studies[15-18]. Especially for a roasting temperature of 350 °C, copper existed in the form of cuprous chloride and there should be valence change during copper oxide chlorination[19].

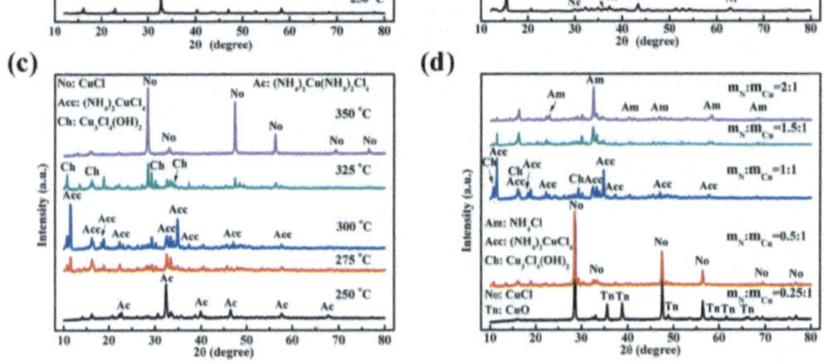

Fig. 5. The X-ray diffraction of pure oxides roasting with ammonium chloride at different temperature and addition for (a) - (b) NiO, (c) - (d) CuO

Based on the preceding analogous experimental results, the main chemical reactions in the roasting can be summarized as follows:

$$CuO + 2NH_4Cl = CuCl_2 + 2NH_3 + H_2O \quad (2)$$
$$CuCl_2 + 2NH_4Cl = (NH_4)_2CuCl_4 \quad (3)$$
$$(NH_4)_2CuCl_4 = NH_4CuCl_3 + NH_3 + HCl \quad (4)$$
$$NH_4CuCl_3 = CuCl_2 + NH_3 + HCl \quad (5)$$
$$CuO + 2NH_3 + CuCl_2 = 2CuCl + N_2 + H_2O \quad (6)$$
$$NiO + 2NH_4Cl = NiCl_2 + 2NH_3 + H_2O \quad (7)$$
$$NiCl_2 + NH_4Cl = NH_4NiCl_3 \quad (8)$$
$$NH_4NiCl_3 = NiCl_2 + NH_3 + HCl \quad (9)$$

Conclusions

The treatment of multi-metallic sulfide concentrate could be performed in a three-step procedure consisting of oxidation roasting, chlorination roasting and water leaching. It was effective to extract nickel and copper from complex bulk concentrate using NH_4Cl as a chlorinating agent under relatively low temperature to obtain high recovery yield. Under the optimum condition, about 94% Ni, 96% Cu and 84% Co were leached into water solution while the majority of iron existed in leaching residues.

The substantial role of ammonium chloride on the chlorination was derived from the breakdown of NH_4Cl into HCl. And hydrogen chloride was conducive to the formation of water-soluable chlorides. But, it must be noted that the ammonium chloride could react with simple chlorides to enhance the leachability directly.

And the ammonia, obtained by the decomposition of ammonium chloride, was absorbed by water and then used to be a complexing agent to promote the metal recovery as well as adjust the pH of the leaching solution.

This process has potential application for the recovery of valuable metals from sulfide or laterite. But the issues such as high amount of the chlorination reagent are required to be further considered.

Acknowledgment

This work was supported financially by Grant 2014CB643403 from the National Key Program for Basic Research of 973 Program.

References

1. Büyükakinci, E. and Y. Topkaya, *Extraction of nickel from lateritic ores at atmospheric pressure with agitation leaching.* Hydrometallurgy, 2009. **97**(1): p. 33-38.
2. Guo, X., et al., *Leaching behavior of metals from a limonitic nickel laterite using a sulfation–roasting–leaching process.* Hydrometallurgy, 2009. **99**(3-4): p. 144-150.

3. Wang, M. and X. Wang, *Extraction of molybdenum and nickel from carbonaceous shale by oxidation roasting, sulphation roasting and water leaching.* Hydrometallurgy, 2010. **102**(1-4): p. 50-54.
4. Nadirov, R.K., et al., *Recovery of value metals from copper smelter slag by ammonium chloride treatment.* International Journal of Mineral Processing, 2013. **124**: p. 145-149.
5. D'yachenko, A.N. and R.I. Kraidenko, *Processing oxide-sulfide copper ores using ammonium chloride.* Russian Journal of Non-Ferrous Metals, 2010. **51**(5): p. 377-381.
6. Turan, M.D. and M.S. Safarzadeh, *Separation of zinc, cadmium and nickel from ZnO–CdO–NiO mixture through baking with ammonium chloride and leaching.* Hydrometallurgy, 2012. **119-120**: p. 1-7.
7. Chi, R., et al., *Recovery of rare earth from bastnasite by ammonium chloride roasting with fluorine deactivation.* Minerals Engineering, 2004. **17**(9-10): p. 1037-1043.
8. Zhang, M., et al., *A study of recovery of copper and cobalt from copper–cobalt oxide ores by ammonium salt roasting.* Hydrometallurgy, 2012. **129-130**: p. 140-144.
9. Andreev, A.A., A.N. D'yachenko, and R.I. Kraidenko, *Processing of oxidized nickel ores with ammonium chloride.* Theoretical Foundations of Chemical Engineering, 2011. **45**(4): p. 521-525.
10. Havlin, J., James Beaton, Samuel Tisdale, and Werner Nelson, *Soil Fertility and Fertilizers 7th edition.* Pearson-Prentice Hall, New Jersey, 2005: p. 138.
11. Mukherjee, T.K., et al., *Chloridizing Roasting Process for a Complex Sulfide Concentrate.* Jom, 1985. **37**(6): p. 29-33.
12. Aneesuddin, M., et al., *Studies on thermal oxidation of chalcopyrite from Chitradurga, Karnataka State, India.* Journal of Thermal Analysis, 1983. **26**(2): p. 205-215.
13. Dunn, J.G. and C.E. Kelly, *A TG/MS and DTA study of the oxidation of pentlandite.* Journal of Thermal Analysis, 1980. **18**(1): p. 147-154.
14. Yu, D. and T.A. Utigard, *TG/DTA study on the oxidation of nickel concentrate.* Thermochimica Acta, 2012. **533**: p. 56-65.
15. Borisov, V.A., A.N. D'yachenko, and R.I. Kraidenko, *Mechanism of reaction between cobalt(II) oxide and ammonium chloride.* Russian Journal of Inorganic Chemistry, 2012. **57**(7): p. 923-926.
16. Grineva, O.V., A.N. D'yachenko, and R.I. Kraidenko, *Chlorination of copper-containing raw material by ammonium chloride.* Russian Journal of Applied Chemistry, 2013. **86**(3): p. 339-342.
17. Willett, R.D., et al., *Crystal Structures of KCuCl3 and NH4CuCl3.* The Journal of Chemical Physics, 1963. **38**(10): p. 2429.
18. Willett, R.D., *Crystal Structure of (NH4)2CuCl4.* The Journal of Chemical Physics, 1964. **41**(8): p. 2243.
19. Vasquez, R.P., *CuCl by XPS.* Surface Science Spectra, 1993. **2**(2): p. 138.

7th International Symposium on High-Temperature Metallurgical Processing
Edited by: Jiann-Yang Hwang, Tao Jiang, P. Chris Pistorius, Gerardo R.F. Alvear F., Onuralp Yücel, Liyuan Cai, Baojun Zhao, Dean Gregurek, and Varadarajan Seshadri
TMS (The Minerals, Metals & Materials Society), 2016

REDUCTION BEHAVIOR OF MAGNETITE PELLETS BY CO-CO_2 MIXTURES USING DIRECT REDUCTION PROCESS

Guihong Han[1], Wenjuan Wang[1], Yanfang Huang[1*], Dianyuan Dang[1], Tao Jiang[2]

[1]School of Chemical Engineering and Energy, Zhengzhou University, 450001, Zhengzhou, P. R. China
[2]Central South University; School of Minerals Processing & Bioengineering, Central South University, Changsha, Hunan 410083, P. R. China

Keywords: "Two-step process", Reduction extent, Swelling index, Reduction kinetics

Abstract

In this study reduction kinetics and swelling behaviors in "two-step process" were investigated by weight-loss technique and visualizing physical changes of pellet. Results showed that reduction degree (r) increased with increasing temperature and time. According to the dynamic calculation, when reduction temperature was lower than 1000℃ the reduction process was controlled by chemical reaction with apparent activation energy of 72.27 kJ/mol. When reduction temperature was higher than 1000℃, the equation of $t/[1-(1-r)^{1/3}]$ against $[1+2(1-r)-3(1-r)^{2/3}]/[1-(1-r)^{1/3}]$ followed straight lines, demonstrating that reduction was mixed-controlled. Swelling experiments indicated that the maximum swelling index increased with the increase of reduction temperature and CO content.

Introduction

Since 1960's, direct reduction iron has been considered as a high quality burdens in electric furnace steelmaking [1]. At present, coal-based direct reduction process, called "two-step process" commonly use cold roasted pellets as the main raw material [2]. Extensive literature studies have been reported on the reduction rate and reduction kinetics of ferrous oxides [3-10]. El-Geassy, et.al concluded that reduction process was controlled by chemical reaction at early stage by employing pure Fe_2O_3 as raw materials and CO as reduction gas [5-7]. Turkdogan and Vintners considered that reduction process was controlled by gas diffusion when reduction degree was more than 60% [8]. Chen carried out research on the reduction kinetics of porous oxidized pellets and obtained the rate constant and activation energy at different reduction stages [9].

Most of the experiments were carried out by using CO as reduction agent [5-7]. In these experiments it was found out that the volume of pellets increased when CO was used for reduction, resulting in swelling. Reduction of swelling could lead to increase of porosity and loss of crushing strength. Obviously swelling could also result in disintegration of pellets [5].

In this paper, an attempt has been made to study the reduction kinetics and swelling behavior of pellets during coal-direct reduction process. For this purpose, experiments were carried out under isothermal conditions in the "two-step process" by using cold oxidized pellets for reduction.

Experimental Procedure

Raw Materials

The chemical analysis of iron ore concentrates used in the present study was given in Table I. The chemical analysis of bentonite was shown in Table II.

Table I. Chemical analysis of magnetite concentrates (wt. %)

TFe	FeO	SiO_2	CaO	MgO	Al_2O_3	K_2O	Na_2O	S	P	LOI
67.15	27.54	4.81	0.20	0.30	1.25	0.043	0.10	0.024	0.098	1.23

Note: loss on ignition (LOI).

Table II. Chemical analysis of bentonite (wt. %)

Fe_2O_3	SiO_2	Al_2O_3	CaO	MgO	K_2O	Na_2O	LOI
7.92	59.09	16.38	1.08	2.42	0.13	3.16	8.75

Preparation and Properties of Oxidized Pellets

Experimental flow-sheet was shown in Figure 1. Green pellets with diameter of 12~15 mm were agglomerated in a laboratory disc pelletizer and then preheated and further roasted in air in an horizontal tube furnace. The properties of cold roasted pellets were listed in Table III.

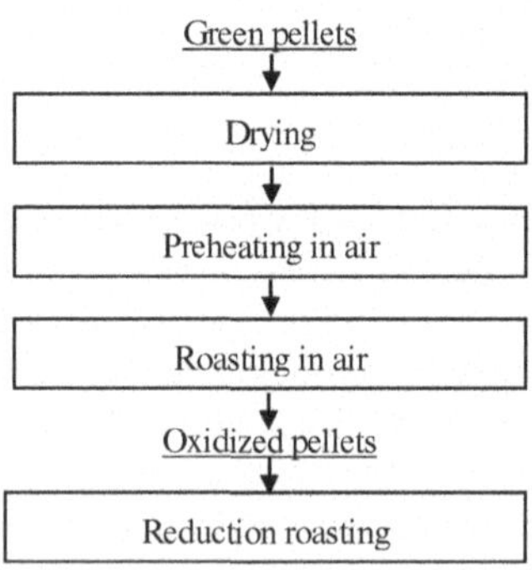

Figure 1. Flow sheet for experiment

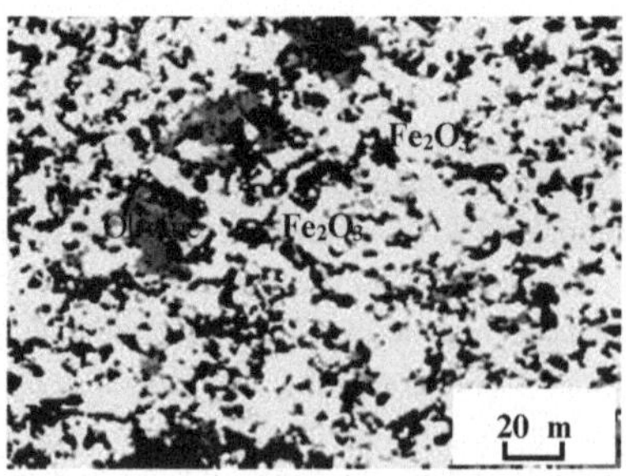

Figure 2. Microstructure of oxidized pellets before reduction. (Fe_2O_3- bright white; olivine-dark gray)

Table III. Main physical and chemical properties of pellets before reduction

TFe(%)	FeO(%)	SiO_2(%)	Na_2O(%)	Al_2O_3(%)	Compression strength (N·per^{-1})	Porosity (%)
65.13	0.26	6.93	0.14	0.78	2772	13.34

As can be seen from Table III, the iron grade of pellets was 65.1 wt. % and the porosity was 13.3 wt. %. The microstructure of roasted pellet was shown in Figure 2. The roasted pellet before reduction was mostly composed of a matrix of hematite grains which were developed by direct bonding between adjacent grains.

Experimental Equipments and Methods

Reduction Behavior

Isothermal gaseous reduction was carried out by the shaft furnace as shown in Figure 3.

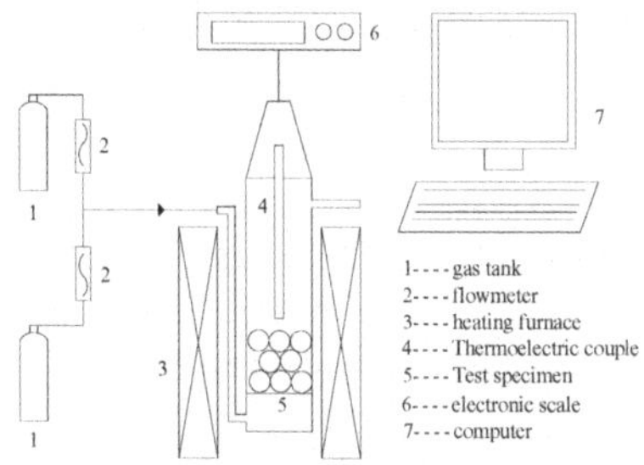

Figure 3. Schematic diagram of experimental equipment

Reduction was investigated up by means of weight-loss technique. The structure of the reduced pellets was examined by reflected light microscope. The weight loss, ΔW_t, which can be obtained from the electronic scale and reduction extent (R, r) was calculated as shown in equations (1) and.(2).

$$R = \frac{\Delta W_t}{\Delta W_T} \times 100 \tag{1}$$

$$r = R/100 = \frac{\Delta W_t}{\Delta W_T} \tag{2}$$

where: ΔW_t refers to the weight loss at the moment of t, while ΔW_T means the total weight loss in theory.

Swelling Behavior

Swelling behavior was examined by visualizing physical changes of pellets in a horizontal tube furnace. The schematic diagram of the experimental setup was shown in Figure 4.

Pellets were located on a specimen holder which could be pushed to a pre-determined location in the furnace. A camera was used to capture the pellets images during reduction. The output from the camera was connected to a computer to record the entire process as a function of time. Images were captured after every 5 min and image processing software was used to calculate frontal cross-sectional area. Swelling index was calculated on the basis of changes in the measured area using equation 3:

$$SI = \frac{V_t - V_0}{V_0} \times 100 = \left[\left(\frac{S_t}{S_0} \right)^{3/2} - 1 \right] \times 100 \quad (3)$$

where: V_0, V_t are the volumes of sample before and after experiment, while S_0, S_t are the corresponding cross-sectional areas.

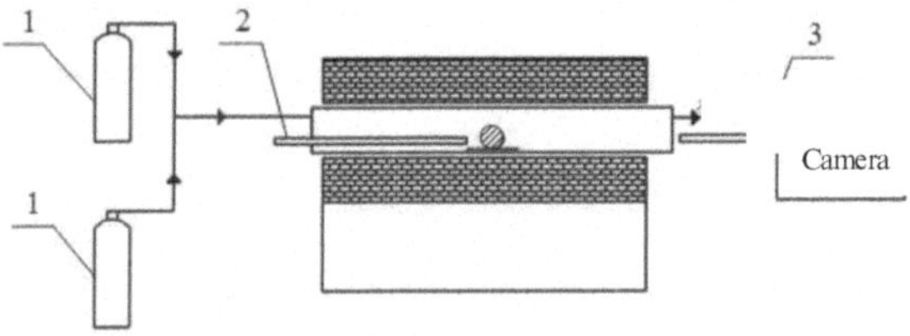

1-Reactant gas; 2-Thermocouple; 3-Camera

Figure 4. Experimental apparatus for the measurement of swelling index

Results and Discussions

Reduction Kinetics of Pellets in "two-step process"

Table III shows pellet porosity of 13.34 %, indicating that reduction process may abide by the un-reacted core model. Kinetics equations of different rate-determining step were different. Literature studies [3, 11] showing the apparent activation energy of different rate-determining step during reduction reaction, are shown in Table V.

Table V. Kinetics equations and apparent activation energies of different rate-determining steps during the reduction reaction of iron oxides

Controlling step	Kinetic Equation	Apparent activation energy (kJ/mol) [3, 11]
Gas internal diffusion	$K_1 t=1-3*(1-r)^{2/3}+2*(1-r)$	10.0 ~ 28.0
Chemical reaction and internal diffusion	$K_3 t=A[1-(1-r)^{1/3}]+B[1-3*(1-r)^{2/3}+2*(1-r)]$	28.0 ~ 50.0
Chemical reaction	$K_2 t=1-(1-r)^{1/3}$	50.0 ~ 75.0

First of all, effects of reduction temperature on the reduction extent of pellets were studied with CO content of 95% in the temperature interval from 900℃to 1100℃. The typical *R-t* curves of pellets at different temperatures were shown in Figure 5.

As can be clearly observed from Figure 5 the reduction rate of pellets was enhanced with the increase of reduction temperature. At different reduction temperatures, the reduction extent increases continuously until it reaches a maximum. However, at the same temperature it will be decreased with increasing the reducing time.

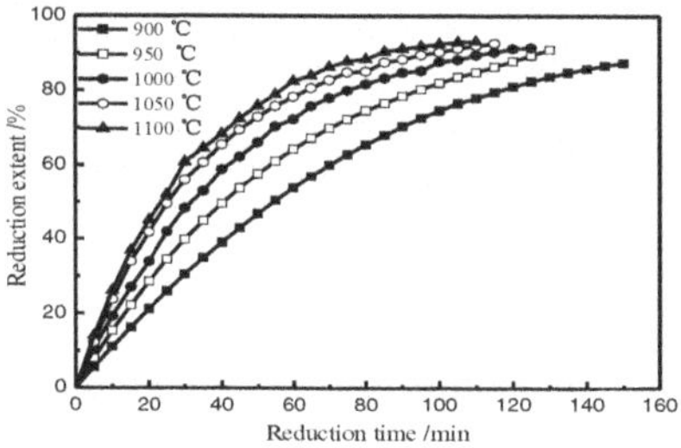

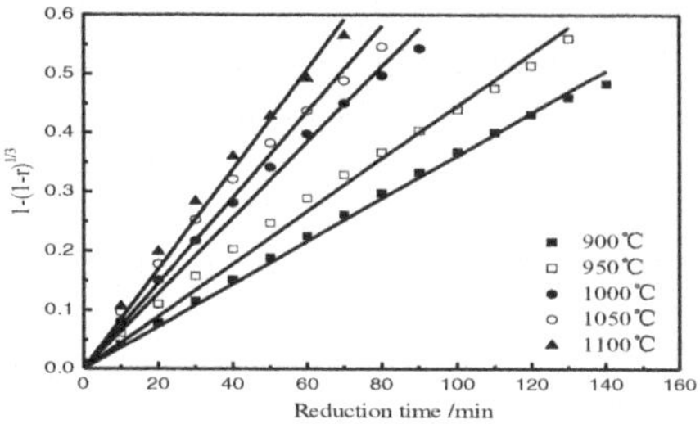

Figure 5. Relationships between R and t   Figure 6. Curves of relationship between $1\text{-}(1\text{-r})^{1/3}$ and t

According to the *R-t* plots in Figure 5, relationships between reduction fraction $1\text{-}(1\text{-r})^{1/3}$ and reduction time (t) were shown in Figure 6. According to Figure 6 and Arrhenius Equation ($k = k_0 \cdot \exp(-E_a / RT)$), the calculated activation energy was 72.27 kJ/mol when reduction temperature was lower than 1000℃

For temperature higher than 1000℃, relationships of $t/(1\text{-}r)^{1/3}$ and $1\text{-}2(1\text{-}r)^{2/3}\text{+}(1\text{-}r)^{1/3}$ were shown in Figure 7. As observed in the literature the equation of $t/[1\text{-}(1\text{-}r)^{1/3}]$ against $[1\text{+}2(1\text{-}r)\text{-}3(1\text{-}r)^{2/3}]/[1\text{-}(1\text{-}r)^{1/3}]$ shows straight lines, which demonstrate that reduction reaction is mixed-controlled when temperature is in the range from 1050℃ to 1100℃.

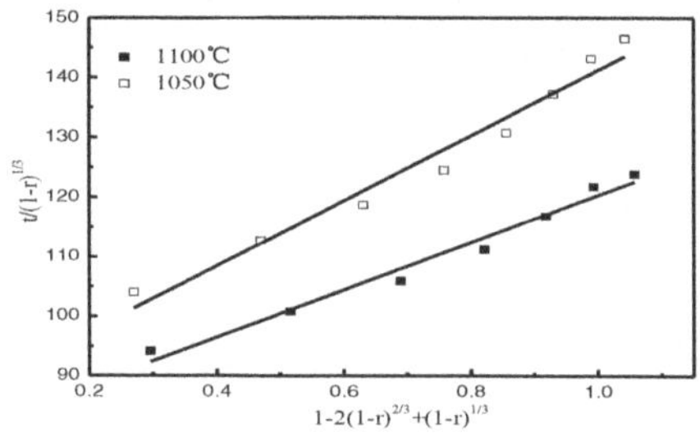

Figure 7. Relationship between $t/[1\text{-}(1\text{-}r)^{1/3}]$ and $[1\text{+}(1\text{-}r]^{2/3}\text{-}2(1\text{-}r)^{1/3}$

According to Table V, the intercept of lines in Figure7 can be explained by the equations 4, and 5:

$$\text{Intercept} = \frac{k_3 \cdot K}{k(1+K)} \tag{4}$$

$$k = \frac{\text{Intercept} \cdot (1+K)}{k_3 \cdot K} \tag{5}$$

By combining the results shown in Figure 7 and equation (5), reduction rate constant were calculated at temperatures 1050℃and 1100℃.According to Arrhenius Equation, apparent activation energy in this temperature range was 21.84 kJ/mol.

According to the above results, reduction process was controlled by chemical reaction with apparent activation energy of 72.27 kJ/mol when temperature was lower than 1000℃, while reduction process was mixed-controlled at higher reduction temperature with apparent activation energy of 21.84 kJ/mol.

Swelling Behavior of Pellets in "Two-Step Process"

Effects of Reduction Time on Swelling Index of Pellets

When reducing temperature was 1050℃and CO content was settled at 95%, effects of reducing time on swelling index of pellets were investigated and the results were shown in Figure 8.

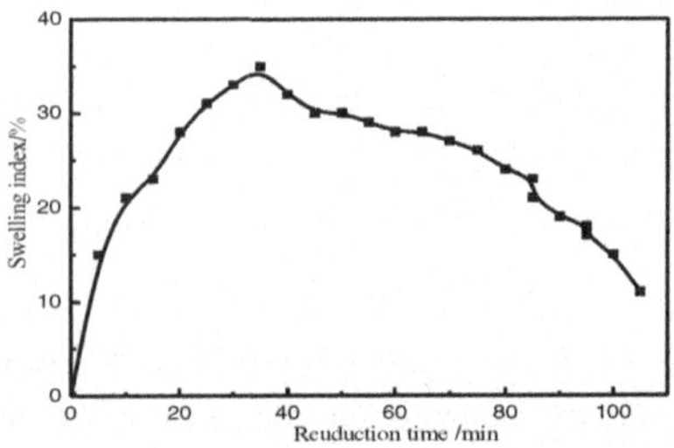

Figure 8. Effects of time on swelling index

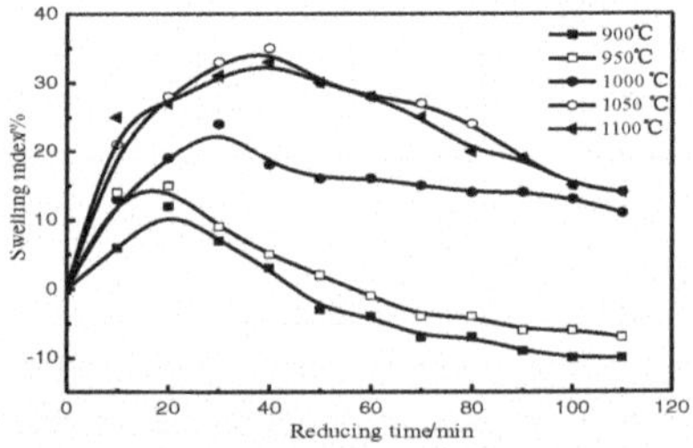

Figure 9. Effects of temperature on swelling index

As can be seen from Figure 8 the swelling index reaches the maximum value with 35% after reduction time of 30 min At reduction time higher than 30 min, the swelling index decreased to 10 %.

Effects of Reduction Temperature on Swelling Index of Pellets

When temperature was changed from 900℃ to 1100℃, effects of temperature on swelling behavior of pellets were investigated when CO content was 95 %. Results were illustrated in Figure 9 and Table VI.

Table VI. Effects of temperature on maximum swelling index

Reduction temperature /℃	900	950	1000	1050	1100
Maximum swelling index /%	12	15	24	35	33

As can be seen from Figure 9 and Table VI, when temperature is lower than 1050℃, swelling index increases with increasing reducing temperature. In case of temperature higher than 1050℃, swelling index is kept as a constant.

Effects of CO Content on Swelling Index of Pellets

Effects of CO content on swelling index were studied when temperature was fixed at 1050℃. The maximum swelling index of pellets during reduction process was shown in Figure 10. At fixed CO content of 85 %, the maximum swelling index was 30 %. With increasing of CO content, the swelling index gradually increased. At CO content of 95 %, the maximum swelling index increased even to 35 %.

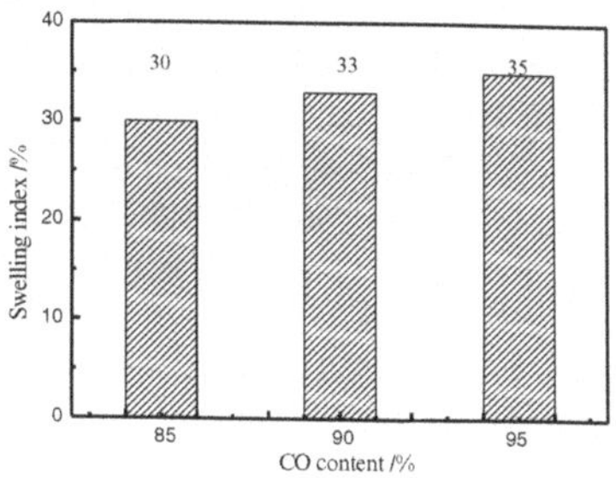

Figure 10. Effects of CO content on the maximum swelling index

Conclusions

From the present study following conclusions can be drawn:

(1) According to the dynamic calculation, at lower temperature (i.e., < 1000℃) the reduction process appears to be controlled by chemical reaction with apparent activation energy of 72.27 kJ/mol. At higher temperature (i.e., > 1000℃), the equation of $t/[1-(1-r)^{1/3}]$ against $[1+2(1-r)-3(1-r)^{2/3}]/[1-(1-r)^{1/3}]$ follows straight line, which demonstrate that reduction is mixed-controlled.

(2) Reduction temperature seems to have a significant effect on the maximum swelling index, which increased with the increasing reduction temperature and CO content.

Acknowledgments

The financial supports from the National Science Fund for Young Scholars of China (No.51404213 and No.51404214) and the Development Fund for Outstanding Young Teachers of Zhengzhou University (No.1421324065) are gratefully acknowledged.

References

[1] S.J. Ripke, "Advances in iron ore pelletizing by understanding bonding and strengthening mechanisms" (Ph.D. thesis, Michigan Technological University, 2002), 1.

[2] L. Jian, "A Study on Mechanism and Process of Direct Reduction of Pellets Made from Concentrate and Composite Binder" (Ph.D. thesis, Central South University, 2007), 2.

[3] M. Bahgata, and M.H. Khedr, "Reduction kinetics, magnetic behavior and morphological changes during reduction of magnetite single crystal," *Materials Science and Engineering B*, 138 (2007), 251–258.

[4] H.W. Kang, W.S. Chung, and T. Murayam, "Effect of iron ore size on kinetics of gaseous reduction," *ISIJ International*, 38 (2) (1998), 109~115.

[5] A.A. El-Geassy, K.A. Shehata, and S.Y. Ezz, "Mechanism of iron oxide reduction with hydrogen/carbon monoxide mixtures," *Transitions ISIJ*, 17 (11) (1977), 629-635.

[6] A.A. El-Geassy, "Gaseous reduction of MgO doped Fe_2O_3 compacts with carbon

monoxide at 1173-1473K," *ISIJ International*, 36 (11) (1996), 1328-1337.
[7] A.A. El-Geassy, "Gaseous reduction of Fe_2O_3 Compacts at 600 to 1050 ℃," *Journal of Materials Science*, 21(11) (1986), 3889-3900.
[8] E.T. Turkdogan, and J.V. Vinters, "Gaseous reduction of iron oxides: Part I. reduction of hematite in hydrogen," *Metallurgical Transactions*, 2 (1971), 3174~3187.
[9] Y.F. Chen, and Z.C Zhang, "Reduction kinetics of porous iron pellets," *Iron and Steel*, 8 (2) (1987), 9-10. (In Chinese)
[10] M. Chang, and L.C. De Jonghe, "Whisker growth in reduction of iron oxides," *Metallurgical Transactions*, 15B (1984), 685.
[11] Q.J. Li, "The study of kinetics of micro/nanoscale iron ore reduction with gases," (Ph.D. thesis, Shanghai University, 2008), 4. (In Chinese)

7th International Symposium on High-Temperature Metallurgical Processing
Edited by: Jiann-Yang Hwang, Tao Jiang, P. Chris Pistorius, Gerardo R.F. Alvear F., Onuralp Yücel, Liyuan Cai, Baojun Zhao, Dean Gregurek, and Varadarajan Seshadri
TMS (The Minerals, Metals & Materials Society), 2016

REFLUX REACTION BEHAVIOR OF PHOSPHORUS UNDER NON-EQUILIBRIUM CONDITION OF CASTING LADLE BETWEEN SLAG AND HOT METAL

MA Jinfang[1], ZHANG Jianliang[1], WANG Zhenyang[1], XING Xiangdong[2]

[1]School of Metallurgical and Ecological Engineering, University of Science and Technology Beijing; 30 xueyuan road, haidian district; Beijing, 100083, People's Republic of China
[2]School of Metallurgical Engineering, Xi'an University of Architecture and Technology; Xi'an, Shanxi, 710055, People's Republic of China.

Keywords: Non-equilibrium, Dephosphorization, Phosphorus capacity, Dephosphorization rate

Abstract

The non-equilibrium reflux reaction behavior of phosphorus was studied with synthetic slags formed by using the AR chemical reagents and changing the initial condition of basicity, FeO, MgO and Al_2O_3. Due to increase of viscosity, the dephosphorization slag with high basicity inhibits the full contact of slag with iron, which leads to the low reaction rate in the early stage. However, since the high basicity improves the activity coefficient of CaO, the dephosphorization rate and phosphorus capacity goes up in late reaction stage. FeO promotes the increase of reaction rate in early dephosphorization stage. But the polarization force of Fe^{2+} is stronger than Ca^{2+}, which hinders the reaction between phosphate ion and CaO and reduces the stability of PO_4^{3-}. Thus, high content of FeO will adversely impact on dephosphorization reaction. In addition, MgO has a promoting effect to the dephosphorization, whereas Al_2O_3 has a negative influence.

Introduction

Phosphorus can affect the steel plasticity, toughness and weldability. The phosphorus atoms are easily enriched on the ferrite grain boundaries and formed phosphides. It can lead to not only solution strength but also toughness drop. Especially in room temperature (for example 20 °C to minus 20 °C) the toughness value reduces a lot. It may promote cold brittleness phenomenon generation and invalidation in the steel loading process. Therefore, the quality control standard is $[P] \leq 0.035\%$ for general steels and $[P] \leq 0.025\%$ for premium steel. As the requirements for special steel are improving gradually, the hot metal with low hazardous elements contents receives more and more attention [1]. However, exhausted resources of high quality iron ore and coal make the clean steel production more difficult. Fortunately, hot metal pretreatment process can eliminate the harmful element and purify the hot metal effectively [2]. At present, the studies about dephosphorization pretreatment are mostly focused on the influence factors such as temperature, slag system *etc.* at balanced reaction conditions [3-6]. But further studies regarding the non-equilibrium conditions process have not been carried out.

Using a high temperature electromagnetic induction furnace to melt iron and a box-type resistance furnace to maintain the temperature, the non-equilibrium dephosphorization was investigated using the reflux contact reaction between slag and hot metal. The initial slag composition, such as the basicity, FeO, MgO and Al_2O_3 contents were changed in the experiment schemes in order to explore their influence on the dephosphorization behavior.

Experimental

Experimental material

The iron used in the experiment was from an iron making plant in north China. Table I shows the composition of this pig iron. The analytical reagents were used to prepare the dephosphorization slag system. The oxide weight used for preparation slags is given in Table II.

Table I The components of iron in the experiments / wt%

C	Si	Mn	P	S
4.1	0.34	0.26	0.098	0.011

Table II The dephosphorization slag proporrtion plan in the experiments

Number	CaO/wt%	SiO_2/wt%	FeO/wt%	Al_2O_3/wt%	MgO/wt%	CaO/SiO_2
a-1	43.6	36.4	20	-	-	1.2
a-2	53.3	26.7	20	-	-	2
a-3	60	20	20	-	-	3
a-4	64	16	20	-	-	4
b-3	52.5	17.5	30	-	-	3
c-3	45	15	40	-	-	3
d-3	37.5	12.5	50	-	-	3
b-3(5%Al_2O_3)	48.75	16.25	30	5	-	3
b-3(5%MgO)	48.75	16.25	30	-	5	3

The basicity, CaO, Al_2O_3 and MgO influence the slag dephosphorization property. a, b, c, d in Table II correspond to 20 wt%, 30 wt%, 40 wt% and 50 wt% FeO respectively. 1, 2, 3, and 4 mean the varying the binary basicity of 1.2, 2, 3, 4. The effect of MgO and Al_2O_3 on phosphorus distribution behaviors have been studied through adding 5wt% Al_2O_3 and 5wt% MgO to the last part of the experiments.

Experimental methods

For preparing the pure FeO, the powders of analytical reagent Fe_2O_3 and Fe were matching at 51 mol% oxygen atom and mixing for 40 min in agate mortar. Then the mixture was put into the iron crucible. The sample was heated into 1050 °C through muffle furnace and kept the temperature for 24 h. The heating process was under argon protective atmosphere. When the scheduled time was reached, the sample was cooled to room temperature in the furnace to get final FeO.

The experimental reagents of CaO, SiO_2, Al_2O_3 and MgO were roasted at 1000 °C in argon atmosphere for 2 h. The carbonate and hydroxide existed in the reagents were decomposed during the heating, which refined the raw materials and reduced experimental error through removing of CO_2 and H_2O. After roasting was finished, the reagents were cooled down to room temperature and put into the sealed bottles for the next experiments.

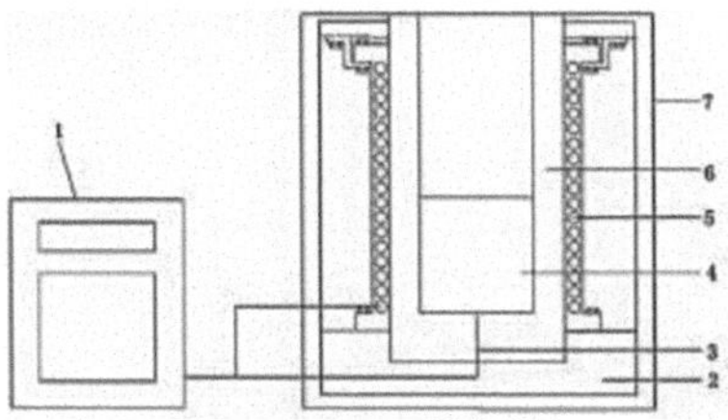

1-Process control system 2-Fire resistant cotton 3-Pt/Pt-Rh thermocouple 4—hot metal
5-60 kW induction coil 6-MgO ramming furnace lining 7-Electromagnetic induction furnace
Fig. 1 The laboratory equipment

The dephosphorization slags were prepared by mixing uniformly in the agate mortar. 200 g slag for each test was put into a 55 mm - diameter - 70 mm - deep graphite crucible. 60 kW electromagnetic induction furnace shown as Fig. 1 was used to melt the pig iron at 1500 °C. Simulating industrial ladle method, the hot metal was poured into the graphite crucible. And then the crucible was immediately taken into the muffle furnace at 1350 °C and kept at the temp for predetermined time as seen in Fig. 2. After that, the samples were rapidly quenching to ice water. The iron and slag were separated after drying and detected the phosphorus contents by chemical neutralization titration.

Results and discussion

Dephosphorization rate under non-equilibrium

Through Equation (1), the relationship between dephosphorization rate and reaction time were given in Fig. 2. The ternary slag phase diagram CaO-SiO_2-FeO at 1350 °C was calculated using Factsage 6.4 software provided by Thermfact and GTT-Technologies, which was labeled with the primary slag composition as shown in Fig. 3.

$$R_{[P]} = \frac{[P_{\text{Before reaction}}]\%-[P_{\text{After reaction}}]\%}{[P_{\text{Before reaction}}]\%}\times 100\% \qquad (1)$$

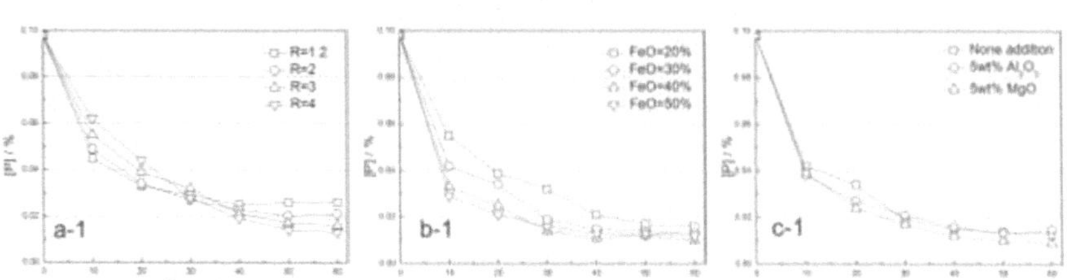

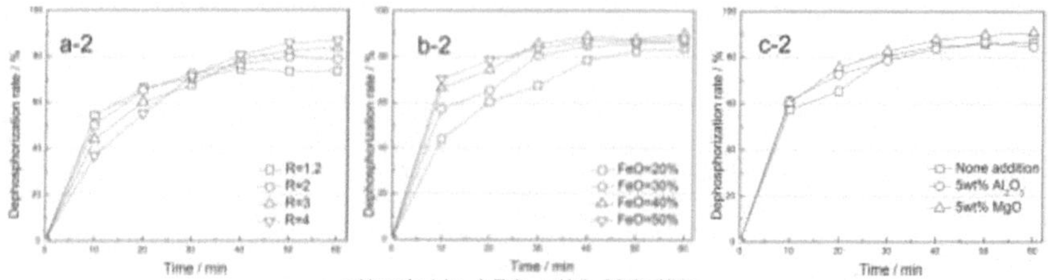

a-binary basicity; b-FeO; c-Al_2O_3, MgO addition

Fig. 2. Dephosphorization rate and P content in hot metal under different initial conditions

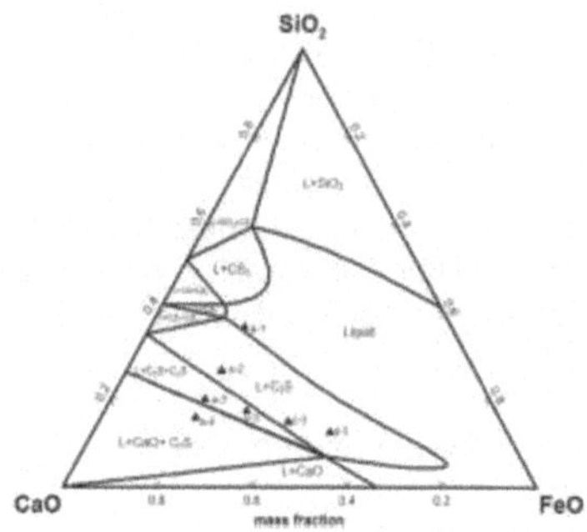

Fig. 3. The CaO-SiO_2-FeO ternary phase diagram at 1350℃ and the initial slag composition

Figure 2 shows that, the dephosphorization rate increases significantly in all instances at the beginning of the reaction and it can reach more than 64% in 10 min. It takes 20 min for the dephosphorization rate to exceed 77%. However, after 30 min, the dephosphorization rate remains stable, which can hold at around 80%.

Based on Fig. 2-a, with the primary slag basicity increases from R=1.2 to R=4, the dephosphorization rate becomes lower in the early stage of the reaction. That is, the high basicity slag reduces the dephosphorization reaction speed. From Fig.3, high basicity slag has more solid solution phase and less liquid phase, which increases the slag viscosity and inhibits the full contact between slag and hot metal. Thus the high basicity slag does not correspond to a high dephosphorization rate in the initial reaction. However, since the increasement of basicity is benefit for the enhancement of CaO activity, the high basicity slag enhances dephosphorization rate in the end of reaction, as seen in Equation (2).

$$3(CaO) + 5(FeO) + 2[P] = (3CaO{\cdot}P_2O_5) + 5[Fe] \quad (2)$$

The dephosphorization rate under different FeO contents is shown in Fig. 2-b. With the FeO content in primary slag increasing, the dephosphorization rate becomes higher in the early stage of the reaction. That is, high FeO primary slag corresponds to a quick dephosphorization rate at beginning. The increasing amount of FeO in primary slag promotes the liquid phase expanding, which is advantageous to the completely contact between slag and hot metal, as present in Fig. 3. Moreover, FeO can increase the oxygen potential in the slag system. Thus, high FeO slags keep the high dephosphorization rate within 30 min. However, as the polarization force of Fe^{2+} is stronger than Ca^{2+}, the reaction between phosphate ion and CaO is hindered and the stability of PO_4^{3-} is reduced. Thus, high content of FeO exceeding a certain limit can adversely impacts on dephosphorization reaction. That is why the dephosphorization rate of 50 wt% FeO slag is less than that of 40 wt% FeO slag after 30 min.

$$3(FeO) + (3CaO \cdot P_2O_5) = (3FeO \cdot P_2O_5) + 3(CaO) \quad (3)$$

Based on Fig. 2-c, after adding 5 wt% Al_2O_3 or 5 wt% MgO into primary slag, the dephosphorization rates are both slightly increasing. The dropping of slag liquidus temperature shortens the transition time from solid-liquid interface reaction to liquid-liquid interface reaction. Therefore, the dephosphorization reaction rate in the early stage is improved. When the reactions are closed to the end, the dephosphorization rate is enhanced of the 5 wt% MgO slag but is declined of the 5 wt% Al_2O_3.

Slag phosphorus capacity

The slag phosphorus capacity is defined by the gaseous phosphorus dissolution in the alkaline oxidation slag, as shown in Equation (4) and Equation (5) [7, 8]:

$$\frac{1}{2}P_{2(g)} + \frac{5}{4}O_{2(g)} + \frac{3}{2}(O^{2-}) = (PO_4^{3-}) \quad (4)$$

$$C_{PO_4^{3-}} = \frac{PO_4^{3-}wt\%}{(\frac{P_{P_2}}{P^\theta})^{\frac{1}{2}} \cdot (\frac{P_{O_2}}{P^\theta})^{\frac{5}{4}}} \quad (5)$$

where PO_4^{3}wt% means the phosphate mass fraction. The reaction interface partial pressure of the phosphorus and oxygen are represented by P_{P2} and P_{O2}. The P_{P2} and P_{O2} can be obtained by Equation (6) to Equation (11):

$$\frac{1}{2}P_{2(g)} = [P] \quad (6)$$

$$K_{[P]}^\theta = \frac{a_{[P]}}{(\frac{P_{P_2}}{P^\theta})^{\frac{1}{2}}} = \frac{[\%P] \cdot f_{[P]}}{(\frac{P_{P_2}}{P^\theta})^{\frac{1}{2}}} \quad (7)$$

$$\lg K_{[P]}^\theta = \frac{6381}{T} + 1.01 \quad (8)$$

$$\frac{1}{2}O_{2(g)} = [O] \quad (9)$$

$$K_{[O]}^\theta = \frac{a_{[O]}}{(\frac{P_{O_2}}{P^\theta})^{\frac{1}{2}}} \quad (10)$$

$$\lg K_{[O]}^\theta = \frac{6118}{T} + 0.151 \quad (11)$$

where [%P] and $f_{[P]}$ express the mass fraction and activity coefficient of phosphorus in hot metal. $K_{[P]}^\theta$ and $K_{[O]}^\theta$ are the chemical reaction equilibrium constants of Equation (6) and Equation (8).

Oxygen activity in hot metal is represented by $a_{[O]}$. Since the phosphorus content in hot metal conforms the Henry's law conditions, $f_{[P]}$ is approximate to 1 for calculation. The oxygen activity in hot metal is measured by zirconia base oxygen-content probe. Based on Fig. 2, the phosphorus contents in hot metal have been no longer changed around 60 min. For calculating the phosphorus capacity under different conditions, the phosphorus mass fractions in final slags are chemically analyzed and converted into PO_4^{3-}wt%. Combined with the phosphorus mass fraction in hot metal reflected by Fig. 2, the slag phosphorus capacities of each group are received in Fig.4.

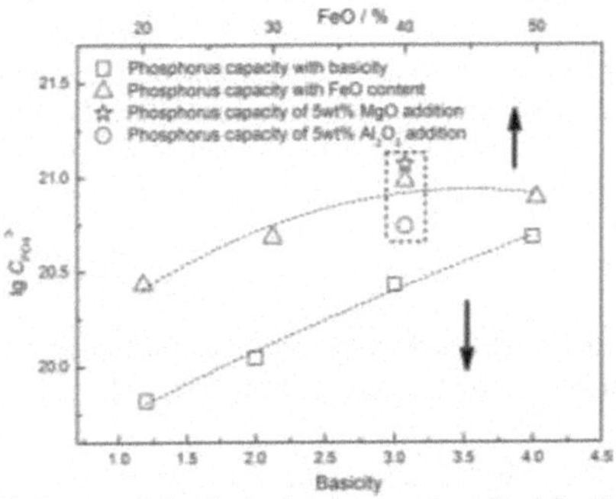

Fig.4. Relationship between dephosphorization slag capacity and basicity, FeO, Al_2O_3, MgO

From Fig. 4, with the slag basicity increasing, the phosphorus capacities are rising. When the basicity increases from 1.2 to 4.0, the phosphorus capacities enhances from $10^{19.82}$ to $10^{20.68}$. After FeO content increases from 20wt% to 40wt%, the phosphorus capacity raises $10^{0.55}$ times. However, as the FeO content increases to 50wt%, there is a turning point presenting the downward trend. At this moment, Fe^{2+} tends to the round of PO_4^{3-} and hinders the phosphorus fixation effect of Ca^{2+}. Finally, the phosphorus capacities show up falling back phenomenon. In addition, the addition of MgO improves the slag phosphorus capacity while that of Al_2O_3 lowers the slag phosphorus capacity in the same conditions of basicity and FeO content.

Conclusions

1) At beginning of the dephosphorization, the reaction rate increases significantly and reaches to 64% at 10 min. It achieves more than 77% before 20 min. And in 30 min, the dephosphorization rates of all slag systems in the experiment are more than 80%. After that, the variations in dephosphorization rate become small. The method of casting ladle reflux reaction is beneficial to improve the dephosphorization dynamics condition.

2) The higher basicity slag has a higher content of solid solution and a lower content of liquid phase, the increasing slag viscosity inhibits the full contact between hot metal and slag and leads to a low dephosphorization rate in the early reaction stage. As the reaction progressing, the dephosphorization rate of high basicity slag increases gradually. The high basicity was advantageous to the increase of CaO activity coefficient. Therefore, high basicity slag enhances the dephosphorization rate and phosphorus capacity when the reaction is nearing completion.

3) FeO has promoted the liquid phase increasing and the fully contacted between slag and hot metal. That is, high FeO primary slag is corresponding to a quick dephosphorization rate at the reaction beginning. But the polarization force of Fe^{2+} is stronger than Ca^{2+}, which tends to be round of PO_4^{3-} and hinders the reaction between phosphate ion and CaO and reduces the stability of PO_4^{3-}. Thus, 50wt% FeO in slag adversely impacts on dephosphorization after reaction proceeding to 30 min. In addition, MgO has a promoting effect to the dephosphorization, whereas Al_2O_3 has a negative influence.

References

[1] I Jeoungkiu et al., "Phosphorus distribution ratios between CaO-SiO_2-Fe_tO slags and carbon-saturated iron at 1573K," *ISIJ*, 36 (5) (1996), 517-521.

[2] P Farshid et al., "Distribution of P_2O_5 between solid solution of 2CaO·SiO_2-3CaO·P_2O_5 and liquid phase," *ISIJ*, 50 (6) (2010), 822-829.

[3] S Kenichi et al., "Distribution of P_2O_5 between solid dicalcium silicate and liquid phases in CaO-SiO_2-Fe_2O_3 system," *ISIJ*, 49 (4) (2009), 505-511.

[4] K Yasunobu et al., "Equilibrium distribution ratios of phosphorus and chromium between BaO-MnO melts and carbon saturated Fe-Cr-Mn alloys at 1573K," *ISIJ*, 34 (10) (1994), 810-814.

[5] D Y Shin et al., "Distribution behavior of Vanadium and phosphorus between slag and molten steel," *Metals and materials international*, 13 (2) (2007), 171-176.

[6] W H Niekerk et al., "Phosphorus distribution between carbon-saturated iron at 1350℃ and lime-based slags containing Na_2O and CaF_2," *Metallurgical and materials transactions B*, 29 (2) (1998), 147-153.

[7] M Ek et al., "Capacity of some CaO-SiO_2-FeO-MnO-MgO slag and slag-soild mixtures in capturing phosphorous," *Ironmaking and steelmaking*, 40 (4) (2013), 305-311.

[8] M D Johnston et al., "Effect of slag basicity and oxygen potential on distribution of boron and phosphorus between slag and silicon," *Journal of non-crystalline solids*, 357 (3) (2011), 970-975.

7th International Symposium on High-Temperature Metallurgical Processing
Edited by: Jiann-Yang Hwang, Tao Jiang, P. Chris Pistorius, Gerardo R.F. Alvear F., Onuralp Yücel, Liyuan Cai, Baojun Zhao, Dean Gregurek, and Varadarajan Seshadri
TMS (The Minerals, Metals & Materials Society), 2016

STUDY ON COMPRESSIVE STRENGTH OF COKE AFTER GASIFIED WITH CO_2 AND STEAM

Wentao Guo[1], Qingguo Xue[1], Xuefeng She[1], Jingsong Wang[1]

State Key Laboratory of Advanced Metallurgy, University of Science and Technology Beijing, 100083, Beijing, P. R. China

Keywords: Coke, Gasification reaction, Carbon species, Tensile strength

Abstract

By the coke gasified reaction device and online measurement high-temperature compressive strength apparatus, the variation of pore structures of coke after gasified reaction with CO_2 and H_2O was investigated. Research illuminating the influences of solution loss rate, temperature, and carbon species on compressive strength of coke was also carried out. The results showed that the total compressive strength was higher than for the reaction with CO_2. The high-temperature compressive strength of coke decreased with the increasing of solution loss ratio or temperature after reacting with CO_2 and H_2O. At the same solution loss ratio, the damage of pore structure to coke strength after gasified reaction with H_2O was relatively smaller and the deformation resistance was stronger when compared with that in CO_2.

Introduction

Coke is utilized in blast furnaces to create heat, reduce metals, and carburize molten iron and columnar skeletons [1]. Injection technology which is commonly applied as a substitute for coke, does not provide the same framework. Injecting coal into a blast furnace generates H_2 at high temperatures, increasing H_2O content as an indirect reduction product [2]. The gasification reaction between coke and CO_2 or H_2O forms and expands pores in the material. During this process, the size difference between gas molecules causes differing gasification reaction rates [3,4]. Therefore, the occurrence of a large number of coke powders affecting the permeability of the blast furnace should be prevented by increasing coke strength. Coke is a porous, brittle material, it strength depends on the matrix strength, crack, and porosity. In addition, the H_2O content of the blast furnace gas was further increased due to a high amount of H_2 produced by the blast furnace [5]. The change of reaction gas composition also influence the structure and strength of the coke [6].

In this study, the influence of pore structure, reduction rate and temperature on the coke tensile strength were studied under CO_2, H_2O and CO_2-H_2O atmosphere.

Experimental

Coke Sample Preparation

Single-coking coal with particle sizes <3 mm, (approximate analysis and properties were listed in Table I) was mixed with 3% organic binder agent additive, and then pressed into coal briquette samples (16 mm in height and diameter) under 20 MPa pressure. Samples were then placed in a container, which was filled with coke powder over the course of 6 hours. The furnace wall temperature was 1273 K.

Table I. Approximate analysis and property of coal

Brand	Approximate analysis wt. %		Elemental analysis wt. %			
	Ash	VM	C	H	O	N
Main coking coal	9.42	23.42	77.67	4.36	4.42	1.35
1/3 coke coal	7.92	19.95	70.34	81.0	4.25	2.51
Gas-fat coal	10.81	20.52	68.67	81.1	4.51	2.13
Lean coal	8.71	10.01	81.29	82.1	4.35	1.04

In order to distinguish the cokes prepared by different coking the main coking coal, the cokes made of 1/3 coke coal, gas coal, lean coal and the blended coal were renamed to Z-coke, 1/3-coke, Q-coke, S-coke and Z+S-coke, respectively.

Gasification Reaction Experiment of Coke

Figure 1 shows the experimental device of coke-gasified reaction. 200 g coke samples were placed into reaction tubes that were 80x800 mm in size, containing either CO_2 or H_2O reactant gas.

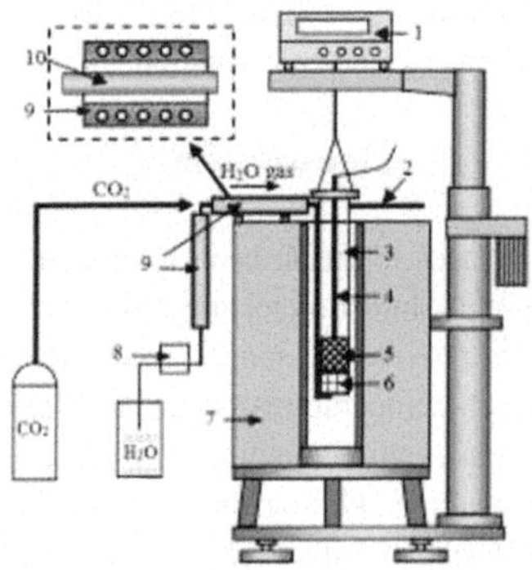

Figure 1. Experimental device of coke gasified reaction.
1.balance; 2.outlet gas; 3.reaction tube; 4.thermocouple; 5.coke; 6.Al_2O_3 ball; 7.heating furnace; 8.peristaltic pump; 9.pipeline heaters; 10.inlet gas

A H_2O gas generator was used for the gasification test. The pipeline heater controlled at temperature of 200±5K evaporated deionized water. During the experiment, deionized water flow was carefully controlled by peristatic pump, which indirectly controlled the gas flow. The gasification experiment proceeded at 1373 K, and CO_2/H_2O flow of 5 L/min. The setting concentrations of a CO_2 and H_2O mixture were 0%, 25%, 50%, 75% and 100%, respectively. Reaction rate was set to 10%, 20%, 30%, or 40%, and controlled by weight loss.

High Temperature Tensile Strength Experiment

The pressure sensor was tensile strength device in the range of 0-6000 N with precision of 1 N. The range of the displacement meter was 25 mm with 0.01 mm precision. Pressure (P) and displacement (u) values were recorded automatically as a function of time. As the coke was crushed, pressure decreased rapidly. Tensile strength measurement theory was shown in Figure 2 [6]. The following relationship exists between tensile strength and pressure:

$$\sigma = \frac{2P}{\pi dl} \tag{1}$$

where σ is the tensile strength (GPa); d is the coke sample diameter (mm); l is the length (mm).

When samples were crushed, the σ value was recorded as coke tensile strength. Each set of experimental conditions was run ten times, and an average of the σ_r values was used to obtain final coke tensile strength measurements.

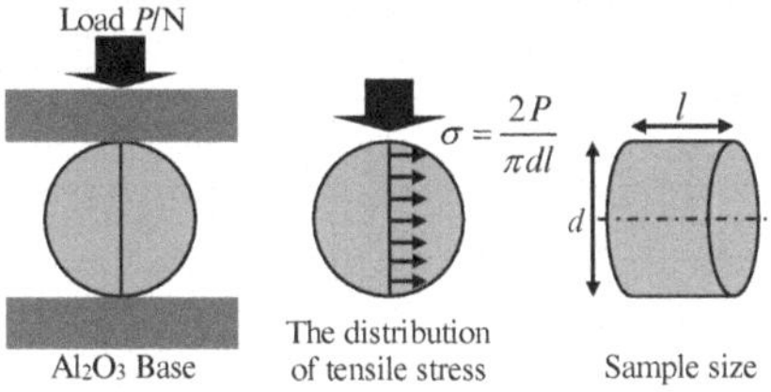

Figure 2. Schematic diagram of tensile strength measurement

Experimental Results and Discussions

Influence of Reaction Rate on the Tensile Strength

The influence of reaction rate on the tensile strength of different cokes as a function of CO_2-H_2O ratio was shown in Figure 3. With the gasification reaction, the coke porosity increased and the wall thickness as well as strength decreased. The tensile strength of the coke decreased with increasing reaction rate. The Z-coke and 1/3-coke showed higher strength than other coke species. With increased reaction rate, the internal structure of coke was destroyed completely. When the reaction rate was above 20%, the tensile strength of 1/3-coke was higher than that of Z-coke. Although the optical microstructure of Q-coke was similar to that of 1/3-coke, it higher porosity and low wall strength, resulted in a significant decrease of tensile strength to be even lower than that of 1/3-coke. Lean coal pyrolysis process produced a small volume of colloid, whereas S-coke showed low tensile strength. The Z+S-coke prepared by mixed lean coal, and main coking coal with the proportion of 50% showed lower tensile strength in comparison with the coke prepared by single coking coal.

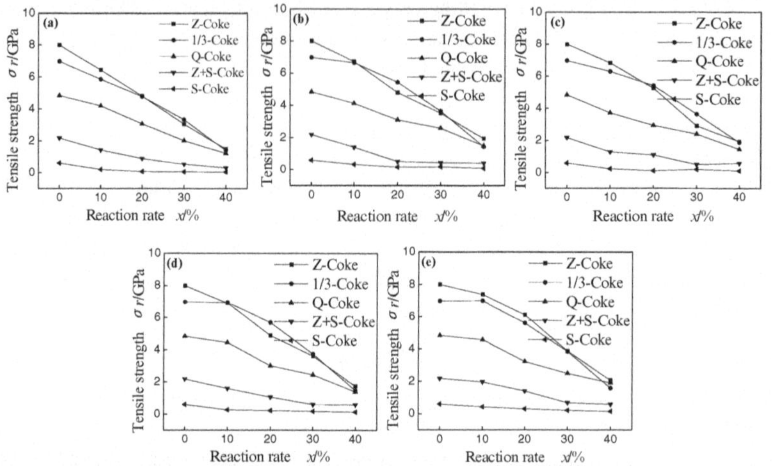

Figure 3. The influence of reaction rate on the tensile strength of coke under different CO_2 or H_2O atmosphere. (a) 100% CO_2; (b) 75% CO_2+25% H_2O; (c) 50% CO_2 + 50% H_2O; (d) 25% CO_2 + 75% H_2O; (e) 100% H_2O

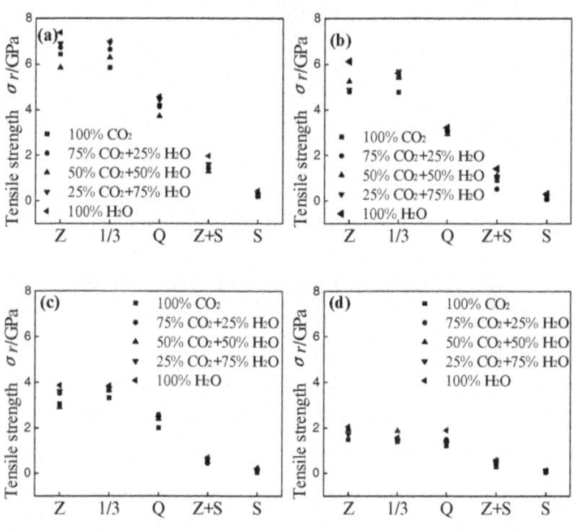

Figure 4. The relationship between different reaction rates of carbon and coke tensile strength. (a) Reaction rate-10%; (b) Reaction rate-20%; (c) Reaction rate-30%; (d) Reaction rate-40%

The different tensile strength is mainly caused by differences in the coke pore structure, as well as microstructure. The relationship between the different reaction rates of carbon and coke tensile strength was shown in Figure 4. The results indicated that the high coking property coal has the high coking capacity and can form the coke substrate with high strength. The tensile strengths of Z-coke, 1/3-coke and Q-coke were higher than the coke made by lean coal with poor coking property. The damage of coke substrate was significant and the structural differences between different kinds of charcoal were smaller with increased reaction rate. Although the H_2O content effects the tensile strength of coke, the variation trend among different coals as a function of H_2O content was basically similar.

The structure of coke leads generally to the different strength which is mainly caused by differences in the reaction mechanism of CO_2 and H_2O with coke. The influence of H_2O content on tensile strength for different cokes were shown in Figure 5. In coke gasification reaction, the reaction rate of coke with mixture gas was accelerated with increased H_2O content.

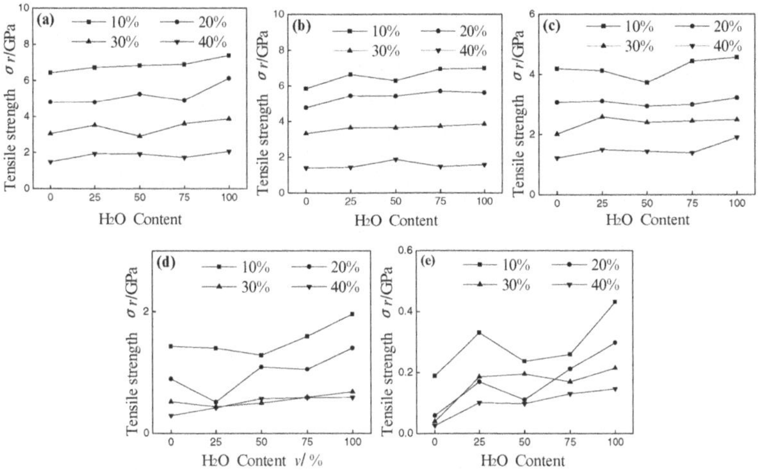

Figure 5. The influence of H_2O content on tensile strength of different cokes.
(a) Z-coke; (b) 1/3-coke; (c) Q-coke; (d) Z+S-coke; (e) S-coke

As shown in Figure 6, under the same reaction rate, the shorter the reaction time, the thinner the reaction layers of coke and the smaller the damage to coke structure. In addition, generation of micropores was promoted by H_2O. Therefore, the tensile strength increased with increasing H_2O content. However, due to the poor coking property the lean coke showed low porosity strength and low tensile strength. As can be seen from Figure 5 (d) and (e) the Z+S-coke and S-coke showed increased tensile strength with higher H_2O content.

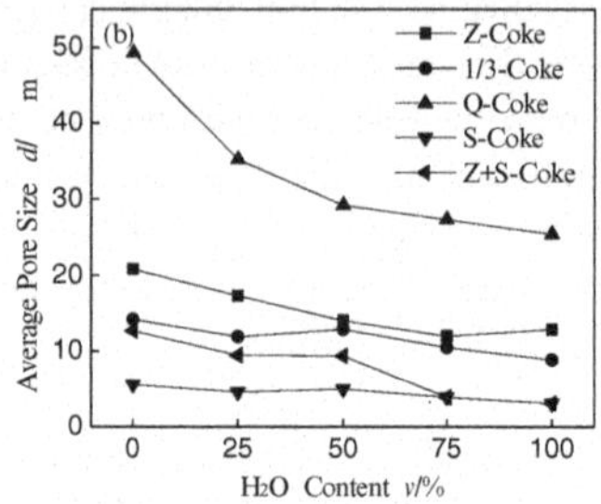

Figure 6. Average pore size of coke as under different CO_2 and H_2O atmospheres

Compared to CO_2, H_2O restrained decrease of coke strength. The inhibitory effect was more obvious with increased H_2O content in the gas mixture. In order to quantitatively evaluate the influence of H_2O content on restrain strength decrease, reduction rate of coke tensile strength was defined by equation (2):

$$Rr = 1 - \frac{\sigma_{r\text{-}x}}{\sigma_{r\text{-}U}} \qquad (2)$$

where R_r is the reduction rate of coke tensile strength, (%); $\sigma_{r\text{-}x}$ is the tensile strength of coke, when the reaction rate is x, (GPa); $\sigma_{r\text{-}U}$ is the tensile strength of unreacted coke.

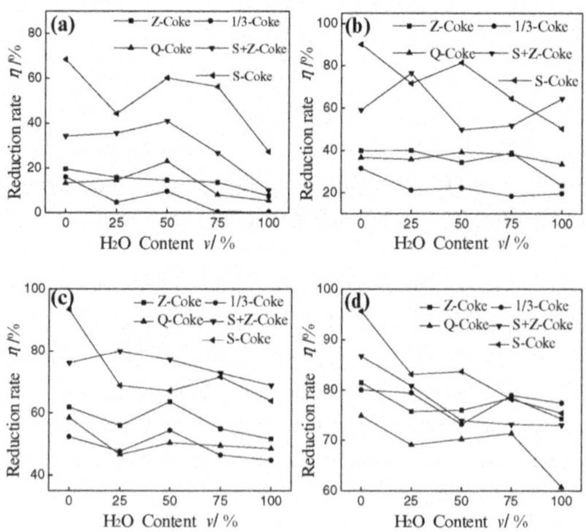

Figure 7. Different reduction rate of coke tensile strength under different CO_2 and H_2O content conditions. (a) Reaction rate-10%; (b) Reaction rate-20%; (c) Reaction rate-30%; (d) Reaction rate-40%

Reduction rate of coke tensile strength under different CO_2 and H_2O content conditions was shown in Figure 7. When the reaction rate was 10%, 20%, 30% and 40%, with increasing H_2O content R_r showed a decreasing trend. Rr from different coke was analyzed and listed from high to low: Z-coke>1/3-coke>Q-coke> Z+S-coke>S-coke. As shown in figure 7 the higher H_2O content resulted in stronger anti-deformability.

As presented in Figure 8, with different reaction rate the reduction rate of coke tensile strength changed. For the five cokes shown in figure 8, the reduction rate of coke tensile strength increased progressively with increased reaction rate.

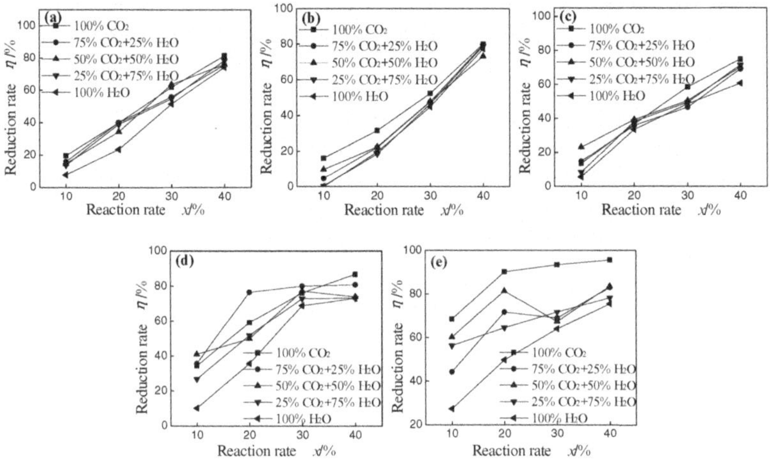

Figure 8. The influence of reaction rate on the reduction rate of coke tensile strength. (a) Z-coke; (b) 1/3-coke; (c) Q-coke; (d) Z+S-coke; (e) S-coke

Under different reaction conditions, variation of coke formed by coking coal and gas-fat coal were consistent. However, the experimental results of S-coke and Z+S-coke formed by lean coal showed that there was significant difference between the value of R_r at same reaction ratio. This was possibly caused due to poor cokeability of lean coal significantly affecting the coke tensile strength by external conditions. As shown in Figure 3, during gasification process, the lower reaction rate resulted in the smaller impact of the pore structure within the coke.

Conclusions

With the increasing reaction rate, coke tensile strength and the tensile strength difference between different cokes gradually decreased, especially at the reaction rate of 40%. At this percentage, the tensile strength of Z-coke, 1/3-coke and Q-coke showed no significant difference.

The tensile strength always ascended with the added H_2O content for different kinds of cokes.

The reduction rate of coke tensile strength was reduced with increasing H_2O content, but increased again with the reaction rate. The higher H_2O content in reaction gas, had a strong influence on the deformation resistance of coke.

References

1. F.L. Shen, S. Gupta and Y. Liu, "Effect of reaction conditions on coke tumbling strength carbon structure and mineralogy," *Fuel*, 111 (2013), 223-228.
2. Y. Ujisawa et al., "Subjects for achievement of blast furnace operation with low reducing agent rate," *Tetsu to Hagane*, 92 (12) (2006), 1015-1021.
3. Y. Kashiwaya and K. Ishii., "The kinds of reactions in coke gasification by H_2O," *Tetsu to Hagane*, 79 (12) (1993), 1305-1310.
4. A. Karimi, N. Semagina and M.R. Gray., "Kinetics of catalytic steam gasification of bitumen coke," *Fuel*, 90 (2011), 1285-1291.
5. Y. Iwanaga., "Degradation behavior of coke at high temperatures in blast furnace," *Ironmaking & Steelmaking*, 16 (2) (1989), 101-109.
6. H. Norio, S. Mototsugu and N. Seiji., "Failure strength of cokes reacted with CO_2," *Tetsu to Hagane*, 96 (5) (2010), 305-312.

7th International Symposium on High-Temperature Metallurgical Processing
Edited by: Jiann-Yang Hwang, Tao Jiang, P. Chris Pistorius, Gerardo R.F. Alvear F., Onuralp Yücel, Liyuan Cai, Baojun Zhao, Dean Gregurek, and Varadarajan Seshadri
TMS (The Minerals, Metals & Materials Society), 2016

STUDYING ON SOFTENING AND MELTING BEHAVIOR OF LUMP ORE IN BLAST FURNACE

Zhennan Qi[1], Shengli Wu[1, 2], Mingyin Kou[1], Xinliang Liu[1], Laixin Wang[1], Yujue Wang[1]

[1]School of Metallurgical and Ecological Engineering, University of Science and Technology Beijing; 30 Xueyuan Road, Haidian District, Beijing 100083, P. R. China
[2]School of Metallurgy and Chemical Engineering, Jiangxi University of Science and Technology; No.156, Kejia Ave.; Ganzhou, Jiangxi, 341000, P. R. China

Keywords: Blast furnace, Lump ore, Softening and melting, Inflation, Shrinkage

Abstract

Softening and melting experiments were conducted on four kinds of imported lump ores named A-D, under different temperature intervals within the self-developed visualization high temperature furnace. By calculating the reduction and shrinking degree under different intervals, softening and melting indexes were determined. These ores all expanded firstly followed by shrinkage with the temperature increasing. The difference of lump ore's own characteristic led to different softening and melting behavior. The results showed that the looser the structure was, the better the reducibility will be. The compact structure and good reducibility led to high expansion rate. The beginning softening temperature declined and the ending softening temperature increased because of high porosity, which resulted in wide softening temperature interval. The charging ratio of lump ore A should be improved. Lump ore B should be distributed in the center of the blast furnace. Lump ores C and D should be used less.

Introduction

The production of Chinese pure iron in 2014 was approximately 712 million, which is about 60% of global whole production. Optimization of primary materials was one of technical measures to deal with ore resource shortage and cost pressure. The softening and melting of iron burden directly influences the shape and location of the cohesive zone, as well as stable and smooth operations of the blast furnace.

In China, iron burdens of blast furnace include sinter, pellet and lump ore. In the past study, considering softening and melting behavior of sinter and pellet, was carried out by Yutaka J, et al [1]. However, the behavior of lump ore was not investigated in detail. Generally the natural lump ore has low beginning softening temperature and wide softening temperature interval [2-3]. Therefore the main focus of this study was to study the softening-melting behavior of four normal imported lump ores to make full use of them.

Experimental

To investigate the softening and melting behavior, four normal imported lump ores in a size range between 12.5 mm~16 mm were sampled. All samples were dried for 24 h at 378 K and stored in a desiccator. Samples used for future investigation were described as A-D.

The self-developed visualization high temperature furnace was used to conduct softening and melting experiments under different temperature intervals. Through simulating reduction condition in the blast furnace, different images of the lump ore as a function of temperature interval were obtained. After the experiments, samples were cooled by water. Then, all samples were taken out and stored in the desiccator for the next test. The chemical analysis was performed by X-ray fluorescence whereas the mineralogical investigation was carried out by X-ray diffraction.

Results and Discussion

Behavioral Analysis

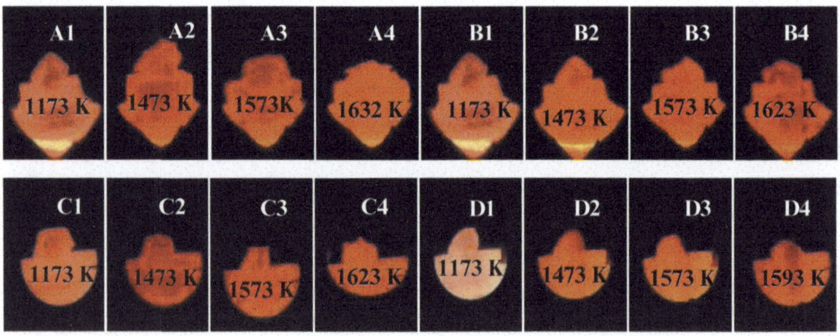

Figure 1. Softening-melting images of the four lump ores

The softening-melting images of the four lump ores at different temperatures were shown in Figure 1. For the interval at 1473 K, samples B and C showed obvious softening shrinking, whereas sample D softens slightly. Sample A kept original shape. For blast furnace, at this stage, there was enough space between lump ore A and lump ores B, C, D, which provided enough gas channels and increased the gas permeability. In the interval of 1573 K, these samples all showed visible softening behavior. In case of sample A, change in softening behavior was relatively smaller than for samples B, C and D. At this condition, the space between lump ore A and lump ores B, C, D was larger than that between any two of B, C, D. Therefore the lump ore A performed as a function of the "coke window", keeping the aeration at a very high level. Lump ores A-C had visible softening and melting at temperature 1623 K, and 1593 K for lump ore D. The volume shrinking ratio of samples B and C is smaller than for samples A and D. In this case, samples B and C acted as the function of the "coke window", improving in a certain extent the gas permeability of the furnace burden.

Softening and melting experiment results demonstrated that due to ore's characteristics, there was a difference in softening and melting behavior.

Shrinkage Analysis

Figure 2 shows the shrink ratio of four lump ores at different temperatures. Positive and negative values represent shrinkage and inflation respectively. Before 1473 K, sample A kept a slight inflating, then shrinked fast. Sample B shrinked after a small extent of inflation at temperature 1173 K. Sample C had a visible inflation followed by shrinking. Sample D inflated during temperature 1173 K and 1323 K. After 1323 K, sample D shrinked slightly. After temperature 1523 K, the shrinkage ratio of sample D increased sharply.

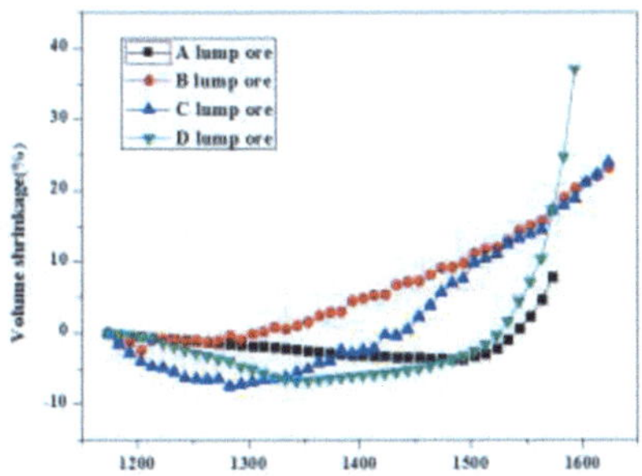

Figure 2. Area shrinkage of four lump ores

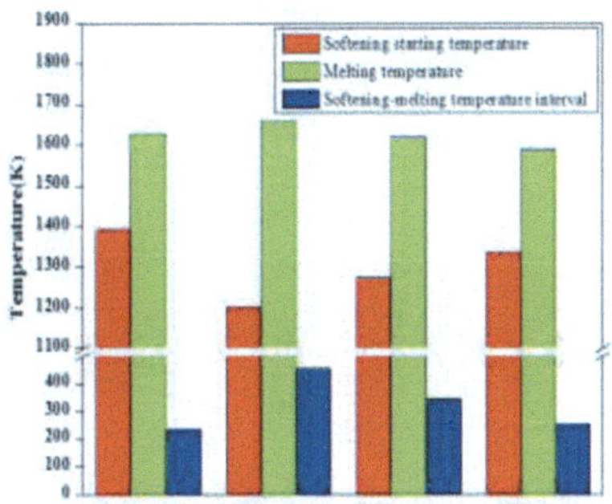

Figure 3. Softening-melting parameters

For the early reduction, the maximum inflation ratio order was:

sample C> sample D> sample A> sample B.

Sample A had the longest inflation time because of the small reduction rate. Sample A's compact structure led to the small reduction rate. Sample B had a slight inflation. This was because sample B had high porosity which adapted it to the inflation caused by crystal transition. Sample C showed a rapid inflation for its compact structure and good reducibility. Sample D's inflation was between samples A and C because of its porous structure.

Softening and Melting Properties Analysis

The softening and melting parameters of four lump ores were shown in Table I and Figure 3. For sample B, its softening starting temperature (1203 K) was the lowest, and its melting temperature (1658 K) was the highest. Thus, its softening and melting temperature interval was the widest. For sample A, softening and melting properties was better than sample B, for higher softening starting temperature (1393 K), lower melting temperature (1628 K) and narrower softening and melting temperature interval (235 K) than for sample B. The softening starting temperatures of samples C and D were 1303 K and 1323 K respectively. The melting temperature (1618 K) of sample C was higher than for sample D (1588 K).

Table I. Softening and Melting Parameters of Four Lump Ores

Lump ore	Softening starting temperature/K	Melting temperature /K	Softening-melting temperature interval/K
A	1393	1628	235
B	1203	1658	455
C	1303	1618	315
D	1323	1588	265

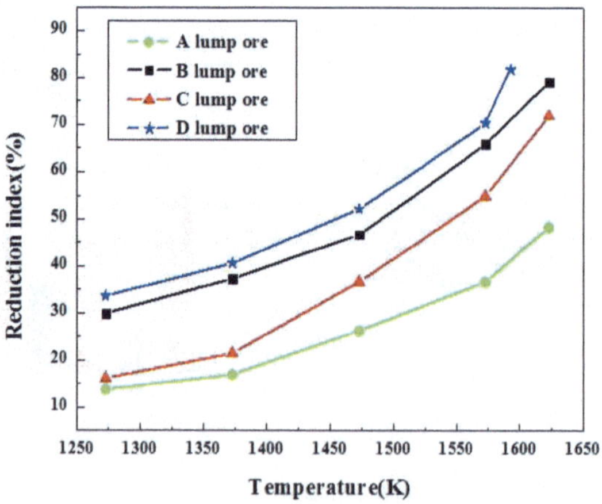

Figure 4. Reduction index of four lump ores

Reducibility Analysis

Research[1-2] show that the reduction index has an important influence on softening-melting and dropping behavior. The reduction index of the four lump ores was shown in Figure 4. At the temperature 1273 K, sample D had the highest reduction degree (33.67%) followed by sample B.

The reducibility of samples A and C were low. The low temperature reducibility of Sample D reached a very high level and was the highest of four lump ores. The reducibility increased with the rising temperature, but the rate of increase differed. In 1623 K, the RI (reduction index) of samples B and D were both about 80%, which were 48.41% and 72.01% for samples A and C respectively.

Hematite was the main metallic mineral of lump ore A with the most compact mineral structure among lump ores A-D. Main hematite and a little limonite were the metallic mineral of lump ore B with a porous mineral structure. Lump ore C, whose main metallic mineral was hematite, had a compact mineral structure. Lump ore D, consisting of limonite, had porous mineral structure. The reducibility of limonite is better than hematite. The results showed that the RI of sample D was the highest followed by sample B. The RI of samples A and C were low because of the existence of water in limonite. After sintering and dehydration, limonite became porous mineral, which contributed to the reduction of iron oxide. Besides, the reducibility of porous ores is better than that of compact ores. The porous structure cuts down the time of gas getting to the reaction interface, leading to easier reducing reaction and better reducibility.

By comparing the mineral structure and metallic mineral, the reducibility order was:

lump ore D> lump ore B> lump ore C> lump ore A.

Increasing the rate of ferrous metals of burden, helps to improve the softening-melting and dropping properties[4].

Chemical Composition and Mineral Component Analysis

X-ray fluorescence(XRF) examination and X-ray diffraction(XRD) analysis of the four lump ores was performed (Table II and Figure 5 respectively). The softening and melting temperature depends on the chemical composition and phase of the lump ore. For sample A, at 1273 K and 1373 K, the content of FeO was low, and the RI was low. The main compound of Fe element was Fe_3O_4, as was shown in XRF results. Fayalite (Fe_2SiO_4), a low melting mineral, is produced by chemical reaction of SiO_2 and FeO. Sample A didn't soften until 1393 K, because of Fe_3O_4 — a high melting point compound. With the increase of RI, the content of Fe_xO and Fe_2SiO_4 rose accordingly, and sample A slightly softened at the same time. Some iron was formed, and the main compounds of Fe element were FeO and iron at 1623 K in XRF results. Sample A had a big shrinkage because of high content of Fe_2SiO_4.

For sample B, a few iron was detected, and the main iron oxide was FeO and few Fe_3O_4 at 1273 K. Sample B softened at relatively low temperature. This was because SiO_2 and FeO reacted and formed some Fe_2SiO_4. With the temperature increasing, the content of iron rose and the content of FeO decreased. Besides, iron and FeO both have high melting point. Then, the content of Fe_2SiO_4 decreased and disappeared with both temperature and iron increase. At 1623 K, iron's content was very high together with a few FeO and few Fe_2SiO_4, resulting in high melting temperature and wide softening and melting temperature interval.

As for sample C, Fe_3O_4 was the main iron oxide and the mineral consisted of Fe_2SiO_4 and high melting point mineral such as pleonaste ($MgFe_2O_4$) and magnesium aluminate spinel ($MgAl_2O_4$) at 1273 K. These high melting point mineral led to high softening starting temperature. Between 1273 K and 1373 K, pleonaste and magnesium aluminate spinel disappeared and sample C softened with the content of Fe_2SiO_4 rising. Its RI improved as the temperature increases. Then, the RI continued rising with the temperature increasing, the content of FeO and Fe_2SiO_4 going down, and the content of metal iron increase.

Table II. X-ray Fluorescence Examination Results

Samples	Test Temperature	Fe	FeO	Fe_3O_4	CaO	SiO_2	Al_2O_3	MgO	MnO	P_2O_5	TiO_2
A	1273 K	-	7.73	51.37	0.13	26.12	13.82	0.12	0.07	0.08	-
	1373 K	-	24.75	74.12	0.08	0.28	0.55	0.05	0.04	0.11	-
	1473 K	-	63.09	30.89	0.12	2.41	3.12	0.07	0.06	0.17	-
	1573 K	5.51	81.04	-	0.09	5.02	7.90	0.09	0.04	0.04	-
	1623 K	31.65	63.34	-	0.60	3.14	0.70	0.20	0.05	0.33	-
B	1273 K	-	81.66	14.94	0.05	2.81	0.35	0.04	0.03	0.10	-
	1373 K	5.08	93.62	-	0.04	0.26	0.55	0.04	--	0.39	-
	1473 K	16.05	81.52	-	0.04	1.51	0.44	0.04	0.02	0.38	-
	1573 K	42.34	56.03	-	0.05	0.62	0.58	0.05	0.11	0.22	-
	1623 K	60.67	35.39	-	0.07	1.50	1.73	0.03	0.04	0.56	-
C	1273 K	-	--	99.32	0.05	0.22	0.14	0.03	0.03	0.20	-
	1373 K	-	47.59	51.68	0.04	0.32	0.24	0.03	0.02	0.08	-
	1473 K	4.45	89.06	-	0.05	2.70	2.05	0.06	0.04	1.60	-
	1573 K	27.04	70.74	-	0.10	1.06	0.45	0.11	0.08	0.43	-
	1623 K	66.61	30.01	-	0.08	1.68	0.78	0.16	0.08	0.61	-
D	1273 K	0.99	94.98	-	0.10	1.44	1.92	0.20	0.07	0.18	0.12
	1373 K	8.82	84.43	-	0.16	2.50	2.57	0.23	1.07	0.15	0.07
	1473 K	16.05	76.37	-	0.14	3.20	3.68	0.19	0.08	0.11	0.18
	1573 K	34.03	33.74	-	0.39	15.20	16.00	0.22	0.08	0.16	0.18
	1593 K	52.26	20.04	-	0.53	12.81	13.46	0.26	0.16	0.20	0.29

At 1273 K, sample D had high reduction index. At the same time, some iron appeared in sample D and the main iron oxide was FeO. Besides, the mineral composition of sample D were the low melting compound (Fe_2SiO_4) and few magnesium aluminate spinel, which contributed to the low softening starting temperature. Between 1273 K and 1373 K, the main iron oxide was FeO, and iron's content rose during softening process. Then, the contents of complex compound changed slightly with iron's content rising. In the process of the temperature reaching 1573 K, the content of the complex compound decreased and SiO_2 and Al_2O_3's content increased during melting process.

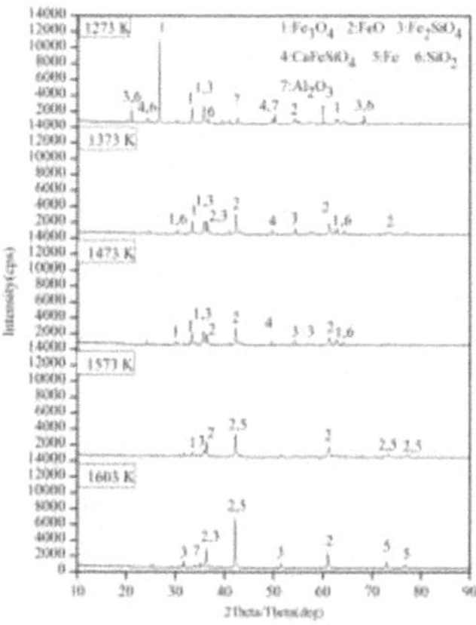

A lump ore

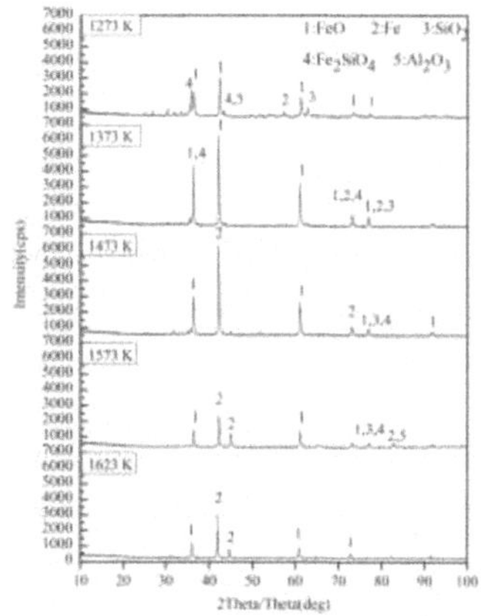

B lump ore

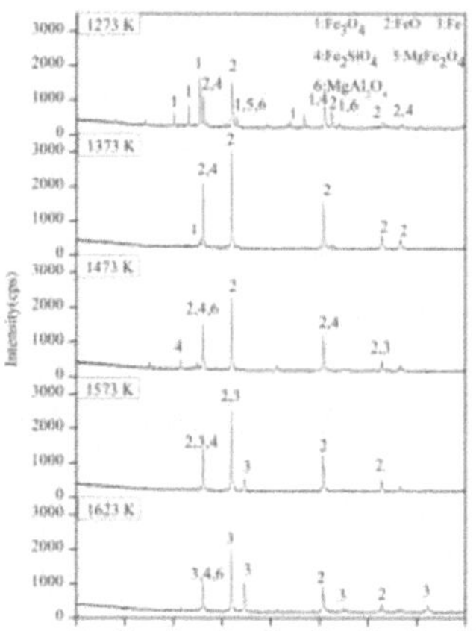

C lump ore

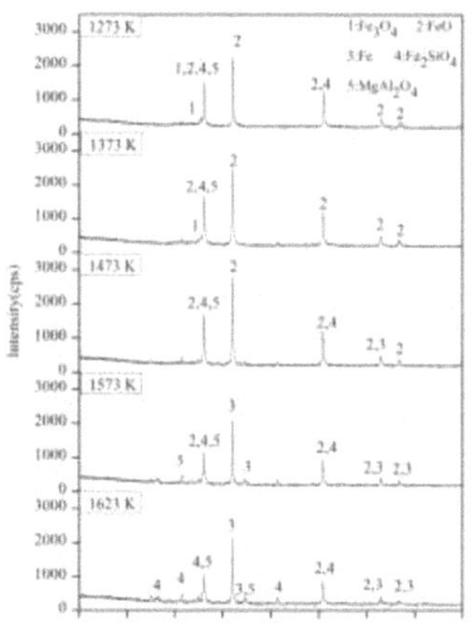

D lump ore

Figure 5. X-ray diffraction analysis results

Conclusions

(1) The lump ore began to shrink the time after the softening starting temperature followed by melting. Sample B had the lowest softening starting temperature (1203 K), the highest melting temperature (1658 K), and the widest softening-melting temperature interval (455 K) among

samples A-D, which was caused by porous characteristics. Sample A had the highest softening starting temperature (1393 K), the highest melting temperature (1628 K), and the narrowest softening-melting temperature interval (235 K).

(2) The lump ore with few extent of softening-melting, acted as "coke window" and improved the gas permeability in the blast furnace, because of the "one after another" property of softening and melting behavior. Before 1573 K, lump ore A acted as this role, which was followed by lump ores B and C during 1573 K and 1623 K.

(3) For samples A and B, the minerals consisted of fayalite and little kirschsteinite which were low melting point materials. For samples C and D, the minerals were main high melting point materials, such as magnesium aluminate spinel.

(4) The reduction index order was:

lump ore D > lump ore B > lump ore C > lump ore A.

The looser the structure was, the better the reducibility will be.

(5) It is suggested that the charging ratio of lump ore A should be improved. Lump ore B should be distributed in the center of the blast furnace. Lump ores C and D should be used less.

References

[1] Yutaka J, et al., "Subjects for Achievement of Blast Furnace Operation With Low Reducing Agent Rate," *ISIJ International*, 45 (10) (2005), 1379.

[2] Ritz V J, Kortmann H A, and Koch K., "Reduction. Softening and Melting Properties of Pellets, Sinters, Lumpy Ore and Mixed Blast Furnace Burden" (Paper presented at the 2nd International Congress on the Science and Technology of Ironmaking and 57th Ironmaking Conference, Toronto, Canada, 1998), 1635.

[3] Nogueira P F, and Fruehan R J. "Fundamental Studies on Blast Furnace Burden Softening and Melting" (Paper presented at the 2nd International Meeting on Ironmaking and the 1st International Symposium on Iron Ore, Rio de Janeiro, Brazil, 2004), 893.

[4] Mousa E, Senk D, and Babich A, "Reduction of Pellets-Nut Coke Mixture Under Simulating Blast Furnace Condition," *Steel Research International*, 81 (9) (2010), 706.

[5] Loo C E, Matthews L T, and O'dea D P, "Lump Ore and Sinter Behavior during Softening and Melting," *ISIJ International*, 51 (6) (2011), 930-938.

[6] Kaushik P., and Fruehan R J, "Mixed Burden Softening and Melting Phenomena in Blast Furnace Operation, Part 3-Mechanism of Burden Interaction and Melt Exudation Phenomenon," *Ironmaking and Steelmaking*, 34 (1) (2007), 10.

AUTHOR INDEX
7th International Symposium on High-Temperature Metallurgical Processing

M

P

Q

SUBJECT INDEX

7th International Symposium on High-Temperature Metallurgical Processing

D

E

F

G

H

T

U

V

W

Z

The manufacturer's authorised representative in the EU is Springer Nature Customer Service Centre GmbH, Europaplatz 3, 69115 Heidelberg, Germany. If you have any concerns regarding our products, please contact ProductSafety@springernature.com

Printed and bound by CPI Group (UK) Ltd, Croydon, CR0 4YY
30/09/2026
02225567-0007